www.brookscole.com

www.brookscole.com is the World Wide Web site for Brooks/Cole and is your direct source to dozens of online resources.

At *www.brookscole.com* you can find out about supplements, demonstration software, and student resources. You can also send email to many of our authors and preview new publications and exciting new technologies.

www.brookscole.com
Changing the way the world learns®

Duxbury Titles of Related Interest

Berger & Maure, *Experimental Design with Applications in Management, Engineering, and the Sciences*

Daniel, *Applied Nonparametric Statistics,* 2nd ed.

Derr, *Statistical Consulting: A Guide to Effective Communication*

Durrett, *Probability: Theory and Examples,* 2nd ed.

Graybill, *Theory and Application of the Linear Model*

Higgins, *An Introduction to Modern Nonparametric Statistics*

Johnson, *Applied Multivariate Methods for Data Analysts*

Kuehl, *Design of Experiments: Statistical Principles of Research Design and Analysis,* 2nd ed.

Larsen, Marx, & Cooil, *Statistics for Applied Problem Solving and Decision Making*

Lattin, Carroll, & Green, *Analyzing Multivariate Data*

Lohr, *Sampling: Design and Analysis*

Lunneborg, *Data Analysis by Resampling: Concepts and Applications*

Minh, *Applied Probability Models*

Minitab, Inc., *MINITAB™ Student Version 12 for Windows*

Myers, *Classical and Modern Regression with Applications,* 2nd ed.

Newton & Harvill, *StatConcepts: A Visual Tour of Statistical Ideas*

Ramsey & Schafer, *The Statistical Sleuth,* 2nd ed.

SAS Institute Inc., *JMP-IN: Statistical Discovery Software*

Savage, *Decision Making with Insight: Includes Insight.XLA 2.0*

Scheaffer, Mendenhall, & Ott, *Elementary Survey Sampling,* 5th ed.

Seila, Ceric, & Tadikamalla, *Applied Simulation Modeling*

Shapiro, *Modeling the Supply Chain*

Trumbo, *Learning Statistics with Real Data*

Winston, *Simulation Modeling Using @RISK*

To order copies, contact your local bookstore or call 1-800-354-9706. For more information, contact Duxbury Press at 10 Davis Drive, Belmont, CA 94002, or go to www.duxbury.com.

A First Course
in Statistical Methods

R. Lyman Ott

Michael T. Longnecker
Texas A&M University

THOMSON
™
BROOKS/COLE

Australia • Canada • Mexico • Singapore • Spain
United Kingdom • United States

THOMSON

™

BROOKS/COLE

Publisher: Curt Hinrichs
Senior Acquisitions Editor: Carolyn Crockett
Development Editor: Cheryll Linthicum
Assistant Editor: Ann Day
Editorial Assistants: Julie Bliss, Rhonda Letts
Technology Project Manager: Burke Taft
Marketing Manager: Joseph Rogove
Marketing Assistant: Jessica Perry
Advertising Project Manager: Tami Strang

Project Manager, Editorial Production: Sandra Craig
Print/Media Buyer: Kristine Waller
Permissions Editor: Elizabeth Zuber
Production, Composition, Illustrations: G & S Typesetters
Copy Editor: Julie Nemer
Cover Designer: Bill Stanton
Cover Image: Jean Louis Batt/GettyImages
Cover Printer: The Lehigh Press, Inc.
Printer: Edwards Brothers, Incorporated

For more information about our products, contact us at:
Thomson Learning Academic Resource Center
1-800-423-0563

For permission to use material from this text, contact us by:
Phone: 1-800-730-2214 **Fax:** 1-800-730-2215
Web: http://www.thomsonrights.com

Library of Congress Control Number: 2003100264
ISBN 0-534-40806-0

Brooks/Cole — Thomson Learning
10 Davis Drive
Belmont, CA 94002
USA

Asia
Thomson Learning
5 Shenton Way #01-01
UIC Building
Singapore 068808

Australia/New Zealand
Thomson Learning
102 Dodds Street
Southbank, Victoria 3006
Australia

Canada
Nelson
1120 Birchmount Road
Toronto, Ontario M1K 5G4
Canada

Europe/Middle East/Africa
Thomson Learning
High Holborn House
50/51 Bedford Row
London WC1R 4LR
United Kingdom

Latin America
Thomson Learning
Seneca, 53
Colonia Polanco
11560 Mexico D.F.
Mexico

Spain/Portugal
Paraninfo
Calle/Magallanes, 25
28015 Madrid, Spain

Contents

Preface

A First Course in Statistical Methods is a concise new text for a one-term course in statistical methods for advanced undergraduate and graduate students from a variety of disciplines. It is based on selected topics and material from our book *An Introduction to Statistical Methods and Data Analysis, Fifth Edition,* and it is intended to prepare students to deal with solving problems encountered in research projects, decision making based on data, and general life experiences beyond the classroom and university setting. We presume students using this book have a minimal mathematical background (high school algebra) and no prior coursework in statistics.

Major Features

Learning from Data   In the broadest sense statistics is a set of principles and techniques that are useful in solving real-life, practical problems based on limited, variable data. As such it has wide applicability in almost all areas of our lives including business, government, science, and our personal lives. In order to help students stay focused on solving problems, we approach the study of statistics by using a four-step process that closely parallels the scientific method:

1. Defining the Problem
2. Collecting Data
3. Summarizing Data
4. Analyzing Data, Interpreting the Analyses, and Communicating Results

Encounters with Real Data   An "Encounters with Real Data" section appears at the end of each chapter, where appropriate. These sections make use of challenging, practical data sets compiled by Bruce Trumbo in his casebook *Learning Statistics with Real Data* (2002) and case studies taken from our book *An Introduction to Statistical Methods and Data Analysis, Fifth Edition* (2001). We hope these data sets, which are also available on the CD that comes with this book, will give the student an appreciation for the broad applicability of statistics and the four-step statistical thought process that we have found helpful and have used throughout many years of teaching, consulting, and R&D management. Here are some examples of the data sets used in those sections:

- **California Earthquakes.**   Data from a network of monitoring stations maintained by government agencies and universities used to better understand the size and frequency of these quakes

- **Percentage of Calories from Fat.**   Assessment and quantification of a person's diet, which is crucial in evaluating the degree of relationship between diet and diseases
- **Effect of an Oil Spill on Plant Growth.**   A study to determine the lingering effects of a ruptured oil pipeline on the plant growth following the cleanup process
- **Smoking and Survival Rates.**   Use of data from a health survey to explore the effects of smoking on the life span of women

With each data set, we illustrate the use of the four-step, learning-from-data process. A discussion of sample size determination, graphical displays of the data, and a summary of the necessary ingredients for a complete report of the statistical findings of the study are included with many of the data "encounters."

Graphical Displays of Data

Throughout the text, we provide a variety of graphical displays that are essential to the evaluation of the critical assumptions underlying the application of such statistical procedures as normal probability plots, boxplots, scatterplots, matrix plots, and residual plots. Furthermore, we emphasize the summarization of data using graphical displays in order to provide tools the reader can use to demonstrate treatment differences. These types of plots provide a method for making a distinction between statistically different treatments and practically different treatments.

For example, in the "Encounters with Real Data" section of Chapter 8, we are interested in evaluating whether the treatment of port-wine stains is more effective for younger children than for older ones. Here we use side-by-side boxplots to summarize the improvement data across the various age groups (Figure 8.8). A normal probability plot is used to assess the normality assumption for the residuals (Figure 8.9).

Examples and Exercises

We have further enhanced the practical nature of statistics by using examples and exercises from journal articles, newspapers, and our many consulting experiences. The many, varied examples and exercises in this book will be a great asset to both

FIGURE 8.8

Boxplots of improvement by age group (means are indicated by solid circles)

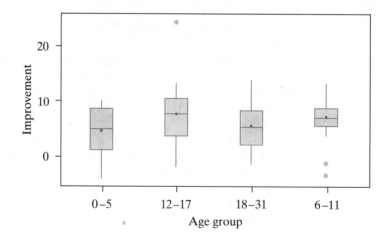

FIGURE 8.9

Normal probability plot of the residuals for the improvement data

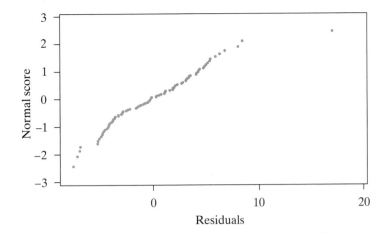

the instructor and the students and will provide students with further evidence of the practical uses of statistics in solving problems that are relevant to their everyday lives. In many of the exercises, we include computer output for the students to use as they solve the exercises. For example, the output for Exercise 8.7 (a portion of which is shown here) provides an AOV table for the comparison of five weight-reducing agents. Also provided are pairwise comparisons of the means, as well as a normal probability plot and boxplot of the residuals. The student is then asked a variety of questions that a researcher would ask when attempting to summarize the results of the study.

```
General Linear Models Procedure
Class Level Information

Class     Levels    Values

AGENT        5      1 2 3 4 S

Number of observations in data set = 50

Dependent Variable: L    WEIGHTLOSS

SOURCE                DF    Sum of Squares    F Value    Pr > F
Model                  4       61.61800000     15.68     0.0001
Error                 45       44.20700000
Corrected Total       49      105.82500000

              R-Square                C.V.              L Mean
              0.582263            9.035093          10.9700000

Source                DF       Type III SS    F Value    Pr > F
AGENT                  4       61.61800000     15.68     0.0001
-------------------------------------------------------------------
```

```
FISHER'S LSD  for variable: WEIGHTLOSS

Alpha= 0.05  df= 45  MSE= 0.982378
Critical Value of T= 2.01
Least Significant Difference= 0.8928

Means with the same letter are not significantly different.

T Grouping              Mean        N  A
          A          12.2400      10  4
          A          12.0500      10  1
          B          11.0200      10  2
          B          10.2700      10  3
          C           9.2700      10  S
------------------------------------------------------------------

Variable=RESIDUAL

  Stem Leaf                       #  Boxplot
    2 9                           1     0
    2 2                           1     0
    1 5                           1     |
    1 00003                       5     |
    0 556679                      6  +-----+
    0 0112233444                 10  |  +  |
   -0 444433321100               12  +-----+
   -0 88765                       5  +-----+
   -1 4331000                     7     |
   -1 98                          2     |
      ----+----+----+----+

                 Normal Probability Plot
   2.75+                                              *
       |                                        *  ++++
       |                                          ++++++
   1.25+                                   +**+**
       |                                +*****
       |                           +****+**
  -0.25+                      ******
       |                   *****
       |                **+****
  -1.75+      *+++*++
       +----+----+----+----+----+----+----+----+----+----+
          -2        -1         0        +1        +2
```

Topics Covered

A brief table of contents for this one-term book is shown here:

Step 1: DEFINING THE PROBLEM
Chapter 1. Statistics and the Scientific Method
Step 2: COLLECTING DATA
Chapter 2. Collecting Data Using Surveys and Scientific Studies

Emphasis on Interpretation, Not Computation

The standard methodology in a first course in statistics is to define a particular statistical procedure and then carry out data analysis by using a computational form of the procedure. However, we find this to be a necessary impediment to the analysis of larger data sets. Furthermore, the students spend too much time in hand calculation with a form of the statistical procedure that is void of the intuitive nature of why the calculations are being done. Therefore, we provide examples and exercises that allow the student to study how to calculate the value of statistical estimators and test statistics using the definitional form of the procedure. After the student becomes comfortable with these aspects of the data the statistical procedure is reflecting, we then emphasize the use of computer software in making computations when analyzing larger data sets. We provide output from several major statistical packages: SAS, Minitab, Stata, JMP, and SPSS. We find that this approach gives the student the experience of computing the value of the procedure using the definition and hence, the student learns the basics behind each procedure. In most situations beyond the statistics course, the student should be using computer software in making the computations for both expedience and quality of calculation. In many exercises and examples, using the computer allows more time for emphasizing the interpretation of results of the computations rather than expending enormous time and effort in the actual computations.

We have demonstrated through examples and exercises the importance of the following aspects of hypothesis testing:

1. The statement of the research hypothesis through the summarization of the researcher's goals into a statement about population parameters.
2. The selection of the most appropriate test statistic, including sample size computations for many procedures.
3. The necessity of considering both Type I and Type II error rates (α and β) when discussing the results of a statistical test of hypotheses.
4. The importance of considering both statistical significance of a test result and the practical significance of the results. Thus, we illustrate the importance of estimating effect sizes and the construction of confidence intervals for population parameters.

5. The statement of the results of the statistical test in nonstatistical jargon that goes beyond the statements "reject H_0" or "fail to reject H_0."

We have provided displays for tests of hypotheses that give the null and alternative hypotheses, the test statistic, rejection region, and the assumptions under which the test is valid. Alternative test statistics are suggested when the stated conditions are not realized in a study. For example, the display shown here is given in Chapter 5.

Summary of a Statistical Test for μ with a Normal Population Distribution (σ Known) or Large Sample Size n

Hypotheses:

Case 1. $H_0: \mu \leq \mu_0$ vs. $H_a: \mu > \mu_0$ (right-tailed test)
Case 2. $H_0: \mu \geq \mu_0$ vs. $H_a: \mu < \mu_0$ (left-tailed test)
Case 3. $H_0: \mu = \mu_0$ vs. $H_a: \mu \neq \mu_0$ (two-tailed test)

T.S.: $z = \dfrac{\bar{y} - \mu_0}{\sigma/\sqrt{n}}$

R.R.:   For a probability α of a Type I error,

Case 1. Reject H_0 if $z \geq z_\alpha$.
Case 2. Reject H_0 if $z \leq -z_\alpha$.
Case 3. Reject H_0 if $|z| \geq z_{\alpha/2}$.

Note: These procedures are appropriate if the population distribution is normally distributed with σ known. In most situations, if $n \geq 30$, then the Central Limit Theorem allows us to use these procedures when the population distribution is nonnormal. Also, if $n \geq 30$, then we can replace σ with the sample standard deviation s. The situation in which $n < 30$ is presented later in this chapter.

Additional Features

- The first chapter provides a discussion of statistics and the scientific method. We provide a discussion of why students should study statistics along with a discussion of several major studies that illustrate the use of statistics in the solution of real-world problems.
- The second chapter draws a distinction between surveys and scientific studies and discusses the designs used for collecting data from surveys and scientific studies.
- The completely randomized design is discussed in Chapter 8 along with an introduction to the analysis of variance and multiple comparisons.
- An introduction to multiple regression is presented in Chapter 12 following linear regression and correlation in Chapter 11.
- Although the last chapter is devoted to communicating and documenting the results of the data analyses, we have incorporated many of these ideas throughout the text in the "Encounters with Real Data" sections and selected exercises.
- The text emphasizes the importance of the assumptions on which statistical procedures are based. We discuss and—through the use of computer simulations—illustrate the robustness or lack of robustness of many estimation and test procedures. The procedures needed to investigate whether

a study satisfies the necessary assumptions are presented and illustrated in many examples and exercises. Furthermore, in many settings, we provide alternative procedures when the assumptions are not met.

- We encourage the use of the computer to make most calculations after an illustration of the computations using the definitional form of the procedure using a small data set.
- Most examples include a graphical display of the data. Computer use greatly facilitates the use of more sophisticated graphical illustrations of statistical results.
- Exercises are grouped into "Basic Techniques" and "Applications."
- Solutions to selected exercises appear at the back of the book.

Ancillaries

- **Student Solutions Manual (0-534-40807-9)** contains the solutions to selected exercises.
- **Instructor's Suite CD-ROM (0-534-40809-5)** contains the Test Bank and complete Solutions Manual.
- **Duxbury Web Site (www.duxbury.com)** offers a rich array of Web-based teaching and learning resources.
- **Book Companion Web Site,** accessible through the Duxbury Web site, includes answers to the review exercises at the end of each chapter.

Acknowledgments

Special thanks and appreciation go to Alex Kugushev for his inspiration, motivation, and guidance in the writing of *An Introduction to Statistical Methods and Data Analysis* and for the initial discussions we had toward revising the manuscript to meet the emerging needs for a one-term course. Carolyn Crockett, our editor at Duxbury, has been our constant companion and guide in making *A First Course in Statistical Methods* a reality. We greatly appreciate the invaluable assistance and support of the numerous editorial, production and marketing people within Duxbury who have impacted the manuscript in so many positive ways. We are also indebted to Mosuk Chow, The Pennsylvania State University; Dale Everson, University of Idaho; and Harry Dean Johnson, Visiting Assistant Professor, University of Idaho, for their insightful and constructive comments during the review process. We also thank once again the reviewers of *An Introduction to Statistical Methods and Data Analysis, Fifth Edition:* Christine Franklin, University of Georgia; Darcy P. Mays, Virginia Commonwealth University; Larry J. Ringer, Texas A&M University; and Deborah J. Rumsey, Ohio State University.

R. Lyman Ott
Michael Longnecker

1 Statistics and the
Scientific Method

CHAPTER 1

Statistics and the Scientific Method

1.1   Introduction

Let's begin with a definition. According to *Merriam-Webster's Collegiate Dictionary,* statistics is "(1) a branch of mathematics dealing with the collection, analysis, interpretation and presentation of masses of numerical data" and "(2) a collection of quantitative data." We'll focus on definition 1 and emphasize that statistics, as a subject matter, is a set of scientific principles and techniques that are useful in reaching conclusions about populations and processes when the available information is both limited and variable; that is, statistics is the science of *learning from data.* Almost everyone—including corporate presidents, marketing representatives, social scientists, engineers, medical researchers, and consumers—deals with data. These data could be in the form of quarterly sales figures, percent increase in juvenile crime, contamination levels in water samples, survival rates for patients undergoing medical therapy, census figures, or input that helps determine which brand of car to purchase. In this text, we approach the study of statistics by considering the four-step process in learning from data: (1) defining the problem, (2) collecting data, (3) summarizing data, and (4) analyzing data, interpreting the analyses, and communicating results. Through the use of these four steps in learning from data, our study of statistics closely parallels the scientific method, which, according to *Merriam-Webster's Collegiate Dictionary,* relates to the "principles and procedures for the scientific pursuit of knowledge involving the recognition and formulation of a problem, the collection of data through observation and experiment, and the formulation and testing of hypotheses."

This book is divided into sections corresponding to the four-step process in learning from data. The relationship among these steps and the chapters of the book is shown in Table 1.1. As you can see from this table, much time is spent discussing how to analyze data using the basic methods presented in Chapters 5–12. However, you must remember that for each data set requiring analysis, someone has defined the problem to be examined (Step 1), developed a plan for collecting data to address the problem (Step 2), and summarized the data and prepared the data for analysis (Step 3). Then following the analysis of the data, the results of the analysis must be interpreted and communicated either verbally or in written form to the intended audience (Step 4).

All four steps are important in learning from data; in fact, unless the problem to be addressed is clearly defined and the data collection carried out properly, the interpretation of the results of the analyses may convey misleading information be-

The Four-Step Process	Chapters
1 Defining the Problem	1 Statistics and the Scientific Method
2 Collecting Data	2 Collecting Data Using Surveys and Scientific Studies
3 Summarizing Data	3 Summarizing Data
	4 Probability and Probability Distributions
4 Analyzing Data, Interpreting the Analyses, and Communicating Results	5 Inferences about Population Central Values
	6 Inferences Comparing Two Population Central Values
	7 Inferences about Population Variances
	8 The Completely Randomized Design
	9 More Complicated Experimental Designs
	10 Categorical Data
	11 Linear Regression and Correlation
	12 Multiple Regression
	13 Communicating and Documenting the Results of Analyses

cause the analyses were based on a data set that did not address the problem or that was incomplete and contained improper information. Throughout the text, we will try to keep you focused on the bigger picture of learning from data through the four-step process. Most chapters will end with a section entitled "Staying Focused," which emphasizes how the material of the chapter fits into the study of statistics—learning from data.

Before we jump into the study of statistics, let's consider four instances in which the application of statistics could help to solve a practical problem.

1. **Problem: Monitoring the ongoing quality of a lightbulb manufacturing facility.** A lightbulb manufacturer produces approximately half a million bulbs per day. The quality department must monitor the defect rate of the bulbs. It could accomplish this task by testing each bulb, but the cost would be substantial and would greatly increase the price per bulb. An alternative approach is to select 1,000 bulbs from the daily production of 500,000 bulbs and test each of the 1,000. The fraction of defective bulbs in the 1,000 tested could be used to estimate the fraction defective in the entire day's production, provided that the 1,000 bulbs were selected in the proper fashion. We will demonstrate in later chapters that the fraction defective in the tested bulbs will probably be quite close to the fraction defective for the entire day's production of 500,000 bulbs.

2. **Problem: Is there a relationship between quitting smoking and gaining weight?** To investigate the claim that people who quit smoking often experience a subsequent weight gain, researchers selected a random sample of 400 participants who had successfully participated in programs to quit smoking. The individuals were weighed at the beginning of the program and again one year later. The average change in weight of the participants was an increase of 5 pounds. The investigators concluded that there was

evidence that the claim was valid. We will develop techniques in later chapters to assess when changes are truly significant changes and not changes due to random chance.

3. **Problem: What effect does nitrogen fertilizer have on wheat production?** For a study of the effects of nitrogen fertilizer on wheat production, a total of 15 fields were available to the researcher. She randomly assigned three fields to each of the five nitrogen rates under investigation. The same variety of wheat was planted in all 15 fields. The fields were cultivated in the same manner until harvest, and the number of pounds of wheat per acre was then recorded for each of the 15 fields. The experimenter wanted to determine the optimal level of nitrogen to apply to *any* wheat field, but, of course, she was limited to running experiments on a limited number of fields. After determining the amount of nitrogen that yielded the largest production of wheat in the study fields, the experimenter then concluded that similar results would hold for wheat fields possessing characteristics somewhat the same as the study fields. Is the experimenter justified in reaching this conclusion?

4. **Problem: Determining public opinion toward a question, issue, product, or candidate.** Similar applications of statistics are brought to mind by the frequent use of the *New York Times/CBS News, Washington Post/ABC News,* CNN, Harris, and Gallup polls. How can these pollsters determine the opinions of more than 195 million Americans who are of voting age? They certainly do not contact every potential voter in the United States. Rather, they sample the opinions of a small number of potential voters, perhaps as few as 1,500, to estimate the reaction of every person of voting age in the country. The amazing result of this process is that the fraction of those persons contacted who hold a particular opinion will closely match the fraction in the total population holding that opinion at a particular time. We will supply convincing supportive evidence of this assertion in subsequent chapters.

These problems illustrate the four-step process in learning from data. First, there was a problem or question to be addressed. Next, for each problem a study or experiment was proposed to collect meaningful data to answer the problem. The quality group had to decide both how many bulbs needed to be tested and how to select the sample of 1,000 bulbs from the total production of bulbs to obtain valid results. The polling groups must decide how many voters to sample and how to select these individuals in order to obtain information that is representative of the population of all voters. Similarly, it was necessary to carefully plan how many participants in the weight gain study were needed and how they were to be selected from the list of all such participants. Furthermore, what variables should the researchers have measured on each participant? Was it necessary to know each participant's age, sex, physical fitness, and other health-related variables, or was weight the only important variable? The results of the study may not be relevant to the general population if many of the participants in the study had a particular health condition. In the wheat experiments, it was important to measure both soil characteristics of the fields and environmental conditions, such as temperature and rainfall, to obtain results that could be generalized to fields not included in the study. The design of a study or experiment is crucial to obtaining results that can be generalized beyond the study.

Finally, having collected, summarized, and analyzed the data, it is important to report the results in unambiguous terms to interested people. For the lightbulb example, management and technical staff need to know the quality of their production batches. Based on this information, they can determine whether adjustments in the process are necessary. Therefore, the results of the statistical analyses cannot be presented in ambiguous terms; decisions must be made from a well-defined knowledge base. The results of the weight gain study are of vital interest to physicians who have patients participating in the smoking-cessation program. If a significant increase in weight was recorded for those individuals who had quit smoking, physicians may have to recommend diets so that the former smokers do not go from one health problem (smoking) to another (elevated blood pressure due to being overweight). It is crucial that a careful description of the participants—that is, age, sex, and other health-related information—be included in the report. In the wheat study, the experiments provide farmers with information that allows them to economically select the optimum amount of nitrogen required for their fields. Therefore, the report must contain information concerning the amount of moisture and types of soils present in the study fields. Otherwise, the conclusions about optimal wheat production may not pertain to farmers' growing wheat under considerably different conditions.

population

sample

To infer validly that the results of a study are applicable to a larger group than just the participants in the study, we must carefully define the **population** (see Definition 1.1) to which inferences are sought and design a study in which the **sample** (see Definition 1.2) has been appropriately selected from the designated population. We will discuss these issues in Chapter 2.

DEFINITION 1.1

A **population** is the set of all measurements of interest to the sample collector. (See Figure 1.1.)

DEFINITION 1.2

A **sample** is any subset of measurements selected from the population. (See Figure 1.1.)

FIGURE 1.1
Population and sample

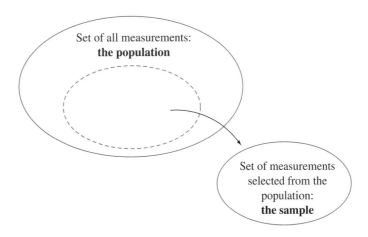

Set of all measurements:
the population

Set of measurements
selected from the
population:
the sample

1.2 Why Study Statistics?

We can think of many reasons for taking an introductory course in statistics. One reason is that you need to know how to evaluate published numerical facts. Every person is exposed to manufacturers' claims for products; to the results of sociological, consumer, and political polls; and to the published results of scientific research. Many of these results are inferences based on sampling. Some inferences are valid; others are invalid. Some are based on samples of adequate size; others are not. Yet all these published results bear the ring of truth. Some people say that statistics can be made to support almost anything. Others say it is easy to lie with statistics. Both statements are true. It is easy, purposely or unwittingly, to distort the truth by using statistics when presenting the results of sampling to the uninformed.

A second reason for studying statistics is that your profession or employment may require you to interpret the results of sampling (surveys or experimentation) or to employ statistical methods of analysis to make inferences in your work. For example, practicing physicians receive large amounts of advertising describing the benefits of new drugs. These advertisements frequently display the numerical results of experiments that compare a new drug with an older one. Do such data really imply that the new drug is more effective, or is the observed difference in results due simply to random variation in the experimental measurements?

Recent trends in the conduct of court trials indicate an increasing use of probability and statistical inference in evaluating the quality of evidence. The use of statistics in the social, biological, and physical sciences is essential because all these sciences make use of observations of natural phenomena, through sample surveys or experimentation, to develop and test new theories. Statistical methods are employed in business when sample data are used to forecast sales and profit. In addition, they are used in engineering and manufacturing to monitor product quality. The sampling of accounts is a useful tool to assist accountants in conducting audits. Thus, statistics plays an important role in almost all areas of science, business, and industry; people employed in these areas need to know the basic concepts, strengths, and limitations of statistics.

1.3 Some Current Applications of Statistics

Defining the Problem: Reducing the Threat of Acid Rain to Our Environment

The accepted causes of acid rain are sulfuric and nitric acids; the sources of these acidic components of rain are hydrocarbon fuels, which spew sulfur and nitric oxide into the atmosphere when burned. Here are some of the many effects of acid rain:

- Acid rain, when present in spring snow melts, invades breeding areas for many fish, which prevents successful reproduction. Forms of life that depend on ponds and lakes contaminated by acid rain begin to disappear.
- In forests, acid rain is blamed for weakening some varieties of trees, making them more susceptible to insect damage and disease.
- In areas surrounded by affected bodies of water, vital nutrients are leached from the soil.

- Human-made structures are also affected by acid rain. Experts from the United States estimate that acid rain has caused nearly $15 billion of damage to buildings and other structures thus far.

Solutions to the problems associated with acid rain will not be easy. The National Science Foundation (NSF) has recommended that we strive for a 50% reduction in sulfur oxide emissions. Perhaps that is easier said than done. High-sulfur coal is a major source of these emissions, but in states dependent on coal for energy, a shift to lower-sulfur coal is not always possible. Instead, better scrubbers must be developed to remove these contaminating oxides from the burning process before they are released into the atmosphere. Fuels for internal combustion engines are also major sources of the nitric and sulfur oxides of acid rain. Clearly, better emission control is needed for automobiles and trucks.

Reducing the oxide emissions from coal-burning furnaces and motor vehicles will require greater use of existing scrubbers and emission control devices as well as the development of new technology to allow us to use available energy sources. Developing alternative, cleaner energy sources is also important if we are to meet NSF's goal. Statistics and statisticians will play a key role in monitoring atmospheric conditions, testing the effectiveness of proposed emission control devices, and developing new control technology and alternative energy sources.

Defining the Problem: Determining the Effectiveness of a New Drug Product

The development and testing of the Salk vaccine for protection against poliomyelitis (polio) provide an excellent example of how statistics can be used in solving practical problems. Most people who grew up before 1954 can recall the panic brought on by the outbreak of polio cases during the summer months. Although relatively few children fell victim to the disease each year, the pattern of outbreak of polio was unpredictable and caused great concern because of the possibility of paralysis or death. The fact that very few of today's youth have even heard of polio demonstrates the great success of the vaccine and the testing program that preceded its release on the market.

It is standard practice in establishing the effectiveness of a particular drug product to conduct an experiment (often called a *clinical trial*) with human participants. For some clinical trials, assignments of participants are made at random, with half receiving the drug product and the other half receiving a solution or tablet (called a *placebo*) that does not contain the medication. One statistical problem concerns the determination of the total number of participants to be included in the clinical trial. This problem was particularly important in the testing of the Salk vaccine because data from previous years suggested that the incidence rate for polio might be less than 50 cases for every 100,000 children. Hence, a large number of participants had to be included in the clinical trial in order to detect a difference in the incidence rates for those treated with the vaccine and those receiving the placebo.

With the assistance of statisticians, it was decided that a total of 400,000 children should be included in the Salk clinical trial begun in 1954, with half of them randomly assigned the vaccine and the remaining children assigned the placebo. No other clinical trial had ever been attempted on such a large group of participants. Through a public school inoculation program, the 400,000 participants were treated and then observed over the summer to determine the number of children contracting polio. Although fewer than 200 cases of polio were reported for the 400,000 participants in the clinical trial, more than three times as many cases appeared in the group receiving the placebo. These results, together with some statistical calculations, were sufficient to indicate the effectiveness of the Salk polio vaccine. However, these conclusions would not have been possible if the statisticians and scientists had not planned for and conducted such a large clinical trial.

The development of the Salk vaccine is not an isolated example of the use of statistics in the testing and developing of drug products. In recent years, the Food and Drug Administration (FDA) has placed stringent requirements on pharmaceutical firms to establish the effectiveness of proposed new drug products. Thus, statistics has played an important role in the development and testing of birth control pills, rubella vaccines, chemotherapeutic agents in the treatment of cancer, and many other preparations.

Defining the Problem: Use and Interpretation of Scientific Data in Our Courts

Libel suits related to consumer products have touched each one of us; you may have been involved as a plaintiff or defendant in a suit or you may know of someone who was involved in such litigation. Certainly we all help to fund the costs of this litigation indirectly through increased insurance premiums and increased costs of goods. The testimony in libel suits concerning a particular product (automobile, drug product, and so on) frequently leans heavily on the interpretation of data from one or more scientific studies involving the product. This is how and why statistics and statisticians have been pulled into the courtroom.

For example, epidemiologists have used statistical concepts applied to data to determine whether there is a statistical "association" between a specific characteristic, such as the leakage in silicone breast implants, and a disease condition, such as an autoimmune disease. An epidemiologist who finds an association should try to determine whether the observed statistical association from the study is due to random variation or whether it reflects an actual association between the characteristic and the disease. Courtroom arguments about the interpretations of these types of associations involve data analyses using statistical concepts as well as a clinical interpretation of the data. Many other examples exist in which statistical models are used in court cases. In salary discrimination cases, a lawsuit is filed claiming that an employer underpays employees on the basis of age, ethnicity, or sex. Statistical models are developed to explain salary differences based on many factors, such as work experience, years of education, and work performance. The adjusted salaries are then compared across age groups or ethnic groups to determine whether significant salary differences exist after adjusting for the relevant work-performance factors.

Defining the Problem: Estimating Bowhead Whale Population Size

Raftery and Zeh (1998) discuss the estimation of the population size and rate of increase in bowhead whales, *Balaena mysticetus*. The importance of such a study derives from the fact that bowheads were the first species of great whale for which commercial whaling was stopped; thus, their status indicates the recovery prospects of other great whales. Also, the International Whaling Commission uses these estimates to determine the aboriginal subsistence whaling quota for Alaskan Eskimos. To obtain the necessary data, researchers conducted a visual and acoustic census off Point Barrow, Alaska. The researchers then applied statistical models and estimation techniques to the data obtained in the census to determine whether the bowhead population had increased or decreased since commercial whaling was stopped. The statistical estimates showed that the bowhead population was increasing at a healthy rate, indicating that stocks of great whales that have been decimated by commercial hunting can recover after hunting is discontinued.

Defining the Problem: Assessing Ozone Exposure in Urban Areas

Ambient ozone pollution in urban areas is one of the nation's most pervasive environmental problems. Whereas the decreasing stratospheric ozone layer may lead to increased instances of skin cancer, high ambient ozone intensity has been shown to cause damage to the human respiratory system as well as to agricultural crops and trees. The Houston, Texas, area has ozone concentrations rated second only to Los Angeles in exceeding the National Ambient Air Quality Standard. Carroll et al. (1997) describe how to analyze the hourly ozone measurements collected in Houston from 1980 to 1993 by 9 to 12 monitoring stations. Besides the ozone level, each station also recorded three meteorological variables: temperature, wind speed, and wind direction.

The statistical aspect of the project had three major goals:

1. Provide information (and/or tools to obtain such information) about the amount and pattern of missing data, as well as about the quality of the ozone and the meteorological measurements.
2. Build a model of ozone intensity to predict the ozone concentration at any given location within Houston at any given time between 1980 and 1993.
3. Apply this model to estimate exposure indices that account for either a long-term exposure or a short-term high-concentration exposure; also, relate census information to different exposure indices to achieve population exposure indices.

The spatial–temporal model the researchers built provided estimates demonstrating that the highest ozone levels occurred at locations with relatively small populations of young children. Also, the model estimated that the exposure of young children to ozone decreased by approximately 20% from 1980 to 1993. An examination of the distribution of population exposure had several policy implications. In particular, it was concluded that the current placement of monitors is not ideal if we are concerned with assessing population exposure. This project involved all four components of learning from data: planning where the monitoring stations should be placed within the city, how often data should be collected, and what variables should be recorded; conducting spatial–temporal graphing of the data; creating spatial–temporal models of the ozone data, meteorological data, and demographic data; and, finally, writing a report that could assist local and federal officials in formulating policy with respect to decreasing ozone levels.

Defining the Problem: Assessing Public Opinion Using Polls

Public opinion, consumer preference, and election polls are commonly used to assess the opinions or preferences of a segment of the public for issues, products, or candidates of interest. We, the American public, are exposed to the results of these polls daily in newspapers, in magazines, on the radio, and on television. For example, the results of polls related to the following subjects were printed in local newspapers over a 2-day period:

- Consumer confidence related to future expectations about the economy
- Preferences for candidates in upcoming elections and caucuses
- Attitudes toward cheating on federal income tax returns
- Preference polls related to specific products (for example, foreign vs. American cars, Coke vs. Pepsi, McDonald's vs. Wendy's)
- Reactions of North Carolina residents toward arguments about the morality of tobacco
- Opinions of voters toward proposed tax decreases and proposed changes in the Defense Department budget

A number of questions can be raised about polls. Suppose we consider a poll on the public's opinion toward a proposed income tax increase in the state of Michigan. *What was the population of interest to the pollster?* Was the pollster interested in all residents of Michigan or just those citizens who currently pay income taxes? *Was the sample in fact selected from this population?* If the population of interest was all people currently paying income taxes, did the pollster make sure that all the individuals sampled were current taxpayers? *What questions were asked and how were the questions phrased?* Was each person asked the same question? Were the questions phrased in such a manner as to bias the responses? Can we believe the results of these polls? Do these results "represent" how the general public *currently* feels about the issues raised in the polls?

Opinion and preference polls are an important, visible application of statistics for the consumer. We will discuss this topic in more detail in Chapter 2. We hope that after studying this material you will have a better understanding of how to interpret the results of these polls.

1.4   A Note to the Student

We think with words and concepts. A study of the discipline of statistics requires us to memorize new terms and concepts (as does the study of a foreign language). Commit these definitions and concepts to memory.

Also, focus on the broader concept of learning from data. Do not let details obscure these broader characteristics of the subject. The teaching objective of this text is to identify and amplify these broader concepts of statistics.

1.5   Staying Focused

The discipline of statistics and those who apply the tools of that discipline deal with learning from data. Medical researchers, social scientists, accountants, agronomists, consumers, and government leaders, knowingly or not, deal with the scientific method and hence the four-step process of (1) defining the problem, (2) collecting data, (3) summarizing data, and (4) analyzing data, interpreting the analyses, and communicating results to address problems. We will delve into all this as we proceed through the book.

Supplementary Exercises

Basic Techniques

Bio. **1.1** Selecting the proper diet for shrimp or other sea animals is an important aspect of sea farming. A researcher wishes to estimate the mean weight of shrimp maintained on a specific diet for a period of 6 months. One hundred shrimp are randomly selected from an artificial pond and each is weighed.
  a. Identify the population of measurements that is of interest to the researcher.
  b. Identify the sample.
  c. What characteristics of the population are of interest to the researcher?
  d. If the sample measurements are used to make inferences about certain characteristics of the population, why is a measure of the reliability of the inferences important?

Env. **1.2** Radioactive waste disposal and the production of radioactive material in some mining operations are creating a serious pollution problem in some areas of the United States. State health officials have decided to investigate the radioactivity levels in one suspect area. Two hundred points in the area are randomly selected and the level of radioactivity is measured at each point. Answer questions (a), (b), (c), and (d) in Exercise 1.1 for this sampling situation.

Soc. **1.3** A social researcher in a particular city wishes to obtain information on the number of children in households that receive welfare support. A random sample of 400 households is selected from the city welfare rolls. A check on welfare recipient data provides the number of children in each household. Answer questions (a), (b), (c), and (d) of Exercise 1.1 for this sample survey.

Pol. Sci. **1.4** Search issues of your local newspaper or news magazine to locate the results of a recent opinion survey.
  a. Identify the items that were observed in order to obtain the sample measurements.
  b. Identify the measurement made on each item.
  c. Clearly identify the population associated with the survey.
  d. What characteristic(s) of the population was (were) of interest to the pollster?
  e. Does the article explain how the sample was selected?
  f. Does the article include the number of measurements in the sample?
  g. What type of inference was made concerning the population characteristics?
  h. Does the article tell you how much faith you can place in the inference about the population characteristic?

Gov. **1.5** Because of a recent increase in the number of neck injuries incurred by high school football players, the Department of Commerce designed a study to evaluate the strength of football helmets worn by high school players in the United States. A total of 540 helmets were collected from the five companies that currently produce helmets. The agency then sent the helmets to an independent testing agency to evaluate the impact cushioning of the helmet and the amount of shock transmitted to the neck when the face mask was twisted.
  a. What is the population of interest?
  b. What is the sample?
  c. What variables should be measured?
  d. What are some of the major limitations of this study in regard to the safety of helmets worn by high school players? For example, is the neck strength of the player related to the amount of shock transmitted to the neck and whether the player will be injured?

Edu. **1.6** The faculty senate at a major university with 35,000 students is considering changing the current grading policy from A, B, C, D, F to a plus and minus system—that is, B+, B, or B− rather than just B. The faculty is interested in the students' opinions concerning this change and will sample 500 students.
  a. What is the population of interest?
  b. What is the sample?
  c. How could the sample be selected?
  d. What type of questions should be included in the questionnaire?

STEP

2

Collecting

Data

2 Collecting Data Using
Surveys and Scientific Studies

CHAPTER 2

Collecting Data Using Surveys and Scientific Studies

2.1 Introduction

As mentioned in Chapter 1, the first step in learning from data is to define the problem. The design of the data collection process is the crucial step in *intelligent data gathering* (Step 2). The process takes a conscious, concerted effort focused on the following steps:

- Specifying the objective of the study, survey, or experiment
- Identifying the variable(s) of interest
- Choosing an appropriate design for the survey or scientific study
- Collecting the data

To specify the objective of the study you must understand the problem being addressed. For example, the transportation department in a large city wants to assess the public's perception of the city's bus system in order to increase the use of buses within the city. Thus, the department needs to determine what aspects of the bus system determine whether a person will ride the bus. The objective of the study is to identify factors that the transportation department can alter to increase the number of people using the bus system.

To identify the variables of interest, you must examine the objective of the study. For the bus system, some major factors can be identified by reviewing studies conducted in other cities and by brainstorming with the bus system employees. Some of the factors may be safety, cost, cleanliness of the buses, whether there is a bus stop close to the person's home or place of employment, and how often the bus fails to be on time. The measurements to be obtained in the study would consist of importance ratings (very important, important, no opinion, somewhat unimportant, very unimportant) of the identified factors. Demographic information, such as age, sex, income, and place of residence, would also be measured. Finally, the measurement of variables related to how frequently a person currently rides the buses would be of importance. Once the objectives are determined and the variables of interest are specified, you must select the most appropriate method to collect the data. Data collection processes include surveys, experiments, and the examination of existing data from business records, censuses, government records, and previous studies. The theory of sample surveys and the theory of experimental designs provide excellent methodology for data collection. Usually surveys are passive. The goal of the survey is to gather data on existing conditions, attitudes, or behaviors. Thus, the transportation department would need to construct a questionnaire and then sample current riders of the buses and people who use other forms of transportation within the city.

Scientific studies, on the other hand, tend to be more active: The person conducting the study varies the experimental conditions to study the effect of the conditions on the outcome of the experiment. For example, the transportation department could decrease the bus fares on a few selected routes and assess whether the use of its buses increased. However, in this example, other factors not under the bus system's control may also have changed during this time period. Thus, an increase in bus use may have taken place because of a strike of subway workers or an increase in gasoline prices. The decrease in fares could be only one of several factors that "caused" the increase in the number of people riding the buses.

In most scientific experiments, as many of the factors that affect the measurements as possible are under the control of the experimenter. Say, a floriculturist wants to determine the effect of a new plant stimulator on the growth of a commercially produced flower. The floriculturist would run the experiments in a greenhouse, where temperature, humidity, moisture levels, and sunlight are controlled. An equal number of plants would be treated with each of the selected quantities of the growth stimulator, including a control—that is, no stimulator applied. At the conclusion of the experiment, the size and health of the plants would be measured. The optimal level of the stimulator for the plant could then be determined because ideally all other factors affecting the size and health of the plants would be the same for all plants in the experiment.

In this chapter, we will consider some sampling designs for surveys and some experimental designs for scientific studies. We will also make a distinction between a scientific study and an observational study.

2.2   Surveys

Information from surveys affects almost every facet of our daily lives. These surveys determine such government policies as the control of the economy and the promotion of social programs. Opinion polls are the basis of much of the news reported by the various news media. Ratings of television shows determine which shows will be available for viewing in the future.

Who conducts surveys? We are all familiar with public opinion polls: the *New York Times/CBS News, Washington Post/ABC News,* MSNBC, Harris, Gallup for *Newsweek,* and CNN polls. However, the vast majority of surveys are conducted for a specific industrial, governmental, administrative, or scientific purpose. For example, auto manufacturers use surveys to find out how satisfied customers are with their cars. Frequently we are asked to complete a survey as part of the warranty registration process following the purchase of a new product. Many important studies involving health issues are determined using surveys—for example, amount of fat in the diet, exposure to secondhand smoke, condom use and the prevention of AIDS, and the prevalence of adolescent depression.

The U.S. Bureau of the Census is required by the U.S. Constitution to enumerate the population every 10 years. With the growing involvement of the government in the lives of its citizens, the Census Bureau has expanded its role beyond just counting the population. An attempt is made to send a census questionnaire in the mail to every household in the United States. Since the 1940 census, in addition to the complete count information, further information has been obtained from representative samples of the population. In the 2000 census, variable sampling rates were employed. For most of the country, approximately five of six households were asked to answer the 14 questions on the short version of the form. The remaining households responded to a longer version of the form containing an additional 45 questions. Many agencies and individuals use the resulting information for many purposes. The federal government uses it to determine allocations of funds to states and cities. Businesses use it to forecast sales, to manage personnel, and to establish future site locations. Urban and regional planners use it to plan land use, transportation networks, and energy consumption. Social scientists use it to study economic conditions, racial balance, and other aspects of the quality of life.

The U.S. Bureau of Labor Statistics (BLS) routinely conducts more than 20 surveys. Some of the best known and most widely used are the surveys that establish the consumer price index (CPI). The CPI is a measure of price change for a fixed market basket of goods and services over time. It is a measure of inflation and serves as an economic indicator for government policies. Businesses tie wage rates and pension plans to the CPI. Federal health and welfare programs, as well as many state and local programs, tie their bases of eligibility to the CPI. Escalator clauses in rents and mortgages are based on the CPI. This one index, determined on the basis of sample surveys, plays a fundamental role in our society.

Many other surveys from the BLS are crucial to society. The monthly Current Population Survey establishes basic information on the labor force, employment, and unemployment. The consumer expenditure surveys collect data on family expenditures for goods and services used in day-to-day living. The Establishment Survey collects information on employment hours and earnings for nonagricultural business establishments. The survey on occupational outlook provides information on future employment opportunities for a variety of occupations, projecting to approximately 10 years ahead. Other activities of the BLS are addressed in the *BLS Handbook of Methods* (U.S. Bureau of Labor Statistics 1982).

Opinion polls are constantly in the news, and the names Gallup and Harris have become well known to everyone. These polls, or sample surveys, reflect the attitudes and opinions of citizens on everything from politics and religion to sports and entertainment. The Nielsen ratings determine the success or failure of TV shows. The Nielsen retail index furnishes up-to-date sales data on foods, cosmetics, pharmaceuticals, beverages, and many other classes of products. The data come from auditing inventories and sales in 1,600 stores across the United States every 60 days.

Businesses conduct sample surveys for their internal operations in addition to using government surveys for crucial management decisions. Auditors estimate account balances and check on compliance with operating rules by sampling accounts. Quality control of manufacturing processes relies heavily on sampling techniques.

Another area of business activity that depends on detailed sampling activities is marketing. Decisions about which products to market, where to market them, and how to advertise them are often made on the basis of sample survey data. The data may come from surveys conducted by the firm that manufactures the product or may be purchased from survey firms that specialize in marketing data.

2.3 Sampling Designs for Surveys

A crucial element in any survey is the manner in which the sample is selected from the population. If the individuals included in the survey are selected based on convenience alone, there may be biases in the sample survey, which would prevent the survey from accurately reflecting the population as a whole. For example, a marketing graduate student developed a new approach to advertising and, to evaluate this new approach, selected the students in a large undergraduate business course to assess whether the new approach was an improvement over standard advertisements. Would the opinions of this class of students be representative of the general population of people to which the new approach to advertising would be applied? The income levels, ethnicity, education levels, and many other socioeconomic characteristics of the students may differ greatly from the population of interest. Furthermore, the students may be coerced into participating in the study by their instructor and hence may not give the most candid answers to questions on a survey. Thus, the manner in which a sample is selected is of utmost importance to the credibility and applicability of the study's results.

simple random sampling

The basic design (**simple random sampling**) consists of selecting a group of *n* units in such a way that each sample of size *n* has the same chance of being selected. Thus, we can obtain a random sample of eligible voters in a bond-issue poll by drawing names from the list of registered voters in such a way that each sample of size *n* has the same probability of selection. The details of simple random sampling are discussed in Section 4.11. At this point, we merely state that a simple random sample will contain as much information on community preference as any other sample survey design, provided all voters in the community have similar socioeconomic backgrounds.

Suppose, however, that the community consists of people in two distinct income brackets, high and low. Voters in the high-income bracket may have opinions on the bond issue that are quite different from the opinions of voters in the low-income bracket. Therefore, to obtain accurate information about the population, we want to sample voters from each bracket. We can divide the population elements into two groups, or strata, according to income and select a simple random sample

stratified random sample

from each group. The resulting sample is called a **stratified random sample.** (See Scheaffer et al., 1996, Chap. 5.) Note that stratification is accomplished by using knowledge of an auxiliary variable, namely, personal income. By stratifying on high

ratio estimation

and low values of income, we increase the accuracy of our estimator. **Ratio estimation** is a second method for using the information contained in an auxiliary variable. Ratio estimators not only use measurements on the response of interest but they also incorporate measurements on an auxiliary variable. Ratio estimation can also be used with stratified random sampling.

Although individual preferences are desired in the survey, a more economical procedure, especially in urban areas, may be to sample specific families, apartment buildings, or city blocks rather than individual voters. Individual preferences

cluster sampling

can then be obtained from each eligible voter within the unit sampled. This technique is called **cluster sampling.** Although we divide the population into groups for both cluster sampling and stratified random sampling, the techniques differ. In stratified random sampling, we take a simple random sample within each group, whereas, in cluster sampling, we take a simple random sample of groups and then sample all items within the selected groups (clusters). (See Scheaffer et al., 1996, Chaps. 8–9 for details.)

Sometimes, the names of people in the population of interest are available in a list, such as a registration list, or on file cards stored in a drawer. For this situation, an economical technique is to draw the sample by selecting one name near the beginning of the list and then selecting every tenth or fifteenth name thereafter. If the sampling is conducted in this manner, we obtain a **systematic sample.** As you might expect, systematic sampling offers a convenient means of obtaining sample information; unfortunately, we do not necessarily obtain the most information for a specified amount of money. (Details are given in Scheaffer et al., 1996, Chap. 7.)

systematic sample

The important point to understand is that there are different kinds of surveys that can be used to collect sample data. For the surveys discussed in this text, we will deal with simple random sampling and methods for summarizing and analyzing data collected in such a manner. More complicated surveys lead to even more complicated problems at the summarizing and analysis stages of statistics.

The American Statistical Association (ASA, at Internet address http:// www .amstat.org or e-mail asainfo@amstat.org) publishes a series of documents on surveys: *What Is a Survey? How to Plan a Survey, How to Collect Survey Data, Judging the Quality of a Survey, How to Conduct Pretesting, What Are Focus Groups?* and *More about Mail Surveys*. These documents describe many of the elements crucial to obtaining a valid and useful survey. They list many of the potential sources of errors commonly found in surveys with guidelines on how to avoid these pitfalls. A discussion of some of the issues raised in these brochures follows.

Problems Associated with Surveys

Even when the sample is selected properly, there may be uncertainty about whether the survey represents the population from which the sample was selected. Two of the major sources of uncertainty are nonresponse, which occurs when a portion of the individuals sampled cannot or will not participate in the survey, and measurement problems, which occur when the respondent's answers to questions do not provide the type of data that the survey was designed to obtain.

survey nonresponse

Survey nonresponse may result in a biased survey because the sample is not representative of the population. It is stated in *Judging the Quality of a Survey* that in surveys of the general population women are more likely to participate than men; that is, the nonresponse rate for males is higher than for females. Thus, a political poll may be biased if the percentage of women in the population in favor of a particular issue is larger than the percentage of men in the population supporting the issue. The poll would overestimate the percentage of the population in favor of the issue because the sample had a larger percentage of women than their percentage in the population. In all surveys, a careful examination of the nonresponse group must be conducted to determine whether a particular segment of the population is either under- or overrepresented in the sample. Some of the remedies for nonresponse are:

1. Offering an inducement for participating in the survey.
2. Sending reminders or making follow-up telephone calls to the individuals who did not respond to the first contact.
3. Using statistical techniques to adjust the survey findings to account for the sample profile differing from the population profile.

measurement problems

Measurement problems are the result of the respondent's not providing the information that the survey seeks. These problems often are due to the specific wording of questions in a survey, the manner in which the respondent answers the survey questions, and the fashion in which an interviewer phrases questions during the interview. Examples of specific problems and possible remedies are as follows:

1. *Inability to recall answers to questions:* The interviewee is asked how many times he or she visited a particular city park during the past year. This type of question often results in an underestimate of the average number of times a family visits the park during a year because people often tend to underestimate the number of occurrences of a common event or an event occurring far from the time of the interview. A possible remedy is to request respondents to use written records or to consult with other family members before responding.
2. *Leading questions:* The fashion in which an opinion question is posed may result in a response that does not truly represent the interviewee's opinion. Thus, the survey results may be biased in the direction in which the question is slanted. For example, a question concerning whether the state should impose a large fine on a chemical company for environmental violations is phrased, "Do you support the state fining the chemical company, which is the major employer of people in our community, considering that this fine may result in their moving to another state?" This type of question tends to elicit a "no" response and thus produces a distorted representation of the community's opinion on the imposition of the fine. The remedy is to write questions carefully in an objective fashion.
3. *Unclear wording of questions:* An exercise club attempted to determine the number of times a person exercises per week. The question asked of the respondent was "How many times in the last week did you exercise?" The word *exercise* has different meanings to different individuals. Allowing different definitions of important words or phrases in survey questions greatly reduces the accuracy of survey results. Several remedies are possible. The questions should be tested on a variety of individuals prior to conducting the survey to determine whether there are any confusing or misleading terms in the questions. During the training of the interviewer, all interviewers should have the "correct" definitions of all key words and be advised to provide these definitions to the respondents.

Many other issues, problems, and remedies are provided in the brochures from the ASA.

Data Collection Techniques

Having chosen a particular sample survey design, how do we actually collect the data? The most commonly used methods of data collection in sample surveys are personal interviews and telephone interviews. These methods, with appropriately trained interviewers and carefully planned callbacks, commonly achieve response rates of 60–75% and sometimes even higher. A mailed questionnaire sent to a specific group of interested people can achieve good results, but generally the response rates for this type of data collection are so low that all reported results are suspect. Frequently, objective information can be found from direct observation rather than from an interview or mailed questionnaire.

personal interviews

Data are frequently obtained by **personal interviews.** For example, we can use personal interviews with eligible voters to obtain a sample of public sentiment toward a community bond issue. The procedure usually requires the interviewer to ask prepared questions and to record the respondent's answers. The primary advantage of these interviews is that people will usually respond when confronted in person. In addition, the interviewer can note specific reactions and eliminate misunderstandings about the questions asked. The major limitations of the personal interview (aside from the cost involved) concern the interviewers. If they are not thoroughly trained, they may deviate from the required protocol, thus introducing a bias into the sample data. Any movement, facial expression, or statement by the interviewer can affect the response obtained. For example, a leading question such as "Are you also in favor of the bond issue?" may tend to elicit a positive response. Finally, errors in recording the responses can lead to erroneous results.

telephone interviews

Information can also be obtained from people in the sample through **telephone interviews.** With the competition among telephone service providers, an interviewer can place any number of calls to specified areas of the country relatively inexpensively. Surveys conducted through telephone interviews are frequently less expensive than personal interviews, owing to the elimination of travel expenses. The investigator can also monitor the interviews to be certain that the specified interview procedure is being followed.

A major problem with telephone surveys is that it is difficult to find a list or directory that closely corresponds to the population. Telephone directories have many numbers that do not belong to households, and many households have unlisted numbers. A few households have no phone service, although lack of phone service is now only a minor problem for most surveys in the United States. A technique that avoids the problem of unlisted numbers is random-digit dialing. In this method, a telephone exchange number (the first three digits of a seven-digit number) is selected, and then the last four digits are dialed randomly until a fixed number of households of a specified type are reached. This technique produces samples from the target population and avoids many of the problems inherent in sampling a telephone directory.

Telephone interviews generally must be kept shorter than personal interviews because responders tend to get impatient more easily when talking over the telephone. With appropriately designed questionnaires and trained interviewers, telephone interviews can be as successful as personal interviews.

self-administered
questionnaire

Another useful method of data collection is the **self-administered questionnaire,** to be completed by the respondent. These questionnaires usually are mailed to the individuals included in the sample, although other distribution methods can be used. The questionnaire must be carefully constructed if it is to encourage participation by the respondents.

The self-administered questionnaire does not require interviewers, and thus its use results in savings in the survey cost. This savings in cost is usually bought at the expense of a lower response rate. Nonresponse can be a problem in any form of data collection, but since we have the least contact with respondents in a mailed questionnaire, we frequently have the lowest rate of response. The low response rate can introduce a bias into the sample because the people who answer questionnaires may not be representative of the population of interest. To eliminate some of the bias, investigators frequently contact the nonrespondents through follow-up letters, telephone interviews, or personal interviews.

direct observation

The fourth method for collecting data is **direct observation.** If we were interested in estimating the number of trucks that use a particular road during the 4–6 P.M. rush hours, we could assign a person to count the number of trucks passing a specified point during this period or electronic counting equipment could be used. The disadvantage in using an observer is the possibility of error in observation.

Direct observation is used in many surveys that do not involve measurements on people. The U.S. Department of Agriculture measures certain variables on crops in sections of fields in order to produce estimates of crop yields. Wildlife biologists may count animals, animal tracks, eggs, or nests to estimate the size of animal populations.

A closely related notion to direct observation is that of getting data from objective sources not affected by the respondents themselves. For example, health information can sometimes be obtained from hospital records and income information from employer's records (especially for state and federal government workers). This approach may take more time but can yield large rewards in important surveys.

EXERCISES **Basic Techniques**

Soc. **2.1** An experimenter wants to estimate the average water consumption per family in a city. Discuss the relative merits of choosing individual families, dwelling units (single-family houses, apartment buildings, etc.), and city blocks as sampling units.

H.R. **2.2** An industry consists of many small plants located throughout the United States. An executive wants to survey the opinions of employees on the industry vacation policy. What would you suggest she sample?

Pol. Sci. **2.3** A political scientist wants to estimate the proportion of adult residents of a state who favor a unicameral legislature. What could be sampled? Also, discuss the relative merits of personal interviews, telephone interviews, and mailed questionnaires as methods of data collection.

Bus. **2.4** Two surveys were conducted to measure the effectiveness of an advertising campaign for a low-fat brand of peanut butter. In one of the surveys, the interviewers visited the home and asked whether the low-fat brand was purchased. In the other survey, the interviewers asked the person to show them the peanut butter container when the interviewee stated he or she had purchased low-fat peanut butter.
a. Do you think the two types of surveys will yield similar results on the percentage of households using the product?
b. What types of biases may be introduced into each of the surveys?

Edu. **2.5** *Time* magazine, in a late-1950s article, stated that "the average Yaleman, class of 1924, makes $25,111 a year," which, in today's dollars, would be over $150,000. *Time*'s estimate was based on replies to a sample survey questionnaire mailed to those members of the Yale class of 1924 whose addresses were on file with the Yale administration in late 1950.

  a. What is the survey's population of interest?
  b. Were the techniques used in selecting the sample likely to produce a sample that was representative of the population of interest?
  c. What are the possible sources of bias in the procedures used to obtain the sample?
  d. Based on the sources of bias, do you believe that *Time*'s estimate of the salary of a 1924 Yale graduate in the late 1950s is too high, too low, or nearly the correct value?

2.4 Scientific Studies

A scientific study may be conducted in many different ways. In some studies, the researcher is interested in collecting information from an undisturbed natural process or setting—for example, a study of the differences in reading scores of second-grade students in public, religious, and private schools. In other studies, the scientist is working a highly controlled laboratory, which may be a completely artificial setting for the study. The study of the effect of humidity and temperature on the length of life cycles of ticks would be conducted in a laboratory because it would be impossible to control the humidity or temperature in the tick's natural environment. This control of the factors under study allows the entomologist to obtain results that can then be more easily attributed to differences in the levels of the temperature and humidity because nearly all other conditions have remained constant throughout the experiment. In a natural setting, however, many other factors vary that may also result in changes in the life cycles of the ticks. Thus, the greater the control is in these artificial settings, the less likely the experiment is to portray the true state of nature. A careful balance between control of conditions and depiction of reality must be maintained for the experiments to be useful. In this section and the next one, we will present some standard designs of experiments and methods for analyzing the data obtained from the experiment.

Terminology

designed experiment

A **designed experiment** is an investigation in which a specified framework is provided to observe, measure, and evaluate groups with respect to a designated response. The researcher controls the elements of the framework during the experiment to obtain data from which statistical inferences can provide valid comparisons of the groups of interest.

We use the following example to illustrate the concepts and terminology that will be defined here.

EXAMPLE 2.1

A researcher is studying the conditions under which commercially raised shrimp reach maximum weight gain. Three water temperatures (25°, 30°, 35°) and four water salinity levels (10%, 20%, 30%, 40%) are selected for study. Shrimp are raised in containers with specified water temperatures and salinity levels. The weight gain of the shrimp in each container is recorded after a 6-week study period. There are many other factors that may affect weight gain, such as density of shrimp in the containers, variety of shrimp, size of shrimp, and type of feeding. The experiment is conducted as follows. Twenty-four containers are available for the study. A specific variety and size of shrimp are selected for study. The density of shrimp in the container is fixed at a given amount. One of the three water temperatures and one of the four salinity levels are randomly assigned to each of the 24 containers. All other identifiable conditions are specified to be maintained at the same level for all 24 containers for the duration of the study. In reality there will be some variation in the levels of these variables. After 6 weeks in the tanks, the shrimp are harvested and weighed.

factors
measurements
observations

treatments

completely randomized design
treatment design
factorial treatment design

There are two types of variables in a designed experiment. Controlled variables called **factors** are selected by the researchers for comparison. Response variables are **measurements** or **observations** that are recorded but not controlled by the researcher. Water temperature and salinity level are the control variables or factors in our example, and shrimp weight is the response variable.

The **treatments** in a designed experiment are the conditions constructed from the factors. The factors are selected by examining the questions raised by the research hypothesis. In some experiments, there may only be a single factor and hence the treatments and levels of the factor would be the same. This type of treatment design is referred to as a **completely randomized design.** In most cases we will have several factors, and the treatments are formed by combining levels of the factors. This type of **treatment design** is called a **factorial treatment design.** For the shrimp study described in Example 2.1, there are two factors: water temperature at three levels (25°, 30°, and 35°) and water salinity at four levels (10%, 20%, 30%, and 40%). We can thus create 3(4) = 12 treatments from the combination of levels of the two factors. These factor–level combinations representing the 12 treatments are shown here:

$$(25°, 10\%) \quad (25°, 20\%) \quad (25°, 30\%) \quad (25°, 40\%)$$
$$(30°, 10\%) \quad (30°, 20\%) \quad (30°, 30\%) \quad (30°, 40\%)$$
$$(35°, 10\%) \quad (35°, 20\%) \quad (35°, 30\%) \quad (35°, 40\%)$$

Twenty-four containers were used in Example 2.1, and containers were randomly assigned to the 12 treatments so that there were two containers for each of the 12 treatments.

In other circumstances, there may be a large number of factors and hence the number of treatments may be so large that only a subset of all possible treatments can be examined in the experiment. For example, suppose we were investigating the effect of the following factors on the yield per acre of soybeans: factor 1—five varieties of soybeans; factor 2—three planting densities; factor 3—four levels of fertilization; factor 4—six locations in Texas; and factor 5—three irrigation rates. From the five factors, we can form (5)(3)(4)(6)(3) = 1,080 distinct treatments. This would make for a very large and expensive experiment. In this type of situation, a subset of the 1,080 possible treatments would be selected for studying the relationship between the five factors and the yield of soybeans. This type of experiment is called a **fractional factorial experiment** because only a fraction of the possible treatments are actually used in the experiment. A great deal of care must be taken in selecting which treatments should be used in the experiment so as to be able to answer as many of the researcher's questions as possible.

fractional factorial experiment

control treatment

A special type of treatment is called the **control treatment.** This treatment is the benchmark to which the effectiveness of the remaining treatments are compared. There are three situations in which a control treatment is particularly necessary. First, the conditions under which the experiments are conducted may prevent generally effective treatments from demonstrating their effectiveness. In this case, the control treatment consisting of no treatment may help to demonstrate that the experimental conditions are not allowing the treatments to demonstrate the differences in their effectiveness. For example, an experiment is conducted to determine the most effective level of nitrogen in a garden that is growing tomatoes. If the soil used in the study already has a high level of fertility prior to adding nitrogen to the soil, all levels of nitrogen will appear to be equally effective. However, if a treatment consisting of adding no nitrogen, the control, is used in the study, the high fertility of the soil will be revealed because the control treatment will be as effective as the nitrogen-added treatments.

A second type of control is the standard method treatment to which all other treatments are compared. In this situation, several new procedures are proposed to replace an already existing, well-established procedure. Finally, the placebo control is used when a response may be obtained from the subject simply by the manipulation of the subject during the experiment. For example, a person may demonstrate a temporary reduction in pain level simply by visiting with the physician and having a treatment prescribed. Thus, in evaluating several different methods of reducing pain level in patients, a treatment with no active ingredients, the placebo, is given to a set of patients without the patients' knowledge. The treatments with active ingredients are then compared to the placebo to determine their true effectiveness.

experimental unit   The **experimental unit** is the physical entity to which the treatment is randomly assigned or the subject that is randomly selected from one of the treatment populations. For the shrimp study of Example 2.1, the experimental unit is the container.

Consider another experiment in which a researcher is testing various dose levels (treatments) of a new drug on laboratory rats. If the researcher randomly assigned a single dose of the drug to each rat, then the experimental unit would be the individual rat. Once the treatment is assigned to an experimental unit, a single

replication   **replication** of the treatment has occurred. In general, we will randomly assign several experimental units to each treatment. We will thus obtain several independent observations on any particular treatment and hence will have several replications of the treatments. In Example 2.1, we had two replications of each treatment.

measurement unit   Distinct from the experimental unit is the **measurement unit.** This is the physical entity on which a measurement is taken. In many experiments, the experimental and measurement unit are identical. In Example 2.1, the measurement unit is the container, the same as the experimental unit. If, however, the individual shrimp were measured in each container, the experimental unit would be the container, because the treatments were applied to the containers, but the measurement unit would be the shrimp.

EXAMPLE 2.2

Consider the following experiment. Four types of protective coatings for frying pans are to be evaluated. Five frying pans are randomly assigned to each of the four coatings. A measure of the abrasive resistance of the coatings is taken at three locations on each of the 20 pans. Identify the following items for this study: experimental design, treatments, replications, experimental unit, measurement unit, and total number of measurements.

Solution

Experimental design: Completely randomized design
Treatments: Four types of protective coatings
Replication: There are five frying pans (replications) for each treatment
Experimental unit: Frying pan, because coatings (treatments) are randomly assigned to the frying pans
Measurement unit: Location in the frying pan
Total number of measurements: $4(5)(3) = 60$

experimental error   The term **experimental error** is used to describe the variation in the responses among experimental units that are assigned the same treatment and are observed under the "same" experimental conditions. The reasons that the experimental error is not zero include: (a) the natural differences in the experimental units prior to their receiving the treatment, (b) the variation in the devices that record the measurements, (c) the variation in setting the treatment conditions, and (d) the effect on the response variable of all extraneous factors other than the treatment factors.

EXAMPLE 2.3

Refer to the previously discussed laboratory experiment in which the researcher randomly assigns a single dose of the drug to each rat and measures the level of drug in the rat's bloodstream after 2 hours. For this experiment, the experimental unit and measurement unit are the same: the rat. Identify the four possible sources of experimental error for this study. [See (a) to (d) in the previous paragraph.]

Solution   We can address these sources as follows:

(a) Natural differences in experimental units prior to receiving the treatment. There will be slight physiological differences among rats, so two rats receiving the exact same dose level (treatment) will have slightly different blood levels after 2 hours.

(b) Variation in the devices used to record the measurement. There will be differences in the responses due to the method by which the quantity of the drug in the rat is determined by the laboratory technician. If several determinations of drug level were made in the blood of the same rat, there may be differences in the amount of drug found due to equipment variation, technician variation, or conditions in the laboratory variation.

(c) Variation in setting the treatment conditions. If there is more than one replication per treatment, the treatment may not be the same from one rat to another. Suppose, for example, that we had ten replications of each dose (treatment); it is highly unlikely that each of the ten rats receives exactly the same dose of drug specified by the treatment. There could be slightly different amounts of drug in the syringes and slightly different amounts could be injected and enter the bloodstream.

(d) The effect on the response (blood level) of all extraneous factors other than the treatment factors. Presumably, the rats are all placed in cages and given the same amount of food and water prior to determining the amount of drug in their blood. However, the temperature, humidity, external stimulation, and other conditions may be somewhat different in the cages. This may have an effect on the responses of the rats.

Thus, these differences and variation in the external conditions within the laboratory during the experiment all contribute to the size of the experimental error in the experiment.

EXAMPLE 2.4

Refer to Example 2.1. Suppose that each treatment is assigned to two containers and that 40 shrimp are placed in the containers. After 6 weeks, the individual shrimp are weighed. Identify the experimental unit, measurement unit, factors, treatments, number of replications, and possible sources of experimental error.

Solution   This is a factorial treatment design with two factors: temperature and salinity level. The treatments are constructed by selecting a temperature and salinity level to be assigned to a particular container. We have a total of $(3)(4) = 12$ possible treatments for this experiment. The 12 treatments are as follows:

(25°, 10%)	(25°, 20%)	(25°, 30%)	(25°, 40%)
(30°, 10%)	(30°, 20%)	(30°, 30%)	(30°, 40%)
(35°, 10%)	(35°, 20%)	(35°, 30%)	(35°, 40%)

We next randomly assigned two containers to each of the 12 treatments. This results in two replications of each treatment. The experimental unit is the container because the treatments were randomly assigned to individual containers. Forty shrimp were placed in the containers, and after 6 weeks the weights of the individual shrimp were recorded. The measurement unit is the individual shrimp because this is the physical entity on which an observation was made. Thus, in this experiment, the experimental and measurement units are different. Several possible sources of experimental error are the difference in the weights of the shrimp prior to being placed in the container, the accuracy at which the temperature and salinity levels are maintained over the 6-week study period, the accuracy with which the shrimp are weighed at the conclusion of the study, the consistency of the amount of food fed to the shrimp (was each shrimp given exactly the same quantity of food over the 6 weeks?), and the variation in any other conditions that may affect shrimp growth.

2.5   Experimental Designs for Scientific Studies

The subject of experimental designs for scientific studies cannot be given much justice at the beginning of a statistical methods course—entire courses at the undergraduate and graduate levels are needed for a comprehensive understanding of the methods and concepts of experimental design. Even so, we will attempt to give you a brief overview of the subject because much data requiring summarizing and analysis arise from scientific studies involving one of a number of experimental designs. We will work by way of examples.

A consumer testing agency decides to evaluate the wear characteristics of four major brands of tires. For this study, the agency selects four cars of a standard car model and four tires of each brand. The tires will be placed on the cars and then driven 30,000 miles on a 2-mile racetrack. The decrease in tread thickness over the 30,000 miles is the variable of interest in this study. Four different drivers will drive the cars, but the drivers are professional drivers with comparable training and experience. The weather conditions, smoothness of track, and the maintenance of the four cars will be essentially the same for all four brands over the study period. All extraneous factors that may affect the tires are nearly the same for all four brands. Thus, the testing agency feels confident that if there is a difference in wear characteristics among the brands at the end of the study, then this is truly a difference in the four brands and not a difference due to the manner in which the study was conducted. The testing agency is interested in recording other factors, such as the cost of the tires, the length of warranty offered by the manufacturer, whether the tires go out of balance during the study, and the evenness of wear across the width of the tires. In this example, we will only consider tread wear. There should be a recorded tread wear for each of the sixteen tires, four tires for each brand. The methods presented in Chapters 8 and 9 will be used to summarize and analyze the sample tread-wear data in order to make comparisons (inferences) among the four tire brands. One possible inference of interest could be the selection of the brand having minimum tread wear. Can the best-performing tire brand in the sample data be expected to provide the best tread wear if the same study is repeated? Are the results of the study applicable to the driving habits of the typical motorist?

Experimental Designs

There are many ways in which the tires can be assigned to the four cars. We will consider one running of the experiment in which we have four tires of each of the four brands. First, we need to decide how to assign the tires to the cars. We could randomly assign a single brand to each car, but this would result in a design having the unit of measurement the total loss of tread for all four tires on the car and not the individual tire loss. Thus, we must randomly assign the sixteen tires to the four cars. One possible arrangement of the tires on the cars is shown in Table 2.1.

In general, a completely randomized design is used when we are interested in comparing t treatments (in our case, $t = 4$; the treatments are brand of tire). For each treatment, we obtain a sample of observations. The sample sizes could be different for the individual treatments. For example, we could test 20 tires from Brands A, B, and C but only 12 tires from Brand D. The sample of observations from a treatment is assumed to be the result of a simple random sample of observations from the hypothetical population of possible values that could have resulted from that treatment. In our example, the sample of four tire-wear thicknesses from Brand A was considered to be the outcome of a simple random sample of four observations selected from the hypothetical population of possible tire-wear thicknesses for standard model cars traveling 30,000 miles using Brand A.

The experimental design could be altered to accommodate the effect of a variable related to how the experiment is conducted. In our example, we assumed

TABLE 2.1

Completely randomized design of tire wear

Car 1	Car 2	Car 3	Car 4
Brand B	Brand A	Brand A	Brand D
Brand B	Brand A	Brand B	Brand D
Brand B	Brand C	Brand C	Brand D
Brand C	Brand C	Brand A	Brand D

that the effect of the different cars, weather, drivers, and various other factors was the same for all four brands. Now, if the wear on tires imposed by Car 4 was less severe than that of the other three cars, would our design take this effect into account? Because Car 4 had all four tires of Brand D placed on it, the wear observed for Brand D may be less than the wear observed for the other three brands because all four tires of Brand D were on the "best" car. In some situations, the objects being observed have existing differences prior to their assignment to the treatments. For example, in an experiment evaluating the effectiveness of several drugs for reducing blood pressure, the age or physical condition of the participants in the study may decrease the effectiveness of the drug. To avoid masking the effectiveness of the drugs, we want to take these factors into account. Also, the environmental conditions encountered during the experiment may reduce the effectiveness of the treatment.

randomized block design

In our example, we want to avoid having the comparison of the tire brands distorted by the differences in the four cars. The experimental design used to accomplish this goal is called a **randomized block design** because we want to "block" out any differences in the four cars to obtain a precise comparison of the four brands of tires. In a randomized block design, each treatment appears in every block. In the blood pressure example, we would group the patients according to the severity of their blood pressure problem and then randomly assign the drugs to the patients within each group. Thus, the randomized block design is similar to a stratified random sample used in surveys. In the tire-wear example, we would use the four cars as the blocks and randomly assign one tire of each brand to each of the four cars, as shown in Table 2.2. Now, if there are any differences in the cars that may affect tire wear, that effect will be equally applied to all four brands.

What happens if the position of the tires on the car affects the wear on the tire? The positions on the car are right front (RF), left front (LF), right rear (RR), and left rear (LR). In Table 2.2, suppose that all four tires from Brand A are placed on the RF position, Brand B on RR, Brand C on LF, and Brand D on LR. Now, if the greatest wear occurs for tires placed on the RF, then Brand A would be at a great disadvantage when compared to the other three brands. In this type of situation we would state that the effect of brand and the effect of position on the car were confounded; that is, using the data in the study, the effects of two or more factors cannot be unambiguously attributed to a single factor. If we observe a large difference in the average wear among the four brands, is this difference due to differences in the brands or differences due to the position of the tires on the car? Using the design given in Table 2.2, this question cannot be answered. Thus, we now need two blocking variables: the "car" the tire is placed on and the "position" on the car. A design having two blocking variables is called a **Latin square design.** A Latin square design for our example is shown in Table 2.3.

Latin square design

TABLE 2.2
Randomized block design of tire wear

Car 1	Car 2	Car 3	Car 4
Brand A	Brand A	Brand A	Brand A
Brand B	Brand B	Brand B	Brand B
Brand C	Brand C	Brand C	Brand C
Brand D	Brand D	Brand D	Brand D

TABLE 2.3
Latin square design
of tire wear

Position	Car 1	Car 2	Car 3	Car 4
RF	Brand A	Brand B	Brand C	Brand D
RR	Brand B	Brand C	Brand D	Brand A
LF	Brand C	Brand D	Brand A	Brand B
LR	Brand D	Brand A	Brand B	Brand C

Note that with this design, each brand is placed in each of the four positions and on each of the four cars. Thus, if position or car has an effect on the wear of the tires, the position effect and/or car effect will be equalized across the four brands. The observed differences in wear can now be attributed to differences in the brand of the car.

The randomized block and Latin square designs are both extensions of the completely randomized design in which the objective is to compare *t* treatments. The analysis of data collected according to a completely randomized design and the inferences made from such analysis are discussed further in Chapters 8 and 9.

Factorial Treatment Structure in a Completely Randomized Design

In this section, we will discuss how treatments are constructed from several factors rather than just being *t* levels of a single factor. These types of experiments are involved with examining the effect of two or more independent variables on a response variable *y*. For example, suppose a company has developed a new adhesive for use in the home and wants to examine the effects of temperature and humidity on the bonding strength of the adhesive. Several treatment design questions arise in any study. First, we must consider what factors (independent variables) are of greatest interest. Second, the number of levels and the actual settings of these levels must be determined for each factor. Third, having separately selected the levels for each factor, we must choose the factor–level combinations (treatments) that will be applied to the experimental units.

The ability to choose the factors and the appropriate settings for each of the factors depends on budget, time to complete the study, and most important, the experimenter's knowledge of the physical situation under study. In many cases, this will involve conducting a detailed literature review to determine the current state of knowledge in the area of interest. Then, assuming that the experimenter has chosen the levels of each independent variable, he or she must decide which factor–level combinations are of greatest interest and are viable. In some situations, certain factor–level combinations will not produce an experimental setting that can elicit a reasonable response from the experimental unit. Certain combinations may not be feasible due to toxicity or practicality issues.

one-at-a-time approach

One approach for examining the effects of two or more factors on a response is the **one-at-a-time approach.** To examine the effect of a single variable, an experimenter changes the levels of this variable while holding the levels of the other independent variables fixed. This process is continued for each variable while holding the other independent variables constant. Suppose that an experimenter is interested in examining the effects of two independent variables, nitrogen and phosphorus, on the yield of a crop. For simplicity we will assume two levels of each variable have been selected for the study: 40 and 60 pounds per plot for nitrogen, and 10 and

TABLE 2.4

Factor–level combinations for a one-at-a-time approach

Combination	Nitrogen	Phosphorus
1	60	10
2	40	10
3	40	20

FIGURE 2.1

Factor–level combinations for a one-at-a-time approach

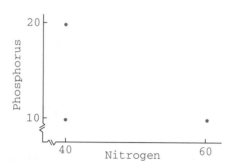

20 pounds per plot for phosphorus. For this study the experimental units are small, relatively homogeneous plots that have been partitioned from the acreage of a farm. For our experiment the factor–level combinations chosen might be as shown in Table 2.4. These factor–level combinations are illustrated in Figure 2.1.

From the graph in Figure 2.1, we see that there is one difference that can be used to measure the effects of nitrogen and phosphorus separately. The difference in response for combinations 1 and 2 would estimate the effect of nitrogen; the difference in response for combinations 2 and 3 would estimate the effect of phosphorus.

Hypothetical yields corresponding to the three factor–level combinations of our experiment are given in Table 2.5. Suppose the experimenter is interested in using the sample information to determine the factor–level combination that will give the maximum yield. From the table, we see that crop yield increases when the nitrogen application is increased from 40 to 60 (holding phosphorus at 10). Yield also increases when the phosphorus setting is changed from 10 to 20 (at a fixed nitrogen setting of 40). Thus, it might seem logical to predict that increasing both the nitrogen and phosphorus applications to the soil will result in a larger crop yield. The fallacy in this argument is that our prediction is based on the assumption that the effect of one factor is the same for both levels of the other factor.

We know from our investigation what happens to yield when the nitrogen application is increased from 40 to 60 for a phosphorus setting of 10. But will the yield also increase by approximately 20 units when the nitrogen application is changed from 40 to 60 at a setting of 20 for phosphorus?

TABLE 2.5

Yields for the three factor–level combinations

Observation (yield)	Nitrogen	Phosphorus
145	60	10
125	40	10
160	40	20
?	60	20

To answer this question, we could apply the factor–level combination of 60 nitrogen–20 phosphorus to another experimental plot and observe the crop yield. If the yield is 180, then the information obtained from the three factor–level combinations would be correct and would have been useful in predicting the factor–level combination that produces the greatest yield. However, suppose the yield obtained from the high settings of nitrogen and phosphorus turns out to be 110. If this happens, the two factors nitrogen and phosphorus are said to **interact.** That is, the effect of one factor on the response does not remain the same for different levels of the second factor, and the information obtained from the one-at-a-time approach would lead to a faulty prediction.

interaction

The two outcomes just discussed for the crop yield at the 60–20 setting are displayed in Figure 2.2, along with the yields at the three initial design points. Figure 2.2 (a) illustrates a situation with no interaction between the two factors; the effect of nitrogen on yield is the same for both levels of phosphorus. In contrast, Figure 2.2 (b) illustrates a case in which the two factors nitrogen and phosphorus do interact.

We have seen that the one-at-a-time approach to investigating the effect of two factors on a response is suitable only for situations in which the two factors do not interact. Although this was illustrated for the simple case in which two factors were investigated at each of two levels, the inadequacies of a one-at-a-time approach are even more salient when we investigate the effects of more than two factors on a response.

factorial treatment structures

Factorial treatment structures are useful for examining the effects of two or more factors on a response *y*, whether or not interaction exists. As before, the choice of the number of levels of each variable and the actual settings of these variables is important. However, assuming that we have made these selections with help

FIGURE 2.2

Yields of the three design points and possible yield at a fourth design point

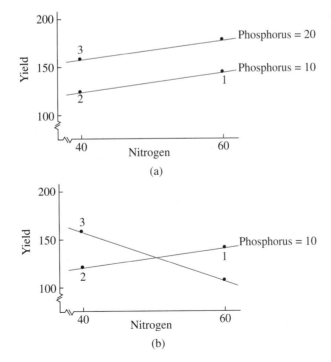

TABLE 2.6

2 × 2 factorial treatment structure for crop yield

x_1	x_2
40	10
40	20
60	10
60	20

TABLE 2.7

2 × 3 factorial treatment structure for crop yield

x_1	x_2
40	10
40	15
40	20
60	10
60	15
60	20

from an investigator knowledgeable in the area being examined, we must decide at what factor–level combinations we will observe y.

Using our previous example, if we are interested in examining the effect of two levels of nitrogen x_1 at 40 and 60 pounds per plot and two levels of phosphorus x_2 at 10 and 20 pounds per plot on the yield of a crop, we can use a completely randomized design where the four factor–level combinations (treatments) of Table 2.6 are assigned at random to the experimental units.

Similarly, if we wish to examine x_1 at the two levels 40 and 60 and x_2 at the three levels 10, 15, and 20, we could use the six factor–level combinations of Table 2.7 as treatments in a completely randomized design.

EXAMPLE 2.5

An auto manufacturer is interested in examining the effect of engine speed x_1, measured in revolutions per minute, and ground speed x_2, measured in miles per hour, on gasoline mileage. The investigators, in consultation with company mechanics and other personnel, decide to consider settings of x_1 at 800, 1,000, and 1,200 and settings of x_2 at 30, 50, and 70. Give the factor–level combinations to be used in a 3 × 3 factorial treatment structure.

Solution   Using all combinations of the factors, we observe gasoline mileage at the following settings of x_1 and x_2:

x_1	800	800	800	1,000	1,000	1,000	1,200	1,200	1,200
x_2	30	50	70	30	50	70	30	50	70

The examples of factorial treatment structures presented in this section have concerned two independent variables. However, the procedure applies to any number of factors and levels per factor. Thus, if we had four different factors x_1, x_2, x_3, and x_4 at two, three, three, and four levels, respectively, we could formulate a 2 × 3 × 3 × 4 factorial treatment structure by considering all $2 \cdot 3 \cdot 3 \cdot 4 = 72$ factor–level combinations. The analysis of data obtained from factorial treatment structures in various experimental designs will be discussed in Chapter 9.

EXERCISES

Engin.   **2.6**   Researchers ran a quality control study to evaluate the quality of plastic irrigation pipes. The study design involved a total of 24 pipes, with 12 pipes randomly selected from each of two manufacturing plants. The pipes were manufactured using one of two water temperatures and one of three types of hardeners. The compressive strength of each pipe was determined for analysis. The experimental conditions are as follows:

Pipe No.	Plant	Temperature (°F)	Hardener
1	1	200	H_1
2	1	175	H_2
3	2	200	H_1
4	2	175	H_2
5	1	200	H_1
6	1	175	H_2
7	2	200	H_1
8	2	175	H_2
9	1	200	H_3
10	1	175	H_3
11	2	200	H_3
12	2	175	H_3
13	1	200	H_3
14	1	175	H_3
15	2	200	H_3
16	2	175	H_3
17	1	200	H_2
18	1	175	H_1
19	2	200	H_2
20	2	175	H_1
21	1	200	H_2
22	1	175	H_1
23	2	200	H_2
24	2	175	H_1

Identify each of the following components of the experimental design.
- **a.** factors
- **b.** factor levels
- **c.** blocks
- **d.** experimental unit
- **e.** measurement unit
- **f.** replications
- **g.** treatments

2.7 In each of the following descriptions of experiments, identify the important features of each design. Include as many of the components from Exercise 2.6 as needed to adequately describe the design.
- **a.** A horticulturalist is measuring the vitamin C concentration in oranges in an orchard on a research farm in south Texas. He is interested in the variation in vitamin C concentration across the orchard, across the productive months, and within each tree. He divides the orchard into eight sections and randomly selects a tree from each section during October through May, the months in which the trees are in production. During each month, from eight trees he selects 10 oranges near the top of the tree, 10 oranges near the middle of the tree, and 10 oranges near the bottom of the tree. The horticulturalist wants to monitor the vitamin C concentration across the productive season and determine whether there is a substantial difference in vitamin C concentration in oranges at various locations in the tree.
- **b.** A medical specialist wants to compare two different treatments (T_1, T_2) for treating a particular illness. She will use eight hospitals for the study. She believes there may be differences in the response among hospitals. Each hospital has four wards of patients. She will

randomly select four patients in each ward to participate in the study. Within each hospital, two wards are randomly assigned to get T_1; the other two wards will receive T_2. All patients in a ward will get the same treatment. A single response variable is measured on each patient.

c. In the design described in (b) make the following change. Within each hospital, the two treatments will be randomly assigned to the patients, with two patients in each ward receiving T_1 and two patients receiving T_2.

d. An experiment is planned to compare three types of schools—public, private-non-parochial, and parochial—all with respect to the reading abilities of students in sixth-grade classes. The researcher selects two large cities in each of five geographical regions of the United States for the study. In each city, she randomly selects one school of each of the three types and randomly selects a single sixth-grade class within each school. The scores on a standardized test are recorded for each of 20 students in each classroom. The researcher is concerned about differences in family income levels among the 30 schools, so she obtains the family income for each of the students who participated in the study.

Bio. **2.8** A research specialist for a large seafood company plans to investigate bacterial growth on oysters and mussels subjected to three different storage temperatures. Nine cold-storage units are available. She plans to use three storage units for each of the three temperatures. One package of oysters and one package of mussels will be stored in each of the storage units for 2 weeks. At the end of the storage period, the packages will be removed and the bacterial count made for two samples from each package. The treatment factors of interest are temperature (levels: 0, 5, 10°C) and seafood (levels: oysters, mussels). She will also record the bacterial count for each package prior to placing seafood in the cooler. Identify each of the following components of the experimental design.

a. factors
b. factor levels
c. blocks
d. experimental unit
e. measurement unit
f. replications
g. treatments

2.9 In each of the following situations, identify whether the design is a completely randomized design, randomized block design, or Latin square. If there is a factorial structure of treatments, specify whether it has a two-factor or three-factor structure. If the experiment's measurement unit is different from the experimental unit, identify both.

a. The 48 treatments comprised 3, 4, and 4 levels of fertilizers N, P, and K, respectively, in all possible combinations. Five peanut farms were randomly selected and the 48 treatments assigned at random at each farm to 48 plots of peanut plants.

b. Ten different software packages were randomly assigned to 30 graduate students. The time to complete a specified task was determined.

c. Four different glazes are applied to clay pots at two different thicknesses. The kiln used in the glazing can hold eight pots at a time, and it takes 1 day to apply the glazes. The experimenter wants eight replications of the experiment. Because the conditions in the kiln vary somewhat from day to day, the experiment is conducted over an 8-day period. Each combination of a thickness and type of glaze is randomly assigned to one pot in the kiln each day.

2.10 A colleague has approached you for help with an experiment she is conducting. The experiment consists of asking a sample of consumers to taste five different recipes for meat loaf. When a consumer tastes a sample he or she will give scores to several characteristics and these scores will be combined into a single overall score. Hence, there will be one value for each recipe for a consumer. The literature indicates that in this kind of study some consumers tend to give low scores to all samples; others tend to give high scores to all samples.

a. There are two possible experimental designs. Design A would use a random sample of 100 consumers. From this group, 20 would be randomly assigned to each of the five recipes, so that each consumer tastes only one recipe. Design B would use a random sample of 100 consumers, with each consumer tasting all five recipes, the recipes being presented in a random order for each consumer. Which design would you recommend? Justify your answer.

b. When asked how the experiment is going, the researcher replies that one recipe smelled so bad that she eliminated it from the analysis. Is this a problem for the analysis if design B was used? Why or why not? Is it a problem if design A was used? Why or why not?

2.6 Observational Studies

observational study

Before leaving the subject of sample data collection, we will draw a distinction between an **observational study** and a scientific study. In experimental designs for scientific studies, the observation conditions are fixed or controlled. For example, with a factorial treatment structure laid off in a completely randomized design, an observation is made at each factor–level combination. Similarly, with a randomized block design, an observation is obtained on each treatment in every block. These "controlled" studies are very different from observational studies, which are sometimes used because it is not feasible to do a proper scientific study. This can be illustrated using an example.

Much research and public interest centers on the effect of cigarette smoking on lung cancer and cardiovascular disease. One possible experimental design would be to randomize a fixed number of individuals (say, 1,000) to each of two groups— one group would be required to smoke cigarettes for the duration of the study (say, 10 years), while those in the second group would not be allowed to smoke throughout the study. At the end of the study, the two groups would be compared for lung cancer and cardiovascular disease. Even if we ignore ethical questions, this type of study would be impossible to do. Because of the long duration, it would be difficult to follow all participants and make certain that they followed the study plan. It would also be difficult to find nonsmoking individuals willing to take the chance of being assigned to the smoking group.

Another possible study would be to sample a fixed number of smokers and a fixed number of nonsmokers to compare the groups for lung cancer and for cardiovascular disease. Assuming we could obtain willing groups of participants, this study could be done in a *much shorter* period of time.

What has been sacrificed? Well, the fundamental difference between an observational study and a scientific study lies in the inferences(s) that can be drawn. For a scientific study comparing smokers to nonsmokers, assuming the two groups of individuals followed the study plan, the observed differences between the smoking and nonsmoking groups could be attributed to the effects of cigarette smoking because individuals were randomized to the two groups; hence, the groups were assumed to be comparable at the outset.

This type of reasoning does not apply to the observational study of cigarette smoking. Differences between the two groups in the observation could not necessarily be attributed to the effects of cigarette smoking because, for example, there may be hereditary factors that predispose people to smoking and lung cancer and/or cardiovascular disease. Thus, differences between the groups might be due to hereditary factors, smoking, or a combination of the two. Typically, the results of an observational study are reported by way of a statement of association. For our example, if the observational study showed a higher frequency of lung cancer and cardiovascular disease for smokers relative to nonsmokers, it would be stated that this study showed that cigarette smoking was associated with an increased frequency of lung cancer and cardiovascular disease. It is a careful rewording; we cannot infer that cigarette smoking *causes* lung cancer and cardiovascular disease.

Often, however, an observational study is the only type of study that can be run. Our job is to make certain that we understand the type of study run and, hence, understand how the data were collected. Then we can critique inferences drawn from an analysis of the study data.

2.7 Data Management: Preparing Data for Summarizing and Analysis (Optional)

In this section, we concentrate on some important data management procedures that are followed between the time the data are gathered and the time they are available in computer-readable form for analysis. This is not a complete manual with all tools required; rather, it is an overview—what a manager should know about these steps. As an example, this section reflects standard procedures in the pharmaceutical industry, which is highly regulated. Procedures may differ somewhat in other industries and settings.

We begin with a discussion of the procedures involved in processing data from a study. In practice, these procedures may consume 75% of the total effort from the receipt of the raw data to the presentation of results from the analysis. What are these procedures, why are they so important, and why are they so time consuming?

To answer these questions, let's list the major data-processing procedures in the cycle, which begins with receipt of the data and ends when the statistical analysis begins. Then we'll discuss each procedure separately.

Procedures in Processing Data for Summarization and Analysis

1. Receive the raw data source.
2. Create the database from the raw data source.
3. Edit the database.
4. Correct and clarify the raw data source.
5. Finalize the database.
6. Create data files from the database.

raw data source

1. *Receiving the raw data source.* For each study that is to be summarized and analyzed, the data arrive in some form, which we'll refer to as the **raw data source.** For a clinical trial, the raw data source is usually case report forms, sheets of $8\frac{1}{2} \times 11$-inch paper that have been used to record study data for each patient entered into the study. For other types of studies, the raw data source may be sheets of paper from a laboratory notebook, a magnetic tape (or other form of machine-readable data), compact disks, hand tabulations, and so on.

data trail

It is important to retain the raw data source, because it is the beginning of the **data trail,** which leads from the raw data to the conclusions drawn from a study. Many consulting operations involved with the analysis and summarization of many different studies keep a log that contains vital information related to the study and raw data source. In a regulated environment such as the pharmaceutical industry, we may have to redo or reproduce data and data analyses based on previous work.

Other situations outside the pharmaceutical industry may also require a retrospective review of what was done in the analysis of a study. In these situations, the study log can be an invaluable source of study information. General information contained in a study log is shown next.

Log for Study Data

1. Date received and from whom
2. Study investigator
3. Statistician (and others) assigned to study
4. Brief description of study
5. Treatments (compounds, preparations, etc.) studied
6. Raw data source
7. Response(s) measured and how measured
8. Reference number for study
9. Estimated (actual) completion date
10. Other pertinent information

Later, when the study has been analyzed and results have been communicated, additional information can be added to the log on how the study results were communicated, where these results are recorded, what data files have been saved, and where these files are stored.

2. *Creating the database from the raw data source.* For most studies that are scheduled for a statistical analysis, a machine-readable database is created. The steps taken to create the database and its eventual form vary from one operation to another, depending on the software systems to be used in the statistical analysis. However, we can give a few guidelines based on the form of the entry system.

When data are to be entered at a terminal, the raw data are first checked for legibility. Any illegible numbers or letters or other problems should be brought to the attention of the study coordinator. Then a coding guide that assigns column numbers and variable names to the data is filled out. Certain codes for missing values (for example, not available) are also defined here. Also, it is helpful to give a brief description of each variable. The data file keyed in at the terminal is referred to as the **machine-readable database.** A listing (printout) of the database should be obtained and checked carefully against the raw data source. Any errors should be corrected at the terminal and verified against an updated listing.

machine-readable database

Sometimes data are received in machine-readable form. For these situations, the magnetic tape or disk file is considered to be the database. However, you must have a coding guide to "read" the database. Using the coding guide, obtain a listing of the database and check it *carefully* to see that all numbers and characters look reasonable and that the proper formats were used to create the file. Any problems that arise must be resolved before proceeding further.

Some data sets are so small that it is not necessary to create a machine-readable data file from the raw data source. Instead, calculations are performed by hand or the data are entered into an electronic calculator. For these situations, check any calculations to see that they make sense. Don't believe everything you see; redoing the calculations is not a bad idea.

3. *Editing the database.* The types of edits done and the completeness of the editing process really depend on the type of study and how concerned you are about the accuracy and completeness of the data prior to the analysis. For example, in using a statistical software package (such as SAS or Minitab), it is wise to examine the minimum, maximum, and frequency distribution for each variable to make certain nothing looks unreasonable.

Certain other checks should be made. Plot the data and look for problems. Also, certain **logic checks** should be done depending on the structure of the data. If, for example, data are recorded for patients at several different visits, then the data recorded for visit 2 can't be earlier than the data for visit 1; similarly, if a patient is lost to follow-up after visit 2, there should not be any data for that patient at later visits.

For small data sets, the data edits can be done by hand, but for large data sets the job may be too time consuming and tedious. If machine editing is required, look for a software system that allows the user to specify certain data edits. Even so, for more complicated edits and logic checks it may be necessary to have a customized edit program written to machine edit the data. This programming chore can be a time-consuming step; plan for this well in advance of the receipt of the data.

4. *Correcting and clarifying the raw data source.* Questions frequently arise concerning the legibility or accuracy of the raw data during any one of the steps from the receipt of the raw data to the communication of the results from the statistical analysis. We have found it helpful to keep a list of these problems or discrepancies in order to define the data trail for a study. If a correction to (or clarification of) the raw data source is required, indicate this on the form and make the appropriate change to the raw data source. If no correction is required, indicate this on the form as well. Keep in mind that the machine-readable database should be changed to reflect any changes made to the raw data source.

5. *Finalizing the database.* You may have been led to believe that all data for a study arrive at one time. This, of course, is not always the case. For example, with a marketing survey, different geographic locations may be surveyed at different times and, hence, those responsible for data processing do not receive all the data at one time. All these subsets of data, however, must be processed through the cycles required to create, edit, and correct the database. Eventually the study is declared complete and the data are processed into the database. At this time, the database should be reviewed again and final corrections made before beginning the analysis because for large data sets, the analysis and summarization chores take considerable human labor and computer time. It's better to agree on a final database analysis than to have to repeat all the analyses on a changed database at a later date.

6. *Creating data files from the database.* Generally there are one or two sets of data files created from the machine-readable database. The first set, referred to as **original files,** reflects the basic structure of the database. A listing of the files is checked against the database listing to verify that the variables have been read with correct formats and missing value codes have been retained. For some studies, the original files are actually used for editing the database.

A second set of data files, called **work files,** may be created from the original files. Work files are designed to facilitate the analysis. They may require restructuring of the original files, a selection of important variables, or the creation or addition of new variables by insertion, computation, or transformation. A listing of the work files is checked against that of the original files to ensure proper restructuring and variable selection. Computed and transformed variables are checked using hand calculations to verify the program code.

If original and work files are SAS data sets, you should use the documentation features provided by SAS. At the time an SAS data set is created, a descriptive label for the data set, up to 40 characters, should be assigned. The label can be

logic checks

original files

work files

stored with the data set and imprinted wherever the contents procedure is used to print the data set's contents. All variables can be given descriptive names, up to 8 characters in length, that are meaningful to those involved in the project. In addition, variable labels up to 40 characters in length can be used to provide additional information. Title statements can be included in the SAS code to identify the project and describe each job. For each file, a listing (proc print) and a dictionary (proc contents) can be retained.

For files created from the database using other software packages, you should use the labeling and documentation features available in the computer program.

Even if appropriate statistical methods are applied to data, the conclusions drawn from the study are only as good as the data on which they are based—so you be the judge. The amount of time you should spend on these data-processing chores before analysis really depends on the nature of the study, the quality of the raw data source, and how confident you want to be about the completeness and accuracy of the data.

2.8 Staying Focused

The first step in learning from data involves defining the problem. This was discussed in Chapter 1. Next we discussed intelligent data gathering (Step 2), which involves specifying the objectives of the data-gathering exercise, identifying the variables of interest, and choosing an appropriate design for the survey or scientific study. In this chapter, we discussed various survey designs and experimental designs for scientific studies. Armed with a basic understanding of some design considerations for conducting surveys or scientific studies, you can address how to collect data on the variables of interest in order to address the stated objectives of the data-gathering exercise.

We also drew a distinction between observational and scientific studies in terms of the inferences (conclusions) that can be drawn from the sample data. Differences found between treatment groups from an observational study are said to be *associated with* the use of the treatments; on the other hand, differences found between treatments in a scientific study are said to be *due to* the treatments. In the next chapter, we will examine the methods for summarizing the data we collect.

CHAPTER 3

Summarizing Data

3.1 Introduction

In the previous chapter, we discussed how to gather data intelligently for an experiment or survey, Step 2 in learning from data. We turn now to Step 3, summarizing the data.

The field of statistics can be divided into two major branches: descriptive statistics and inferential statistics. In both branches, we work with a set of measurements. For situations in which data description is our major objective, the set of measurements available to us is frequently the entire population. For example, suppose that we wish to describe the distribution of annual incomes for all families registered in the 2000 census. Because all these data are recorded and are available on computer tapes, we do not need to obtain a random sample from the population; the complete set of measurements is at our disposal. Our major problem is in organizing, summarizing, and describing these data—that is, making sense of the data. Similarly, vast amounts of monthly, quarterly, and yearly data of medical costs are available for the managed health care industry (HMOs). These data are broken down by type of illness, age of patient, inpatient or outpatient care, prescription costs, and out-of-region reimbursements, along with many other types of expenses. However, in order to present such data in formats useful to HMO managers, congressional staffs, doctors, and the consuming public, it is necessary to organize, summarize, and describe the data. Good descriptive statistics enable us to make sense of the data by reducing a large set of measurements to a few summary measures that provide a good, rough picture of the original measurements.

In situations in which we are concerned with statistical inference, a sample is usually the only set of measurements available to us. We use information in the sample to draw conclusions about the population from which the sample was drawn. Of course, in the process of making inferences, we also need to organize, summarize, and describe the sample data.

For example, the tragedy surrounding isolated incidents of product tampering has brought about federal legislation requiring tamper-resistant packaging for certain drug products sold over the counter. These same incidents also brought about increased industry awareness of the need for rigid standards of product and packaging quality that must be maintained while delivering these products to the store shelves. In particular, one company is interested in determining the proportion of packages out of total production that are improperly sealed or have been damaged in transit. Obviously, it would be impossible to inspect all packages at all stores where the product is sold, but a random sample of the production could be obtained, and the proportion defective in the sample could be used to estimate the actual proportion of improperly sealed or damaged packages.

Similarly, in developing an economic forecast of new housing starts for the next year, it is necessary to use sample data from various economic indicators to make such a prediction (inference).

A third situation involves an experiment in which a food scientist wants to study the effect of two factors on the specific volume of bread loaves. The factors are type of fat and type of surfactant. (A surfactant is a substance that is mixed into the bread dough to lower the surface tension of the dough and thus produce loaves with increased specific volumes.) The experiment involves mixing a type of fat with a type of surfactant into the bread dough prior to baking the bread. The specific volume of the bread is then measured. This experiment is repeated several times for each of the fat–surfactant combinations. In this experiment, the scientist wants to make inferences from the results of the experiment to the commercial production of bread. In many such experiments, the use of proper graphical displays adds to the insight the scientist obtains from the data.

Whether we are describing an observed population or using sampled data to draw an inference from the sample to the population, an insightful description of the data is an important step in drawing conclusions from it. No matter what our objective, statistical inference or population description, we must first adequately describe the set of measurements at our disposal.

The two major methods for describing a set of measurements are graphical techniques and numerical descriptive techniques. Section 3.3 deals with graphical methods for describing data on a single variable. In Sections 3.4, 3.5, and 3.6, we discuss numerical techniques for describing data. The final topics on data description are presented in Section 3.7, in which we consider a few techniques for describing (summarizing) data on more than one variable.

3.2 Calculators, Computers, and Software Systems

Electronic calculators can be great aids in performing some of the calculations mentioned later in this chapter, especially for small data sets. For larger data sets, even hand-held calculators are of little use because of the time required to enter data. A computer can help in these situations. Specific programs or more general software systems can be used to perform statistical analyses almost instantaneously even for very large data sets after the data are entered into the computer from a terminal, a magnetic tape, or disk storage. It is not necessary to know computer programming to make use of specific programs or software systems for planned analyses—most have user's manuals that give detailed directions for their use. Others, developed for use at a terminal, provide program prompts that lead the user through the analysis of choice.

Many statistical software packages are available for use on computers. Three of the more commonly used systems are Minitab, SAS, and SPSS. Each is available in a mainframe version as well as in a personal computer version. Because a software system is a group of programs that work together, it is possible to obtain plots, data descriptions, and complex statistical analyses in a single job. Most people find that they can use any particular system easily, although they may be frustrated by minor errors committed on the first few tries. The ability of such packages to perform complicated analyses on large amounts of data more than repays the initial investment of time and irritation.

In general, to use a system you need to learn about only the programs in which you are interested. Typical steps in a job involve describing your data to the software system, manipulating your data if they are not in the proper format or if you want a subset of your original data set, and then calling the appropriate set of programs or procedures using the key words particular to the software system you are using. The results obtained from calling a program are then displayed at your terminal or sent to your printer.

If you have access to a computer and are interested in using it, find out how to obtain an account, what programs and software systems are available for doing statistical analyses, and where to obtain instruction on data entry for these programs and software systems.

Because computer configurations, operating systems, and text editors vary from site to site, it is best to talk to someone knowledgeable about gaining access to a software system. Once you have mastered the commands to begin executing programs in a software system, you will find that running a job within a given software system is similar from site to site.

Because this isn't a book on computer use, we won't spend additional time and space on the mechanics, which are best learned by doing. Our main interest is in interpreting the output from these programs. The designers of these programs tend to include in the output everything that a user could conceivably want to know; as a result, in any particular situation, some of the output is irrelevant. When reading computer output, look for the values you want; if you don't need or don't understand an output statistic, don't worry. Of course, as you learn more about statistics, more of the output will be meaningful. In the meantime, look for what you need and disregard the rest.

There are dangers in using such packages carelessly. A computer is a mindless beast and will do anything asked of it, no matter how absurd the result might be. For instance, suppose that the data include age, gender (1 = female, 2 = male), religion (1 = Catholic, 2 = Jewish, 3 = Protestant, 4 = other or none), and monthly income of a group of people. If we asked the computer to calculate averages we would get averages for the variables gender and religion, as well as for age and monthly income, even though these averages are meaningless. Used intelligently, these packages are convenient, powerful, and useful—but be sure to examine the output from any computer run to make certain the results make sense. Did anything go wrong? Was something overlooked? In other words, be *skeptical*. One of the important acronyms of computer technology still holds; namely, GIGO: garbage in, garbage out.

Throughout the textbook, we will use computer software systems to do some of the more tedious calculations of statistics *after* we have explained how the calculations can be done. Used in this way, computers (and associated graphical and statistical analysis packages) will enable us to spend additional time on interpreting the results of the analyses rather than on doing the analyses.

3.3 Summarizing Data on a Single Variable: Graphical Methods

After the measurements of interest have been collected, ideally the data are organized, displayed, and examined by using various graphical techniques. As a general rule, the data should be arranged into categories so that *each measurement is classified into one, and only one, of the categories.* This procedure eliminates any ambiguity that might otherwise arise when categorizing measurements. For example, suppose a sex discrimination lawsuit is filed. The law firm representing the plaintiffs needs to summarize the salaries of all employees in a large corporation. To examine possible inequities in salaries, the law firm decides to summarize the 1997 yearly income rounded to the nearest dollar for all female employees into the following categories:

Income Level	Salary
1	less than $20,000
2	$20,000 to $39,999
3	$40,000 to $59,999
4	$60,000 to $79,999
5	$80,000 to $99,999
6	$100,000 or more

The yearly salary of each female employee falls into one, and only one, income category. However, if the income categories had been defined as

Income Level	Salary
1	less than $20,000
2	$20,000 to $40,000
3	$40,000 to $60,000
4	$60,000 to $80,000
5	$80,000 to $100,000
6	$100,000 or more

then there would be confusion as to which category should be checked. For example, an employee earning $40,000 could be placed in either category 2 or 3. To reiterate: If the data are organized into categories, it is important to define the categories so that a measurement can be placed into only one category.

Recruitment From	Number	Percentage
Corporate	501	37.2
Public-interest	683	50.8
Government	94	7.0
Other	67	5.0

Note: Includes trustees of private colleges and universities, directors of large private foundations, senior partners of top law firms, and directors of certain large cultural and civic organizations.

Source: Thomas R. Dye and L. Harmon Zeigler, *The Irony of Democracy,* 5th ed. (Pacific Grove, CA: Duxbury Press, 1981), p. 130.

FIGURE 3.1

Pie chart for the data of Table 3.1

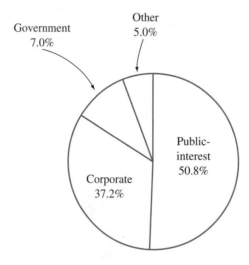

pie chart

When data are organized according to this guideline, there are several ways to display the data graphically. The first and simplest graphical procedure for data organized in this manner is the **pie chart.** It is used to display the percentage of the total number of measurements falling into each of the categories of the variable by partitioning a circle (similar to slicing a pie).

The data in Table 3.1 represent a summary of a study to determine paths to authority for individuals occupying top positions of responsibility in key public-interest organizations. Using biographical information, each of 1,345 individuals was classified according to how she or he was recruited for the current elite position.

Although you can scan the data in Table 3.1, the results are more easily interpreted by using a pie chart. From Figure 3.1 we can make certain inferences about channels to positions of authority. For example, more people were recruited for elite positions from public-interest organizations (approximately 51%) than from elite positions in other organizations.

Other variations of the pie chart are shown in Figures 3.2 and 3.3. Clearly, from Figure 3.2, cola soft drinks have gained in popularity from 1980 to 1990 at the expense of some of the other types of soft drinks. Also, it's evident from Figure 3.3 that the loss of a major food chain account affected fountain sales for PepsiCo, Inc. In summary, the pie chart can be used to display percentages associated with each

FIGURE 3.2

Approximate market share of soft drinks by type, 1980 and 1990

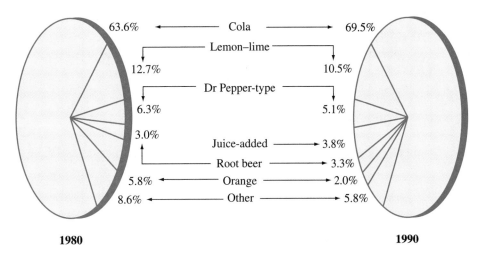

1980   **1990**

FIGURE 3.3

Estimated U.S. market share before and after switch in accounts*

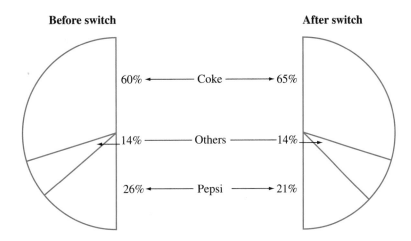

category of the variable. The following guidelines should help you to obtain clarity of presentation in pie charts.

Guidelines for Constructing Pie Charts

1. Choose a small number (five or six) of categories for the variable because too many make the pie chart difficult to interpret.
2. Whenever possible, construct the pie chart so that percentages are in either ascending or descending order.

bar chart

A second graphical technique for data organized according to the recommended guideline is the **bar chart,** or bar graph. Figure 3.4 displays the number of workers in the Cincinnati, Ohio, area for the largest five foreign investors. The estimated total workforce is 680,000. There are many variations of the bar chart. Sometimes the bars are displayed horizontally, as in Figures 3.5(a) and (b). They can also be used to display data across time, as in Figure 3.6. Bar charts are relatively easy to construct if you use the guidelines given.

———

*A major fast-food chain switched its account from Pepsi to Coca-Cola for fountain sales.

FIGURE 3.4

Number of workers by major foreign investors

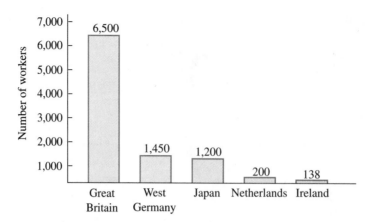

FIGURE 3.5

Greatest per capita consumption by country

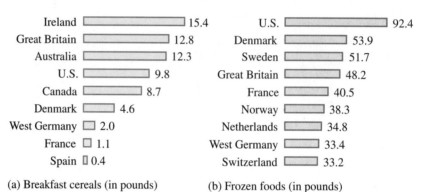

(a) Breakfast cereals (in pounds)   (b) Frozen foods (in pounds)

FIGURE 3.6

Estimated direct and indirect costs for developing a new drug by selected years

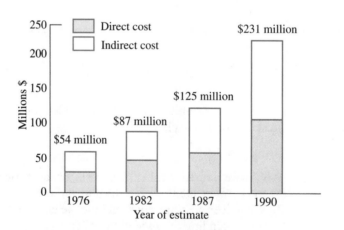

Guidelines for Constructing Bar Charts

1. Label frequencies on one axis and categories of the variable on the other axis.
2. Construct a rectangle at each category of the variable with a height equal to the frequency (number of observations) in the category.
3. Leave a space between each category to connote distinct, separate categories and to clarify the presentation.

frequency histogram, relative frequency histogram

The next two graphical techniques that we discuss are the **frequency histogram** and the **relative frequency histogram.** Both of these graphical techniques are

TABLE 3.2
Weight gains for chicks (grams)

3.7	4.2	4.4	4.4	4.3	4.2	4.4	4.8	4.9	4.4
4.2	3.8	4.2	4.4	4.6	3.9	4.3	4.5	4.8	3.9
4.7	4.2	4.2	4.8	4.5	3.6	4.1	4.3	3.9	4.2
4.0	4.2	4.0	4.5	4.4	4.1	4.0	4.0	3.8	4.6
4.9	3.8	4.3	4.3	3.9	3.8	4.7	3.9	4.0	4.2
4.3	4.7	4.1	4.0	4.6	4.4	4.6	4.4	4.9	4.4
4.0	3.9	4.5	4.3	3.8	4.1	4.3	4.2	4.5	4.4
4.2	4.7	3.8	4.5	4.0	4.2	4.1	4.0	4.7	4.1
4.7	4.1	4.8	4.1	4.3	4.7	4.2	4.1	4.4	4.8
4.1	4.9	4.3	4.4	4.4	4.3	4.6	4.5	4.6	4.0

applicable only to quantitative (measured) data. As with the pie chart, we must organize the data before constructing a graph.

An animal scientist is carrying out an experiment to investigate whether adding an antibiotic to the diet of chicks will promote growth over the standard diet without the antibiotic. The scientist determines that 100 chicks will provide sufficient information to validate the results of the experiment. (In Chapter 5, we will present techniques for determining the proper sample size for a study to achieve specified goals.) From previous research studies, the average weight gain of chicks fed the standard diet over an 8-week period is 3.9 grams. The scientist wants to compare the weight gain of the chicks in the study to the standard value of 3.9 grams. To eliminate the effects of many other factors that could affect weight gain, the scientist rears and feeds the 100 chicks in the same building with individual feeders for each chick. The weight gains for the 100 chicks are recorded in Table 3.2.

An initial examination of the weight gain data reveals that the largest weight gain is 4.9 grams and the smallest is 3.6 grams. Although we might examine the table very closely to determine whether the weight gains of the chicks are substantially greater than 3.9 grams, it is difficult to describe how the measurements are distributed along the interval 3.6 to 4.9. Are most of the measurements greater than 3.9, concentrated near 3.6, concentrated near 4.9, or evenly distributed along the interval? One way to obtain the answers to these questions is to organize the data in a
frequency table.

frequency table

class intervals

To construct a frequency table, we begin by dividing the range from 3.6 to 4.9 into an arbitrary number of subintervals called **class intervals.** The number of subintervals chosen depends on the number of measurements in the set, but we generally recommend using from 5 to 20 class intervals. The more data we have, the larger the number of classes we tend to use. The guidelines given here can be used for constructing the appropriate class intervals.

Guidelines for Constructing Class Intervals

1. Divide the *range* of the measurements (the difference between the largest and the smallest measurements) by the approximate number of class intervals desired. Generally, we want to have from 5 to 20 class intervals.
2. After dividing the range by the desired number of subintervals, round the resulting number to a convenient (easy to work with) unit. This unit represents a common width for the class intervals.
3. Choose the first class interval so that it contains the smallest measurement. It is also advisable to choose a starting point for the first interval

so that no measurement falls on a point of division between two subintervals, which eliminates any ambiguity in placing measurements into the class intervals. (One way to do this is to choose boundaries to one more decimal place than the data.)

For the data in Table 3.2,

$$\text{range} = 4.9 - 3.6 = 1.3$$

Assume that we want to have approximately 10 subintervals. Dividing the range by 10 and rounding to a convenient unit, we have $1.3/10 = .13 \approx .1$. Thus, the class interval width is .1.

It is convenient to choose the first interval to be 3.55–3.65, the second to be 3.65–3.75, and so on. Note that the smallest measurement, 3.6, falls in the first interval and that no measurement falls on the endpoint of a class interval. (See Table 3.3.)

Having determined the class interval, we construct a frequency table for the data. The first column labels the classes by number and the second column indicates the class intervals. We then examine the 100 measurements of Table 3.2, keeping a tally of the number of measurements falling in each interval. The number of mea-
class frequency surements falling in a given class interval is called the **class frequency.** These data are recorded in the third column of the frequency table. (See Table 3.3.)

relative frequency The **relative frequency** of a class is defined to be the frequency of the class divided by the total number of measurements in the set (total frequency). Thus, if we let f_i denote the frequency for class i and n denote the total number of measurements, the relative frequency for class i is f_i/n. The relative frequencies for all the classes are listed in the fourth column of Table 3.3.

The data of Table 3.2 have been organized into a frequency table, which can
histogram now be used to construct a *frequency histogram* or a *relative frequency* **histogram.**

TABLE 3.3
Frequency table for the chick data

Class	Class Interval	Frequency f_i	Relative frequency f_i/n
1	3.55–3.65	1	.01
2	3.65–3.75	1	.01
3	3.75–3.85	6	.06
4	3.85–3.95	6	.06
5	3.95–4.05	10	.10
6	4.05–4.15	10	.10
7	4.15–4.25	13	.13
8	4.25–4.35	11	.11
9	4.35–4.45	13	.13
10	4.45–4.55	7	.07
11	4.55–4.65	6	.06
12	4.65–4.75	7	.07
13	4.75–4.85	5	.05
14	4.85–4.95	4	.04
Totals		$n = 100$	1.00

FIGURE 3.7(a)

Frequency histogram for the chick data of Table 3.3

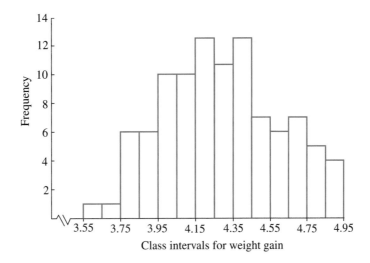

FIGURE 3.7(b)

Relative frequency histogram for the chick data of Table 3.3

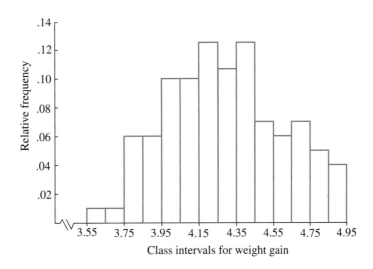

To construct a frequency histogram, draw two axes: a horizontal axis labeled with the class intervals and a vertical axis labeled with the frequencies. Then construct a rectangle over each class interval with a height equal to the number of measurements falling in a given subinterval. The frequency histogram for the data of Table 3.3 is shown in Figure 3.7(a).

The relative frequency histogram is constructed in much the same way as a frequency histogram. In the relative frequency histogram, however, the vertical axis is labeled as relative frequency, and a rectangle is constructed over each class interval with a height equal to the class relative frequency (the fourth column of Table 3.3). The relative frequency histogram for the data of Table 3.3 is shown in Figure 3.7(b). Clearly, the two histograms of Figures 3.7(a) and (b) are of the same shape and would be identical if the vertical axes were equivalent. We will frequently refer to either one as simply a histogram.

There are several comments that should be made concerning histograms. First, the distinction between bar charts and histograms is based on the distinction between *qualitative* and *quantitative* variables. Values of qualitative variables vary in kind but not degree and hence are not measurements. For example, the variable political party affiliation can be categorized as Republican, Democrat, or other, and, although we could label the categories as one, two, or three, these values are only codes and have no quantitative interpretation. In contrast, quantitative variables have actual units of measure. For example, the variable yield (in bushels) per acre of corn can assume specific values. *Pie charts and bar charts are used to display frequency data from qualitative variables; histograms are appropriate for displaying frequency data for quantitative variables.*

Second, the histogram is the most imporant graphical technique we will present because of the role it plays in statistical inference, a subject we will discuss in later chapters. Third, if we had an extremely large set of measurements, and if we constructed a histogram using many class intervals, each with a very narrow width, the histogram for the set of measurements would be, for all practical purposes, a smooth curve. Fourth, the fraction of the total number of measurements in an interval is equal to the fraction of the total area under the histogram over the interval. For example, if we consider the intervals containing weights greater than 3.9 for the chick data in Table 3.3, we see that there are exactly 86 of the 100 measurements in those intervals. Thus, .86, the proportion of the total measurements falling in those intervals, is equal to the proportion of the total area under the histogram over those intervals.

Fifth, if a single measurement is selected at random from the set of sample measurements, the chance, or **probability,** that it lies in a particular interval is equal to the fraction of the total number of sample measurements falling in that interval. This same fraction is used to estimate the probability that a measurement randomly selected from the population lies in the interval of interest. For example, from the sample data of Table 3.2, the chance or probability of selecting a chick with a weight gain greater than 3.9 grams is .86. The number .86 is an approximation of the proportion of all chickens fed the diet containing the antibiotic that would yield a weight gain greater than 3.9 grams, the value obtained from the standard diet.

Because of the arbitrariness in the choice of number of intervals, starting value, and length of intervals, histograms can be made to take on different shapes for the same set of data, especially for small data sets. Histograms are most useful for describing data sets when the number of data points is fairly large, say 50 or more. In Figures 3.8(a)–(d), a set of histograms for the chick data constructed using 5, 10, 14, and 18 class intervals illustrates the problems that can be encountered in attempting to construct a histogram. These graphs were obtained using the Minitab software program.

When the number of data points is relatively small and the number of intervals is large, the histogram has several intervals in which there are no data values; see Figure 3.8(d). This results in a graph that is not a realistic depiction of the histogram for the whole population. When the number of class intervals is too small, most of the patterns or trends in the data are not displayed; see Figure 3.8(a). In the set of graphs in Figure 3.8, the histogram with 14 class intervals appears to be the most appropriate graph.

Finally, because we use proportions rather than frequencies in a relative frequency histogram, we can compare two different samples (or populations) by examining their relative frequency histograms even if the samples (populations) are of different sizes. When describing relative frequency histograms and comparing the plots from a number of samples, we examine the overall shape in the histogram. Figure 3.9 depicts many of the common shapes for relative frequency histograms.

FIGURE 3.8   Histograms for the chick data

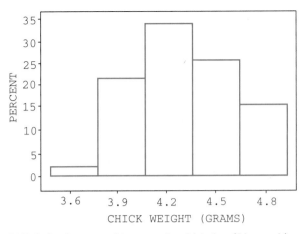

(a) Relative frequency histogram for chick data (5 intervals)

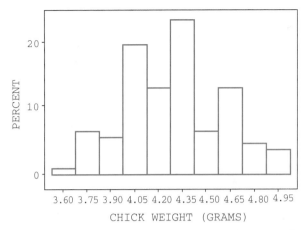

(b) Relative frequency histogram for chick data (10 intervals)

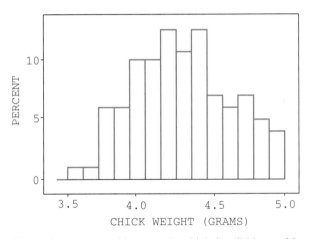

(c) Relative frequency histogram for chick data (14 intervals)

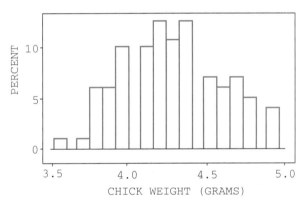

(d) Relative frequency histogram for chick data (18 intervals)

unimodal

bimodal

uniform

symmetric

skewed to the right

skewed to the left

A histogram with one major peak is called **unimodal,** see Figures 3.9(b), (c), and (d). When the histogram has two major peaks, such as in Figures 3.9(e) and (f), we say that the histogram is **bimodal.** In many instances, bimodal histograms are an indication that the sampled data are in fact from two distinct populations. Finally, when every interval has essentially the same number of observations, the histogram is called a **uniform** histogram, see Figure 3.9(a).

A histogram is **symmetric** in shape if the right and left sides have essentially the same shape. Thus, Figures 3.9(a), (b), and (e) have symmetric shapes. When the right side of the histogram, containing the larger half of the observations in the data, extends a greater distance than the left side, the histogram is referred to as **skewed to the right;** see Figure 3.9(c). The histogram is **skewed to the left** when its left side extends a much larger distance than the right side; see Figure 3.9(d). We will see later in the text that knowing the shape of the distribution will help us choose the appropriate measures to summarize the data (Sections 3.4–3.7) and the methods for analyzing the data (Chapter 5 and beyond).

FIGURE 3.9 Some common shapes of distributions

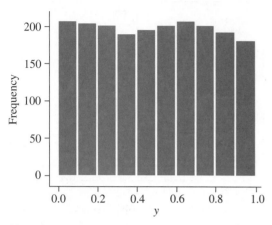

(a) Uniform distribution

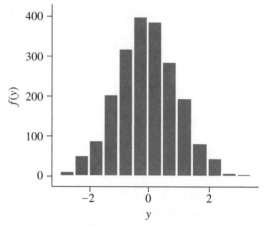

(b) Symmetric, unimodal (normal) distribution

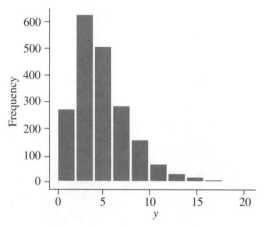

(c) Right-skewed distribution

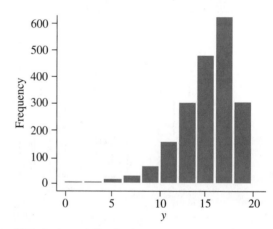

(d) Left-skewed distribution

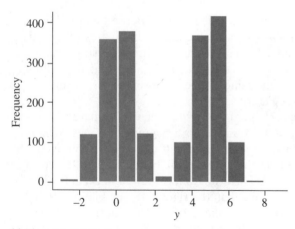

(e) Bimodal distribution

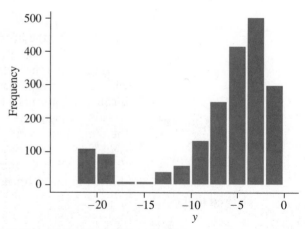

(f) Bimodal distribution skewed to left

exploratory data
analysis (EDA)

stem-and-leaf plot

The next graphical technique presented in this section is a display technique taken from an area of statistics called **exploratory data analysis (EDA).** Professor John Tukey (1977) has been the leading proponent of this practical philosophy of data analysis aimed at exploring and understanding data.

The **stem-and-leaf plot** is a clever, simple device for constructing a histogram-like picture of a frequency distribution. It allows us to use the information contained in a frequency distribution to show the range of scores, where the scores are concentrated, the shape of the distribution, whether there are any specific values or scores not represented, and whether there are any stray or extreme scores. The stem-and-leaf plot does not follow the organization principles stated previously for histograms. We will use the data shown in Table 3.4 to illustrate how to construct a stem-and-leaf plot.

The original scores in Table 3.4 are either three- or four-digit numbers. We will use the first, or leading, digit of each score as the stem (see Figure 3.10) and the

TABLE 3.4

Violent crime rates for 90 standard metropolitan statistical areas selected from the north, south, and west

South	Rate	North	Rate	West	Rate
Albany, GA	876	Allentown, PA	189	Abilene, TX	570
Anderson, SC	578	Battle Creek, MI	661	Albuquerque, NM	928
Anniston, AL	718	Benton Harbor, MI	877	Anchorage, AK	516
Athens, GA	388	Bridgeport, CT	563	Bakersfield, CA	885
Augusta, GA	562	Buffalo, NY	647	Brownsville, TX	751
Baton Rouge, LA	971	Canton, OH	447	Denver, CO	561
Charleston, SC	698	Cincinnati, OH	336	Fresno, CA	1,020
Charlottesville, VA	298	Cleveland, OH	526	Galveston, TX	592
Chattanooga, TN	673	Columbus, OH	624	Houston, TX	814
Columbus, GA	537	Dayton, OH	605	Kansas City, MO	843
Dothan, AL	642	Des Moines, IA	496	Lawton, OK	466
Florence, SC	856	Dubuque, IA	296	Lubbock, TX	498
Fort Smith, AR	376	Gary, IN	628	Merced, CA	562
Gadsden, AL	508	Grand Rapids, MI	481	Modesto, CA	739
Greensboro, NC	529	Janesville, WI	224	Oklahoma City, OK	562
Hickery, NC	393	Kalamazoo, MI	868	Reno, NV	817
Knoxville, TN	354	Lima, OH	804	Sacramento, CA	690
Lake Charles, LA	735	Madison, WI	210	St. Louis, MO	720
Little Rock, AR	811	Milwaukee, WI	421	Salinas, CA	758
Macon, GA	504	Minneapolis, MN	435	San Diego, CA	731
Monroe, LA	807	Nassau, NY	291	Santa Ana, CA	480
Nashville, TN	719	New Britain, CT	393	Seattle, WA	559
Norfolk, VA	464	Philadelphia, PA	605	Sioux City, IA	505
Raleigh, NC	410	Pittsburgh, PA	341	Stockton, CA	703
Richmond, VA	491	Portland, ME	352	Tacoma, WA	809
Savannah, GA	557	Racine, WI	374	Tucson, AZ	706
Shreveport, LA	771	Reading, PA	267	Victoria, TX	631
Washington, DC	685	Saginaw, MI	684	Waco, TX	626
Wilmington, DE	448	Syracuse, NY	685	Wichita Falls, TX	639
Wilmington, NC	571	Worcester, MA	460	Yakima, WA	585

Note: Rates represent the number of violent crimes (murder, forcible rape, robbery, and aggravated assault) per 100,000 inhabitants, rounded to the nearest whole number.

Source: Department of Justice, Uniform Crime Reports for the United States, 1990.

FIGURE 3.10

Stem-and-leaf plot for violent crime rates of Table 3.4

1	89
2	98 96 24 10 91 67
3	88 76 93 54 36 93 41 52 74
4	64 10 91 48 47 96 81 21 35 60 66 98 80
5	78 62 37 08 29 04 57 71 63 26 70 16 61 92 62 62 59 05 85
6	98 73 42 85 61 47 24 05 28 05 84 85 90 31 26 39
7	18 35 19 71 51 39 20 58 31 03 06
8	76 56 11 07 77 68 04 85 14 43 17 09
9	71 28
10	20

FIGURE 3.11

Stem-and-leaf plot with ordered leaves

1	89
2	10 24 67 91 96 98
3	36 41 52 54 74 76 88 93 93
4	10 21 35 47 48 60 64 66 80 81 91 96 98
5	04 05 08 16 26 29 37 57 59 61 62 62 62 63 70 71 78 85 92
6	05 05 24 26 28 31 39 42 47 61 73 84 85 85 90 98
7	03 06 18 19 20 31 35 39 51 58 71
8	04 07 09 11 14 17 43 56 68 76 77 85
9	28 71
10	20

trailing digits as the leaf. For example, the violent crime rate in Albany is 876. The leading digit is 8 and the trailing digits are 76. In the case of Fresno, the leading digits are 10 and the trailing digits are 20. If our data consisted of six-digit numbers such as 104,328, we might use the first two digits as stem numbers and the second two digits as leaf numbers and ignore the last two digits.

For the data on violent crime, the smallest rate is 189, the largest is 1,020, and the leading digits are 1, 2, 3, . . . , 10. In the same way that a class interval determines where a measurement is placed in a frequency table, the leading digit (stem of a score) determines the row in which a score is placed in a stem-and-leaf plot. The trailing digits for the score are then written in the appropriate row. In this way, each score is recorded in the stem-and-leaf plot, as in Figure 3.10 for the violent crime data.

We can see that each stem defines a class interval and the limits of each interval are the largest and smallest possible scores for the class. The values represented by each leaf must be between the lower and upper limits of the interval.

Note that a stem-and-leaf plot is a graph that looks much like a histogram turned sideways, as in Figure 3.10. The plot can be made a bit more useful by ordering the data (leaves) within a row (stem) from lowest to highest (Figure 3.11). The advantage of such a graph over the histogram is that it reflects not only frequencies, concentration(s) of scores, and shapes of the distribution but also the actual scores. The disadvantage is that for large data sets, the stem-and-leaf plot can be more unwieldy than the histogram.

Guidelines for Constructing Stem-and-Leaf Plots

1. Split each score or value into two sets of digits. The first or leading set of digits is the stem and the second or trailing set of digits is the leaf.
2. List all possible stem digits from lowest to highest.
3. For each score in the mass of data, write the leaf values on the line labeled by the appropriate stem number.

4. If the display looks too cramped and narrow, stretch the display by using two lines per stem so that, for example, leaf digits 0, 1, 2, 3, and 4 are placed on the first line of the stem and leaf digits 5, 6, 7, 8, and 9 are placed on the second line.
5. If too many digits are present, such as in a six- or seven-digit score, drop the right-most trailing digit(s) to maximize the clarity of the display.
6. The rules for developing a stem-and-leaf plot are somewhat different from the rules governing the establishment of class intervals for the traditional frequency distribution and for a variety of other procedures that we will consider in later sections of the text. Class intervals for stem-and-leaf plots are, then, in a sense slightly atypical.

The following stem-and-leaf plot is obtained from Minitab. The data consist of the number of employees in the wholesale and retail trade industries in Wisconsin measured each month for a 5-year period.

Data Display

```
Trade
    322   317   319   323   327   328   325   326   330   334
    337   341   322   318   320   326   332   334   335   336
    335   338   342   348   330   326   329   337   345   350
    351   354   355   357   362   368   348   345   349   355
    362   367   366   370   371   375   380   385   361   354
    357   367   376   381   381   383   384   387   392   396
```

Character Stem-and-Leaf Display

```
Stem-and-leaf of Trade    N = 60
Leaf Unit = 1.0
      31  789
      32  0223
      32  5666789
      33  00244
      33  556778
      34  12
      34  55889
      35  0144
      35  5577
      36  122
      36  6778
      37  01
      37  56
      38  01134
      38  57
      39  2
      39  6
```

Note that most of the stems are repeated twice, with the leaf digits split into two groups: 0 to 4 and 5 to 9.

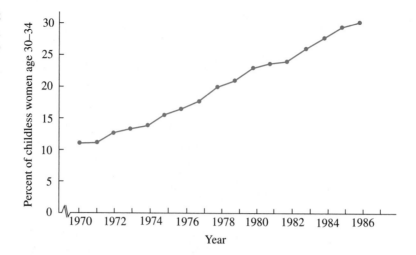

FIGURE 3.12
Percentage of childless women
age 30 to 34, 1970–1986

The last graphical technique to be presented in this section deals with how certain variables change over time. For macroeconomic data such as disposable income and microeconomic data such as weekly sales data of one particular product at one particular store, plots of data over time are fundamental to business management. Similarly, social researchers are often interested in showing how variables change over time. They might be interested in changes over time in attitudes toward various racial and ethnic groups, changes in the rate of savings in the United States, or changes in crime rates for various cities. A pictorial method of presenting changes in a variable over time is called a **time series.** Figure 3.12 is a time series showing the percentage of white women age 30 to 34 who did not have any children during the years 1970 to 1986.

time series

Usually, time points are labeled chronologically across the horizontal axis (abscissa), and the numerical values (frequencies, percentages, rates, etc.) of the variable of interest are labeled along the vertical axis (ordinate). Time can be measured in days, months, years, or whichever unit is most appropriate. As a rule of thumb, a time series should consist of no fewer than four or five time points; typically, these time points are equally spaced. Many more time points than this are desirable, however, in order to show a more complete picture of changes in a variable over time.

How we display the time axis in a time series frequently depends on the time intervals at which data are available. For example, the U.S. Census Bureau reports average family income in the United States only on a yearly basis. When information about a variable of interest is available in different units of time, we must decide which unit or units are most appropriate for the research. In an election year, a political scientist would most likely examine weekly or monthly changes in candidate preferences among registered voters. On the other hand, a manufacturer of machine-tool equipment might keep track of sales (in dollars and number of units) on a monthly, quarterly, and yearly basis. Figure 3.13 shows the quarterly sales (in thousands of units) of a machine-tool product from 1998 to 2000. Note that from this time series, it is clear that the company has experienced a gradual but steady growth in the number of units sold over the 3-year period.

Time series plots are useful for examining general trends and seasonal or cyclic patterns. For example, the "Money and Investing" section of the *Wall Street Journal* gives the daily workday values for the Dow Jones Industrials, Utilities, and

FIGURE 3.13

Quarterly sales (in thousands), 1998–2000

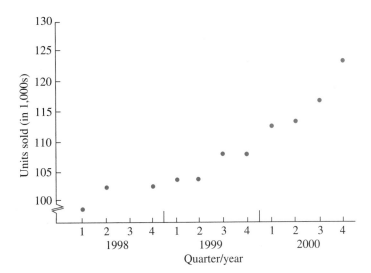

Transportation Averages for a 6-month period. These are displayed in Figure 3.14 for a typical period. An examination of these plots reveals that for the Industrial and Transportation indices in 1998 there is a somewhat increasing trend from June to mid-July, followed by a downward trend through September, with a sharply increasing trend from October through November. In contrast to these two indices, the Utilities index is fairly flat from June to September, followed by an increase through early October, at which time the index stays fairly flat through November. To detect seasonal or cyclical patterns, it is necessary to have weekly or monthly data over a number of years.

Sometimes it is important to compare trends over time in a variable for two or more groups. Figure 3.15 reports the values of two ratios from 1976 to 1988: the ratio of the median family income of African Americans to the median income of Anglo-Americans and the ratio of the median income of Hispanics to the median income of Anglo-Americans.

Median family income represents the income amount that divides family incomes into two groups—the top half and the bottom half. In 1987, the median family income for African Americans was $18,098, meaning that 50% of all African American families had incomes above $18,098, and 50% had incomes below $18,098. The median, one of several measures of central tendency, is discussed more fully later in this chapter.

Figure 3.15 shows that the ratio of African American or Hispanic to Anglo-American family income fluctuated between 1976 and 1988, but the overall trend in both ratios indicates that they declined over this time. A social researcher would interpret these trends to mean that the income of African American and Hispanic families generally declined relative to the income of Anglo-American families.

Sometimes information is not available in equal time intervals. For example, polling organizations such as Gallup or the National Opinion Research Center do not necessarily ask the American public the same questions about their attitudes or behavior on a yearly basis. Sometimes there is a time gap of more than 2 years before a question is asked again.

When information is not available in equal time intervals, it is important for the interval width between time points (the horizontal axis) to reflect this fact. If, for example, a social researcher is plotting values of a variable for 1985, 1986, 1987,

FIGURE 3.14

Time series plots for the
Dow Jones Industrials,
Utilities, and Transportation
Averages

Source: *Wall Street Journal*,
30 November 1998. Copyright
1998 by Dow Jones & Co., Inc.
Reproduced with permission
of Dow Jones & Co., Inc., via
Copyright Clearance Center.

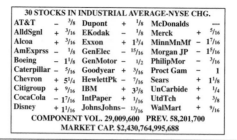

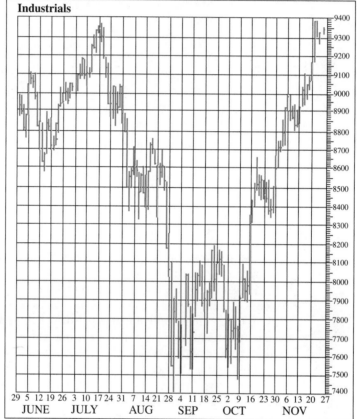

and 1990, the interval width between 1987 and 1990 on the horizontal axis should be three times the width of that between the other years. If these interval widths were spaced evenly, the resulting trend line could be seriously misleading. Other examples of graphic distortion are discussed in Chapter 13.

Figure 3.16 presents the trend in church attendance among American Catholics and Protestants from 1954 to 1988. The width of the intervals between time points reflects the fact that Catholics were not asked about their church attendance every year.

Before leaving graphical methods for describing data, there are several general guidelines that can be helpful in developing graphs with an impact. These guidelines pay attention to the design and presentation techniques and should help you make better, more informative graphs.

FIGURE 3.14

(continued)

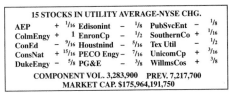

15 STOCKS IN UTILITY AVERAGE-NYSE CHG.

AEP	+ $^{1}/_{16}$	Edisonint – $^{1}/_{8}$	PubSvcEnt	–	$^{1}/_{8}$
ColmEngy	+ 1	EnronCp – $^{1}/_{2}$	SouthernCo	+	$^{1}/_{16}$
ConEd	– $^{9}/_{16}$	Houstnind – $^{5}/_{16}$	Tex Util	–	$^{1}/_{2}$
ConsNat	+ $^{15}/_{16}$	PECO Engy – $^{7}/_{16}$	UnicomCp	+	$^{3}/_{16}$
DukeEngy	– $^{5}/_{8}$	PG&E – $^{3}/_{8}$	WillmsCos	+	$^{3}/_{8}$

COMPONENT VOL. 3,283,900   PREV. 7,217,700
MARKET CAP. $175,964,191,750

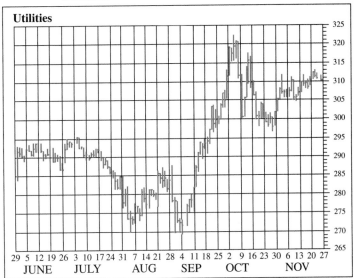

20 STOCKS IN TRANSPORTATION AVERAGE-NYSE CHG.

AMR	+ $^{1}/_{4}$	FDX Cp + $^{1}/_{16}$	SowestAir	+	$^{5}/_{16}$
AirbrnFrt	– $^{1}/_{16}$	GATX + $^{7}/_{8}$	UAL Cp	+	$^{5}/_{8}$
AlxBldwn*	– $^{1}/_{8}$	HuntJB* + $^{5}/_{16}$	UnPacific	–	$^{1}/_{8}$
BurlNthSF	+ $^{3}/_{8}$	NorflkSo – 1	US Airways	+	1$^{5}/_{8}$
CNF Trnsp	....	NowestAir* + $^{7}/_{16}$	USFrght*	+	$^{3}/_{8}$
CSX	+ $^{1}/_{16}$	RoadwEx* + $^{9}/_{16}$	YellowCp*	+	$^{1}/_{4}$
DeltaAir	+ $^{1}/_{8}$	RyderSys – $^{5}/_{16}$	* Nasdaq		NMS

COMPONENT VOL. 3,598,500   PREV. 9,572,100
MARKET CAP. $107,744,107,125

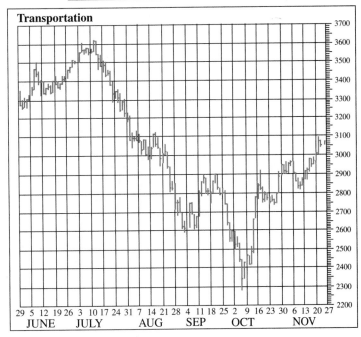

FIGURE 3.15

Ratio of African American and Hispanic median family income to Anglo-American median family income, 1976–1988

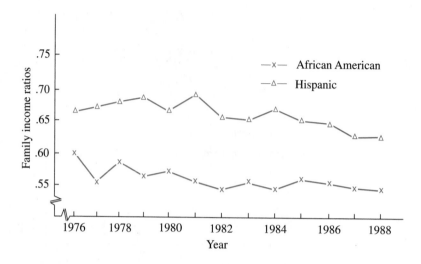

FIGURE 3.16

Church attendance of American Protestants and Catholics in a typical week, 1954–1988

Source: 1954–1988 The Gallup Organization. All rights reserved, used with permission.

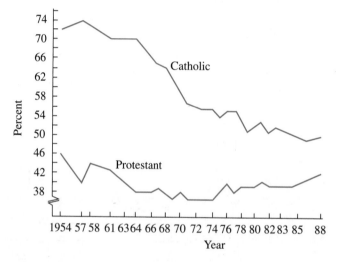

General Guidelines for Successful Graphs

1. Before constructing a graph, set your priorities. What messages should the viewer obtain from the graph?
2. Choose the type of graph (pie chart, bar graph, histogram, and so on).
3. Pay attention to the title. One of the most important aspects of a graph is its title. The title should immediately inform the viewer of the point of the graph and draw the eye toward the most important elements of the graph.
4. Fight the urge to use many type sizes, styles, and color changes. The indiscriminate and excessive use of different type sizes, styles, and colors will confuse the viewer. Generally, we recommend using only two typefaces; color changes and italics should be used in only one or two places.
5. Convey the tone of your graph by using colors and patterns. Intense, warm colors (yellows, oranges, reds) are more dramatic than blues and purples and help to stimulate enthusiasm in the viewer. On the other hand, pastels (particularly grays) convey a conservative, businesslike tone. Similarly, simple patterns convey a conservative tone, whereas busier patterns stimulate more excitement.

6. Don't underestimate the effectiveness of a simple, straightforward graph.
7. Practice making graphs frequently. As with almost anything, practice improves skill.

EXERCISES **Basic Techniques**

Edu. **3.1** University officials periodically review the distribution of undergraduate majors within the colleges of the university to help determine a fair allocation of resources to departments within the colleges. At one such review, the following data were obtained:

College	Number of Majors
Agriculture	1,500
Arts and Sciences	11,000
Business Administration	7,000
Education	2,000
Engineering	5,000

a. Construct a pie chart for these data.
b. Use the same data to construct a bar graph.

Bus. **3.2** Because the import of basic materials is an indication of the strength of the U.S. economy, the Commerce Department monitors the importation of steel. The following data are the level of steel imports (in millions of tons) for the years 1985 to 1996:

Year	1985	1986	1987	1988	1989	1990	1991	1992	1993	1994	1995	1996
Import	27.6	22.7	21.9	20.4	19.7	21.9	20.2	21.9	21.8	32.7	27.3	32.1

a. Would a pie chart be an appropriate graphical method for describing the data?
b. Construct a bar graph for the data.
c. Do you observe any trends in steel imports?

3.3 A large study of employment trends, based on a survey of 45,000 businesses, was conducted by Ohio State University. Assuming an unemployment rate of 5% or less, the study predicted that 2.1 million job openings would be created between 1980 and 1990. This employment growth is shown by major industry groups.

Industry Group	Employment Growth Percentage, 1980–1990
Service	33.2
Manufacturing	25.0
Retail trade	17.9
Finance, insurance, real estate	6.6
Wholesale trade	4.8
Construction	4.6
Transportation	3.9
Government	2.7
Other	1.3

Construct a pie chart to display these data.

Bus. **3.4** From the same study described in Exercise 3.3, data were obtained on the job openings between 1980 and 1990. Use the data to construct a bar chart.

Occupational Groups	Percentage of Job Openings, 1980–1990
Clerical workers	20.9
Sales	7.3
Managers	9.5
Professional and technical	16.3
Laborers	3.7
Service workers	18.1
Operatives	13.1
Craft and kindred workers	11.1

Gov. **3.5** The National Highway Traffic Safety Administration has studied the use of rear-seat automobile lap and shoulder seat belts. The number of lives potentially saved with the use of lap and shoulder seat belts is shown for various percentages of use.

	Lives Saved Wearing	
Percentage of Use	Lap Belt Only	Lap and Shoulder Belt
100	529	678
80	423	543
60	318	407
40	212	271
20	106	136
10	85	108

Suggest several different ways to graph these data. Which one seems more appropriate and why?

Med. **3.6** The survival times (in months) for two treatments for patients with severe chronic left-ventricular heart failure are given in the following tables.

Standard Therapy						
4	15	24	10	1	27	31
14	2	16	32	7	13	36
29	6	12	18	14	15	18
6	13	21	20	8	3	24

New Therapy						
5	20	29	15	7	32	36
17	15	19	35	10	16	39
27	14	10	16	12	13	16
9	18	33	30	29	31	27

a. Construct separate relative frequency histograms for the survival times of both the therapies.
b. Compare the two histograms. Does the new therapy appear to generate a longer survival time? Explain your answer.

3.7 Combine the data from the separate therapies in Exercise 3.6 into a single data set and construct a relative frequency histogram for this combined data set. Does the plot indicate that the data are from two separate populations? Explain your answer.

Gov. **3.8** Many public interest groups assert that the federal government expends an increasing ฯ tion of the nation's resources on the nation's defense. The following table contains the outlays (iน billions of dollars) for the Defense Department since 1980 and the outlays as a percentage of gross national product (% GNP).

Year	80	81	82	83	84	85	86	87	88	89	90	91	92	93	94	95	96	97
Outlay	134	158	185	210	227	253	273	282	290	304	299	273	298	291	282	272	266	267
% GNP	4.9	5.2	5.8	6.1	6.0	6.2	6.2	6.1	5.9	5.7	5.3	4.7	4.9	4.5	4.1	3.8	3.6	3.4

Source: *Statistical Abstract of the United States, 1997.*

a. Plot the outlay time series data and describe any trends in the outlays.
b. Plot the % GNP time series data and describe any trends in the % GNP.
c. Do the two time series have similar trends? Do either of the plots support the public interest group's contention?

Edu. **3.9** Educational researchers study trends in SAT scores to assess claimed differences between male and female performance on the exams.

Gender/Type	1967	1970	1975	1980	1985	1990	1993	1994	1995	1996
Male/Verbal	540	536	515	506	514	505	504	501	505	507
Female/Verbal	545	538	509	498	503	496	497	497	502	503
Male/Math	535	531	518	515	522	521	524	523	525	527
Female/Math	495	493	479	473	480	483	484	487	490	492

(Year spans the data columns.)

Source: *Statistical Abstract of the United States, 1997.*

a. Plot the four separate time series and describe any trends in the separate time series.
b. Do the trends appear to be the same for males and females?
c. What differences do you observe in the plots for males and females?

Soc. **3.10** The following table presents the homeownership rates, in percentages, by state for the years 1985 and 1996. These values represent the proportion of homes owned by the occupant to the total number of occupied homes.

State	1985	1996	State	1985	1996
Alabama	70.4	71.0	Montana	66.5	68.6
Alaska	61.2	62.9	Nebraska	68.5	66.8
Arizona	64.7	62.0	Nevada	57.0	61.1
Arkansas	66.6	66.6	New Hampshire	65.5	65.0
California	54.2	55.0	New Jersey	62.3	64.6
Colorado	63.6	64.5	New Mexico	68.2	67.1
Connecticut	69.0	69.0	New York	50.3	52.7
Delaware	70.3	71.5	North Carolina	68.0	70.4
Dist. of Columbia	37.4	40.4	North Dakota	69.9	68.2
Florida	67.2	67.1	Ohio	67.9	69.2
Georgia	62.7	69.3	Oklahoma	70.5	68.4
Hawaii	51.0	50.6	Oregon	61.5	63.1

(*continued*)

State	1985	1996	State	1985	1996
Idaho	71.0	71.4	Pennsylvania	71.6	71.7
Illinois	60.6	68.2	Rhode Island	61.4	56.6
Indiana	67.6	74.2	South Carolina	72.0	72.9
Iowa	69.9	72.8	South Dakota	67.6	67.8
Kansas	68.3	67.5	Tennessee	67.6	68.8
Kentucky	68.5	73.2	Texas	60.5	61.8
Louisiana	70.2	64.9	Utah	71.5	72.7
Maine	73.7	76.5	Vermont	69.5	70.3
Maryland	65.6	66.9	Virginia	68.5	68.5
Massachusetts	60.5	61.7	Washington	66.8	63.1
Michigan	70.7	73.3	West Virginia	75.9	74.3
Minnesota	70.0	75.4	Wisconsin	63.8	68.2
Mississippi	69.6	73.0	Wyoming	73.2	68.0
Missouri	69.2	70.2			

Source: U.S. Bureau of the Census, Internet site: http://www.census.gov/ftp/pub/hhes/www/hvs.html.

    a. Construct a relative frequency histogram plot for the homeownership data given in the table for the years 1985 and 1996.

    b. What major differences exist between the plots for the two years?

    c. Why do you think the plots have changed over these 11 years?

    d. How could Congress use the information in these plots for writing tax laws that allow major tax deductions for homeownership?

3.11 Construct a stem-and-leaf plot for the data in Exercise 3.10.

3.12 Describe the shape of the stem-and-leaf plot and histogram for the homeownership data in Exercises 3.10 and 3.11, using the terms *modality, skewness,* and *symmetry* in your description.

Bus.   **3.13** A supplier of high-quality audio equipment for automobiles accumulates monthly sales data on speakers and receiver–amplifier units for 5 years. The data (in thousands of units per month) are shown in a table. Plot the sales data. Do you see any overall trend in the data? Do there seem to be any cyclic or seasonal effects?

Year	J	F	M	A	M	J	J	A	S	O	N	D
1	101.9	93.0	93.5	93.9	104.9	94.6	105.9	116.7	128.4	118.2	107.3	108.6
2	109.0	98.4	99.1	110.7	100.2	112.1	123.8	135.8	124.8	114.1	114.9	112.9
3	115.5	104.5	105.1	105.4	117.5	106.4	118.6	130.9	143.7	132.2	120.8	121.3
4	122.0	110.4	110.8	111.2	124.4	112.4	124.9	138.0	151.5	139.5	127.7	128.0
5	128.1	115.8	116.0	117.2	130.7	117.5	131.8	145.5	159.3	146.5	134.0	134.2

3.4 Summarizing Data on a Single Variable: Measures of Central Tendency

Numerical descriptive measures are commonly used to convey a mental image of pictures, objects, and other phenomena. There are two main reasons for this. First, graphical descriptive measures are inappropriate for statistical inference because it is difficult to describe the similarity of a sample frequency histogram and the corresponding population frequency histogram. The second reason for using numerical descriptive measures is one of expediency—we never seem to carry the appropriate graphs or histograms with us, and so must resort to our powers of verbal communication to convey the appropriate picture. We seek several numbers, called *numerical descriptive measures,* that will create a mental picture of the frequency distribution for a set of measurements.

central tendency

variability

The two most common numerical descriptive measures are measures of **central tendency** and measures of **variability;** that is, we seek to describe the center of the distribution of measurements and also how the measurements vary about the center of the distribution. We will draw a distinction between numerical descriptive measures for a population, called **parameters,** and numerical descriptive measures for a sample, called **statistics.** In problems requiring statistical inference, we will not be able to calculate values for various parameters, but we will be able to compute corresponding statistics from the sample and use these quantities to estimate the corresponding population parameters.

parameters

statistics

In this section, we will consider various measures of central tendency, followed in Section 3.5 by a discussion of measures of variability.

mode

The first measure of central tendency we consider is the **mode.**

DEFINITION 3.1

> The **mode** of a set of measurements is defined to be the measurement that occurs most often (with the highest frequency).

We illustrate the use and determination of the mode in an example.

EXAMPLE 3.1

Slaughter weights (in pounds) for a sample of 15 Herefords, each with a frame size of 3 (on a 1–7 scale), are shown here.

962	1,005	1,033
980	965	1,030
975	989	955
1,015	1,000	970
1,042	1,005	995

Determine the modal slaughter weight.

Solution For these data, the weight 1,005 occurs twice and all other weights occur once. Hence, the mode is 1,005.

Identification of the mode for Example 3.1 was quite easy because we were able to count the number of times each measurement occurred. When dealing with grouped data—data presented in the form of a frequency table—we can define the modal interval to be the class interval with the highest frequency. However, because we do not know the actual measurements but only how many measurements fall into each interval, the mode is taken as the midpoint of the modal interval; it is an approximation to the mode of the actual sample measurements.

The mode is also commonly used as a measure of popularity that reflects central tendency or opinion. For example, we might talk about the most preferred stock, a most preferred model of washing machine, or the most popular candidate.

In each case, we would be referring to the mode of the distribution. In Figure 3.9 of the previous section, frequency histograms (b), (c), and (d) had a single mode with the mode located at the center of the class having the highest frequency. Thus, the modes would be −.25 for histogram (b), 3 for histogram (c), and 17 for histogram (d). It should be noted that some distributions have more than one measurement that occurs with the highest frequency. Thus, we might encounter bimodal, trimodal, and so on, distributions. The relative frequency histogram for the chick diet data given in Figure 3.8(c) would be a bimodal distribution with modes at 4.2 grams and 4.4 grams. In Figure 3.9, both histograms (e) and (f) are bimodal.

median   The second measure of central tendency we consider is the **median.**

DEFINITION 3.2   The **median** of a set of measurements is defined to be the middle value when the measurements are arranged from lowest to highest.

The median is most often used to measure the midpoint of a large set of measurements. For example, we may read about the median wage increase won by union members, the median age of people receiving Social Security benefits, and the median weight of cattle prior to slaughter during a given month. Each of these situations involves a large set of measurements, and the median reflects the central value of the data—that is, the value that divides the set of measurements into two groups, with an equal number of measurements in each group.

However, we may use the definition of median for small sets of measurements by using the following convention. The median for an even number of measurements is the average of the two middle values when the measurements are arranged from lowest to highest. For an odd number of measurements, the median is still the middle value. Thus, whether there are an even or odd number of measurements, there are an equal number of measurements above and below the median.

EXAMPLE 3.2

Each of 10 children in the second grade was given a reading aptitude test. The scores were as follows:

95   86   78   90   62   73   89   92   84   76

Determine the median test score.

Solution First we must arrange the scores in order of magnitude.

62   73   76   78   84   86   89   90   92   95

Because there are an even number of measurements, the median is the average of the two midpoint scores.

$$\text{median} = \frac{84 + 86}{2} = 85$$

EXAMPLE 3.3

An experiment was conducted to measure the effectiveness of a new procedure for pruning grapes. Each of 13 workers was assigned the task of pruning an acre of grapes. The productivity, measured in worker-hours/acre, is recorded for each person.

$$4.4 \quad 4.9 \quad 4.2 \quad 4.4 \quad 4.8 \quad 4.9 \quad 4.8 \quad 4.5 \quad 4.3 \quad 4.8 \quad 4.7 \quad 4.4 \quad 4.2$$

Determine the mode and median productivity for the group.

Solution First arrange the measurements in order of magnitude:

$$4.2 \quad 4.2 \quad 4.3 \quad 4.4 \quad 4.4 \quad 4.4 \quad 4.5 \quad 4.7 \quad 4.8 \quad 4.8 \quad 4.8 \quad 4.9 \quad 4.9$$

For these data, we have two measurements appearing three times each. Hence, the data are bimodal, with modes of 4.4 and 4.8. The median for the odd number of measurements is the middle score, 4.5.

The third, and last, measure of central tendency we will discuss in this text is
mean the arithmetic mean, known simply as the **mean.**

DEFINITION 3.3   The **arithmetic mean,** or **mean,** of a set of measurements is defined to be the sum of the measurements divided by the total number of measurements.

When people talk about an "average," they quite often are referring to the mean. It is the balancing point of the data set. Because of the important role that the mean will play in statistical inference in later chapters, we give special symbols to the population mean and the sample mean. The *population mean* is denoted by the Greek
μ letter μ (read "mu"), and the *sample mean* is denoted by the symbol $\bar{y}$ (read "y-bar").
$\bar{y}$ As indicated in Chapter 1, a population of measurements is the complete set of measurements of interest to us; a sample of measurements is a subset of measurements selected from the population of interest. If we let $y_1, y_2, \ldots, y_n$ denote the measurements observed in a sample of size n, then the sample mean $\bar{y}$ can be written as

$$\bar{y} = \frac{\sum_i y_i}{n}$$

where the symbol appearing in the numerator, $\sum_i y_i$, is the notation used to designate a sum of n measurements, y_i:

$$\sum_i y_i = y_1 + y_2 + \cdots + y_n$$

The corresponding population mean is μ.

In most situations, we will not know the population mean; the sample will be used to make inferences about the corresponding unknown population mean. For example, the accounting department of a large department store chain is conducting an examination of its overdue accounts. The store has thousands of such accounts, which would yield a population of overdue values having a mean value, μ. The value of μ could only be determined by conducting a large-scale audit that would take several days to complete. The accounting department monitors the overdue accounts on a daily basis by taking a random sample of n overdue accounts and computing the sample mean, $\bar{y}$. The sample mean, $\bar{y}$, is then used as an estimate of the mean value, μ, in *all* overdue accounts for that day. The accuracy of the estimate and approaches for determining the appropriate sample size will be discussed in Chapter 5.

EXAMPLE 3.4

A sample of $n = 15$ overdue accounts in a large department store yields the following amounts due:

$55.20	$ 4.88	$271.95
18.06	180.29	365.29
28.16	399.11	807.80
44.14	97.47	9.98
61.61	56.89	82.73

a. Determine the mean amount due for the 15 accounts sampled.
b. If there are a total of 150 overdue accounts, use the sample mean to predict the total amount overdue for all 150 accounts.

Solution

a. The sample mean is computed as follows:

$$\bar{y} = \frac{\Sigma_i y_i}{15} = \frac{55.20 + 18.06 + \cdots + 82.73}{15} = \frac{2,483.56}{15} = \$165.57$$

b. From part (a) we found that the 15 accounts sampled averaged $165.57 overdue. Using this information, we predict, or estimate, the total amount overdue for the 150 accounts to be $150(165.57) = \$24,835.50$.

The mean is a useful measure of the central value of a set of measurements, but it is subject to distortion due to the presence of one or more extreme values in **outliers**   the set. In these situations, the extreme values (called **outliers**) pull the mean in the direction of the outliers to find the balancing point, thus distorting the mean as a **trimmed mean**   measure of the central value. A variation of the mean, called a **trimmed mean,** drops the highest and lowest extreme values and averages the rest. For example, a 5% trimmed mean drops the highest 5% and the lowest 5% of the measurements and averages the rest. Similarly, a 10% trimmed mean drops the highest and the lowest 10% of the measurements and averages the rest. In Example 3.4, a 10% trimmed mean would drop the smallest and largest accounts, resulting in a mean of

$$\bar{y} = \frac{2,483.56 - 4.88 - 807.80}{13} = \$128.53$$

By trimming the data, we are able to reduce the impact of very large (or small) values on the mean, and thus get a more reliable measure of the central value of the set. This will be particularly important when the sample mean is used to predict the corresponding population central value.

Note that in a limiting sense the median is a 50% trimmed mean. Thus, the median is often used in place of the mean when there are extreme values in the data set. In Example 3.4, the value $807.80 is considerably larger than the other values in the data set. This results in 10 of the 15 accounts having values less than the mean and only 5 larger. The median value for the 15 accounts is $61.61. There are 7 accounts less than the median and 7 accounts greater than the median. Thus, in

FIGURE 3.17

Relation among the mean μ, the trimmed mean TM, the median M_d, and the mode M_o

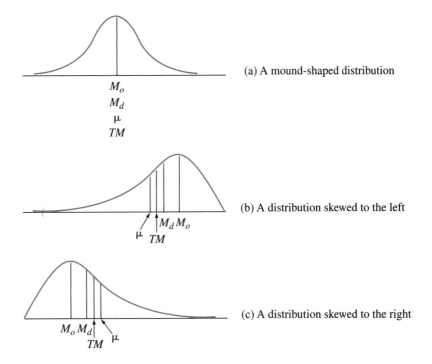

(a) A mound-shaped distribution

(b) A distribution skewed to the left

(c) A distribution skewed to the right

selecting a typical overdue account, the median is a more appropriate value than the mean. However, if we want to estimate the total amount overdue in all 150 accounts, we would want to use the mean and not the median. When estimating the sum of all measurements in a population, we do not want to exclude the extremes in the sample. Suppose a sample contains a few extremely large values. If the extremes are trimmed, then the population sum will be grossly underestimated using the sample trimmed mean or sample median in place of the sample mean.

In this section, we discussed the mode, median, mean, and trimmed mean. How are these measures of central tendency related for a given set of measurements? The answer depends on the **skewness** of the data. If the distribution is mound-shaped and symmetrical about a single peak, the mode (M_o), median (M_d), mean (μ), and trimmed mean (TM) will all be the same. This is shown using a smooth curve and population quantities in Figure 3.17(a). If the distribution is skewed, having a long tail in one direction and a single peak, the mean is pulled in the direction of the tail; the median falls between the mode and the mean; and depending on the degree of trimming, the trimmed mean usually falls between the median and the mean. Figures 3.17(b) and (c) illustrate this for distributions skewed to the left and to the right.

The important thing to remember is that we are not restricted to using only one measure of central tendency. For some data sets, it will be necessary to use more than one of these measures to provide an accurate descriptive summary of central tendency for the data.

skewness

Major Characteristics of Each Measure of Central Tendency

Mode

1. It is the most frequent or probable measurement in the data set.
2. There can be more than one mode for a data set.

3. It is not influenced by extreme measurements.
4. Modes of subsets cannot be combined to determine the mode of the complete data set.
5. For grouped data, its value can change depending on the categories used.
6. It is applicable for both qualitative and quantitative data.

Median

1. It is the central value; 50% of the measurements lie above it and 50% fall below it.
2. There is only one median for a data set.
3. It is not influenced by extreme measurements.
4. Medians of subsets cannot be combined to determine the median of the complete data set.
5. For grouped data, its value is rather stable even when the data are organized into different categories.
6. It is applicable to quantitative data only.

Mean

1. It is the arithmetic average of the measurements in a data set.
2. There is only one mean for a data set.
3. Its value is influenced by extreme measurements; trimming can help to reduce the degree of influence.
4. Means of subsets can be combined to determine the mean of the complete data set.
5. It is applicable to quantitative data only.

Measures of central tendency do not provide a complete mental picture of the frequency distribution for a set of measurements. In addition to determining the center of the distribution, we must have some measure of the spread of the data. In the next section, we discuss measures of variability, or dispersion.

EXERCISES   Basic Techniques

3.14 Compute the mean, median, and mode for the following data:

11   17   18   10   22   23   15   17
14   13   10   12   18   18   11   14

3.15 Refer to the data in Exercise 3.14. Replace the measurements 22 and 23 with 42 and 43. Recompute the mean, median, and mode. Discuss the impact of these extreme measurements on the three measures of central tendency.

3.16 Refer to Exercises 3.14 and 3.15. Compute a 10% trimmed mean for both data sets. Do the extreme values affect the 10% trimmed mean? Is a 5% trimmed mean affected?

Env. **3.17** The ratio of DDE (related to DDT) to PCB concentrations in bird eggs has been shown to have had a number of biological implications. The ratio is used as an indication of the movement of contamination through the food chain. The paper "The ratio of DDE to PCB concentrations in Great Lakes herring gull eggs and its use in interpreting contaminants data" [*Journal of Great Lakes Research* (1998) 24(1): 12–31] reports the following ratios for eggs collected at 13 study sites from the five Great Lakes. The eggs were collected from both terrestrial- and aquatic-feeding birds.

	DDE to PCB Ratio									
Terrestrial feeders	76.50	6.03	3.51	9.96	4.24	7.74	9.54	41.70	1.84	2.50	1.54
Aquatic feeders	0.27	0.61	0.54	0.14	0.63	0.23	0.56	0.48	0.16	0.18

   a. Compute the mean and median for the 21 ratios, ignoring the type of feeder.
   b. Compute the mean and median separately for each type of feeder.
   c. Using your results from parts (a) and (b), comment on the relative sensitivity of the mean and median to extreme values in a data set.
   d. Which measure, mean or median, do you recommend as the most appropriate measure of the DDE to PCB level for both types of feeders? Explain your answer.

Med.   **3.18** A study of the survival times, in days, of skin grafts on burn patients was examined in Woolson and Lachenbruch [*Biometrika* (1980) 67:597–606]. Two of the patients left the study prior to the failure of their grafts. The survival time for these individuals is some number greater than the reported value.

   Survival time (days):   37, 19, 57*, 93, 16, 22, 20, 18, 63, 29, 60*

(The "*" indicates that the patient left the study prior to failure of the graft; values given are for the day the patient left the study.)
   a. Calculate the measures of center (if possible) for the 11 patients.
   b. If the survival times of the two patients who left the study were obtained, how would these new values change the values of the summary statistics calculated in (a)?

Engin.   **3.19** A study of the reliability of diesel engines was conducted on 14 engines. The engines were run in a test laboratory. The time (in days) until the engine failed is given here. The study was terminated after 300 days. For those engines that did not fail during the study period, an asterisk is placed by the number 300. Thus, for these engines, the time to failure is some value greater than 300.

   Failure time (days):   130, 67, 300*, 234, 90, 256, 87, 120, 201, 178, 300*, 106, 289, 74

   a. Calculate the measures of center for the 14 engines.
   b. What are the implications of computing the measures of center when some of the exact failure times are not known?

Gov.   **3.20** Effective tax rates (per $100) on residential property for three groups of large cities, ranked by residential property tax rate, are shown in the following table.

Group 1	Rate	Group 2	Rate	Group 3	Rate
Detroit, MI	4.10	Burlington, VT	1.76	Little Rock, AR	1.02
Milwaukee, WI	3.69	Manchester, NH	1.71	Albuquerque, NM	1.01
Newark, NJ	3.20	Fargo, ND	1.62	Denver, CO	.94
Portland, OR	3.10	Portland, ME	1.57	Las Vegas, NV	.88
Des Moines, IA	2.97	Indianapolis, IN	1.57	Oklahoma City, OK	.81
Baltimore, MD	2.64	Wilmington, DE	1.56	Casper, WY	.70
Sioux Falls, IA	2.47	Bridgeport, CT	1.55	Birmingham, AL	.70
Providence, RI	2.39	Chicago, IL	1.55	Phoenix, AZ	.68
Philadelphia, PA	2.38	Houston, TX	1.53	Los Angeles, CA	.64
Omaha, NE	2.29	Atlanta, GA	1.50	Honolulu, HI	.59

Source: Government of the District of Columbia, Department of Finance and Revenue, *Tax Rates and Tax Burdens in the District of Columbia: A Nationwide Comparison,* annual.

a. Compute the mean, median, and mode separately for the three groups.
b. Compute the mean, median, and mode for the complete set of 30 measurements.
c. What measure or measures best summarize the center of these distributions? Explain.

3.21 Refer to Exercise 3.20. Average the three group means, the three group medians, and the three group modes, and compare your results to those of part (b). Comment on your findings.

3.5 Summarizing Data on a Single Variable: Measures of Variability

It is not sufficient to describe a data set using only measures of central tendency, such as the mean or the median. For example, suppose we are monitoring the production of plastic sheets that have a nominal thickness of 3 mm. If we randomly select 100 sheets from the daily output of the plant and find that the average thickness of the 100 sheets is 3 mm, does this indicate that all 100 sheets have the desired thickness of 3 mm? We may have a situation in which 50 sheets have a thickness of 1 mm and the remaining 50 sheets have a thickness of 5 mm. This would result in an average thickness of 3 mm, but none of the 100 sheets would have a thickness close to the specified 3 mm. Thus, we need to determine how dispersed the sheet thicknesses are about the mean of 3 mm.

variability
Graphically, we can observe the need for some measure of variability by examining the relative frequency histograms in Figure 3.18. All the histograms have the same mean, but each has a different spread, or **variability,** about the mean. For illustration, we have shown the histograms as smooth curves. Suppose the three histograms represent the amount of PCB (in ppb) found in a large number of 1-liter samples taken from three lakes that are close to chemical plants. The average amount of PCB, μ, in a 1-liter sample is the same for all three lakes. However, the variability in the PCB quantity is considerably different. Thus, the lake with PCB quantity depicted in histogram (a) would have fewer samples containing very small or large quantities of PCB as compared to the lake with PCB values depicted in

FIGURE 3.18

Relative frequency histograms with different variabilities but the same mean

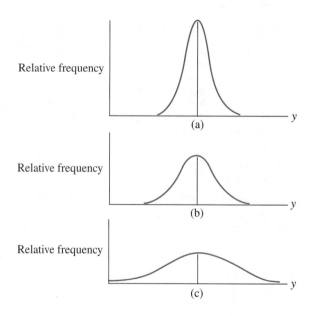

histogram (c). Knowing only the mean PCB quantity in the three lakes would mislead the investigator concerning the level of PCB present in all three lakes.

range   The simplest but least useful measure of data variation is the **range,** which we alluded to in Section 3.2. We now present its definition.

DEFINITION 3.4

> The **range** of a set of measurements is defined to be the difference between the largest and the smallest measurements of the set.

EXAMPLE 3.5

Determine the range of the 15 overdue accounts of Example 3.4.

Solution The smallest measurement is $4.88 and the largest is $807.80. Hence, the range is

$$807.80 - 4.88 = \$802.92$$

Although the range is easy to compute, it is sensitive to outliers because it depends on the most extreme values. It does not give much information about the pattern of variability. Referring to the situation described in Example 3.4, if in the current budget period the 15 overdue accounts consisted of 10 accounts having a value of $4.88, 3 accounts of $807.80, and 1 account of $11.36, then the mean value would be $165.57 and the range would be $802.92. The mean and range would be identical to the mean and range calculated for the data in Example 3.4. However, the data in the current budget period are more spread out about the mean than the data in the earlier budget period. What we seek is a measure of variability that discriminates among data sets having different degrees of concentration of the data about the mean.

percentile   A second measure of variability involves the use of **percentiles.**

DEFINITION 3.5

> The *p*th **percentile** of a set of n measurements arranged in order of magnitude is that value that has at most $p\%$ of the measurements below it and at most $(100 - p)\%$ above it.

For example, Figure 3.19 illustrates the 60th percentile of a set of measurements. Percentiles are frequently used to describe the results of achievement test scores and the ranking of a person in comparison to the rest of the people taking an

FIGURE 3.19

The 60th percentile of a set of measurements

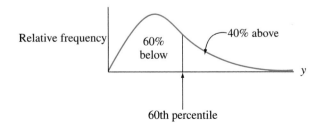

FIGURE 3.20

Quartiles of a distribution

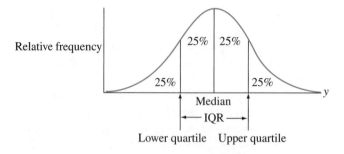

examination. Specific percentiles of interest are the 25th, 50th, and 75th percentiles, often called the *lower quartile,* the *middle quartile* (median), and the *upper quartile,* respectively (see Figure 3.20).

The computation of percentiles is accomplished as follows. Each data value corresponds to a percentile for the percentage of the data values that are less than or equal to it. Let $y_{(1)}, y_{(2)}, \ldots, y_{(n)}$ denote the ordered observations for a data set; that is,

$$y_{(1)} \leq y_{(2)} \leq \cdots \leq y_{(n)}$$

The *j*th ordered observation, $y_{(j)}$, corresponds to the $100(j - .5)/n$ percentile. We use this formula in place of assigning the percentile $100j/n$ so that we avoid assigning the 100th percentile to $y_{(n)}$, which would imply that the largest possible data value in the population was observed in the data set, an unlikely happening. For example, a study of serum total cholesterol (mg/l) levels recorded the levels given in Table 3.5 for 20 adult patients. Thus, each ordered observation is a data percentile corresponding to a multiple of the fraction $100(j - .5)/n = 100(2j - 1)/2n = 100(2j - 1)/40$.

TABLE 3.5

Observation (*j*)	Cholesterol (mg/l)	Percentile
1	133	2.5
2	137	7.5
3	148	12.5
4	149	17.5
5	152	22.5
6	167	27.5
7	174	32.5
8	179	37.5
9	189	42.5
10	192	47.5
11	201	52.5
12	209	57.5
13	210	62.5
14	211	67.5
15	218	72.5
16	238	77.5
17	245	82.5
18	248	87.5
19	253	92.5
20	257	97.5

The 22.5th percentile is 152 (mg/l). Thus, 22.5% of the people in the study have a serum cholesterol less than or equal to 152. Also, the median of the data set, which is the 50th percentile, is halfway between 192 and 201; that is, median = (192 + 201)/2 = 196.5. Thus, approximately one-half of the people in the study have a serum cholesterol level less than 196.5 and one-half have one greater than 196.5.

When dealing with large data sets, the percentiles are generalized to quantiles, where a quantile, denoted $Q(u)$, is a number that divides a sample of n data values into two groups so that the specified fraction u of the data values is less than or equal to the value of the quantile, $Q(u)$. Plots of the quantiles $Q(u)$ versus the data fraction u provide a method of obtaining estimated quantiles for the population from which the data were selected. We can obtain a quantile plot using the following steps:

1. Place a scale on the horizontal axis of a graph covering the interval (0, 1).
2. Place a scale on the vertical axis covering the range of the observed data, y_1 to y_n.
3. Plot $y_{(i)}$ versus $u_i = (i - .5)/n = (2i - 1)/2n$, for $i = 1, \ldots, n$.

Using the Minitab software, we obtain the plot shown in Figure 3.21 for the cholesterol data. Note that, with Minitab, the vertical axis is labeled $Q(u)$ rather than $y_{(i)}$. We plot $y_{(i)}$ versus u to obtain a quantile plot. Specific quantiles can be read from the plot.

We can obtain the quantile, $Q(u)$, for any value of u as follows. First, place a smooth curve through the plotted points in the quantile plot and then read the value off the graph corresponding to the desired value of u.

To illustrate the calculations, suppose we want to determine the 80th percentile for the cholesterol data—that is, the cholesterol level such that 80% of the people in the population have a cholesterol level less than this value, $Q(.80)$.

Referring to Figure 3.21, locate the point $u = .8$ on the horizontal axis and draw a perpendicular line up to the quantile plot and then a horizontal line over to the vertical axis. The point where this line touches the vertical axis is our estimate of the 80th quantile. (See Figure 3.22.) Roughly 80% of the population have a cholesterol level less than 243.

interquartile range   The second measure of variability, the **interquartile range,** is now defined.

FIGURE 3.21

Quantile plot of cholesterol data

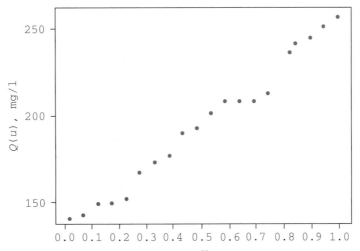

FIGURE 3.22

80th quantile of cholesterol data

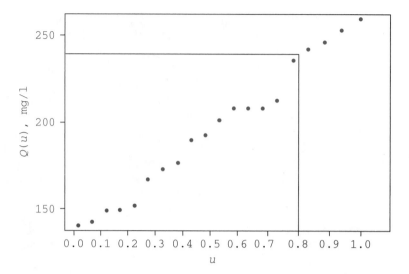

DEFINITION 3.6

The **interquartile range (IQR)** of a set of measurements is defined to be the difference between the upper and lower quartiles; that is,

IQR = 75th percentile − 25th percentile

The interquartile range, although more sensitive to data pileup about the midpoint than the range, is still not sufficient for our purposes. In fact, the IQR can be very misleading when the data set is highly concentrated about the median. For example, suppose we have a sample consisting of 10 data values:

20, 50, 50, 50, 50, 50, 50, 50, 50, 80

The mean, median, lower quartile, and upper quartile all equal 50. Thus, IQR equals 50 − 50 = 0. This is very misleading because a measure of variability equal to 0 should indicate that the data consist of *n identical* values, which is not the case in our example. The IQR ignores the extremes in the data set completely. In fact, the IQR only measures the distance needed to cover the middle 50% of the data values and, hence, totally ignores the spread in the lower and upper 25% of the data. In summary, the IQR does not provide a lot of useful information about the variability of a single set of measurements, but can be quite useful when comparing the variabilities of two or more data sets. This is especially true when the data sets have some skewness. The IQR will be discussed further as part of the boxplot (Section 3.6).

In most data sets, we typically need a minimum of five summary values to provide a minimal description of the data set: smallest value, $y_{(1)}$; lower quartile, $Q(.25)$; median; upper quartile, $Q(.75)$; and the largest value, $y_{(n)}$. When the data set has a unimodal, bell-shaped, and symmetric relative frequency histogram, just the sample mean and a measure of variability, the sample variance, can represent the data set. We will now develop the sample variance.

We seek now a sensitive measure of variability, not only for comparing the variabilities of two sets of measurements but also for interpreting the variability of a single set of measurements. To do this, we work with the **deviation** $y - \overline{y}$ of a measurement y from the mean $\overline{y}$ of the set of measurements.

deviation

FIGURE 3.23

Dot diagram of the
percentages of registered
voters in five cities

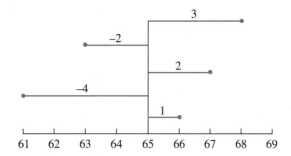

To illustrate, suppose we have five sample measurements $y_1 = 68$, $y_2 = 67$, $y_3 = 66$, $y_4 = 63$, and $y_5 = 61$, which represent the percentages of registered voters in five cities who exercised their right to vote at least once during the past year. These measurements are shown in the dot diagram of Figure 3.23. Each measurement is located using a dot above the horizontal axis of the diagram. We use the sample mean

$$\bar{y} = \frac{\Sigma_i y_i}{n} = \frac{325}{5} = 65$$

to locate the center of the set and we construct horizontal lines in Figure 3.23 to represent the deviations of the sample measurements from their mean. The deviations of the measurements are computed by using the formula $y - \bar{y}$. The five measurements and their deviations are shown in Figure 3.23.

A data set with very little variability has most of the measurements located near the center of the distribution. Deviations from the mean for a more variable set of measurements are relatively large.

Many different measures of variability can be constructed by using the deviations $y - \bar{y}$. A first thought is to use the mean deviation, but this will always equal zero, as it does for our example. A second possibility is to ignore the minus signs and compute the average of the absolute values. However, a more easily interpreted function of the deviations involves the sum of the squared deviations of the measurements from their mean. This measure is called the **variance.**

variance

DEFINITION 3.7

> The **variance** of a set of n measurements $y_1, y_2, \ldots, y_n$ with mean $\bar{y}$ is the sum of the squared deviations divided by $n - 1$:
>
> $$\frac{\Sigma_i(y_i - \bar{y})^2}{n - 1}$$

s^2

σ^2

As with the sample and population means, we have special symbols to denote the sample and population variances. The symbol s^2 represents the sample variance, and the corresponding population variance is denoted by the symbol σ^2.

The definition for the variance of a set of measurements depends on whether the data are regarded as a sample or population of measurements. The definition we have given here assumes we are working with the sample, because the population measurements usually are not available. Many statisticians define the sample variance to be the average of the squared deviations, $\Sigma_i(y_i - \bar{y})^2/n$. However, the use of $(n - 1)$ as the denominator of s^2 is not arbitrary. This definition of the sample variance makes it an *unbiased estimator* of the population variance σ^2. This means roughly that, if we were to draw a very large number of samples, each of size n, from the population of interest and if we computed s^2 for each sample, the average sample variance would equal the population variance σ^2. Had we divided by n in the definition of the sample variance s^2, the average sample variance computed from a large number of samples would be less than the population variance; hence, s^2 would tend to underestimate σ^2.

standard deviation   Another useful measure of variability, the **standard deviation,** involves the square root of the variance. One reason for defining the standard deviation is that it yields a measure of variability having the same units of measurement as the original data, whereas the units for variance are the square of the measurement units.

DEFINITION 3.8   The **standard deviation** of a set of measurements is defined to be the positive square root of the variance.

s
σ

We then have *s* denoting the sample standard deviation and *σ* denoting the corresponding population standard deviation.

EXAMPLE 3.6

The time between an electric light stimulus and a bar press to avoid a shock was noted for each of five conditioned rats. Use the given data to compute the sample variance and standard deviation.

shock avoidance times (seconds):   5, 4, 3, 1, 3

Solution The deviations and the squared deviations are shown in Table 3.6. The sample mean $\bar{y}$ is 3.2.

TABLE 3.6

	y_i	$y_i - \bar{y}$	$(y_i - \bar{y})^2$
	5	1.8	3.24
	4	.8	.64
	3	−.2	.04
	1	−2.2	4.84
	3	−.2	.04
Totals	16	0	8.80

Using the total of the squared deviations column, we find the sample variance to be

$$s^2 = \frac{\Sigma_i(y_i - \bar{y})^2}{n - 1} = \frac{8.80}{4} = 2.2$$

We have now discussed several measures of variability, each of which can be used to compare the variabilities of two or more sets of measurements. The standard deviation is particularly appealing for two reasons: (1) we can compare the variabilities of *two or more* sets of data using the standard deviation, and (2) we can also use the results of the rule that follows to interpret the standard deviation of a single set of measurements. This rule applies to data sets with roughly a "mound-shaped" histogram—that is, a histogram that has a single peak, is symmetrical, and tapers off gradually in the tails. Because so many data sets can be classified as mound-shaped, the rule has wide applicability. For this reason, it is called the *Empirical Rule*.

EMPIRICAL RULE

Given a set of n measurements possessing a mound-shaped histogram, then

the interval $\bar{y} \pm s$ contains approximately 68% of the measurements
the interval $\bar{y} \pm 2s$ contains approximately 95% of the measurements
the interval $\bar{y} \pm 3s$ contains approximately 99.7% of the measurements

EXAMPLE 3.7

The yearly report from a particular stockyard gives the average daily wholesale price per pound for steers as $.61, with a standard deviation of $.07. What conclusions can we reach about the daily steer prices for the stockyard? Because the original daily price data are not available, we are not able to provide much further information about the daily steer prices. However, from past experience it is known that the daily price measurements have a mound-shaped relative frequency histogram. Applying the Empirical Rule, what conclusions can we reach about the distribution of daily steer prices?

Solution   Applying the Empirical Rule, the interval

$.61 \pm .07$ or $.54 to $.68

contains approximately 68% of the measurements. The interval

$.61 \pm .14$ or $.47 to $.75

contains approximately 95% of the measurements. The interval

$.61 \pm .21$ or $.40 to $.82

contains approximately 99.7% of the measurements.

In English, approximately $\frac{2}{3}$ of the steers sold for between $.54 and $.68 per pound and 95% sold for between $.47 and $.75 per pound, with minimum and maximum prices being approximately $.40 and $.82.

To increase our confidence in the Empirical Rule, let us see how well it describes the five frequency distributions of Figure 3.24. We calculated the mean and standard deviation for each of the five data sets (not given), and these are shown next to each frequency distribution. Figure 3.24(a) shows the frequency distribution for measurements made on a variable that can take values $y = 0, 1, 2, \ldots, 10$. The mean and standard deviation $\bar{y} = 5.50$ and $s = 1.49$ for this symmetric mound-shaped distribution were used to calculate the interval $\bar{y} \pm 2s$, which is marked below the horizontal axis of the graph. We found 94% of the measurements falling in this interval—that is, lying within two standard deviations of the mean. Note that this percentage is very close to the 95% specified in the Empirical Rule. We also calculated the percentage of measurements lying within one standard deviation of the mean. We found this percentage to be 60%, a figure that is not too far from the

FIGURE 3.24   A demonstration of the utility of the Empirical Rule

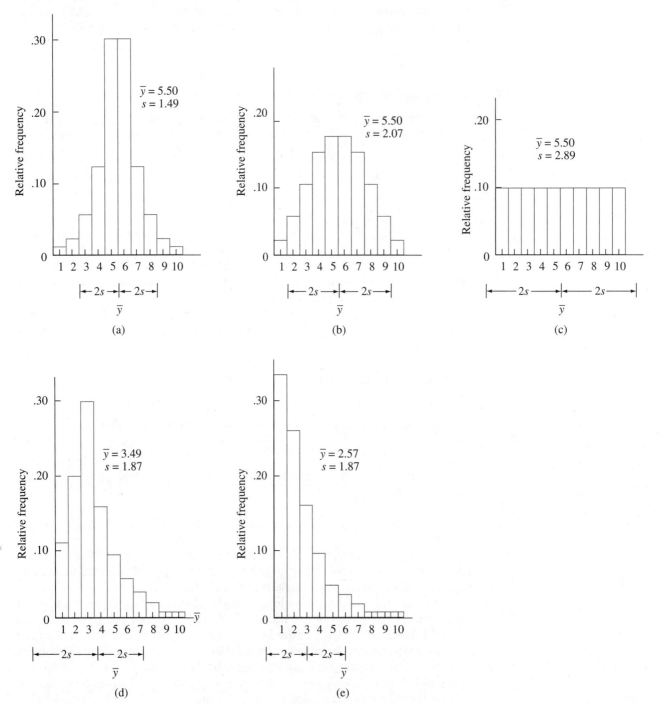

68% specified by the Empirical Rule. Consequently, we think the Empirical Rule provides an adequate description for Figure 3.24(a).

Figure 3.24(b) shows another mound-shaped frequency distribution, but one that is less peaked than the distribution of Figure 3.24(a). The mean and standard deviation for this distribution, shown to the right of the figure, are 5.50 and 2.07, respectively. The percentages of measurements lying within one and two standard deviations of the mean are 64% and 96%, respectively. Once again, these percentages agree very well with the Empirical Rule.

Now let us look at three other distributions. The distribution in Figure 3.24(c) is perfectly flat, while the distributions of Figures 3.24(d) and (e) are nonsymmetric and skewed to the right. The percentages of measurements that lie within two standard deviations of the mean are 100%, 96%, and 95%, respectively, for these three distributions. All these percentages are reasonably close to the 95% specified by the Empirical Rule. The percentages that lie within one standard deviation of the mean are 60%, 75%, and 87% respectively. Only the value for Figure 3.24(e) shows a large disagreement with the 68% of the Empirical Rule.

To summarize, you can see that the Empirical Rule accurately forecasts the percentage of measurements falling within two standard deviations of the mean for all five distributions of Figure 3.24, even for the distributions that are flat, as in Figure 3.24(c), or highly skewed to the right, as in Figure 3.24(e). The Empirical Rule is less accurate in forecasting the percentages within one standard deviation of the mean, but the forecast, 68%, compares reasonably well for the three distributions that might be called mound-shaped, Figures 3.24(a), (b), and (d).

The results of the Empirical Rule enable us to obtain a quick approximation to the sample standard deviation s. The Empirical Rule states that approximately 95% of the measurements lie in the interval $\bar{y} \pm 2s$. The length of this interval is, therefore, $4s$. Because the range of the measurements is approximately $4s$, we obtain an **approximate value for s** by dividing the range by 4:

approximating s

$$\text{approximate value of } s = \frac{\text{range}}{4}$$

Some people might wonder why we did not equate the range to $6s$, because the interval $\bar{y} \pm 3s$ should contain almost all the measurements. This procedure yields an approximate value for s that is smaller than the one obtained by the preceding procedure. If we are going to make an error (as we are bound to do with any approximation), it is better to overestimate the sample standard deviation so that we are not led to believe there is less variability than may be the case.

EXAMPLE 3.8

The following data represent the percentages of family income allocated to groceries for a sample of 30 shoppers:

26	28	30	37	33	30
29	39	49	31	38	36
33	24	34	40	29	41
40	29	35	44	32	45
35	26	42	36	37	35

For these data, $\Sigma y_i = 1,043$ and $\Sigma(y_i - \bar{y})^2 = 1,069.3667$. Compute the mean, variance, and standard deviation of the percentage of income spent on food. Check your calculation of s.

Solution The sample mean is

$$\bar{y} = \frac{\Sigma_i y_i}{30} = \frac{1,043}{30} = 34.77$$

The corresponding sample variance and standard deviation are

$$s^2 = \frac{1}{n-1}\Sigma(y_i - \bar{y})^2$$

$$= \frac{1}{29}(1{,}069.3667) = 36.8747$$

$$s = \sqrt{36.8747} = 6.07$$

We can check our calculation of s by using the range approximation. The largest measurement is 49 and the smallest is 24. Hence, an approximate value of s is

$$s \approx \frac{\text{range}}{4} = \frac{49 - 24}{4} = 6.25$$

Note how close the approximation is to our computed value.

Although there will not always be the close agreement found in Example 3.8, the range approximation provides a useful and quick check on the calculation of s.

EXERCISES

Basic Techniques

Engin.

3.22 Pushing economy and wheelchair propulsion technique were examined for eight wheel-chair racers on a motorized treadmill in a paper by Goosey and Campbell [*Adapted Physical Activity Quarterly* (1998) 15:36–50]. The eight racers had the following years of racing experience:

Racing experience (years): 6, 3, 10, 4, 4, 2, 4, 7

a. Verify that the mean years' experience is 5 years. Does this value appear to adequately represent the center of the data set?
b. Verify that $\Sigma_i(y - \bar{y})^2 = \Sigma_i(y - 5)^2 = 46$.
c. Calculate the sample variance and standard deviation for the experience data. How would you interpret the value of the standard deviation relative to the sample mean?

3.23 In the study described in Exercise 3.22, the researchers also recorded the ages of the eight racers.

Age (years): 39, 38, 31, 26, 18, 36, 20, 31

a. Calculate the sample standard deviation of the eight racers' ages.
b. Why would you expect the standard deviation of the racers' ages to be larger than the standard deviation of their years of experience?

3.24 For the racing experience data in Exercise 3.22, estimate the standard deviation by dividing the range by 4. How close is this estimate to the value you calculated in Exercise 3.22? Repeat this exercise for the data in Exercise 3.23.

Applications

Bus.

3.25 *Consumer Reports* in its June 1998 issue reports on the typical daily room rate at six luxury and nine budget hotels. The room rates are given in the following table.

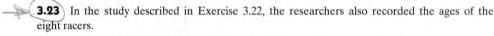

| Luxury Hotel | $175 | $180 | $120 | $150 | $120 | $125 | | | |
| Budget Hotel | $50 | $50 | $49 | $45 | $36 | $45 | $50 | $50 | $40 |

a. Compute the mean and standard deviation of the room rates for both luxury and budget hotels.

   b. Verify that luxury hotels have a more variable room rate than budget hotels.
   c. Give a practical reason why the luxury hotels are more variable than the budget hotels.
   d. Could another measure of variability be better to compare luxury and budget hotel rates? Explain.

Env. **3.26** Many marine phanerogam species are highly sensitive to changes in environmental conditions. In the article "*Posidonia oceanica:* A biological indicator of past and present mercury contamination in the Mediterranean Sea" [*Marine Environmental Research*, 45:101–111], the researchers report the mercury concentrations over a period of about 20 years at several locations in the Mediterranean Sea. Samples of *Posidonia oceanica* were collected by scuba diving at a depth of 10 m. For each site, 45 orthotropic shoots were sampled and the mercury concentration was determined. The average mercury concentration is recorded in the following table for each of the sampled years.

	Mercury Concentration (ng/g dry weight)	
Year	**Site 1 Calvi**	**Site 2 Marseilles-Coriou**
1992	14.8	70.2
1991	12.9	160.5
1990	18.0	102.8
1989	8.7	100.3
1988	18.3	103.1
1987	10.3	129.0
1986	19.3	156.2
1985	12.7	117.6
1984	15.2	170.6
1983	24.6	139.6
1982	21.5	147.8
1981	18.2	197.7
1980	25.8	262.1
1979	11.0	123.3
1978	16.5	363.9
1977	28.1	329.4
1976	50.5	542.6
1975	60.1	369.9
1974	96.7	705.1
1973	100.4	462.0
1972	*	556.1
1971	*	461.4
1970	*	628.8
1969	*	489.2

   a. Generate a time series plot of the mercury concentrations and place lines for both sites on the same graph. Comment on any trends in the lines across the years of data. Are the trends similar for both sites?
   b. Select the most appropriate measure of center for the mercury concentrations. Compare the center for the two sites.
   c. When comparing the center and variability of the two sites, should the years 1969–1972 be used for site 2?

3.6   **The Boxplot**

boxplot

As mentioned earlier in this chapter, a stem-and-leaf plot provides a graphical representation of a set of scores that can be used to examine the shape of the distribution, the range of scores, and where the scores are concentrated. The **boxplot,** which builds on the information displayed in a stem-and-leaf plot, is more concerned with the symmetry of the distribution and incorporates numerical measures of central tendency and location to study the variability of the scores and the concentration of scores in the tails of the distribution.

Before we show how to construct and interpret a boxplot, we need to introduce several new terms that are peculiar to the language of exploratory data analysis (EDA). We are familiar with the definitions for the first, second (median), and third quartiles of a distribution presented earlier in this chapter. The boxplot uses the median and **quartiles** of a distribution.

quartiles

We can now illustrate a *skeletal boxplot* using an example.

EXAMPLE 3.9

Use the stem-and-leaf plot in Figure 3.25 for the 90 violent crime rates of Table 3.4 in Section 3.3 to construct a skeletal boxplot.

FIGURE 3.25
Stem-and-leaf plot

```
 1   89
 2   10 24 67 91 96 98
 3   36 41 52 54 74 76 88 93 93
 4   10 21 35 47 48 60 64 66 80 81 91 96 98
 5   04 05 08 16 26 29 37 57 59 61 62 62 62 63 70 71 78 85 92
 6   05 05 24 26 28 31 39 42 47 61 73 84 85 85 90 98
 7   03 06 18 19 20 31 35 39 51 58 71
 8   04 07 09 11 14 17 43 56 68 76 77 85
 9   28 71
10   20
```

Solution When the scores are ordered from lowest to highest, the median is computed by averaging the 45th and 46th scores. For these data, the 45th score (counting from the lowest to the highest in Figure 3.25) is 571 and the 46th is 578; hence, the median is

$$M = \frac{571 + 578}{2} = 574.5$$

To find the lower and upper quartiles for this distribution of scores, we need to determine the 25th and 75th percentiles. We can use the method given in Section 3.5 to compute $Q(.25)$ and $Q(.75)$. A quick method that yields essentially the same values for the two quartiles consists of the following steps:

1. Order the data from smallest to largest value.
2. Divide the ordered data set into two data sets using the median as the dividing value.
3. Let the lower quartile be the median of the set of values consisting of the smaller values.
4. Let the upper quartile be the median of the set of values consisting of the larger values.

In the example, the data set has 90 values. Thus, we create two data sets, one containing the $90/2 = 45$ smallest values and the other containing the 45 largest values. The lower quartile is the $(45 + 1)/2 = 23$rd smallest value and the upper quartile is the 23rd value counting from the largest value in the data set. The 23rd-lowest and 23rd-highest scores are 464 and 719.

lower quartile, $Q_1 = 464$
upper quartile, $Q_3 = 719$

skeletal boxplot

These three descriptive measures and the smallest and largest values in a data set are used to construct a skeletal boxplot (see Figure 3.26). The **skeletal boxplot** is constructed by drawing a box between the lower and upper quartiles with a solid line drawn across the box to locate the median. A straight line is then drawn connecting the box to the largest value; a second line is drawn from the box to the smallest value. These straight lines are sometimes called whiskers, and the entire graph is called a **box-and-whiskers plot.**

box-and-whiskers plot

FIGURE 3.26

Skeletal boxplot for the data of Figure 3.25

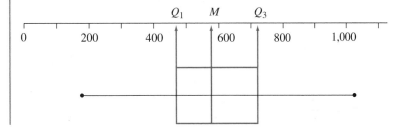

With a quick glance at a skeletal boxplot, it is easy to obtain an impression about the following aspects of the data:

1. The lower and upper quartiles, Q_1 and Q_3
2. The interquartile range (IQR), the distance between the lower and upper quartiles
3. The most extreme (lowest and highest) values
4. The symmetry or asymmetry of the distribution of scores

If we were presented with Figure 3.26 without having seen the original data, we would have observed that

$$Q_1 \approx 475$$
$$Q_3 \approx 725$$
$$IQR \approx 725 - 475 = 250$$
$$M \approx 575$$

most extreme values: 175 and 1,025

Also, because the median is closer to the lower quartile than the upper quartile and because the upper whisker is a little longer than the lower whisker, the distribution is slightly asymmetrical. To see that this conclusion is true, construct a frequency histogram for these data.

The skeletal boxplot can be expanded to include more information about extreme values in the tails of the distribution. To do so, we need the following additional quantities:

lower inner fence: $Q_1 - 1.5(\text{IQR})$
upper inner fence: $Q_3 + 1.5(\text{IQR})$
lower outer fence: $Q_1 - 3(\text{IQR})$
upper outer fence: $Q_3 + 3(\text{IQR})$

Any score beyond an inner fence on either side is called a *mild outlier,* and a score beyond an outer fence on either side is called an *extreme outlier.*

EXAMPLE 3.10

Compute the inner and outer fences for the data of Example 3.9. Identify any mild and extreme outliers.

Solution For these data, we found the lower and upper quartiles to be 464 and 719, respectively; $\text{IQR} = 719 - 464 = 255$. Then

lower inner fence $= 464 - 1.5(255) = 81.5$
upper inner fence $= 719 + 1.5(255) = 1{,}101.5$
lower outer fence $= 464 - 3(255) = -301$
upper outer fence $= 719 + 3(255) = 1{,}484$

Also, from the stem-and-leaf plot we see that the lower and upper adjacent values are 189 and 1,020. Because the upper and lower fences are 1,101.5 and 81.5, respectively, there are no observations beyond the inner fences. Hence, there are no mild or extreme outliers.

We now have all the quantities necessary for constructing a boxplot.

Steps in Constructing a Boxplot

1. As with a skeletal boxplot, mark off a box from the lower quartile to the upper quartile.
2. Draw a solid line across the box to locate the median.
3. Mark the location of the upper and lower adjacent values with an x.
4. Draw a dashed line between each quartile and its adjacent value.
5. Mark each outlier with the symbol *.
6. Mark each extreme outlier with the symbol o.

EXAMPLE 3.11

Construct a boxplot for the data of Example 3.9.

Solution The boxplot is shown in Figure 3.27.

FIGURE 3.27
The boxplot for the data of Example 3.9

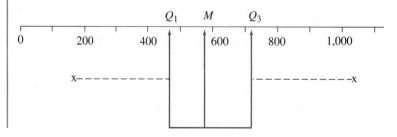

What information can be drawn from a boxplot? First, the center of the distribution of scores is indicated by the median line in the boxplot. Second, a measure of the variability of the scores is given by the interquartile range, the length of the box. Recall that the box is constructed between the lower and upper quartiles so it contains the middle 50% of the scores in the distribution, with 25% on either side of the median line inside the box. Third, by examining the relative position of the median line, we can gauge the symmetry of the middle 50% of the scores. For example, if the median line is closer to the lower quartile than the upper, there is a greater concentration of scores on the lower side of the median within the box than on the upper side; for a symmetric distribution of scores the median line is located in the center of the box. Fourth, additional information about skewness is obtained from the lengths of the whiskers; the longer one whisker is relative to the other, the more skewness there is in the tail with the longer whisker. Fifth, a general assessment can be made about the presence of outliers by examining the number of scores classified as mild outliers and the number classified as extreme outliers.

Boxplots provide a powerful graphical technique for comparing samples from several different treatments or populations. We can illustrate these concepts using the following example. Several new filtration systems have been proposed for use in small city water systems. The three systems under consideration have very similar initial and operating costs, and will be compared on the basis of the amount of impurities that remain in the water after passing through the system. After careful assessment, it is determined that monitoring 20 days of operation will provide sufficient information to determine any significant difference among the three systems. Water samples are collected on a hourly basis. The amount of impurities (in ppm) remaining in the water after the water passes through the filter is recorded. The average daily values for the three systems are plotted using a side-by-side boxplot, as presented in Figure 3.28.

An examination of the boxplots in Figure 3.28 reveals the shapes of the relative frequency histograms for the three types of filters based on their boxplots. Filter A has approximately a symmetric distribution, filter B is skewed to the right, and filter C is skewed to the left. Filters A and B have nearly equal medians. However, filter B is much more variable than both filters A and C. Filter C has a larger median than both filters A and B but smaller variability than A with the exception of the two very small values obtained using filter C. The extreme values obtained by filters C and B, identified by *, would be examined to make sure that they are valid measurements. These measurements could be either recording errors or

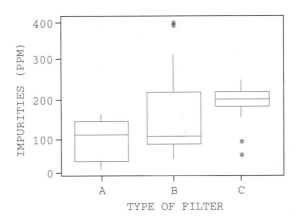

FIGURE 3.28

Removing impurities using three filter types

operational errors. They must be carefully checked because they have such a large influence on the summary statistics. Filter A produces a more consistent filtration than filter B. Filter A generally filters the water more thoroughly than filter C. We will introduce statistical techniques in Chapter 8 that will provide us with ways to differentiate among the three filter types.

EXERCISES   Applications

Med.   **3.27** The number of people who volunteered to give a pint of blood at a central donor center was recorded for each of 20 successive Fridays. The data are shown here:

| 320 | 370 | 386 | 334 | 325 | 315 | 334 | 301 | 270 | 310 |
| 274 | 308 | 315 | 368 | 332 | 260 | 295 | 356 | 333 | 250 |

a. Construct a stem-and-leaf plot.
b. Construct a boxplot and describe the shape of the distribution of the number of people donating blood.

Bus.   **3.28** *Consumer Reports* in its May 1998 issue provides cost per daily feeding for 28 brands of dry dog food and 23 brands of canned dog food. Using the Minitab computer program, the following side-by-side boxplot for these data was created.

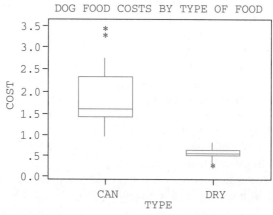

a. From these graphs, determine the median, lower quartile, and upper quartile for the daily costs of both dry and canned dog food.
b. Comment on the similarities and differences in the distributions of daily costs for the two types of dog food.

3.7 Summarizing Data from More than One Variable

In the previous sections, we've discussed graphical methods and numerical descriptive methods for summarizing data from a single variable. Frequently, more than one variable is being studied at the same time, and we might be interested in summarizing the data on each variable separately and also in studying relations among the variables. For example, we might be interested in the prime interest rate and in the consumer price index, as well as in the relation between the two. In this section, we'll discuss a few techniques for summarizing data from two (or more) variables. Material in this section will provide a brief preview and introduction to chi-square methods (Chapter 10), analysis of variance (Chapters 8 and 9), and regression (Chapters 11 and 12).

TABLE 3.7

Data from a survey of television viewing

Network Preference	Residence			
	Urban	Suburban	Rural	Total
ABC	144	180	90	414
CBS	135	240	96	471
NBC	108	225	54	387
Other	63	105	60	228
Total	450	750	300	1,500

TABLE 3.8

Comparing the distribution of residences for each network

Network Preference	Residence			
	Urban	Suburban	Rural	Total
ABC	34.8	43.5	21.7	$100 (n = 414)$
CBS	28.7	50.9	20.4	$100 (n = 471)$
NBC	27.9	58.1	14.0	$100 (n = 387)$
Other	27.6	46.1	26.3	$100 (n = 228)$

contingency table

Consider first the problem of summarizing data from two qualitative variables. Cross-tabulations can be constructed to form a **contingency table.** The rows of the table identify the categories of one variable, and the columns identify the categories of the other variable. The entries in the table are the number of times each value of one variable occurs with each possible value of the other. For example, a television viewing survey was conducted on 1,500 individuals. Each individual surveyed was asked to state his or her place of residence and network preference for national news. The results of the survey are shown in Table 3.7. As you can see, 144 urban residents preferred ABC, 135 urban residents preferred CBS, and so on.

The simplest method for looking at relations between variables in a contingency table is a percentage comparison based on the row totals, the column totals, or the overall total. If we calculate percentages within each row of Table 3.7, we can compare the distribution of residences within each network preference. A percentage comparison such as this, based on the row totals, is shown in Table 3.8.

Except for ABC, which has the highest urban percentage among the networks, the differences among the residence distributions are in the suburban and rural categories. The percentage of suburban preferences rises from 43.5% for ABC to 58.1% for NBC. Corresponding shifts downward occur in the rural category. In Chapter 10, we will use chi-square methods to explore further relations between two (or more) qualitative variables.

stacked bar graph

An extension of the bar graph provides a convenient method for displaying data from a pair of qualitative variables. Figure 3.29 is a **stacked bar graph,** which displays the data in Table 3.8.

The graph represents the distribution of television viewers of each of the major network's news programs based on the location of the viewer's residence. This type of information is often used by advertisers to determine on which networks' programs they will place their commercials.

A second extension of the bar graph provides a convenient method for displaying the relationship between a single quantitative and a qualitative variable. A

FIGURE 3.29

Comparison of distribution of residences for each network

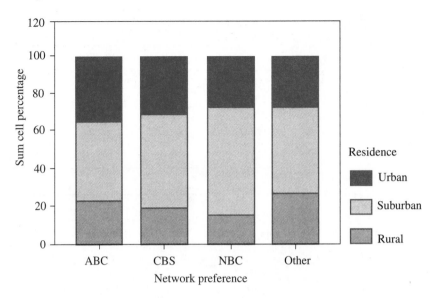

TABLE 3.9

Descriptive statistics with the dependent variable, specific volume

Fat	Surfactant	Mean	Standard Deviation	N
1	1	5.567	1.206	3
	2	6.200	.794	3
	3	5.900	.458	3
	Total	5.889	.805	9
2	1	6.800	.794	3
	2	6.200	.849	2
	3	6.000	.606	4
	Total	6.311	.725	9
3	1	6.500	.849	2
	2	7.200	.668	4
	3	8.300	1.131	2
	Total	7.300	.975	8
Total	1	6.263	1.023	8
	2	6.644	.832	9
	3	6.478	1.191	9
	Total	6.469	.997	26

food scientist is studying the effects of combining different types of fats with different surfactants on the specific volume of baked bread loaves. The experiment is designed with three levels of surfactant and three levels of fat, a 3×3 factorial experiment with a varying number of loaves baked from each of the nine treatments. She bakes bread from dough mixed from the nine different combinations of the types of fat and types of surfactants and then measures the specific volume of the bread. The data and summary statistics are displayed in Table 3.9.

In this experiment, the scientist wants to make inferences from the results of the experiment to the commercial production process. Figure 3.30 is a **cluster bar**

cluster bar graph

graph from the baking experiment. This type of graph allows the experimenter to examine the simultaneous effects of two factors, type of fat and type of surfactant,

FIGURE 3.30
Specific volumes from
baking experiment

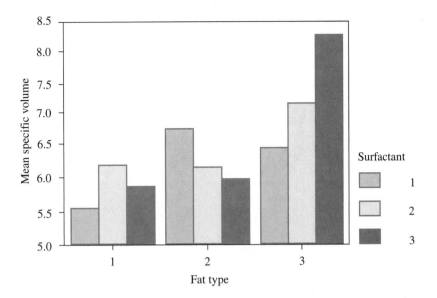

on the specific volume of the bread. Thus, the researcher can examine the differences in the specific volumes of the nine different ways in which the bread was formulated.

We can also construct data plots to summarize the relation between two quantitative variables. Consider the following example. A manager of a small machine shop examined the starting hourly wage y offered to machinists with x years of experience. The data are shown here:

y (dollars)	8.90	8.70	9.10	9.00	9.79	9.45	10.00	10.65	11.10	11.05
x (years)	1.25	1.50	2.00	2.00	2.75	4.00	5.00	6.00	8.00	12.00

scatterplot

Is there a relationship between hourly wage offered and years of experience? One way to summarize these data is to use a **scatterplot**, as shown in Figure 3.31. Each point on the plot represents a machinist with a particular starting wage and years of experience. The smooth curve fitted to the data points, called the *least squares line*, represents a summary of the relationship between y and x. This line allows the prediction of hourly starting wages for a machinist having years of experience not represented in the data set. How this curve is obtained will be discussed in Chapters 11 and 12. In general, the fitted curve indicates that, as the years of experience x increases, the hourly starting wage increases to a point and then levels off.

Finally, we can construct data plots for summarizing the relation between several quantitative variables. Consider the following example. Thall and Vail (1990) described a study to evaluate the effectiveness of the anti-epileptic drug progabide as an adjuvant to standard chemotherapy. A group of 59 epileptics was selected to be used in the clinical trial. The patients suffering from simple or complex partial seizures were randomly assigned to receive either the anti-epileptic drug progabide or a placebo. At each of four successive postrandomization clinic visits, the number of seizures occurring over the previous 2 weeks was reported. The measured variables were y_i ($i = 1, 2, 3, 4$—the seizure counts recorded at the four clinic visits); Trt (x_1) (0 is the placebo, 1 is progabide); Base (x_2), the baseline seizure rate; Age (x_3),

FIGURE 3.31

Scatterplot of starting hourly wage and years of experience

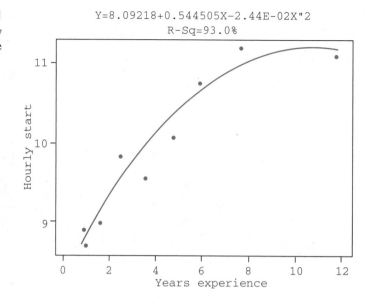

the patient's age in years. The data and summary statistics are given in Tables 3.10 and 3.11.

side-by-side boxplots

The first plots are **side-by-side boxplots** that compare the base number of seizures and ages of the treatment patients to the patients assigned to the placebo. These plots provide a visual assessment of whether the treatment patients and placebo patients had similar distributions of age and base seizure counts prior to the start of the clinical trials. An examination of Figure 3.32(a) reveals that the number of seizures prior to the beginning of the clinical trials has similar patterns for the two groups of patients. There is a single patient with a base seizure count greater than 100 in both groups. The base seizure count for the placebo group is somewhat more variable than for the treatment group—its box is wider than the box for the treatment group. The descriptive statistics table contradicts this observation. The sample standard deviation is 26.10 for the placebo group and 27.37 for the treatment group. This seemingly inconsistent result occurs due to the large base count for a single patient in the treatment group. The median number of base seizures is higher for the treatment group than for the placebo group. The means are nearly

TABLE 3.10

Data for epilepsy study: successive 2-week seizure counts for 59 epileptics. Covariates are adjuvant treatment (0 = placebo, 1 = progabide), 8-week baseline seizure counts, and age (in years)

ID	y_1	y_2	y_3	y_4	Trt	Base	Age
104	5	3	3	3	0	11	31
106	3	5	3	3	0	11	30
107	2	4	0	5	0	6	25
114	4	4	1	4	0	8	36
116	7	18	9	21	0	66	22
118	5	2	8	7	0	27	29
123	6	4	0	2	0	12	31
126	40	20	23	12	0	52	42
130	5	6	6	5	0	23	37
135	14	13	6	0	0	10	28
141	26	12	6	22	0	52	36

TABLE 3.10
(*continued*)

ID	y_1	y_2	y_3	y_4	Trt	Base	Age
145	12	6	8	4	0	33	24
201	4	4	6	2	0	18	23
202	7	9	12	14	0	42	36
205	16	24	10	9	0	87	26
206	11	0	0	5	0	50	26
210	0	0	3	3	0	18	28
213	37	29	28	29	0	111	31
215	3	5	2	5	0	18	32
217	3	0	6	7	0	20	21
219	3	4	3	4	0	12	29
220	3	4	3	4	0	9	21
222	2	3	3	5	0	17	32
226	8	12	2	8	0	28	25
227	18	24	76	25	0	55	30
230	2	1	2	1	0	9	40
234	3	1	4	2	0	10	19
238	13	15	13	12	0	47	22
101	11	14	9	8	1	76	18
102	8	7	9	4	1	38	32
103	0	4	3	0	1	19	20
108	3	6	1	3	1	10	30
110	2	6	7	4	1	19	18
111	4	3	1	3	1	24	24
112	22	17	19	16	1	31	30
113	5	4	7	4	1	14	35
117	2	4	0	4	1	11	27
121	3	7	7	7	1	67	20
122	4	18	2	5	1	41	22
124	2	1	1	0	1	7	28
128	0	2	4	0	1	22	23
129	5	4	0	3	1	13	40
137	11	14	25	15	1	46	33
139	10	5	3	8	1	36	21
143	19	7	6	7	1	38	35
147	1	1	2	3	1	7	25
203	6	10	8	8	1	36	26
204	2	1	0	0	1	11	25
207	102	65	72	63	1	151	22
208	4	3	2	4	1	22	32
209	8	6	5	7	1	41	25
211	1	3	1	5	1	32	35
214	18	11	28	13	1	56	21
218	6	3	4	0	1	24	41
221	3	5	4	3	1	16	32
225	1	23	19	8	1	22	26
228	2	3	0	1	1	25	21
232	0	0	0	0	1	13	36
236	1	4	3	2	1	12	37

TABLE 3.11

Descriptive statistics:
Minitab output for
epilepsy example

0=PLACEBO
1=TREATED

Variable	TREATMENT	N	Mean	Median	Tr Mean	StDev	SE Mean
Y1	0	28	9.36	5.00	8.54	10.14	1.92
	1	31	8.58	4.00	5.26	18.24	3.28
Y2	0	28	8.29	4.50	7.81	8.16	1.54
	1	31	8.42	5.00	6.37	11.86	2.13
Y3	0	28	8.79	5.00	6.54	14.67	2.77
	1	31	8.13	4.00	5.63	13.89	2.50
Y4	0	28	7.96	5.00	7.46	7.63	1.44
	1	31	6.71	4.00	4.78	11.26	2.02
BASE	0	28	30.79	19.00	28.65	26.10	4.93
	1	31	31.61	24.00	27.37	27.98	5.03
AGE	0	28	29.00	29.00	28.88	6.00	1.13
	1	31	27.74	26.00	27.52	6.60	1.19

Variable	TREATMENT	Min	Max	Q1	Q3
Y1	0	0.00	40.00	3.00	12.75
	1	0.00	102.00	2.00	8.00
Y2	0	0.00	29.00	3.00	12.75
	1	0.00	65.00	3.00	10.00
Y3	0	0.00	76.00	2.25	8.75
	1	0.00	72.00	1.00	8.00
Y4	0	0.00	29.00	3.00	11.25
	1	0.00	63.00	2.00	8.00
BASE	0	6.00	111.00	11.00	49.25
	1	7.00	151.00	13.00	38.00
AGE	0	19.00	42.00	24.25	32.00
	1	18.00	41.00	22.00	33.00

FIGURE 3.32

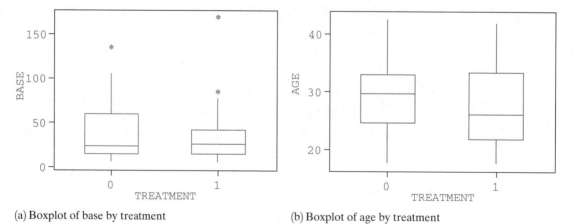

(a) Boxplot of base by treatment

(b) Boxplot of age by treatment

identical for the two groups. The means are in greater agreement than are the medians due to the skewed-to-the-right distribution of the middle 50% of the data for the placebo group, whereas the treatment group is nearly symmetric for the middle 50% of its data. Figure 3.32(b) displays the nearly identical distribution of age for the two treatment groups; the only difference is that the treatment group has a slightly smaller median age and is slightly more variable than the placebo group. Thus, the two groups appear to have similar age and base-seizure distributions prior to the start of the clinical trials.

EXERCISES
Soc.

3.29 In the paper "Demographic implications of socioeconomic transition among the tribal populations of Manipur, India" [*Human Biology* (1998) 70(3): 597–619], the authors describe the tremendous changes that have taken place in all the tribal populations of Manipur, India, since the beginning of the twentieth century. The tribal populations of Manipur are in the process of socio-economic transition from a traditional subsistence economy to a market-oriented economy. The following table displays the relation between literacy level and subsistence group for a sample of 614 married men and women in Manipur, India.

	Literacy Level		
Subsistence Group	Illiterate	Primary Schooling	At Least Middle School
Shifting cultivators	114	10	45
Settled agriculturists	76	2	53
Town dwellers	93	13	208

a. Graphically depict the data in the table using a stacked bar graph.
b. Do a percentage comparison based on the row and column totals. What conclusions do you reach with respect to the relation between literacy and subsistence group?

Engin.

3.30 In the manufacture of soft contact lenses, the power (the strength) of the lens needs to be very close to the target value. In the paper "An ANOM-type test for variances from normal populations" [*Technometrics* (1997) 39:274–283], a comparison of several suppliers is made relative to the consistency of the power of the lens. The following table contains the deviations from the target power value of lenses produced using materials from three different suppliers:

Supplier	Deviations from Target Power Value							
1	189.9	191.9	190.9	183.8	185.5	190.9	192.8	188.4	189.0
2	156.6	158.4	157.7	154.1	152.3	161.5	158.1	150.9	156.9
3	218.6	208.4	187.1	199.5	202.0	211.1	197.6	204.4	206.8

a. Compute the mean and standard deviation for the deviations of each supplier.
b. Plot the sample deviation data.
c. Describe the deviation from specified power for the three suppliers.
d. Which supplier appears to provide material that produces lenses having power closest to the target value?

Bus.

3.31 The federal government keeps a close watch on money growth versus targets that have been set for that growth. We list two measures of the money supply in the United States, M2 (private checking deposits, cash, and some savings) and M3 (M2 plus some investments), which are given here for 20 consecutive months.

Month	Money Supply (in trillions of dollars)		Month	Money Supply (in trillions of dollars)	
	M2	M3		M2	M3
1	2.25	2.81	11	2.43	3.05
2	2.27	2.84	12	2.42	3.05
3	2.28	2.86	13	2.44	3.08
4	2.29	2.88	14	2.47	3.10
5	2.31	2.90	15	2.49	3.10
6	2.32	2.92	16	2.51	3.13
7	2.35	2.96	17	2.53	3.17
8	2.37	2.99	18	2.53	3.18
9	2.40	3.02	19	2.54	3.19
10	2.42	3.04	20	2.55	3.20

    a. Would a scatterplot describe the relation between M2 and M3?

    b. Construct a scatterplot. Is there an obvious relation?

3.32 Refer to Exercise 3.31. What other data plot might be used to describe and summarize these data? Make the plot and interpret your results.

3.8   Encounters with Real Data: California Earthquakes

Defining the Problem (1)

The crust of Earth is made up of several tectonic plates that are in constant motion relative to one another. On a geologic time scale of millions of years, these movements are grand indeed. For example, the Atlantic Ocean now lies between land masses that were formerly adjacent. On the smaller time scale of recorded human history, the interactions between plates have sometimes been dramatically experienced as earthquakes. Most earthquakes in California are caused by the northward slippage of the Pacific plate against the North American plate. These quakes occur along a family of faults, of which the San Andreas fault is perhaps the largest and best known.

    A network of monitoring stations, maintained by government agencies and universities, records the waves of seismic energy released by earthquakes. From this information, it is possible to deduce the location (epicenter), depth, and magnitude of individual quakes. Magnitude is measured on a logarithmic scale of the energy released. Quakes with magnitudes less than 3 are seldom felt by humans. Those with magnitudes around 5 are usually felt beyond the immediate vicinity where they occur and can cause considerable local damage and injury. Great quakes, with magnitudes of 7 or larger, usually cause massive damage over a wide area. Injuries and deaths from great quakes can be widespread and numerous, depending on population density, the construction standards of buildings, the time of day, and so on.

    In the very early morning local time (0837 GMT) of September, 3, 2000, an earthquake of magnitude somewhat above 5 occurred near Yountville, California, in the Napa Valley, an area just northwest of San Francisco famous for its production of fine wines. This quake injured about 125 people, several of them seriously.

Property damage in the area was estimated to be in the $50 million range, and the governor declared a state of emergency, qualifying some of those affected for federal loans and other government assistance.

Collecting Data (2)

The dataset CAQUAKES contains information on 460 earthquakes of magnitude 0.5 and greater occurring between August 28 and September 9, 2000, in California (including a few in western Nevada and northern Baja California). The complete measured variables for CAQUAKES follow.

```
c1   Date    Text variable: Date of earthquake in format YYYY/MM/DD
c2   GMT     Time of quake, Greenwich Mean Time. (Pacific Daylight Time + 7 hours)
c3   Lat     Latitude of quake location (positive values for North)
c4   Lon     Longitude (negative values for West)
c5   Depth   Depth in kilometers
c6   Mag     Magnitude
c7   BefAft  1 = Before Yountville quake (230 quakes), 2 = After (229). * = Yountville
```

A portion of this dataset is displayed here, showing the Yountville earthquake and ten others that occurred at about the same time.

Partial Listing of Dataset CAQUAKES (Rows 225–235, including the Yountville earthquake)

Date	GMT	Lat	Lon	Depth	Mag	BefAft
2000/09/03	3:45:26.42	37.4822	-118.838	4.43	1.42	1
2000/09/03	3:52:01.15	36.3540	-120.933	4.53	1.30	1
2000/09/03	4:00:08.63	38.9082	-122.997	5.18	1.14	1
2000/09/03	4:52:39.67	38.8267	-122.805	2.36	0.96	1
2000/09/03	5:29:33.92	38.7610	-122.716	2.59	1.62	1
2000/09/03	8:26:46.98	38.8037	-122.780	0.95	1.40	1
2000/09/03	8:36:30.09	38.3788	-122.413	10.21	5.17	*
2000/09/03	8:51:43.71	35.1000	-119.074	29.72	2.06	2
2000/09/03	9:11:15.97	38.3922	-122.403	6.10	1.64	2
2000/09/03	9:18:23.43	38.7645	-122.743	2.17	1.87	2
2000/09/03	10:04:46.53	38.3787	-122.409	6.01	1.41	2

A complete listing of the data is found in the file CAQUAKES on the CD that came with your book.

Summarizing Data (3)

A histogram of these earthquakes shows a distribution that is skewed to the right (see Figure 3.33). Even in this skewed distribution, the magnitude of the Yountville earthquake stands out as extraordinary. According to the usual rules for declaring outliers, it is an extreme outlier. A few of the other relatively large quakes are also tagged as outliers.

Sometimes it is important to think about the mean or median of a distribution, but here it is the extremes that require attention. As far as the general public is concerned, the only earthquakes that matter are included among the outliers.

FIGURE 3.33

Histogram of California earthquakes, August 28– September 9, 2002

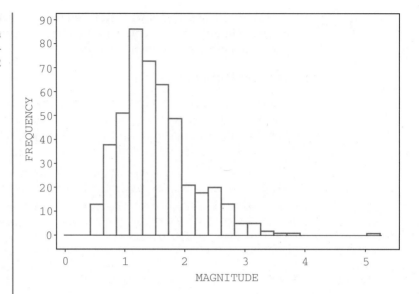

FIGURE 3.33

Histogram of California earthquakes, August 28– September 9, 2002

Even among the outliers, only the few largest were strong enough to be felt (except possibly by very alert people who happened to be very near the epicenters). The vast majority of the 460 earthquakes in our dataset were detectable only by sensitive scientific instruments.

The histogram shows *detected* quakes. As the tectonic plates creep in opposite directions along their faults, they cause very many tiny quakes—much greater in number than quakes of moderate magnitudes. But the smallest ones are quite difficult to detect with the technology used to collect our data.

This difficulty in measuring small quakes leads us to look at the lower end of the distribution, to the left of the mode located near magnitude 1.2. Quakes below that magnitude often have to be near a monitoring station to be identified with confidence. Some seismic events between magnitudes 0.5 and 2 have been eliminated from this dataset because experts strongly suspected them to be explosions in gravel quarries rather than earthquakes. Very few seismic events below magnitude 0.5 were reported in the database we used, and we have eliminated all of them as being of questionable origin. (The tiniest of them might have competed for recognition against the background noise of big trucks going over bumps, jackhammers digging up sewers, trees falling in windstorms, and so on.)

Thus, the downslope at the left of the histogram is due to the difficulty in detecting and identifying quakes with magnitudes below a small threshold value, not to the fact that small quakes are sparse. Imagining *all* quakes in the designated region and time period, we suppose that a true histogram of their magnitudes may have no mode, continuing to increase in height as magnitudes decrease to 0 (that is, toward the left end of the scale shown).

A record of earthquake magnitudes such as we show here can be viewed as a sample from an ongoing random process. Except that our time span was deliberately chosen to include the relatively large Yountville earthquake, the histogram of this sample looks very much the same as histograms for California earthquakes recorded over other time periods of equal length.

3.9 Staying Focused

This chapter deals with the third step in learning from data. After defining the problem (Step 1) and collecting the data (Step 2), we are faced with the job of summarizing the data (Step 3). We considered graphical and numerical ways to summarize data. The pie chart and bar graph are particularly appropriate for graphically displaying data obtained from a qualitative variable. The frequency and relative frequency histograms and stem-and-leaf plots are graphical techniques applicable only to quantitative data.

Numerical descriptive measures of data are used to convey a mental image of the distribution of measurements. Measures of central tendency include the mode, the median, and the arithmetic mean. Measures of variability include the range, the interquartile range, the variance, and the standard deviation of a set of measurements.

We extended the concept of data description to summarize the relations between two qualitative variables. Here cross-tabulations were used to develop percentage comparisons. We examined plots for summarizing the relations between quantitative and qualitative variables and between two quantitative variables. Material presented here (namely, summarizing relations among variables) will be discussed and expanded in later chapters on chi-square methods, on the analysis of variance, and on regression.

Key Formulas

1. Sample mean

$$\bar{y} = \frac{\sum_i y_i}{n}$$

2. Sample variance

$$s^2 = \frac{1}{n-1} \sum_i (y_i - \bar{y})^2$$

3. Sample standard deviation

$$s = \sqrt{s^2}$$

Supplementary Exercises

Env.   **3.33**   To control the risk of severe core damage during a commercial nuclear power station blackout accident, the reliability of the emergency diesel generators to start on demand must be maintained at a high level. The paper "Empirical Bayes estimation of the reliability of nuclear-power emergency diesel generators" [*Technometrics* (1996) 38:11–23] contains data on the failure history of seven nuclear power plants. The following data are the number of successful demands between failures for the diesel generators at one of these plants from 1982 to 1988.

| 28 | 50 | 193 | 55 | 4 | 7 | 147 | 76 | 10 | 0 | 10 | 84 | 0 | 9 | 1 | 0 | 62 |
| 26 | 15 | 226 | 54 | 46 | 128 | 4 | 105 | 40 | 4 | 273 | 164 | 7 | 55 | 41 | 26 | 6 |

(*Note:* The failure of the diesel generator does not necessarily result in damage to the nuclear core because all nuclear power plants have several emergency diesel generators.)

a. Calculate the mean and median of the successful demands between failures.

> **b.** Which measure appears to best represent the center of the data?
>
> **c.** Calculate the range and standard deviation, s.
>
> **d.** Use the range approximation to estimate s. How close is the approximation to the true value?
>
> **e.** Construct the intervals
>
> $$\bar{y} \pm s \qquad \bar{y} \pm 2s \qquad \bar{y} \pm 3s$$
>
> Count the number of demands between failures falling in each of the three intervals. Convert these numbers to percentages and compare your results to the Empirical Rule.
>
> **f.** Why do you think the Empirical Rule and your percentages do not match well?

Bus.   **3.34**   The February 1998 issue of *Consumer Reports* provides data on the price of 24 brands of paper towels. The prices are given in both cost per roll and cost per sheet because the brands had varying numbers of sheets per roll.

Brand	Price per Roll	Number of Sheets per Roll	Cost per Sheet
1	1.59	50	.0318
2	0.89	55	.0162
3	0.97	64	.0152
4	1.49	96	.0155
5	1.56	90	.0173
6	0.84	60	.0140
7	0.79	52	.0152
8	0.75	72	.0104
9	0.72	80	.0090
10	0.53	52	.0102
11	0.59	85	.0069
12	0.89	80	.0111
13	0.67	85	.0079
14	0.66	80	.0083
15	0.59	80	.0074
16	0.76	80	.0095
17	0.85	85	.0100
18	0.59	85	.0069
19	0.57	78	.0073
20	1.78	180	.0099
21	1.98	180	.0110
22	0.67	100	.0067
23	0.79	100	.0079
24	0.55	90	.0061

> **a.** Compute the standard deviation for both the price per roll and the price per sheet.
>
> **b.** Which is more variable, price per roll or price per sheet?

3.35   For the data in Exercise 3.34, use a scatterplot to plot the price per roll and number of sheets per roll.

> **a.** Do the 24 points appear to fall on a straight line?
>
> **b.** If not, is there any other relation between the two prices?
>
> **c.** What factors may explain why the ratio of price per roll to number of sheets is not a constant?

3.36   For the data in Exercise 3.34, construct boxplots for both price per roll and number of sheets per roll. Are there any "unusual" brands in the data?

Env.   **3.37** The paper "Conditional simulation of waste-site performance" [*Technometrics* (1994) 36: 129–161] discusses the evaluation of a pilot facility for demonstrating the safe management, storage, and disposal of defense-generated, radioactive, transuranic waste. Researchers have determined that one potential pathway for release of radionuclides is through contaminant transport in groundwater. Recent focus has been on the analysis of transmissivity, a function of the properties and the thickness of an aquifer that reflects the rate at which water is transmitted through the aquifer. The following table contains 41 measurements of transmissivity, T, made at the pilot facility.

9.354	6.302	24.609	10.093	0.939	354.81	15399.27	88.17	1253.43	0.75	312.10
1.94	3.28	1.32	7.68	2.31	16.69	2772.68	0.92	10.75	0.000753	
1.08	741.99	3.23	6.45	2.69	3.98	2876.07	12201.13	4273.66	207.06	
2.50	2.80	5.05	3.01	462.38	5515.69	118.28	10752.27	956.97	20.43	

a. Draw a relative frequency histogram for the 41 values of T.
b. Describe the shape of the histogram.
c. When the relative frequency histogram is highly skewed to the right, the Empirical Rule may not yield very accurate results. Verify this statement for the data given.
d. Data analysts often find it easier to work with mound-shaped relative frequency histograms. A transformation of the data will sometimes achieve this shape. Replace the given 41 T values with the logarithm base 10 of the values and reconstruct the relative frequency histogram. Is the shape more mound-shaped than the original data? Apply the Empirical Rule to the transformed data and verify that it yields more accurate results than it did with the original data.

3.38 Compute the mean, median, and standard deviation for the homeownership data in Exercise 3.10.
a. Compare the mean and median for the 1996 data. Which value is most appropriate for this data set? Explain your answer.
b. Is there a substantial difference between the summary statistics for the two years? What conclusions can you draw about the change in homeownership during the 11 years using these summary statistics?

3.39 Construct boxplots for the two years of homeownership data in Exercise 3.10.
a. Has homeownership percentage changed over the years?
b. Are there any states that have extremely low homeownership?
c. Are there any states that have extremely high homeownership?
d. What similarities exist for the states classified as having low homeownership? High ownership?

3.40 Using the homeownership data in Exercise 3.10, construct a quantile plot for both years.
a. Find the 20th percentile for the homeownership percentage and interpret this value for the 1996 data.
b. Congress wants to designate those states that have the highest homeownership percentage in 1996. Which states fall into the upper 10th percentile of homeownership rates?
c. Similarly identify those states that fall into the upper 10th percentile of homeownership rates during 1985. Are these states different from the states in this group during 1996?

Engin.   **3.41** The Insurance Institute for Highway Safety published data on the total damage suffered by compact automobiles in a series of controlled, low-speed collisions. The data (in dollars) with brand names removed are as follows:

361	393	430	543	566	610	763	851
886	887	976	1,039	1,124	1,267	1,328	1,415
1,425	1,444	1,476	1,542	1,544	2,048	2,197	

a. Draw a histogram of the data using six or seven categories.
b. On the basis of the histogram, what would you guess the mean to be?
c. Calculate the median and mean.

d. What does the relation between the mean and median indicate about the shape of the data?

3.42   Production records for an automobile manufacturer show the following figures for production per shift (maximum production is 720 cars per shift):

| 688 | 711 | 625 | 701 | 688 | 667 | 694 | 630 | 547 | 703 | 688 | 697 | 703 |
| 656 | 677 | 700 | 702 | 688 | 691 | 664 | 688 | 679 | 708 | 699 | 667 | 703 |

a. Would the mode be a useful summary statistic for these data?
b. Find the median.
c. Find the mean.
d. What does the relationship between the mean and median indicate about the shape of the data?

3.43   Draw a stem-and-leaf plot of the data in Exercise 3.42. The stems should include (from highest to lowest) 71, 70, 69, ... Does the shape of the stem-and-leaf display confirm your judgment in part (d) of Exercise 3.42?

3.44   Refer to Exercise 3.43.
a. Find the median and IQR.
b. Find the inner and outer fences. Are there any outliers?
c. Draw a boxplot of the data.

3.45   A company revised a long-standing policy to eliminate the time clocks and cards for non-exempt employees. Along with this change, all employees (exempt and nonexempt) were expected to account for their own time on the job as well as absences due to sickness, vacation, holidays, and so on. The previous policy of allocating a certain number of sick days was eliminated—if an employee was sick, he or she was given time off with pay; otherwise, he or she was expected to be working.

In order to see how well the new program was working, the records of a random sample of 15 employees were examined to determine the number of sick days this year (under the new plan) and the corresponding number for the preceding year. The data are shown here:

Employee	This Year (new policy)	Preceding Year (old policy)
1	0	2
2	0	2
3	0	3
4	0	4
5	2	5
6	1	2
7	1	6
8	3	8
9	1	5
10	0	4
11	5	5
12	6	12
13	1	3
14	2	4
15	12	4

a. Obtain the mean and standard deviation for each column.
b. Based on the sample data, what might you conclude (infer) about the new policies? Explain your reason(s).

3.46 Refer to Exercise 3.45. What happens to $\bar{y}$ and s for each column if we eliminate the two 12s and substitute the value 7? Are the ranges for the old and new policies affected by these substitutions?

Bus.

3.47 The most widely reported index of the performance of the New York Stock Exchange (NYSE) is the Dow Jones Industrial Average (DJIA). This index is computed from the stock prices of 30 companies. When the DJIA was invented in 1896, the index was the average price of 12 stocks. The index was modified over the years as new companies were added and dropped from the index and was also altered to reflect when a company splits its stock. The closing NYSE prices for the 30 components (as of May 1999) of the DJIA are given in the following table.

    a. Compute the average price of the 30 stock prices in the DJIA.

    b. Compute the range of the 30 stock prices in the DJIA.

    c. The DJIA is no longer an average; the name includes the word *average* only for historical reasons. The index is computed by summing the stock prices and dividing by a constant, which is changed as stocks are added or removed from the index and when stocks split.

$$DJIA = \frac{\sum_{i=1}^{30} y_i}{C}$$

where y_i is the closing price for stock i, and $C = .211907$. Using the stock prices given, compute the DJIA for May 20, 1999.

    d. The DJIA is a summary of data. Does the DJIA provide information about a population using sampled data? If so, which population? Is the sample a random sample?

Components of DJIA		
Company	Percent of DJIA	NYSE Stock Price (5/18/99)
Alcoa	2.532	58.3125
Allied-Signal	2.640	60.8125
American Express	5.357	123.3750
AT&T	2.570	59.1875
Boeing	1.948	44.8750
Caterpillar	2.559	58.9375
Chevron	4.060	93.5000
Citigroup	2.996	69.0000
Coca-Cola	2.950	67.9375
DuPont	2.999	69.0625
Eastman Kodak	3.281	75.5625
Exxon	3.474	80.0000
General Electric	4.619	106.3750
General Motors	3.479	80.1250
Goodyear	2.578	59.3750
Hewlett-Packard	4.090	94.1875
IBM	2.380	232.5000
International Paper	2.380	54.8125
J.P. Morgan	6.049	139.3125
Johnson & Johnson	4.125	95.0000
McDonald's	1.739	40.0625
Merck	3.126	72.0000

(continued)

Components of DJIA

Company	Percent of DJIA	NYSE Stock Price (5/18/99)
Minnesota Mining	3.884	89.4375
Phillip Morris	1.745	40.1875
Procter & Gamble	4.198	96.6875
Sears, Roebuck	2.125	48.9375
Union Carbide	2.421	55.7500
United Technologies	2.640	60.8125
Wal-Mart Stores	1.935	44.5625
Walt Disney	1.294	29.8125

H.R. **3.48** As one part of a review of middle-manager selection procedures, a study was made of the relation between hiring source (promoted from within, hired from related business, hired from unrelated business) and the 3-year job history (additional promotion, same position, resigned, dismissed). The data for 120 middle managers follow.

	Source			
Job History	Within Firm	Related Business	Unrelated Business	Total
Promoted	13	4	10	27
Same position	32	8	18	58
Resigned	9	6	10	25
Dismissed	3	3	4	10
Total	57	21	42	120

a. Calculate job-history percentages within each source.
b. Would you say that there is a strong dependence between source and job history?

Env. **3.49** A survey was taken of 150 residents of major coal-producing states, 200 residents of major oil and natural gas–producing states, and 450 residents of other states. Each resident chose a most preferred national energy policy. The results are shown in the following SPSS printout.

```
                            STATE
            COUNT
            ROW PCT  COAL    OIL AND   OTHER    ROW
            COL PCT          GAS                TOTAL
            TOT PCT
OPINION               62       25       102      189
   COAL ENCOURAGED   32.8     13.2      54.0     23.6
                     41.3     12.5      22.7
                      7.8      3.1      12.8

                       3       12        26       41
   FUSION DEVELOP      7.3     29.3      63.4      5.1
                       2.0      6.0       5.8
                       0.4      1.5       3.3

                       8        6        22       36
   NUCLEAR DEVELOP    22.2     16.7      61.1      4.5
                       5.3      3.0       4.9
                       1.0      0.8       2.8
```

```
                          19       79       53      151
      OIL DEREGULATION   12.6     52.3     35.1     18.9
                         12.7     39.5     11.8
                          2.4      9.9      6.6

                          58       78      247      383
      SOLAR DEVELOP      15.1     20.4     64.5     47.9
                         38.7     39.0     54.9
                          7.3      9.8     30.9

          COLUMN         150      200      450      800
          TOTAL         18.8     25.0     56.3    100.0
CHI SQUARE = 106.19406 WITH 8 DEGREES OF FREEDOM SIGNIFICANCE = 0.0000
CRAMER'S V = 0.25763
CONTINGENCY COEFFICIENT = 0.34233
LAMBDA = 0.01199 WITH OPINION DEPENDENT, = 0.07429 WITH STATE DEPENDENT.
```

a. Interpret the values 62, 32.8, 41.3, and 7.8 in the upper left cell of the cross-tabulation. Note the labels COUNT, ROW PCT, COL PCT, and TOT PCT at the upper left corner.

b. Which of the percentage calculations seems most meaningful to you?

c. According to the percentage calculations you prefer, does there appear to be a strong dependence between state and opinion?

Bus. **3.50** A municipal workers' union that represents sanitation workers in many small midwestern cities studied the contracts that were signed in the previous years. The contracts were subdivided into those settled by negotiation without a strike, those settled by arbitration without a strike, and all those settled after a strike. For each contract, the first-year percentage wage increase was determined. Summary figures follow.

Contract Type	Negotation	Arbitration	Poststrike
Mean percentage wage increase	8.20	9.42	8.40
Variance	0.87	1.04	1.47
Standard deviation	0.93	1.02	1.21
Sample size	38	16	6

Does there appear to be a relationship between contract type and mean percent wage increase? If you were management rather than union affiliated, which posture would you take in future contract negotiations?

3.51 Refer to the California earthquake data in Section 3.8, found in the file CAQUAKES on the CD that came with your book. Minitab programs are shown here to help with solving the problems.

a. Make a boxplot of the earthquake magnitudes. What is the approximate magnitude of the smallest outlier in the right-hand tail?

b. It doesn't make sense to talk about the median (or mean) magnitude of all California earthquakes in the period of time covered by our dataset. Why? Throughout the region specified, it seems reasonable to suppose that all quakes of magnitude 2.5 or greater are detected and correctly located. How many of these bigger quakes are there? Find their median magnitude.

```
MTB > Copy 'Mag' 'Biggrmag';
SUBC> Use 'Mag' 2.5:10.
MTB > Describe 'BiggrMag'
```

c. Can you discern an important difference in the magnitudes of California earthquakes before and after the time of the Yountville quake?

d. Sometimes an earthquake triggers subsequent ones in its immediate vicinity—called aftershocks. Restrict the dataset to a small rectangle centered on the Yountville quake of magnitude 5.17. (A tenth of a degree of latitude and longitude in either direction would be appropriate.)

```
MTB > Copy 'Lat' 'Lon' 'Mag' 'Ylat' 'Ylon' 'Ymag';
SUBC>  Use 'Lat' 38.28:38.48.
MTB > Copy 'Ylat' 'Ylon' 'Ymag' 'Ylat' 'Ylon' 'YMag';
SUBC>  Use 'Ylon' -122.513:-122.313.
```

Explore this restricted dataset with an appropriate scatterplot (showing locations of the quakes), time series plot, and so on. Within the time span of our dataset and within this small rectangle, how many quakes occurred before the one of magnitude 5.17? After it? Were aftershocks located all around the initial quake or in a particular direction from it?

CHAPTER 4

Probability and Probability Distributions

4.1 How Probability Can Be Used in Making Inferences

We stated in Chapter 1 that a scientist uses inferential statistics to make statements about a population based on information contained in a sample of units selected from that population. Graphical and numerical descriptive techniques were presented in Chapter 3 as a means to summarize and describe a sample. However, a sample is not identical to the population from which it was selected. We need to assess the degree of accuracy to which the sample mean, sample standard deviation, or sample proportion represent the corresponding population values.

Most management decisions must be made in the presence of uncertainty. Prices and designs for new automobiles must be selected on the basis of shaky forecasts of consumer preference, national economic trends, and competitive actions. The size and allocation of a hospital staff must be decided with limited information on patient load. The inventory of a product must be set in the face of uncertainty about demand. Probability is the language of uncertainty. Now let us examine probability, the mechanism for making inferences.

For example, *Newsweek,* in its June 20, 1998, issue, asks the question, "Who Needs Doctors? The Boom in Home Testing." The article discusses the dramatic increase in medical screening tests for home use. The home-testing market has expanded beyond the two most frequently used tests, pregnancy and diabetes glucose monitoring, to a variety of diagnostic tests that were previously used only by doctors and certified laboratories. There is a DNA test to determine whether twins are fraternal or identical, a test to check cholesterol level, a screening test for colon cancer, and tests to determine whether your teenager is a drug user. However, the major question that needs to be addressed is how reliable are the testing kits? When a test indicates that a woman is not pregnant, what is the chance that the test is incorrect and the woman is truly pregnant? This type of incorrect result from a home test could translate into a woman not seeking the proper prenatal care in the early stages of her pregnancy.

Suppose a company states in its promotional materials that its pregnancy test provides correct results in 75% of its applications by pregnant women. We want to evaluate the claim, and so we select 20 women who have been determined by their physicians, using the best possible testing procedures, to be pregnant. The test is taken by each of the 20 women, and for all 20 women the test result is negative, indicating that none of the 20 is pregnant. What do you conclude about the company's claim on the reliability of its test? Suppose you are further assured that each of the 20 women was in fact pregnant, as was determined several months after the test was taken.

If the company's claim of 75% reliability is correct, we expect somewhere near 75% of the tests in the sample to be positive. However, none of the test results was positive. Thus, we conclude that the company's claim is probably false. Why did we fail to state with certainty that the company's claim was false? Consider the possible setting. Suppose we have a large population consisting of millions of units, and 75% of the units are Ps for positives and 25% of the units are Ns for negatives. We randomly select 20 units from the population and count the number of units in the sample that are Ps. Is it possible to obtain a sample consisting of 0 Ps and 20 Ns? Yes, it is possible, *but* it is highly *improbable.* Later in this chapter we will compute the probability of such a sample occurrence.

To obtain a better view of the role that probability plays in making inferences from sample results to conclusions about populations, suppose the 20 tests result in 14 tests being positive—that is, a 70% correct response rate. Would you consider this result highly improbable and reject the company's claim of a 75% correct response rate? How about 12 positives and 8 negatives, or 16 positives and 4 negatives? At what point do we decide that the result of the observed sample is so improbable, assuming the company's claim is correct, that we disagree with its claim? To answer this question, we must know how to find the probability of obtaining a particular sample outcome. Knowing this probability, we can then determine whether we agree or disagree with the company's claim. Probability is the tool that enables us to make an inference. Later in this chapter we will discuss in detail how the FDA and private companies determine the reliability of screening tests.

Because probability is the tool for making inferences, we need to define probability. In the preceding discussion, we used the term *probability* in its everyday sense. Let us examine this idea more closely.

Observations of phenomena can result in many different outcomes, some of which are more likely than others. Numerous attempts have been made to give a precise definition of the probability of an outcome. We will cite three of these.

classical interpretation

The first interpretation of probability, called the **classical interpretation of probability,** arose from games of chance. Typical probability statements of this type are, for example, "the probability that a flip of a balanced coin will show heads is 1/2" and "the probability of drawing an ace when a single card is drawn from a standard deck of 52 cards is 4/52." The numerical values for these probabilities arise from the nature of the games. A coin flip has two possible outcomes (a head or a tail); the probability of a head should then be 1/2 (1 out of 2). Similarly, there are 4 aces in a standard deck of 52 cards, so the probability of drawing an ace in a single draw is 4/52, or 4 out of 52.

outcome

event

In the classical interpretation of probability, each possible distinct result is called an **outcome;** an **event** is identified as a collection of outcomes. The probability of an event E under the classical interpretation of probability is computed by taking the ratio of the number of outcomes, N_e, favorable to event E to the total number N of possible outcomes:

$$P(\text{event } E) = \frac{N_e}{N}$$

The applicability of this interpretation depends on the assumption that all outcomes are equally likely. If this assumption does not hold, the probabilities indicated by the classical interpretation of probability will be in error.

relative frequency interpretation

A second interpretation of probability is called the **relative frequency concept of probability;** this is an empirical approach to probability. If an experiment is repeated a large number of times and event E occurs 30% of the time, then .30 should be a very good approximation of the probability of event E. Symbolically, if an experiment is conducted n different times and if event E occurs on n_e of these trials, then the probability of event E is approximately

$$P(\text{event } E) \approx \frac{n_e}{n}$$

We say "approximately" because we think of the actual probability $P(\text{event } E)$ as the relative frequency of the occurrence of event E over a very large number of observations or repetitions of the phenomenon. The fact that we can check probabilities that have a relative frequency interpretation (by simulating many repetitions of the experiment) makes this interpretation very appealing and practical.

The third interpretation of probability can be used for problems in which it is difficult to imagine a repetition of an experiment. These are one-shot situations. For example, the director of a state welfare agency who estimates the probability that a proposed revision in eligibility rules will be passed by the state legislature is not thinking in terms of a long series of trials. Rather, the director uses a **personal** or

subjective interpretation

subjective probability to make a one-shot statement of belief regarding the likelihood of passage of the proposed legislative revision. The problem with subjective probabilities is that they can vary from person to person and they cannot be checked.

Of the three interpretations presented, the relative frequency concept seems to be the most reasonable one because it provides a practical interpretation of the probability for most events of interest. Even though we will never run the necessary repetitions of the experiment to determine the exact probability of an event, the fact that we can check the probability of an event gives meaning to the relative frequency concept. Throughout the remainder of this text we will lean heavily on this interpretation of probability.

EXERCISES

Applications

4.1 Indicate which interpretation of the probability statement seems most appropriate.

a. The National Angus Association has stated that there is a 60/40 chance that wholesale beef prices will rise by the summer—that is, a .60 probability of an increase and a .40 probability of a decrease.

b. The quality control section of a large chemical manufacturing company has undertaken an intensive process-validation study. From this study, the QC section claims that the probability that the shelf life of a newly released batch of chemical will exceed the minimal time specified is .998.

c. A new blend of coffee is being contemplated for release by the marketing division of a large corporation. Preliminary marketing survey results indicate that 550 of a random sample of 1,000 potential users rated this new blend better than a brand-name competitor. The probability of this happening is approximately .001, assuming that there is actually no difference in consumer preference for the two brands.

d. The probability that a customer will receive a package the day after it was sent by a business using an overnight delivery service is .92.

e. The sportscaster in College Station, Texas, states that the probability that the Aggies will win their football game against the University of Florida is .75.

f. The probability of a nuclear power plant having a meltdown on a given day is .00001.

g. If a customer purchases a single ticket for the Texas lottery, the probability of that ticket being the winning ticket is 1/15,890,700.

4.2 Give your own personal probability for each of the following situations. It would be instructive to tabulate these probabilities for the entire class. In which cases did you have large disagreements?

a. The federal income tax will be eliminated.

b. You will receive an A in this course.

c. Two or more individuals in the classroom have the same birthday.

d. An asteroid will strike the planet Earth in the next year.

e. A woman will be elected as vice president or president of the United States in the next presidential election.

4.2 Finding the Probability of an Event

In the preceding section, we discussed three different interpretations of probability. In this section, we will use the classical interpretation and the relative frequency concept to illustrate the computation of the probability of an outcome or event. Consider an experiment that consists of tossing two coins, a penny and then a dime, and observing the upturned faces. There are four possible outcomes:

TT: tails for both coins
TH: a tail for the penny, a head for the dime
HT: a head for the penny, a tail for the dime
HH: heads for both coins

What is the probability of observing the event exactly one head from the two coins?

This probability can be obtained easily if we can assume that all four outcomes are equally likely. In this case, that seems quite reasonable. There are $N = 4$ possible outcomes, and $N_e = 2$ of these are favorable for the event of interest, observing exactly one head. Hence, by the classical interpretation of probability,

$$P(\text{exactly 1 head}) = \frac{2}{4} = \frac{1}{2}$$

Because the event of interest has a relative frequency interpretation, we could also obtain this same result empirically, using the relative frequency concept. To demonstrate how relative frequency can be used to obtain the probability of an event, we will use simulation. Simulation is a technique that produces outcomes having the same probability of occurrence as the real situation events. The computer is a convenient tool for generating these outcomes. Suppose we wanted to simulate 500 tosses of the two coins. We can use a computer program such as SAS or Minitab to simulate the tossing of a pair of coins. The program has a random number generator. We designate an even number as H and an odd number as T. Because there are five even and five odd single-digit numbers, the probability of obtaining an even number is $5/10 = .5$, which is the same as the probability of obtaining an odd number. Thus, we can request 500 pairs of single-digit numbers. This set of 500 pairs of numbers represents 500 tosses of the two coins, with the first digit representing the outcome of tossing the penny and the second digit representing the outcome of tossing the dime. For example, the pair (3, 6) represents a tail for the penny and a head for the dime. Using version 13 of Minitab, we produced the 500 pairs of numbers shown here.

25	32	70	15	96	87	80	43	15	77	89	51	08	36	29	55	42	86	45	93	68	72	49	99	37
82	81	58	50	85	27	99	41	10	31	42	35	50	02	68	33	50	93	73	62	15	15	90	97	24
46	86	89	82	20	23	63	59	50	40	32	72	59	62	58	53	01	85	49	27	31	48	53	07	78
15	81	39	83	79	21	88	57	35	33	49	37	85	42	28	38	50	43	82	47	01	55	42	02	52
66	44	15	40	29	73	11	06	79	81	49	64	32	06	07	31	07	78	73	07	26	36	39	20	14
48	20	27	73	53	21	44	16	00	33	43	95	21	08	19	60	68	30	99	27	22	74	65	22	05
26	79	54	64	94	01	21	47	86	94	24	41	06	81	16	07	30	34	99	54	68	37	38	71	79
86	12	83	09	27	60	49	54	21	92	64	57	07	39	04	66	73	76	74	93	50	56	23	41	23
18	87	21	48	75	63	09	97	96	86	85	68	65	35	92	40	57	87	82	71	04	16	01	03	45
52	79	14	12	94	51	39	40	42	17	32	94	42	34	68	17	39	32	38	03	75	56	79	79	57
07	40	96	46	22	04	12	90	80	71	46	11	18	81	54	95	47	72	06	07	66	05	59	34	81
66	79	83	82	62	20	75	71	73	79	48	86	83	74	04	13	36	87	96	11	39	81	59	41	70
21	47	34	02	05	73	71	57	64	58	05	16	57	27	66	92	97	68	18	52	09	45	34	80	57
87	22	18	65	66	18	84	31	09	38	05	67	10	45	03	48	52	48	33	36	00	49	39	55	35
70	84	50	37	58	41	08	62	42	64	02	29	33	68	87	58	52	39	98	78	72	13	13	15	96
57	32	98	05	83	39	13	39	37	08	17	01	35	13	98	66	89	40	29	47	37	65	86	73	42
85	65	78	05	24	65	24	92	03	46	67	48	90	60	02	61	21	12	80	70	35	15	40	52	76
29	11	45	22	38	33	32	52	17	20	03	26	34	18	85	46	52	66	63	30	84	53	76	47	21
42	97	56	38	41	87	14	43	30	35	99	06	76	67	00	47	83	32	52	42	48	51	69	15	18
08	30	37	89	17	89	23	58	13	93	17	44	09	08	61	05	35	44	91	89	35	15	06	39	27

TABLE 4.1
Simulation of tossing a penny
and a dime 500 times

Event	Outcome of Simulation	Frequency	Relative Frequency
TT	(Odd, Odd)	129	129/500 = .258
TH	(Odd, Even)	117	117/500 = .234
HT	(Even, Odd)	125	125/500 = .250
HH	(Even, Even)	129	129/500 = .258

The summary of the simulation of the 500 tosses is shown in Table 4.1. Note that this approach yields simulated probabilities that are nearly in agreement with our intuition; that is, intuitively we might expect these outcomes to be equally likely. Thus, each of the four outcomes should occur with a probability equal to 1/4, or .25. This assumption was made for the classical interpretation. We will show in Chapter 10 that in order to be 95% certain that the simulated probabilities are within .01 of the true probabilities, the number of tosses should be at least 7,200 and not 500, as we used previously.

If we wish to find the probability of tossing two coins and observing exactly one head, we have, from Table 4.1,

$$P(\text{exactly 1 head}) \approx \frac{117 + 125}{500} = .484$$

This is very close to the theoretical probability, which we have shown to be .5.

Note that we could easily modify our example to accommodate the tossing of an unfair coin. Suppose we are tossing a penny that is weighted so that the probability of a head occurring in a toss is .70 and the probability of a tail is .30. We could designate an H outcome whenever one of the random digits 0, 1, 2, 3, 4, 5, or 6 occurs and a T outcome whenever one of the digits 7, 8, or 9 occurs. The same simulation program can be run as before, but we would interpret the output differently.

EXERCISES   **Applications**

Edu.   **4.3**   Suppose an exam consists of 20 true-or-false questions. A student takes the exam by guessing the answer to each question. What is the probability that the student correctly answers 15 or more of the questions? [*Hint:* Use a simulation approach. Generate a large number (2,000 or more sets) of 20 single-digit numbers. Each number represents the answer to one of the questions on the exam, with even digits representing correct answers and odd digits representing wrong answers. Determine the relative frequency of the sets having 15 or more correct answers.]

Med.   **4.4**   The example in Section 4.1 considered the reliability of a screening test. Suppose we wanted to simulate the probability of observing at least 15 positive results and 5 negative results in a set of 20 results, when the probability of a positive result was claimed to be .75. Use a random number generator to simulate the running of 20 screening tests.

   a. Let a two-digit number represent an individual running of the screening test. Which numbers represent a positive outcome of the screening test? Which numbers represent a negative outcome?

   b. If we generate 2,000 sets of 20 two-digit numbers, how can the outcomes of this simulation be used to approximate the probability of obtaining at least 15 positive results in the 20 runnings of the screening test?

4.3 Basic Event Relations and Probability Laws

The probability of an event, say event A, will always satisfy the property

$$0 \leq P(A) \leq 1$$

That is, the probability of an event lies anywhere in the interval from 0 (the occurrence of the event is impossible) to 1 (the occurrence of an event is a sure thing).

either A or B occurs   Suppose A and B represent two experimental events and you are interested in a new event, the event that **either A or B occurs.** For example, suppose that we toss a pair of dice and define the following events:

> A: A total of 7 shows
> B: A total of 11 shows

Then the event "either A or B occurs" is the event that you toss a total of either 7 or 11 with the pair of dice.

mutually exclusive   Note that, for this example, the events A and B are **mutually exclusive;** that is, if you observe event A (a total of 7), you could not at the same time observe event B (a total of 11). Thus, if A occurs, B cannot occur (and vice versa).

DEFINITION 4.1

Two events A and B are said to be **mutually exclusive** if (when the experiment is performed a single time) the occurrence of one of the events excludes the possibility of the occurrence of the other event.

The concept of mutually exclusive events is used to specify a second property that the probabilities of events must satisfy. When two events are mutually exclusive, then the probability that either one of the events will occur is the sum of the event probabilities.

DEFINITION 4.2

If two events, A and B, are mutually exclusive, the **probability** that either event occurs is $P(\text{either } A \text{ or } B) = P(A) + P(B)$.

Definition 4.2 is a special case of the union of two events, which we will soon define.

The definition of additivity of probabilities for mutually exclusive events can be extended beyond two events. For example, when we toss a pair of dice, the sum S of the numbers appearing on the dice can assume any one of the values $S = 2, 3, 4, \ldots, 11, 12$. On a single toss of the dice, we can observe only one of these values. Therefore, the values $2, 3, \ldots, 12$ represent mutually exclusive events. If we want to find the probability of tossing a sum less than or equal to 4, this probability is

$$P(S \leq 4) = P(2) + P(3) + P(4)$$

For this particular experiment, the dice can fall in 36 different equally likely ways. We can observe a 1 on die 1 and a 1 on die 2, denoted by the symbol $(1, 1)$. We can observe a 1 on die 1 and a 2 on die 2, denoted by $(1, 2)$. In other words, for this experiment, the possible outcomes are

(1, 1)	(2, 1)	(3, 1)	(4, 1)	(5, 1)	(6, 1)
(1, 2)	(2, 2)	(3, 2)	(4, 2)	(5, 2)	(6, 2)
(1, 3)	(2, 3)	(3, 3)	(4, 3)	(5, 3)	(6, 3)
(1, 4)	(2, 4)	(3, 4)	(4, 4)	(5, 4)	(6, 4)
(1, 5)	(2, 5)	(3, 5)	(4, 5)	(5, 5)	(6, 5)
(1, 6)	(2, 6)	(3, 6)	(4, 6)	(5, 6)	(6, 6)

As you can see, only one of these events, (1, 1), will result in a sum equal to 2. Therefore, we would expect a 2 to occur with a relative frequency of 1/36 in a long series of repetitions of the experiment, and we let $P(2) = 1/36$. The sum $S = 3$ will occur if we observe either of the outcomes (1, 2) or (2, 1). Therefore, $P(3) = 2/36 = 1/18$. Similarly, we find $P(4) = 3/36 = 1/12$. It follows that

$$P(S \leq 4) = P(2) + P(3) + P(4) = \frac{1}{36} + \frac{1}{18} + \frac{1}{12} = \frac{1}{6}$$

complement   A third property of event probabilities concerns an event and its **complement.**

DEFINITION 4.3   The **complement** of an event A is the event that A *does not* occur. The complement of A is denoted by the symbol $\overline{A}$.

Thus, if we define the complement of an event A as a new event—namely, "A does not occur"—it follows that

$$P(A) + P(\overline{A}) = 1$$

For an example, refer again to the two-coin-toss experiment. If, in many repetitions of the experiment, the proportion of times you observe event A, "two heads show," is 1/4, then it follows that the proportion of times you observe the event $\overline{A}$, "two heads do not show," is 3/4. Thus, $P(A)$ and $P(\overline{A})$ will always sum to 1.

We can summarize the three properties that the probabilities of events must satisfy as follows:

Properties of Probabilities   If A and B are any two mutually exclusive events associated with an experiment, then $P(A)$ and $P(B)$ must satisfy the following properties:

1. $0 \leq P(A) \leq 1$ and $0 \leq P(B) \leq 1$
2. $P(\text{either } A \text{ or } B) = P(A) + P(B)$
3. $P(A) + P(\overline{A}) = 1$ and $P(B) + P(\overline{B}) = 1$

union
intersection   We can now define two additional event relations: the **union** and the **intersection** of two events.

DEFINITION 4.4   The **union** of two events A and B is the set of all outcomes that are included in either A or B (or both). The union is denoted as $A \cup B$.

DEFINITION 4.5   The **intersection** of two events A and B is the set of all outcomes that are included in both A and B. The intersection is denoted as $A \cap B$.

These definitions along with the definition of the complement of an event formalize some simple concepts. The event $\overline{A}$ occurs when A *does not*; $A \cup B$ occurs when either A or B occurs; $A \cap B$ occurs when A *and* B occur.

The additivity of probabilities for mutually exclusive events, called the *addition law for mutually exclusive events,* can be extended to give the general addition law.

DEFINITION 4.6   Consider two events A and B; the **probability of the union** of A and B is

$$P(A \cup B) = P(A) + P(B) - P(A \cap B)$$

EXAMPLE 4.1

Events and event probabilities are shown in the Venn diagram in Figure 4.1. Use this diagram to determine the following probabilities:

    a. $P(A), P(\overline{A})$
    b. $P(B), P(\overline{B})$
    c. $P(A \cap B)$
    d. $P(A \cup B)$

FIGURE 4.1

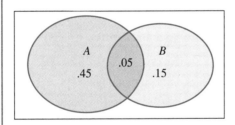

Solution From the Venn diagram, we are able to determine the following probabilities:

    a. $P(A) = .5$, therefore $P(\overline{A}) = 1 - .5 = .5$
    b. $P(B) = .2$, therefore $P(\overline{B}) = 1 - .2 = .8$
    c. $P(A \cap B) = .05$
    d. $P(A \cup B) = P(A) + P(B) - P(A \cap B) = .5 + .2 - .05 = .65$

4.4 Conditional Probability and Independence

Consider the following situation: The examination of a large number of insurance claims, categorized according to type of insurance and whether the claim was fraudulent, produced the results shown in Table 4.2. Suppose you are responsible for checking insurance claims—in particular, for detecting fraudulent claims—and you examine the next claim that is processed. What is the probability of the event F, "the claim is fraudulent"? To answer the question, you examine Table 4.2 and note

TABLE 4.2
Categorization of
insurance claims

Category	Type of Policy (%)			Total %
	Fire	**Auto**	**Other**	
Fraudulent	6	1	3	10
Nonfraudulent	14	29	47	90
Total	20	30	50	100

that 10% of all claims are fraudulent. Thus, assuming that the percentages given in the table are reasonable approximations to the true probabilities of receiving specific types of claims, it follows that $P(F) = .10$. Would you say that the risk that you face a fraudulent claim has probability .10? We think not, because you have additional information that may affect the assessment of $P(F)$. This additional information concerns the type of policy you were examining (fire, auto, or other).

Suppose that you have the additional information that the claim was associated with a fire policy. Checking Table 4.2, we see that 20% (or .20) of all claims are associated with a fire policy and that 6% (or .06) of all claims are fraudulent fire policy claims. Therefore, it follows that the probability that the claim is fraudulent, given that you know the policy is a fire policy, is

$$P(F|\text{fire policy}) = \frac{\text{proportion of claims that are fraudulent fire policy claims}}{\text{proportion of claims that are against fire policies}}$$

$$= \frac{.06}{.20} = .30$$

conditional probability   This probability, $P(F|\text{fire policy})$, is called a **conditional probability** of the event F—that is, the probability of event F given the fact that the event "fire policy" has already occurred. This tells you that 30% of all fire policy claims are fraudulent. The vertical bar in the expression $P(F|\text{fire policy})$ represents the phrase *given that*, or simply *given*. Thus, the expression is read, "the probability of the event F given the event fire policy."

unconditional probability   The probability $P(F) = .10$, called the **unconditional** or **marginal probability** of the event F, gives the proportion of times a claim is fraudulent—that is, the proportion of times event F occurs in a very large (infinitely large) number of repetitions of the experiment (receiving an insurance claim and determining whether the claim is fraudulent). In contrast, the conditional probability of F, given that the claim is for a fire policy, $P(F|\text{fire policy})$, gives the proportion of fire policy claims that are fraudulent. Clearly, the conditional probabilities of F, given the types of policies, will be of much greater assistance in measuring the risk of fraud than the unconditional probability of F.

DEFINITION 4.7   Consider two events A and B with nonzero probabilities, $P(A)$ and $P(B)$. The **conditional probability** of event A given event B is

$$P(A|B) = \frac{P(A \cap B)}{P(B)}$$

The conditional probability of event B given event A is

$$P(B|A) = \frac{P(A \cap B)}{P(A)}$$

This definition for conditional probabilities gives rise to what is referred to as the *multiplication law*.

DEFINITION 4.8

The **probability of the intersection** of two events A and B is

$$P(A \cap B) = P(A)P(B|A)$$
$$= P(B)P(A|B)$$

The only difference between Definitions 4.7 and 4.8, both of which involve conditional probabilities, relates to which probabilities are known and which needs to be calculated. When the intersection probability $P(A \cap B)$ and the individual probability $P(A)$ are known, we can compute $P(B|A)$. When we know $P(A)$ and $P(B|A)$, we can compute $P(A \cap B)$.

EXAMPLE 4.2

Two supervisors are to be selected as safety representatives within the company. Given that there are six supervisors in research and four in development, and each group of two supervisors has the same chance of being selected, find the probability of choosing both supervisors from research.

Solution   Let A be the event that the first supervisor selected is from research, and let B be the event that the second supervisor is also from research. Clearly, we want $P(A \text{ and } B) = P(A \cap B) = P(B|A)P(A)$.
For this example,

$$P(A) = \frac{\text{number of research supervisors}}{\text{number of supervisors}} = \frac{6}{10}$$

and

$$P(B|A) = \frac{\begin{array}{c}\text{number of research supervisors after}\\\text{one research supervisor was selected}\end{array}}{\text{number of supervisors after one supervisor was selected}} = \frac{5}{9}$$

Thus,

$$P(A \cap B) = P(A)P(B|A) = \frac{6}{10}\left(\frac{5}{9}\right) = \frac{30}{90} = .333$$

Thus the probability of choosing both supervisors from research is .333, assuming that each group of two has the same chance of being selected.

Suppose that the probability of event A is the same whether event B has or has not occurred; that is, suppose

$$P(A|B) = P(A)$$

Then we say that the occurrence of event A is not dependent on the occurrence of event B or, simply, that A and B are **independent events.** When $P(A|B) \neq P(A)$, the occurrence of A depends on the occurrence of B, and events A and B are said to be **dependent events.**

independent events

dependent events

DEFINITION 4.9

Two events A and B are **independent events** if

$$P(A|B) = P(A) \quad \text{or} \quad P(B|A) = P(B)$$

(Note: You can show that if $P(A|B) = P(A)$, then $P(B|A) = P(B)$, and vice versa.)

Definition 4.9 leads to a special case of $P(A \cap B)$. When events A and B are independent, it follows that

$$P(A \cap B) = P(A)P(B|A) = P(A)P(B)$$

The concept of independence is of particular importance in sampling. Later in the book, we will discuss drawing samples from two (or more) populations to compare the population means, variances, or some other population parameters. For most of these applications, we will select samples in such a way that the observed values in one sample are independent of the values that appear in another sample. **independent samples** We call these **independent samples.**

EXERCISES   **Basic Techniques**

4.5   A coin is to be flipped three times. List the possible outcomes in the form (result on toss 1, result on toss 2, result on toss 3).

4.6   In Exercise 4.5, assume that each one of the outcomes has probability 1/8 of occurring. Find the probability of
a. A: observing exactly 1 head
b. B: observing 1 or more heads
c. C: observing no heads

4.7   For Exercise 4.6:
a. Compute the probability of the complement of event A, event B, and event C.
b. Determine whether events A and B are mutually exclusive.

4.8   Determine the following conditional probabilities for the events of Exercise 4.6.
a. $P(A|B)$ **b.** $P(A|C)$ **c.** $P(B|C)$

4.9   Refer to Exercise 4.8. Are events A and B independent? Why or why not? What about A and C? What about B and C?

4.10   Consider the following outcomes for an experiment:

Outcome	1	2	3	4	5
Probability	.20	.25	.15	.10	.30

Let event A consist of outcomes 1, 3, and 5 and event B consist of outcomes 4 and 5.
a. Find $P(A)$ and $P(B)$.
b. Find P(both A and B occur).
c. Find P(either A or B occurs).

4.11   Refer to Exercise 4.10. Does P(either A or B occurs) $= P(A) + P(B)$? Why or why not?

Applications

Engin.   **4.12**   The emergency room of a hospital has two backup generators, either of which can supply enough electricity for basic hospital operations. We define events A and B as follows:

event A: generator 1 works properly
event B: generator 2 works properly

Describe the following events in words:

    a. complement of A **b.** $B|A$ **c.** either A or B

Bus.   **4.13**   An institutional investor is considering a large investment in two of five companies. Suppose that, unknown to the investor, two of the five firms are on shaky ground with regard to the development of new products.

    a. List the possible outcomes for this situation.
    b. Determine the probability of choosing two of the three firms that are on better ground.
    c. What is the probability of choosing one of two firms on shaky ground?
    d. What is the probability of choosing the two shakiest firms?

Soc.   **4.14**   A survey of workers in two manufacturing sites of a firm included the following question: How effective is management in responding to legitimate grievances of workers? The results are shown here.

	Number Surveyed	Number Responding "Poor"
Site 1	192	48
Site 2	248	80

Let A be the event the worker comes from Site 1 and B be the event the response is "poor." Compute $P(A)$, $P(B)$, and $P(A \cap B)$.

H.R.   **4.15**   A large corporation has spent considerable time developing employee performance rating scales to evaluate an employee's job performance on a regular basis, so major adjustments can be made when needed and employees who should be considered for a fast track can be isolated. Keys to this latter determination are ratings on the ability of an employee to perform to his or her capabilities and on his or her formal training for the job.

	Formal Training			
Workload Capacity	None	Little	Some	Extensive
Low	.01	.02	.02	.04
Medium	.05	.06	.07	.10
High	.10	.15	.16	.22

The probabilities for being placed on a fast track are as indicated for the 12 categories of workload capacity and formal training. The following three events (A, B, and C) are defined:

    A: an employee works at the high-capacity level
    B: an employee falls into the highest (extensive) formal training category
    C: an employee has little or no formal training and works below high capacity

    a. Find $P(A)$, $P(B)$, and $P(C)$.
    b. Find $P(A|B)$, $P(A|\overline{B})$, and $P(\overline{B}|C)$.
    c. Find $P(A \cup B)$, $P(A \cap C)$, and $P(B \cap C)$

Bus.   **4.16**   The utility company in a large metropolitan area finds that 70% of its customers pay a given monthly bill in full.

    a. Suppose two customers are chosen at random from the list of all customers. What is the probability that both customers will pay their monthly bill in full?
    b. What is the probability that at least one of them will pay in full?

4.17   Refer to Exercise 4.16. A more detailed examination of the company records indicates that 95% of the customers who pay one monthly bill in full will also pay the next monthly bill in full; only 10% of those who pay less than the full amount one month will pay in full the next month.

a. Find the probability that a customer selected at random will pay two consecutive months in full.

b. Find the probability that a customer selected at random will pay neither of two consecutive months in full.

c. Find the probability that a customer chosen at random will pay exactly one month in full.

4.5   Bayes' Formula

In this section, we will show how Bayes' Formula can be used to update conditional probabilities by using sample data when available. These updated conditional probabilities are useful in decision making. A particular application of these techniques involves the evaluation of diagnostic tests. Suppose a meat inspector must decide whether a randomly selected meat sample contains *Escherichia coli* bacteria. The inspector conducts a diagnostic test. Ideally, a positive result (Pos) means that the meat sample actually has *E. coli,* and a negative result (Neg) means that the meat sample is free of *E. coli.* However, the diagnostic test is occasionally in error. The

false positive   results of the test may be a **false positive,** for which the test's indication of *E. coli*
false negative   presence is incorrect, or a **false negative,** for which the test's conclusion of *E. coli* absence is incorrect. Large-scale screening tests are conducted to evaluate the accuracy of a given diagnostic test. For example, *E. coli* (E) is placed in 10,000 meat samples, and the diagnostic test yields a positive result for 9,500 samples and a negative result for 500 samples; that is, there are 500 false negatives out of the 10,000 tests. Another 10,000 samples have all traces of *E. coli* (NE) removed, and the diagnostic test yields a positive result for 100 samples and a negative result for 9,900 samples. There are 100 false positives out of the 10,000 tests. We can summarize the results in the following table:

Diagnostic Test Result	Meat Sample Status	
	E	NE
Positive	9,500	100
Negative	500	9,900
Total	10,000	10,000

Evaluation of test results is as follows:

$$\text{True positive rate} = P(\text{Pos}|\text{E}) = \frac{9,500}{10,000} = .95$$

$$\text{False positive rate} = P(\text{Pos}|\text{NE}) = \frac{100}{10,000} = .01$$

$$\text{True negative rate} = P(\text{Neg}|\text{NE}) = \frac{9,900}{10,000} = .99$$

$$\text{False negative rate} = P(\text{Neg}|\text{E}) = \frac{500}{10,000} = .05$$

sensitivity
specificity

The **sensitivity** of the diagnostic test is the true positive rate, and the **specificity** of the diagnostic test is the true negative rate.

The primary question facing the inspector is to evaluate the probability of *E. coli* being present in the meat sample when the test yields a positive result; that is, the inspector needs to know $P(E|\text{Pos})$. Bayes' formula answers this question, as the following calculations show. To make this calculation, we need to know the *rate* of *E. coli* in the type of meat being inspected. For this example, suppose that *E. coli* is present in 4.5% of all meat samples; that is, *E. coli* has prevalence $P(E) = .045$. We can then compute $P(E|\text{Pos})$ as follows:

$$P(E|\text{Pos}) = \frac{P(E \cap \text{Pos})}{P(\text{Pos})} = \frac{P(E \cap \text{Pos})}{P(E \cap \text{Pos}) + P(NE \cap \text{Pos})}$$

$$= \frac{P(\text{Pos}|E)P(E)}{P(\text{Pos}|E)P(E) + P(\text{Pos}|NE)P(NE)}$$

$$= \frac{(.95)(.045)}{(.95)(.045) + (.01)(1 - .045)} = .817$$

Thus, *E. coli* is truly present in 81.7% of the tested samples in which a positive test result occurs. Also, we can conclude that 18.3% of the tested samples indicated *E. coli* was present when in fact there was no *E. coli* in the meat sample.

EXAMPLE 4.3

A book club classifies members as heavy, medium, or light purchasers, and separate mailings are prepared for each of these groups. Overall, 20% of the members are heavy purchasers, 30% medium, and 50% light. A member is not classified into a group until 18 months after joining the club, but a test is made of the feasibility of using the first 3 months' purchases to classify members. The following percentages are obtained from existing records of individuals classified as heavy, medium, or light purchasers:

First 3 Months' Purchases	Group (%)		
	Heavy	Medium	Light
0	5	15	60
1	10	30	20
2	30	40	15
3+	55	15	5

If a member purchases no books in the first 3 months, what is the probability that the member is a light purchaser? (*Note:* This table contains conditional percentages for each column.)

Solution Using the conditional probabilities in the table, the underlying purchase probabilities, and Bayes' Formula, we can compute this conditional probability.

$$P(\text{light}|0)$$

$$= \frac{P(0|\text{light})P(\text{light})}{P(0|\text{light})P(\text{light}) + P(0|\text{medium})P(\text{medium}) + P(0|\text{heavy})P(\text{heavy})}$$

$$= \frac{(.60)(.50)}{(.60)(.50) + (.15)(.30) + (.05)(.20)}$$

$$= .845$$

These examples indicate the basic idea of Bayes' Formula. There is some number k of possible, mutually exclusive, underlying events $A_1, \ldots, A_k$, which are sometimes called the **states of nature.** Unconditional probabilities $P(A_1), \ldots, P(A_k)$, often called **prior probabilities,** are specified. There are m possible, mutually exclusive, **observable events** $B_1, \ldots, B_m$. The conditional probabilities of each observable event given each state of nature, $P(B_i|A_i)$, are also specified, and these probabilities are called **likelihoods.** The problem is to find the **posterior probabilities** $P(A_i|B_i)$. *Prior* and *posterior* refer to probabilities before and after observing an event B_i.

states of nature

prior probabilities

observable events

likelihoods

posterior probabilities

Bayes' Formula

If $A_1, \ldots, A_k$ are mutually exclusive states of nature, and if $B_1, \ldots, B_m$ are m possible mutually exclusive observable events, then

$$P(A_i|B_j) = \frac{P(B_j|A_i)P(A_i)}{P(B_j|A_1)P(A_1) + P(B_j|A_2)P(A_2) + \cdots + P(B_j|A_k)P(A_k)}$$

$$= \frac{P(B_j|A_i)P(A_i)}{\Sigma_i P(B_j|A_i)P(A_i)}$$

EXAMPLE 4.4

In the manufacture of circuit boards, there are three major types of defective boards. The types of defects, along with the percentage of all circuit boards having these defects, are (1) improper electrode coverage (D_1), 2.8%; (2) plating separation (D_2), 1.2%; and (3) etching problems (D_3), 3.2%. A circuit board will contain at most one of the three defects. Defects can be detected with certainty using destructive testing of the finished circuit boards; however, this is not a very practical method for inspecting a large percentage of the circuit boards. A nondestructive inspection procedure has been developed, which has the following outcomes: A_1, which indicates the board has only defect D_1; A_2, which indicates the board has only defect D_2; A_3, which indicates the board has only defect D_3; and A_4, which indicates the board has no defects. The respective likelihoods for the four outcomes of the nondestructive test determined by evaluating a large number of boards known to have exactly one of the three types of defects are given in the following table:

Test Outcome	Type of Defect			
	D_1	D_2	D_3	None
A_1	.90	.06	.02	.02
A_2	.05	.80	.06	.01
A_3	.03	.05	.82	.02
A_4 (no defects)	.02	.09	.10	.95

If a circuit board is tested using the nondestructive test and the outcome indicates no defects (A_4), what are the probabilities that the board has no defect or a D_1, D_2, or D_3 type of defect?

Let D_4 represent the situation in which the circuit board has no defects.

$$P(D_1|A_4) = \frac{P(A_4|D_1)P(D_1)}{P(A_4|D_1)P(D_1) + P(A_4|D_2)P(D_2) + P(A_4|D_3)P(D_3) + P(A_4|D_4)P(D_4)}$$

$$= \frac{(.02)(.028)}{(.02)(.028) + (.09)(.012) + (.10)(.032) + (.95)(.928)} = \frac{.00056}{.88644} = .00063$$

$$P(D_2|A_4) = \frac{P(A_4|D_2)P(D_2)}{P(A_4|D_1)P(D_1) + P(A_4|D_2)P(D_2) + P(A_4|D_3)P(D_3) + P(A_4|D_4)P(D_4)}$$

$$= \frac{(.09)(.012)}{(.02)(.028) + (.09)(.012) + (.10)(.032) + (.95)(.928)} = \frac{.00108}{.88644} = .00122$$

$$P(D_3|A_4) = \frac{P(A_4|D_3)P(D_3)}{P(A_4|D_1)P(D_1) + P(A_4|D_2)P(D_2) + P(A_4|D_3)P(D_3) + P(A_4|D_4)P(D_4)}$$

$$= \frac{(.10)(.032)}{(.02)(.028) + (.09)(.012) + (.10)(.032) + (.95)(.928)} = \frac{.0032}{.88644} = .00361$$

$$P(D_4|A_4) = \frac{P(A_4|D_4)P(D_4)}{P(A_4|D_1)P(D_1) + P(A_4|D_2)P(D_2) + P(A_4|D_3)P(D_3) + P(A_4|D_4)P(D_4)}$$

$$= \frac{(.95)(.928)}{(.02)(.028) + (.09)(.012) + (.10)(.032) + (.95)(.928)} = \frac{.8816}{.88644} = .99454$$

Thus, if the new test indicates that none of the three types of defects is present in the circuit board, there is a very high probability, .99454, that the circuit board in fact is free of defects.

EXERCISES   **Applications**

Bus.   **4.18**   Of a finance company's loans, 1% are defaulted (not completely repaid). The company routinely runs credit checks on all loan applicants. It finds that 30% of defaulted loans went to poor risks, 40% to fair risks, and 30% to good risks. Of the nondefaulted loans, 10% went to poor risks, 40% to fair risks, and 50% to good risks. Use Bayes' Formula to calculate the probability that a poor-risk loan will be defaulted.

4.19   Refer to Exercise 4.18. Show that the posterior probability of default, given a fair risk, equals the prior probability of default. Explain why this is a reasonable result.

Bus.   **4.20**   An underwriter of home insurance policies studies the problem of home fires resulting from wood-burning furnaces. Of all homes having such furnaces, 30% own a type 1 furnace, 25% a type 2 furnace, 15% a type 3, and 30% other types. Over 3 years, 5% of type 1 furnaces, 3% of type 2, 2% of type 3, and 4% of other types have resulted in fires. If a fire occurs in a particular home, what is the probability that a type 1 furnace is in the home?

Med.   **4.21**   In a January 15, 1998, article, the *New England Journal of Medicine* reported on the utility of using computerized tomography (CT) as a diagnostic test for patients with clinically suspected appendicitis. In at least 20% of patients with appendicitis, the correct diagnosis was not made. On the other hand, the appendix was normal in 15% to 40% of patients who underwent emergency appendectomy. A study was designed to determine the prospective effectiveness of using CT as a diagnostic test to improve the treatment of these patients. The study examined 100 consecutive patients suspected of having acute appendicitis who presented to the emergency department or were referred there from a physician's office. The 100 patients underwent a CT scan, and the surgeon made an assessment of the presence of appendicitis for each of the patients. The final clinical outcomes were determined at surgery and by pathological examination of the appendix after appendectomy or by clinical follow-up at least 2 months after CT scanning.

Radiologic Determination	Presence of Appendicitis	
	Confirmed (C)	Ruled Out (RO)
Definitely appendicitis (DA)	50	1
Equivocally appendicitis (EA)	2	2
Definitely not appendicitis (DNA)	1	44
Total	53	47

The 1996 rate of occurrence of appendicitis was approximately $P(C) = .00108$.

   a. Find the sensitivity and specificity of the radiological determination of appendicitis.
   b. Find the probability that a patient truly had appendicitis given that the radiological determination was definite appendicitis (DA).
   c. Find the probability that a patient truly did not have appendicitis given that the radiological determination was definite appendicitis (DA).
   d. Find the probability that a patient truly did not have appendicitis given that the radiological determination was definitely not appendicitis (DNA).

4.6   Variables: Discrete and Continuous

The basic language of probability deals with many different kinds of events. We are interested in calculating the probabilities associated with both quantitative and qualitative events. For example, we developed techniques that could be used to determine the probability that a machinist selected at random from the workers in a large automotive plant would suffer an accident during an 8-hour shift. These same techniques are also applicable to finding the probability that a machinist selected at random would work more than 80 hours without suffering an accident.

These qualitative and quantitative events can be classified as events (or outcomes) associated with qualitative and quantitative variables. For example, in the automotive plant accident study, the randomly selected machinist's accident report would consist of checking one of the following: No Accident, Minor Accident, or Major Accident. Thus, the data on 100 machinists in the study would be observations on a qualitative variable because the possible responses are the different categories of accident and are not different in any measurable, numerical amount. Because we cannot predict with certainty what type of accident a particular machinist will suffer, the variable is classified as a **qualitative random variable.** Other examples of qualitative random variables that are commonly measured are political party affiliation, socioeconomic status, the species of insect discovered on an apple leaf, and the brand preferences of customers. There are a finite (and typically quite small) number of possible outcomes associated with any qualitative variable. Using the methods of this chapter, it is possible to calculate the probabilities associated with these events.

qualitative random variable

Many times the events of interest in an experiment are quantitative outcomes associated with a **quantitative random variable** because the possible responses vary in numerical magnitude. For example, in the automotive plant accident study, the number of consecutive 8-hour shifts between accidents for a randomly selected

quantitative random variable

machinist is an observation on a quantitative random variable and events of interest, such as the number of 8-hour shifts between accidents for a randomly selected machinist, are observations on a quantitative random variable. Other examples of quantitative random variables are the change in earnings per share of a stock over the next quarter, the length of time a patient is in remission after a cancer treatment, the yield per acre of a new variety of wheat, and the number of people voting for the incumbent in an upcoming election. The methods of this chapter can be applied to calculate the probability associated with any particular event.

There are major advantages to dealing with quantitative random variables. The numerical yardstick underlying a quantitative variable makes the mean and standard deviation (for instance) sensible. With qualitative random variables the methods of this chapter can be used to calculate the probabilities of various events, and that's about all. With quantitative random variables, we can do much more: we can average the resulting quantities, find standard deviations, and assess probable

random variable   errors, among other things. Hereafter, we use the term **random variable** to mean quantitative random variable.

Most events of interest result in numerical observations or measurements. If a quantitative variable measured (or observed) in an experiment is denoted by the symbol y, we are interested in the values that y can assume. These values are called *numerical outcomes*. The number of different plant species per acre in a coal strip mine after a reclamation project is a numerical outcome. The percentage of registered voters who cast ballots in a given election is also a numerical outcome. The quantitative variable y is called a *random variable* because the value that y assumes in a given experiment is a chance or random outcome.

DEFINITION 4.10

When observations on a quantitative random variable can assume only a countable number of values, the variable is called a **discrete random variable.**

Examples of discrete variables are these:

1. Number of diseased trees per acre on an apple orchard
2. Change in the number of accidents per month at an intersection after a new signaling device has been installed
3. Number of dead people voting in the last mayoral election in a major midwest city

Note that it is possible to count the number of values that each of these random variables can assume.

DEFINITION 4.11

When observations on a quantitative random variable can assume any one of the uncountable number of values in a line interval, the variable is called a **continuous random variable.**

For example, the daily maximum temperature in Rochester, New York, can assume any of the infinitely many values on a line interval. It can be 89.6, 89.799, or 89.7611114. Typical continuous random variables are temperature, pressure, height, weight, and distance.

The distinction between **discrete** and **continuous random variables** is pertinent when we are seeking the probabilities associated with specific values of a random variable. The need for the distinction will be apparent when probability distributions are discussed in later sections of this chapter.

4.7 Probability Distributions for Discrete Random Variables

probability distribution

As previously stated, we need to know the probability of observing a particular sample outcome in order to make an inference about the population from which the sample was drawn. To do this, we need to know the probability associated with each value of the variable y. Viewed as relative frequencies, these probabilities generate a distribution of theoretical relative frequencies called the **probability distribution** of y. Probability distributions differ for discrete and continuous random variables. For discrete random variables, we compute the probability of specific individual values occurring. For continuous random variables, the probability of an interval of values is the event of interest.

The *probability distribution for a discrete random variable* displays the probability $P(y)$ associated with each value of y. This display can be presented as a table, a graph, or a formula. To illustrate, consider the tossing of two coins in Section 4.2 and let y be the number of heads observed. Then y can take the values 0, 1, or 2. From the data in Table 4.1, we can determine the approximate probability for each value of y, as given in Table 4.3. We point out that the relative frequencies in the table are very close to the theoretical relative frequencies (probabilities), which can be shown to be .25, .50, and .25 using the classical interpretation of probability. If we had employed 2,000,000 tosses of the coins instead of 500, the relative frequencies for $y = 0, 1,$ and 2 would be indistinguishable from the theoretical probabilities.

The probability distribution for y, the number of heads in the toss of two coins, is shown in Table 4.4 and is presented graphically as a *probability histogram* in Figure 4.2.

The probability distribution for this simple discrete random variable illustrates three important properties of discrete random variables.

TABLE 4.3		
Empirical sampling results for y: the number of heads in 500 tosses of two coins		

y	Frequency	Relative Frequency
0	129	.258
1	242	.484
2	129	.258

TABLE 4.4	
Probability distribution for the number of heads when two coins are tossed	

y	$P(y)$
0	.25
1	.50
2	.25

FIGURE 4.2

Probability distribution for the number of heads when two coins are tossed

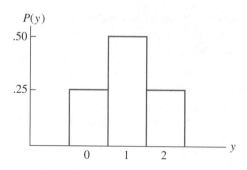

Properties of Discrete Random Variables

1. The probability associated with every value of y lies between 0 and 1.
2. The sum of the probabilities for all values of y is equal to 1.
3. The probabilities for a discrete random variable are additive. Hence, the probability that $y = 1$ or 2 is equal to $P(1) + P(2)$.

The relevance of the probability distribution to statistical inference will be emphasized when we discuss the probability distribution for the binomial random variable.

4.8 A Useful Discrete Random Variable: The Binomial

Many populations of interest to business people and scientists can be viewed as large sets of 0s and 1s. For example, consider the set of responses of all adults in the United States to the question, "Do you favor the development of nuclear energy?" If we disallow "no opinion," the responses will constitute a set of "yes" responses and "no" responses. If we assign a 1 to each yes and a 0 to each no, the population will consist of a set of 0s and 1s, and the sum of the 1s will equal the total number of people favoring the development. The sum of the 1s divided by the number of adults in the United States will equal the proportion of people who favor the development.

Gallup and Harris polls are examples of the sampling of 0, 1 populations. People are surveyed, and their opinions are recorded. Based on the sample responses, Gallup and Harris estimate the proportions of people in the population who favor some particular issue or possess some particular characteristic.

Similar surveys are conducted in the biological sciences, engineering, and business, but they may be called experiments rather than polls. For example, experiments are conducted to determine the effect of new drugs on small animals, such as rats or mice, before progressing to larger animals and, eventually, to human participants. Many of these experiments bear a marked resemblance to a poll in that the experimenter records only whether the drug was effective. Thus, if 300 rats are injected with a drug and 230 show a favorable response, the experimenter has conducted a poll—a poll of rat reaction to the drug, 230 "in favor" and 70 "opposed."

Similar polls are conducted by most manufacturers to determine the fraction of a product that is of good quality. Samples of industrial products are collected before shipment and each item in the sample is judged "defective" or "acceptable" according to criteria established by the company's quality control department. Based on the number of defectives in the sample, the company can decide whether the product is suitable for shipment. Note that this example, as well as those preceding, has the practical objective of making an inference about a population based on information contained in a sample.

The public opinion poll, the consumer preference poll, the drug-testing experiment, and the industrial sampling for defectives are all examples of a common, frequently conducted sampling situation known as a *binomial experiment*. The binomial experiment is conducted in all areas of science and business and only differs from one situation to another in the nature of objects being sampled (people, rats, electric lightbulbs, oranges). Thus, it is useful to define its characteristics. We can then apply our knowledge of this one kind of experiment to a variety of sampling experiments.

For all practical purposes the binomial experiment is identical to the coin-tossing example of previous sections. Here, n different coins are tossed (or a single coin is tossed n times), and we are interested in the number of heads observed. We assume that the probability of tossing a head on a single trial is π (π may equal .50, as it would for a balanced coin, but in many practical situations π will take some other value between 0 and 1). We also assume that the outcome for any one toss is unaffected by the results of any preceding tosses. These characteristics can be summarized as shown here.

DEFINITION 4.12

A **binomial experiment** is one that has the following properties:

1. The experiment consists of n identical trials.
2. Each trial results in one of two outcomes. We will label one outcome a success and the other a failure.
3. The probability of success on a single trial is equal to π and π remains the same from trial to trial.*
4. The trials are independent; that is, the outcome of one trial does not influence the outcome of any other trial.
5. The random variable y is the number of successes observed during the n trials.

EXAMPLE 4.5

An article in the March 5, 1998, issue of *The New England Journal of Medicine* discussed a large outbreak of tuberculosis. One person, called the index patient, was diagnosed with tuberculosis in 1995. The 232 co-workers of the index patient were given a tuberculin screening test. The number of co-workers recording a positive reading on the test was the random variable of interest. Did this study satisfy the properties of a binomial experiment?

Solution To answer the question, we check each of the five characteristics of the binomial experiment to determine whether they were satisfied.

1. Were there n identical trials? Yes. There were $n = 232$ workers who had approximately equal contact with the index patient.
2. Did each trial result in one of two outcomes? Yes. Each co-worker recorded either a positive or negative reading on the test.
3. Was the probability of success the same from trial to trial? Yes, if the co-workers had equivalent risk factors and equal exposures to the index patient.
4. Were the trials independent? Yes. The outcome of one screening test was unaffected by the outcome of the other screening tests.
5. Was the random variable of interest to the experimenter the number of successes y in the 232 screening tests? Yes. The number of co-workers who obtained a positive reading on the screening test was the variable of interest.

---

*Some books and computer programs use the letter p rather than π. We have chosen π to avoid confusion with p-values, discussed in Chapter 5.

All five characteristics were satisfied, so the tuberculin screening test represented a binomial experiment.

EXAMPLE 4.6

An economist interviews 75 students in a class of 100 to estimate the proportion of students who expect to obtain a C or better in the course. Is this a binomial experiment?

Solution Check this experiment against the five characteristics of a binomial experiment.

1. Are there identical trials? Yes. Each of 75 students is interviewed.
2. Does each trial result in one of two outcomes? Yes. Each student either does or does not expect to obtain a grade of C or higher.
3. Is the probability of success the same from trial to trial? No. If we let success denote a student expecting to obtain a C or higher, then the probability of success can change considerably from trial to trial. For example, unknown to the professor, suppose that 75 of the 100 students expect to obtain a grade of C or higher. Then π, the probability of success for the first student interviewed, is $75/100 = .75$. If the student is a failure (does not expect a C or higher), the probability of success for the next student is $75/99 = .76$. Suppose that after 70 students are interviewed, 60 are successes and 10 are failures. Then the probability of success for the next (71st) student is $15/30 = .50$.

This example shows how the probability of success can change substantially from trial to trial in situations in which the sample size is a relatively large portion of the total population size. This experiment does not satisfy the properties of a binomial experiment.

Note that very few real-life situations satisfy perfectly the requirements stated in Definition 4.12, but for many the lack of agreement is so small that the binomial experiment still provides a very good model for reality.

Having defined the binomial experiment and suggested several practical applications, we now examine the probability distribution for the binomial random variable y, the number of successes observed in n trials. Although it is possible to approximate $P(y)$, the probability associated with a value of y in a binomial experiment, by using a relative frequency approach, it is easier to use a general formula for binomial probabilities.

Formula for Computing $P(y)$ in a Binomial Experiment

The probability of observing y successes in n trials of a binomial experiment is

$$P(y) = \frac{n!}{y!(n-y)!}\pi^{y}(1-\pi)^{n-y}$$

where

$$n = \text{number of trials}$$
$$\pi = \text{probability of success on a single trial}$$
$$1 - \pi = \text{probability of failure on a single trial}$$
$$y = \text{number of successes in } n \text{ trials}$$
$$n! = n(n-1)(n-2)\cdots(3)(2)(1)$$

As indicated in the box, the notation $n!$ (referred to as n factorial) is used for the product

$$n! = n(n-1)(n-2)\cdots(3)(2)(1)$$

For $n = 3$,

$$n! = 3! = (3)(3-1)(3-2) = (3)(2)(1) = 6$$

Similarly, for $n = 4$,

$$4! = (4)(3)(2)(1) = 24$$

We also note that $0!$ is defined to be equal to 1.

To see how the formula for binomial probabilities can be used to calculate the probability for a specific value of y, consider the following examples.

EXAMPLE 4.7

A new variety of turf grass has been developed for use on golf courses, with the goal of obtaining a germination rate of 85%. To evaluate the grass, 20 seeds are planted in a greenhouse so that each seed will be exposed to identical conditions. If the 85% germination rate is correct, what is the probability that 18 or more of the 20 seeds will germinate?

$$P(y) = \frac{n!}{y!(n-y)!}\pi^y(1-\pi)^{n-y}$$

and substituting for $n = 20$, $\pi = .85$, $y = 18$, 19, and 20, we obtain

$$P(y = 18) = \frac{20!}{18!(20-18)!}(.85)^{18}(1-.85)^{20-18} = 190(.85)^{18}(.15)^2 = .229$$

$$P(y = 19) = \frac{20!}{19!(20-19)!}(.85)^{19}(1-.85)^{20-19} = 20(.85)^{19}(.15)^1 = .137$$

$$P(y = 20) = \frac{20!}{20!(20-20)!}(.85)^{20}(1-.85)^{20-20} = (.85)^{20} = .0388$$

$$P(y \geq 18) = P(y = 18) + P(y = 19) + P(y = 20) = .405$$

The calculations in Example 4.7 entail a considerable amount of effort even though n was only 20. For those situations involving a large value of n, we can use computer software to make the exact calculations. An approach that yields fairly accurate results in many situations and does not require the use of a computer will be discussed later in this chapter.

EXAMPLE 4.8

Suppose that a sample of households is randomly selected from all the households in the city in order to estimate the percentage in which the head of the household is unemployed. To illustrate the computation of a binomial probability, suppose that the unknown percentage is actually 10% and that a sample of $n = 5$ (we select a small sample to make the calculation manageable) is selected from the population. What is the probability that all five heads of the households are employed?

Solution   We must carefully define which outcome we wish to call a success. For this example, we define a success as being employed. Then the probability of success when one person is selected from the population is $\pi = .9$ (because the proportion unemployed is .1). We wish to find the probability that $y = 5$ (all five are employed) in five trials.

$$P(y = 5) = \frac{5!}{5!(5-5)!}(.9)^5(.1)^0$$

$$= \frac{5!}{5!0!}(.9)^5(.1)^0$$

$$= (.9)^5 = .590$$

The binomial probability distribution for $n = 5$, $\pi = .9$ is shown in Figure 4.3. The probability of observing five employed in a sample of five is shaded in the figure.

FIGURE 4.3
The binomial probability distribution for $n = 5$, $\pi = .9$

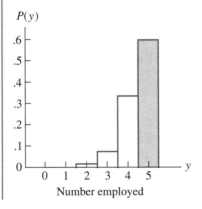

EXAMPLE 4.9

Refer to Example 4.8 and calculate the probability that exactly one person in the sample of five households is unemployed. What is the probability of one or fewer being unemployed?

Solution Because y is the number of employed in the sample of five, one unemployed person corresponds to four employed ($y = 4$). Then,

$$P(4) = \frac{5!}{4!(5-4)!}(.9)^4(.1)^1$$

$$= \frac{(5)(4)(3)(2)(1)}{(4)(3)(2)(1)(1)}(.9)^4(.1)$$

$$= 5(.9)^4(.1)$$

$$= .328$$

Thus, the probability of selecting four employed heads of households in a sample of five is .328, or, roughly, one chance in three.

The outcome "one or fewer unemployed" is the same as the outcome "4 or 5 employed." Because y represents the number employed, we seek the probability that $y = 4$ or 5. Because the values associated with a random variable represent mutually exclusive events, the probabilities for discrete random variables are additive. Thus, we have

$$P(y = 4 \text{ or } 5) = P(4) + P(5)$$
$$= .328 + .590$$
$$= .918$$

Thus, the probability that a random sample of five households will yield either four or five employed heads of households is .918. This high probability is consistent with our intuition: we could expect the number of employed in the sample to be large if 90% of all heads of households in the city are employed.

Like any relative frequency histogram, a binomial probability distribution possesses a mean, μ, and a standard deviation, σ. Although we omit the derivations, we give the formulas for these parameters.

Mean and Standard Deviation of the Binomial Probability Distribution

$$\mu = n\pi \quad \text{and} \quad \sigma = \sqrt{n\pi(1 - \pi)}$$

where π is the probability of success in a given trial and n is the number of trials in the binomial experiment.

If we know π and the sample size, n, we can calculate μ and σ to locate the center and describe the variability for a particular binomial probability distribution. Thus, we can quickly determine those values of y that are probable and those that are improbable.

EXAMPLE 4.10

We consider the turf grass seed example to illustrate the calculation of the mean and standard deviation. Suppose the company producing the turf grass takes a sample of 20 seeds on a regular basis to monitor the quality of the seeds. If the germination rate of the seeds stays constant at 85%, then the average number of seeds that will germinate in the sample of 20 seeds is

$$\mu = n\pi = 20(.85) = 17$$

with a standard deviation of

$$\sigma = \sqrt{n\pi(1 - \pi)} = \sqrt{20(.85)(1 - .85)} = 1.60$$

Suppose we examine the germination records of a large number of samples of 20 seeds each. If the germination rate has remained constant at 85%, then the average number of seeds that germinate should be close to 17 per sample. If in a particular sample of 20 seeds we determine that only 12 have germinated, does the germination rate of 85% seem consistent with our results? Using a computer software program, we can generate the probability distribution for the number of seeds that germinate in the sample of 20 seeds, as shown in Figure 4.4.

Although the distribution is tending toward left skewness (see Figure 4.4), the Empirical Rule should work well for this relatively mound-shaped distribution. Thus, $y = 12$ seeds is more than 3 standard deviations less than the mean number of seeds, $\mu = 17$; it is highly improbable that in 20 seeds we would obtain only 12 germinated seeds if π really is equal to .85. The germination rate is most likely a value considerably less than .85.

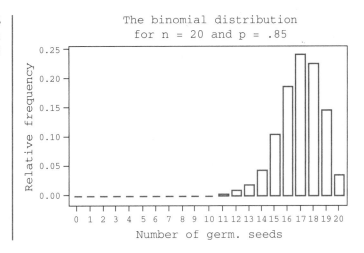

FIGURE 4.4
The binomial distribution
for $n = 20$ and $p = .85$

EXAMPLE 4.11

A poll shows that 516 of 1,218 voters favor the reelection of a particular political candidate. Do you think that the candidate will win?

Solution   To win the election, the candidate will need at least 50% of the votes. Let us see whether $y = 516$ is too small a value of y to imply a value of π (the proportion of voters favoring the candidate) equal to .5 or larger. If $\pi = .5$,

$$\mu = n\pi = (1,218)(.5) = 609$$

$$\sigma = \sqrt{n\pi(1 - \pi)} = \sqrt{(1,218)(.5)(.5)}$$

$$= \sqrt{304.5} = 17.45$$

and $3\sigma = 52.35$.

You can see from Figure 4.5 that $y = 516$ is more than 3σ, or 52.35, away from $\mu = 609$. In fact, if you wish to check, you will see that $y = 516$ is more than 5σ away from $\mu = 609$, the value of μ if π were really equal to .5. Thus, it appears that the number of voters in the sample who favor the candidate is much too small if the candidate does, in fact, possess a majority favoring reelection. Consequently, we conclude that he or she will lose. (Note that this conclusion is based on the assumption that the set of voters from which the sample was drawn is the same as the set who will vote. We also must assume that the opinions of the voters will not change between the time of sampling and the date of the election.)

FIGURE 4.5
Location of the observed value
of y ($y = 516$) relative to μ

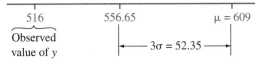

The purpose of this section is to present the binomial probability distribution so you can see how binomial probabilities are calculated and so you can calculate them for small values of n, if you wish. In practice, n is usually large (in national surveys, sample sizes as large as 1,500 are common), and the computation of the binomial probabilities is tedious. Later in this chapter, we will present a simple procedure for obtaining approximate values to the probabilities we need in making inferences. We can also use some very rough procedures for evaluating probabilities by using the mean and standard deviation of the binomial random variable y along with the Empirical Rule.

   We refer the interested reader to Hildebrand and Ott (1998) and Devore (2000) for more information about discrete random variables. In the next section, we discuss probability distributions with emphasis on the normal distribution.

EXERCISES   **Basic Techniques**

4.22 Let y be a binomial random variable; compute $P(y)$ for each of the following situations:
   a. $n = 10, \pi = .2, y = 3$ **b.** $n = 4, \pi = .4, y = 2$ **c.** $n = 16, \pi = .7, y = 12$

4.23 Let y be a binomial random variable with $n = 8$ and $\pi = .4$. Find the following values:
   a. $P(y \leq 4)$ **b.** $P(y > 4)$ **c.** $P(y \leq 7)$ **d.** $P(y > 6)$

Applications

Bus.   **4.24** An appliance store has the following probabilities for y, the number of major appliances sold on a given day:

y	$P(y)$
0	.100
1	.150
2	.250
3	.140
4	.090
5	.080
6	.060
7	.050
8	.040
9	.025
10	.015

   a. Construct a graph of $P(y)$.
   b. Find $P(y \leq 2)$.
   c. Find $P(y \geq 7)$.
   d. Find $P(1 \leq y \leq 5)$.

Bus.   **4.25** The weekly demand for copies of a popular word-processing program at a computer store has the probability distribution shown here.

y	$P(y)$
0	.06
1	.14
2	.16
3	.14
4	.12
5	.10
6	.08
7	.07
8	.06
9	.04
10	.03

a. What is the probability that three or more copies will be needed in a particular week?

b. What is the probability that the demand will be for at least two but no more than six copies?

c. If the store has eight copies of the program available at the beginning of each week, what is the probability the demand will exceed the supply in a given week?

Bio. **4.26** A biologist randomly selects 10 portions of water, each equal to .1 cm^3 in volume, from the local reservoir and counts the number of bacteria present in each portion. The biologist then totals the number of bacteria for the 10 portions to obtain an estimate of the number of bacteria per cubic centimeter present in the reservoir water. Is this a binomial experiment?

Pol. Sci. **4.27** Examine the accompanying newspaper clipping. Does this sampling appear to satisfy the characteristics of a binomial experiment?

Poll Finds Opposition to Phone Taps

New York—People surveyed in a recent poll indicated they are 81% to 13% against having their phones tapped without a court order.

The people in the survey, by 68% to 27%, were opposed to letting the government use a wiretap on citizens suspected of crimes, except with a court order.

The survey was conducted for 1,495 households and also found the following results:

—The people surveyed are 80% to 12% against the use of any kind of electronic spying device without a court order.

—Citizens are 77% to 14% against allowing the government to open their mail without court orders.

—They oppose, by 80% to 12%, letting the telephone company disclose records of long-distance phone calls, except by court order.

For each of the questions, a few of those in the survey had no responses.

Env. **4.28** In an inspection of automobiles in Los Angeles, 60% of all automobiles had emissions that do not meet EPA regulations. For a random sample of 10 automobiles, compute the following probabilities:

a. All 10 automobiles failed the inspection.

b. Exactly 6 of the 10 failed the inspection.

c. Six or more failed the inspection.

d. All 10 passed the inspection.

Use the following Minitab output to answer the questions. Note that with Minitab, the binomial probability π is denoted by p and the binomial variable y is represented by x.

```
Binomial Distribution with n = 10 and p = 0.6

     x          P(X = x)         P(X <= x)
  0.00           0.0001           0.0001
  1.00           0.0016           0.0017
  2.00           0.0106           0.0123
  3.00           0.0425           0.0548
  4.00           0.1115           0.1662
  5.00           0.2007           0.3669
  6.00           0.2508           0.6177
  7.00           0.2150           0.8327
  8.00           0.1209           0.9536
  9.00           0.0403           0.9940
 10.00           0.0060           1.0000
```

4.29   Refer to Exercise 4.28.
  a. Compute the probabilities for parts (a)–(d) if $\pi = .3$.
  b. Indicate how you would compute $P(y \leq 100)$ for $n = 1,000$ and $\pi = .3$.

Bus.   **4.30** Over a long period of time in a large multinational corporation, 10% of all sales trainees are rated as outstanding, 75% are rated as excellent/good, 10% are rated as satisfactory, and 5% are considered unsatisfactory. Find the following probabilities for a sample of 10 trainees selected at random:
  a. Two are rated as outstanding.
  b. Two or more are rated as outstanding.
  c. Eight of the 10 are rated either outstanding or excellent/good.
  d. None of the trainees is rated as unsatisfactory.

Med.   **4.31** A new technique, balloon angioplasty, is being widely used to open clogged heart valves and vessels. The balloon is inserted via a catheter and is inflated, opening the vessel; thus, no surgery is required. Left untreated, 50% of the people with heart-valve disease die within about 2 years. If experience with this technique suggests that approximately 70% live for more than 2 years, would the next five patients treated with balloon angioplasty at a hospital constitute a binomial experiment with $n = 5$, $\pi = .70$? Why?

Bus.   **4.32** A random sample of 50 price changes is selected from the many listed for a large supermarket during a reporting period. If the probability that a price change is posted correctly is .93,
  a. Write an expression for the probability that three or fewer changes are posted incorrectly.
  b. What assumptions were made for part (a)?

4.9 Probability Distributions for Continuous Random Variables

Discrete random variables (such as the binomial) have possible values that are distinct and separate, such as 0 or 1 or 2 or 3. Other random variables are most usefully considered to be *continuous:* their possible values form a whole interval (or range, or continuum). For instance, the 1-year return per dollar invested in a common stock could range from 0 to some quite large value. In practice, virtually all random variables assume a discrete set of values; the return per dollar of a million-dollar common-stock investment could be $1.06219423 or $1.06219424 or $1.06219425. However, when there are many possible values for a random variable, it is sometimes mathematically useful to treat the random variable as continuous.

Theoretically, then, a continuous random variable is one that can assume values associated with infinitely many points in a line interval. We state, without elaboration, that it is impossible to assign a small amount of probability to each value of y (as was done for a discrete random variable) and retain the property that the probabilities sum to 1.

To overcome this difficulty, we revert to the concept of the relative frequency histogram of Chapter 3, where we talked about the probability of y falling in a given interval. Recall that the relative frequency histogram for a population containing a large number of measurements will almost be a smooth curve because the number of class intervals can be made large and the width of the intervals can be decreased. Thus, we envision a smooth curve that provides a model for the population relative frequency distribution generated by repeated observation of a continuous random variable. This will be similar to the curve shown in Figure 4.6(a).

FIGURE 4.6
Probability distribution for a
continuous random variable

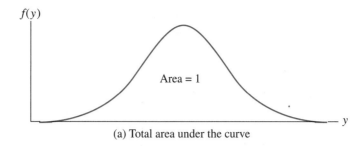

Area = 1

(a) Total area under the curve

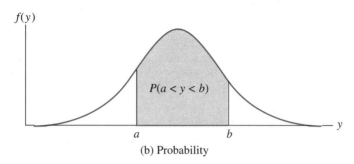

$P(a < y < b)$

a b

(b) Probability

Recall that the histogram relative frequencies are proportional to areas over the class intervals and that these areas possess a probabilistic interpretation. Thus, if a measurement is randomly selected from the set, the probability that it will fall in an interval is proportional to the histogram area above the interval. Because a population is the whole (100%, or 1), we want the total area under the probability curve to equal 1. If we let the total area under the curve equal 1, then the areas over intervals are exactly equal to the corresponding probabilities.

The graph for the probability distribution for a continuous random variable is shown in Figure 4.7. The ordinate (height of the curve) for a given value of y is denoted by the symbol $f(y)$. Many people are tempted to say that $f(y)$, like $P(y)$ for the binomial random variable, designates the probability associated with the continuous random variable y. However, as we mentioned before, it is impossible to assign a probability to each of the infinitely many possible values of a continuous random variable. Thus, all we can say is that $f(y)$ represents the height of the probability distribution for a given value of y.

The probability that a continuous random variable falls in an interval, say between two points a and b, follows directly from the probabilistic interpretation given to the area over an interval for the relative frequency histogram (Section 3.3)

FIGURE 4.7
Hypothetical probability
distribution for student
examination scores

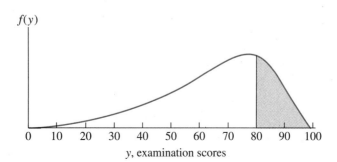

0   10   20   30   40   50   60   70   80   90   100

y, examination scores

and is equal to the area under the curve over the interval a to b, as shown in Figure 4.6(b). This probability is written $P(a < y < b)$.

There are curves of many shapes that can be used to represent the population relative frequency distribution for measurements associated with a continuous random variable. Fortunately, the areas for many of these curves have been tabulated and are ready for use. Thus, if we know that student examination scores possess a particular probability distribution, as in Figure 4.7, and if areas under the curve have been tabulated we can find the probability that a particular student will score more than 80% by looking up the tabulated area, which is shaded in Figure 4.7.

Figure 4.8 depicts four important probability distributions that will be used extensively in the following chapters. Which probability distribution we use in a particular situation is very important because probability statements are determined by the area under the curve. As can be seen in Figure 4.8, we obtain very different answers depending on which distribution is selected. For example, the probability that the random variable takes on a value less than 5.0 is essentially 1.0 for the probability distributions in Figures 4.8(a) and (b) but is 0.584 and 0.947 for the probability distributions in Figures 4.8(c) and (d), respectively. In some situations, we will not know exactly the distribution for the random variable in a particular

FIGURE 4.8

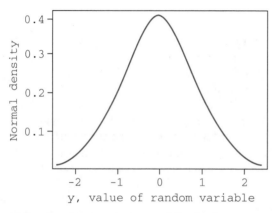

(a) Density of the standard normal distribution

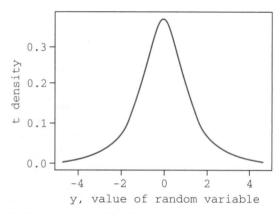

(b) Density of the t (df = 3) distribution

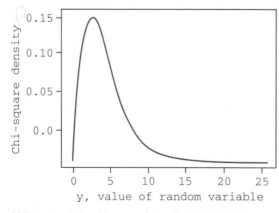

(c) Density of the chi-square (df = 5) distribution

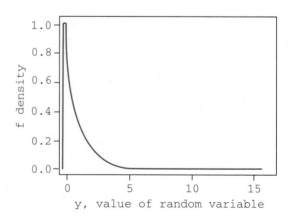

(d) Density of the F (df = 2, 6) distribution

study. In these situations, we can use the observed values for the random variable to construct a relative frequency histogram, which is a sample estimate of the true probability frequency distribution. As far as statistical inferences are concerned, the selection of the *exact* shape of the probability distribution for a continuous random variable is not crucial in many cases because most of our inference procedures are insensitive to the exact specification of the shape.

We will find that data collected on continuous variables often possess a nearly bell-shaped frequency distribution, such as depicted in Figure 4.8(a). A continuous variable (the normal) and its probability distribution (bell-shaped curve) provide a good model for these types of data. The normally distributed variable is also very important in statistical inference. We will study the normal distribution in detail in the next section.

4.10   A Useful Continuous Random Variable: The Normal Distribution

normal curve

Many variables of interest, including several statistics to be discussed in later sections and chapters, have mound-shaped frequency distributions that can be approximated by using a **normal curve.** For example, the distribution of total scores on the Brief Psychiatric Rating Scale for outpatients having a current history of repeated aggressive acts is mound-shaped. Other practical examples of mound-shaped distributions are social perceptiveness scores of preschool children selected from a particular socioeconomic background, psychomotor retardation scores for patients with circular-type manic-depressive illness, milk yields for cattle of a particular breed, and perceived anxiety scores for residents of a community. Each of these mound-shaped distributions can be approximated with a normal curve.

Because the normal distribution has been well tabulated, areas under a normal curve—which correspond to probabilities—can be used to approximate probabilities associated with the variables of interest in our experimentation. Thus, the normal random variable and its associated distribution play an important role in statistical inference.

The relative frequency histogram for the normal random variable, called the *normal curve* or *normal probability distribution,* is a smooth bell-shaped curve. Figure 4.9(a) shows a normal curve. If we let y represent the normal random variable, then the height of the probability distribution for a specific value of y is represented by $f(y)$.* The probabilities associated with a normal curve form the basis for the Empirical Rule.

As we see from Figure 4.9(a), the normal probability distribution is bell shaped and symmetrical about the mean μ. Although the normal random variable y may theoretically assume values from $-\infty$ to $+\infty$, we know from the Empirical Rule that approximately all the measurements are within 3 standard deviations (3σ) of μ. From the Empirical Rule, we also know that if we select a measurement at random from a population of measurements that possesses a mound-shaped distribution, the probability is approximately .68 that the measurement will lie within 1 standard

————

*For the normal distribution, $f(y) = \dfrac{1}{\sqrt{2\pi}\sigma}e^{-(y-\mu)^2/2\sigma^2}$, where μ and σ are the mean and standard deviation, respectively, of the population of y-values.

FIGURE 4.9

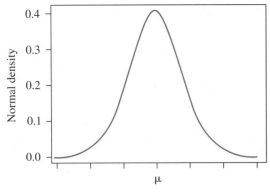

(a) Density of the normal distribution

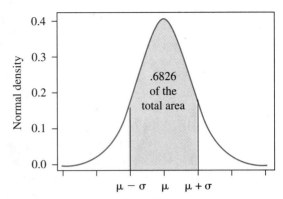

(b) Area under normal curve within 1 standard deviation of mean

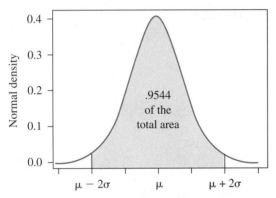

(c) Area under normal curve within 2 standard deviations of mean

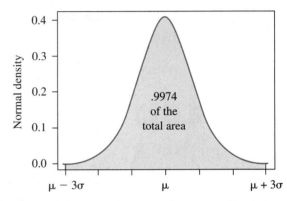

(d) Area under normal curve within 3 standard deviations of mean

deviation of its mean [see Figure 4.9(b)]. Similarly, we know that the probability is approximately .954 that a value will lie in the interval $\mu \pm 2\sigma$ and .997 in the interval $\mu \pm 3\sigma$ [see Figures 4.9(c) and (d)]. What we do not know, however, is the probability that the measurement will be within 1.65 standard deviations of its mean or within 2.58 standard deviations of its mean. The procedure we will discuss in this section will enable us to calculate the probability that a measurement falls within any distance of the mean μ for a normal curve.

Because there are many different normal curves (depending on the parameters μ and σ), it might seem to be an impossible task to tabulate areas (probabilities) for all normal curves, especially if each curve requires a separate table. Fortunately, this is not the case. By specifying the probability that a variable y lies within a certain number of standard deviations of its mean (just as we did in using the Empirical Rule), we need only one table of probabilities.

area under a normal curve   Table 1 in the Appendix gives the **area under a normal curve** to the left of a value y that is z standard deviations ($z\sigma$) away from the mean (see Figure 4.10). The area shown by the shading in Figure 4.10 is the probability listed in Table 1 in the Appendix. Values of z to the nearest tenth are listed along the left-hand column of the table, with z to the nearest hundredth along the top of the table. To find the

FIGURE 4.10

Area under a normal curve as given in Appendix Table 1

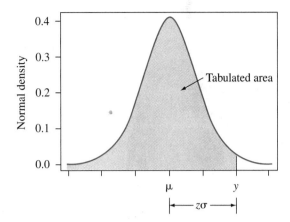

FIGURE 4.11

Area under a normal curve from μ to a point 1.65 standard deviations above the mean

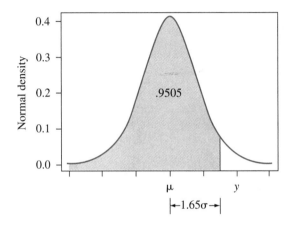

probability that a normal random variable lies to the left of a point 1.65 standard deviations above the mean, we look up the table entry corresponding to $z = 1.65$. This probability is .9505 (see Figure 4.11).

To determine the probability that a measurement will be less than some value y, we first calculate the number of standard deviations that y lies away from the mean by using the formula

$$z = \frac{y - \mu}{\sigma}$$

z score

The value of z computed using this formula is sometimes referred to as the **z score** associated with the y-value. Using the computed value of z, we determine the appropriate probability by using Table 1 in the Appendix. Note that we are merely coding the value y by subtracting μ and dividing by σ. (In other words, $y = z\sigma + \mu$.) Figure 4.12 illustrates the values of z corresponding to specific values of y. Thus, a value of y that is 2 standard deviations below (to the left of) μ corresponds to $z = -2$.

FIGURE 4.12

Relationship between specific
values of y and $z = (y - \mu)/\sigma$

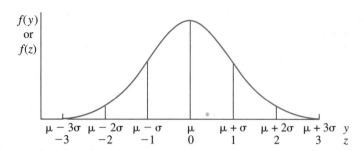

EXAMPLE 4.12

Consider a normal distribution with $\mu = 20$ and $\sigma = 2$. Determine the probability that a measurement is less than 23.

Solution   When first working problems of this type, it is a good idea to draw a picture so that you can see the area in question, as we have in Figure 4.13.

FIGURE 4.13

Area less than $y = 23$
under normal curve,
with $\mu = 20$, $\sigma = 2$

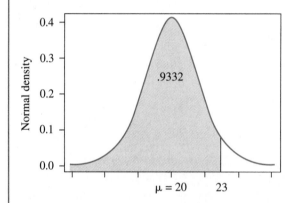

To determine the area under the curve to the left of the value $y = 23$, we first calculate the number of standard deviations $y = 23$ lies away from the mean.

$$z = \frac{y - \mu}{\sigma} = \frac{23 - 20}{2} = 1.5$$

Thus, $y = 23$ lies 1.5 standard deviations above $\mu = 20$. Referring to Table 1 in the Appendix, we find the area corresponding to $z = 1.5$ to be .9332. This is the probability that a measurement is less than 23.

EXAMPLE 4.13

For the normal distribution in Example 4.12 with $\mu = 20$ and $\sigma = 2$, find the probability that y is less than 16.

Solution   In determining the area to the left of 16, we use

$$z = \frac{y - \mu}{\sigma} = \frac{16 - 20}{2.} = -2$$

We find the appropriate area from Table 1 to be .0228; thus, .0228 is the probability that a measurement is less than 16. The area is shown in Figure 4.14.

FIGURE 4.14
Area less than $y = 16$
under normal curve,
with $\mu = 20$, $\sigma = 2$

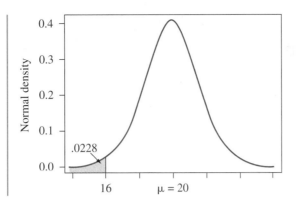

EXAMPLE 4.14

The mean daily milk production of a herd of Guernsey cows has a normal distribution with $\mu = 70$ pounds and $\sigma = 13$ pounds.

a. What is the probability that the milk production for a cow chosen at random will be less than 60 pounds?
b. What is the probability that the milk production for a cow chosen at random will be greater than 90 pounds?
c. What is the probability that the milk production for a cow chosen at random will be between 60 pounds and 90 pounds?

Solution   We begin by drawing pictures of the areas we are looking for [Figures 4.15(a)–(c)]. To answer part (a) we must compute the z values corresponding to the value of 60. The value $y = 60$ corresponds to a z score of

$$z = \frac{y - \mu}{\sigma} = \frac{60 - 70}{13} = -.77$$

From Appendix Table 1, the area to the left of 60 is .2206 [see Figure 4.15(a)].

FIGURE 4.15(a)
Area less than $y = 60$
under normal curve,
with $\mu = 70$, $\sigma = 13$

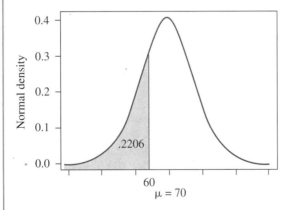

To answer part (b), the value $y = 90$ corresponds to a z score of

$$z = \frac{y - \mu}{\sigma} = \frac{90 - 70}{13} = 1.54$$

so from Appendix Table 1 we obtain .9382, the tabulated area less than 90. Thus, the area greater than 90 must be $1 - .9382 = .0618$ because the total area under the curve is 1 [see Figure 4.15(b)].

FIGURE 4.15(b)

Area greater than $y = 90$ under normal curve, with $\mu = 70$, $\sigma = 13$

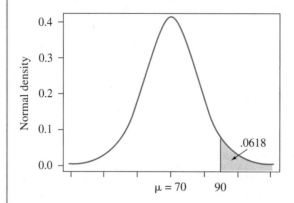

To answer part (c), we can use our results from (a) and (b). The area between two values y_1 and y_2 is determined by finding the difference between the areas to the left of the two values [see Figure 4.15(c)]. We have the area less than 60 is .2206, and the area less than 90 is .9382. Hence, the area between 60 and 90 is $.9382 - .2206 = .7176$. We can thus conclude that 22.06% of cow production is less than 60 pounds, 6.18% is greater than 90 pounds, and 71.76% is between 60 and 90 pounds.

FIGURE 4.15(c)

Area between 60 and 90 under normal curve, with $\mu = 70$, $\sigma = 13$

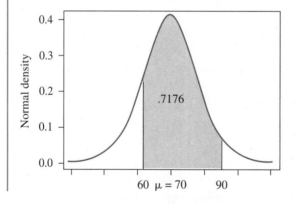

100pth percentile

An important aspect of the normal distribution is that we can easily find the percentiles of the distribution. The **100pth percentile** of a distribution is that value, y_p, such that $100p\%$ of the population values fall below y_p and $100(1 - p)\%$ are above y_p. For example, the median of a population is the 50th percentile, $y_{.50}$, and the quartiles are the 25th and 75th percentiles. The normal distribution is symmetric, so the median and the mean are the same value, $y_{.50} = \mu$ [see Figure 4.16(a)].

To find the percentiles of the standard normal distribution, we reverse our use of Appendix Table 1. To find the 100pth percentile, z_p, we find the probability p in Table 1 and then read out its corresponding number, z_p, along the margins of the table. For example, to find the 80th percentile, $z_{.80}$, we locate the probability, $p = .8000$ in Table 1. The value nearest to .8000 is .7995, which corresponds to a z score of 0.84. Thus, $z_{.80} = 0.84$ [see Figure 4.16(b)]. Now, to find the 100pth per-

FIGURE 4.16

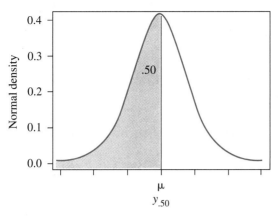

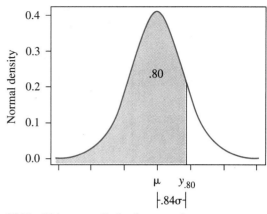

(a) For the normal curve, the mean and median agree

(b) The 80th percentile for the normal curve

centile, y_p, of a normal distribution with mean μ and standard deviation σ, we need to apply the reverse of our standardization formula,

$$y_p = \mu + z_p\sigma$$

Suppose we wanted to determine the 80th percentile of a population having a normal distribution with $\mu = 55$ and $\sigma = 3$. We have determined that $z_{.80} = 0.84$; thus, the 80th percentile for the population is $y_{.80} = 55 + (.84)(3) = 57.52$.

EXAMPLE 4.15

The Scholastic Aptitude Test (SAT) is an exam used to measure a person's readiness for college. The mathematics scores are scaled to have a normal distribution with mean 500 and standard deviation 100. What proportion of the people taking the SAT will score below 350? To identify a group of students needing remedial assistance, we want to determine the lower 10% of all scores; that is, we want to determine the 10th percentile, $y_{.10}$.

Solution   To find the proportion of scores below 350 [see Figure 4.17(a)], we need to find the area below 350:

$$z = \frac{y - \mu}{\sigma} = \frac{350 - 500}{100} = -1.5$$

FIGURE 4.17(a)

Area less than 350 under normal curve, with $\mu = 500$, $\sigma = 100$

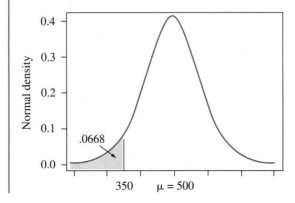

In a normal distribution, the area to the left of a value -1.5 standard deviations from the mean is, from Appendix Table 1, .0668. Hence, 6.68%, or approximately 7%, of the people taking the exam score below 350. The score of 350 is approximately the seventh percentile, $y_{.07}$, of the population of all scores.

To find the tenth percentile [see Figure 4.17(b)], we first find $z_{.10}$ in Appendix Table 1. Because .1003 is the value nearest .1000 and its corresponding z-value is -1.28, we take $z_{.10} = -1.28$. We then compute

$$y_{.10} = \mu + z_{.10}\sigma = 500 + (-1.28)(100) = 500 - 128 = 372$$

Thus, 10% of the scores on the SAT are less than 372.

FIGURE 4.17(b)

The 10th percentile for a normal curve, with $\mu = 500$, $\sigma = 100$

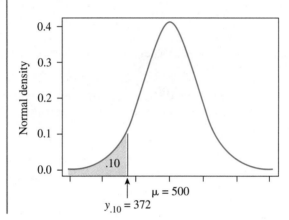

EXAMPLE 4.16

An analysis of income tax returns from the previous year indicates that for a given income classification, the amount of money owed to the government over and above the amount paid in the estimated tax vouchers for the first three payments is approximately normally distributed with a mean of \$530 and a standard deviation of \$205. Find the 75th percentile for this distribution of measurements. The government wants to target that group of returns having the largest 25% of amounts owed.

Solution   We need to determine the 75th percentile, $y_{.75}$ (Figure 4.18). From Appendix Table 1, we find $z_{.75} = .67$ because the probability nearest .7500 is .7486, which corresponds to a z score of .67. We then compute

$$y_{.75} = \mu + z_{.75}\sigma = 530 + (.67)(205) = 667.35$$

FIGURE 4.18

The 75th percentile for a normal curve, with $\mu = 530$, $\sigma = 205$

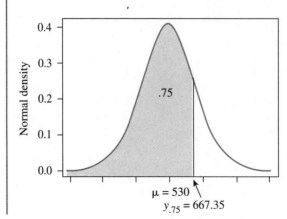

Thus, 25% of the tax returns in this classification exceed $667.35 in the amount owed the government.

EXERCISES   **Basic Techniques**

4.33   Use Table 1 of the Appendix to find the area under the normal curve between these values:
   a. $z = 0$ and $z = 1.3$
   b. $z = 0$ and $z = -1.9$

4.34   Repeat Exercise 4.33 for these values:
   a. $z = 0$ and $z = .7$
   b. $z = 0$ and $z = -1.2$

4.35   Repeat Exercise 4.33 for these values:
   a. $z = -.21$ and $z = 1.35$
   b. $z = .37$ and $z = 1.20$

4.36   Find a value for z, say z_0, such that $P(z > z_0) = .5$.

4.37   Find a value for z, say z_0, such that $P(z > z_0) = .025$.

4.38   Find a value for z, say z_0, such that $P(z > z_0) = .0089$.

4.39   Find a value for z, say z_0, such that $P(z > z_0) = .05$.

4.40   Find a value for z, say z_0, such that $P(-z_0 < z < z_0) = .95$.

4.41   Let y be a normal random variable with $\mu = 500$ and $\sigma = 100$. Find the following probabilities:
   a. $P(500 < y < 696)$
   b. $P(y > 696)$
   c. $P(304 < y < 696)$
   d. k such that $P(500 - k < y < 500 + k) = .60$

4.42   Suppose that y is a normal random variable with $\mu = 100$ and $\sigma = 15$.
   a. Show that $y < 130$ is equivalent to $z < 2$.
   b. Convert $y > 82.5$ to the z-score equivalent.
   c. Find $P(y < 130)$ and $P(y > 82.5)$.
   d. Find $P(y > 106)$, $P(y < 94)$, and $P(94 < y < 106)$.
   e. Find $P(y < 70)$, $P(y > 130)$, and $P(70 < y < 130)$.

4.43   Find the value of z for these areas.
   a. an area .01 to the right of z
   b. an area .10 to the left of z

4.44   Find the probability of observing a value of z greater than these values.
   a. 1.96
   b. 2.21
   c. 2.86
   d. 0.73

Applications

Gov.   **4.45**   Records maintained by the office of budget in a particular state indicate that the amount of time elapsed between the submission of travel vouchers and the final reimbursement of funds has approximately a normal distribution with a mean of 39 days and a standard deviation of 6 days.
   a. What is the probability that the elapsed time between submission and reimbursement will exceed 50 days?
   b. If you had a travel voucher submitted more than 55 days ago, what might you conclude?

Edu.   **4.46**   The College Boards, which are administered each year to many thousands of high school students, are scored so as to yield a mean of 500 and a standard deviation of 100. These scores are close to being normally distributed. What percentage of the scores can be expected to satisfy each condition?

a. greater than 600
b. greater than 700
c. less than 450
d. between 450 and 600

Bus. **4.47** Monthly sales figures for a particular food industry tend to be normally distributed with mean of 150 (in thousands of dollars) and a standard deviation of 35 (in thousands of dollars). Compute the following probabilities:
a. $P(y > 200)$
b. $P(y > 220)$
c. $P(y < 120)$
d. $P(100 < y < 200)$

4.48 Refer to Exercise 4.46. An exclusive club wishes to invite those scoring in the top 10% on the College Boards to join.
a. What score is required to be invited to join the club?
b. What score separates the top 60% of the population from the bottom 40%? What do we call this value?

4.11 Random Sampling

We have discussed random samples and introduced various sampling schemes in Chapter 2. What is the importance of random sampling? We must know how the sample was selected so we can determine probabilities associated with various sample outcomes. The probabilities of samples selected *in a random manner* can be determined, and we can use these probabilities to make inferences about the population from which the sample was drawn.

Sample data selected in a nonrandom fashion are frequently distorted by a *selection bias.* A selection bias exists whenever there is a systematic tendency to overrepresent or underrepresent some part of the population. For example, a survey of households conducted during the week entirely between the hours of 9 A.M. and 5 P.M. would be severely biased toward households with at least one member at home. Hence, any inferences made from the sample data would be biased toward the attributes or opinions of those families with at least one member at home and may not be truly representative of the population of households in the region.

random sample Now we turn to a definition of a **random sample** of n measurements selected from a population containing N measurements ($N > n$). (*Note:* This is a simple random sample as discussed in Chapter 2. Because most of the random samples discussed in this book are simple random samples, we'll drop the adjective unless needed for clarification.)

DEFINITION 4.13

A sample of n measurements selected from a population is said to be a **random sample** if every different sample of size n from the population has an equal probability of being selected.

EXAMPLE 4.17

A study of crimes related to handguns is being planned for the 10 largest cities in the United States. The study will randomly select 2 of the 10 largest cities for an in-depth study following the preliminary findings. The population of interest is the 10 largest cities $\{C_1, C_2, C_3, C_4, C_5, C_6, C_7, C_8, C_9, C_{10}\}$. List all possible different samples consisting of two cities that could be selected from the population of 10 cities. Give the probability associated with each sample in a random sample of $n = 2$ cities selected from the population.

Solution   All possible samples are listed next.

Sample	Cities	Sample	Cities	Sample	Cities
1	C_1, C_2	16	C_2, C_9	31	C_5, C_6
2	C_1, C_3	17	C_2, C_{10}	32	C_5, C_7
3	C_1, C_4	18	C_3, C_4	33	C_5, C_8
4	C_1, C_5	19	C_3, C_5	34	C_5, C_9
5	C_1, C_6	20	C_3, C_6	35	C_5, C_{10}
6	C_1, C_7	21	C_3, C_7	36	C_6, C_7
7	C_1, C_8	22	C_3, C_8	37	C_6, C_8
8	C_1, C_9	23	C_3, C_9	38	C_6, C_9
9	C_1, C_{10}	24	C_3, C_{10}	39	C_6, C_{10}
10	C_2, C_3	25	C_4, C_5	40	C_7, C_8
11	C_2, C_4	26	C_4, C_6	41	C_7, C_9
12	C_2, C_5	27	C_4, C_7	42	C_7, C_{10}
13	C_2, C_6	28	C_4, C_8	43	C_8, C_9
14	C_2, C_7	29	C_4, C_9	44	C_8, C_{10}
15	C_2, C_8	30	C_4, C_{10}	45	C_9, C_{10}

Now, let us suppose that we select a random sample of $n = 2$ cities from the 45 possible samples. The sample selected is called a *random* sample if every sample has an equal probability, 1/45, of being selected.

random number table

One of the simplest and most reliable ways to select a random sample of n measurements from a population is to use a table of random numbers (see Table 2 in the Appendix). **Random number tables** are constructed in such a way that, no matter where you start in the table and no matter in which direction you move, the digits occur randomly and with equal probability. Thus, if we wish to choose a random sample of $n = 10$ measurements from a population containing 100 measurements, we can label the measurements in the population from 0 to 99 (or 1 to 100). Then by referring to Table 2 in the Appendix and choosing a random starting point, the next 10 two-digit numbers going across the page indicate the labels of the particular measurements to be included in the random sample. Similarly, by moving up or down the page, we can also obtain a random sample.

This listing of all possible samples is feasible only when both the sample size n and the population size N are small. We can determine the number, M, of distinct samples of size n that can be selected from a population of N measurements using the following formula:

$$M = \frac{N!}{n!(N-n)!}$$

In Example 4.17, we had $N = 10$ and $n = 2$. Thus,

$$M = \frac{10!}{2!(10-2)!} = \frac{10!}{2!8!} = 45$$

The value of M becomes very large even when N is fairly small. For example, if $N = 50$ and $n = 5$, then $M = 2{,}118{,}760$. Thus, it would be very impractical to list all 2,118,760 possible samples consisting of $n = 5$ measurements from a population of $N = 50$ measurements and then randomly select one of the samples. In practice, we construct a list of elements in the population by assigning a number from 1 to N to each element in the population, called the *sampling frame*. We then randomly select n integers from the integers $(1, 2, \ldots, N)$ by using a table of random numbers (see Table 2 in the Appendix) or by using a computer program. Most statistical software programs contain routines for randomly selecting n integers from the integers $(1, 2, \ldots, N)$, where $N > n$. Exercise 4.52 contains the necessary commands for using Minitab to generate the random sample.

EXAMPLE 4.18

A small community consists of 850 families. We wish to obtain a random sample of 20 families to ascertain public acceptance of a wage and price freeze. Refer to Table 2 in the Appendix to determine which families should be sampled.

Solution   Assuming that a list of all families in the community is available (such as a telephone directory), we can label the families from 0 to 849 (or, equivalently, from 1 to 850). Then, referring to Table 2 in the Appendix, we choose a starting point. Suppose we have decided to start at line 1, column 3. Going down the page we choose the first 20 three-digit numbers between 000 and 849. From Table 2, we have

015	110	482	333
255	564	526	463
225	054	710	337
062	636	518	224
818	533	524	055

These 20 numbers identify the 20 families that are to be included in our sample.

A telephone directory is not always the best source for names, especially in surveys related to economics or politics. In the 1936 presidential campaign, Franklin Roosevelt was running as the Democratic candidate against the Republican candidate, Governor Alfred Landon of Kansas. This was a difficult time for the nation; the country had not yet recovered from the Great Depression of the 1930s, and there were still 9 million people unemployed.

The *Literary Digest* set out to sample the voting public and predict the winner of the election. Using names and addresses taken from telephone books and club memberships, the *Literary Digest* sent out 10 million questionnaires and got 2.4 million back. Based on the responses to the questionnaire, the *Digest* predicted a Landon victory by 57 to 43%.

At this time, George Gallup was starting his survey business. He conducted two surveys. The first one, based on 3,000 people, predicted the results of the *Digest* survey long before the *Digest* results were published; the second survey, based on 50,000, was used to forecast *correctly* the Roosevelt victory.

How did Gallup correctly predict the *Literary Digest* survey results and then, with another survey, correctly predict the outcome of the election? Where did the *Literary Digest* go wrong? The first problem was a severe selection bias. By taking the names and addresses from telephone directories and club memberships, its survey systematically excluded the poor. Unfortunately for the *Digest,* the vote was split along economic lines; the poor gave Roosevelt a large majority, whereas the rich tended to vote for Landon. A second reason for the error could be due to a *nonresponse bias.* Because only 20% of the 10 million people returned their surveys, and approximately half of those responding favored Landon, we might suspect that maybe the nonrespondents had different preferences than did the respondents. This was, in fact, true.

How, then do we achieve a random sample? Careful planning and a certain amount of ingenuity are required to have even a chance to approximate random sampling. This is especially true when the universe of interest involves people. People can be difficult to work with; they have a tendency to discard mail questionnaires and refuse to participate in personal interviews. Unless we are very careful, the data we obtain may be full of biases having unknown effects on the inferences we are attempting to make.

We do not have sufficient time to explore the topic of random sampling further in this book; entire courses at the undergraduate and graduate levels are devoted to sample survey research methodology. The important point to remember is that data from a random sample will provide the foundation for making statistical inferences in later chapters. Random samples are not easy to obtain, but with care we can avoid many potential biases that could affect the inferences we make.

EXERCISES   **Applications**

Soc.   **4.49** City officials want to sample the opinions of the homeowners in a community regarding the desirability of increasing local taxes to improve the quality of the public schools. If a random number table is used to identify the homes to be sampled and a home is discarded if the homeowner is not home when visited by the interviewer, is it likely this process will approximate random sampling? Explain.

4.50 A local TV network wants to run an informal survey of individuals from a local voting station to ascertain early results on a proposal to raise funds to move the city-owned historical museum to a new location. How might the network sample voters to approximate random sampling?

4.51 A psychologist was interested in studying women who are in the process of obtaining a divorce to determine whether the women experience significant attitudinal changes after the divorce has been finalized. Existing records from the geographic area in question show that 798 couples have recently filed for divorce. Assume that a sample of 25 women is needed for the study, and use Table 2 in the Appendix to determine which women should be asked to participate in the study. (*Hint:* Begin in column 2, row 1, and proceed down.)

4.52 Suppose you have been asked to run a public opinion poll related to an upcoming election. There are 230 precincts in the city, and you need to randomly select 50 registered voters from each precinct. Suppose that each precinct has 1,000 registered voters and it is possible to obtain a list of these people. You assign the numbers 1–1,000 to the 1,000 people on each list, with 1 to the first person on the list and 1,000 to the last person. You need to next obtain a random sample of 50 numbers from the numbers 1 to 1,000. The names on the sampling frame corresponding to these 50 numbers will be the 50 persons selected for the poll. A Minitab program is shown here for purposes of illustration. Note that you need to run this program 230 separate times to obtain a new random sample for each of the 230 precincts.

Follow these steps:

1. Click on **Calc.**
2. Click on **Random Data.**
3. Click on **Integer.**
4. Type **5** in the **Generate rows of data** box.

5. Type **c1-c10** in the **Store in Column(s): box.**
6. Type **1** in the **Minimum value:** box.
7. Type **1000** in the **Maximum value:** box.
8. Click on **OK.**
9. Click on **File.**
10. Click on **Print Worksheet.**

	C1	C2	C3	C4	C5	C6	C7	C8	C9	C10
1	340	701	684	393	313	312	834	596	321	739
2	783	877	724	498	315	282	175	611	725	571
3	862	625	971	30	766	256	40	158	444	546
4	974	402	768	593	980	536	483	244	51	201
5	232	742	1	861	335	129	409	724	340	218

a. Using either a random number table or a computer program, generate a second random sample of 50 numbers from the numbers 1 to 1,000.

b. Give several reasons why you need to generate a different set of random numbers for each of the precincts. Why not use the same set of 50 numbers for all 230 precincts?

4.12 Sampling Distributions

We discussed several different measures of central tendency and variability in Chapter 3 and distinguished between numerical descriptive measures of a population (parameters) and numerical descriptive measures of a sample (statistics). Thus, μ and σ are parameters, whereas $\bar{y}$ and s are statistics.

The numerical value of a sample statistic cannot be predicted exactly in advance. Even if we knew that the population mean μ was $216.37 and that the population standard deviation σ was $32.90—even if we knew the complete population distribution—we could not say that the sample mean $\bar{y}$ would be exactly equal to $216.37. A sample statistic is a random variable; it is subject to random variation because it is based on a random sample of measurements selected from the population of interest. Also, like any other random variable, a sample statistic has a probability distribution. We call the probability distribution of a sample statistic the *sampling distribution* of that statistic. Stated differently, the sampling distribution of a statistic is the population of all possible values for that statistic.

The actual mathematical derivation of sampling distributions is one of the basic problems of mathematical statistics. We will illustrate how the sampling distribution for $\bar{y}$ can be obtained for a simplified population. Later in the chapter, we will present several general results.

EXAMPLE 4.19

The sample $\bar{y}$ is to be calculated from a random sample of size two taken from a population consisting of 10 values (2, 3, 4, 5, 6, 7, 8, 9, 10, 11). Find the sampling distribution of $\bar{y}$, based on a random sample of size two.

Solution One way to find the sampling distribution is by counting. There are 45 possible samples of two items selected from the 10 items. These are shown here:

Sample	Value of $\bar{y}$	Sample	Value of $\bar{y}$	Sample	Value of $\bar{y}$
2, 3	2.5	3, 10	6.5	6, 7	6.5
2, 4	3	3, 11	7	6, 8	7
2, 5	3.5	4, 5	4.5	6, 9	7.5
2, 6	4	4, 6	5	6, 10	8
2, 7	4.5	4, 7	5.5	6, 11	8.5
2, 8	5	4, 8	6	7, 8	7.5
2, 9	5.5	4, 9	6.5	7, 9	8
2, 10	6	4, 10	7	7, 10	8.5
2, 11	6.5	4, 11	7.5	7, 11	9
3, 4	3.5	5, 6	5.5	8, 9	8.5
3, 5	4	5, 7	6	8, 10	9
3, 6	4.5	5, 8	6.5	8, 11	9.5
3, 7	5	5, 9	7	9, 10	9.5
3, 8	5.5	5, 10	7.5	9, 11	10
3, 9	6	5, 11	8	10, 11	10.5

Assuming each sample of size two is equally likely, it follows that the sampling distribution for $\bar{y}$ based on $n = 2$ observations selected from the population {2, 3, 4, 5, 6, 7, 8, 9, 10, 11} is as indicated here.

$\bar{y}$	$P(\bar{y})$	$\bar{y}$	$P(\bar{y})$
2.5	1/45	7	4/45
3	1/45	7.5	4/45
3.5	2/45	8	3/45
4	2/45	8.5	3/45
4.5	3/45	9	2/45
5	3/45	9.5	2/45
5.5	4/45	10	1/45
6	4/45	10.5	1/45
6.5	5/45		

The sampling distribution is shown as a graph in Figure 4.19. Note that the distribution is symmetric, with a mean of 6.5 and a standard deviation of approximately 2.0 (the range divided by 4).

FIGURE 4.19
Sampling distribution for $\bar{y}$

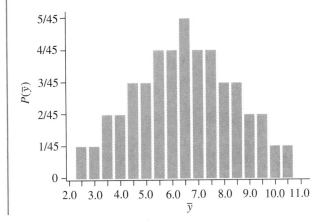

Example 4.19 illustrates for a very small population that we could in fact enumerate every possible sample of size two selected from the population and then compute all possible values of the sample mean. The next example will illustrate the properties of the sample mean, $\bar{y}$, when sampling from a larger population. This example will illustrate that the behavior of $\bar{y}$ as an estimator of μ depends on the sample size, n. Later in this chapter, we will illustrate the effect of the shape of the population distribution on the sampling distribution of $\bar{y}$.

EXAMPLE 4.20

In this example, the population values are known and, hence, we can compute the exact values of the population mean, μ, and population standard deviation, σ. We will then examine the behavior of $\bar{y}$ based on samples of size $n = 5$, 10, and 25 selected from the population.

The population consists of 500 pennies from which we compute the age of each penny: Age = 2000 − Date on penny. The histogram of the 500 ages is displayed in Figure 4.20(a). The shape is skewed to the right with a very long right tail. The mean and standard deviation are computed to be $\mu = 13.468$ years and $\sigma = 11.164$ years. In order to generate the sampling distribution of $\bar{y}$ for $n = 5$, we need to generate all possible samples of size $n = 5$ and then compute the $\bar{y}$ from each of these samples. This would be an enormous task because there are 255,244,687,600 possible samples of size 5 that could be selected from a population of 500 elements. The number of possible samples of size 10 or 25 is so large it makes even the national debt look small. Thus, we will use a computer program, S-plus, to select 25,000 samples of size 5 from the population of 500 pennies. For example, the first sample consists of pennies with ages 4, 12, 26, 16, and 9. The sample mean $\bar{y} = (4 + 12 + 26 + 16 + 9)/5 = 13.4$. We repeat 25,000 times the process of selecting 5 pennies, recording their ages, y_1, y_2, y_3, y_4, y_5, and then computing $\bar{y} = (y_1 + y_2 + y_3 + y_4 + y_5)/5$. The 25,000 values for $\bar{y}$ are then plotted in a frequency histogram, called the *sampling distribution* of $\bar{y}$ for $n = 5$. A similar procedure is followed for samples of size $n = 10$ and 25. The sampling distributions obtained are displayed in Figures 4.20(b)–(d).

Note that all three sampling distributions have nearly the same central value, approximately 13.5. (See Table 4.5.) The mean values of $\bar{y}$ for the three samples are nearly the same as the population mean, $\mu = 13.468$. In fact, if we had generated all possible samples for all three values of n, the mean of the possible values of $\bar{y}$ would agree exactly with μ.

The next characteristic to notice about the three histograms is their shape. All three are somewhat symmetric in shape, achieving a nearly normal distribution shape when $n = 25$. However, the histogram for $\bar{y}$ based on samples of size $n = 5$ is more spread out than the histogram based on $n = 10$, which, in turn, is more spread

TABLE 4.5

Means and standard deviations for the sampling distributions of $\bar{y}$

Sample Size	Mean of $\bar{y}$	Standard Deviation of $\bar{y}$	$\dfrac{11.1638}{\sqrt{n}}$
1 (Population)	13.468 (μ)	11.1638 (σ)	11.1638
5	13.485	4.9608	4.9926
10	13.438	3.4926	3.5303
25	13.473	2.1766	2.2328

FIGURE 4.20

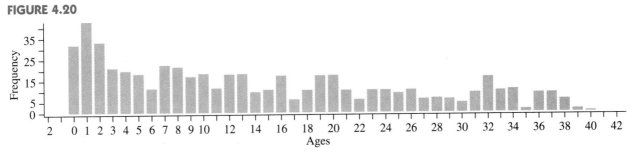

(a) Histogram of ages for 500 pennies

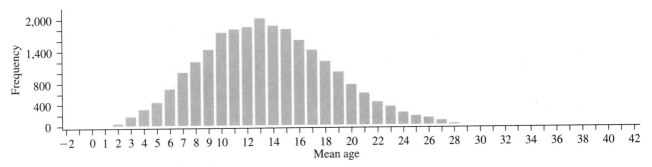

(b) Sampling distribution of $\bar{y}$ for $n = 5$

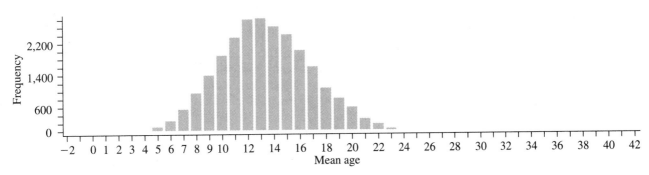

(c) Sampling distribution of $\bar{y}$ for $n = 10$

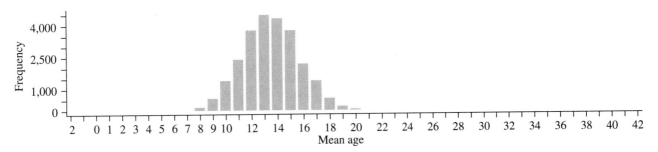

(d) Sampling distribution of $\bar{y}$ for $n = 25$

out than the histogram based on $n = 25$. When n is small, we are much more likely to obtain a value of $\bar{y}$ far from μ than when n is larger. What causes this increased dispersion in the values of $\bar{y}$? A single extreme y, either large or small relative to μ, in the sample has a greater influence on the size of $\bar{y}$ when n is small than when n is large. Thus, sample means based on small n are less accurate in their estimation of μ than their large-sample counterparts.

Table 4.5 contains summary statistics for the sampling distribution of $\bar{y}$. The sampling distribution of $\bar{y}$ has mean $\mu_{\bar{y}}$ and standard deviation $\sigma_{\bar{y}}$, which are related to the population mean μ and standard deviation σ by the following relationship:

$$\mu_{\bar{y}} = \mu \qquad \sigma_{\bar{y}} = \frac{\sigma}{\sqrt{n}}$$

standard error of $\bar{y}$

From Table 4.5, we note that the three sampling deviations have means that are approximately equal to the population mean. Also, the three sampling deviations have standard deviations that are approximately equal to $\sigma/\sqrt{n}$. If we had generated all possible values of $\bar{y}$, then the standard deviation of $\bar{y}$ would equal $\sigma/\sqrt{n}$ exactly. This quantity, $\sigma_{\bar{y}} = \sigma/\sqrt{n}$, is called the **standard error of $\bar{y}$**.

Quite a few of the more common sample statistics, such as the sample median and the sample standard deviation, have sampling distributions that are nearly normal for moderately sized values of n. We can observe this behavior by computing the sample median and sample standard deviation from each of the three sets of 25,000 sample ($n = 5, 10, 25$) selected from the population of 500 pennies. The resulting sampling distributions are displayed in Figures 4.21(a)–(d), for the sample median, and Figures 4.22(a)–(d), for the sample standard deviation. The sampling distribution of both the median and the standard deviation are more highly skewed in comparison to the sampling distribution of the sample mean. In fact, the value of n at which the sampling distributions of the sample median and standard deviation have a nearly normal shape is much larger than the value required for the

Central Limit Theorems

sample mean. A series of theorems in mathematical statistics called the **Central Limit Theorems** provide theoretical justification for our approximating the true sampling distribution of many sample statistics with the normal distribution. We will discuss one such theorem for the sample mean. Similar theorems exist for the sample median, sample standard deviation, and the sample proportion.

THEOREM 4.1

Central Limit Theorem for $\bar{y}$

Let $\bar{y}$ denote the sample mean computed from a random sample of n measurements from a population having a mean, μ, and finite standard deviation, σ. Let $\mu_{\bar{y}}$ and $\sigma_{\bar{y}}$ denote the mean and standard deviation of the sampling distribution of $\bar{y}$, respectively. Based on repeated random samples of size n from the population, we can conclude the following:

1. $\mu_{\bar{y}} = \mu$
2. $\sigma_{\bar{y}} = \sigma/\sqrt{n}$
3. When n is large, the sampling distribution of $\bar{y}$ is approximately normal (with the approximation becoming more precise as n increases).
4. When the population distribution is normal, the sampling distribution of $\bar{y}$ is exactly normal for any sample size n.

FIGURE 4.21

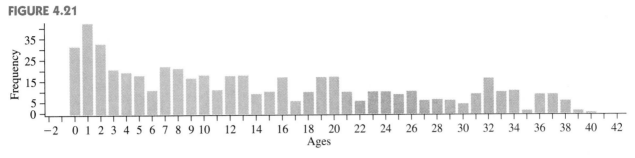

(a) Histogram of ages for 500 pennies

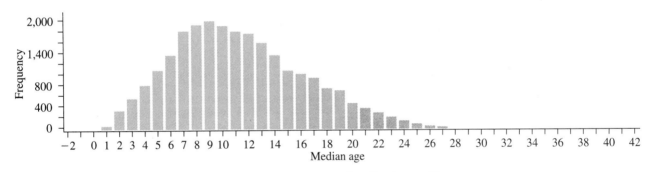

(b) Sampling distribution of median for $n = 5$

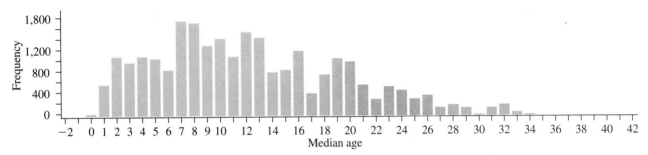

(c) Sampling distribution of median for $n = 10$

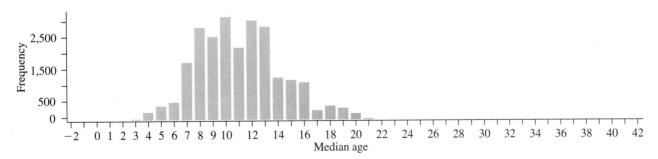

(d) Sampling distribution of median for $n = 25$

FIGURE 4.22

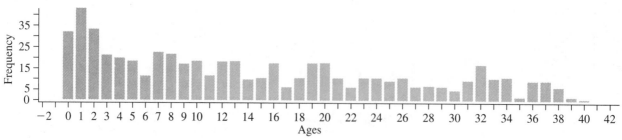

(a) Histogram of ages for 500 pennies

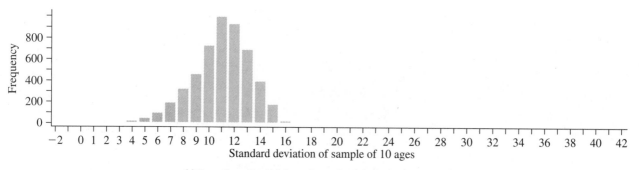

(b) Sampling distribution of standard deviation for $n = 5$

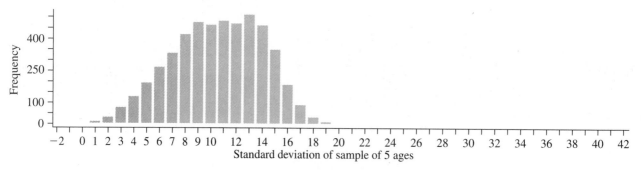

(c) Sampling distribution of standard deviation for $n = 10$

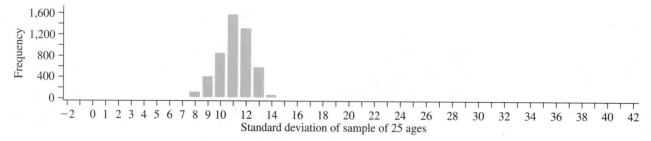

(d) Sampling distribution of standard deviation for $n = 25$

Figure 4.20 illustrates the Central Limit Theorem. Figure 4.20(a) displays the distribution of the measurements y in the population from which the samples are to be drawn. No specific shape was required for these measurements for the Central Limit Theorem to be validated. Figures 4.20(b)–(d) illustrate the sampling distribution for the sample mean $\bar{y}$ when n is 5, 10, and 25, respectively. We note that even for a very small sample size, $n = 10$, the shape of the sampling distribution of $\bar{y}$ is very similar to that of a normal distribution. This is not true in general. If the population distribution had many extreme values or several modes, the sampling distribution of $\bar{y}$ would require n to be considerably larger in order to achieve a symmetric bell shape.

We have seen that the sample size n has an effect on the shape of the sampling distribution of $\bar{y}$. The shape of the distribution of the population measurements also affects the shape of the sampling distribution of $\bar{y}$. Figures 4.23 and 4.24 illustrate the effect of the population shape on the shape of the sampling distribution of $\bar{y}$. In Figure 4.23, the population measurements have a normal distribution and the sampling distribution of $\bar{y}$ is *exactly* a normal distribution for all values of n, as is illustrated for $n = 5, 10,$ and 25. When the population distribution is nonnormal, as depicted in Figure 4.24, the sampling distribution of $\bar{y}$ will not have a normal shape for small n (see Figure 4.24 with $n = 5$); however, for $n = 10$ and 25, the sampling distributions are nearly normal in shape.

It is very unlikely that the exact shape of the population distribution will be known. Thus, the exact shape of the sampling distribution of $\bar{y}$ will not be known

FIGURE 4.23
Sampling distribution of $\bar{y}$ for $n = 5, 10, 25$ when sampling from a normal distribution

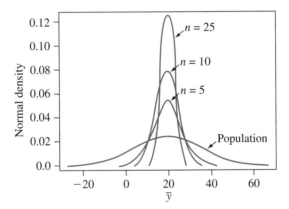

FIGURE 4.24
Sampling distribution of $\bar{y}$ for $n = 5, 10, 25$ when sampling from a skewed distribution

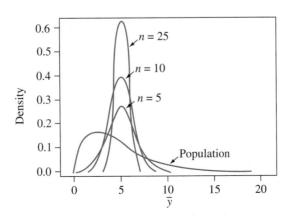

either. The important point to remember is that the sampling distribution of $\bar{y}$ will be approximately normally distributed with a mean $\mu_{\bar{y}} = \mu$, the population mean, and a standard deviation $\sigma_{\bar{y}} = \sigma/\sqrt{n}$. The approximation will be more precise as n, the sample size for each sample, increases and as the shape of the population distribution becomes more like the shape of a normal distribution.

An obvious question is: How large should the sample size be for the Central Limit Theorem to hold? Numerous simulation studies have been conducted over the years and the results of these studies suggest that, in general, the Central Limit Theorem holds for $n > 30$. However, we should not apply this rule blindly. If the population is heavily skewed, the sampling distribution for $\bar{y}$ will still be skewed even for $n > 30$. On the other hand, if the population is symmetric, the Central Limit Theorem holds for $n < 30$.

Therefore, take a look at the data. If the sample histogram is clearly skewed, then the population will also probably be skewed. Consequently, a value of n much higher than 30 may be required to have the sampling distribution of $\bar{y}$ be approximately normal. Any inference based on the normality of $\bar{y}$ for $n = 30$ under this condition should be examined carefully.

As demonstrated in Figures 4.21 and 4.22, the Central Limit Theorem can be extended to many different sample statistics. The form of the Central Limit Theorem for the sample median and sample standard deviation is somewhat more complex than for the sample mean. Many of the statistics that we will encounter in later chapters will be either averages or sums of variables. The Central Limit Theorem for sums can be easily obtained from the Central Limit Theorem for the sample mean. Suppose we have a random sample of n measurements, $y_1, \ldots, y_n$, from a population and let $\sum y = y_1 + \cdots + y_n$.

THEOREM 4.2

> **Central Limit Theorem for $\sum y$**
>
> Let $\sum y$ denote the sum of a random sample of n measurements from a population having a mean μ and finite standard deviation σ. Let $\mu_{\sum y}$ and $\sigma_{\sum y}$ denote the mean and standard deviation of the sampling distribution of $\sum y$, respectively. Based on repeated random samples of size n from the population, we can conclude the following:
>
> **1.** $\mu_{\sum y} = n\mu$
> **2.** $\sigma_{\sum y} = \sqrt{n}\sigma$
> **3.** When n is large, the sampling distribution of $\sum y$ will be approximately normal (with the approximation becoming more precise as n increases).
> **4.** When the population distribution is normal, the sampling distribution of $\sum y$ is exactly normal for any sample size n.

Usually, a sample statistic is used as an estimate of a population parameter. For example, a sample mean $\bar{y}$ can be used to estimate the population mean μ from which the sample was selected. Similarly, a sample median and sample standard deviation estimate the corresponding population median and standard deviation. The sampling distribution of a sample statistic is then used to determine how accurate the estimate is likely to be. In Example 4.19, the population mean μ is known to be 6.5. Obviously, we do not know μ in any practical study or experiment. However, we can use the sampling distribution of $\bar{y}$ to determine the probability that the value of $\bar{y}$ for a random sample of $n = 2$ measurements from the population will be more than three units from μ. Using the data in Example 4.19, this probability is

$$P(2.5) + P(3) + P(10) + P(10.5) = \frac{4}{45}$$

In general, we would use the normal approximation from the Central Limit Theorem in making this calculation because the sampling distribution of a sample statistic is seldom known. This type of calculation will be developed in Chapter 5. Because a sample statistic is used to make inferences about a population parameter, the sampling distribution of the statistic is crucial in determining the accuracy of the inference.

interpretations of a sampling distribution   **Sampling distributions** can be **interpreted** in at least two ways. One way uses the long-run relative frequency approach. Imagine taking repeated samples of a fixed size from a given population and calculating the value of the sample statistic for each sample. In the long run, the relative frequencies for the possible values of the sample statistic will approach the corresponding sampling distribution probabilities. For example, if we took a large number of samples from the population distribution corresponding to the probabilities of Example 4.19 and, for each sample, computed the sample mean, approximately 9% would have $\bar{y} = 5.5$.

The other way to interpret a sampling distribution makes use of the classical interpretation of probability. Imagine listing all possible samples that could be drawn from a given population. The probability that a sample statistic will have a particular value (say, that $\bar{y} = 5.5$) is then the proportion of all possible samples that yield that value. In Example 4.19, $P(\bar{y} = 5.5) = 4/45$ corresponds to the fact that 4 of the 45 samples have a sample mean equal to 5.5. Both the repeated-sampling and the classical approaches to finding probabilities for a sample statistic are legitimate.

In practice, however, a sample is taken only once, and only one value of the sample statistic is calculated. A sampling distribution is not something you can see in practice; it is not an empirically observed distribution. Rather, it is a theoretical concept, a set of probabilities derived from assumptions about the population and about the sampling method.

There's an unfortunate similarity between the phrase *sampling distribution,* meaning the theoretically derived probability distribution of a statistic, and the phrase *sample distribution,* which refers to the histogram of individual values actually observed in a particular sample. The two phrases mean very different things.

sample histogram   To avoid confusion, we will refer to the distribution of sample values as the **sample histogram** rather than as the sample distribution.

EXERCISES   **Basic Techniques**

4.53   A random sample of 16 measurements is drawn from a population with a mean of 60 and a standard deviation of 5. Describe the sampling distribution of $\bar{y}$, the sample mean. Within which interval do you expect $\bar{y}$ to lie approximately 95% of the time?

4.54   Refer to Exercise 4.53. Describe the sampling distribution for the sample sum $\sum y_i$ Is it unlikely (improbable) that $\sum y_i$ would be more than 70 units away from 960? Explain.

Applications

Psy.   **4.55**   Psychomotor retardation scores for a large group of manic–depressive patients were approximately normal, with a mean of 930 and a standard deviation of 130. What fraction of the patients scored

  a. Between 800 and 1,100?
  b. Less than 800?
  c. Greater than 1,200?

Soc.   **4.56** Federal resources have been tentatively approved for the construction of an outpatient clinic. In order to design a facility that will handle patient load requirements and stay within a limited budget, the designers studied patient demand. From studying a similar facility in the area, they found that the distribution of the number of patients requiring hospitalization during a week could be approximated by a normal distribution with a mean of 125 and a standard deviation of 32.
  a. Use the Empirical Rule to describe the distribution of y, the number of patients requesting service in a week.
  b. If the facility is built with a 160-patient capacity, what fraction of the weeks might the clinic be unable to handle the demand?

Env.   **4.57** The level of a particular pollutant, nitrogen oxide, in the exhaust of a hypothetical model of car, the Polluter, when driven in city traffic has approximately a normal distribution with a mean level of 2.1 grams per mile (g/m) and a standard deviation of 0.3 g/m.
  a. If the EPA mandates that a nitrogen oxide level of 2.7 g/m cannot be exceeded, what proportion of Polluters will be in violation of the mandate?
  b. At most, 25% of Polluters exceed which nitrogen oxide level value (that is, find the 75th percentile)?
  c. The company producing the Polluter must reduce the nitrogen oxide level so that at most 5% of its cars exceed the EPA level of 2.7 g/m. If the standard deviation remains 0.3 g/m, to what value must the mean level be reduced so that at most 5% of Polluters exceed 2.7 g/m?

4.58 Refer to Exercise 4.57. A company has a fleet of 150 Polluters used by its sales staff. Describe the distribution of the total amount (in g/m) of nitrogen oxide produced in the exhaust of this fleet. What are the mean and standard deviation of the total amount (in g/m) of nitrogen oxide in the exhaust for the fleet? (*Hint:* The total amount of nitrogen oxide can be represented as $\sum_{i=1}^{150} W_i$, where W_i is the amount of nitrogen oxide in the exhaust of the ith car. Thus, the Central Limit Theorem for sums is applicable.)

Soc.   **4.59** The baggage limit for an airplane is set at 100 pounds per passenger. Thus, for an airplane with 200 passenger seats there is a limit of 20,000 pounds. The weight of the baggage of an individual passenger is a random variable with a mean of 95 pounds and a standard deviation of 35 pounds. If all 200 seats are sold for a particular flight, what is the probability that the total weight of the passengers' baggage will exceed the 20,000-pound limit?

4.13   Normal Approximation to the Binomial

A binomial random variable y was defined earlier to be the number of successes observed in n independent trials of a random experiment in which each trial resulted in either a success (S) or a failure (F) and $P(S) = \pi$ for all n trials. We will now demonstrate how the Central Limit Theorem for sums enables us to calculate probabilities for a binomial random variable by using an appropriate normal curve as an approximation to the binomial distribution. We said in Section 4.8 that probabilities associated with values of y can be computed for a binomial experiment for any values of n or π, but the task becomes more difficult when n gets large. For example, suppose a sample of 1,000 voters is polled to determine sentiment toward the consolidation of city and county government. What is the probability of observing 460 or fewer favoring consolidation if we assume that 50% of the entire population favor the change? Here we have a binomial experiment with $n = 1,000$ and π, the probability of selecting a person favoring consolidation, equal to .5. To determine the probability of observing 460 or fewer favoring consolidation in the random sample of 1,000 voters, we can compute $P(y)$ using the binomial formula for $y = 460, 459, \ldots, 0$. The desired probability is then

$$P(y = 460) + P(y = 459) + \cdots + P(y = 0)$$

There are 461 probabilities to calculate, with each one being somewhat difficult because of the factorials. For example, the probability of observing 460 favoring consolidation is

$$P(y = 460) = \frac{1000!}{460!540!}(.5)^{460}(.5)^{540}$$

A similar calculation is needed for all other values of y.

To justify the use of the Central Limit Theorem, we need to define n random variables, $I_1, \ldots, I_n$, by

$$I_i = \begin{cases} 1 & \text{if the } i\text{th trial results in a success} \\ 0 & \text{if the } i\text{th trial results in a failure} \end{cases}$$

The binomial random variable y is the number of successes in the n trials. Now, consider the sum of the random variable $I_1, \ldots, I_n$, $\sum_{i=1}^{n} I_i$. A 1 is placed in the sum for each S that occurs and a 0 for each F that occurs. Thus, $\sum_{i=1}^{n} I_i$ is the number of S's that occurred during the n trials. Hence, we conclude that $y = \sum_{i=1}^{n} I_i$. Because the binomial random variable y is the sum of independent random variables, each having the same distribution, we can apply the Central Limit Theorem for sums to y. Thus, the normal distribution can be used to approximate the binomial distribution when n is of an appropriate size. The normal distribution that will be used has a mean and standard deviation given by the following formulas:

$$\mu = n\pi \qquad \sigma = \sqrt{n\pi(1 - \pi)}$$

These are the mean and standard deviation of the binomial random variable y.

EXAMPLE 4.21

Use the normal approximation to the binomial to compute the probability of observing 460 or fewer in a sample of 1,000 favoring consolidation if we assume that 50% of the entire population favor the change.

Solution The normal distribution used to approximate the binomial distribution will have

$$\mu = n\pi = 1,000(.5) = 500$$

$$\sigma = \sqrt{n\pi(1 - \pi)} = \sqrt{1,000(.5)(.5)} = 15.8$$

The desired probability is represented by the shaded area shown in Figure 4.25. We calculate the desired area by first computing

$$z = \frac{y - \mu}{\sigma} = \frac{460 - 500}{15.8} = -2.53$$

Referring to Table 1 in the Appendix, we find that the area under the normal curve to the left of 460 (for $z = -2.53$) is .0057. Thus, the probability of observing 460 or fewer favoring consolidation is approximately .0057.

FIGURE 4.25

Approximating normal
distribution for the binomial
distribution, $\mu = 500$ and
$\sigma = 15.8$

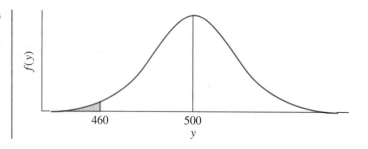

The normal approximation to the binomial distribution can be unsatisfactory if $n\pi < 5$ or $n(1 - \pi) < 5$. If π, the probability of success, is small, and n, the sample size, is modest, the actual binomial distribution is seriously skewed to the right. In such a case, the symmetric normal curve will give an unsatisfactory approximation. If π is near 1, so $n(1 - \pi) < 5$, the actual binomial will be skewed to the left, and again the normal approximation will not be very accurate. The normal approximation, as described, is quite good when $n\pi$ and $n(1 - \pi)$ exceed about 20. In the middle zone, $n\pi$ or $n(1 - \pi)$ between 5 and 20, a modification called a **continuity correction** makes a substantial contribution to the quality of the approximation.

continuity correction

The point of the continuity correction is that we are using the continuous normal curve to approximate a discrete binomial distribution. A picture of the situation is shown in Figure 4.26.

The binomial probability that $y \leq 5$ is the sum of the areas of the rectangles above 5, 4, 3, 2, 1, and 0. This probability (area) is approximated by the area under the superimposed normal curve to the left of 5. Thus, the normal approximation ignores half of the rectangle above 5. The continuity correction simply includes the area between $y = 5$ and $y = 5.5$. For the binomial distribution with $n = 20$ and $\pi = .30$ (pictured in Figure 4.26), the correction is to take $P(y \leq 5)$ as $P(y \leq 5.5)$. Instead of

$$P(y \leq 5) = P[z \leq (5 - 20(.3))/\sqrt{20(.3)(.7)}] = P(z \leq -.49) = .3121$$

use

$$P(y \leq 5.5) = P[z \leq (5.5 - 20(.3))/\sqrt{20(.3)(.7)}] = P(z \leq -.24) = .4052$$

The actual binomial probability can be shown to be .4164. The general idea of the continuity correction is to add or subtract .5 from a binomial value before using normal probabilities. The best way to determine whether to add or subtract is to draw a picture like Figure 4.26.

FIGURE 4.26

Normal approximation
to binomial

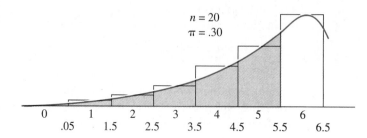

Normal Approximation to the Binomial Probability Distribution

For large n and π not too near 0 or 1, the distribution of a binomial random variable y may be approximated by a normal distribution with $\mu = n\pi$ and $\sigma = \sqrt{n\pi(1 - \pi)}$. This approximation should be used only if $n\pi \geq 5$ and $n(1 - \pi) \geq 5$. A continuity correction will improve the quality of the approximation in cases in which n is not overwhelmingly large.

EXAMPLE 4.22

A large drug company has 100 potential new prescription drugs under clinical test. About 20% of all drugs that reach this stage are eventually licensed for sale. What is the probability that at least 15 of the 100 drugs will be eventually licensed? Assume that the binomial assumptions are satisfied, and use a normal approximation with continuity correction.

Solution The mean of y is $\mu = 100(.2) = 20$; the standard deviation is $\sigma = \sqrt{100(.2)(.8)} = 4.0$. The desired probability is that 15 or more drugs are approved. Because $y = 15$ is included, the continuity correction is to take the event as y greater than or equal to 14.5.

$$P(y \geq 14.5) = P\left(z \geq \frac{14.5 - 20}{4.0}\right) = P(z \geq -1.38) = 1 - P(z < -1.38)$$

$$= 1 - .0838 = .9162$$

4.14 Minitab Instructions (Optional)

Generating Random Numbers

To generate 1,000 random numbers from the set $[0, 1, \ldots, 9]$,

1. Click on **Calc,** then **Random Data,** then **Integer.**
2. Type the number of rows of data: **Generate 20 rows of data.**
3. Type the columns in which the data are to be stored: **Store in column(s): c1-c50.**
4. Type the first number in the list: **Minimum value: 0.**
5. Type the last number in the list: **Maximum value: 9.**
6. Click on **OK.**

Note that we have generated $(20)(50) = 1,000$ random numbers.

Calculating Binomial Probabilities

To calculate binomial probabilities when $n = 10$ and $\pi = 0.6$,

1. Enter the values of x in column c1: **0, 1, 2, 3, 4, 5, 6, 7, 8, 9, 10.**
2. Click on **Calc,** then **Probability Distributions,** then **Binomial.**
3. Select either **Probability** [to compute $P(X = x)$] or **Cumulative probability** [to compute $P(X \leq x)$].

4. Type the value of n: **Number of trials: 10.**
5. Type the value of π: **Probability of success: 0.6.**
6. Click on **Input column.**
7. Type the column number where values of x are located: **C1.**
8. Click on **Optional storage.**
9. Type the column number to store probability: **C2.**
10. Click on **OK.**

Calculating Normal Probabilities

To calculate $P(X \le 18)$ when X is normally distributed with $\mu = 23$ and $\sigma = 5$,

1. Click on **Calc,** then **Probability Distributions,** then **Normal.**
2. Click on **Cumulative probability.**
3. Type the value of μ: **Mean: 23.**
4. Type the value of σ: **Standard deviation: 5.**
5. Click on **Input constant.**
6. Type the value of x: **18.**
7. Click on **OK.**

Generating Sampling Distribution of $\bar{y}$

To create the sampling distribution of $\bar{y}$ based on 500 samples of size $n = 16$ from a normal distribution with $\mu = 60$ and $\sigma = 5$,

1. Click on **Calc,** then **Random Data,** then **Normal.**
2. Type the number of samples: **Generate 500 rows.**
3. Type the sample size n in terms of number of columns: **Store in column(s) c1-c16.**
4. Type in the value of μ: **Mean: 60.**
5. Type in the value of σ: **Standard deviation: 5.**
6. Click on **OK.** There are now 500 rows in columns c1–c16, 500 samples of 16 values each to generate 500 values of $\bar{y}$.
7. Click on **Calc,** then **Row Statistics,** then **mean.**
8. Type in the location of data: **Input Variables c1-c16.**
9. Type in the column in which the 500 means will be stored: **Store Results in c17.**
10. To obtain the mean of the 500 $\bar{y}$s, click on **Calc,** then **Column Statistics,** then **mean.**
11. Type in the location of the 500 means: **Input Variables c17.**
12. Click on **OK.**
13. To obtain the standard deviation of the 500 $\bar{y}$s, click on **Calc,** then **Column Statistics,** then **standard deviation.**
14. Type in the location of the 500 means: **Input Variables c17.**
15. Click on **OK.**
16. To obtain the sampling distribution of $\bar{y}$, click **Graph,** then **Histogram.**
17. Type **c17** in the Graph box.
18. Click on **OK.**

4.15   Staying Focused

In this chapter, we presented a brief introduction to probability, probability distributions, and sampling distributions as part of Step 3, summarizing data. Knowledge of the probabilities of sample outcomes is vital to a statistical inference. Three different interpretations of the probability of an outcome were given: the classical, relative frequency, and subjective interpretations. Although each has a place in statistics, the relative frequency approach has the most intuitive appeal because it can be checked.

Quantitative random variables are classified as either discrete or continuous random variables. The probability distribution for a discrete random variable y is a display of the probability $P(y)$ associated with each value of y. This display may be presented in the form of a histogram, table, or formula.

The binomial is a very important and useful discrete random variable. Many experiments that scientists conduct are similar to a coin-tossing experiment in which dichotomous (yes–no) type data are accumulated. The binomial experiment frequently provides an excellent model for computing probabilities of various sample outcomes.

Probabilities associated with a continuous random variable correspond to areas under the probability distribution. Computations of such probabilities were illustrated for areas under the normal curve. The importance of this exercise is borne out by the Central Limit Theorem: Any random variable that is expressed as a sum or average of a random sample from a population having a finite standard deviation will have a normal distribution for a sufficiently large sample size. Direct application of the Central Limit Theorem gives the sampling distribution for the sample mean. Because many sample statistics are either sums or averages of random variables, the application of the Central Limit Theorem provides us with information about probabilities of sample outcomes. These probabilities are vital for the statistical inferences we wish to make as part of Step 4 of learning from data.

Key Formulas

1. Binomial probability distribution

$$P(y) = \frac{n!}{y!(n-y)!}\pi^y(1-\pi)^{n-y}$$

2. Sampling distribution for $\bar{y}$

Mean: μ

Standard error: $\sigma_{\bar{y}} = \sigma/\sqrt{n}$

3. Normal approximation to the binomial

$$\mu = n\pi \qquad \sigma = \sqrt{n\pi(1-\pi)}$$

provided that $n\pi$ and $n(1-\pi)$ are greater than or equal to 5 or, equivalently, if

$$n \geq \frac{5}{\min(\pi, 1-\pi)}$$

Supplementary Exercises

Bus. **4.60** One way to audit expense accounts for a large consulting firm is to sample all reports dated the last day of each month. Comment on whether such a sample constitutes a random sample.

Bus. **4.61** Critical key-entry errors in the data processing operation of a large district bank occur approximately .1% of the time. If a random sample of 10,000 entries is examined, determine the following:

   a. The expected number of errors
   b. The probability of observing fewer than five errors
   c. The probability of observing fewer than two errors

Engin. **4.62** The breaking strengths for 1-foot-square samples of a particular synthetic fabric are approximately normally distributed with a mean of 2,250 pounds per square inch (psi) and a standard deviation of 10.2 psi.

   a. Find the probability of selecting a 1-foot-square sample of material at random that on testing would have a breaking strength in excess of 2,265 psi.
   b. Describe the sampling distribution for $\bar{y}$ based on random samples of 15 1-foot sections.

4.63 Refer to Exercise 4.62. Suppose that a new synthetic fabric has been developed that may have a different mean breaking strength. A random sample of 15 1-foot sections is obtained and each section is tested for breaking strength. If we assume that the population standard deviation for the new fabric is identical to that for the old fabric, give the standard deviation for the sampling distribution of $\bar{y}$ using the new fabric.

4.64 Refer to Exercise 4.63. Suppose that the mean breaking strength for the sample of 15 1-foot sections of the new synthetic fabric is 2,268 psi. What is the probability of observing a value of $\bar{y}$ equal to or greater than 2,268, assuming that the mean breaking strength for the new fabric is 2,250, the same as that for the old?

4.65 Based on your answer in Exercise 4.64, do you believe the new fabric has the same mean breaking strength as the old? (Assume $\sigma = 10.2$.)

4.66 In Figure 4.19, we visually inspected the relative frequency histogram for sample means based on two measurements and noted its bell shape. Another way to determine whether a set of measurements is bell shaped (normal) is to construct a **normal probability plot** of the sample data. If the plotted points are nearly a straight line, we say the measurements were selected from a normal population. We can generate a normal probability plot using the following Minitab code. If the plotted points fall within the curved dotted lines, we consider the data to be a random sample from a normal distribution.

Minitab code:

1. Enter the 45 measurements into C1 of the data spreadsheet.
2. Click on **Graph,** then **Probability Plot.**
3. Type **c1** in the box labeled **Variables:.**
4. Click on **OK.**

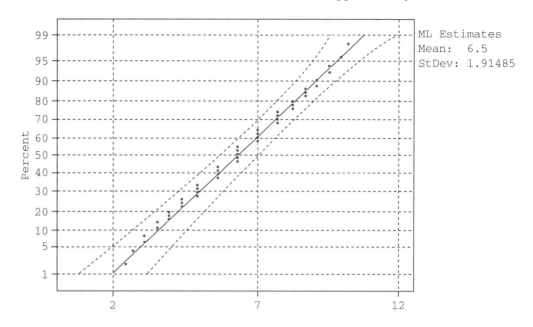

ML Estimates
Mean: 6.5
StDev: 1.91485

The 45 data values clearly fall within the two curved lines and fairly close to the straight line. Thus, we conclude there is a strong indication that the data values follow a normal distribution.

  a. Suppose our population was the 10 measurements (2, 3, 6, 8, 9, 12, 25, 29, 39, 50). Generate the 45 sample means based on $n = 2$ observations per sample and determine whether the sampling distribution of the sample mean is approximately normally distributed by constructing a histogram and normal probability plot of the 45 sample means.

  b. Why do you think the plots for the means from the population (2, 3, 6, 8, 9, 12, 25, 29, 39, 50) differ greatly from the plots obtained for the means from the population (2, 3, 4, 5, 6, 7, 8, 9, 10, 11)?

Gov.   **4.67**   Suppose that you are a regional director of the IRS office and that you are charged with sampling 1% of the returns with gross income levels above $15,000. How might you go about this? Would you use random sampling? How?

Bus.   **4.68**   Many firms are using or exploring the possibility of using telemarketing techniques—that is, marketing their products via the telephone to supplement the more traditional marketing strategies. Assume a firm finds that approximately 1 in every 100 calls yields a sale.

  a. Find the probability the first sale will occur somewhere in the first 5 calls.

  b. Find the probability the first sale will occur sometime after 10 calls.

Bus.   **4.69**   Marketing analysts have determined that a particular advertising campaign should make at least 20% of the adult population aware of the advertised product. After a recent campaign, 25 of 400 adults sampled indicated that they had seen the ad and were aware of the new product.

  a. Find the approximate probability of observing $y \leq 25$ given that 20% of the population is aware of the product through the campaign.

  b. Based on your answer to part (a), does it appear the ad was successful? Explain.

Med.   **4.70**   One or more specific minor birth defects occur with probability .0001 (that is, 1 in 10,000 births). If 20,000 babies are born in a given geographic area in a given year, can we calculate the probability of observing at least one of the minor defects using the binomial or normal approximation to the binomial? Explain.

4.71   Random samples of size 5, 20, and 80 are drawn from a population with mean $\mu = 100$ and standard deviation $\sigma = 15$.

a. Give the mean of the sampling distribution of $\bar{y}$ for each of the sample sizes 5, 20, and 80.
b. Give the standard deviation of the sampling distribution of $\bar{y}$ for each of the sample sizes 5, 20, and 80.
c. Based on the results obtained in parts (a) and (b), what do you conclude about the accuracy of using the sample mean $\bar{y}$ as an estimate of population mean μ?

4.72 Refer to Exercise 4.71. To evaluate how accurately the sample mean $\bar{y}$ estimates the population mean μ, we need to know the chance of obtaining a value of $\bar{y}$ that is far from μ. Suppose it is important that the sample mean $\bar{y}$ is within 5 units of the population mean μ. Find the following probabilities for each of the three sample sizes and comment on the accuracy of using $\bar{y}$ to estimate μ.

a. $P(\bar{y} \geq 105)$
b. $P(\bar{y} \leq 95)$
c. $P(95 \leq \bar{y} \leq 105)$

STEP

4

Analyzing Data,

Interpreting the

Analyses, and

Communicating

Results

CHAPTER 5

Inferences about Population Central Values

5.1 Introduction

Inference, specifically decision making and prediction, is centuries old and plays a very important role in our lives. Each of us faces daily personal decisions and situations that require predictions concerning the future. The U.S. government is concerned with the balance of trade with countries in Europe and Asia. An investment advisor wants to know whether inflation will be rising in the next 6 months. A metallurgist would like to use the results of an experiment to determine whether a new lightweight alloy possesses the strength characteristics necessary for use in automobile manufacturing. A veterinarian investigates the effectiveness of a new chemical for treating heartworm in dogs. The inferences that these individuals make should be based on relevant facts, which we call observations or data.

In many practical situations, the relevant facts are abundant, seemingly inconsistent, and, in many respects, overwhelming. As a result, a careful decision or prediction is often little better than an outright guess. You need only refer to the Market Views section of the *Wall Street Journal* or one of the financial news shows on cable TV to observe the diversity of expert opinion concerning future stock market behavior. Similarly, a visual analysis of data by scientists and engineers often yields conflicting opinions regarding conclusions to be drawn from an experiment.

Many individuals tend to feel that their own built-in inference-making equipment is quite good. However, experience suggests that most people are incapable of using large amounts of data, mentally weighing each bit of relevant information, and arriving at a good inference. (You may test your own inference-making ability by using the exercises in Chapters 5 through 10. Scan the data and make an inference before you use the appropriate statistical procedure. Then compare the results.) The statistician, rather than relying on his or her own intuition, uses statistical results to aid in making inferences. Although we touched on some of the notions involved in statistical inference in preceding chapters, we will now collect our ideas in a presentation of some of the basic ideas involved in statistical inference.

The objective of statistics is to make inferences about a population based on information contained in a sample. Populations are characterized by numerical descriptive measures called *parameters*. Typical population parameters are the mean μ, the median M, the standard deviation σ, and a proportion π. Most inferential problems can be formulated as an inference about one or more parameters of a population. For example, a study is conducted by the Wisconsin Education Department to assess the reading ability of children in the primary grades. The population consists of the scores on a standard reading test of all children in the primary grades in Wisconsin. We are interested in estimating the value of the population mean score μ and the proportion π of scores below a standard, which designates that a student needs remedial assistance.

Methods for making inferences about parameters fall into one of two categories. Either we will **estimate** (predict) the value of the population parameter of interest or we will **test a hypothesis** about the value of the parameter. These two methods of statistical inference—estimation and hypothesis testing—involve different procedures, and, more important, they answer two different questions about the parameter. In estimating a population parameter, we are answering the question, "What is the value of the population parameter?" In testing a hypothesis, we are answering the question, "Is the parameter value equal to this specific value?"

estimation

hypothesis testing

Consider a study in which an investigator wishes to examine the effectiveness of a drug product in reducing anxiety levels of anxious patients. The investigator uses a screening procedure to identify a group of anxious patients. After the patients are admitted into the study, each one's anxiety level is measured on a rating scale immediately before he or she receives the first dose of the drug and then at the end of 1 week of drug therapy. These sample data can be used to make inferences about the population from which the sample was drawn either by estimation or by a statistical test:

Estimation: Information from the sample can be used to estimate (or predict) the mean decrease in anxiety ratings for the set of all anxious patients who may conceivably be treated with the drug.

Statistical test: Information from the sample can be used to determine whether the population mean decrease in anxiety ratings is greater than zero.

Notice that the inference related to estimation is aimed at answering the question, "What is the mean decrease in anxiety ratings for the population?" In contrast, the statistical test attempts to answer the question, "Is the mean drop in anxiety ratings greater than zero?"

EXERCISES   Basic Techniques

Pol. Sci.   **5.1**   A researcher is interested in estimating the percentage of registered voters in her state who have voted in at least one election over the past 2 years.

a. Identify the population of interest to the researcher.

b. How might you select a sample of voters to gather this information?

Engin.   **5.2**   A manufacturer claims that the average lifetime of a particular fuse is 1,500 hours. Information from a sample of 35 fuses shows that the average lifetime is 1,380 hours. What can be said about the manufacturer's claim?

   a. Identify the population of interest to us.

   b. Would an answer to the question posed involve estimation or testing a hypothesis?

5.3   Refer to Exercise 5.2. How might you select a sample of fuses from the manufacturer to test the claim?

5.2   Estimation of μ

The first step in statistical inference is point estimation, in which we compute a single value (statistic) from the sample data to estimate a population parameter. Suppose that we are interested in estimating a population mean and that we are willing to assume the underlying population is normal. One natural statistic that could be used to estimate the population mean is the sample mean, but we also could use the median and the trimmed mean. Which sample statistic should we use?

A whole branch of mathematical statistics deals with problems related to developing point estimators (the formulas for calculating specific point estimates from sample data) of parameters from various underlying populations and determining whether a particular point estimator has certain desirable properties. Fortunately, we will not have to derive these point estimators—they'll be given to us for each parameter. When we know which point estimator (formula) to use for a given parameter, we can develop confidence intervals (interval estimates) for these same parameters.

In this section, we deal with point and interval estimation of a population mean μ. Tests of hypotheses about μ are covered in Section 5.4.

For most problems in this book, we will use sample mean $\bar{y}$ as a point estimate of μ; we also use it to form an interval estimate for the population mean μ. From the Central Limit Theorem for the sample mean given in Chapter 4, we know that for large n (crudely, $n \geq 30$), $\bar{y}$ will be approximately normally distributed, with a mean μ and a standard error $\sigma_{\bar{y}}$. Then from our knowledge of the Empirical Rule and areas under a normal curve, we know that the interval $\mu \pm 2\sigma_{\bar{y}}$, or more precisely, the interval $\mu \pm 1.96\sigma_{\bar{y}}$, includes 95% of the $\bar{y}$s in repeated sampling, as shown in Figure 5.1.

From Figure 5.1 we can observe that the sample mean $\bar{y}$ may not be very close to the population mean μ, the quantity it is supposed to estimate. Thus, when the value of $\bar{y}$ is reported, we should also provide an indication of how accurately $\bar{y}$ estimates μ. We will accomplish this by considering an interval of possible values for

FIGURE 5.1

Sampling distribution for $\bar{y}$

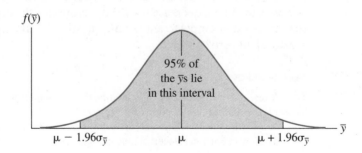

$f(\bar{y})$

95% of the $\bar{y}$s lie in this interval

$\mu - 1.96\sigma_{\bar{y}}$ μ $\mu + 1.96\sigma_{\bar{y}}$ $\bar{y}$

FIGURE 5.2

When the observed value of $\bar{y}$ lies in the interval $\mu \pm 1.96\sigma_{\bar{y}}$, the interval $\bar{y} \pm 1.96\sigma_{\bar{y}}$ contains the parameter μ

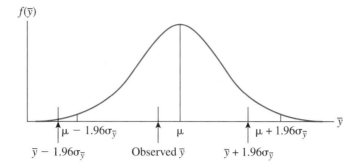

FIGURE 5.3

Fifty interval estimates of the population mean (27)

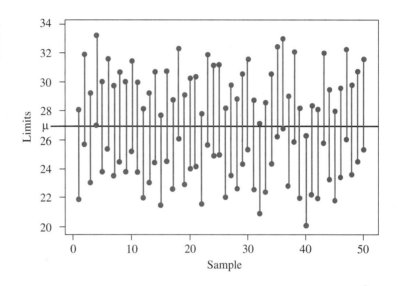

μ in place of using just a single value $\bar{y}$. Consider the interval $\bar{y} \pm 1.96\sigma_{\bar{y}}$. Any time $\bar{y}$ falls in the interval $\mu \pm 1.96\sigma_{\bar{y}}$, the interval $\bar{y} \pm 1.96\sigma_{\bar{y}}$ will contain the parameter μ (see Figure 5.2). The probability of $\bar{y}$ falling in the interval $\mu \pm 1.96\sigma_{\bar{y}}$ is .95, so we state that $\bar{y} \pm 1.96\sigma_{\bar{y}}$ is an **interval estimate** of μ with **level of confidence** .95.

interval estimate

level of confidence

confidence coefficient

We evaluate the goodness of an interval estimation procedure by examining the fraction of times in repeated sampling that interval estimates encompass the parameter to be estimated. This fraction, called the **confidence coefficient,** is .95 when using the formula $\bar{y} \pm 1.96\sigma_{\bar{y}}$; that is, 95% of the time in repeated sampling, intervals calculated using the formula $\bar{y} \pm 1.96\sigma_{\bar{y}}$ will contain the mean μ.

This idea is illustrated in Figure 5.3. Suppose we want to study a commercial process that produces shrimp for sale to restaurants. The shrimp are monitored for size by randomly selecting 40 shrimp from the tanks and measuring their length. We will consider a simulation of the shrimp monitoring. Suppose that the distribution of shrimp length in the tank has a normal distribution with a mean $\mu = 27$ cm and a standard deviation $\sigma = 10$ cm. Fifty samples of size $n = 40$ are drawn from the shrimp population. From each of these samples we compute the interval estimate $\bar{y} \pm 1.96\sigma_{\bar{y}} = \bar{y} \pm 1.96(10/\sqrt{40})$, because $\sigma_{\bar{y}} = \sigma/\sqrt{n}$. (See Table 5.1.) Note that although the intervals vary in location, only 2 of the 50 intervals failed to capture the population mean μ. The fact that two samples produced intervals that did

TABLE 5.1

Fifty 95% interval estimates
of the population mean (27)

	Sample Mean	Lower Limit	Upper Limit	Interval Contains Population Mean
1	25.0080	21.9089	28.1070	Yes
2	28.8373	25.7382	31.9363	Yes
3	26.1587	23.0597	29.2578	Yes
4	30.1301	27.0310	33.2291	No
5	26.9420	23.8430	30.0411	Yes
6	28.5148	25.4158	31.6139	Yes
7	26.6456	23.5465	29.7446	Yes
8	27.6168	24.5178	30.7158	Yes
9	26.9287	23.8297	30.0278	Yes
10	28.3338	25.2348	31.4329	Yes
11	26.9008	23.8017	29.9998	Yes
12	25.0978	21.9988	28.1969	Yes
13	26.1866	23.0875	29.2856	Yes
14	27.6062	24.5072	30.7053	Yes
15	24.6580	21.5590	27.7570	Yes
16	27.6427	24.5437	30.7418	Yes
17	25.7136	22.6146	28.8127	Yes
18	29.2075	26.1084	32.3065	Yes
19	26.0411	22.9421	29.1402	Yes
20	27.1937	24.0947	30.2928	Yes
21	27.2848	24.1858	30.3838	Yes
22	24.7320	21.6330	27.8310	Yes
23	28.8036	25.7046	31.9026	Yes
24	28.0333	24.9343	31.1324	Yes
25	28.1065	25.0075	31.2055	Yes
26	25.1701	22.0711	28.2691	Yes
27	26.7039	23.6049	29.8029	Yes
28	25.7587	22.6597	28.8577	Yes
29	27.4835	24.3844	30.5825	Yes
30	28.5009	25.4019	31.5999	Yes
31	25.7142	22.6152	28.8133	Yes
32	24.0557	20.9567	27.1547	Yes
33	25.5259	22.4269	28.6249	Yes
34	27.5036	24.4046	30.6026	Yes
35	29.3654	26.2664	32.4645	Yes
36	29.9348	26.8358	33.0338	Yes
37	25.9826	22.8835	29.0816	Yes
38	29.0128	25.9138	32.1118	Yes
39	25.1266	22.0276	28.2256	Yes
40	23.2452	20.1462	26.3442	No
41	25.3202	22.2212	28.4192	Yes
42	25.0905	21.9914	28.1895	Yes
43	28.9345	25.8354	32.0335	Yes
44	26.4079	23.3089	29.5069	Yes
45	24.9458	21.8467	28.0448	Yes
46	26.5274	23.4284	29.6265	Yes
47	29.1770	26.0780	32.2760	Yes
48	26.7114	23.6123	29.8104	Yes
49	27.6640	24.5650	30.7631	Yes
50	28.5054	25.4063	31.6044	Yes

not contain μ is not an indication that the procedure for producing intervals is faulty. Because our level of confidence is 95%, we would expect that, in a large collection of 95% confidence intervals, approximately 5% of the intervals would fail to include μ. Thus, in 50 intervals we would expect two or three intervals (5% of 50) to not contain μ. It is crucial to understand that even when experiments are properly conducted, a number of the experiments will yield results that in some sense are in error. This occurs when we run only a small number of experiments or select only a small subset of the population. In our example, we randomly selected 40 observations from the population and then constructed a 95% confidence interval for the population mean μ. If this process were repeated a very large number of times —for example, 10,000 times instead of the 50 in our example—the proportion of intervals not containing μ would be very close to 5%.

EXAMPLE 5.1

In a random sample of $n = 36$ parochial schools throughout the south, the average number of pupils per school is 379.2, with a standard deviation of 124. Use the sample to construct a 95% confidence interval for μ, the mean number of pupils per school for all parochial schools in the south.

Solution   The sample data indicate that $\bar{y} = 379.2$ and $s = 124$. The appropriate 95% confidence interval is then computed by using the formula

$$\bar{y} \pm 1.96\sigma_{\bar{y}}$$

where $\sigma_{\bar{y}} = \sigma/\sqrt{n}$. In Section 5.7 we present a procedure for obtaining a confidence interval for μ when σ is unknown. However, for all practical purposes, if the sample size is 30 or more, we can estimate the population standard deviation σ with s in the confidence interval formula. Also, based on the results from the Central Limit Theorem, if the population distribution is not too nonnormal and the sample size is large, the level of confidence of our interval will be approximately the same as if we were sampling from a normal distribution. With s replacing σ, our interval is

$$379.2 \pm 1.96\frac{124}{\sqrt{36}} \quad \text{or} \quad 379.2 \pm 40.51$$

The interval from 338.69 to 419.71 forms a 95% confidence interval for μ. In other words, we are 95% sure that the average number of pupils per school for parochial schools throughout the south lies between 338.69 and 419.71.

There are many different confidence intervals for μ, depending on the confidence coefficient we choose. For example, the interval $\mu \pm 2.58\sigma_{\bar{y}}$ includes 99% of the values of $\bar{y}$ in repeated sampling, and the interval $\bar{y} \pm 2.58\sigma_{\bar{y}}$ forms a **99% confidence interval** for μ.

99% confidence interval

$(1 - \alpha)$ = confidence coefficient

We can state a general formula for a confidence interval for μ with a **confidence coefficient of $(1 - \alpha)$,** where α (Greek letter alpha) is between 0 and 1. For a specified value of $(1 - \alpha)$, a $100(1 - \alpha)\%$ confidence interval for μ is given by the following formula. Here we assume that σ is known or that the sample size is large enough to replace σ with s.

Confidence Interval for μ, σ Known

$$\bar{y} \pm z_{\alpha/2}\sigma_{\bar{y}}, \text{ where } \sigma_{\bar{y}} = \sigma/\sqrt{n}$$

FIGURE 5.4

Interpretation of $z_{\alpha/2}$ in the confidence interval formula

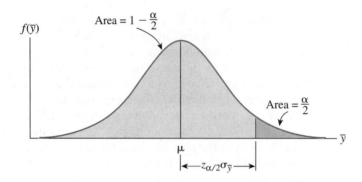

TABLE 5.2

Common values of the confidence coefficient $(1 - \alpha)$ and the corresponding z-value, $z_{\alpha/2}$

Confidence Coefficient $(1 - \alpha)$	Value of $\alpha/2$	Area in Table 1 $1 - \alpha/2$	Corresponding z-Value, $z_{\alpha/2}$
.90	.05	.95	1.645
.95	.025	.975	1.96
.98	.01	.99	2.33
.99	.005	.995	2.58

$z_{\alpha/2}$ The quantity $z_{\alpha/2}$ is a value of z having a tail area of $\alpha/2$ to its right. In other words, at a distance of $z_{\alpha/2}$ standard deviations to the right of μ, there is an area of $\alpha/2$ under the normal curve. Values of $z_{\alpha/2}$ can be obtained from Table 1 in the Appendix by looking up the z-value corresponding to an area of $1 - (\alpha/2)$ (see Figure 5.4). Common values of the confidence coefficient $(1 - \alpha)$ and $z_{\alpha/2}$ are given in Table 5.2.

EXAMPLE 5.2

A forester wishes to estimate the average number of count trees per acre (trees larger than a specified size) on a 2,000-acre plantation. She can then use this information to determine the total timber volume for trees in the plantation. A random sample of $n = 50$ 1-acre plots is selected and examined. The average (mean) number of count trees per acre is found to be 27.3, with a standard deviation of 12.1. Use this information to construct a 99% confidence interval for μ, the mean number of count trees per acre for the entire plantation.

Solution   We use the general confidence interval with confidence coefficient equal to .99 and a $z_{\alpha/2}$-value equal to 2.58 (see Table 5.2). Substituting into the formula $\bar{y} \pm 2.58 \, \sigma_{\bar{y}}$ and replacing σ with s in $\sigma_{\bar{y}} = \sigma/\sqrt{n}$, we have

$$27.3 \pm 2.58 \frac{12.1}{\sqrt{50}}$$

This corresponds to the confidence interval 27.3 ± 4.41—that is, the interval from 22.89 to 31.71. Thus, we are 99% sure that the average number of count trees per acre is between 22.89 and 31.71.

The discussion in this section has included one rather unrealistic assumption —namely, that the population standard deviation is known. In practice, it's difficult to find situations in which the population mean is unknown but the standard devi-

ation is known. Usually both the mean and the standard deviation must be estimated from the sample. Because σ is estimated by the sample standard deviation s,
the actual standard error of the mean, $\sigma/\sqrt{n}$, is naturally estimated by $s/\sqrt{n}$. This
estimation introduces another source of random error (s will vary randomly, from
substituting s for σ   sample to sample, around σ) and, strictly speaking, invalidates our confidence interval formula. Fortunately, the formula is still a very good approximation for large
sample sizes. As a very rough rule, we can use this formula when n is larger than 30;
a better way to handle this issue is described in Section 5.7.

Statistical inference-making procedures differ from ordinary procedures in
that we not only make an inference, but also provide a measure of how good that
inference is. For interval estimation, the width of the confidence interval and the
confidence coefficient measure the goodness of the inference. For a given value of
the confidence coefficient, the smaller the width of the interval, the more precise
the inference. The confidence coefficient, on the other hand, is set by the experimenter to express how much assurance he or she places on whether the interval estimate encompasses the parameter of interest. For a fixed sample size, increasing
the level of confidence will result in an interval of greater width. Thus, the experimenter will generally express a desired level of confidence and specify the desired
width of the interval. Next we will discuss a procedure to determine the appropriate
sample size to meet these specifications.

EXERCISES   **Basic Techniques**

Engin.   **5.4**   A cereal company randomly selects 25 12-ounce boxes of corn flakes every 10 minutes and
weighs the boxes. Suppose the weights have a normal distribution with $\sigma = 0.2$ ounces. One such
sample yields $\bar{y} = 12.3$ oz.
  a. Calculate a 95% confidence interval for the mean weight μ of the packages produced
  during the period of time from which the sample was selected.
  b. Give a careful nonstatistical jargon interpretation of the confidence interval.

5.5   The process engineer at the cereal company is concerned that the confidence intervals for μ
are too wide to be of practical use.
  a. If we double the sample size from 25 to 50, what is the impact on the width of the 95%
  confidence intervals?
  b. If we increase the level of confidence from 95 to 99%, what is the impact on the width of
  the confidence intervals?

Applications

Bus.   **5.6**   The Chamber of Commerce in a city wants to estimate the gross profit margin of small businesses (under \$500,000 in sales) in their city. A random sample of the year-end statements of
10 small businesses shows the mean gross profit margin to be 5.2% (of sales) with a standard deviation of 7.5%.
  a. Construct a 99% confidence interval for the mean gross profit margin μ of all small businesses in the city.
  b. What are some limitations in using the confidence interval that you constructed in (a)?
  For example, since the sample size is small, do you think that the data come from a normal distribution? Is it valid to replace σ with s?

Engin.   **5.7**   As a result of the massive shift from full-service to self-service gas stations, a consumer's
group is concerned that many cars are being driven on underinflated tires. This results in excessive tire wear, and unsafe steering and braking of the car. A tire is considered to be seriously underinflated if its tire pressure is more than 10 psi under its recommended level. A random sample
of 400 cars is selected and the mean underinflation is $\bar{y} = 10.4$ psi, with a standard deviation of
$s = 4.2$ psi.

a. Construct a 99% confidence interval for the mean underinflation μ.
b. Based on your confidence interval, would you recommend that the consumer group issue a report that the mean tire pressure is seriously underinflated? Explain your answer.
c. Would your answer in (b) change if a 90% confidence interval was used to reach the decision?

Soc.   **5.8**   A social worker is interested in estimating the average length of time spent outside of prison for first offenders who later commit a second crime and are sent to prison again. A random sample of $n = 150$ prison records in the county courthouse indicates that the average length of prison-free life between first and second offenses is 3.2 years, with a standard deviation of 1.1 years. Use the sample information to estimate μ, the mean prison-free life between first and second offenses for all prisoners on record in the county courthouse. Construct a 95% confidence interval for μ. Assume that σ can be replaced by s.

Ag.   **5.9**   The rust mite, a major pest of citrus in Florida, punctures the cells of the leaves and fruit. Damage by rust mites is readily recognizable because the injured fruit displays a brownish (rust) color and is somewhat reduced in size depending on the severity of the attack. If the rust mites are not controlled, the affected groves have a substantial reduction in both the fruit yield and the fruit quality. In either case, the citrus grower suffers financially because the produce is of a lower grade and sells for less on the fresh-fruit market. This year, more and more citrus growers have gone to a program of preventive maintenance spraying for rust mites. In evaluating the effectiveness of the program, a random sample of 60 10-acre plots, one plot from each of 60 groves, is selected. These show an average yield of 850 boxes, with a standard deviation of 100 boxes. Give a 95% confidence interval for μ, the average (10-acre) yield for all groves using such a maintenance spraying program. Assume that σ can be replaced by s.

Gov.   **5.10**   A problem of interest to the United States, other governments, and world councils concerned with the critical shortage of food throughout the world is finding a method to estimate the total amount of grain crops that will be produced throughout the world in a particular year.

One method of predicting total crop yields is based on satellite photographs of Earth's surface. Because a scanning device reads the total acreage of a particular type of grain with error, it is necessary to have the device read many equal-sized plots of a particular planting to calibrate the reading on the scanner with the actual acreage. Satellite photographs of 100 50-acre plots of wheat are read by the scanner and give a sample average and standard deviation

$$\bar{y} = 3.27 \qquad s = .23$$

Find a 95% confidence interval for the mean scanner reading for the population of all 50-acre plots of wheat. Explain the meaning of this interval.

5.3   Choosing the Sample Size for Estimating μ

How can we determine the number of observations to include in the sample? The implications of such a question are clear. Data collection costs money. If the sample is too large, time and talent are wasted. Conversely, it is wasteful if the sample is too small, because inadequate information has been purchased for the time and effort expended. Also, it may be impossible to increase the sample size at a later time. Hence, the number of observations to be included in the sample will be a compromise between the desired accuracy of the sample statistic as an estimate of the population parameter and the required time and cost to achieve this degree of accuracy.

Suppose we want to estimate the average amount for accident claims filed against an insurance company. To decide how many claims must be examined, we would have to determine how accurate the company wants to be. For example, the company might indicate that the tolerable error is to be 10 units (± 5 units) or less. Then we would want the confidence interval to be of the form $\bar{y} \pm 5$.

There are two considerations in determining the appropriate sample size for estimating μ using a confidence interval. First, the tolerable error establishes the desired width of the interval. The second consideration is the level of confidence. In selecting our specifications, we need to consider that, if the confidence interval of μ is too wide, then our estimation of μ will be imprecise and not very informative. Similarly, a very low level of confidence (say 50%) will yield a confidence interval that very likely will be in error—that is, fail to contain μ. However, to obtain a confidence interval having a narrow width and a high level of confidence may require a large value for the sample size and hence be unreasonable in terms of cost and/or time.

What constitutes reasonable certainty? In most situations, the confidence level is set at 95% or 90%, partly because of tradition and partly because these levels represent (to some people) a reasonable level of certainty. The 95% (or 90%) level translates into a long-run chance of 1 in 20 (or 1 in 10) of not covering the population parameter. This seems reasonable and is comprehensible, whereas 1 chance in 1,000 or 1 in 10,000 is too small.

The tolerable error depends heavily on the context of the problem, and only someone who is familiar with the situation can make a reasonable judgment about its magnitude.

When considering a confidence interval for a population mean μ, the plus-or-minus term of the confidence interval is $z_{\alpha/2}\sigma_{\bar{y}}$, where $\sigma_{\bar{y}} = \sigma/\sqrt{n}$. Three quantities determine the value of the plus-or-minus term: the desired confidence level (which determines the z-value used), the standard deviation (σ), and the sample size (which together with σ determines the standard error $\sigma_{\bar{y}}$). Usually, a guess must be made about the size of the population standard deviation. (Sometimes an initial sample is taken to estimate the standard deviation; this estimate provides a basis for determining the additional sample size that is needed.) For a given tolerable error, once the confidence level is specified and an estimate of σ supplied, the required sample size can be calculated using the formula shown here.

Suppose we want to estimate μ using a $100(1 - \alpha)\%$ confidence interval having tolerable error W. Our interval will be of the form $\bar{y} \pm E$, where $E = W/2$. Note that W is the width of the confidence interval. To determine the sample size n, we solve the equation

$$E = z_{\alpha/2}\sigma_{\bar{y}} = z_{\alpha/2}\sigma/\sqrt{n}$$

for n. This formula for n is shown here.

Sample Size Required for a 100(1 − α)% Confidence Interval for μ of the Form $\bar{y} \pm E$

$$n = \frac{(z_{\alpha/2})^2\sigma^2}{E^2}$$

Note that determining a sample size to estimate μ requires knowledge of the population variance σ^2 (or standard deviation σ). We can obtain an approximate sample size by estimating σ^2, using one of these two methods:

1. Employ information from a prior experiment to calculate a sample variance s^2. This value is used to approximate σ^2.
2. Use information on the range of the observations to obtain an estimate of σ.

We then substitute the estimated value of σ^2 in the sample-size equation to determine an approximate sample size n.

We illustrate the procedure for choosing a sample size with two examples.

EXAMPLE 5.3

Union officials are concerned about reports of inferior wages paid to a company's employees under their jurisdiction. It is decided to take a random sample of n wage sheets from the company to estimate the average hourly wage. If it is known that wages in the company have a range of $10 per hour, determine the sample size required to estimate the average hourly wage μ using a 95% confidence interval with width equal to $1.20.

Solution   Because we want a 95% confidence interval with width $1.20, $E = \$.60$. The value that we use to substitute for σ is range/4 = 2.50. Substituting into the formula for n we have

$$n = \frac{(1.96)^2(2.5)^2}{(.60)^2} = 66.69$$

To be on the safe side, we round this number up to the next integer. A sample size of 67 should give a 95% confidence interval with the desired width of $1.20.

EXAMPLE 5.4

A federal agency has decided to investigate the advertised weight printed on cartons of a certain brand of cereal. The company in question periodically samples cartons of cereal coming off the production line to check their weight. A summary of 1,500 of the weights made available to the agency indicates a mean weight of 11.80 ounces per carton and a standard deviation of .75 ounce. Use this information to determine the number of cereal cartons the federal agency must examine to estimate the average weight of cartons being produced now, using a 99% confidence interval of width .50.

Solution   The federal agency has specified that the width of the confidence interval is to be .50, so $E = .25$. Assuming that the weights made available to the agency by the company are accurate, we can take $\sigma = .75$. The required sample size with $z_{\alpha/2} = 2.58$ is

$$n = \frac{(2.58)^2(.75)^2}{(.25)^2} = 59.91$$

Thus, the federal agency must obtain a random sample of 60 cereal cartons to estimate the mean weight to within $\pm.25$.

EXERCISES   **Basic Techniques**

5.11   Refer to Example 5.3. Suppose we continue to estimate σ with $\hat{\sigma} = 2.5$.
   a. If the level of confidence remains at 95% but the tolerable width is 1, how large a sample size is required?
   b. If the level of confidence increases to 99% but the specified width remains at 1.2, how large a sample size is required?

c. If the level of confidence decreases to 90% but the specified width remains at 1.2, how large a sample size is required?

5.12 In general, if we keep the level of confidence fixed, how much do you need to increase the sample size to cut the width in half?

Applications

Bio.   **5.13** A biologist wishes to estimate the effect of an antibiotic on the growth of a particular bacterium by examining the mean amount of bacteria present per plate of culture when a fixed amount of the antibiotic is applied. Previous experimentation with the antibiotic on this type of bacteria indicates that the standard deviation of the amount of bacteria present is approximately 13 cm^2. Use this information to determine the number of observations (cultures that must be developed and then tested) to estimate the mean amount of bacteria present, using a 99% confidence interval with a half width of 3 cm^2.

Soc.   **5.14** The city housing department wants to estimate the average rent for rent-controlled apartments. They need to determine the number of renters to include in the survey in order to estimate the average rent to within $50 using a 95% confidence interval. From past results, the rent for controlled apartments ranged from $200 to $1500 per month. How many renters are needed in the survey to meet the requirements?

5.15 Refer to Exercise 5.14. Suppose the mayor has reviewed the proposed survey and decides on some changes.
a. If the level of confidence is increased to 99% with the average rent estimated to within $25, what sample size is required?
b. Suppose the budget for the project will not support both increasing the level of confidence and reducing the width of the interval. Explain to the mayor the impact on the estimation of the average rent of not raising the level of confidence from 95% to 99%.

5.4   A Statistical Test for μ

The second type of inference-making procedure is statistical testing (or hypothesis testing). As with estimation procedures, we will make an inference about a population parameter, but here the inference will be of a different sort. With point and interval estimates there was no supposition about the actual value of the parameter prior to collecting the data. Using sampled data from the population, we are simply attempting to determine the value of the parameter. In hypothesis testing, there is a preconceived idea about the value of the population parameter. For example, in studying the antipsychotic properties of an experimental compound, we might ask whether the average shock-avoidance response of rats treated with a specific dose of the compound is greater than 60, $\mu > 60$, the value that has been observed after extensive testing using a suitable standard drug. Thus, there are two theories or hypotheses involved in a statistical study. The first is the hypothesis being proposed by the person conducting the study, called the **research hypothesis,** $\mu > 60$ in our example. The second theory is the negation of this hypothesis, called the **null hypothesis,** $\mu \leq 60$ in our example. The goal of the study is to decide whether the data tend to support the research hypothesis.

A **statistical test** is based on the concept of proof by contradiction and is composed of the five parts listed here.

1. Research hypothesis (also called the alternative hypothesis), denoted by H_a.
2. Null hypothesis, denoted by H_0.
3. Test statistics, denoted by T.S.
4. Rejection region, denoted by R.R.
5. Checking assumptions and drawing conclusions.

For example, the Texas A&M agricultural extension service wants to determine whether the mean yield per acre (in bushels) for a particular variety of soybeans has increased during the current year over the mean yield in the previous 2 years when μ was 52 bushels per acre. The first step in setting up a statistical test is determining the proper specification of H_0 and H_a. The following guidelines will be helpful:

1. The statement that μ equals a specific value will always be included in H_0. The particular value specified for μ is called its null value and is denoted μ_0.
2. The statement about μ that the researcher is attempting to support or detect with the data from the study is the research hypothesis, H_a.
3. The negation of H_a is the null hypothesis, H_0.
4. The null hypothesis is presumed correct unless there is overwhelming evidence in the data that the research hypothesis is supported.

In our example, μ_0 is 52. The research statement is that yield in the current year has increased above 52; that is, $H_a: \mu > 52$. (Note that we will include 52 in the null hypothesis.) Thus, the null hypothesis, the negation of H_a, is $H_0: \mu \leq 52$.

To evaluate the research hypothesis, we take the information in the sample data and attempt to determine whether the data support the research hypothesis or the null hypothesis, but we will give the benefit of the doubt to the null hypothesis.

After stating the null and research hypotheses, we then obtain a random sample of 1-acre yields from farms throughout the state. The decision to state whether or not the data support the research hypothesis is based on a quantity

test statistic   computed from the sample data called the **test statistic.** If the population distribution is determined to be mound shaped, a logical choice as a test statistic for μ is $\overline{y}$ or some function of $\overline{y}$.

If we select $\overline{y}$ as the test statistic, we know that the sampling distribution of $\overline{y}$ is approximately normal with a mean μ and standard deviation $\sigma_{\overline{y}} = \sigma/\sqrt{n}$, provided the population distribution is normal or the sample size is fairly large. We are attempting to decide between $H_a: \mu > 52$ or $H_0: \mu \leq 52$. The decision will be to either reject H_0 or fail to reject H_0. In developing our decision rule, we will assume that $\mu = 52$, the null value of μ. We will now determine the values of $\overline{y}$, called the

rejection region   **rejection region,** that we are very unlikely to observe if $\mu = 52$ (or if μ is any other value in H_0). The rejection region contains the values of $\overline{y}$ that support the research hypothesis and contradict the null hypothesis, hence the region of values for $\overline{y}$ that reject the null hypothesis. The rejection region will be the values of $\overline{y}$ in the upper tail of the null distribution ($\mu = 52$) of $\overline{y}$. See Figure 5.5.

FIGURE 5.5

Assuming that H_0 is true, contradictory values of $\bar{y}$ are in the upper tail

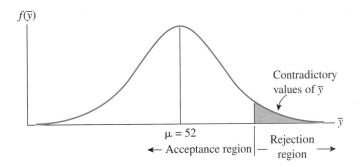

Type I error

Type II error

As with any two-way decision process, we can make an error by falsely rejecting the null hypothesis or by falsely accepting the null hypothesis. We give these errors the special names **Type I error** and **Type II error.**

DEFINITION 5.1

A **Type I error** is committed if we reject the null hypothesis when it is true. The probability of a Type I error is denoted by the symbol α.

DEFINITION 5.2

A **Type II error** is committed if we accept the null hypothesis when it is false and the research hypothesis is true. The probability of a Type II error is denoted by the symbol β (Greek letter beta).

The two-way decision process is shown in Table 5.3 with corresponding probabilities associated with each situation.

Although it is desirable to determine the acceptance and rejection regions to simultaneously minimize both α and β, this is not possible. The probabilities associated with Type I and Type II errors are inversely related. For a fixed sample size n, as we change the rejection region to decrease α, β increases, and vice versa.

To alleviate what appears to be an impossible bind, the experimenter specifies a tolerable probability for a Type I error of the statistical test. Thus, the experimenter may choose α to be .01, .05, .10, and so on. Specification of a value for α then locates the rejection region. The determination of the associated probability of a Type II error is more complicated and will be delayed until later in the chapter.

Let us now see how the choice of α locates the rejection region. Returning to our soybean example, we will reject the null hypothesis for large values of the

TABLE 5.3

Two-way decision process

Decision	Null Hypothesis	
	True	**False**
Reject H_0	Type I error α	Correct $1 - \beta$
Accept H_0	Correct $1 - \alpha$	Type II error β

FIGURE 5.6

Rejection region
for the soybean example
when $\alpha = .025$

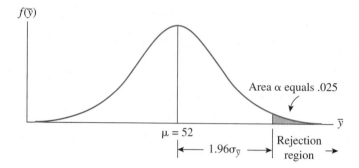

$f(\bar{y})$

Area α equals .025

$\mu = 52$

$\leftarrow 1.96\sigma_{\bar{y}} \rightarrow$

Rejection
region

$\bar{y}$

sample mean $\bar{y}$. Suppose we have decided to take a sample of $n = 36$ 1-acre plots, and from these data we compute $\bar{y} = 56$ and $s = 9.6$. Can we conclude that the mean yield for all farms is above 52?

specifying α

Before answering this question we must **specify α**. If we are willing to take the risk that 1 time in 40 we would incorrectly reject the null hypothesis, then $\alpha = 1/40 = .025$. An appropriate rejection region can be specified for this value of α by referring to the sampling distribution of $\bar{y}$. Assuming that $\mu = 52$ and that σ can be replaced by s, then $\bar{y}$ is normally distributed, with $\mu = 52$ and $\sigma_{\bar{y}} = 9.6/\sqrt{36} = 1.6$. Because the shaded area of Figure 5.6 corresponds to α, locating a rejection region with an area of .025 in the right tail of the distribution of $\bar{y}$ is equivalent to determining the value of z that has an area .025 to its right. Referring to Table 1 in the Appendix, this value of z is 1.96. Thus, the rejection region for our example is located 1.96 standard errors $(1.96\sigma_{\bar{y}})$ above the mean $\mu = 52$. If the observed value of $\bar{y}$ is greater than 1.96 standard errors above $\mu = 52$, we reject the null hypothesis, as shown in Figure 5.6.

EXAMPLE 5.5

Set up all the parts of a statistical test for the soybean example and use the sample data to reach a decision on whether to accept or reject the null hypothesis. Set $\alpha = .025$. Assume that σ can be estimated by s.

Solution   The five parts of the test are as follows:

H_0: $\mu \leq 52$
H_a: $\mu > 52$
T.S.: $\bar{y}$
R.R.: For $\alpha = .025$, reject the null hypothesis if $\bar{y}$ lies more than 1.96 standard errors above $\mu = 52$.

The computed value of $\bar{y}$ is 56. To determine the number of standard errors that $\bar{y}$ lies above $\mu = 52$, we compute a z score for $\bar{y}$ using the formula

$$z = \frac{\bar{y} - \mu_0}{\sigma_{\bar{y}}}$$

where $\sigma_{\bar{y}} = \sigma/\sqrt{n}$. Substituting into the formula,

$$z = \frac{\bar{y} - \mu_0}{\sigma_{\bar{y}}} = \frac{56 - 52}{9.6/\sqrt{36}} = 2.5$$

Checking assumptions and drawing conclusions: With a sample size of $n = 36$, the Central Limit Theorem should hold for the distribution of $\bar{y}$. Because the observed value of $\bar{y}$ lies more than 1.96 standard errors above the hypothesized mean $\mu = 52$, we reject the null hypothesis in favor of the research hypothesis and conclude that the average soybean yield per acre is greater than 52.

one-tailed test   The statistical test conducted in Example 5.5 is called a **one-tailed test** because the rejection region is located in only one tail of the distribution of $\bar{y}$. If our research hypothesis is $H_a: \mu < 52$, small values of $\bar{y}$ indicate rejection of the null hypothesis. This test is also one-tailed, but the rejection region is located in the lower tail of the distribution of $\bar{y}$. Figure 5.7 displays the rejection region for the alternative hypothesis $H_a: \mu < 52$ when $\alpha = .025$.

two-tailed test   We can formulate a **two-tailed test** for the research hypothesis $H_a: \mu \neq 52$, where we are interested in detecting whether the mean yield per acre of soybeans is greater or less than 52. Clearly both large and small values of $\bar{y}$ would contradict the null hypothesis, and we locate the rejection region in both tails of the distribution of $\bar{y}$. A two-tailed rejection region for $H_a: \mu \neq 52$ and $\alpha = .05$ is shown in Figure 5.8.

FIGURE 5.7

Rejection region for $H_a: \mu < 52$ when $\alpha = .025$ for the soybean example

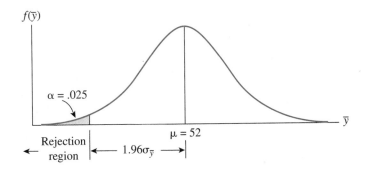

FIGURE 5.8

Two-tailed rejection region for $H_a: \mu \neq 52$ when $\alpha = .05$ for the soybean example

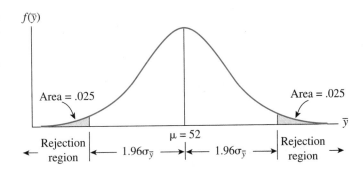

EXAMPLE 5.6

A corporation maintains a large fleet of company cars for its salespeople. To check the average number of miles driven per month per car, a random sample of $n = 40$ cars is examined. The mean and standard deviation for the sample are 2,752 miles and 350 miles, respectively. Records for previous years indicate that the average number of miles driven per car per month was 2,600. Use the sample data to test the research hypothesis that the current mean μ differs from 2,600. Set $\alpha = .05$ and assume that σ can be estimated by s.

FIGURE 5.9

Rejection region for
$H_a: \mu \neq 2{,}600$ when $\alpha = .05$

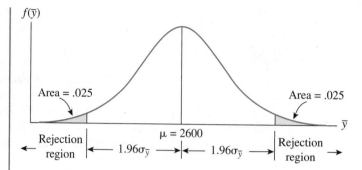

Solution   The research hypothesis for this statistical test is $H_a: \mu \neq 2{,}600$ and the null hypothesis is $H_0: \mu = 2{,}600$. Using $\alpha = .05$, the two-tailed rejection region for this test is located as shown in Figure 5.9.

   With a sample size of $n = 40$, the Central Limit Theorem should hold for $\bar{y}$. To determine how many standard errors our test statistic $\bar{y}$ lies away from $\mu = 2{,}600$, we compute

$$z = \frac{\bar{y} - \mu_0}{\sigma/\sqrt{n}} = \frac{2{,}752 - 2{,}600}{350/\sqrt{40}} = 2.75$$

The observed value for $\bar{y}$ lies more than 1.96 standard errors above the mean, so we reject the null hypothesis in favor of the alternative $H_a: \mu \neq 2{,}600$. We conclude that the mean number of miles driven is different from 2,600.

   The mechanics of the statistical test for a population mean can be greatly simplified if we use z rather than $\bar{y}$ as a test statistic. Using

$$H_0: \quad \mu \leq \mu_0 \text{ (where } \mu_0 \text{ is some specified value)}$$
$$H_a: \quad \mu > \mu_0$$

and the test statistic

$$z = \frac{\bar{y} - \mu_0}{\sigma/\sqrt{n}}$$

then for $\alpha = .025$ we reject the null hypothesis if $z \geq 1.96$—that is, if $\bar{y}$ lies more than 1.96 standard errors above the mean. Similarly, for $\alpha = .05$ and $H_a: \mu \neq \mu_0$, we reject the null hypothesis if the computed value of $z \geq 1.96$ or the computed value of $z \leq -1.96$. This is equivalent to rejecting the null hypothesis if the computed value of $|z| \geq 1.96$.

test for a population mean   The statistical **test for a population mean** μ is summarized next. Three different sets of hypotheses are given with their corresponding rejection regions. In a given situation, you will choose only one of the three alternatives with its associated rejection region. The tests given are appropriate only when the population distribution is normal with known σ. The rejection region will be approximately the correct region even when the population distribution is nonnormal provided the sample size is large; in most cases, $n \geq 30$ is sufficient. We can then apply the results from the Central Limit Theorem with the sample standard deviation s replacing σ to conclude that the sampling distribution of $z = (\bar{y} - \mu_0)/(s/\sqrt{n})$ is approximately normal.

Summary of a Statistical Test for μ with a Normal Population Distribution (σ Known) or Large Sample Size n

Hypotheses:

Case 1. $H_0: \mu \leq \mu_0$ vs. $H_a: \mu > \mu_0$ (right-tailed test)
Case 2. $H_0: \mu \geq \mu_0$ vs. $H_a: \mu < \mu_0$ (left-tailed test)
Case 3. $H_0: \mu = \mu_0$ vs. $H_a: \mu \neq \mu_0$ (two-tailed test)

T.S.: $z = \dfrac{\bar{y} - \mu_0}{\sigma/\sqrt{n}}$

R.R.: For a probability α of a Type I error,

Case 1. Reject H_0 if $z \geq z_\alpha$.
Case 2. Reject H_0 if $z \leq -z_\alpha$.
Case 3. Reject H_0 if $|z| \geq z_{\alpha/2}$.

Note: These procedures are appropriate if the population distribution is normally distributed with σ known. In most situations, if $n \geq 30$, then the Central Limit Theorem allows us to use these procedures when the population distribution is nonnormal. Also, if $n \geq 30$, then we can replace σ with the sample standard deviation s. The situation in which $n < 30$ is presented later in this chapter.

EXAMPLE 5.7

As a part of her evaluation of municipal employees, the city manager audits the parking tickets issued by city parking officers to determine the number of tickets that were contested by the car owner and found to be improperly issued. In past years, the number of improperly issued tickets per officer had a normal distribution with mean $\mu = 380$ and $\sigma = 35.2$. Because there has recently been a change in the city's parking regulations, the city manager suspects that the mean number of improperly issued tickets has increased. An audit of 50 randomly selected officers is conducted to test whether there has been an increase in improper tickets. Use the sample data given here and $\alpha = .01$ to test the research hypothesis that the mean number of improperly issued tickets is greater than 380. The audit generates the following data: $n = 50$ and $\bar{y} = 390$.

Solution   Using the sample data with $\alpha = .01$, the five parts of a statistical test are as follows:

H_0: $\mu \leq 380$
H_a: $\mu > 380$
T.S.: $z = \dfrac{\bar{y} - \mu_0}{\sigma/\sqrt{n}} = \dfrac{390 - 380}{35.2/\sqrt{50}} = \dfrac{10}{35.2/7.07} = 2.01$
R.R.:   For $\alpha = .01$ and a right-tailed test, we reject H_0 if $z \geq z_{.01}$, where $z_{.01} = 2.33$.

Checking assumptions and drawing conclusions: Because the observed value of z, 2.01, does not exceed 2.33, we might be tempted to accept the null hypothesis that $\mu \leq 380$. The only problem with this conclusion is that we do not know β, the probability of incorrectly accepting the null hypothesis. To hedge somewhat in situations in which z does not fall in the rejection region and β has not been calculated, we recommend stating that there is insufficient evidence to reject the null hypothesis. To reach a conclusion about whether to accept H_0, the experimenter would have to compute β. If β is small for reasonable alternative values of μ, then H_0 is accepted. Otherwise, the experimenter should conclude that there is insufficient evidence to reject the null hypothesis.

computing β

We can illustrate the **computation of β**, the probability of a Type II error, using the data in Example 5.7. If the null hypothesis is H_0: $\mu \leq 380$, the probability of incorrectly accepting H_0 will depend on how close the actual mean is to 380. For example, if the actual mean number of improperly issued tickets is 400, we would expect β to be much smaller than if the actual mean is 387. The closer the actual mean is to μ_0 the more likely we are to obtain data having a value $\bar{y}$ in the acceptance region. The whole process of determining β for a test is a "what-if" type of process. In practice, we compute the value of β for a number of values of μ in the alterna-

OC curve

tive hypothesis H_a and plot β versus μ in a graph called the **OC curve.** Alternatively, tests of hypotheses are evaluated by computing the probability that the test rejects

power

false null hypotheses, called the **power** of the test. We note that power $= 1 - \beta$. The

power curve

plot of power versus the value of μ is called the **power curve.** We attempt to design tests that have large values of power and hence small values for β.

Let us suppose that the actual mean number of improper tickets is 395 per officer. What is β? With the null and research hypotheses as before,

$$H_0: \quad \mu \leq 380$$
$$H_a: \quad \mu > 380$$

and with $\alpha = .01$, we use Figure 5.10(a) to display β. The shaded portion of Figure 5.10(a) represents β, as this is the probability of $\bar{y}$ falling in the acceptance region when the null hypothesis is false and the actual value of μ is 395. The power of the test for detecting that the actual value of μ is 395 is $1 - \beta$, the area in the rejection region.

Let us consider two other possible values for μ—namely, 387 and 400. The corresponding values of β are shown as the shaded portions of Figures 5.10(b) and (c), respectively; power is the unshaded portion in the rejection region of Figure 5.10(b) and (c). The three situations illustrated in Figure 5.10 confirm what we alluded to earlier; that is, the probability of a Type II error β decreases (and hence power increases) the further μ lies away from the hypothesized means under H_0.

The following notation will facilitate the calculation of β. Let μ_0 denote the null value of μ and let μ_a denote the value of the mean in H_a. Let $\beta(\mu_a)$ be the probability of a Type II error if the actual value of the mean is μ_a and let $\text{PWR}(\mu_a)$ be the power at μ_a. Note that $\text{PWR}(\mu_a)$ equals $1 - \beta(\mu_a)$. Although we never really know the actual mean, we select feasible values of μ and determine β for each of these values. This will allow us to determine the probability of a Type II error occurring if one of these feasible values happens to be the actual value of the mean. The decision whether or not to accept H_0 depends on the magnitude of β for one or more reasonable values for μ_a. Alternatively, researchers calculate the power curve for a test of hypotheses. Recall, that the power of the test at μ_a, $\text{PWR}(\mu_a)$ is the probability the test will detect that H_0 is false when the actual value of μ is μ_a. Hence, we want tests of hypotheses in which $\text{PWR}(\mu_a)$ is large when μ_a is far from μ_0.

FIGURE 5.10

The probability β of a Type II error

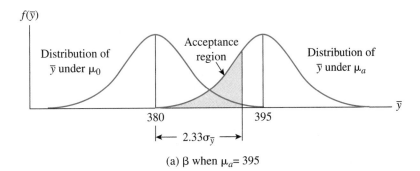

(a) β when $\mu_a = 395$

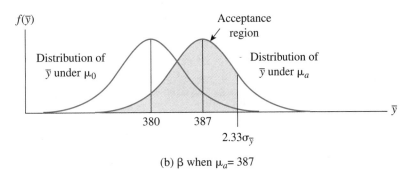

(b) β when $\mu_a = 387$

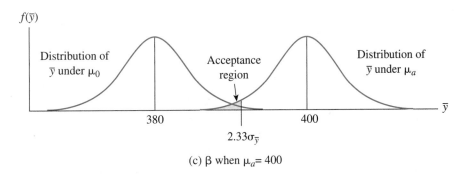

(c) β when $\mu_a = 400$

For a one-tailed test, $H_0: \mu \le \mu_0$ or $H_0: \mu \ge \mu_0$, the value of β at μ_a is the probability that z is less than

$$z_\alpha - \frac{|\mu_0 - \mu_a|}{\sigma_{\bar{y}}}$$

This probability is written as

$$\beta(\mu_a) = P\left[z < z_\alpha - \frac{|\mu_0 - \mu_a|}{\sigma_{\bar{y}}}\right]$$

The value of $\beta(\mu_a)$ is found by looking up the probability corresponding to the number $z_\alpha - |\mu_0 - \mu_a|/\sigma_{\bar{y}}$ in Table 1 in the Appendix.

Formulas for β are given here for one- and two-tailed tests. Examples using these formulas follow.

Calculation of β for a One- or Two-Tailed Test about μ

1. One-tailed test:

$$\beta(\mu_a) = P\left(z \le z_\alpha - \frac{|\mu_0 - \mu_a|}{\sigma_{\bar{y}}}\right) \qquad \text{PWR}(\mu_a) = 1 - \beta(\mu_a)$$

2. Two-tailed test:

$$\beta(\mu_a) \approx P\left(z \le z_{\alpha/2} - \frac{|\mu_0 - \mu_a|}{\sigma_{\bar{y}}}\right) \qquad \text{PWR}(\mu_a) = 1 - \beta(\mu_a)$$

EXAMPLE 5.8

Compute β and power for the test in Example 5.7 if the actual mean number of improperly issued tickets is 395.

Solution The research hypothesis for Example 5.7 was $H_a: \mu > 380$. Using $\alpha = .01$ and the computing formula for β with $\mu_0 = 380$ and $\mu_a = 395$, we have

$$\beta(395) = P\left[z < z_{.01} - \frac{|\mu_0 - \mu_a|}{\sigma_{\bar{y}}}\right] = P\left[z < 2.33 - \frac{|380 - 395|}{35.2/\sqrt{50}}\right]$$

$$= P[z < 2.33 - 3.01] = P[z < -.68]$$

Referring to Table 1 in the Appendix, the area corresponding to $z = -.68$ is .2483. Hence, $\beta(395) = .2483$ and $\text{PWR}(395) = 1 - .2483 = .7517$.

Previously, when $\bar{y}$ did not fall in the rejection region, we concluded that there was insufficient evidence to reject H_0 because β was unknown. Now when $\bar{y}$ falls in the acceptance region, we can compute β corresponding to one (or more) alternative values for μ that appear reasonable in light of the experimental setting. Then provided we are willing to tolerate a probability of falsely accepting the null hypothesis equal to the computed value of β for the alternative value(s) of μ considered, our decision is to accept the null hypothesis. Thus, in Example 5.8, if the actual mean number of improperly issued tickets is 395, then there is about a .25 probability (1 in 4 chance) of accepting the hypothesis that μ is less than or equal to 380 when in fact μ equals 395. The city manager would have to analyze the consequence of making such a decision. If the risk was acceptable then she could state that the audit has determined that the mean number of improperly issued tickets has not increased. If the risk is too great, then the city manager would have to expand the audit by sampling more than 50 officers. In the next section, we will describe how to select the proper value for n.

EXAMPLE 5.9

Prospective salespeople for an encyclopedia company are now being offered a sales training program. Previous data indicate that the average number of sales per month for those who do not participate in the program is 33. To determine whether the training program is effective, a random sample of 35 new employees is given the sales training and then sent out into the field. One month later, the mean and standard deviation for the number of sets of encyclopedias sold are 35 and 8.4, respectively. Do the data present sufficient evidence to indicate that the training program enhances sales? Use $\alpha = .05$.

Solution The five parts to our statistical test are as follows:

$$H_0: \quad \mu \le 33$$
$$H_a: \quad \mu > 33$$

T.S.: $\quad z = \dfrac{\overline{y} - \mu_0}{\sigma_{\overline{y}}} \approx \dfrac{35 - 33}{8.4/\sqrt{35}} = 1.41$

R.R.: For $\alpha = .05$ we will reject the null hypothesis if $z \ge z_{.05} = 1.645$.

Checking assumptions and drawing conclusions: With $n = 35$, the Central Limit Theorem should hold. Because the observed value of z does not fall into the rejection region, we reserve judgment on accepting H_0 until we calculate β. In other words, we conclude that there is insufficient evidence to reject the null hypothesis that people in the sales program have the same or a smaller mean number of sales per month as those not in the program.

EXAMPLE 5.10

Refer to Example 5.9. Suppose that the encyclopedia company thinks that the cost of financing the sales program will be offset by increased sales if those in the program average 38 sales per month. Compute β for $\mu_a = 38$ and, based on the value of $\beta(38)$, indicate whether you accept the null hypothesis.

Solution Using the computational formula for β with $\mu_0 = 33$, $\mu_a = 38$, and $\alpha = .05$, we have

$$\beta(38) = P\left[z \le z_{.05} - \frac{|\mu_0 - \mu_a|}{\sigma_{\overline{y}}} \right] = P\left[z \le 1.645 - \frac{|33 - 38|}{8.4/\sqrt{35}} \right]$$
$$= P[z \le -1.88]$$

The area corresponding to $z = -1.88$ in Table 1 of the Appendix is .0301. Hence,

$$\beta(38) = .0301 \qquad \text{PWR}(38) = 1 - .0301 = .9699$$

Because β is relatively small, we accept the null hypothesis and conclude that the training program has not increased the average sales per month above the point at which increased sales would offset the cost of the training program.

There are several factors that influence the magnitude of a Type II error β and the corresponding power $(1 - \beta)$ in a given test. For a given standard deviation σ, we will examine separately how the value of μ_a, the sample size, and the size of the Type I error α influence the value of β and power.

Influence of μ_a

We will use the situation described in Examples 5.9 and 5.10. The encyclopedia company wants to compute the chance of a Type II error for several other values of μ in H_a so they will have a reasonable idea of their chance of making a Type II error based on the data collected in the random sample of new employees. Repeating the calculations for obtaining $\beta(38)$ for different values of μ, we obtain the values in Table 5.4.

Figure 5.11 is a plot of the $\beta(\mu)$ values in Table 5.4 with a smooth curve through the points. Note that as the value of μ increases, the probability of Type II

TABLE 5.4

Probability of Type II error and power for values of μ in H_a

μ	33	34	35	36	37	38	39	40	41
$\beta(\mu)$	.9500	.8266	.5935	.3200	.1206	.0301	.0049	.0005	.0000
PWR(μ)	.0500	.1734	.4065	.6800	.8794	.9699	.9951	.9995	.9999

FIGURE 5.11

Probability of Type II error: effect of μ_a

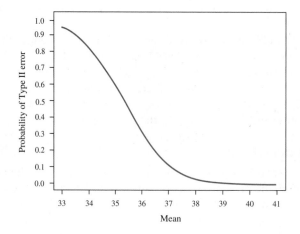

error decreases to 0 and the corresponding power value increases to 1.0. The company could examine this curve to determine whether the chances of Type II error are reasonable for values of μ in H_a that are important to the company. From Table 5.4 or Figure 5.11 we observe that $\beta(38) = .0301$, a relatively small number. Based on the results from Example 5.10, we find that the test statistic does not fall in the rejection region. Because $\beta(38)$ is small, we can now state that we accept the null hypothesis and conclude that the training program has not increased the average sales per month above the point at which increased sales offsets the cost of the training program.

Influence of Sample Size

In Section 5.2, we discussed how we measure the effectiveness of interval estimates. The effectiveness of a statistical test can be measured by the magnitudes of the Type I and Type II errors, α and $\beta(\mu)$. When α is preset at a tolerable level by the experimenter, $\beta(\mu_a)$ is a function of the sample size for a fixed value of μ_a. The larger the sample size n, the more information we have concerning μ, the less likely we are to make a Type II error, and, hence, the smaller the value of $\beta(\mu_a)$. To illustrate this idea, suppose we are testing the hypotheses H_0: $\mu \le 84$ against H_a: $\mu > 84$, where μ is the mean of a population having a normal distribution with $\sigma = 1.4$. If we take $\alpha = .05$, then the probability of Type II errors is plotted in Figure 5.12 for three possible sample sizes, $n = 10, 18,$ and 25. Note that $\beta(84.6)$ becomes smaller as we increase n from 10 to 25.

Influence of Type I Error

Another relationship of interest is that between α and $\beta(\mu)$. For a fixed sample size n, if we change the rejection region to increase the value of α, the value of $\beta(\mu_a)$ will decrease. This relationship can be observed in Figure 5.13. Here we have fixed the

FIGURE 5.12
Probability of Type II error:
effect of sample size,
$\alpha = .05, n = 10, 18, 25$

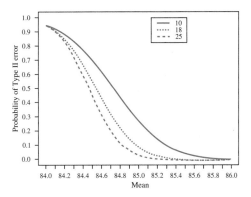

FIGURE 5.13
Probability of Type II error:
effect of Type I error,
$n = 25, \alpha = .05, .01, .001$

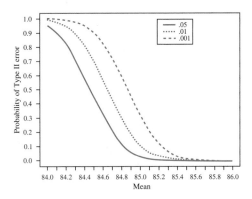

sample size at 25 and plotted $\beta(\mu)$ for three different values of $\alpha = .05, .01, .001$. We observe that $\beta(84.6)$ becomes smaller as α increases from .001 to .05.

A similar set of graphs can be obtained for the power of the test by simply plotting $\text{PWR}(\mu) = 1 - \beta(\mu)$ vs. μ. Here the relationships described would be reversed; that is, for fixed α increasing the value of the sample size would increase the value of $\text{PWR}(\mu)$ and, for fixed sample size, increasing the value of α would increase the value of $\text{PWR}(\mu)$.

We will consider now the problem of designing an experiment for testing hypotheses about μ when α is specified and $\beta(\mu_a)$ is preset for a fixed value μ_a. This problem reduces to determining the sample size needed to achieve the fixed values of α and $\beta(\mu_a)$. Note that in those cases in which the determined value of n is too large for the initially specified values of α and β, we can increase our specified value of α and achieve the desired value of $\beta(\mu_a)$ with a smaller sample size.

5.5 Choosing the Sample Size for μ

The quantity of information available for a statistical test about μ is measured by the magnitudes of the Type I and II error probabilities, α and $\beta(\mu)$ for various values of μ in the alternative hypothesis H_a. Suppose that we are interested in testing $H_0: \mu \le \mu_0$ against the alternative $H_a: \mu > \mu_0$. First, we must specify the value of α. Next we must determine a value of μ in the alternative, μ_1, such that if the actual value of the mean is larger than μ_1, then the consequences of making a Type II error would be substantial. Finally, we must select a value for $\beta(\mu_1)$, β. Note that for any value of μ larger than μ_1, the probability of Type II error will be smaller than $\beta(\mu_1)$; that is,

$$\beta(\mu) < \beta(\mu_1), \text{ for all } \mu > \mu_1$$

Let $\Delta = \mu_1 - \mu_0$. The sample size necessary to meet these requirements is

$$n = \sigma^2 \frac{(z_\alpha + z_\beta)^2}{\Delta^2}$$

Note: If σ^2 is unknown, substitute an estimated value from previous studies or a pilot study to obtain an approximate sample size.

The same formula applies when testing $H_0: \mu \geq \mu_0$ against the alternative $H_a: \mu < \mu_0$, with the exception that we want the probability of a Type II error to be of magnitude β or less when the actual value of μ is *less* than μ_1, a value of the mean in H_a; that is,

$$\beta(\mu) < \beta, \text{ for all } \mu < \mu_1$$

with $\Delta = \mu_0 - \mu_1$.

EXAMPLE 5.11

A cereal manufacturer produces cereal in boxes having a labeled weight of 16 ounces. The boxes are filled by machines that are set to have a mean fill per box of 16.37 ounces. Because the actual weight of a box filled by these machines has a normal distribution with a standard deviation of approximately .225 ounces, the percentage of boxes having weight less than 16 ounces is 5% using this setting. The manufacturer is concerned that one of its machines is underfilling the boxes and wants to sample boxes from the machine's output to determine whether the mean weight μ is less than 16.37—that is, to test

$$H_0: \quad \mu \geq 16.37$$
$$H_a: \quad \mu < 16.37$$

with $\alpha = .05$. If the true mean weight is 16.27 or less, the manufacturer needs the probability of failing to detect this underfilling of the boxes with a probability of at most .01 or risk incurring a civil penalty from state regulators. Thus, we need to determine the sample size n such that our test of H_0 versus H_a has $\alpha = .05$ and $\beta(\mu)$ less than .01 whenever μ is less than 16.27 ounces.

Solution   We have $\alpha = .05$, $\beta = .01$, $\Delta = 16.37 - 16.27 = .1$, and $\sigma = .225$. Using our formula with $z_{.05} = 1.645$ and $z_{.01} = 2.33$, we have

$$n = \frac{(.225)^2(1.645 + 2.33)^2}{(.1)^2} = 79.99 \approx 80$$

Thus, the manufacturer must obtain a random sample of $n = 80$ boxes to conduct this test under the specified conditions.

Suppose that after obtaining the sample, we compute $\bar{y} = 16.35$ ounces. The computed value of the test statistic is

$$z = \frac{\bar{y} - 16.37}{\sigma_{\bar{y}}} = \frac{16.35 - 16.37}{.225/\sqrt{80}} = -.795$$

Because the rejection region is $z < -1.645$, the computed value of z does not fall in the rejection region. What is our conclusion? In similar situations in previous sections, we concluded that there was insufficient evidence to reject H_0. Now, however, knowing that $\beta(\mu) \le .01$ when $\mu \le 16.27$, we feel safe in our conclusion to accept H_0: $\mu \ge 16.37$. Thus, the manufacturer is somewhat secure in concluding that the mean fill from the examined machine is at least 16.37 ounces.

With a slight modification of the sample size formula for the one-tailed tests, we can test

$$H_0: \quad \mu = \mu_0$$
$$H_a: \quad \mu \ne \mu_0$$

for a specified α, β, and Δ, where

$$\beta(\mu) \le \beta, \text{ whenever} |\mu - \mu_0| \ge \Delta$$

Thus, the probability of Type II error is at most β whenever the actual mean differs from μ_0 by at least Δ. A formula for an approximate sample size n when testing a two-sided hypothesis for μ is presented here.

Approximate Sample Size for a Two-Sided Test of H_0: $\mu = \mu_0$

$$n = \frac{\sigma^2}{\Delta^2}(z_{\alpha/2} + z_\beta)^2$$

Note: If σ^2 is unknown, substitute an estimated value to get an approximate sample size.

EXERCISES

Basic Techniques

5.16 A researcher wanted to test the hypotheses H_0: $\mu \le 38$ against H_a: $\mu > 38$ with $\alpha = .05$. A random sample of 50 measurements from a population yielded $\bar{y} = 40.1$ and $s = 5.6$.

  a. What conclusions can you make about the hypotheses based on the sample information?
  b. Could you have made a Type II error in this situation? Explain.
  c. Calculate the probability of a Type II error if the actual value of μ is at least 39.

5.17 For the data of Exercise 5.16, sketch the power curve for rejecting H_0: $\mu \le 38$ by determining $PWR(\mu_a)$ for the following values of μ in the alternative hypothesis: 39, 40, 41, 42, 43, and 44.

  a. Interpret the values on your curve.
  b. Without actually recalculating the values for $PWR(\mu)$, sketch the power curve for $\alpha = .025$ and $n = 50$.
  c. Without actually recalculating the values for $PWR(\mu)$, sketch the power curve for $\alpha = .05$ and $n = 20$.

Applications

Bus.   **5.18** The administrator of a nursing home would like to do a time-and-motion study of staff time spent per day performing nonemergency tasks. Prior to the introduction of some efficiency measures, the average worker-hours per day spent on these tasks was $\mu = 16$. The administrator wants to test whether the efficiency measures have reduced the value of μ. How many days must be sampled to test the proposed hypothesis if she wants a test having $\alpha = .05$ and the probability of a Type II error of at most .10 when the actual value of μ is 12 hours or less (at least a 25% decrease from prior to the efficiency measures being implemented)? Assume $\sigma = 7.64$.

Med.   **5.19** A study was conducted of 90 adult male patients following a new treatment for congestive heart failure. One of the variables measured on the patients was the increase in exercise capacity (in minutes) over a 4-week treatment period. The previous treatment regime had produced an average increase of $\mu = 2$ minutes. The researchers wanted to evaluate whether the new treatment had increased the value of μ in comparison to the previous treatment. The data yielded $\bar{y} = 2.17$ and $s = 1.05$.

accept

a. Using $\alpha = .05$, what conclusions can you draw about the research hypothesis?

b. What is the probability of making a Type II error if the actual value of μ is 2.1?

5.20 Refer to Exercise 5.19. Compute the power of the test PWR(μ_a) at $\mu_a = 2.1, 2.2, 2.3, 2.4,$ and 2.5. Sketch a smooth curve through a plot of PWR(μ_a) versus μ_a.

a. If α is reduced from .05 to .01, what is the effect on the power curve?

b. If the sample size is reduced from 90 to 50, what is the effect on the power curve?

Edu. **5.21** To evaluate the success of a 1-year experimental program designed to increase the mathematical achievement of underprivileged high school seniors, a random sample of participants in the program will be selected and their mathematics scores will be compared with the previous year's statewide average of 525 for underprivileged seniors. The researchers want to determine whether the experimental program has increased the mean achievement level over the previous year's statewide average. If $\alpha = .05$, what sample size is needed to have a probability of Type II error of at most .025 if the actual mean is increased to 550? From previous results, $\sigma \approx 80$.

5.22 Refer to Exercise 5.21. Suppose a random sample of 100 students is selected yielding *reject* $\bar{y} = 542$ and $s = 76$. Is there sufficient evidence to conclude that the mean mathematics achievement level has been increased? Explain.

Env. **5.23** The vulnerability of inshore environments to contamination due to urban and industrial expansion in Mombasa is discussed in the paper "Metals, petroleum hydrocarbons and organochlorines in inshore sediments and waters on Mombasa, Kenya" [*Marine Pollution Bulletin* 34 (1997): 570–577]. A geochemical and oceanographic survey of the inshore waters of Mombasa, Kenya was undertaken during the period from September 1995 to January 1996. In the survey, suspended particulate matter and sediment were collected from 48 stations within Mombasa's estuarine creeks. The concentrations of major oxides and 13 trace elements were determined for a varying number of cores at each of the stations. In particular, the lead concentrations in suspended particulate matter (in mg kg^{-1} dry weight) were determined at 37 stations. The researchers were interested in determining whether the average lead concentration was greater than 30 mg kg^{-1} dry weight. The data are given in the following table along with summary statistics and a normal probability plot (discussed briefly in Exercise 4.66 of Chapter 4).

Lead concentrations (mg kg^{-1} dry weight) from 37 stations in Kenya

48	53	44	55	52	39	62	38	23	27
41	37	41	46	32	17	32	41	23	12
3	13	10	11	5	30	11	9	7	11
77	210	38	112	52	10	6			

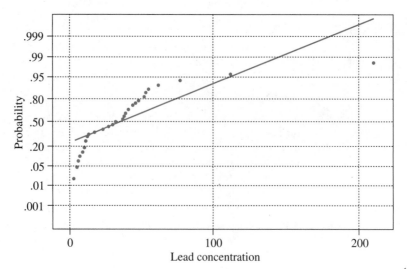

a. Is there sufficient evidence ($\alpha = .05$) in the data that the mean lead concentration exceeds 30 mg kg^{-1} dry weight?

b. What is the probability of a Type II error if the actual mean concentration is 50?

c. Do the data appear to have a normal distribution?

d. Based on your answer in (c), is the sample size large enough for the test procedures to be valid? Explain.

5.6 The Level of Significance of a Statistical Test

In Section 5.4, we introduced hypothesis testing along rather traditional lines: we defined the parts of a statistical test along with the two types of errors and their associated probabilities α and $\beta(\mu_a)$. The problem with this approach is that if other researchers want to apply the results of your study using a different value for α then they must compute a new rejection region before reaching a decision concerning H_0 and H_a. An alternative approach to hypothesis testing follows the following steps: specify the null and alternative hypotheses, specify a value for α, collect the sample data, and determine the weight of evidence for rejecting the null hypothesis. This weight, given in terms of a probability, is called the **level of significance** (or **level of significance** *p*-**value**) of the statistical test. More formally, the level of significance is defined as follows: *the probability of obtaining a value of the test statistic that is as likely or more likely to reject H_0 as the actual observed value of the test statistic. This probability is computed assuming that the null hypothesis is true.* Thus, if the level of significance is a small value, then the sample data fail to support H_0 and our decision is to reject H_0. On the other hand, if the level of significance is a large value, then we fail to reject H_0. We must next decide what is a large or small value for the level of significance. The following decision rule yields results that will always agree with the testing procedures we introduced in Section 5.5.

Decision Rule for Hypothesis Testing Using the *p*-Value

1. If the *p*-value $\leq \alpha$, then reject H_0.
2. If the *p*-value $> \alpha$, then fail to reject H_0.

We illustrate the calculation of a level of significance with several examples.

EXAMPLE 5.12

Refer to Example 5.7.

a. Determine the level of significance (*p*-value) for the statistical test and reach a decision concerning the research hypothesis using $\alpha = .01$.

b. If the preset value of α is .05 instead of .01, does your decision concerning H_a change?

Solution

a. The null and alternative hypotheses are

$$H_0: \quad \mu \leq 380$$
$$H_a: \quad \mu > 380$$

From the sample data, the computed value of the test statistic is

$$z = \frac{\bar{y} - 380}{s/\sqrt{n}} = \frac{390 - 380}{35.2/\sqrt{50}} = 2.01$$

The level of significance for this test (i.e., the weight of evidence for rejecting H_0) is the probability of observing a value of $\bar{y}$ greater than or equal to 390 assuming that the null hypothesis is true; that is, $\mu = 380$. This value can be computed by using the z-value of the test statistic, 2.01, because

$$p\text{-value} = P(\bar{y} \geq 390, \text{ assuming } \mu = 380) = P(z \geq 2.01)$$

Referring to Table 1 in the Appendix, $P(z \geq 2.01) = 1 - P(z < 2.01) = 1 - .9778 = .0222$. This value is shown by the shaded area in Figure 5.14. Because the p-value is greater than α (.0222 > .01), we fail to reject H_0 and conclude that the data do not support the research hypothesis.

FIGURE 5.14

Level of significance
for Example 5.12

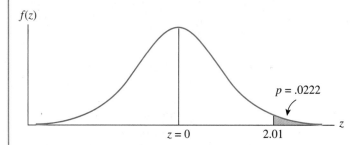

$f(z)$

$p = .0222$

$z = 0$ 2.01

z

b. Another person examines the same data but with a preset value for $\alpha = .05$. This person is willing to support a higher risk of a Type I error; hence the decision is to reject H_0 because the p-value is less than α (.0222 < .05). It is important to emphasize that the value of α used in the decision rule is *preset* and not selected after calculating the p-value.

As we can see from Example 5.12, the level of significance represents the probability of observing a sample outcome more contradictory to H_0 than the observed sample result. *The smaller the value of this probability, the heavier the weight of the sample evidence against H_0.* For example, a statistical test with a level of significance of $p = .01$ shows more evidence for the rejection of H_0 than does another statistical test with $p = .20$.

EXAMPLE 5.13

Refer to Example 5.11. Using a preset value of $\alpha = .05$, is there sufficient evidence in the data to support the research hypothesis?

Solution   The null and alternative hypotheses are

$$H_0: \quad \mu \geq 16.37$$
$$H_a: \quad \mu < 16.37$$

From the sample data, the computed value of the test statistic is

$$z = \frac{\bar{y} - \mu_0}{s/\sqrt{n}} = \frac{16.35 - 16.37}{.225/\sqrt{80}} = -.80$$

The level of significance for this test statistic is computed by determining which values of $\bar{y}$ are more extreme to H_0 than the observed $\bar{y}$. Because H_a specifies μ less than 16.37, the values of $\bar{y}$ that would be more extreme to H_0 are those values less than 16.35, the observed value. Thus,

$$p\text{-value} = P(\bar{y} \leq 16.35, \text{ assuming } \mu = 16.37) = P(z \leq -.80) = .2119$$

There is considerable evidence to support H_0. More precisely, p-value $= .2119 > .05 = \alpha$, and hence we fail to reject H_0. Thus, we conclude that there is insufficient evidence (p-value $= .2119$) to support the research hypothesis. Note that this is exactly the same conclusion reached using the traditional approach.

For two-tailed tests, $H_a: \mu \neq \mu_0$, we still determine the level of significance by computing the probability of obtaining a sample having a value of the test statistic that is more contradictory to H_0 than the observed value of the test statistic. However, for two-tailed research hypotheses, we compute this probability in terms of the magnitude of the distance from $\bar{y}$ to the null value of μ because both values of $\bar{y}$ much less than μ_0 and values of $\bar{y}$ much larger than μ_0 contradict $\mu = \mu_0$. Thus, the level of significance is written as

$$p\text{-value} = P(|\bar{y} - \mu_0| \geq \text{observed}|\bar{y} - \mu_0|) = P(|z| \geq |\text{computed } z|)$$

$$= 2P(z \geq |\text{computed } z|)$$

To summarize, the level of significance (p-value) can be computed as

Case 1	Case 2	Case 3
$H_0: \mu \leq \mu_0$	$H_0: \mu \geq \mu_0$	$H_0: \mu = \mu_0$
$H_a: \mu > \mu_0$	$H_a: \mu < \mu_0$	$H_a: \mu \neq \mu_0$
p-value: $P(z \geq \text{computed } z)$	$P(z \leq \text{computed } z)$	$2P(z \geq |\text{computed } z|)$

EXAMPLE 5.14

Refer to Example 5.6. Using a preset value of $\alpha = .01$, is there sufficient evidence in the data to support the research hypothesis?

Solution The null and alternative hypotheses are

$$H_0: \quad \mu = 2,600$$
$$H_a: \quad \mu \neq 2,600$$

From the sample data, the computed value of the test statistic is

$$z = \frac{\bar{y} - \mu_0}{s/\sqrt{n}} = \frac{2,752 - 2,600}{350/\sqrt{40}} = 2.75$$

The level of significance for this test statistic is computed using the formula for a two-tailed test.

$$p\text{-value} = 2P(z \geq |\text{computed } z|) = 2P(z \geq |2.75|) = 2P(z \geq 2.75)$$

$$= 2(1 - .9970) = .006$$

Because the p-value is very small, there is very little evidence to support H_0. More precisely, p-value $= .006 \leq .05 = \alpha$, and hence we reject H_0. Thus, there is sufficient evidence (p-value $= .006$) to support the research hypothesis and conclude that the mean number of miles driven is different from 2,600. Note that this is exactly the same conclusion reached using the traditional approach.

There is much to be said in favor of this approach to hypothesis testing. Rather than reaching a decision directly, the statistician (or person performing the statistical test) presents the experimenter with the weight of evidence for rejecting the null hypothesis. The experimenter can then draw his or her own conclusion. Some experimenters reject a null hypothesis if $p \leq .10$, whereas others require $p \leq .05$ or $p \leq .01$ for rejecting the null hypothesis. The experimenter is left to make the decision based on what he or she believes is enough evidence to indicate rejection of the null hypothesis.

Many professional journals have followed this approach by reporting the results of a statistical test in terms of its level of significance. Thus, we might read that a particular test was significant at the $p = .05$ level or perhaps the $p < .01$ level. By reporting results this way, the reader is left to draw his or her own conclusion.

One word of warning is needed here. The p-value of .05 has become a magic level, and many seem to feel that a particular null hypothesis should not be rejected unless the test achieves the .05 level or lower. This has resulted in part from the decision-based approach with α preset at .05. Try not to fall into this trap when reading journal articles or reporting the results of your statistical tests. After all, statistical significance at a particular level does not dictate importance or practical significance. Rather, it means that a null hypothesis can be rejected with a specified low risk of error. For example, suppose that a company is interested in determining whether the average number of miles driven per car per month for the sales force has risen above 2,600. Sample data from 400 cars show that $\overline{y} = 2,640$ and $s = 35$. For these data, the z test statistic for $H_0: \mu = 2,600$ is $z = 22.86$ based on $\sigma = 35$; the level of significance is $p < .0000000001$. Thus, even though there has only been a 1.5% increase in the average monthly miles driven for each car, the result is (highly) statistically significant. Is this increase of any practical significance? Probably not. What we have proved *conclusively* is that the mean μ has increased slightly.

Throughout the book we will conduct statistical tests from both the decision-based approach and from the level-of-significance approach to familiarize you with both avenues of thought. For either approach, remember to consider the practical significance of your findings after drawing conclusions based on the statistical test.

EXERCISES

Basic Techniques

5.24   The sample data for a statistical test concerning μ yielded $n = 50$, $\overline{y} = 48.2$, $s = 12.57$. Determine the level of significance for testing $H_0: \mu \leq 45$ versus $H_a: \mu > 45$. Is there significant evidence in the data to support the claim that μ is greater than 45 using $\alpha = .05$?

5.25   Refer to Exercise 5.24. If the researcher used $\alpha = .025$ in place of $\alpha = .05$, would the conclusion about μ change? Explain how the same data can be used to reach a different conclusion about μ.

Applications

Med.   **5.26**   A tobacco company advertises that the average nicotine content of its cigarettes is at most 14 milligrams. A consumer protection agency wants to determine whether the average nicotine content is in fact greater than 14. A random sample of 300 cigarettes of the company's brand yield an average nicotine content of 14.6 and a standard deviation of 3.8 milligrams. Determine the level of significance of the statistical test of the agency's claim that μ is greater than 14. If $\alpha = .01$, is there significant evidence that the agency's claim has been supported by the data?

Psy.   **5.27** A psychological experiment was conducted to investigate the length of time (time delay) between the administration of a stimulus and the observation of a specified reaction. A random sample of 36 people was subjected to the stimulus and the time delay was recorded. The sample mean and standard deviation were 2.2 and .57 seconds, respectively. Is there significant evidence that the mean time delay for the hypothetical population of all people who may be subjected to the stimulus differs from 1.6 seconds? Use $\alpha = .05$. What is the level of significance of the test?

5.7   Inferences about μ for a Normal Population, σ Unknown

The estimation and test procedures about μ presented earlier in this chapter were based on the assumption that the population variance was known or that we had enough observations to allow s to be a reasonable estimate of σ. In this section, we present a test that can be applied when σ is unknown, no matter what the sample size, provided the population distribution is approximately normal. In Section 5.8, we will provide inference techniques for the situation where the population distribution is nonnormal. Consider the following example. Researchers would like to determine the average concentration of a drug in the bloodstream 1 hour after it is given to patients suffering from a rare disease. For this situation, it might be impossible to obtain a random sample of 30 or more observations at a given time. What test procedure could be used in order to make inferences about μ?

W. S. Gosset faced a similar problem around the turn of the twentieth century. As a chemist for Guinness Breweries, he was asked to make judgments on the mean quality of various brews, but was not supplied with large sample sizes to reach his conclusions.

Gosset thought that when he used the test statistic

$$z = \frac{\bar{y} - \mu_0}{\sigma/\sqrt{n}}$$

with σ replaced by s for small sample sizes, he was falsely rejecting the null hypothesis $H_0: \mu = \mu_0$ at a slightly higher rate than that specified by α. This problem intrigued him, and he set out to derive the distribution and percentage points of the test statistic

$$\frac{\bar{y} - \mu_0}{s/\sqrt{n}}$$

for $n < 30$.

For example, suppose an experimenter sets α at a nominal level—say, .05. Then he or she expects falsely to reject the null hypothesis approximately 1 time in 20. However, Gosset proved that the actual probability of a Type I error for this test was somewhat higher than the nominal level designated by α. He published the results of his study under the pen name Student, because at that time it was against company policy for him to publish his results in his own name. The quantity

$$\frac{\bar{y} - \mu_0}{s/\sqrt{n}}$$

Student's t   is called the t statistic and its distribution is called the *Student's t distribution* or, simply, **Student's t**. (See Figure 5.15.)

FIGURE 5.15

Two *t* distributions and a
standard normal distribution

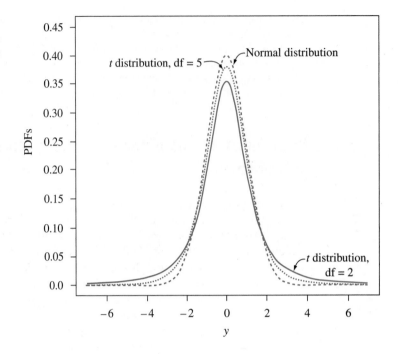

Although the quantity

$$\frac{\bar{y} - \mu_0}{s/\sqrt{n}}$$

possesses a *t* distribution only when the sample is selected from a normal population, the *t* distribution provides a reasonable approximation to the distribution of

$$\frac{\bar{y} - \mu_0}{s/\sqrt{n}}$$

when the sample is selected from a population with a mound-shaped distribution. We summarize the properties of *t* here.

**Properties of Student's
t Distribution**

1. There are many different *t* distributions. We specify a particular one by a parameter called the degrees of freedom (df). See Figure 5.15.
2. The *t* distribution is symmetrical about 0 and hence has mean equal to 0, the same as the *z* distribution.
3. The *t* distribution has variance $\text{df}/(\text{df} - 2)$, and hence is more variable than the *z* distribution, which has variance equal to 1. See Figure 5.15.
4. As the df increases, the *t* distribution approaches the *z* distribution. [Note that as df increases, the variance $\text{df}/(\text{df} - 2)$ approaches 1.]
5. Thus, with

$$t = \frac{\bar{y} - \mu_0}{s/\sqrt{n}}$$

we conclude that *t* has a *t* distribution with $\text{df} = n - 1$, and, as *n* increases, the distribution of *t* approaches the distribution of *z*.

FIGURE 5.16

Illustration of area tabulated in Table 3 in the Appendix for the t distribution

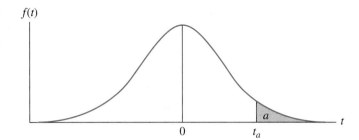

The phrase *degrees of freedom* sounds mysterious now, but the idea will eventually become second nature to you. The technical definition requires advanced mathematics, which we will avoid; on a less technical level, the basic idea is that degrees of freedom are pieces of information for estimating σ using s. The standard deviation s for a sample of n measurements is based on the deviations $y_i - \bar{y}$. Because $\Sigma(y_i - \bar{y}) = 0$ always, if $n - 1$ of the deviations are known, the last $(n$th$)$ is fixed mathematically to make the sum equal 0. It is therefore noninformative. Thus, in a sample of n measurements there are $n - 1$ pieces of information (degrees of freedom) about σ. A second method of explaining degrees of freedom is to recall that σ measures the dispersion of the population values about μ, so prior to estimating σ we must first estimate μ. Hence, the number of pieces of information (degrees of freedom) in the data that can be used to estimate σ is $n - 1$, the number of original data values minus the number of parameters estimated prior to estimating σ.

Because of the symmetry of t, only upper-tail percentage points (probabilities or areas) of the distribution of t have been tabulated; these appear in Table 3 in the Appendix. The degrees of freedom (df) are listed along the left column of the page.

t_a An entry in the table specifies a value of t, say t_a, such that an area a lies to its right. See Figure 5.16. Various values of a appear across the top of Table 3 in the Appendix. Thus, for example, with df $= 7$, the value of t with an area .05 to its right is 1.895 (found in the $a = .05$ column and df $= 7$ row). Because the t distribution approaches the z distribution as df approaches ∞, the values in the last row of Table 3 are the same as z_a. Thus, we can quickly determine z_α by using values in the last row of Table 3.

We can use the t distribution to make inferences about a population mean μ. The sample test concerning μ is summarized next. The only difference between the z test discussed earlier in this chapter and the test given here is that s replaces σ. The t test (rather than the z test) should be used any time σ is unknown and the distribution of y-values is mound-shaped.

Summary of a Statistical Test for μ with a Normal Population Distribution (σ Unknown)

Hypotheses:

Case 1. $H_0: \mu \leq \mu_0$ vs. $H_a: \mu > \mu_0$ (right-tailed test)
Case 2. $H_0: \mu \geq \mu_0$ vs. $H_a: \mu < \mu_0$ (left-tailed test)
Case 3. $H_0: \mu = \mu_0$ vs. $H_a: \mu \neq \mu_0$ (two-tailed test)

T.S.: $t = \dfrac{\bar{y} - \mu_0}{s/\sqrt{n}}$

R.R.: For a probability α of a Type I error and df $= n - 1$,

> **Case 1.** Reject H_0 if $t \geq t_\alpha$.
> **Case 2.** Reject H_0 if $t \leq -t_\alpha$.
> **Case 3.** Reject H_0 if $|t| \geq t_{\alpha/2}$.

Level of significance (p-value):

> **Case 1.** p-value $= P(t \geq \text{computed } t)$
> **Case 2.** p-value $= P(t \leq \text{computed } t)$
> **Case 3.** p-value $= 2P(t \geq |\text{computed } t|)$

Recall that a denotes the area in the tail of the t distribution. For a one-tailed test with the probability of a Type I error equal to α, we locate the rejection region using the value from Table 3 in the Appendix, for $a = \alpha$ and df $= n - 1$. However, for a two-tailed test we would use the t-value from Table 3 corresponding to $a = \alpha/2$ and df $= n - 1$.

Thus, for a one-tailed test we reject the null hypothesis if the computed value of t is greater than the t-value from Table 3 in the Appendix, and $a = \alpha$ and df $= n - 1$. Similarly, for a two-tailed test we reject the null hypothesis if $|t|$ is greater than the t-value from Table 3 for $a = \alpha/2$ and df $= n - 1$.

EXAMPLE 5.15

A massive multistate outbreak of food-borne illness was attributed to *Salmonella enteritidis*. Epidemiologists determined that the source of the illness was ice cream. They sampled nine production runs from the company that had produced the ice cream to determine the level of *S. enteritidis* in the ice cream. These levels (MPN/g) are as follows:

.593　.142　.329　.691　.231　.793　.519　.392　.418

Use these data to determine whether the average level of *S. enteritidis* in the ice cream is greater than .3 MPN/g, a level that is considered to be very dangerous. Set $\alpha = .01$.

Solution The null and research hypotheses for this example are

H_0: $\mu \leq .3$
H_a: $\mu > .3$

Because the sample size is small, we need to examine whether the data appear to have been randomly sampled from a normal distribution. Figure 5.17 is a normal probability plot of the data values. All nine points fall nearly on the straight line. We conclude that the normality condition appears to be satisfied. Before setting up the rejection region and computing the value of the test statistic, we must first compute the sample mean and standard deviation. You can verify that

$\bar{y} = .456$　and　$s = .2128$

FIGURE 5.17

Normal probability plot for *Salmonella* data

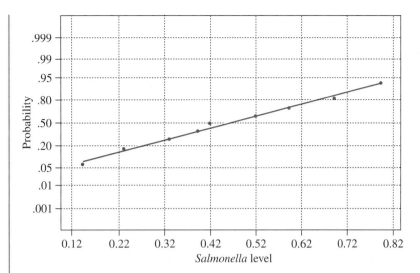

The rejection region with $\alpha = .01$ is

R.R.:   Reject H_0 if $t > 2.896$,

where, from Table 3 in the Appendix, the value of $t_{.01}$ with df $= 9 - 1 = 8$ is 2.896. The computed value of t is

$$t = \frac{\bar{y} - \mu_0}{s/\sqrt{n}} = \frac{.456 - .3}{.2128/\sqrt{9}} = 2.21$$

The observed value of t is not greater than 2.896, so we have insufficient evidence to indicate that the average level of *S. enteritidis* in the ice cream is greater than .3 MPN/g. The level of significance of the test is given by

$$p\text{-value} = P(t > \text{computed } t) = P(t > 2.21)$$

The t tables have only a few areas (a) for each value of df. The best we can do is bound the p-value. From Table 3 with df $= 8$, $t_{.05} = 1.860$ and $t_{.025} = 2.306$. Because computed $t = 2.21$, $.025 < p\text{-value} < .05$. However, with $\alpha = .01 < .025 < p\text{-value}$, we can still conclude that $p\text{-value} > \alpha$, and hence fail to reject H_0. The output from Minitab given here shows that the $p\text{-value} = .029$.

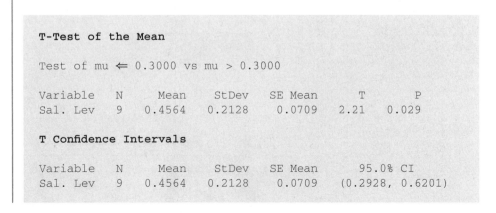

As we commented previously, in order to state that the level of *S. enteritidis* is less than or equal to .3, we need to calculate the probability of Type II error for some crucial values of μ in H_a. These calculations are somewhat more complex than the calculations for the z test. We will use a set of graphs to determine $\beta(\mu_a)$. The value of $\beta(\mu_a)$ depends on three quantities, df $= n - 1$, α, and the distance d from μ_a to μ_0 in σ units,

$$d = \frac{|\mu_a - \mu_0|}{\sigma}$$

Thus, to determine $\beta(\mu_a)$, we must specify α, μ_a, and provide an estimate of σ. Then with the calculated d and df $= n - 1$, we locate $\beta(\mu_a)$ on the graph. Table 4 in the Appendix provides graphs of $\beta(\mu_a)$ for $\alpha = .01$ and .05 for both one-sided and two-sided hypotheses for a variety of values for d and df.

EXAMPLE 5.16

Refer to Example 5.15. We have $n = 9$, $\alpha = .01$, and a one-sided test. Thus, df $= 8$ and if we estimate $\sigma \approx .25$, we can compute the values of d corresponding to selected values of μ_a. The values of $\beta(\mu_a)$ can then be determined using the graphs in Table 4 in the Appendix. Figure 5.18 is the necessary graph for this example. To illustrate the calculations, let $\mu_a = .45$. Then

$$d = \frac{|\mu_a - \mu_0|}{\sigma} = \frac{|.45 - .3|}{.25} = .6$$

We draw a vertical line from $d = .6$ on the horizontal axis to the line labeled 8, our df. We then locate the value on the vertical axis at the height of the intersection, .79.

FIGURE 5.18

Probability of Type II error curves $\alpha = .01$, one-sided

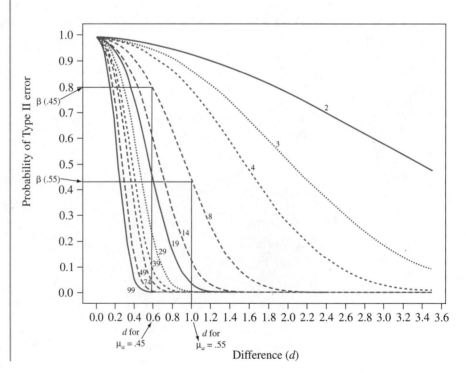

TABLE 5.5
Probability of Type II errors

μ_a	.35	.4	.45	.5	.55	.6	.65	.7	.75	.8
d	.2	.4	.6	.8	1.0	1.2	1.4	1.6	1.8	2.0
$\beta(\mu_a)$	.97	.91	.79	.63	.43	.26	.13	.05	.02	.00

Thus, $\beta(.45) = .79$. Similarly, to determine $\beta(.55)$, first compute $d = 1.0$, draw a vertical line from $d = 1.0$ to the line labeled 8, and locate .43 on the vertical axis. Thus, $\beta(.55) = .43$. Table 5.5 contains values of $\beta(\mu_a)$ for several values of μ_a. Because the values of $\beta(\mu_a)$ are large for values of μ_a that are considerably larger than $\mu_0 = .3$— for example, $\beta(.6) = .26$—we will not state that μ is less than or equal to .3, but will only state that the data fail to support the contention that μ is larger than .3.

In addition to being able to run a statistical test for μ when σ is unknown, we can construct a confidence interval using t. The confidence interval for μ with σ unknown is identical to the corresponding confidence interval for μ when σ is known, with z replaced by t and σ replaced by s.

100(1 − α)% Confidence Interval for μ, σ Unknown

$$\bar{y} \pm t_{\alpha/2}\frac{s}{\sqrt{n}}$$

Note: df $= n - 1$ and the confidence coefficient is $(1 - \alpha)$.

EXAMPLE 5.17

An airline wants to evaluate the depth perception of its pilots over the age of 50. A random sample of $n = 14$ airline pilots over the age of 50 are asked to judge the distance between two markers placed 20 feet apart at the opposite end of the laboratory. The sample data listed here are the pilots' error (recorded in feet) in judging the distance.

2.7   2.4   1.9   2.6   2.4   1.9   2.3
2.2   2.5   2.3   1.8   2.5   2.0   2.2

Use the sample data to place a 95% confidence interval on μ, the average error in depth perception for the company's pilots over the age of 50.

Solution   Before setting up a 95% confidence interval on μ, we must first assess the normality assumption by plotting the data in a normal probability plot or a boxplot. Figure 5.19 is a boxplot of the 14 data values. The median line is near the center of the box, the right and left whiskers are approximately the same length, and there are no outliers. The data appear to be a sample from a normal distribution. Thus, it is appropriate to construct the confidence interval based on the t distribution. You can verify that

$$\bar{y} = 2.26 \quad \text{and} \quad s = .28$$

Referring to Table 3 in the Appendix, the t-value corresponding to $a = .025$ and df $= 13$ is 2.160. Hence, the 95% confidence interval for μ is

$$\bar{y} \pm t_{\alpha/2}\frac{s}{\sqrt{n}} \quad \text{or} \quad 2.26 \pm 2.160\frac{.28}{\sqrt{14}}$$

FIGURE 5.19
Boxplot of distance
(with 95% *t* confidence
interval for the mean)

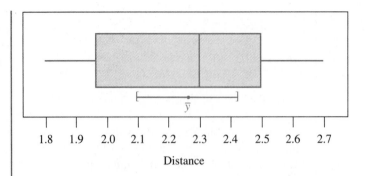

which is the interval 2.26 ± .16, or 2.10 to 2.42. Thus, we are 95% confident that the average error in the pilots' judgment of the distance is between 2.10 and 2.42 feet.

In this section, we have made the formal mathematical assumption that the population is normally distributed. *In practice, no population has exactly a normal distribution.* How does nonnormality of the population distribution affect inferences based on the *t* distribution?

There are two issues to consider when populations are assumed to be non-normal. First, what kind of nonnormality is assumed? Second, what possible effects do these specific forms of nonnormality have on the *t*-distribution procedures? The most important deviations from normality are **skewed distributions** and **heavy-tailed distributions.** Heavy-tailed distributions are roughly symmetric but have outliers relative to a normal distribution. Figure 5.20 displays four such distributions:

skewed distributions

heavy-tailed distributions

FIGURE 5.20

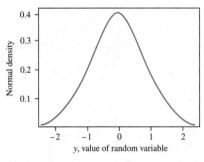

(a) Density of the standard normal distribution.

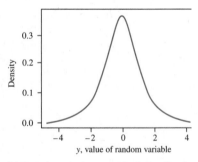

(b) Density of a heavy-tailed distribution.

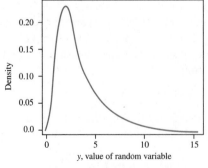

(c) Density of a lightly skewed distribution.

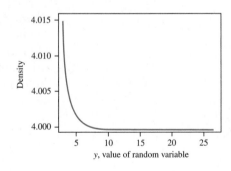

(d) Density of a highly skewed distribution.

(a) is the standard normal distribution, (b) is a heavy-tailed distribution (a t distribution with df = 3), (c) is a distribution mildly skewed to the right, and (d) is a distribution highly skewed to the right.

To evaluate the effect of nonnormality as exhibited by skewness or heavy tails, we will consider whether the t-distribution procedures are still approximately correct for these forms of nonnormality and whether there are other more efficient procedures. For example, even if a test procedure for μ based on the t distribution gave nearly correct results for, say, a heavy-tailed population distribution, it might be possible to obtain a test procedure with a more accurate probability of Type I error and greater power if we test hypotheses about the population median in place of the population μ. Also, in the case of heavy-tailed or highly skewed population distributions, the median rather than μ is a more appropriate representation of the population center.

The question of approximate correctness of t procedures has been studied extensively. In general, probabilities specified by the t procedures, particularly the confidence level for confidence intervals and the Type I error for statistical tests, have been found to be fairly accurate, even when the population distribution is heavy-tailed. However, when the population is very heavy-tailed, as is the case in Figure 5.20(b), the tests of hypotheses tend to have probability of Type I errors smaller than the specified level, which leads to a test having much lower power and hence greater chances of committing Type II errors. Skewness, particularly with small sample sizes, can have an even greater effect on the probability of both Type I and Type II errors. When we are sampling from a population distribution that is normal, the sampling distribution of a t statistic is symmetric. However, when we are sampling from a population distribution that is highly skewed, the sampling distribution of a t statistic is skewed, not symmetric. Although the degree of skewness decreases as the sample size increases, there is no procedure for determining the sample size at which the sampling distribution of the t statistic becomes symmetric.

As a consequence, the level of a nominal $\alpha = .05$ test may actually have a level of .01 or less when the sample size is less than 20 and the population distribution looks like that of Figure 5.20(b), (c), or (d). Furthermore, the power of the test will be considerably less than when the population distribution is a normal distribution, thus causing an increase in the probability of Type II errors. A simulation study of the effect of skewness and heavy-tailedness on the level and power of the t test yielded the results given in Table 5.6. The values in the table are the power values for a level $\alpha = .05$ t test of $H_0: \mu \leq \mu_0$ versus $H_a: \mu > \mu_0$. The power values are calculated for shifts of size $d = |\mu_a - \mu_0|/\sigma$ for values of $d = 0, .2, .6, .8$. Three different

TABLE 5.6
Level and power values
for t test

Population Distribution	$n = 10$ Shift d				$n = 15$ Shift d				$n = 20$ Shift d			
	0	**.2**	**.6**	**.8**	**0**	**.2**	**.6**	**.8**	**0**	**.2**	**.6**	**.8**
Normal	.05	.145	.543	.754	.05	.182	.714	.903	.05	.217	.827	.964
Heavy tailed	.035	.104	.371	.510	.049	.115	.456	.648	.045	.163	.554	.736
Light skewness	.025	.079	.437	.672	.037	.129	.614	.864	.041	.159	.762	.935
Heavy skewness	.007	.055	.277	.463	.006	.078	.515	.733	.011	.104	.658	.873

sample sizes were used: $n = 10, 15$, and 20. When $d = 0$, the level of the test is given for each type of population distribution. We want to compare these values to .05. The values when $d > 0$ are compared to the corresponding values when sampling from a normal population. We observe that when sampling from the lightly skewed distribution and the heavy-tailed distribution, the levels are somewhat less than .05 with values nearly equal to .05 when using $n = 20$. However, when sampling from a heavily skewed distribution, even with $n = 20$ the level is only .011. The power values for the heavy-tailed and heavily skewed populations are considerably less than the corresponding values when sampling from a normal distribution. Thus, the test is much less likely to correctly detect that the alternative hypothesis H_a is true. This reduced power is present even when $n = 20$. When sampling from a lightly skewed population distribution, the power values are very nearly the same as the values for the normal distribution.

robust methods

Because the t procedures have reduced power when sampling from skewed populations with small sample sizes, procedures have been developed that are not as affected by the skewness or extreme heavy-tailedness of the population distribution. These procedures are called **robust methods** of estimation and inference. Two robust procedures, the sign test and Wilcoxon signed rank test, will be considered in Section 5.8 and Chapter 6, respectively. They are both more efficient than the t test when the population distribution is very nonnormal in shape. Also, they maintain the selected α level of the test unlike the t test, which, when applied to very nonnormal data, has a true α value much different from the selected α value. The same comments can be made with respect to confidence intervals for the mean. When the population distribution is highly skewed, the coverage probability of a nominal $100(1 - \alpha)$ confidence interval is considerably less than $100(1 - \alpha)$.

So what is a nonexpert to do? First, examine the data through graphs. A boxplot or normal probability plot will reveal any gross skewness or extreme outliers. If the plots do not reveal extreme skewness or many outliers, the nominal t-distribution probabilities should be reasonably correct. Thus, the level and power calculations for tests of hypotheses and the coverage probability of confidence intervals should be reasonably accurate. If the plots reveal severe skewness or heavy-tailedness, the test procedures and confidence intervals based on the t distribution will be highly suspect. In these situations, the median is a more appropriate measure of the center of the population than is the mean. In Section 5.8, we will develop a test of hypotheses and confidence intervals for the median of a population. These procedures will maintain the nominal coverage probability of confidence intervals and the stated α level of tests of hypotheses when the population distribution is highly skewed or heavy-tailed.

EXERCISES   **Basic Techniques**

5.28   Set up the rejection region based on t for the following conditions with $\alpha = .05$:
   a. $H_a: \mu < \mu_0, n = 15$
   b. $H_a: \mu \neq \mu_0, n = 23$
   c. $H_a: \mu > \mu_0, n = 6$

5.29   Repeat Exercise 5.28 with $\alpha = .01$.

Applications

Edu.   **5.30**   A new reading program was being evaluated in the fourth grade at an elementary school. A random sample of 20 students was thoroughly tested to determine reading speed and reading comprehension. Based on a fixed-length standardized test reading passage, the following speeds (in minutes) and comprehension scores (based on a 100-point scale) were obtained.

Student	1	2	3	4	5	6	7	8	9	10	11	12	13	14	15	16	17	18	19	20	n	$\bar{y}$	s
Speed	5	7	15	12	8	7	10	11	9	13	10	6	11	8	10	8	7	6	11	8	20	9.10	2.573
Comprehension	60	76	76	90	81	75	95	98	88	73	90	66	91	83	100	85	76	69	91	78	20	82.05	10.88

a. Use the reading-speed data to place a 95% confidence interval on μ, the average reading speed, for all fourth-grade students in the large school from which the sample was selected.

b. Plot the reading-speed data using a normal probability plot or boxplot to evaluate whether the data set has any extreme outliers and/or gross skewness.

c. Interpret the interval estimate in part (a).

d. How would your inference change if you used a 98% confidence interval in place of the 95% confidence interval?

5.31 Refer to Exercise 5.30. Using the reading-comprehension data, is there significant evidence that the mean comprehension for all fourth graders is greater than 80, the statewide average for comparable students during the previous year? Give the level of significance for your test. Interpret your findings.

Bus. **5.32** A consumer testing agency wants to evaluate the claim made by a manufacturer of discount tires. The manufacturer claims that their tires can be driven at least 35,000 miles before wearing out. To determine the average number of miles that can be obtained from the manufacturer's tires, the agency randomly selects 60 tires from the manufacturer's warehouse and places the tires on 15 cars driven by test drivers on a 2-mile oval track. The number of miles driven (in thousands of miles) until the tires are determined to be worn out is given in the following table.

Car	1	2	3	4	5	6	7	8	9	10	11	12	13	14	15	n	$\bar{y}$	s
Miles driven	25	27	35	42	28	37	40	31	29	33	30	26	31	28	30	15	31.47	5.04

a. Place a 99% confidence interval on the average number of miles driven, μ, prior to the tires wearing out.

b. Is there significant evidence ($\alpha = .01$) that the manufacturer's claim is false? What is the level of significance of your test? Interpret your findings.

5.33 Refer to Exercise 5.32. Using the Minitab output given, compare your results to the results given by the computer program.

a. Does the normality assumption appear to be valid?

b. How close to the true value were your bounds on the p-value?

c. Is there a contradiction between the interval estimate of μ and the conclusion reached by your test of the hypotheses?

```
Test of mu = 35.00 vs mu < 35.00

Variable   N    Mean    StDev   SE Mean     T        P          99.0 % CI
Miles      15   31.47   5.04      1.30    -2.71   0.0084   (   27.59,   35.3
```

Boxplot of tire data

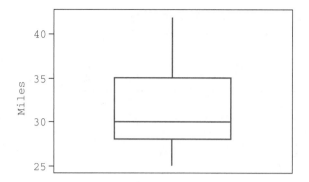

Test of normality for tire data

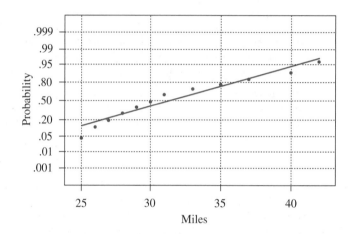

Env.   **5.34**   The amount of sewage and industrial pollutants dumped into a body of water affects the health of the water by reducing the amount of dissolved oxygen available for aquatic life. Over a 2-month period, 8 samples were taken from a river at a location 1 mile downstream from a sewage treatment plant. The amount of dissolved oxygen in the samples was determined and is reported in the following table. The current research asserts that the mean dissolved oxygen level must be at least 5.0 parts per million (ppm) for fish to survive.

Sample	1	2	3	4	5	6	7	8	n	$\bar{y}$	s
Oxygen (ppm)	5.1	4.9	5.6	4.2	4.8	4.5	5.3	5.2	8	4.95	.45

   a. Place a 95% confidence on the mean dissolved oxygen level during the 2-month period.
   b. Using the confidence interval from (a), does the mean oxygen level appear to be less than 5 ppm?
   c. Test the research hypothesis that the mean oxygen level is less than 5 ppm. What is the level of significance of your test? Interpret your findings.

Ag.   **5.35**   Commercial growers of ornamental shrubs often want to retard the growth of the shrubs so that they do not become too large prior to being sold. A growth retardant, dikegulac, was evaluated on kalanchoe, an ornamental shrub. The paper "Dikegulac alters growth and flowering of kalanchoe" [*HortScience* (1985) 20:722–724] describes the results of these experiments. Ten shrubs were treated with dikegulac and another ten shrubs were untreated so the effect of the dikegulac on plant growth could be determined. The heights (in cm) of the 20 plants were measured 13 weeks after the treatment date and the summary statistics are given in the following table.

	n	$\bar{y}$	s
Untreated	10	43.6	5.7
Treated	10	36.1	4.9

a. Construct 90% confidence intervals on the average height of the treated and untreated shrubs. Interpret these intervals.

b. Do the two confidence intervals overlap? What conclusions can you make about the effectiveness of dikegulac as a growth retardant?

5.8   Inferences about the Median

When the population distribution is highly skewed or very heavily tailed, the median is more appropriate than the mean as a representation of the center of the population. Furthermore, as was demonstrated in Section 5.7, the t procedures for constructing confidence intervals and for tests of hypotheses for the mean are not appropriate when applied to random samples from such populations with small sample sizes. In this section, we will develop a test of hypotheses and a confidence interval for the population median that will be appropriate for all types of population distributions.

The estimator of the population median M is based on the order statistics that were discussed in Chapter 3. Recall that if the measurements from a random sample of size n are given by $y_1, y_2, \ldots, y_n$, then the order statistics are these values ordered from smallest to largest. Let $y_{(1)} \leq y_{(2)} \leq \ldots \leq y_{(n)}$ represent the data in ordered fashion. Thus, $y_{(1)}$ is the smallest data value and $y_{(n)}$ is the largest data value. The estimator of the population median is the sample median $\hat{M}$. Recall that $\hat{M}$ is computed as follows.

If n is an odd number, then $\hat{M} = y_{(m)}$, where $m = (n + 1)/2$.
If n is an even number, then $\hat{M} = (y_{(m)} + y_{(m+1)})/2$, where $m = n/2$.

To take into account the variability of $\hat{M}$ as an estimator of M, we next construct a confidence interval for M. A confidence interval for the population median M may be obtained by using the binomial distribution with $\pi = 0.5$.

100(1 − α)% Confidence Interval for the Median

A confidence interval for M with level of confidence at least $100(1 - \alpha)\%$ is given by

$$(M_L, M_U) = (y_{(L_{\alpha/2})}, y_{(U_{\alpha/2})})$$

where

$$L_{\alpha/2} = C_{\alpha(2),n} + 1$$
$$U_{\alpha/2} = n - C_{\alpha(2),n}$$

Table 5 in the Appendix contains values for $C_{\alpha(2),n}$, which are percentiles from a binomial distribution with $\pi = 0.5$.

Because the confidence limits are computed using the binomial distribution, which is a discrete distribution, the level of confidence of (M_L, M_U) will generally be somewhat larger than the specified $100(1 - \alpha)\%$. The exact level of confidence is given by

$$\text{Level} = 1 - 2Pr[Bin(n, .5) \le C_{\alpha(2),n}]$$

The following example will demonstrate the construction of the interval.

EXAMPLE 5.18

The sanitation department of a large city wants to investigate ways to reduce the amount of recyclable materials that are placed in the city's landfill. By separating the recyclable material from the remaining garbage, the city could prolong the life of the landfill site. More important, the number of trees needed to be harvested for paper products and aluminum needed for cans could be greatly reduced. From an analysis of recycling records from other cities, it is determined that if the average weekly amount of recyclable material is more than 5 pounds per household a commercial recycling firm could make a profit collecting the material. To determine the feasibility of the recycling plan, a random sample of 25 households is selected. The weekly weight of recyclable material (in pounds/week) for each household is given here.

14.2	5.3	2.9	4.2	1.2	4.3	1.1	2.6	6.7	7.8	25.9	43.8	2.7
5.6	7.8	3.9	4.7	6.5	29.5	2.1	34.8	3.6	5.8	4.5	6.7	

Determine an appropriate measure of the amount of recyclable waste from a typical household in the city.

Solution A boxplot and normal probability of the recyclable waste data reveal the extreme right skewness of the data. Thus, the mean is not an appropriate representation of the typical household's potential recyclable material. The sample median and a confidence interval on the population are given by the following computations. First, we order the data from smallest value to largest value:

1.1	1.2	2.1	2.6	2.7	2.9	3.6	3.9	4.2	4.3	4.5	4.7	5.3
5.6	5.8	6.5	6.7	6.7	7.8	7.8	14.2	25.9	29.5	34.8	43.8	

Boxplot of recyclable wastes

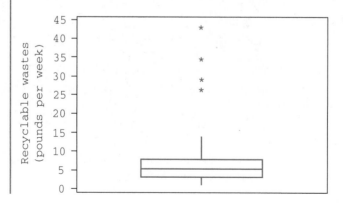

Normal probability plot
of recyclable wastes

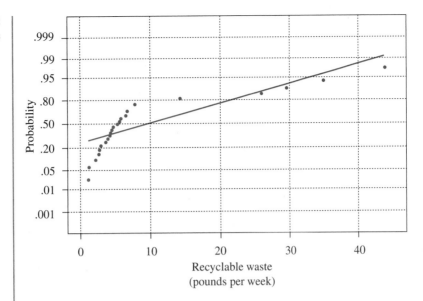

The number of values in the data set is an odd number, so the sample median is given by

$$\hat{M} = y_{((25+1)/2)} = y_{(13)} = 5.3$$

The sample mean is calculated to be $\bar{y} = 9.53$. Thus, we have that for 20 of the 25 households the weekly recyclable wastes are less than the sample mean. Note that 12 of the 25 waste values are less and 12 of the 25 are greater than the sample median. Thus, the sample median is more representative of the typical household's recyclable waste than is the sample mean. Next we will construct a 95% confidence interval for the population median.

From Appendix Table 5, we find

$$C_{\alpha(2),n} = C_{.05,25} = 7$$

Thus,

$$L_{.025} = C_{.05,25} + 1 = 8$$
$$U_{.025} = n - C_{.05,n} = 25 - 7 = 18$$

The 95% confidence interval for the population median is given by

$$(M_L, M_U) = (y_{(8)}, y_{(18)}) = (3.9, 6.7)$$

Using the binomial distribution, the exact level of coverage is given by

$$1 - 2Pr[Bin(25, .5) \le 7] = .957$$

which is slightly larger than the desired level 95%. Thus, we are at least 95% confident that the median amount of recyclable waste per household is between 3.9 and 6.7 pounds per week.

Large-Sample Approximation

When the sample size n is large, we can apply the normal approximation to the binomial distribution to obtain approximations to $C_{\alpha(2),n}$. The approximate value is given by

$$C_{\alpha(2),n} \approx \frac{n}{2} - z_{\alpha/2}\sqrt{\frac{n}{4}}$$

Because this approximate value for $C_{\alpha(2),n}$ is generally not an integer, we set $C_{\alpha(2),n}$ to be the largest integer that is less than or equal to the approximate value.

EXAMPLE 5.19

Using the data in Example 5.18, find a 95% confidence interval for the median using the approximation to $C_{\alpha(2),n}$.

Solution   We have $n = 25$ and $\alpha = .05$. Thus, $z_{.05/2} = 1.96$, and

$$C_{\alpha(2),n} \approx \frac{n}{2} - z_{\alpha/2}\sqrt{\frac{n}{4}} = \frac{25}{2} - 1.96\sqrt{\frac{25}{4}} = 7.6$$

Thus, we set $C_{\alpha(2),n} = 7$, and our confidence interval is identical to the interval constructed in Example 5.18. If n is larger than 30, the approximate and the exact values of $C_{\alpha(2),n}$ will often be the same integer.

In Example 5.18, the city wanted to determine whether the median amount of recyclable material was more than 5 pounds per household per week. We constructed a confidence interval for the median but we still have not answered the question of whether the median is greater than 5. Thus, we need to develop a test of hypotheses for the median.

We will use the ideas developed for constructing a confidence interval for the median in our development of the testing procedures for hypotheses concerning a population median. In fact, a $100(1 - \alpha)\%$ confidence interval for the population median M can be used to test two-sided hypotheses about M. If we want to test H_0: $M = M_0$ vs. H_a: $M \neq M_0$ at level α, then we construct a $100(1 - \alpha)\%$ confidence interval for M. If M_0 is contained in the confidence interval, then we fail to reject H_0. If M_0 is outside the confidence interval, then we reject H_0.

sign test   For testing one-sided hypotheses about M, we will use the binomial distribution to determine the rejection region. The testing procedure is called the **sign test** and is constructed as follows. Let $y_1, \ldots, y_n$ be a random sample from a population having median M. Let the null value of M be M_0 and define $W_i = y_i - M_0$. The sign test statistic B is the number of positive W_is. Note that B is simply the number of y_is that are greater than M_0. Because M is the population median, 50% of the data values are greater than M and 50% are less than M. Now, if $M = M_0$, then there is a 50% chance that y_i is greater than M_0 and hence a 50% chance that W_i is positive. Because the W_is are independent, each W_i has a 50% chance of being positive whenever $M = M_0$, and B counts the number of positive W_is under H_0. B is a binomial random variable with $\pi = .5$ and the percentiles from the binomial distribution with $\pi = .5$ given in Table 5 in the Appendix can be used to construct the rejection region for the test of hypothesis.

test for a population   The statistical **test for a population median** M is summarized next. Three dif-
median M   ferent sets of hypotheses are given with their corresponding rejection regions. The tests given are appropriate for any population distribution.

Summary of a Statistical Test for the Median M

Hypotheses:

    Case 1. H_0: $M \le M_0$ vs. H_a: $M > M_0$ (right-tailed test)
    Case 2. H_0: $M \ge M_0$ vs. H_a: $M < M_0$ (left-tailed test)
    Case 3. H_0: $M = M_0$ vs. H_a: $M \ne M_0$ (two-tailed test)

T.S.: Let $W_i = y_i - M_0$ and B = number of positive W_is.
R.R.: For a probability α of a Type I error,

    Case 1. Reject H_0 if $B \ge n - C_{\alpha(1),n}$.
    Case 2. Reject H_0 if $B \le C_{\alpha(1),n}$.
    Case 3. Reject H_0 if $B \le C_{\alpha(2),n}$ or $B \ge n - C_{\alpha(2),n}$.

The following example will illustrate the test of hypotheses for the population median.

EXAMPLE 5.20

Refer to Example 5.18. The Sanitation Department wanted to determine whether the median household recyclable wastes was greater than 5 pounds per week. Test this research hypothesis at level $\alpha = .05$ using the data from Exercise 5.18.

Solution The set of hypotheses are

    H_0: $M \le 5$ versus H_a: $M > 5$

The data set consisted of a random sample of $n = 25$ households. From Table 5 in the Appendix, we find $C_{\alpha(1),n} = C_{.05,25} = 7$. Thus, we will reject H_0: $M \le 5$ if $B \ge n - C_{\alpha(1),n} = 25 - 7 = 18$. Let $W_i = y_i - M_0 = y_i - 5$, which yields

−3.9	−3.8	−2.9	−2.4	−2.3	−2.1	−1.4	−1.1	−0.8
−0.7	−0.5	−0.3	0.3	0.6	0.8	1.5	1.7	1.7
2.8	2.8	9.2	20.9	24.5	29.8	38.8		

The 25 values of W_i contain 13 positive values. Thus, $B = 13$, which is not greater than 18. We conclude the data set fails to demonstrate that the median household level of recyclable waste is greater than 5 pounds.

Large-Sample Approximation

When the sample size n is larger than the values given in Table 5 in the Appendix, we can use the normal approximation to the binomial distribution to set the rejection region. The standardized version of the sign test is given by

$$B_{ST} = \frac{B - (n/2)}{\sqrt{n/4}}$$

When M equals M_0, B_{ST} has approximately a standard normal distribution. Thus, we have the following decision rules for the three different research hypotheses.

    Case 1. Reject H_0: $M \le M_0$ if $B_{ST} \ge z_\alpha$, with p-value = $Pr(z \ge B_{ST})$
    Case 2. Reject H_0: $M \ge M_0$ if $B_{ST} \le -z_\alpha$, with p-value = $Pr(z \le B_{ST})$
    Case 3. Reject H_0: $M = M_0$ if $|B_{ST}| \ge z_{\alpha/2}$, with p-value = $2Pr(z \ge |B_{ST}|)$

where z_α is the standard normal percentile.

EXAMPLE 5.21

Using the information in Example 5.20, construct the large-sample approximation to the sign test, and compare your results to those obtained using the exact sign test.

Solution   Refer to Example 5.20, in which we had $n = 25$ and $B = 13$. We conduct the large-sample approximation to the sign test as follows. We will reject H_0: $M \leq 5$ in favor of H_a: $M > 5$ if $B_{ST} \geq z_{.05} = 1.96$.

$$B_{ST} = \frac{B - (n/2)}{\sqrt{n/4}} = \frac{13 - (25/2)}{\sqrt{25/4}} = 0.2$$

Because B_{ST} is not greater than 1.96, we fail to reject H_0. The p-value $= Pr(z \geq 0.2)$ $= 1 - Pr(z < 0.2) = 1 - .5793 = .4207$ using Table 1 in the Appendix. Thus, we reach the same conclusion as was obtained using the exact sign test.

In Section 5.7, we observed that the performance of the t test deteriorated when the population distribution was either very heavily tailed or highly skewed. In Table 5.7, we compute the level and power of the sign test and compare these values to the comparable values for the t test for the four population distributions depicted in Figure 5.20 in Section 5.7. Ideally, the level of the test should remain the same for all population distributions. Also, we want tests having the largest possible power values because the power of a test is its ability to detect false null hypotheses. When the population distribution is either heavy tailed or highly skewed, the level of the t test changes from its stated value of .05. In these situations, the level of the sign test stays the same because the level of the sign test is the same for all distributions. The power of the t test is greater than the power of the sign test when sampling from a population having a normal distribution. However, the power of the sign test is greater than the power of the t test when sampling from very heavily tailed distributions or highly skewed distributions.

TABLE 5.7   Level and power values of the t test versus the sign test

Population Distribution	Test Statistic	n = 10 $(M_a - M_0)/\sigma$				n = 15 $(M_a - M_0)/\sigma$				n = 20 $(M_a - M_0)/\sigma$			
		Level	.2	.6	.8	Level	.2	.6	.8	Level	.2	.6	.8
Normal	t	.05	.145	.543	.754	.05	.182	.714	.903	.05	.217	.827	.964
	Sign	.055	.136	.454	.642	.059	.172	.604	.804	.058	.194	.704	.889
Heavy Tailed	t	.035	.104	.371	.510	.049	.115	.456	.648	.045	.163	.554	.736
	Sign	.055	.209	.715	.869	.059	.278	.866	.964	.058	.325	.935	.990
Lightly Skewed	t	.055	.140	.454	.631	.059	.178	.604	.794	.058	.201	.704	.881
	Sign	.025	.079	.437	.672	.037	.129	.614	.864	.041	.159	.762	.935
Highly Skewed	t	.007	.055	.277	.463	.006	.078	.515	.733	.011	.104	.658	.873
	Sign	.055	.196	.613	.778	.059	.258	.777	.912	.058	.301	.867	.964

EXERCISES   **Basic Techniques**

5.36   Suppose we have a random sample of n measurements from a population having median M. We want to place a 90% confidence interval on M.
     a. If $n = 20$, find $L_{\alpha/2}$ and $U_{\alpha/2}$ using Table 5 in the Appendix.
     b. Use the large-sample approximation to find $L_{\alpha/2}$ and $U_{\alpha/2}$ and compare these values to the values found in part (a).

5.37 Suppose we have a random sample of 30 measurements from a population having median M. We want to test $H_0: M \le M_0$ versus $H_a: M > M_0$ at level $\alpha = .05$. Set up the rejection region for testing these hypotheses using the values in Table 5 of the Appendix.

5.38 Refer to Exercise 5.37. Use the large-sample approximation to set up the rejection region and compare your results to the rejection region obtained in Exercise 5.37.

Applications

Bus.

5.39 The amount of money spent on health care is an important issue for workers because many companies provide health insurance that only partially covers many medical procedures. The director of employee benefits at a midsize company wants to determine the amount spent on health care by the typical hourly worker in the company. A random sample of 25 workers is selected and the amount they spent on their families' health care needs during the past year is given here.

| 400 | 345 | 248 | 1,290 | 398 | 218 | 197 | 342 | 208 | 223 | 531 | 172 | 4,321 |
| 143 | 254 | 201 | 3,142 | 219 | 276 | 326 | 207 | 225 | 123 | 211 | 108 | |

  a. Graph the data using a boxplot or normal probability plot and determine whether the population has a normal distribution.
  b. Based on your answer to part (a), is the mean or the median cost per household a more appropriate measure of what the typical worker spends on health care needs?
  c. Place a 95% confidence interval on the amount spent on health care by the typical worker. Explain what the confidence interval is telling us about the amount spent on health care needs.
  d. Does the typical worker spend more than $400 per year on health care needs? Use $\alpha = .05$.

Gov.

5.40 Many states have attempted to reduce the blood-alcohol level at which a driver is declared to be legally drunk. There has been resistance to this change in the law by certain business groups who have argued that the current limit is adequate. A study was conducted to demonstrate the effect on reaction time of a blood-alcohol level of .1%, the current limit in many states. A random sample of 25 people of legal driving age had their reaction time recorded in a standard laboratory test procedure before and after drinking a sufficient amount of alcohol to raise their blood alcohol to a .1% level. The difference (After − Before) in their reaction times in seconds was recorded as follows:

| .01 | .02 | .04 | .05 | .07 | .09 | .11 | .26 | .27 | .27 | .28 | .28 | .29 |
| .29 | .30 | .31 | .31 | .32 | .33 | .35 | .36 | .38 | .39 | .39 | .40 | |

  a. Place a 99% confidence interval on both the mean and median difference in reaction times of drivers who have a blood-alcohol level of .1%.
  b. Is there sufficient evidence that a blood-alcohol level of .1% causes any increase in the mean reaction time?
  c. Is there sufficient evidence that a blood-alcohol level of .1% causes any increase in the median reaction time?
  d. Which summary of reaction time differences seems more appropriate, the mean or median? Justify your answer.

5.41 Refer to Exercise 5.40. The lobbyist for the business group has their expert examine the experimental equipment and determines that there may be measurement errors in recording the reaction times. Unless the difference in reaction time is at least .25 seconds, the expert claims that the two times are essentially equivalent.
  a. Is there sufficient evidence that the median difference in reaction time is greater than .25 seconds?
  b. What other factors about the drivers are important in attempting to decide whether moderate consumption of alcohol affects reaction time?

5.9 **Encounters with Real Data: Percentage of Calories from Fat**

Defining the Problem (1)

Many studies have proposed relationships between diet and many diseases. For example, the percentage of calories from fat in the diet may be related to the incidence of certain types of cancer and heart disease. The assessment and quantification of a person's usual diet is crucial in evaluating the degree of relationship between diet and diseases. This is a very difficult task, but it is important in an effort to monitor dietary behavior among individuals. Rosner, Willett, and Spiegelman (1989) describe the Nurses' Health Study, which examined the diet of a large sample of women.

Collecting Data (2)

One of the objectives of the study was to determine the percentage of calories from fat in the diet of a population of women. There are many dietary assessment methodologies. The most commonly used method in large nutritional epidemiology studies is the food frequency questionnaire (FFQ), which uses a carefully designed series of questions to determine the dietary intakes of participants in the study. In the Nurses' Health Study, a sample of 168 women who represented a random sample from a population of female nurses completed a single FFQ. From the information gathered from the questionnaire, the percentage of calories from fat (PCF) was computed. The parameters of interest were the average value of PCF μ for the population of nurses, the standard deviation σ in PCF for the population of nurses, the proportion π of nurses having PCF greater than 50%, as well as other parameters. The complete data set, which contains the ages of the women and several other variables, may be found in the file PERCENTF on the CD that came with your book.

  The number of people needed in the study was determined by specifying the degree of accuracy in the estimation of the parameters μ, σ, and π. For this study, it was decided that a sample of 168 participants would be adequate.

Summarizing Data (3)

The researchers needed to carefully examine the data from the questionnaires to determine whether the responses were recorded correctly. The data were then transferred to computer files and prepared for analysis following the steps outlined in Section 2.7.

Analyzing Data, Interpreting the Analyses, and Communicating Results (4)

The next step in the study was to summarize the data through plots and summary statistics. The PCF values for the 168 women are displayed in Figure 5.21 in a stem-and-leaf diagram, along with a table of summary statistics.

FIGURE 5.21   The percentage of calories from fat (PCF) for 168 women in a dietary study

```
1  5
2  0 0 4 4
2  5 5 6 6 6 6 7 7 8 8 8 8 9 9 9 9 9 9 9 9
3  0 0 0 0 0 1 1 1 1 1 1 1 2 2 2 2 2 2 2 3 3 3 3 3 3 3 3 3 3 3 4 4 4 4 4 4 4 4 4
3  5 5 5 5 5 5 5 5 5 5 5 5 5 5 5 6 6 6 6 6 6 6 7 7 7 7 7 7 7 7 7 7 8 8 8 8 8 8 8 8 8 8 8 8 8 9 9 9 9 9 9 9 9 9
4  0 0 0 0 0 0 0 1 1 1 1 1 1 1 1 1 1 1 1 1 1 1 2 2 2 2 2 2 3 3 3 4 4 4 4 4
4  5 5 5 5 5 6 6 6 7 7 8 9 9
5  0 3 4
5  5 7
```

Descriptive Statistics for Percentage Calories from Fat Data

Variable	N	Mean	Median	TrMean	St Dev	SE Mean
PCF	168	36.919	36.473	36.847	6.728	0.519

Variable	Minimum	Maximum	Q1	Q3		
PCF	15.925	57.847	32.766	41.295		

From the stem-and-leaf plot, it appears that the data are nearly normally distributed with PCF values ranging from 15 to 57%. The proportion of the women that have PCF greater than 50% is $\hat{\pi} = 4/168 = 2.4\%$. From the table of summary statistics, the sample mean is $\bar{y} = 36.919$ and sample standard deviation is $s = 6.728$. The researchers wanted to draw inferences from the random sample of 168 women to the population from which they were selected. Thus, we need to place bounds on our point estimates to reflect our degree of confidence in their estimation of the population values. Also, the researchers were interested in testing hypotheses about the size of the population mean PCF μ or variance σ^2. For example, many nutritional experts recommend that a person's daily diet have no more than 30% of total calories a day from fat. Thus, we want to test the statistical hypothesis that μ is greater than 30 to determine whether the average value of PCF for this population exceeds the recommended value. The normal probability plot in Figure 5.22

FIGURE 5.22

Normal probability plot for percentage of calories from fat (PCF)

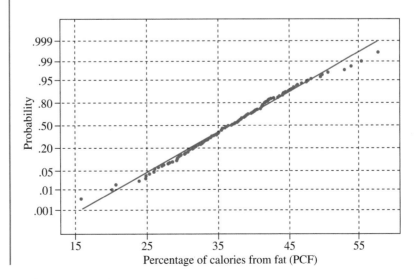

confirms what we observed with the stem-and-leaf plot, namely, that the data appear to follow a normal distribution.

The mean and standard deviation of the PCF data were given by $\bar{y} = 36.92$ and $s = 6.73$. We can next construct a 95% confidence interval for the mean PCF for the population of nurses as follows:

$$36.92 \pm t_{.025,167}\frac{6.73}{\sqrt{168}}, \ 36.92 \pm 1.974\frac{6.73}{\sqrt{168}}, \text{ or } 36.92 \pm 1.02$$

We are 95% confident that the mean PCF in the population of nurses is between 35.90 and 37.94. Thus, there is evidence that the mean PCF for the population of nurses exceeds the recommended value of 30. We next formally test the hypotheses

$$H_0: \mu \le 30 \text{ versus } H_a: \mu > 30$$

Because the data appear to be normally distributed and, in any case, the sample size is large, we can use the t test with rejection region as follows:

R.R.: For a one-tailed t test with $\alpha = 0.5$, we reject H_0 if

$$t = \frac{\bar{y} - 30}{s/\sqrt{168}} \ge t_{.05,167} = 1.654$$

Because $t = (36.92 - 30)/(6.73/\sqrt{168}) = 13.33$, we reject H_0. The p-value of the test is essentially 0, so we can conclude that the mean PCF value is significantly greater than 30.

Based on the confidence interval and backed up by the statistical test, there is strong evidence that the population of nurses has an average PCF larger than the recommended value of 30. The experts in this field would have to determine the practical consequences of having a PCF value between 5.90 and 7.94 units higher than the recommended value.

A report summarizing our findings from the study would include the following items:

1. Statement of the objective for the study
2. Description of the study design and data collection procedures
3. Numerical and graphical summaries of data sets
4. Description of all inference methodologies:
   - t tests
   - t-based confidence interval on the population mean
   - Verification that all necessary conditions for using inference techniques were satisfied
5. Discussion of results and conclusions
6. Interpretation of findings relative to previous studies
7. Recommendations for future studies
8. Listing of the data set

5.10 Staying Focused

A population mean or median can be estimated using point or interval estimation. The selection of the median in place of the mean as a representation of the center of a population depends on the shape of the population distribution. The performance of an interval estimate is determined by the width of the interval and the confidence coefficient. The formulas for a $100(1 - \alpha)\%$ confidence interval for the mean μ and median M were given. A formula was provided for determining the necessary sample size in a study so that a confidence interval for μ will have a predetermined width and level of confidence.

Following the traditional approach to hypothesis testing, a statistical test consists of five parts: research hypothesis, null hypothesis, test statistic, rejection region, and checking assumptions and drawing conclusions. A statistical test employs the technique of proof by contradiction. We conduct experiments and studies to gather data to verify the research hypothesis through the contradiction of the null hypothesis H_0. As with any two-decision process based on variable data, there are two types of errors that can be committed. A Type I error is the rejection of H_0 when H_0 is true and a Type II error is the acceptance of H_0 when the alternative hypothesis H_a is true. The probability for a Type I error is denoted by α. For a given value of the mean μ_a in H_a, the probability of a Type II error is denoted by $\beta(\mu_a)$. The value of $\beta(\mu_a)$ decreases as the distance from μ_a to μ_0 increases. The power of a test of hypothesis is the probability that the test will reject H_0 when the value of μ resides in H_a. Thus, the power at μ_a equals $1 - \beta(\mu_a)$.

We also demonstrated that for a given sample size and value of the mean μ_a, α and $\beta(\mu_a)$ are inversely related; as α increases, $\beta(\mu_a)$ decreases, and vice versa. If we specify the sample size n and α for a given test procedure, we can compute $\beta(\mu_a)$ for values of the mean μ_a in the alternative hypothesis. In many studies, we need to determine the necessary sample size n to achieve a testing procedure having a specified value for α and a bound on $\beta(\mu_a)$. A formula is provided to determine n such that a level α test has $\beta(\mu_a) \leq \beta$ whenever μ_a is a specified distance beyond μ_0.

We developed an alternative to the traditional decision-based approach for a statistical test of hypotheses. Rather than relying on a preset level of α, we compute the weight of evidence in the data for rejecting the null hypothesis. This weight, expressed in terms of a probability, is called the level of significance for the test. Most professional journals summarize the results of a statistical test using the level of significance. We discussed how the level of significance can be used to obtain the same results as the traditional approach.

We also considered inferences about μ when σ is unknown (which is the usual situation). Through the use of the t distribution, we can construct both confidence intervals and a statistical test for μ. The t-based tests and confidence intervals do not have the stated levels or power when the population distribution is highly skewed or very heavy tailed and the sample size is small. In these situations, we may use the median in place of the mean to represent the center of the population. Procedures were provided to construct confidence intervals and tests of hypothesis for the population median.

Key Formulas

Estimation and tests for μ and the median

1. $100(1 - \alpha)\%$ confidence interval for μ (σ known) when sampling from a normal population or n large

$$\bar{y} \pm z_{\alpha/2}\sigma_{\bar{y}}, \text{ where } \sigma_{\bar{y}} = \sigma/\sqrt{n}$$

2. $100(1 - \alpha)\%$ confidence interval for μ (σ unknown) when sampling from a normal population or n large

$$\bar{y} \pm t_{\alpha/2}s/\sqrt{n}, \qquad df = n - 1$$

3. Sample size for estimating μ with a $100(1 - \alpha)\%$ confidence interval, $\bar{y} \pm E$

$$n = \frac{(z_{\alpha/2})^2\sigma^2}{E^2}$$

4. Statistical test for μ (σ known) when sampling from a normal population or n large

$$\text{Test statistic: } z = \frac{\bar{y} - \mu_0}{s/\sqrt{n}}$$

5. Statistical test for μ (σ unknown) when sampling from a normal population or n large

$$\text{Test statistic: } t = \frac{\bar{y} - \mu_0}{\sigma/\sqrt{n}}, \qquad df = n - 1$$

6. Calculation of $\beta(\mu_a)$ (and equivalent power) for a test on μ (σ known) when sampling from a normal population or n large

  a. One-tailed level α test

$$\beta(\mu_a) = P\left(z < z_\alpha - \frac{|\mu_0 - \mu_a|}{\sigma_{\bar{y}}}\right)$$

  where $\sigma_{\bar{y}} = \sigma/\sqrt{n}$

  b. Two-tailed level α test

$$\beta(\mu_a) \approx P\left(z < z_{\alpha/2} - \frac{|\mu_0 - \mu_a|}{\sigma_{\bar{y}}}\right)$$

  where $\sigma_{\bar{y}} = \sigma/\sqrt{n}$

7. Calculation of $\beta(\mu_a)$ (and equivalent power) for a test on μ (σ unknown) when sampling from a normal population or n large: Use Table 4 in the Appendix.

8. Sample size n for a statistical test on μ (σ known) when sampling from a normal population

  a. One-tailed level α test

$$n = \frac{\sigma^2}{\Delta^2}(z_\alpha + z_\beta)^2$$

  b. Two-tailed level α test

$$n \approx \frac{\sigma^2}{\Delta^2}(z_{\alpha/2} + z_{\beta/2})^2$$

9. $100(1 - \alpha)\%$ confidence interval for the population median M

$$(y_{(L_{\alpha/2})}, y_{(U_{\alpha/2})}), \qquad \text{where } L_{\alpha/2} = C_{\alpha(2),n} + 1 \quad \text{and} \quad U_{\alpha/2} = n - C_{\alpha(2),n}$$

10. Statistical test for median

Test statistic:

Let $W_i = y_i - M_0$ and B = number of positive W_is

Supplementary Exercises

H.R.   **5.42**   An office manager has implemented an incentive plan that she thinks will reduce the mean time required to handle a customer complaint. The mean time for handling a complaint was 30 minutes prior to implementing the incentive plan. After the plan was in place for several months, a random sample of the records of 38 customers who had complaints revealed a mean time of 28.7 minutes with a standard deviation of 3.8 minutes.
   a. Give a point estimate of the mean time required to handle a customer complaint.
   b. What is the standard deviation of the point estimate given in (a)?
   c. Construct a 95% confidence on the mean time to handle a complaint after implementing the plan. Interpret the confidence interval for the office manager.
   d. Is there sufficient evidence that the incentive plan has reduced the mean time to handle a complaint?

Env.   **5.43**   The concentration of mercury in a lake has been monitored for a number of years. Measurements taken on a weekly basis yielded an average of 1.20 mg/m^3 (milligrams per cubic meter) with a standard deviation of .32 mg/m^3. Following an accident at a smelter on the shore of the lake, 15 measurements produced the following mercury concentrations.

1.60	1.77	1.61	1.08	1.07	1.79	1.34	1.07
1.45	1.59	1.43	2.07	1.16	0.85	2.11	

   a. Give a point estimate of the mean mercury concentration after the accident.
   b. Construct a 95% confidence interval on the mean mercury concentration after the accident. Interpret this interval.
   c. Is there sufficient evidence that the mean mercury concentration has increased since the accident? Use $\alpha = .05$.
   d. Assuming that the standard deviation of the mercury concentration is .32 mg/m^3, calculate the power of the test to detect mercury concentrations of 1.28, 1.32, 1.36, and 1.40.

Med.   **5.44**   Over the years, projected due dates for expectant mothers have been notoriously bad at a large metropolitan hospital. The physicians attended an in-service program to develop techniques to improve their projections. In a recent survey of 100 randomly selected mothers who had delivered a baby at the hospital since the in-service, the average number of days to birth beyond the projected due date was 9.2 days with a standard deviation of 12.4 days.
   a. Describe how to select the random sample of 100 mothers.
   b. Estimate the mean number of days to birth beyond the due date using a 95% confidence interval. Interpret this interval.
   c. If the mean number of days to birth beyond the due date was 13 days prior to the in-service, is there substantial evidence that the mean has been reduced? What is the level of significance of the test?
   d. What factors may be important in explaining why the doctors' projected due dates are not closer to the actual delivery dates?

Bus.   **5.45**   Statistics has become a valuable tool for auditors, especially where large inventories are involved. It would be costly and time consuming for an auditor to inventory each item in a large operation. Thus, the auditor frequently resorts to obtaining a random sample of items and using the sample results to check the validity of a company's financial statement. For example, a hospital financial statement claims an inventory that averages $300 per item. An auditor's random sample of 20 items yielded a mean and standard deviation of $160 and $90, respectively. Do the data contradict the hospital's claimed mean value per inventoried item and indicate that the average is less than $300? Use $\alpha = .05$.

Bus. **5.46** Over the past 5 years, the mean time for a warehouse to fill a buyer's order has been 25 minutes. Officials of the company believe that the length of time has increased recently, either due to a change in the workforce or due to a change in customer purchasing policies. The processing time (in minutes) was recorded for a random sample of 15 orders processed over the past month.

28	25	27	31	10
26	30	15	55	12
24	32	28	42	38

Do the data present sufficient evidence to indicate that the mean time to fill an order has increased?

Engin. **5.47** If a new process for mining copper is to be put into full-time operation, it must produce an average of more than 50 tons of ore per day. A 15-day trial period gave the results shown in the accompanying table.

Day	1	2	3	4	5	6	7	8	9	10	11	12	13	14	15
Yield (tons)	57.8	58.3	50.3	38.5	47.9	157.0	38.6	140.2	39.3	138.7	49.2	139.7	48.3	59.2	49.7

a. Estimate the typical amount of ore produced by the mine using both a point estimate and a 95% confidence interval.
b. Is there significant evidence that on a typical day the mine produces more than 50 tons of ore? Test by using $\alpha = .05$.

Env. **5.48** The board of health of a particular state was called to investigate claims that raw pollutants were being released into the river flowing past a small residential community. By applying financial pressure, the state was able to get the violating company to make major concessions toward the installation of a new water purification system. In the interim, different production systems were to be initiated to help reduce the pollution level of water entering the river. To monitor the effect of the interim system, a random sample of 50 water specimens was taken throughout the month at a location downstream from the plant. If $\bar{y} = 5.0$ and $s = .70$, use the sample data to determine whether the mean dissolved oxygen count of the water (in ppm) is less than 5.2, the average reading at this location over the previous year.
a. List the five parts of the statistical test, using $\alpha = .05$.
b. Conduct the statistical test and state your conclusion.

Env. **5.49** The search for alternatives to oil as a major source of fuel and energy will inevitably bring about many environmental challenges. These challenges will require solutions to problems in such areas as strip mining and many others. Let us focus on one. If coal is considered as a major source of fuel and energy, we will have to consider ways to keep large amounts of sulfur dioxide (SO_2) and particulates from getting into the air. This is especially important at large government and industrial operations. Here are some possibilities.

1. Build the smokestack extremely high.
2. Remove the SO_2 and particulates from the coal prior to combustion.
3. Remove the SO_2 from the gases after the coal is burned but before the gases are released into the atmosphere. This is accomplished by using a scrubber.

A new type of scrubber has been recently constructed and is set for testing at a power plant. Over a 15-day period, samples are obtained three times daily from gases emitted from the stack. The amounts of SO_2 emissions (in pounds per million BTU) are given here:

Time	\multicolumn{15}{c}{Day}
	1	2	3	4	5	6	7	8	9	10	11	12	13	14	15
6 A.M.	.158	.129	.176	.082	.099	.151	.084	.155	.163	.077	.116	.132	.087	.134	.179
2 P.M.	.066	.135	.096	.174	.179	.149	.164	.122	.063	.111	.059	.118	.134	.066	.104
10 P.M.	.128	.172	.106	.165	.163	.200	.228	.129	.101	.068	.100	.119	.125	.182	.138

a. Estimate the average amount of SO_2 emissions during each of the three time periods using 95% confidence intervals.
b. Does there appear to be a significant difference in average SO_2 emissions over the three time periods?
c. Combining the data over the entire day, is the average SO_2 emissions using the new scrubber less than .145, the average daily value for the old scrubber?

Soc. **5.50** As part of an overall evaluation of training methods, an experiment was conducted to determine the average exercise capacity of healthy male army inductees. To do this, each male in a random sample of 35 healthy army inductees exercised on a bicycle ergometer (a device for measuring work done by the muscles) under a fixed work load until he tired. Blood pressure, pulse rates, and other indicators were carefully monitored to ensure that no one's health was in danger. The exercise capacities (mean time, in minutes) for the 35 inductees are listed here.

```
23  19  36  12  41  43  19
28  14  44  15  46  36  25
35  25  29  17  51  33  47
42  45  23  29  18  14  48
21  49  27  39  44  18  13
```

a. Use these data to construct a 95% confidence interval for μ, the average exercise capacity for healthy male inductees. Interpret your findings.
b. How would your interval change using a 99% confidence interval?

5.51 Using the data in Exercise 5.50, determine the number of sample observations that would be required to estimate μ to within 1 minute, using a 95% confidence interval. (*Hint:* Substitute $s = 12.36$ for σ in your calculations.)

H.R. **5.52** Faculty members in a state university system who resign within 10 years of initial employment are entitled to receive the money paid into a retirement system, plus 4% per year. Unfortunately, experience has shown that the state is extremely slow in returning this money. Concerned about such a practice, a local teachers' organization decides to investigate. From a random sample of 50 employees who resigned from the state university system over the previous 5 years, the average time between the termination date and reimbursement was 75 days, with a standard deviation of 15 days. Use the data to estimate the mean time to reimbursement, using a 95% confidence interval.

5.53 Refer to Exercise 5.52. After a confrontation with the teachers' union, the state promised to make reimbursements within 60 days. Monitoring of the next 40 resignations yields an average of 58 days, with a standard deviation of 10 days. If we assume that these 40 resignations represent a random sample of the state's future performance, estimate the mean reimbursement time, using a 99% confidence interval.

Bus. **5.54** Improperly filled orders are a costly problem for mail-order houses. To estimate the mean loss per incorrectly filled order, a large firm plans to sample n incorrectly filled orders and to determine the added cost associated with each one. The firm estimates that the added cost is between $40 and $400. How many incorrectly filled orders must be sampled to estimate the mean additional cost using a 95% confidence interval of width $20?

5.55 Refer to Section 5.9, Encounters with Real Data.

a. 1. What is the population of interest?
    2. What dietary variables other than PCF might affect a person's health?
    3. What characteristics of the nurses other than dietary intake might be important in studying the nurses' health condition?
    4. Describe a method for randomly selecting which nurses participate in the study.
    5. State several hypotheses that may be of interest to the researchers.

b. Compile a report of this study, including the eight terms identified as appropriate for communicating the results.

CHAPTER 6

Inferences Comparing Two Population Central Values

6.1 Introduction

The inferences we have made so far have concerned a parameter from a single population. Quite often we are faced with an inference involving a comparison of parameters from different populations. We might wish to compare the mean corn crop yield for two different varieties of corn, the mean annual income for two ethnic groups, the mean nitrogen content of two different lakes, or the mean length of time between administration and eventual relief for two different antivertigo drugs.

In many sampling situations, we will select independent random samples from two populations to compare the populations' parameters. The statistics used to make these inferences will, in many cases, be the difference between the corresponding sample statistics. Suppose we select independent random samples of n_1 observations from one population and n_2 observations from a second population. We use the difference between the sample means, $(\bar{y}_1 - \bar{y}_2)$, to make an inference about the difference between the population means, $(\mu_1 - \mu_2)$.

The following theorem will help in finding the sampling distribution for the difference between sample statistics computed from independent random samples.

THEOREM 6.1

If two independent random variables y_1 and y_2 are normally distributed with means and variances (μ_1, σ_1^2) and (μ_2, σ_2^2), respectively, the difference $(y_1 - y_2)$ between the random variables is normally distributed with mean equal to $(\mu_1 - \mu_2)$ and variance equal to $(\sigma_1^2 + \sigma_2^2)$. Similarly, the sum $(y_1 + y_2)$ of the random variables is also normally distributed with mean $(\mu_1 + \mu_2)$ and variance $(\sigma_1^2 + \sigma_2^2)$.

Theorem 6.1 can be applied directly to find the sampling distribution of the difference between two independent sample means or two independent sample proportions. The Central Limit Theorem (discussed in Chapter 4) implies that if two random samples of sizes n_1 and n_2 are independently selected from two populations 1 and 2, then, where n_1 and n_2 are large, the sampling distributions of $\bar{y}_1$ and $\bar{y}_2$ will be approximately normal, with means and variances $(\mu_1, \sigma_1^2/n_1)$ and $(\mu_2, \sigma_2^2/n_2)$, respectively. Consequently, because $\bar{y}_1$ and $\bar{y}_2$ are independent, normally distributed random variables, it follows from Theorem 6.1 that the sampling distribution for the difference in the sample means, $(\bar{y}_1 - \bar{y}_2)$, is approximately normal, with a mean

$$\mu_{\bar{y}_1 - \bar{y}_2} = \mu_1 - \mu_2$$

variance

$$\sigma_{\bar{y}_1 - \bar{y}_2}^2 = \sigma_{\bar{y}_1}^2 + \sigma_{\bar{y}_2}^2 = \frac{\sigma_1^2}{n_1} + \frac{\sigma_2^2}{n_2}$$

and a standard error

$$\sigma_{\bar{y}_1 - \bar{y}_2} = \sqrt{\frac{\sigma_1^2}{n_1} + \frac{\sigma_2^2}{n_2}}$$

The sampling distribution of the difference between two independent, normally distributed sample means is shown in Figure 6.1.

Properties of the Sampling Distribution for the Difference between Two Sample Means, $(\bar{y}_1 - \bar{y}_2)$

1. The sampling distribution of $(\bar{y}_1 - \bar{y}_2)$ is approximately normal for large samples.
2. The mean of the sampling distribution, $\mu_{\bar{y}_1 - \bar{y}_2}$, is equal to the difference between the population means, $(\mu_1 - \mu_2)$.
3. The standard error of the sampling distribution is

$$\sigma_{\bar{y}_1 - \bar{y}_2} = \sqrt{\frac{\sigma_1^2}{n_1} + \frac{\sigma_2^2}{n_2}}$$

The sampling distribution for the difference between two sample means, $(\bar{y}_1 - \bar{y}_2)$, can be used to answer the same types of questions as we asked about the

FIGURE 6.1

Sampling distribution for the difference between two sample means

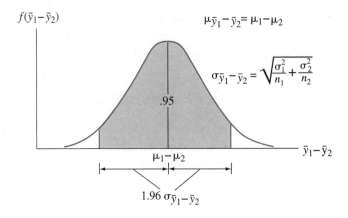

sampling distribution for $\bar{y}$ in Chapter 4. Because sample statistics are used to make inferences about corresponding population parameters, we can use the sampling distribution of a statistic to calculate the probability that the statistic will be within a specified distance of the population parameter. For example, we could use the sampling distribution of the difference in sample means to calculate the probability that $(\bar{y}_1 - \bar{y}_2)$ will be within a specified distance of the unknown difference in population means $(\mu_1 - \mu_2)$. Inferences (estimations or tests) about $(\mu_1 - \mu_2)$ will be discussed in succeeding sections of this chapter.

6.2   Inferences about $\mu_1 - \mu_2$: Independent Samples

In situations where we are making inferences about $\mu_1 - \mu_2$ based on random samples independently selected from two populations, we will consider three cases:

> **Case 1.** Both population distributions are normally distributed with $\sigma_1 = \sigma_2$.
> **Case 2.** Both sample sizes n_1 and n_2 are large.
> **Case 3.** The sample sizes n_1 or n_2 are small and the population distributions are nonnormal.

In this section, we will consider the situation in which we are independently selecting random samples from two populations that have normal distributions with different means μ_1 and μ_2 but identical standard deviations $\sigma_1 = \sigma_2 = \sigma$. The data will be summarized into the statistics: the sample means $\bar{y}_1$ and $\bar{y}_2$, and sample standard deviations s_1 and s_2. We will compare the two populations by constructing appropriate graphs, confidence intervals for $\mu_1 - \mu_2$, and tests of hypotheses concerning the difference $\mu_1 - \mu_2$.

A logical point estimate for the difference in population means is the sample difference $\bar{y}_1 - \bar{y}_2$. The standard error for the difference in sample means is more complicated than for a single sample mean, but the confidence interval has the same form: point estimate $\pm t_{\alpha/2}$ (standard error). A general confidence interval for $\mu_1 - \mu_2$ with confidence level of $(1 - \alpha)$ is given here.

Confidence Interval for $\mu_1 - \mu_2$, Independent Samples

$$(\bar{y}_1 - \bar{y}_2) \pm t_{\alpha/2} s_p \sqrt{\frac{1}{n_1} + \frac{1}{n_2}}$$

where

$$s_p = \sqrt{\frac{(n_1 - 1)s_1^2 + (n_2 - 1)s_2^2}{n_1 + n_2 - 2}} \quad \text{and} \quad df = n_1 + n_2 - 2$$

The sampling distribution of $\bar{y}_1 - \bar{y}_2$ is a normal distribution, with standard deviation

$$\sigma_{\bar{y}_1 - \bar{y}_2} = \sqrt{\frac{\sigma_1^2}{n_1} + \frac{\sigma_2^2}{n_2}} = \sqrt{\frac{\sigma^2}{n_1} + \frac{\sigma^2}{n_2}} = \sigma \sqrt{\frac{1}{n_1} + \frac{1}{n_2}}$$

because we require that the two populations have the same standard deviation σ. If we knew the value of σ, then we could use $z_{\alpha/2}$ in the formula for the confidence

s_p^2, a weighted average

interval. Because σ is unknown in most cases, we must estimate its value. This estimate is denoted by s_p and is formed by combining (pooling) the two independent estimates of σ, s_1, and s_2. In fact, s_p^2 is **a weighted average** of the sample variances s_1^2 and s_2^2. We have to estimate the standard deviation of the point estimate of $\mu_1 - \mu_2$, so we must use the percentile from the t distribution $t_{\alpha/2}$ in place of the normal percentile, $z_{\alpha/2}$. The degrees of freedom for the t-percentile are df $= n_1 + n_2 - 2$, because we have a total of $n_1 + n_2$ data values and two parameters μ_1 and μ_2 that must be estimated prior to estimating the standard deviation σ. Remember that we use $\bar{y}_1$ and $\bar{y}_2$ in place of μ_1 and μ_2, respectively, in the formulas for s_1^2 and s_2^2.

Recall that we are assuming that the two populations from which we draw the samples have normal distributions with a common variance σ^2. If the confidence interval presented were valid only when these assumptions were met exactly, the estimation procedure would be of limited use. Fortunately, the confidence coefficient remains relatively stable if both distributions are mound-shaped and the sample sizes are approximately equal. For situations in which these conditions do not hold, we will discuss alternative procedures in this section and in Section 6.3.

EXAMPLE 6.1

Company officials were concerned about the length of time a particular drug product retained its potency. A random sample of $n_1 = 10$ bottles of the product was drawn from the production line and analyzed for potency.

A second sample of $n_2 = 10$ bottles was obtained and stored in a regulated environment for a period of 1 year. The readings obtained from each sample are given in Table 6.1.

TABLE 6.1
Potency reading for two samples

Fresh		Stored	
10.2	10.6	9.8	9.7
10.5	10.7	9.6	9.5
10.3	10.2	10.1	9.6
10.8	10.0	10.2	9.8
9.8	10.6	10.1	9.9

Suppose we let μ_1 denote the mean potency for all bottles that might be sampled coming off the production line and μ_2 denote the mean potency for all bottles that might be retained for a period of 1 year. Estimate $\mu_1 - \mu_2$ by using a 95% confidence interval.

Solution   The potency readings for the fresh and stored bottles are plotted in Figures 6.2(a) and (b) in normal probability plots to assess the normality assumption. We find that the plotted points in both plots fall very close to a straight line, and hence the normality condition appears to be satisfied for both types of bottles. The summary statistics for the two samples are presented next.

Fresh Bottles	**Stored Bottles**
$n_1 = 10$	$n_2 = 10$
$\bar{y}_1 = 10.37$	$\bar{y}_2 = 9.83$
$s_1 = .3234$	$s_2 = .2406$

FIGURE 6.2
(a) Normal probability plot: potency of fresh bottles;
(b) Normal probability plot: potency of stored bottles

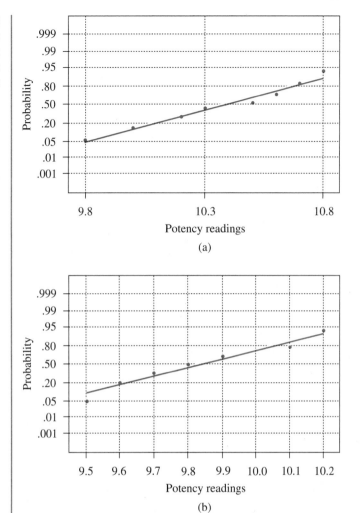

(a)

(b)

In Chapter 7, we will provide a test of equality for two population variances. For the data here, the computed sample standard deviations are approximately equal, considering the small sample sizes. Thus, the required conditions necessary to construct a confidence interval on $(\mu_1 - \mu_2)$—that is, normality, equal variances, and independent random samples—appear to be satisfied. The estimate of the common standard deviation σ is

$$s_p = \sqrt{\frac{(n_1 - 1)s_1^2 + (n_2 - 1)s_2^2}{n_1 + n_2 - 2}} = \sqrt{\frac{9(.3234)^2 + 9(.2406)^2}{18}} = .285$$

The t-value based on df $= n_1 + n_2 - 2 = 18$ and $a = .025$ is 2.101. A 95% confidence interval for the difference in mean potencies is

$$(10.37 - 9.83) \pm 2.101(.285)\sqrt{1/10 + 1/10}$$

$$.54 \pm .268 \quad \text{or} \quad (.272, .808)$$

We estimate that the difference in mean potencies for the bottles from the production line and those stored for 1 year, $\mu_1 - \mu_2$, lies in the interval .272 to .808. Company officials then had to evaluate whether a decrease in mean potency of size between .272 and .808 has a practical impact on the useful potency of the drug.

EXAMPLE 6.2

A school district decided that the number of students attending their high school was nearly unmanageable, so it was split into two districts, with District 1 students going to the old high school and District 2 students going to a newly constructed building. A group of parents became concerned with how the two districts were constructed relative to income levels. A study was thus conducted to determine whether people in suburban District 1 had a different mean income from those in District 2. A random sample of 20 homeowners was taken in District 1. Although 20 homeowners were to be interviewed in District 2 also, one person refused to provide the information requested, even though the researcher promised to keep the interview confidential. Thus, only 19 observations were obtained from District 2. The data (in thousands of dollars) produced sample means and variances as shown in Table 6.2. Use these data to construct a 95% confidence interval for $(\mu_1 - \mu_2)$.

TABLE 6.2

Income data for Example 6.2

	District 1	District 2
Sample Size	20	19
Sample Mean	48.27	46.78
Sample Variance	8.74	6.58

Solution A preliminary analysis using histograms plotted for the two samples suggests that the two populations are mound-shaped (near normal). Also, the sample variances are very similar. The difference in the sample means is

$$\bar{y}_1 - \bar{y}_2 = 48.27 - 46.78 = 1.49$$

The estimate of the common standard deviation σ is

$$s_p = \sqrt{\frac{(n_1 - 1)s_1^2 + (n_2 - 1)s_2^2}{n_1 + n_2 - 2}} = \sqrt{\frac{19(8.74) + 18(6.58)}{20 + 19 - 2}} = 2.77$$

The t-percentile for $a = \alpha/2 = .025$ and df $= 20 + 19 - 2 = 37$ is not listed in Table 3 of the Appendix, but taking the labeled value for the nearest df less than 37 (df $= 35$), we have $t_{.025} = 2.030$. A 95% confidence interval for the difference in mean incomes for the two districts is of the form

$$\bar{y}_1 - \bar{y}_2 \pm t_{\alpha/2} s_p \sqrt{\frac{1}{n_1} + \frac{1}{n_2}}$$

Substituting into the formula, we obtain

$$1.49 \pm 2.030(2.77)\sqrt{\frac{1}{20} + \frac{1}{19}} \quad \text{or} \quad 1.49 \pm 1.80$$

Thus, we estimate the difference in mean incomes to lie somewhere in the interval from $-.31$ to 3.29. If we multiply these limits by \$1,000, the confidence interval for the difference in mean incomes is $-\$310$ to \$3,290. This interval includes both positive and negative values for $\mu_1 - \mu_2$, so we are unable to determine whether the mean income for District 1 is larger or smaller than the mean income for District 2.

We can also test a hypothesis about the difference between two population means. As with any test procedure, we begin by specifying a research hypothesis for the difference in population means. Thus, we might, for example, specify that the difference $\mu_1 - \mu_2$ is greater than some value D_0. (*Note:* D_0 will often be 0.) The entire test procedure is summarized here.

A Statistical Test for $\mu_1 - \mu_2$, Independent Samples

The assumptions under which the test will be valid are the same as were required for constructing the confidence interval on $\mu_1 - \mu_2$: population distributions are normal with equal variances and the two random samples are independent.

H_0:   **1.** $\mu_1 - \mu_2 \le D_0$ (D_0 is a specified value, often 0)
     2. $\mu_1 - \mu_2 \ge D_0$
     3. $\mu_1 - \mu_2 = D_0$

H_a: **1.** $\mu_1 - \mu_2 > D_0$
     2. $\mu_1 - \mu_2 < D_0$
     3. $\mu_1 - \mu_2 \ne D_0$

T.S.: $t = \dfrac{(\bar{y}_1 - \bar{y}_2) - D_0}{s_p\sqrt{\dfrac{1}{n_1} + \dfrac{1}{n_2}}}$

R.R.:   For a level α, Type I error rate and with df $= n_1 + n_2 - 2$,
     1. Reject H_0 if $t \ge t_\alpha$.
     2. Reject H_0 if $t \le -t_\alpha$.
     3. Reject H_0 if $|t| \ge t_{\alpha/2}$.

Check assumptions and draw conclusions.

EXAMPLE 6.3

An experiment was conducted to evaluate the effectiveness of a treatment for tapeworm in the stomachs of sheep. A random sample of 24 worm-infected lambs of approximately the same age and health was randomly divided into two groups. Twelve of the lambs were injected with the drug and the remaining 12 were left untreated. After a 6-month period, the lambs were slaughtered and the following worm counts were recorded:

Drug-Treated Sheep	18	43	28	50	16	32	13	35	38	33	6	7
Untreated Sheep	40	54	26	63	21	37	39	23	48	58	28	39

a. Test whether the mean number of tapeworms in the stomachs of the treated lambs is less than the mean for untreated lambs. Use an $\alpha = .05$ test.
b. What is the level of significance for this test?
c. Place a 95% confidence interval on $\mu_1 - \mu_2$ to assess the size of the difference in the two means.

Solution

a. Boxplots of the worm counts for the treated and untreated lambs are displayed in Figure 6.3. From the plots, we can observe that the data for the untreated lambs are symmetric with no outliers and the data for the treated lambs are slightly skewed to the left with no outliers. Also, the widths of the two boxes are approximately equal. Thus, the condition that the population distributions are normal with equal variances appears to be satisfied. The condition of independence of the worm counts both between and within the two groups is evaluated by considering how the lambs were selected, assigned to the two groups, and cared for during the 6-month experiment. Because the 24 lambs were randomly selected from a representative herd of infected lambs, were randomly assigned to the treated and untreated groups, and were properly separated and cared for during the 6-month period of the experiment, the 24 worm counts are presumed to be independent random samples from the two populations. Finally, we can observe from the boxplots that the untreated lambs appear to have higher worm counts than the treated lambs because the median line is higher for the untreated group. The following test confirms our observation. The data for the treated and untreated sheep are summarized next.

FIGURE 6.3

Boxplots of worm counts for treated (1) and untreated (2) sheep

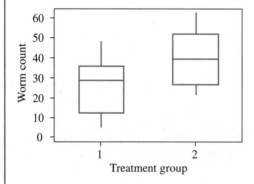

Drug-Treated Lambs   **Untreated Lambs**

$n_1 = 12$ $n_2 = 12$
$\bar{y}_1 = 26.58$ $\bar{y}_2 = 39.67$
$s_1 = 14.36$ $s_2 = 13.86$

The sample standard deviations are of a similar size, so from this and from our observation from the boxplot, the pooled estimate of the common population standard deviation σ is now computed:

$$s_p = \sqrt{\frac{(n_1 - 1)s_1^2 + (n_2 - 1)s_2^2}{n_1 + n_2 - 2}} = \sqrt{\frac{11(14.36)^2 + 11(13.86)^2}{22}} = 14.11$$

The test procedure for evaluating the research hypothesis that the treated lambs have mean tapeworm count (μ_1) less than the mean level (μ_2) for untreated lambs is as follows:

H_0: $\mu_1 - \mu_2 \geq 0$ (that is, the drug does not reduce mean worm count)

H_a: $\mu_1 - \mu_2 < 0$ (that is, the drug reduces mean worm count)

T.S.: $t = \dfrac{(\bar{y}_1 - \bar{y}_2) - D_0}{s_p\sqrt{\dfrac{1}{n_1} + \dfrac{1}{n_2}}} = \dfrac{(26.58 - 39.67) - 0}{14.11\sqrt{\dfrac{1}{12} + \dfrac{1}{12}}} = -2.272$

R.R.:   For $\alpha = .05$, the critical t-value for a one-tailed test with df $= n_1 + n_2 - 2 = 22$ is obtained from Table 3 in the Appendix, using $\alpha = .05$. We will reject H_0 if $t \leq -1.717$.

Conclusion: Because the observed value of $t = -2.272$ is less than -1.717 and hence is in the rejection region, we have sufficient evidence to conclude that the drug treatment does reduce the mean worm count.

b. Using Table 3 in the Appendix with $t = -2.272$ and df $= 22$, we can bound the level of significance in the range $.01 < p$-value $< .025$. From the following computed output, we can observe that the exact level of significance is p-value $= .017$.

```
Two-Sample T-Test and Confidence Interval

Two-sample T for Treated vs Untreated

                N      Mean     StDev     SEMean
Treated        12      26.6     14.4       4.1
Untreated      12      39.7     13.9       4.0


95% CI for mu Treated - mu Untreated: ( -25.0,  -1.1)
T-Test mu Treated = mu Untreated (vs <): T  = -2.27    P = 0.017    DF = 22
Both use Pooled StDev = 14.1
```

c. A 95% confidence interval on $\mu_1 - \mu_2$ provides the experimenter with an estimate of the size of the reduction in mean tapeworm count obtained by using the drug. This interval can be computed as follows:

$$(\bar{y}_1 - \bar{y}_2) \pm t_{.025}s_p\sqrt{\dfrac{1}{n_1} + \dfrac{1}{n_2}}$$

$$(26.58 - 39.67) \pm (2.074)(14.11)\sqrt{\dfrac{1}{12} + \dfrac{1}{12}}, \quad \text{or} \quad -13.09 \pm 11.95$$

Thus, we are 95% certain that the reduction in tapeworm count through the use of the drug is between 1.1 and 25.0 worms.

The confidence interval and test procedures for comparing two population means presented in this section require that three conditions be satisfied. The first and most critical condition is that the two random samples are independent. Practically, we mean that the two samples are randomly selected from two distinct populations and that the elements of one sample are statistically independent of those of the second sample. Two types of dependencies (data are not independent) commonly occur in experiments and studies. The data may have a *cluster effect,* which often results when the data have been collected in subgroups. For example, 50 children are selected from five different classrooms for an experiment to compare the effectiveness of two tutoring techniques. The children are randomly assigned to one of the two techniques. Because children from the same classroom have a common teacher and hence may tend to be more similar in their academic achievement than children from different classrooms, the condition of independence between participants in the study may be lacking.

A second type of dependence is the result of *serial or spatial correlation.* When measurements are taken over time, observations that are closer together in time tend to be more similar than observations collected at greatly different times, serially correlated. A similar dependence occurs when the data are collected at different locations—for example, water samples taken at various locations in a lake to assess whether a chemical plant is discharging pollutants into the lake. Measurements that are physically closer to one another are more likely to be similar than measurements taken farther apart. This type of dependence is *spatial correlation.* When the data are dependent, the procedures based on the t distribution produce confidence intervals having coverage probabilities different from the intended values and tests of hypotheses having Type I error rates different from the stated values. There are appropriate statistical procedures for handling this type of data, but they are more advanced. A book on longitudinal or repeated measures data analysis or the analysis of spatial data can provide the details for the analysis of dependent data.

When the population distributions are either very heavy tailed or highly skewed, the coverage probability for confidence intervals and the level and power of the t test will differ greatly from the stated values. A nonparametric alternative to the t test is presented in the next section; this test does not require normality.

The third and final assumption is that the two population variances σ_1^2 and σ_2^2 are equal. For now, just examine the sample variances to see that they are approximately equal; later (in Chapter 7), we'll give a test for this assumption. Many efforts have been made to investigate the effect of deviations from the equal variance assumption on the t methods for independent samples. The general conclusion is that for equal sample sizes, the population variances can differ by as much as a factor of 3 (for example, $\sigma_1^2 = 3\sigma_2^2$) and the t methods will still apply.

To illustrate the effect of unequal variances, a computer simulation was performed in which two independent random samples were generated from normal populations having the same means but unequal variances: $\sigma_1 = k\sigma_2$ with $k =$ .25, .5, 1, 2, and 4. For each combination of sample sizes and standard deviations, 1,000 simulations were run. For each simulation, a level .05 test was conducted. The proportion of the 1,000 tests that incorrectly rejected H_0 are presented in Table 6.3. If the pooled t test is unaffected by the unequal variances, we expect the proportions to be close to .05, the intended level, in all cases.

From the results in Table 6.3, we can observe that when the sample sizes are equal the proportion of Type I errors remains close to .05 (ranging from .042 to .065). When the sample sizes are different, the proportion of Type I errors deviates greatly from .05. The more serious case is when the smaller sample size is associated with the larger variance. In this case, the error rates are much larger than .05. For example, when $n_1 = 10, n_2 = 40$, and $\sigma_1 = 4\sigma_2$, the error rate is .307. However, when $n_1 = 10, n_2 = 10$, and $\sigma_1 = 4\sigma_2$, the error rate is .063, much closer to .05. This is remarkable and provides a convincing argument to use equal sample sizes.

TABLE 6.3

The effect of unequal variances on the Type I error rates of the pooled t test

		$\sigma_1 = k\sigma_2$				
n_1	n_2	$k = .25$	.50	1	2	4
10	10	.065	.042	.059	.045	.063
10	20	.016	.017	.049	.114	.165
10	40	.001	.004	.046	.150	.307
15	15	.053	.043	.056	.060	.060
15	30	.007	.023	.066	.129	.174
15	45	.004	.010	.069	.148	.250

In the situation in which the sample variances (s_1^2 and s_2^2) suggest that $\sigma_1^2 \neq \sigma_2^2$, there is an approximate t test using the test statistic

$$t' = \frac{\bar{y}_1 - \bar{y}_2 - D_0}{\sqrt{\dfrac{s_1^2}{n_1} + \dfrac{s_2^2}{n_2}}}$$

Welch (1938) showed that percentage points of a t distribution with modified degrees of freedom, known as Satterthwaite's approximation, can be used to set the rejection region for t'. This approximate t test is summarized here.

Approximate t Test for Independent Samples, Unequal Variance

H_0:   **1.** $\mu_1 - \mu_2 \leq D_0$ H_a: **1.** $\mu_1 - \mu_2 > D_0$
   2. $\mu_1 - \mu_2 \geq D_0$ **2.** $\mu_1 - \mu_2 < D_0$
   3. $\mu_1 - \mu_2 = D_0$ **3.** $\mu_1 - \mu_2 \neq D_0$

T.S.: $t' = \dfrac{(\bar{y}_1 - \bar{y}_2) - D_0}{\sqrt{\dfrac{s_1^2}{n_1} + \dfrac{s_2^2}{n_2}}}$

R.R.:   For a level α, Type 1 error rate and with df $= n_1 + n_2 - 2$,
   1. Reject H_0 if $t' \geq t_\alpha$
   2. Reject H_0 if $t' \leq -t_\alpha$
   3. Reject H_0 if $|t'| \geq t_{\alpha/2}$

where

$$\text{df} = \frac{(n_1 - 1)(n_2 - 1)}{(1 - c)^2(n_1 - 1) + c^2(n_2 - 1)}, \qquad \text{with } c = \frac{s_1^2/n_1}{\dfrac{s_1^2}{n_1} + \dfrac{s_2^2}{n_2}}$$

Note: If the computed value of df is not an integer, *round down* to the nearest integer.

The test based on the t' statistic is sometimes referred to as the *separate-variance t test* because we use the separate sample variances s_1^2 and s_2^2 rather than a pooled sample variance.

When there is a large difference between σ_1 and σ_2, we must also modify the confidence interval for $\mu_1 - \mu_2$. The following formula is developed from the separate-variance t test.

Approximate Confidence Interval for $\mu_1 - \mu_2$, Independent Samples with $\sigma_1 \neq \sigma_2$

$$(\bar{y}_1 - \bar{y}_2) \pm t_{\alpha/2}\sqrt{\frac{s_1^2}{n_1} + \frac{s_2^2}{n_2}}$$

where the t-percentile has

$$df = \frac{(n_1 - 1)(n_2 - 1)}{(1 - c)^2(n_1 - 1) + c^2(n_2 - 1)}, \quad \text{with } c = \frac{s_1^2/n_1}{\dfrac{s_1^2}{n_1} + \dfrac{s_2^2}{n_2}}$$

To illustrate that the separate-variance t test is less affected by unequal variances than the pooled t test, the data from the computer simulation reported in Table 6.3 were analyzed using the separate-variance t test. The proportion of the 1,000 tests that incorrectly rejected H_0 is presented in Table 6.4. If the separate-variance t test is unaffected by the unequal variances, we expect the proportions to be close to .05, the intended level, in all cases.

From the results in Table 6.4, we can observe that the separate-variance t test has a Type I error rate that is consistently very close to .05 in all the cases considered. On the other hand, the pooled t test had Type I error rates very different from .05 when the sample sizes were unequal and we sampled from populations having very different variances.

In this section, we developed pooled-variance t methods based on the requirement of independent random samples from normal populations with equal population variances. For situations where the variances are not equal, we introduced the separate-variance t' statistic. Confidence intervals and hypothesis tests based on these procedures (t or t') need not give identical results. Standard computer packages often report the results of both t and t' tests. Which of these results should you use in your report?

If the sample sizes are equal and the population variances are equal, the separate-variance t test and the pooled t test give algebraically identical results; that is, the computed t equals the computed t'. Thus, why not always use t' in place of t when $n_1 = n_2$? The reason we would select t over t' is that the df for t are nearly always larger than the df for t', and hence the power of the t test is greater than the power of the t' test when the variances are equal. When the sample sizes and variances are very unequal, the results of the t and t' procedures may differ greatly. The evidence in such cases indicates that the separate-variance methods are somewhat more reliable and more conservative than the results of the pooled t methods.

TABLE 6.4

The effect of unequal variances on the Type I error rates of the separate-variance t test

				$\sigma_1 = k\sigma_2$		
n_1	n_2	$k = .25$	.50	1	2	4
10	10	.055	.040	.056	.038	.052
10	20	.055	.044	.049	.059	.051
10	40	.049	.047	.043	.041	.055
15	15	.044	.041	.054	.055	.057
15	30	.052	.039	.051	.043	.052
15	45	.058	.042	.055	.050	.058

However, if the populations have both different means and different variances, an examination of just the size of the difference in their means $\mu_1 - \mu_2$ will be an inadequate description of how the populations differ. We should always examine the size of the differences in both the means and the standard deviations of the populations being compared. In Chapter 7, we will discuss procedures for examining the difference in the standard deviations of two populations.

EXERCISES

Basic Techniques

6.1 Set up the rejection regions for testing H_0: $\mu_1 - \mu_2 = 0$ for the following conditions:
  a. H_a: $\mu_1 - \mu_2 \neq 0$, $n_1 = 12$, $n_2 = 14$, and $\alpha = .05$
  b. H_a: $\mu_1 - \mu_2 > 0$, $n_1 = n_2 = 8$, and $\alpha = .01$
  c. H_a: $\mu_1 - \mu_2 < 0$, $n_1 = 6$, $n_2 = 4$, and $\alpha = .05$

What assumptions must be made prior to applying a two-sample t test?

6.2 Conduct a test of H_0: $\mu_1 - \mu_2 \geq 0$ against the alternative hypothesis H_a: $\mu_1 - \mu_2 < 0$ for the sample data shown here. Use $\alpha = .05$.

	Population	
	1	**2**
Sample size	16	13
Sample mean	71.5	79.8
Sample variance	68.35	70.26

6.3 Refer to the data of Exercise 6.2. Give the level of significance for your test.

Applications

Med.

6.4 In an effort to link cold environments with hypertension in humans, a preliminary experiment was conducted to investigate the effect of cold on hypertension in rats. Two random samples of six rats each were exposed to different environments. One sample of rats was held in a normal environment at 26°C. The other sample was held in a cold 5°C environment. Blood pressures and heart rates were measured for rats for both groups. The blood pressures for the 12 rats are shown in the accompanying table.
  a. Do the data provide sufficient evidence that rats exposed to a 5°C environment have a higher mean blood pressure than rats exposed to a 26°C environment? Use $\alpha = .05$.
  b. Evaluate the three conditions required for the test used in part (a).
  c. Provide a 95% confidence interval on the difference in the two population means.

26°C		5°C	
Rat	**Blood Pressure**	**Rat**	**Blood Pressure**
1	152	7	384
2	157	8	369
3	179	9	354
4	182	10	375
5	176	11	366
6	149	12	423

Env.   **6.5** A pollution-control inspector suspected that a riverside community was releasing semi-treated sewage into a river and this, as a consequence, was changing the level of dissolved oxygen of the river. To check this, he drew 15 randomly selected specimens of river water at a location above the town and another 15 specimens below. The dissolved oxygen readings, in parts per million, are given in the accompanying table.

| **Above Town** | 5.2 | 4.8 | 5.1 | 5.0 | 4.9 | 4.8 | 5.0 | 4.7 | 4.7 | 5.0 | 4.7 | 5.1 | 5.0 | 4.9 | 4.9 |
| **Below Town** | 4.2 | 4.4 | 4.7 | 4.9 | 4.6 | 4.8 | 4.9 | 4.6 | 5.1 | 4.3 | 5.5 | 4.7 | 4.9 | 4.8 | 4.7 |

Use the computer output shown here to answer the following questions.

```
Two-Sample T-Test and Confidence Interval

Two-sample T for Above Town vs Below Town

              N       Mean      StDev    SE Mean
Above To     15       4.92      0.157      0.042
Below To     15       4.74      0.320      0.084

95% CI for mu Above To - mu Below To: ( -0.013,   0.378)
T-Test mu Above To = mu Below To (vs not =): T = 1.95   P = 0.065   DF = 20
```

Boxplots of above- and below-town specimens (means are indicated by solid circles)

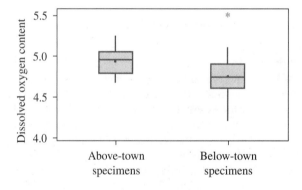

a. Do the data provide sufficient evidence to indicate a difference in mean oxygen content between locations above and below the town? Use $\alpha = .05$.
b. Was the pooled t test or the separate-variance t test used in the computer output?
c. Do the required conditions to use the test in (a) appear to be valid for this study? Justify your answer.
d. How large is the difference between the mean oxygen content above and below the town?

Engin.   **6.6** An industrial engineer conjectures that a major difference between successful and unsuccessful companies is the percentage of their manufactured products returned because of defectives. In a study to evaluate this conjecture, the engineer surveyed the quality control departments of 50 successful companies (identified by the annual profit statement) and 50 unsuccessful companies. The companies in the study all produced products of a similar nature and cost. The percentage of the total output returned by customers in the previous year is summarized in the following table and graphs.

```
Two-Sample T-Test and Confidence Interval

Two-sample T for Unsuccessful vs Successful

                 N       Mean     StDev    Se Mean
Unsuccessful 50          9.08     1.97        0.28
Successful   50          5.40     2.88        0.41

95% CI for mu Unsuccessful - mu Successful: ( 2.70,  4.66)
T-Test mu Unsuccessful = mu Successful (vs >): T = 7.46   P = 0.0000   DF = 86
```

Boxplots of unsuccessful and successful businesses (means are indicated by solid circles)

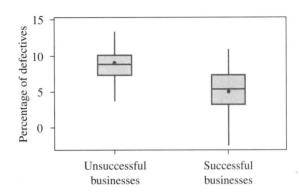

a. Do the data provide sufficient evidence that successful companies have a lower percentage of their products returned by customers? Use $\alpha = .05$.

b. Was the pooled t test or the separate-variance t test used in the computer output?

c. Do the required conditions to use the test in (a) appear to be valid for this study? Justify your answer.

d. How large is the difference between the percentage of returns for successful and unsuccessful companies?

Soc.   **6.7**   The number of households currently receiving a daily newspaper has decreased over the last 10 years, and many people state they obtain information about current events through television news and the Internet. To test whether people who receive a daily newspaper have a greater knowledge of current events than people who don't, a sociologist gave a current events test to 25 randomly selected people who subscribe to a daily newspaper and to 30 randomly selected people who do not receive a daily newspaper. The following stem-and-leaf graphs give the scores for the two groups. Does it appear that people who receive a daily newspaper have a greater knowledge of current events? Be sure to evaluate all necessary conditions for your procedures to be valid.

```
Character Stem-and-Leaf Display

Stem-and-leaf of No Newspaper Deliver    Stem-and-leaf of Newspaper Subscribers
N=30                                     N=25
Leaf Unit = 1.0                          Leaf Unit = 1.0

      0 000
      0
      1 3
```

```
1 59
2 334                    2 2
2 57                     2 99
3 00234                  3 2
3 5589                   3 66889
4 00124                  4 000112333
4 5                      4 55666

5 0                      5 2
5 55                     5 9
6 2
```

```
Two-Sample T-Test and Confidence Interval

Two-sample T for No Newspaper vs Newspaper

                N      Mean    StDev   SE Mean
No Newspaper   30      32.0    16.0      2.9
Newspaper      25      40.91   7.48      1.5

95% CI for mu No Newspaper - mu Newspaper: ( -15.5,   -2.2)
T-Test mu No Newspaper = mu Newspaper (vs <) : T = -2.70  P = 0.0049  DF = 42
```

Env. **6.8** The study of concentrations of atmospheric trace metals in isolated areas of the world has received considerable attention because of the concern that humans might somehow alter the climate of Earth by changing the amount and distribution of trace metals in the atmosphere. Consider a study at the South Pole, where at 10 different sampling periods throughout a 2-month period, 10,000 standard cubic meters (scm) of air were obtained and analyzed for metal concentrations. The results associated with magnesium and europium are listed here. (*Note:* Magnesium results are in units of 10^{-9} g/scm; europium results are in units of 10^{-15} g/scm.) Note that $s > \bar{y}$ for the magnesium data. Would you expect the data to be normally distributed? Explain.

	Sample Size	Sample Mean	Sample Standard Deviation
Magnesium	10	1.0	2.21
Europium	10	17.0	12.65

6.9 Refer to Exercise 6.8. Could we run a *t* test comparing the mean metal concentrations for magnesium and europium? Why?

Env. **6.10** PCBs have been in use since 1929, mainly in the electrical industry, but it was not until the 1960s that they were found to be a major environmental contaminant. In the paper, "The ratio of DDE to PCB concentrations in Great Lakes herring gull eggs and its use in interpreting contaminants data" [*Journal of Great Lakes Research* (1998) 24(1):12–31], researchers report on the following study. Thirteen study sites from the five Great Lakes were selected. At each site, 9 to 13 herring gull eggs were collected randomly each year for several years. Following collection, the PCB content was determined. The mean PCB content at each site is reported in the following table for the years 1982 and 1996.

							Site						
Year	1	2	3	4	5	6	7	8	9	10	11	12	13
1982	61.48	64.47	45.50	59.70	58.81	75.86	71.57	38.06	30.51	39.70	29.78	66.89	63.93
1996	13.99	18.26	11.28	10.02	21.00	17.36	28.20	7.30	12.80	9.41	12.63	16.83	22.74

a. Legislation was passed in the 1970s restricting the production and use of PCBs. Thus, the active input of PCBs from current local sources has been severely curtailed. Do the data provide evidence that there has been a significant decrease in the mean PCB content of herring gull eggs?

b. Estimate the size of the decrease in mean PCB content from 1982 to 1996, using a 95% confidence interval.

c. Evaluate the conditions necessary to validly test hypotheses and construct confidence intervals using the collected data.

d. Does the independence condition appear to be violated?

6.11 Refer to Exercise 6.10. There appears to be a large variation in the mean PCB content across the 13 sites. How could we reduce the effect of variation in PCB content due to site differences on the evaluation of the difference in the mean PCB content between the two years?

H.R. **6.12** A firm has a generous but rather complicated policy concerning end-of-year bonuses for its lower-level managerial personnel. The policy's key factor is a subjective judgment of "contribution to corporate goals." A personnel officer took samples of 24 female and 36 male managers to see whether there was any difference in bonuses, expressed as a percentage of yearly salary. The data are listed here:

Gender				Bonus Percentage					
F	9.2	7.7	11.9	6.2	9.0	8.4	6.9	7.6	7.4
	8.0	9.9	6.7	8.4	9.3	9.1	8.7	9.2	9.1
	8.4	9.6	7.7	9.0	9.0	8.4			
M	10.4	8.9	11.7	12.0	8.7	9.4	9.8	9.0	9.2
	9.7	9.1	8.8	7.9	9.9	10.0	10.1	9.0	11.4
	8.7	9.6	9.2	9.7	8.9	9.2	9.4	9.7	8.9
	9.3	10.4	11.9	9.0	12.0	9.6	9.2	9.9	9.0

```
Two-Sample T-Test and Confidence Interval

Two-sample T for Female vs Male

              N      Mean    StDev   SE Mean
Female       24      8.53    1.19     0.24
Male         36      9.68    1.00     0.17

95% CI for mu Female - mu Male: ( -1.74,   -0.56)
T-Test mu Female = mu Male (vs <): T = -3.90  P = 0.0002  DF = 43

95% CI for mu Female - mu Male: ( -1.72,   -0.58)
T-Test mu Female = mu Male (vs <): T =  -4.04  P = 0.0001  DF = 58
Both use Pooled StDev = 1.08
```

Boxplots of females' and males' bonuses (means are indicated by solid circles)

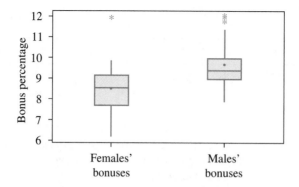

a. Identify the value of the pooled-variance t statistic (the usual t test based on the equal variance assumption).
b. Identify the value of the t' statistic.
c. Use both statistics to test the research hypothesis of unequal means at $\alpha = .05$ and at $\alpha = .01$. Does the conclusion depend on which statistic is used?

6.3   A Nonparametric Alternative: The Wilcoxon Rank Sum Test

Wilcoxon rank sum test

The two-sample t test of the previous section was based on several conditions: independent samples, normality, and equal variances. When the conditions of normality and equal variances are not valid but the sample sizes are large, the results using a t (or t') test are approximately correct. There is, however, an alternative test procedure that requires less stringent conditions. This procedure, called the **Wilcoxon rank sum test,** is discussed here.

The assumptions for this test are that we have independent random samples taken from two populations whose distributions are identical except that one distribution may be shifted to the right of the other distribution, as shown in Figure 6.4. The Wilcoxon rank sum test does not require that populations have normal distributions. Thus, we have removed one of the three conditions that were required of the t-based procedures. The other conditions, equal variances and independence of the random samples, are still required for the Wilcoxon rank sum test. Because the two population distributions are assumed to be identical under the null hypothesis, independent random samples from the two populations should be similar

FIGURE 6.4

Skewed population distributions identical in shape but shifted

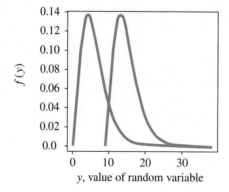

if the null hypothesis is true. Because we are now allowing the population distributions to be nonnormal, the rank sum procedure must deal with the possibility of extreme observations in the data. One way to handle samples containing extreme values is to replace each data value with its rank (from lowest to highest) in the combined sample — that is, the sample consisting of the data from both populations. The smallest value in the combined sample is assigned the rank of 1 and the largest value is assigned the rank of $N = n_1 + n_2$. The ranks are not affected by how far the smallest (largest) data value is from the next smallest (largest) data value. Thus, extreme values in data sets do not have a strong effect on the rank sum statistic, as they do in the t-based procedures.

The calculation of the rank sum statistic consists of the following steps:

ranks

1. List the data values for both samples from smallest to largest.
2. In the next column, assign the numbers 1 to N to the data values with 1 to the smallest value and N to the largest value. These are the **ranks** of the observations.
3. If there are ties — that is, duplicated values — in the combined data set, the ranks for the observations in a tie are taken to be the average of the ranks for those observations.
4. Let T denote the sum of the ranks for the observations from population 1.

If the null hypothesis of identical population distributions is true, the n_1 ranks from population 1 are just a random sample from the N integers $1, \ldots, N$. Thus, under the null hypothesis, the distribution of the sum of the ranks T depends only on the sample sizes, n_1 and n_2, and does not depend on the shape of the population distributions. Under the null hypothesis, the sampling distribution of T has mean and variance given by

$$\mu_T = \frac{n_1(n_1 + n_2 + 1)}{2} \qquad \sigma_T^2 = \frac{n_1 n_2}{12}(n_1 + n_2 + 1)$$

Intuitively, if T is much smaller (or larger) than μ_T, we have evidence that the null hypothesis is false and in fact the population distributions are not equal. The rejection region for the rank sum test specifies the size of the difference between T and μ_T for the null hypothesis to be rejected. Because the distribution of T under the null hypothesis does not depend on the shape of the population distributions, Appendix Table 6 provides the critical values for the test regardless of the shape of the population distribution. The Wilcoxon rank sum test is summarized here.

Wilcoxon Rank Sum Test*

H_0:   The two populations are identical.

H_a: 1. Population 1 is shifted to the right of population 2.
        2. Population 1 is shifted to the left of population 2.
        3. Populations 1 and 2 are shifted from each other.

$(n_1 \leq 10, n_2 \leq 10)$

T.S.: T, the sum of the ranks in sample 1

---

*This test is equivalent to the Mann–Whitney U test, Conover (1998).

R.R.: For $\alpha = .05$, use Table 6 in the Appendix to find critical values for T_U and T_L;
1. Reject H_0 if $T > T_U$.
2. Reject H_0 if $T < T_L$.
3. Reject H_0 if $T > T_U$ or $T < T_L$.

Check assumptions and draw conclusions.

EXAMPLE 6.4

Many states are considering lowering the blood-alcohol level at which a driver is designated as driving under the influence (DUI) of alcohol. An investigator for a legislative committee designed the following test to study the effect of alcohol on reaction time. Ten participants consumed a specified amount of alcohol. Another group of 10 participants consumed the same amount of a nonalcoholic drink, a placebo. The two groups did not know whether they were receiving alcohol or the placebo. The 20 participants' average reaction times (in seconds) to a series of simulated driving situations are reported in the following table. Does it appear that alcohol consumption increases reaction time?

Placebo	0.90	0.37	1.63	0.83	0.95	0.78	0.86	0.61	0.38	1.97
Alcohol	1.46	1.45	1.76	1.44	1.11	3.07	0.98	1.27	2.56	1.32

a. Why is the t test inappropriate for analyzing the data in this study?
b. Use the Wilcoxon rank sum test to test the hypotheses:

H_0: The distributions of reaction times for the placebo and alcohol populations are identical.

H_a: The distribution of reaction times for the placebo consumption population is shifted to the left of the distribution for the alcohol population. (Larger reaction times are associated with the consumption of alcohol.)

c. Place 95% confidence intervals on the median reaction times for the two groups.
d. Compare the results from (b) to the results from Minitab.

Solution

a. A boxplot of the two samples is given here. The plots indicate that the population distributions are skewed to the right, because 10% of the data values are large outliers and the upper whiskers are longer than the lower whiskers. The sample sizes are both small, and hence the t test may be inappropriate for analyzing this study.

b. The Wilcoxon rank sum test will be conducted to evaluate whether alcohol consumption increases reaction time. Table 6.5 contains the ordered data for the combined samples, along with their associated ranks. We will designate observations from the placebo group as 1 and from the alcohol group as 2.

Boxplots of placebo and alcohol populations (means are indicated by solid circles)

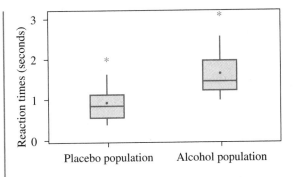

TABLE 6.5

Ordered reaction times and ranks

	Ordered Data	Group	Ranks		Ordered Data	Group	Ranks
1	0.37	1	1	11	1.27	2	11
2	0.38	1	2	12	1.32	2	12
3	0.61	1	3	13	1.44	2	13
4	0.78	1	4	14	1.45	2	14
5	0.83	1	5	15	1.46	2	15
6	0.86	1	6	16	1.63	1	16
7	0.90	1	7	17	1.76	2	17
8	0.95	1	8	18	1.97	1	18
9	0.98	2	9	19	2.56	2	19
10	1.11	2	10	20	3.07	2	20

For $\alpha = .05$, reject H_0 if $T < 83$, using Table 6 in the Appendix with $\alpha = .05$, one-tailed, and $n_1 = n_2 = 10$. The value of T is computed by summing the ranks from group 1: $T = 1 + 2 + 3 + 4 + 5 + 6 + 7 + 8 + 16 + 18 = 70$. Because 70 is less than 83, we reject H_0 and conclude there is significant evidence that the placebo population has smaller reaction times than the population of alcohol consumers.

c. Because we have small sample sizes and the population distributions appear to be skewed to the right, we will construct confidence intervals on the median reaction times in place of confidence intervals on the mean reaction times. Using the methodology from Section 5.8, from Table 5 in the Appendix, we find

$$C_{\alpha(2),n} = C_{.05,10} = 1$$

Thus,

$$L_{.025} = C_{.05,10} + 1 = 2$$

and

$$U_{.025} = n - C_{.05,10} = 10 - 1 = 9$$

The 95% confidence intervals for the population medians are given by

$$(M_L, M_U) = (y_{(2)}, y_{(9)})$$

Thus, a 95% confidence interval is (.38, 1.63) for the placebo population median and (1.11, 2.56) for the alcohol population median. Because
the sample sizes are very small, the confidence intervals are not very
informative.

d. The output from Minitab is given here.

```
Mann-Whitney Confidence Interval and Test

PLACEBO  N = 10      Median =       0.845
ALCOHOL  N = 10      Median =       1.445
Point estimate for ETA1-ETA2 is   -0.600
95.5 Percent CI for ETA1-ETA2 is (-1.080, -0.250)
W = 70.0
Test of ETA1 = ETA2 vs ETA1 < ETA2 is significant at 0.0046
```

Minitab refers to the test statistic as the Mann–Whitney test. This test is
equivalent to the Wilcoxon test statistic. In fact, the value of the test statistic $W = 70$ is identical to the Wilcoxon $T = 70$. The output provides the
p-value $= 0.0046$ and a 95.5% confidence interval on the difference in the
population medians, $(-1.08, -0.25)$.

When both sample sizes are more than 10, the sampling distribution of T is
approximately normal; this allows us to use a z statistic in place of T when using the
Wilcoxon rank sum test:

$$z = \frac{T - \mu_T}{\sigma_T}$$

The theory behind the Wilcoxon rank sum test requires that the population distributions be continuous, so the probability that any two data values are equal is zero.
Because in most studies we record data values only to a few decimal places, we will
often have ties—that is, observations with the same value. For these situations, each
observation in a set of tied values receives a rank score equal to the average of the
ranks for the set of values. When there are ties, the variance of T must be adjusted.
The adjusted value of σ_T^2 is shown here.

$$\sigma_T^2 = \frac{n_1 n_2}{12}\left((n_1 + n_2 + 1) - \frac{\sum_{j=1}^{k} t_j(t_j^2 - 1)}{(n_1 + n_2)(n_1 + n_2 - 1)}\right)$$

where k is the number of tied groups and t_j denotes the number of tied observations
in the jth group. Note that when there are no tied observations, $t_j = 1$ for all j, which
results in

$$\sigma_T^2 = \frac{n_1 n_2}{12}(n_1 + n_2 + 1)$$

From a practical standpoint, unless there are many ties, the adjustment will result
in very little change to σ_T^2. The normal approximation to the Wilcoxon rank sum
test is summarized here.

Wilcoxon Rank Sum Test:
Normal Approximation

$n_1 > 10$ and $n_2 > 10$

H_0: The two populations are identical.

H_a: **1.** Population 1 is shifted to the right of population 2.
2. Population 1 is shifted to the left of population 2.
3. Population 1 and 2 are shifted from each other.

T.S.: $z = \dfrac{T - \mu_T}{\sigma_T}$, where T denotes the sum of the ranks in sample 1.

R.R.: For a specified value of α,
1. Reject H_0 if $z \geq z_\alpha$.
2. Reject H_0 if $z \leq -z_\alpha$.
3. Reject H_0 if $|z| \geq z_{\alpha/2}$.

Check assumptions and draw conclusions.

EXAMPLE 6.5

Environmental engineers were interested in determining whether a cleanup project on a nearby lake was effective. Prior to initiation of the project, they obtained 12 water samples at random from the lake and analyzed the samples for the amount of dissolved oxygen (in ppm). Due to diurnal fluctuations in the dissolved oxygen, all measurements were obtained at the 2 P.M. peak period. The before and after data are presented in Table 6.6.

a. Use $\alpha = .05$ to test the following hypotheses:

H_0: The distributions of measurements for before cleanup and 6 months after the cleanup project began are identical.

H_a: The distribution of dissolved oxygen measurements before the cleanup project is shifted to the right of the corresponding distribution of measurements for 6 months after initiating the cleanup project. (Note that a cleanup project has been effective in one sense if the dissolved oxygen level drops over a period of time.)

For convenience, the data are arranged in ascending order in Table 6.6.
b. Has the correction for ties made much of a difference?

TABLE 6.6

Dissolved oxygen measurements (in ppm)

Before Cleanup		After Cleanup	
11.0	11.6	10.2	10.8
11.2	11.7	10.3	10.8
11.2	11.8	10.4	10.9
11.2	11.9	10.6	11.1
11.4	11.9	10.6	11.1
11.5	12.1	10.7	11.3

TABLE 6.7

Dissolved oxygen measurements and ranks

Before Cleanup		After Cleanup	
11.0	(10)	10.2	(1)
11.2	(14)	10.3	(2)
11.2	(14)	10.4	(3)
11.2	(14)	10.6	(4.5)
11.4	(17)	10.6	(4.5)
11.5	(18)	10.7	(6)
11.6	(19)	10.8	(7.5)
11.7	(20)	10.8	(7.5)
11.8	(21)	10.9	(9)
11.9	(22.5)	11.1	(11.5)
11.9	(22.5)	11.1	(11.5)
12.1	(24)	11.3	(16)
	$T = 216$		

Solution

a. First we must jointly rank the combined sample of 24 observations by assigning the rank of 1 to the smallest observation, the rank of 2 to the next smallest, and so on. When two or more measurements are the same, we assign all of them a rank equal to the average of the ranks they occupy. The sample measurements and associated ranks (shown in parentheses) are listed in Table 6.7.

Because n_1 and n_2 are both greater than 10, we will use the test statistic z. If we are trying to detect a shift to the left in the distribution after the cleanup, we expect the sum of the ranks for the observations in sample 1 to be large. Thus, we will reject H_0 for large values of $z = (T - \mu_T)/\sigma_T$.

Grouping the measurements with tied ranks, we have 18 groups. These groups are listed next with the corresponding values of t_j, the number of tied ranks in the group.

Rank(s)	Group	t_j	Rank(s)	Group	t_j
1	1	1	14, 14, 14	10	3
2	2	1	16	11	1
3	3	1	17	12	1
4.5, 4.5	4	2	18	13	1
6	5	1	19	14	1
7.5, 7.5	6	2	20	15	1
9	7	1	21	16	1
10	8	1	22.5, 22.5	17	2
11.5, 11.5	9	2	24	18	1

For all groups with $t_j = 1$, there is no contribution for

$$\frac{\sum_j t_j(t_j^2 - 1)}{(n_1 + n_2)(n_1 + n_2 - 1)}$$

in σ_T^2, because $t_j^2 - 1 = 0$. Thus, we will need only $t_j = 2, 3$.

Substituting our data into the formulas, we obtain

$$\mu_T = \frac{n_1(n_1 + n_2 + 1)}{2} = \frac{12(12 + 12 + 1)}{2} = 150$$

$$\sigma_T^2 = \frac{n_1 n_2}{12}\left[(n_1 + n_2 + 1) - \frac{\sum t_j(t_j^2 - 1)}{(n_1 + n_2)(n_1 + n_2 - 1)}\right]$$

$$= \frac{12(12)}{12}\left[25 - \frac{6 + 6 + 6 + 24 + 6}{24(23)}\right]$$

$$= 12(25 - .0870) = 298.956$$

$$\sigma_T = 17.29$$

The computed value of z is

$$z = \frac{T - \mu_T}{\sigma_T} = \frac{216 - 150}{17.29} = 3.82$$

This value exceeds 1.645, so we reject H_0 and conclude that the distribution of before-cleanup measurements is shifted to the right of the corresponding distribution of after-cleanup measurements; that is, the after-cleanup measurements on dissolved oxygen tend to be smaller than the corresponding before-cleanup measurements.

b. The value of σ_T^2, without correcting for ties is

$$\sigma_T^2 = \frac{12(12)(25)}{12} = 300 \quad \text{and} \quad \sigma_T = 17.32$$

For this value of σ_T, $z = 3.81$ rather than 3.82, which was found by applying the correction. This should help you understand how little effect the correction has on the final result unless there are a large number of ties.

The Wilcoxon rank sum test is an alternative to the two-sample t test, with the rank sum test requiring fewer conditions than the t test. In particular, Wilcoxon's test does not require the two populations to have normal distributions; it only requires that the distributions be identical except possibly that one distribution is shifted from the other distribution. When both distributions are normal, the t test is more likely to detect an existing difference; that is, the t test has greater power than the rank sum test. This is logical because the t test uses the magnitudes of the observations rather than just their relative magnitudes (ranks) as is done in the rank sum test. However, when the two distributions are nonnormal, the Wilcoxon rank sum test has greater power; that is, it is more likely to detect a shift in the population distributions. Also, the level or probability of a Type I error for the Wilcoxon rank sum test will be equal to the stated level for all population distributions. The t test's *actual* level will deviate from its stated value when the population distributions are nonnormal. This is particularly true when the nonnormality of the population distributions is present in the form of severe skewness or extreme outliers.

Applications

Bus.   **6.13**   A plumbing contractor was interested in making her operation more efficient by cutting down on the average distance between service calls while still maintaining at least the same level of business activity. One plumber (plumber 1) was assigned a dispatcher who monitored all his incoming requests for service and outlined a service strategy for that day. Plumber 2 was to continue as she had in the past, by providing service in roughly sequential order for stacks of service calls received. The total daily mileages for these two plumbers are recorded here for a total of 18 days (3 workweeks).

Plumber 1	88.2	94.7	101.8	102.6	89.3	95.7
	78.2	80.1	83.9	86.1	89.4	71.4
	92.4	85.3	87.5	94.6	92.7	84.6
Plumber 2	105.8	117.6	119.5	126.8	108.2	114.7
	90.2	95.6	110.1	115.3	109.6	112.4
	104.6	107.2	109.7	102.9	99.1	115.5

a. Plot the sample data for each plumber and compute $\bar{y}$ and s.
b. Based on your findings in part (a), which procedure appears more appropriate for comparing the distributions?

Med.   **6.14**   The paper, "Serum beta-2-microglobulin (SB2M) in patients with multiple myeloma treated with alpha interferon" [*Journal of Medicine* (1997) 28:311–318] reports on the influence of alpha interferon administration in the treatment of patients with multiple myeloma (MM). Twenty newly diagnosed patients with MM were entered into the study. The researchers randomly assigned the 20 patients into the two groups. Ten patients were treated with both intermittent melphalan and sumiferon (treatment group), whereas the remaining 10 patients were treated only with intermittent melphalan. The SB2M levels were measured before and at days 3, 8, 15, and months 1, 3, and 6 from the start of therapy. The measurement of SB2M was performed using a radioimmunoassay method. The measurements before treatment are given here.

Treatment Group	2.9	2.7	3.9	2.7	2.1	2.6	2.2	4.2	5.0	0.7
Control Group	3.5	2.5	3.8	8.1	3.6	2.2	5.0	2.9	2.3	2.9

a. Plot the sample data for both groups using boxplots or normal probability plots.
b. Based on your findings in part (a), which procedure appears more appropriate for comparing the distributions of SB2M?
c. Is there significant evidence that there is a difference in the distribution of SB2M for the two groups?
d. Discuss the implications of your findings in part (c) on the evaluation of the influence of alpha interferon.

Bus.   **6.15**   Refer to Exercise 6.13. A second study was done the following year. A treatment group of 18 plumbers was assigned a dispatcher who monitored all the plumbers' incoming requests for service and outlined a service strategy for that day's activities. A control group of 18 plumbers was to conduct their activities as they did in the past, by providing service in roughly sequential order for stacks of service calls received. The average daily mileages for these 36 plumbers were computed over a 30-day period and are recorded here.

Treatment Group	62.2	79.3	83.2	82.2	84.1	89.3
	95.8	87.9	91.5	96.6	90.1	98.6
	85.2	87.9	86.7	99.7	101.1	88.6
Control Group	87.1	70.2	94.6	182.9	85.6	89.5
	109.5	101.7	99.7	193.2	105.3	92.9
	63.9	88.2	99.1	95.1	92.4	87.3

a. The sample data are plotted next. Based on these plots, which procedure appears more appropriate for comparing the distributions of the two groups?

b. Computer output is shown here for a *t* test and a Wilcoxon rank sum test (which is equivalent to the Mann–Whitney test shown in the output). Compare the results for these two tests and draw a conclusion about the effectiveness of the dispatcher program.

c. Comment on the appropriateness or inappropriateness of the *t* test based on the plots of the data and the computer output.

d. Does it matter which test is used here? Might it be reasonable to run both tests in certain situations? Why or why not?

```
Two-Sample T-Test and Confidence Interval Based on Pooled Variances

Two-sample T for Treatment vs Control

                 N      Mean     StDev   SE Mean
Treatment       18     88.33      9.06     2.1
Control         18     102.1     33.2      7.8


95% CI for mu Treatment - mu Control: ( -30.3,    2.7)
T-Test mu Treatment = mu Control (vs <): T =   -1.70   P = 0.049   DF = 34
Both use Pooled StDev = 24.3
```

```
Two-Sample T-Test and Confidence Interval Based on Separate Variances

Two-sample T for Treatment vs Control

                 N      Mean     StDev   SE Mean
Treatment       18     88.33      9.06     2.1
Control         18     102.1     33.2      7.8


95% CI for mu Treatment - mu Control: ( -30.8,    3.2)
T-Test mu Treatment = mu Control (vs <): T =   -1.70   P = 0.053   DF = 19
```

```
Mann-Whitney Confidence Interval and Test

Treatment   N =   18        Median =        88.25
Control     N =   18        Median =        93.75
Point estimate for ETA1-ETA2 is          -5.20
95.2 Percent CI for ETA1-ETA2 is (-12.89, 0.81)
W = 278.5
Test of ETA1 = ETA2 vs ETA1 < ETA2 is significant at 0.0438
The test is significant at 0.0438 (adjusted for ties)
```

Normal probability plot
for control group

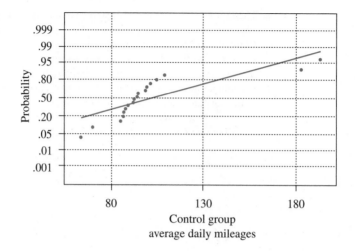

Control group
average daily mileages

Normal probability plot
for treatment group

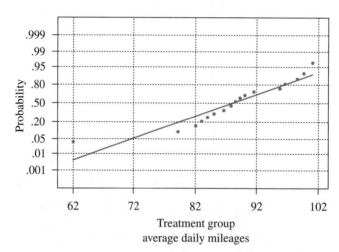

Treatment group
average daily mileages

Boxplots of treatment and
control groups (means are
indicated by solid circles)

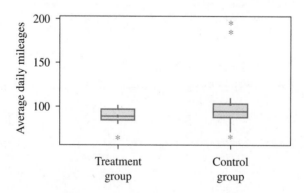

6.4   Inferences about $\mu_1 - \mu_2$: Paired Data

The methods we presented in the preceding three sections were appropriate for situations in which independent random samples are obtained from two populations. These methods are not appropriate for studies or experiments in which each measurement in one sample is *matched* or *paired* with a particular measurement in the other sample. In this section, we will deal with methods for analyzing paired data. We begin with an example.

EXAMPLE 6.6

Insurance adjusters are concerned about the high estimates they are receiving for auto repairs from garage I compared to garage II. To verify their suspicions, each of 15 cars recently involved in an accident was taken to both garages for separate estimates of repair costs. The estimates from the two garages are given in Table 6.8. A preliminary analysis of the data used a two-sample t test.

TABLE 6.8

Repair estimates (in hundreds of dollars)

Car	Garage I	Garage II
1	17.6	17.3
2	20.2	19.1
3	19.5	18.4
4	11.3	11.5
5	13.0	12.7
6	16.3	15.8
7	15.3	14.9
8	16.2	15.3
9	12.2	12.0
10	14.8	14.2
11	21.3	21.0
12	22.1	21.0
13	16.9	16.1
14	17.6	16.7
15	18.4	17.5
	$\bar{y}_1 = 16.85$	$\bar{y}_2 = 16.23$
	$s_1 = 3.20$	$s_1 = 2.94$

Solution Computer output for these data is shown here.

```
Two-Sample T-Test and Confidence Interval

Two-sample T for Garage I vs Garage II

             N      Mean    StDev   SE Mean
Garage I    15     16.85     3.20     0.83
Garage II   15     16.23     2.94     0.76

95% CI for mu Garage I - mu Garage II: (-1.69, 2.92)
T-Test mu Garage I = mu Garage II (vs not =): T = 0.55  P = 0.59  DF = 27
```

From the output, we see there is a consistent difference in the sample means ($\bar{y}_1 - \bar{y}_2 = .62$). However, this difference is rather small considering the variability of the measurements ($s_1 = 3.20$, $s_2 = 2.94$). In fact, the computed t-value (.55) has a p-value of .59, indicating very little evidence of a difference in the average claim estimates for the two garages.

A closer glance at the data in Table 6.8 indicates that something about the conclusion in Example 6.6 is inconsistent with our intuition. For all but one of the 15 cars, the estimate from garage I was higher than that from garage II. From our knowledge of the binomial distribution, the probability of observing garage I estimates higher in $y = 14$ or more of the $n = 15$ trials, assuming no difference ($\pi = .5$) for garages I and II, is

$$P(y = 14 \text{ or } 15) = P(y = 14) + P(y = 15)$$

$$= \binom{15}{14}(.5)^{14}(.5) + \binom{15}{15}(.5)^{15} = .000488$$

Thus, if the two garages in fact have the same distribution of estimates, there is approximately a 5 in 10,000 chance of having 14 or more estimates from garage I higher than those from garage II. Using this probability, we would argue that the observed estimates are highly contradictory to the null hypothesis of equality of distribution of estimates for the two garages. Why are there such conflicting results from the t test and the binomial calculation?

The explanation of the difference in the conclusions from the two procedures is that one of the required conditions for the t test, two samples being independent of one another, has been violated by the manner in which the study was conducted. The adjusters obtained a measurement from both garages for each car. For the two samples to be independent, the adjusters would have to take a random sample of 15 cars to garage I and a *different* random sample of 15 to garage II.

As can be observed in Figure 6.5, the repair estimates for a given car are about the same value, but there is a large variability in the estimates from each garage. The large variability *among* the 15 estimates from each garage diminishes the relative size of any difference *between* the two garages. When designing the study, the adjusters recognized that the large differences in the amount of damage suffered by the cars would result in a large variability in the 15 estimates at both garages. By having both garages give an estimate on each car, the adjusters could calculate the difference between the estimates from the garages and hence reduce the large car-to-car variability.

FIGURE 6.5

Repair estimates from two garages

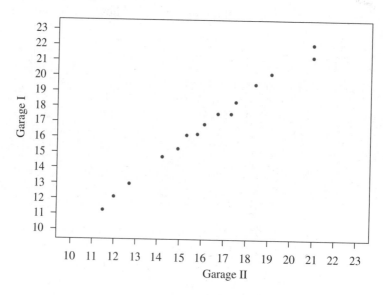

This example illustrates a general design principle. In many situations, the available experimental units may be considerably different prior to their random assignment to the treatments with respect to characteristics that may affect the experimental responses. These differences will often then mask true treatment differences. In the previous example, the cars had large differences in the amount of damage suffered during the accident and hence would be expected to have large differences in their repair estimates no matter what garage gave the repair estimate. When comparing two treatments or groups in which the available experimental units have important differences prior to their assignment to the treatments or groups, the samples should be paired. There are many ways to design experiments to yield paired data. One method involves having the same group of experimental units receive both treatments, as was done in the repair estimates example. A second method involves having measurements taken before and after the treatment is applied to the experimental units. For example, suppose we want to study the effect of a new medicine proposed to reduce blood pressure. We would record the blood pressure of participants before they received the medicine and then after receiving the medicine. A third design procedure uses naturally occurring pairs such as twins or husbands and wives. A final method pairs the experimental units with respect to factors that may mask differences in the treatments. For example, a study is proposed to evaluate two methods for teaching remedial reading. The participants could be paired based on a pretest of their reading ability. After pairing the participants, the two methods are randomly assigned to the participants within each pair.

A proper analysis of paired data needs to take into account the lack of independence between the two samples. The sampling distribution for the difference in the sample means, $(\bar{y}_1 - \bar{y}_2)$, will have mean and standard error

$$\mu_{\bar{y}_1 - \bar{y}_2} = \mu_1 - \mu_2 \quad \text{and} \quad \sigma_{\bar{y}_1 - \bar{y}_2} = \sqrt{\frac{\sigma_1^2 + \sigma_2^2 - 2\sigma_1\sigma_2\rho}{n}}$$

where ρ measures the amount of dependence between the two samples. When the two samples produce similar measurements, ρ is positive and the standard error of $\bar{y}_1 - \bar{y}_2$ is smaller than would be obtained using two independent samples. This was the case in the repair estimates data. The size and sign of ρ can be determined by examining the plot of the paired data values. The magnitude of ρ is large when the plotted points are close to a straight line. The sign of ρ is positive when the plotted points follow an increasing line and negative when plotted points follow a decreasing line. From Figure 6.5, we observe that the estimates are close to an increasing line and thus ρ will be positive. The use of paired data in the repair estimate study will reduce the variability in the standard error of the difference in the sample means in comparison to using independent samples.

The actual analysis of paired data requires us to compute the differences in the n pairs of measurements, $d_i = y_{1i} - y_{2i}$, and obtain $\bar{d}$, s_d, the mean and standard deviations in the d_is. Also, we must formulate the hypotheses about μ_1 and μ_2 into hypotheses about the mean of the differences, $\mu_d = \mu_1 - \mu_2$. The conditions required to develop a t procedure for testing hypotheses and constructing confidence intervals for μ_d are

1. The sampling distribution of the d_is is a normal distribution.
2. The d_is are independent; that is, the pairs of observations are independent.

A summary of the test procedure is given here.

Paired *t* test

H_0: **1.** $\mu_d \leq D_0$ (D_0 is a specified value, often 0)
    2. $\mu_d \geq D_0$
    3. $\mu_d = D_0$

H_a: **1.** $\mu_d > D_0$
    2. $\mu_d < D_0$
    3. $\mu_d \neq D_0$

T.S.: $t = \dfrac{\bar{d} - D_0}{s_d/\sqrt{n}}$

R.R.: For a level α Type I error rate and with df $= n - 1$
    1. Reject H_0 if $t \geq t_\alpha$.
    2. Reject H_0 if $t \leq -t_\alpha$.
    3. Reject H_0 if $|t| \geq t_{\alpha/2}$

Check assumptions and draw conclusions.

The corresponding $100(1 - \alpha)\%$ confidence interval on $\mu_d = \mu_1 - \mu_2$ based on the paired data is shown here.

$100(1 - \alpha)\%$ Confidence Interval for μ_d Based on Paired Data

$$\bar{d} \pm t_{\alpha/2}\frac{s_d}{\sqrt{n}}$$

where n is the number of pairs of observations (and hence the number of differences) and df $= n - 1$.

EXAMPLE 6.7

Refer to the data of Example 6.6 and perform a paired t test. Draw a conclusion based on $\alpha = .05$.

Solution   For these data, the parts of the statistical test are

H_0: $\mu_d = \mu_1 - \mu_2 \leq 0$
H_a: $\mu_d > 0$

T.S.: $t = \dfrac{\bar{d}}{s_d/\sqrt{n}}$

R.R.: For df $= n - 1 = 14$, reject H_0 if $t \geq t_{.05}$.

Before computing t, we must first calculate $\bar{d}$ and s_d. For the data of Table 6.8, we have the differences $d_i =$ garage I estimate $-$ garage II estimate.

Car	1	2	3	4	5	6	7	8	9	10	11	12	13	14	15
d_i	.3	1.1	1.1	−.2	.3	.5	.4	.9	.2	.6	.3	1.1	.8	.9	.9

The mean and standard deviation are given here.

$$\bar{d} = .61 \quad \text{and} \quad s_d = .394$$

Substituting into the test statistic t, we have

$$t = \frac{\bar{d} - 0}{s_d/\sqrt{n}} = \frac{.61}{.394/\sqrt{15}} = 6.00$$

Indeed, $t = 6.00$ is far beyond all tabulated t values for df $= 14$, so the p-value is less than .005; in fact, the p-value is .000016. We conclude that the mean repair estimate for garage I is greater than that for garage II. This conclusion agrees with our intuitive finding based on the binomial distribution.

The point of all this discussion is not to suggest that we typically have two or more analyses that may give *very* conflicting results for a given situation. Rather, the point is that the analysis must fit the experimental situation; and for this experiment, the samples are dependent, demanding we use an analysis appropriate for dependent (paired) data.

After determining that there is a *statistically significant* difference in the means, we should estimate the size of the difference. A 95% confidence interval for $\mu_1 - \mu_2 = \mu_d$ will provide an estimate of the size of the difference in the average repair estimate between the two garages.

$$\bar{d} \pm t_{\alpha/2}\frac{s_d}{\sqrt{n}}$$

$$.61 \pm 2.145\frac{.394}{\sqrt{15}} \quad \text{or} \quad .61 \pm .22$$

Thus, we are 95% confident that the mean repair estimates differ by a value between $390 and $830. The insurance adjusters determined that a difference of this size is of practical significance.

The reduction in standard error of $\bar{y}_1 - \bar{y}_2$ by using the differences d_is in place of the observed values y_{1i}s and y_{2i}s will produce a t test having greater power and confidence intervals having smaller width. Is there any loss in using paired data experiments? Yes, the t procedures using the d_is have df $= n - 1$, whereas the t procedures using the individual measurements have df $= n_1 + n_2 - 2 = 2(n - 1)$. Thus, when designing a study or experiment, the choice between using an independent samples experiment and a paired data experiment will depend on how much difference exists in the experimental units prior to their assignment to the treatments. If there are only small differences, then the independent samples design is more efficient. If the differences in the experimental units are extreme, then the paired data design is more efficient.

EXERCISES   Applications

Engin.   **6.16**   Researchers are studying two existing coatings used to prevent corrosion in pipes that transport natural gas. The study involves examining sections of pipe that had been in the ground at least 5 years. The effectiveness of the coating depends on the pH of the soil, so the researchers recorded the pH of the soil at all 20 sites at which the pipe was buried prior to measuring the amount of corrosion on the pipes. The pH readings are given here. Describe how the researchers could conduct the study to reduce the effect of the differences in the pH readings on the evaluation of the difference in the two coatings' corrosion protection.

pH Readings at 20 Research Sites									
Coating A 3.2	4.9	5.1	6.3	7.1	3.8	8.1	7.3	5.9	8.9
Coating B 3.7	8.2	7.4	5.8	8.8	3.4	4.7	5.3	6.8	7.2

Med.   **6.17** Suppose you are a participant in a project to study the effectiveness of a new treatment for high cholesterol. The new treatment will be compared to a current treatment by recording the change in cholesterol readings over a 10-week treatment period. The effectiveness of the treatment may depend on the participant's age, body fat percentage, diet, and general health. The study will involve at most 30 participants because of cost considerations.

a. Describe how you would conduct the study using independent samples.
b. Describe how you would conduct the study using paired samples.
c. How would you decide which method, paired or independent samples, would be more efficient in evaluating the change in cholesterol readings?

Med.   **6.18** The paper "Effect of long-term blood pressure control on salt sensitivity" [*Journal of Medicine* (1997) 28:147–156] presents the results of a study that evaluated salt sensitivity (SENS) after a period of antihypertensive treatment. Ten hypertensive patients (diastolic blood pressure between 90 and 115 mmHg) were studied after at least 18 months on antihypertensive treatment. SENS readings, which were obtained before and after the patients were placed on an antihypertensive treatment, are given here.

	Patient									
	1	**2**	**3**	**4**	**5**	**6**	**7**	**8**	**9**	**10**
Before treatment	22.86	7.74	15.49	9.97	1.44	9.39	11.40	1.86	−6.71	6.42
After treatment	6.11	−4.02	8.04	3.29	−0.77	6.99	10.19	2.09	11.40	10.70

a. Is there significant evidence that the mean SENS value decreased after the patient received antihypertensive treatment?
b. Estimate the size of the change in the mean SENS value.
c. Do the conditions required for using the t procedures appear to be valid for these data? Justify your answer.

Edu.   **6.19** A study was designed to measure the effect of home environment on academic achievement of 12-year-old students. Because genetic differences may also contribute to academic achievement, the researcher wanted to control for this factor. Thirty sets of identical twins were identified who had been adopted prior to their first birthday, with one twin placed in a home in which academics were emphasized (Academic) and the other twin placed in a home in which academics were not emphasized (Nonacademic). The final grades (based on 100 points) for the 60 students are given here.

Set of Twins	Academic	Non-academic	Set of Twins	Academic	Non-academic
1	78	71	9	98	92
2	75	70	10	52	56
3	68	66	11	67	63
4	92	85	12	55	52
5	55	60	13	49	48
6	74	72	14	66	67
7	65	57	15	75	70
8	80	75	16	90	88

Set of Twins	Academic	Non-academic	Set of Twins	Academic	Non-academic
17	89	80	24	70	62
18	73	65	25	68	73
19	61	60	26	74	73
20	76	74	27	85	75
21	81	76	28	97	88
22	89	78	29	95	94
23	82	78	30	78	75

a. Use the following computer output to evaluate whether there is a difference in the mean final grade between the students in an academically oriented home environment and those in a nonacademic home environment.

b. Estimate the size of the difference in the mean final grades of the students in academic and nonacademic home environments.

c. Do the conditions for using the *t* procedures appear to be satisfied for these data?

d. Does it appear that using twins in this study to control for variation in final scores was effective as compared to taking a random sample of 30 students in both types of home environments? Justify your answer.

```
Paired T-Test and Confidence Interval

Paired T for Academic - Nonacademic

                 N       Mean     StDev    SE Mean
Academic        30      75.23     13.29      2.43
Nonacademic     30      71.43     11.42      2.09
Difference      30      3.800      4.205     0.768

95% CI for mean difference: (2.230,  5.370)
T-Test of mean difference = 0 (vs not = 0): T-value = 4.95  P-Value = 0.000
```

```
Two-Sample T-Test and Confidence Interval

Two-sample T for Academic vs Nonacademic

               N      Mean     StDev    SE Mean
Academic      30      75.2      13.3       2.4
Nonacademic   30      71.4      11.4       2.1

95% CI for mu Academic - mu Nonacademic: ( -2.6,  10.2)
T-Test mu Academic = mu Nonacademic (vs not = ): T = 1.19   P = 0.24   DF = 56
```

Boxplot of differences (with H_0 and 95% t confidence interval for the mean)

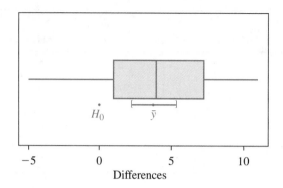

Normal probability plot of differences

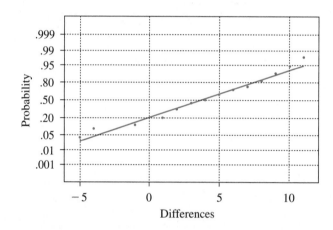

6.5 A Nonparametric Alternative: The Wilcoxon Signed-Rank Test

The Wilcoxon signed-rank test, which makes use of the sign and the magnitude of the rank of the differences between pairs of measurements, provides an alternative to the paired t test when the population distribution of the differences is nonnormal. The Wilcoxon signed-rank test requires that the population distribution of differences be symmetric about the unknown median M. Let D_0 be a specified hypothesized value of M. The test evaluates shifts in the distribution of differences to the right or left of D_0; in most cases, D_0 is 0. The computation of the signed-rank test involves the following steps:

1. Calculate the differences in the n pairs of observations.
2. Subtract D_0 from all the differences.
3. Delete all zero values. Let n be the number of nonzero values.
4. List the *absolute values* of the differences in increasing order, and assign them the ranks $1, \ldots, n$ (or the average of the ranks for ties).

We define the following notation before describing the Wilcoxon signed-rank test:

n = the number of pairs of observations with a nonzero difference
T_+ = the sum of the positive ranks; if there are no positive ranks, $T_+ = 0$
T_- = the sum of the negative ranks; if there are no negative ranks, $T_- = 0$
T = the smaller of T_+ and T_-

μ_T $$\boldsymbol{\mu_T} = \frac{n(n+1)}{4}$$

σ_T $$\boldsymbol{\sigma_T} = \sqrt{\frac{n(n+1)(2n+1)}{24}}$$

g groups   If we group together all differences assigned the same rank, and there are g such **groups,** the variance of T is

$$\sigma_T^2 = \frac{1}{24}\left[n(n+1)(2n+1) - \frac{1}{2}\sum_{j=1}^{g} t_j(t_j - 1)(t_j + 1) \right]$$

t_j where t_j is the number of tied ranks in the jth group. Note that if there are no tied ranks, $g = n$, and $t_j = 1$ for all groups. The formula then reduces to

$$\sigma_T^2 = \frac{n(n+1)(2n+1)}{24}$$

The Wilcoxon signed-rank test is presented here.

Wilcoxon Signed-Rank Test

H_0:   The distribution of differences is symmetrical around D_0. (D_0 is specified; usually D_0 is 0.)

H_a: **1.** The differences tend to be larger than D_0.
        2. The differences tend to be smaller than D_0.
        3. Either 1 or 2 is true (two-sided H_a).

($n \le 50$)

T.S.:   **1.** $T = T_-$
        2. $T = T_+$
        3. T = smaller of T_+ and T_-

R.R.:   For a specified value of α (one-tailed .05, .025, .01, or .005; two-tailed .10, .05, .02, .01) and fixed number of nonzero differences n, reject H_0 if the value of T is less than or equal to the appropriate entry in Table 7 in the Appendix.

($n > 50$)

T.S.:   Compute the test statistic

$$z = \frac{T - \dfrac{n(n+1)}{4}}{\sqrt{\dfrac{n(n+1)(2n+1)}{24}}}$$

R.R.:   For cases 1 and 2, reject H_0 if $z < -z_\alpha$; for case 3, reject H_0 if
$z < -z_{\alpha/2}$.

Check assumptions and draw conclusions.

EXAMPLE 6.8

A city park department compared a new fertilizer, brand A, to the previously used fertilizer, brand B, on each of 20 different softball fields. Each field was divided in half, with brand A randomly assigned to one-half of the field and brand B to the other. Sixty pounds of fertilizer per acre were then applied to the fields. The effect of the fertilizer on the grass grown at each field was measured by the weight (in pounds) of grass clippings produced by mowing the grass at the fields over a 1-month period. Evaluate whether brand A tends to produce more grass than brand B. The data are given here.

Field	Brand A	Brand B	Difference	Field	Brand A	Brand B	Difference
1	211.4	186.3	25.1	11	208.9	183.6	25.3
2	204.4	205.7	−1.3	12	208.7	188.7	20.0
3	202.0	184.4	17.6	13	213.8	188.6	25.2
4	201.9	203.6	−1.7	14	201.6	204.2	−2.6
5	202.4	180.4	22.0	15	201.8	181.6	20.1
6	202.0	202.0	0	16	200.3	208.7	−8.4
7	202.4	181.5	20.9	17	201.8	181.5	20.3
8	207.1	186.7	20.4	18	201.5	208.7	−7.2
9	203.6	205.7	−2.1	19	212.1	186.8	25.3
10	216.0	189.1	26.9	20	203.4	182.9	20.5

Solution   Plots of the differences in grass yields for the 20 fields are given in Figures 6.6 (a) and (b). The differences appear to not follow a normal distribution and appear to form two distinct clusters. Thus, we will apply the Wilcoxon signed-rank test to evaluate the differences in grass yields from brand A and brand B. The null hypothesis is that the distribution of differences is symmetrical about 0 against the alternative that the differences tend to be greater than 0. First we must rank (from smallest to largest) the absolute values of the $n = 20 - 1 = 19$ nonzero differences. These ranks appear in Table 6.9.

The sums of the positive and negative ranks are

$$T_- = 1 + 2 + 3 + 4 + 5 + 6 = 21$$

and

$$T_+ = 7 + 8 + 9 + 10 + 11 + 12 + 13 + 14 + 15 + 16 + 17.5 + 17.5 + 19$$
$$= 169$$

Thus, T, the smaller of T_+ and T_-, is 21. For a one-sided test with $n = 19$ and $\alpha = .05$, we see from Table 7 in the Appendix that we will reject H_0 if T is less than or equal to 53. Thus, we reject H_0 and conclude that brand A fertilizer tends to produce more grass than brand B.

FIGURE 6.6

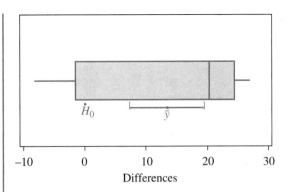

(a) Boxplot of differences (with H_0 and 95% t confidence interval for the mean)

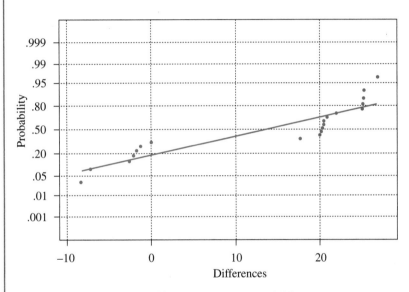

(b) Normal probability of differences

TABLE 6.9

Rankings of grass yield data

Field	Difference	Rank of Absolute Difference	Sign of Difference	Field	Difference	Rank of Absolute Difference	Sign of Difference
1	25.1	15	Positive	11	25.3	17.5	Positive
2	−1.3	1	Negative	12	20.0	8	Positive
3	17.6	7	Positive	13	25.2	16	Positive
4	−1.7	2	Negative	14	−2.6	4	Negative
5	22.0	14	Positive	15	20.1	9	Positive
6	0	None	Positive	16	−8.4	6	Negative
7	20.9	13	Positive	17	20.3	10	Positive
8	20.4	11	Positive	18	−7.2	5	Negative
9	−2.1	3	Negative	19	25.3	17.5	Positive
10	26.9	19	Positive	20	20.5	12	Positive

The choice of an appropriate paired-sample test depends on examining different types of deviations from normality. Because the level of the Wilcoxon signed-rank does not depend on the population distribution, its level is the same as the stated value for all symmetric distributions. The level of the paired t test may be different from its stated value when the population distribution is very nonnormal. First, plot the data and attempt to determine whether the population distribution is very heavy-tailed or very skewed. In such cases, use a Wilcoxon rank-based test. When the plots are not definitive in their detection of nonnormality, perform both tests. If the results from the different tests yield different conclusions, carefully examine the data to identify any peculiarities to understand why the results differ. If the conclusions agree and there are no blatant violations of the required conditions, you should be very confident in your conclusions. This particular hedging strategy is appropriate not only for paired data but also for many situations in which there are several alternative analyses.

EXERCISES

Basic Techniques

6.20 Refer to Exercise 6.18.

    a. Using the data in the table, run a Wilcoxon signed-rank test. Give the p-value and draw a conclusion.

    b. Compare your conclusions here to those in Exercise 6.18. Does it matter which test (t or signed-rank) is used?

Applications

Bio.   **6.21** The effect of Benzedrine on the heart rate of dogs (in beats per minute) was examined in an experiment on 14 dogs chosen for the study. Each dog was to serve as its own control, with half of the dogs assigned to receive Benzedrine during the first study period and the other half assigned to receive a placebo (saline solution). All dogs were examined to determine the heart rates after 2 hours on the medication. After 2 weeks in which no medication was given, the regimens for the dogs were switched for the second study period. The dogs previously on Benzedrine were given the placebo and the others received Benzedrine. Again heart rates were measured after 2 hours.

    The following sample data are not arranged in the order in which they were taken but have been summarized by regimen. Use these data to test the research hypothesis that the distribution of heart rates for the dogs when receiving Benzedrine is shifted to the right of that for the same animals when on the placebo. Use a one-tailed Wilcoxon signed-rank test with $\alpha = .05$.

Dog	Placebo	Benzedrine	Dog	Placebo	Benzedrine
1	250	258	8	296	305
2	271	285	9	301	319
3	243	245	10	298	308
4	252	250	11	310	320
5	266	268	12	286	293
6	272	278	13	306	305
7	293	280	14	309	313

Soc.   **6.22** A study was conducted to determine whether automobile repair charges are higher for female customers than for male customers. Ten auto repair shops were randomly selected from the telephone book. Two cars of the same age, brand, and engine problem were used in the study. For each repair shop, the two cars were randomly assigned to a man and woman participant and then taken to the shop for an estimate of repair cost. The repair costs (in dollars) are given here.

Repair Shop	1	2	3	4	5	6	7	8	9	10	11	12	13	14	15	16	17	18	19	20
Female customers	871	684	795	838	1,033	917	1,047	723	1,179	707	817	846	975	868	1,323	791	1,157	932	1,089	770
Male customers	792	765	511	520		618	447		548	720		899	788	927	657	851	702		918	528	884	702		839	878

a. Which procedure, t or Wilcoxon, is more appropriate in this situation? Why?
b. Are repair costs generally higher for female customers than for male customers? Use $\alpha = .05$.

Boxplot of differences (with H_0 and 95% t confidence interval for the mean)

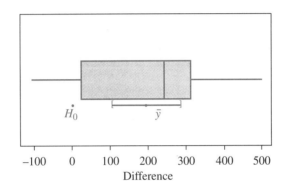

Normal probability plot of differences in cost

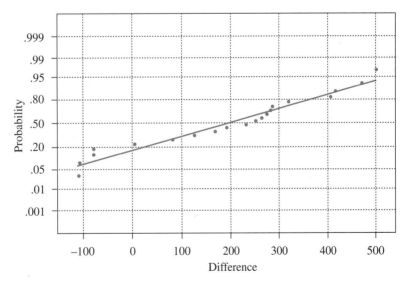

6.6   Choosing Sample Sizes for Inferences about $\mu_1 - \mu_2$

Sections 5.3 and 5.5 were devoted to sample-size calculations to obtain a confidence interval about μ with a fixed width and specified degree of confidence or to conduct a statistical test concerning μ with predefined levels for α and β. Similar calculations can be made for inferences about $\mu_1 - \mu_2$ with either independent samples or with paired data. Determining the sample size for a $100(1 - \alpha)\%$ confidence interval about $\mu_1 - \mu_2$ of width $2E$ based on independent samples is possible by solving the following expression for n. We will assume that both samples are the same size.

$$z_{\alpha/2}\sigma\sqrt{\frac{1}{n} + \frac{1}{n}} = E$$

Note that, in this formula, σ is the common population standard deviation and that we have assumed equal sample sizes.

Sample Sizes for a 100(1 − α)% Confidence Interval for $\mu_1 - \mu_2$ of the Form $\bar{y}_1 - \bar{y}_2 \pm E$, Independent Samples

$$n = \frac{2z_{\alpha/2}^2\sigma^2}{E^2}$$

(*Note:* If σ is unknown, substitute an estimated value to get an approximate sample size.)

The sample sizes obtained using this formula are usually approximate because we have to substitute an estimated value of σ, the common population standard deviation. This estimate will probably be based on an educated guess from information from a previous study or from the range of population values.

Corresponding sample sizes for one- and two-sided tests of $\mu_1 - \mu_2$ based on specified values of α and β, where we desire a level α test having the probability of a Type II error $\beta(\mu_1 - \mu_2) \leq \beta$ whenever $|\mu_1 - \mu_2| \geq \Delta$, are shown here.

Sample Sizes for Testing $\mu_1 - \mu_2$, Independent Samples

One-sided test: $n = 2\sigma^2\dfrac{(z_\alpha + z_\beta)^2}{\Delta^2}$

Two-sided test: $n = 2\sigma^2\dfrac{(z_{\alpha/2} + z_\beta)^2}{\Delta^2}$

where $n_1 = n_2 = n$ and the probability of a Type II error is to be $\leq\beta$ when the true difference $|\mu_1 - \mu_2| \geq \Delta$. (*Note:* If σ is unknown, substitute an estimated value to obtain an approximate sample size.)

EXAMPLE 6.9

An experiment was done to determine the effect on dairy cattle of a diet supplemented with liquid whey. Whereas no differences were noted in milk production measurements among cattle given a standard diet (7.5 kg of grain plus hay by choice) with water and those on the standard diet and liquid whey only, a considerable difference between the groups was noted in the amount of hay ingested. Suppose that we test the null hypothesis of no difference in mean hay consumption for the two diet groups of dairy cattle. For a two-tailed test with $\alpha = .05$, determine the approximate number of dairy cattle that should be included in each group if we want $\beta \leq .10$ for $|\mu_1 - \mu_2| \geq .5$. Previous experimentation has shown σ to be approximately .8.

Solution From the description of the problem, we have $\alpha = .05$, $\beta \le .10$ for $\Delta = |\mu_1 - \mu_2| \ge .5$, and $\sigma = .8$. Table 1 in the Appendix gives us $z_{.025} = 1.96$ and $z_{0.10} = 1.28$. Substituting into the formula, we have

$$n \approx \frac{2(.8)^2(1.96 + 1.28)^2}{(.5)^2} = 53.75, \text{ or } 54$$

Thus, we need 54 cattle per group to run the desired test.

Sample-size calculations can also be performed when the desired sample sizes are unequal, $n_1 \ne n_2$. Let n_2 be some multiple m of n_1; that is, $n_2 = mn_1$. For example, we may want n_1 three times as large as n_2; hence, $n_2 = \frac{1}{3}n_1$. The displayed formulas can still be used, but we must substitute $(m + 1)/m$ for 2 and n_1 for n in the sample-size formulas. After solving for n_1, we have $n_2 = mn_1$.

EXAMPLE 6.10

Refer to Example 6.9. Suppose the experimenters wanted more information about the diet with liquid whey (group II) than about the diet with water (group I). In particular, the experimenters wanted 40% more cattle in group II than in group I; that is, $n_2 = (1.4)n_1$. All other specifications are as specified in Example 6.9.

Solution We replace 2 in our sample-size formula with $\dfrac{(1.4 + 1)}{1.4}$. We then have

$$n_1 \approx \frac{(m + 1)\sigma^2}{m}\frac{(z_{\alpha/2} + z_\beta)^2}{\Delta^2} = \frac{(1.4 + 1)(.8)^2}{1.4}\frac{(1.96 + 1.28)^2}{(.5)^2} = 46.07, \text{ or } 47$$

That is, we need 47 cattle in group I and $(1.4)47 = 65.8$, or 66 in group II.

Sample sizes for estimating μ_d and conducting a statistical test for μ_d based on paired data (differences) are found using the formulas of Chapter 5 for μ. The only change is that we are working with a single sample of differences rather than a single sample of y values. For convenience, the appropriate formulas are shown here.

Sample Size for a 100(1 − α)% Confidence Interval for $\mu_1 - \mu_2$ of the Form $\bar{d} \pm E$, Paired Samples

$$n = \frac{z_{\alpha/2}^2 \sigma_d^2}{E^2}$$

(*Note:* If σ_d is unknown, substitute an estimated value to obtain approximate sample size.)

Sample Sizes for Testing $\mu_1 - \mu_2$, Paired Samples

One-sided test: $n = \dfrac{\sigma_d^2(z_\alpha + z_\beta)^2}{\Delta^2}$

Two-sided test: $n = \dfrac{\sigma_d^2(z_{\alpha/2} + z_\beta)^2}{\Delta^2}$

where the probability of a Type II error is β or less if the true difference $\mu_d \ge \Delta$. (*Note:* If σ_d is unknown, substitute an estimated value to obtain an approximate sample size.)

6.7 Encounters with Real Data:
Effect of an Oil Spill on Plant Growth

Defining the Problem (1)

On January 7, 1992, an underground oil pipeline ruptured and caused the contamination of a marsh along the Chiltipin Creek in San Patricio County, Texas. The cleanup process consisted of burning the contaminated regions in the marsh. To evaluate the influence of the oil spill on the flora, researchers designed a study of plant growth 1 year after the burning. In an unpublished Texas A&M University dissertation, Newman (1998) describes the researchers' findings with respect to *Distichlis spicata,* a flora of particular importance to the area of the spill.

Collecting Data (2)

The researchers needed to determine the important characteristics of the flora that may be affected by the spill. Here are some of the questions to be answered before starting the study:

1. What are the factors that determine the viability of the flora?
2. How did the oil spill affect these factors?
3. Are there data on the important flora factors prior to the spill?
4. How should the researchers measure the flora factors in the oil spill region?
5. How many observations are necessary to confirm that the flora have undergone a change after the oil spill?
6. What type of experimental design or study is needed?
7. What statistical procedures are valid for making inferences about the change in flora parameters after the oil spill?
8. What types of information should be included in a final report to document any changes observed in the flora parameters?

After lengthy discussion, reading of the relevant literature, and searching many databases about similar sites and flora, the researchers found there was no specific information on the flora in this region prior to the oil spill. They determined that the flora parameters of interest were the average density μ of *Distichlis spicata* after burning the spill region, the variability σ in flora density, and the proportion π of the spill region in which the flora density was essentially zero. Because there was no relevant information on flora density in the spill region before the spill, it was necessary to evaluate the flora density in unaffected areas of the marsh to determine whether the plant density had changed after the oil spill. The researchers located several regions that had not been contaminated by the oil spill. The spill region and the unaffected regions were divided into tracts of nearly the same size. The number of tracts needed in the study was determined by specifying how accurately the parameters μ, σ, and π needed to be estimated to achieve a level of precision as specified by the width of 95% confidence intervals and by the power of tests of hypotheses. From these calculations and within budget and time limitations, it was decided that 40 tracts from both the spill and unaffected areas would be used in the study. Forty tracts of exactly the same size were randomly selected in these locations and the *Distichlis spicata* density was recorded. Similar measurements were taken within the spill area of the marsh.

Summarizing Data (3)

The data consist of 40 measurements of flora density in the uncontaminated (control) sites and 40 density measurements in the spill (burned) sites. (The data are in the FLORADEN file of the CD that came with your book.) The researchers next carefully examined the data from the fieldwork to determine whether the measurements were recorded correctly. The data was then transferred to computer files and prepared for analysis following the steps outlined in Section 2.7. The next step in the study was to summarize the data through plots and summary statistics. The data are displayed in Figure 6.7 with summary statistics given in Table 6.10. A boxplot of the data displayed in Figure 6.8 indicates that the control sites have a somewhat greater plant density than the oil spill sites. From the summary statistics, we have that the average flora density in the control sites is $\bar{y}_{Con} = 38.48$ with a standard deviation of $s_{Con} = 16.37$. The sites within the spill region have an average density of $\bar{y}_{Spill} = 26.93$ with a standard deviation of $s_{Spill} = 9.88$. Thus, the control sites have a larger average flora density and a greater variability in flora density than the sites within the spill region. To determine whether these observed differences in flora density reflect similar differences in all the sites and not just the ones in the study will require an analysis of the data.

FIGURE 6.7

Number of plants observed in tracts at oil spill and control sites. The data are displayed in stem-and-leaf plots

Control Tracts				Oil Spill Tracts		
Mean:	38.48	000	0		Mean:	26.93
Median:	41.50	7	0	59	Median:	26.00
St. Dev:	16.37	1	1	14	St. Dev.	9.88
n:	40	6	1	77799	*n*:	40
		4	2	2223444		
		9	2	555667779		
		0	3	11123444		
		55678	3	5788		
		000111222233	4	1		
		57	4			
		0112344	5	02		
		67789	5			

TABLE 6.10

Summary statistics for oil spill data

			Descriptive Statistics			
Variable	**Site Type**	**N**	**Mean**	**Median**	**Tr. Mean**	**St. Dev.**
No. plants	Control	40	38.48	41.50	39.50	16.37
	Oil spill	40	26.93	26.00	26.69	9.88
Variable	**Site Type**	**SE Mean**	**Minimum**	**Maximum**	**Q1**	**Q3**
No. plants	Control	2.59	0.00	59.00	35.00	51.00
	Oil spill	1.56	5.00	52.00	22.00	33.75

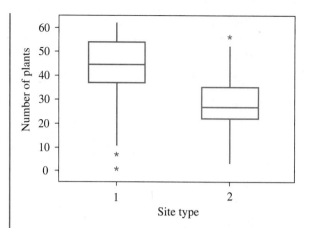

Analyzing Data, Interpreting the Analyses, and Communicating the Results (4)

The researchers hypothesized that the oil spill sites would have a lower plant density than the control sites. Thus, we will construct confidence intervals on the mean plant density in the control plots μ_1 and in the oil spill plots μ_2 to assess their average plant density. Also, we can construct confidence intervals on the difference $\mu_1 - \mu_2$ and test the research hypothesis that μ_1 is greater than μ_2. From Figure 6.8, the data from the oil spill area appear to have a normal distribution, whereas the data from the control area appear to be skewed to the left. The normal probability plots are given in Figure 6.9 (a) and (b) to assess further whether the population distributions are in fact normal in shape. We observe that the data from the oil spill tracts appear to follow a normal distribution, but that the data from the control tracts do not because their plotted points do not fall close to the straight line. Also, the variability in plant density is higher in control sites than in the oil spill sites. Thus, the approximate t procedures will be the most appropriate inference procedures. The sample data yielded the following values:

Control Plots	Oil Spill Plots
$n_1 = 40$	$n_2 = 40$
$\bar{y}_1 = 38.48$	$\bar{y}_2 = 26.93$
$s_1 = 16.37$	$s_2 = 9.88$

The research hypothesis is that the mean plant density for the control plots exceeds that for the oil spill plots. Thus, our approximate t test is set up as follows:

$$H_0: \mu_1 \leq \mu_2 \quad \text{versus} \quad H_a: \mu_1 > \mu_2$$

That is,

$$H_0: \quad \mu_1 - \mu_2 \leq 0$$

$$H_a: \quad \mu_1 - \mu_2 > 0$$

$$\text{T.S.:} \quad t' = \frac{(\bar{y}_1 - \bar{y}_2) - D_0}{\sqrt{\dfrac{s_1^2}{n_1} + \dfrac{s_2^2}{n_2}}} = \frac{(38.48 - 26.93) - 0}{\sqrt{\dfrac{(16.37)^2}{40} + \dfrac{(9.88)^2}{40}}} = 3.82$$

FIGURE 6.9

(a) Normal probability plot
for control sites; (b) Normal
probability plot for
oil spill sites

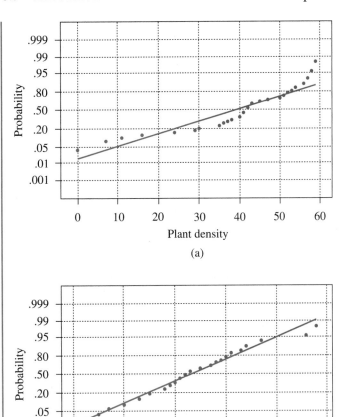

(a)

(b)

To compute the rejection region and p-value, we need to compute the approximate df for t':

$$c = \frac{s_1^2/n_1}{\dfrac{s_1^2}{n_1} + \dfrac{s_2^2}{n_2}} = \frac{(16.37)^2/40}{(16.37)^2/40 + (9.88)^2/40} = .73$$

$$\mathrm{df} = \frac{(n_1-1)(n_2-1)}{(1-c)^2(n_1-1) + c^2(n_2-1)} = \frac{(39)(39)}{(1-.73)^2(39) + (.73)^2(39)}$$

$$= 64.38, \text{ which is rounded to } 64$$

Table 3 in the Appendix does not have df = 64, so we will use df = 60. In fact, the difference is very small when df becomes large: $t_{.05} = 1.671$ and 1.669 for df = 60 and 64, respectively.

R.R.:   For $\alpha = .05$ and df = 60, reject H_0 if $t' > 1.671$.

Because $t' = 3.82$ is greater than 1.671, we reject H_0. We can bound the p-value using Table 3 in the Appendix with df $= 60$. With $t' = 3.82$, the level of significance is p-value $< .001$. Thus we can conclude that there is significant (p-value $< .001$) evidence that μ_1 is greater than μ_2. Although we have determined that there is a statistically significant difference between the mean plant densities at the control and oil spill sites, the question remains whether these differences have *practical* significance. We can estimate the size of the difference in the means by placing a 95% confidence interval on $\mu_1 - \mu_2$.

The appropriate 95% confidence interval for $\mu_1 - \mu_2$ is computed by using the following formula with df $= 64$, the same as was used for the R.R.

$$(\bar{y}_1 - \bar{y}_2) \pm t_{\alpha/2}\sqrt{\frac{s_1^2}{n_1} + \frac{s_2^2}{n_2}} \quad \text{or} \quad (38.48 - 26.93) \pm 2.0\sqrt{\frac{(16.37)^2}{40} + \frac{(9.88)^2}{40}}$$

$$\text{or} \quad 11.55 \pm 6.05$$

Thus, we are 95% confident that the mean plant densities differ by an amount between 5.5 and 17.6. The plant scientists then evaluated whether a difference in this range was of practical importance.

Formal communication and documentation of the results require a report summarizing our findings from the study. The report should include the following items:

1. Statement of the objective for the study
2. Description of the study design and data collection procedures
3. Numerical and graphical summaries of the data sets

   - Table of means, medians, standard deviations, quartiles, range
   - Boxplots
   - Stem-and-leaf plots

4. Description of all inference methodologies:

   - Approximate t tests of differences in means
   - Approximate t-based confidence interval on population means
   - Verification that all necessary conditions for using inference techniques were satisfied using boxplots, normal probability plots

5. Discussion of results and conclusions
6. Interpretation of findings relative to previous studies
7. Recommendations for future studies
8. Listing of the data set

6.8   Staying Focused

In this chapter, we have considered inferences about $\mu_1 - \mu_2$. The first set of methods was based on independent random samples being selected from the populations of interest. We learned how to sample data to run a statistical test or to construct a confidence interval for $\mu_1 - \mu_2$ using t methods. Wilcoxon's rank sum test, which does not require normality of the underlying populations, was presented as an alternative to the t test.

The second major set of procedures can be used to make comparisons between two populations when the sample measurements are paired. In this situation, we no longer have independent random samples, and hence the procedures of Sections 6.2 and 6.3 (t methods and Wilcoxon's rank sum) are inappropriate. The test and estimation methods for paired data are based on the sample differences for the paired measurements or the ranks of the differences. The paired t test and corresponding confidence interval based on the difference measurements were introduced and found to be identical to the single-sample t methods of Chapter 5. The nonparametric alternative to the paired t test is Wilcoxon's signed-rank test.

The material presented in Chapters 5 and 6 lays the foundation for statistical inference (estimation and testing) for the remainder of the book. Review the material in this chapter periodically as new topics are introduced so that you retain the basic elements of statistical inference.

Key Formulas

1. $100(1 - \alpha)\%$ confidence interval for $\mu_1 - \mu_2$, independent samples; y_1 and y_2 approximately normal; $\sigma_1^2 = \sigma_2^2$

$$\bar{y}_1 - \bar{y}_2 \pm t_{\alpha/2} s_p \sqrt{\frac{1}{n_1} + \frac{1}{n_2}}$$

where

$$s_p = \sqrt{\frac{(n_1 - 1)s_1^2 + (n_2 - 1)s_2^2}{n_1 + n_2 - 2}} \quad \text{and} \quad df = n_1 + n_2 - 2$$

2. t test for $\mu_1 - \mu_2$, independent samples; y_1 and y_2 approximately normal; $\sigma_1^2 = \sigma_2^2$

$$\text{T.S.:} \quad t = \frac{(\bar{y}_1 - \bar{y}_2) - D_0}{s_p \sqrt{1/n_1 + 1/n_2}} \quad df = n_1 + n_2 - 2$$

3. t' test for $\mu_1 - \mu_2$, unequal variances; independent samples; y_1 and y_2 approximately normal;

$$\text{T.S.:} \quad t' = \frac{(\bar{y}_1 - \bar{y}_2) - D_0}{\sqrt{\dfrac{s_1^2}{n_1} + \dfrac{s_2^2}{n_2}}} \quad df = \frac{(n_1 - 1)(n_2 - 1)}{(1 - c)^2(n_1 - 1) + c^2(n_2 - 1)}$$

where

$$c = \frac{s_1^2/n_1}{\dfrac{s_1^2}{n_1} + \dfrac{s_2^2}{n_2}}$$

4. $100(1 - \alpha)\%$ confidence interval for $\mu_1 - \mu_2$, unequal variances; independent samples; y_1 and y_2 approximately normal;

$$(\bar{y}_1 - \bar{y}_2) \pm t_{\alpha/2} \sqrt{\frac{s_1^2}{n_1} + \frac{s_2^2}{n_2}}$$

where the t percentile has

$$df = \frac{(n_1 - 1)(n_2 - 1)}{(1 - c)^2(n_1 - 1) + c^2(n_2 - 1)},$$

with

$$c = \frac{s_1^2/n_1}{\dfrac{s_1^2}{n_1} + \dfrac{s_2^2}{n_2}}$$

5. Wilcoxon's rank sum test, independent samples

H_0:   The two populations are identical.

$(n_1 \leq 10, n_2 \leq 10)$

T.S.: T, the sum of the ranks in sample 1

$(n_1, n_2 > 10)$

T.S.: $z = \dfrac{T - \mu_T}{\sigma_T}$

where T denotes the sum of the ranks in sample 1

$$\mu_T = \frac{n_1(n_1 + n_2 + 1)}{2} \quad \text{and} \quad \sigma_T = \sqrt{\frac{n_1 n_2}{12}(n_1 + n_2 + 1)}$$

provided there are no tied ranks.

6. Paired t test; difference approximately normal

T.S.: $t = \dfrac{\bar{d} - D_0}{s_d/\sqrt{n}}$ $df = n - 1$

where n is the number of differences.

7. $100(1 - \alpha)\%$ confidence interval for μ_d, paired data; differences approximately normal

$$\bar{d} \pm t_{\alpha/2}s_d/\sqrt{n}$$

8. Wilcoxon's signed-rank test, paired data

H_0:   The distribution of differences is symmetrical about D_0.
T.S.: $(n \leq 50)$ T_-, or T_+ or smaller of T_+ and T_- depending on the form of H_a.
T.S.: $n > 50$

$$z = \frac{T - \mu_T}{\sigma_T}$$

where

$$\mu_T = \frac{n(n + 1)}{4} \quad \text{and} \quad \sigma_T = \sqrt{\frac{n(n + 1)(2n + 1)}{24}}$$

provided there are no tied ranks.

9. Independent samples: sample sizes for estimating $\mu_1 - \mu_2$ with a $100(1 - \alpha)\%$ confidence interval, of the form $\bar{y}_1 - \bar{y}_2 \pm E$

$$n = \frac{2z_{\alpha/2}^2\sigma^2}{E^2}$$

10. Independent samples: sample sizes for a test of $\mu_1 - \mu_2$
   a. One-sided test:

$$n = \frac{2\sigma^2(z_\alpha + z_\beta)^2}{\Delta^2}$$

   b. Two-sided test:

$$n = \frac{2\sigma^2(z_{\alpha/2} + z_\beta)^2}{\Delta^2}$$

11. Paired samples: sample size for estimating $\mu_1 - \mu_2$ with $100(1 - \alpha)\%$ confidence interval, of the form $\bar{d} \pm E$

$$n = \frac{z_{\alpha/2}^2\sigma_d^2}{E^2}$$

12. Paired samples: sample size for a test of $\mu_1 - \mu_2$
   a. One-sided test:

$$n = \frac{2\sigma_d^2(z_\alpha + z_\beta)^2}{\Delta^2}$$

   b. Two-sided test:

$$n = \frac{2\sigma_d^2(z_{\alpha/2} + z_\beta)^2}{\Delta^2}$$

Supplementary Exercises

Med.   **6.23** Long-distance runners have contended that moderate exposure to ozone increases lung capacity. To investigate this possibility, a researcher exposed 12 rats to ozone at the rate of 2 parts per million for a period of 30 days. The lung capacity of the rats was determined at the beginning of the study and again after the 30 days of ozone exposure. The lung capacities (in mL) are given here.

Rat	1	2	3	4	5	6	7	8	9	10	11	12
Before exposure	8.7	7.9	8.3	8.4	9.2	9.1	8.2	8.1	8.9	8.2	8.9	7.5
After exposure	9.4	9.8	9.9	10.3	8.9	8.8	9.8	8.2	9.4	9.9	12.2	9.3

   a. Is there sufficient evidence to support the conjecture that ozone exposure increases lung capacity? Use $\alpha = .05$. Report the *p*-value of your test.
   b. Estimate the size of the increase in lung capacity after exposure to ozone using a 95% confidence interval.
   c. After completion of the study, the researcher claimed that ozone causes increased lung capacity. Is this statement supported by this experiment?

Env.   **6.24**   In an environmental impact study for a new airport, the noise level of various jets was measured just seconds after their wheels left the ground. The jets were either widebodied or narrow-
bodied. The noise levels in decibels (dB) are recorded here for 15 widebodied jets and 12 narrow-
bodied jets.

Wide-Bodied Jet	109.5	107.3	105.0	117.3	105.4	113.7	121.7	109.2	108.1	106.4	104.6	110.5	110.9	111.0	112.4
Narrow-Bodied Jet	131.4	126.8	114.1	126.9	108.2	122.0	106.9	116.3	115.5	111.6	124.5	116.2			

a. Do the two types of jets have different mean noise levels? Report the level of significance of the test.
b. Estimate the size of the difference in mean noise level between the two types of jets using a 95% confidence interval.
c. How would you select the jets for inclusion in this study?

Env.   **6.25**   Following the March 24, 1989, grounding of the tanker *Exxon Valdez* in Alaska, approximately 35,500 tons of crude oil were released into Prince William Sound. The paper, "The deep
benthos of Prince William Sound, Alaska, 16 months after the *Exxon Valdez* oil spill" [*Marine Pollution Bulletin* (1998) 36:118–130] reports on an evaluation of deep benthic infauna after the spill.
Thirteen sites were selected for study. Seven of the sites were within the oil trajectory and six were
outside the oil trajectory. Collection of environmental and biological data at two depths, 40 m and
100 m, occurred in the period July 1–23, 1990. One of the variables measured was population
abundance (individuals per square meter). The values are given in the following table.

Site	Within Oil Trajectory							Outside Oil Trajectory					
	1	2	3	4	5	6	7	1	2	3	4	5	6
Depth 40 m	5,124	2,904	3,600	2,880	2,578	4,146	1,048	1,336	394	7,370	6,762	744	1,874
Depth 100 m	3,228	2,032	3,256	3,816	2,438	4,897	1,346	1,676	2,008	2,224	1,234	1,598	2,182

a. After combining the data from the two depths, does there appear to be a difference in
population mean abundance between the sites within and outside the oil trajectory?
Use $\alpha = .05$.
b. Estimate the size of the difference in population mean abundance at the two types of
sites using a 95% confidence interval.
c. What are the required conditions for the techniques used in parts (a) and (b)?
d. Check to see whether the required conditions are satisfied.

6.26   Refer to Exercise 6.25. Answer the following questions using the combined data for both
depths.
a. Use the Wilcoxon rank sum test to assess whether there is a difference in population
abundance between the sites within and outside the oil trajectory. Use $\alpha = .05$.
b. What are the required conditions for the techniques used in part (a)?
c. Are the required conditions satisfied?
d. Discuss any differences in the conclusions obtained using the *t*-procedures and the
Wilcoxon rank sum test.

6.27   Refer to Exercise 6.25. The researchers also examined the effect of depth on population
abundance.
a. Plot the four data sets using side-by-side boxplots to demonstrate the effect of depth on
population abundance.
b. Separately for each depth, evaluate differences between the sites within and outside the
oil trajectory. Use $\alpha = .05$.
c. Are your conclusions at 40 m consistent with your conclusions at 100 m?

6.28 Refer to Exercises 6.25 – 6.27.

a. Discuss the veracity of the statement, "The oil spill did not adversely affect the population abundance; in fact, it appears to have increased the population abundance."

b. A possible criticism of the study is that the six sites outside the oil trajectory were not comparable in many aspects to the seven sites within the oil trajectory. Suppose that the researchers had data on population abundance at the seven within sites prior to the oil spill. What type of analysis could be used on these data to evaluate the effect of the oil spill on population abundance? What are some advantages to using this data rather than the data in Exercise 6.25?

c. What are some possible problems with using the before and after oil spill data in assessing the effect of the spill on population abundance?

Bio.   **6.29** A study was conducted to evaluate the effectiveness of an antihypertensive product. Three groups of 20 rats each were randomly selected from a strain of hypertensive rats. The 20 rats in the first group were treated with a low dose of an antihypertensive product, the second group with a higher dose of the same product, and the third group with an inert control. Note that negative values represent increases in blood pressure. The accompanying computer output can be used to answer the following questions.

Boxplot of
blood pressure data

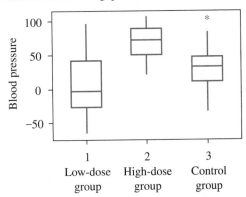

Row	Low Dose	High Dose	Control
1	-45.1	54.2	18.2
2	-59.8	89.1	17.2
3	58.1	89.6	34.8
4	-23.7	98.8	3.2
5	64.9	107.3	42.9
6	12.1	65.1	-27.2
7	10.5	75.6	42.6
8	42.5	52.0	10.0
9	48.5	50.2	102.3
10	-1.7	80.9	61.0
11	-65.4	92.6	-33.1
12	-17.5	55.3	55.1
13	22.1	103.2	84.6
14	-15.4	45.4	40.3
15	96.5	70.9	30.5
16	-27.7	29.7	18.5
17	-16.7	40.3	29.3
18	39.5	73.3	-19.7
19	-4.2	21.0	37.2
20	-41.3	73.2	48.8

```
Two-sample T for Low Dose vs Control

             N    Mean   StDev   SE Mean
Low Dose    20     3.8    44.0       9.8
Control     20    29.8    34.0       7.6

95% CI for mu Low Dose - mu Control: (-51.3,  -0.8)
T-Test mu Low Dose = mu Control (vs not =): T = -2.09  P = 0.044  DF = 35


-------------------------------------------------------------------------------

Two-sample T for High Dose vs Control

             N    Mean   StDev   SE Mean
High Dose   20    68.4    24.5       5.5
Control     20    29.8    34.0       7.6

95% CI for mu High Dose - mu Control: (19.5, 57.6)
T-Test mu High Dose = mu Control (vs not =): T = 4.12  P = 0.0002  DF = 34


-------------------------------------------------------------------------------

Two-sample T for Low Dose vs High Dose

             N    Mean   StDev   SE Mean
Low Dose    20     3.8    44.0       9.8
High Dose   20    68.4    24.5       5.5

95% CI for mu Low Dose - mu High Dose: (-87.6,  -41.5)
T-Test mu Low Dose = mu High Dose (vs not =): T =  -5.73  P = 0.0000  DF = 29
```

a. Compare the mean drop in blood pressure for the high-dose group and the control group. Use $\alpha = .05$ and report the level of significance.

b. Estimate the size of the difference in the mean drop for the high-dose and control groups using a 95% confidence interval.

c. Do the conditions required for the statistical techniques used in (a) and (b) appear to be satisfied? Justify your answer.

6.30   Refer to Exercise 6.29.

a. Compare the mean drop in blood pressure for the low-dose group and the control group. Use $\alpha = .05$ and report the level of significance.

b. Estimate the size of the difference in the mean drop for the low-dose and control groups using a 95% confidence interval.

c. Do the conditions required for the statistical techniques used in (a) and (b) appear to be satisfied? Justify your answer.

6.31   Refer to Exercise 6.29.

a. Compare the mean drop in blood pressure for the low-dose group and the high-dose group. Use $\alpha = .05$ and report the level of significance.

b. Estimate the size of the difference in the mean drop for the low-dose and high-dose groups using a 95% confidence interval.

c. Do the conditions required for the statistical techniques used in (a) and (b) appear to be satisfied? Justify your answer.

6.32 In Exercises 6.29–6.31, we tested three sets of hypotheses using portions of the same data sets in each of the sets of hypotheses. Let the experiment-wide Type I error rate be defined as the probability of making at least one Type I error in testing any set of hypotheses using the data from the experiment.

    a. If we tested each of the three sets of hypotheses at the .05 level, estimate the experiment-wide Type I error rate.

    b. Suggest a procedure by which we could be ensured that the experiment-wide Type I error rate would be at most .05.

Edu.

6.33 To assess whether degreed nurses received a more comprehensive training than registered nurses, a study was designed to compare the two groups. The state nursing licensing board randomly selected 50 nurses from each group for evaluation. They were given the state licensing board examination and their scores are summarized in the following tables and graphs.

Boxplots of degreed and registered nurses (means are indicated by solid circles)

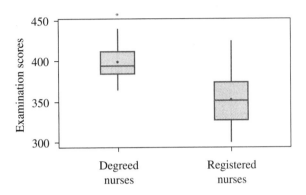

```
Two-Sample T-Test and Confidence Interval

Two-sample T for Degreed vs RN

              N     Mean   StDev   SE Mean
Degreed      50    399.9   17.2      2.4
RN           50    354.7   30.9      4.4


95% CI for mu Degreed - mu RN: (35.3, 55.2)
T-Test mu Degreed = mu RN (vs >): T = 9.04   P = 0.0000   DF = 76
```

    a. Can the licensing board conclude that the mean score of nurses who receive a BS in nursing is higher than the mean score of registered nurses? Use $\alpha = .05$.

    b. Report the approximated p-value for your test.

    c. Estimate the size of the difference in the mean scores of the two groups of nurses using a 95% confidence interval.

    d. The mean test scores are considered to have a meaningful difference only if they differ by more than 40 points. Is the observed difference in the mean scores a meaningful one?

Pol. Sci.

6.34 All people running for public office must report the amount of money spent during their campaign. Political scientists have contended that female candidates generally find it difficult to raise money and therefore spend less in their campaign than male candidates. Suppose the accompanying data represent the campaign expenditures of a randomly selected group of male and female candidates for the state legislature. Do the data support the claim that female candidates generally spend less in their campaigns for public office than male candidates?

Campaign Expenditures (in thousands of dollars)

Candidate	1	2	3	4	5	6	7	8	9	10	11	12	13	14	15	16	17	18	19	20
Female	169	206	257	294	252	283	240	207	230	183	298	269	256	277	300	126	318	184	252	305
Male	289	334	278	268	336	438	388	388	394	394	425	386	356	342	305	365	355	312	209	458

a. State the null and alternative hypotheses in
  i. plain English
  ii. statistical terms or symbols
b. Estimate the size of the difference in campaign expenditures for female and male candidates.
c. Is the difference statistically significant at the .05 level?
d. Is the difference of practical significance?

6.35 Refer to Exercise 6.34. What conditions must be satisfied in order to use the t procedures to analyze the data? Use the accompanying summary data and plot to determine whether these conditions have been satisfied for the data in Exercise 6.34.

```
Two-Sample T-Test and Confidence Interval

Two-sample T for Female vs Male

          N    Mean    StDev   SE Mean
Female   20   245.4    52.1      12
Male     20   350.9    61.9      14

95% CI for mu Female - mu Male: (-142, -69)
T-Test mu Female = mu Male (vs not =): T =  -5.83   P = 0.0000   DF = 38
Both use Pooled StDev = 57.2
```

Boxplots of female and male candidates (means are indicated by solid circles)

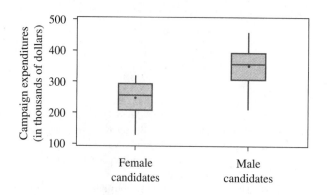

Med.

6.36 Company officials are concerned about the length of time a particular drug retains its potency. A random sample (sample 1) of 10 bottles of the product is drawn from current production and analyzed for potency. A second sample (sample 2) of 10 bottles is obtained, stored for 1 year, and then analyzed. The readings obtained are as follows:

Sample 1	10.2	10.5	10.3	10.8	9.8	10.6	10.7	10.2	10.0	10.6
Sample 2	9.8	9.6	10.1	10.2	10.1	9.7	9.5	9.6	9.8	9.9

The data are analyzed by a standard program package (SAS). The relevant output is shown here:

```
                              TTEST PROCEDURE


Variable: POTENCY

 SAMPLE    N          Mean       Std Dev     Std Error  Variances     T      DF Prob>|T|
------------------------------------------------------  -------------------------------
    1     10    10.37000000    0.32335052   0.10225241  Unequal    4.2368  16.6  0.0006
    2     10     9.83000000    0.24060110   0.07608475  Equal      4.2368  18.0  0.0005

For H0: Variances are equal, F' = 1.81    DF = (9,9)    Prob>F' = 0.3917
```

    a. What is the research hypothesis?
    b. What are the values of the t and t' statistics? Why are they equal for this data set?
    c. What are the p-values for t and t' statistics? Why are they different?
    d. Are the conclusions concerning the research hypothesis the same for the two tests if we use $\alpha = .05$?
    e. Which test, t or t', is more appropriate for this data set?

Bus.   **6.37** Many people purchase sports utility vehicles (SUVs) because they think they are sturdier and hence safer than regular cars. However, preliminary data have indicated that the costs for repairs of SUVs are higher than for midsize cars when both vehicles are in an accident. A random sample of 8 new SUVs and 8 midsize cars are tested for front impact resistance. The amounts of damage (in hundreds of dollars) to the vehicles when crashed at 20 mph head on into a stationary barrier are recorded in the following table.

Car	1	2	3	4	5	6	7	8
SUV	14.23	12.47	14.00	13.17	27.48	12.42	32.59	12.98
Midsize	11.97	11.42	13.27	9.87	10.12	10.36	12.65	25.23

    a. Plot the data to determine whether the conditions required for the t procedures are valid.
    b. Do the data support the conjecture that the mean damage is greater for SUVs than for midsize vehicles? Use $\alpha = .05$ with both the t test and Wilcoxon test.
    c. Which test appears to be the more appropriate procedure for this data set?
    d. Do you reach the same conclusions from both procedures? Why?

Boxplots of midsize and SUV damage amounts (means are indicated by solid circles)

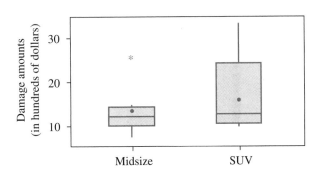

```
Two-Sample T-Test and Confidence Interval

Two-sample T for Midsize vs SUV

            N      Mean     StDev   SE Mean
Midsize  8      13.11      5.05       1.8
SUV      8      17.42      7.93       2.8

95% CI for mu Midsize - mu SUV: ( -11.4, 2.8)
T-Test mu Midsize = mu SUV (vs <): T = -1.30   P = 0.11   DF = 14
Both use Pooled StDev = 6.65
```

```
Mann-Whitney Confidence Interval and Test

Midsize    N =     8        Median =        11.69
SUV        N =     8        Median =        13.59
Point estimate for ETA1-ETA2 is              -2.32
95.9 Percent CI for ETA1-ETA2 is (-14.83,  -0.33)
W = 48.0
Test of ETA1 = ETA2 vs ETA1 < ETA2 is significant at 0.0203
```

6.38 Refer to Exercise 6.37. The small number of vehicles in the study has led to criticism of the results. A new study is to be conducted with a larger sample size. Assume that the populations of damages are both normally distributed with a common $\sigma = \$700$.

a. Determine the sample size so that we are 95% confident that the estimate of the difference in mean repair cost is within \$500 of the true difference.

b. For the research hypothesis H_a: $\mu_{SUV} > \mu_{MID}$, determine the sample size required to obtain a test having $\alpha = .05$ and $\beta(\mu_d) < .05$ when $\mu_{SUV} - \mu_{MID} \geq \$500$.

Law **6.39** The following memorandum opinion on statistical significance was issued by the judge in a trial involving many scientific issues. The opinion has been stripped of some legal jargon and has been taken out of context. Still, it can give us an understanding of how others deal with the problem of ascertaining the meaning of statistical significance. Read this memorandum and comment on the issues raised regarding statistical significance.

Memorandum Opinion

This matter is before the Court upon two evidentiary issues that were raised in anticipation of trial. First, it is essential to determine the appropriate level of statistical significance for the admission of scientific evidence.

With respect to statistical significance, no statistical evidence will be admitted during the course of the trial unless it meets a confidence level of 95%.

Every relevant study before the court has employed a confidence level of at least 95%. In addition, plaintiffs concede that social scientists routinely utilize a 95% confidence level. Finally, all legal authorities agree that statistical evidence is inadmissable unless it meets the 95% confidence level required by statisticians. Therefore, because plaintiffs advance no reasonable basis to alter the accepted approach of mathematicians to the test of statistical significance, no statistical evidence will be admitted at trial unless it satisfies the 95% confidence level.

Env. **6.40** **Defining the Problem (1).** Lead is an environmental pollutant especially worthy of attention because of its damaging effects on the neurological and intellectual development of children. Morton et al. (1982) collected data on lead absorption by children whose parents worked at a factory in Oklahoma where lead was used in the manufacture of batteries. The concern was that children might be exposed to lead inadvertently brought home on the bodies or clothing of their parents. Levels of lead (in micrograms per deciliter) were measured in blood samples taken from 33 children who might have been exposed in this way. They constitute the Exposed group.

Collecting Data (2). The researchers formed a Control group by making matched pairs. For each of the 33 children in the Exposed group they selected a matching child of the same age, living in the same neighborhood, and with parents employed at a place where lead is not used.

The dataset LEADKIDS contains three variables, each with 33 cases. All involve measurements of lead in micrograms per deciliter of blood.

c1	Exposed	Lead(μg/dl of whole blood) for children of workers in the battery factory
c2	Control	Lead(μg/dl of whole blood) for matched controls
c3	Diff	The differences: 'Exposed' - 'Control'.

These data are listed next.

Complete Listing of Dataset LEADKIDS			Complete Listing of Dataset LEADEXP		
Exposed	Control	Diff	Lead	JobExp	JobHyg
38	16	22	14	3	1
23	18	5	13	3	1
41	18	23	25	3	1
18	24	-6	39	2	1
37	19	18	41	3	2
36	11	25	18	3	2
23	10	13	49	3	2
62	15	47	29	2	2
31	16	15	16	1	2
34	18	16	38	3	3
24	18	6	23	3	3
14	13	1	37	3	3
21	19	2	62	3	3
17	10	7	24	3	3
16	16	0	45	3	3
20	16	4	39	3	3
15	24	-9	48	3	3
10	13	-3	44	3	3
45	9	36	35	3	3
39	14	25	43	3	3
22	21	1	34	3	3
35	19	16	73	3	3
49	7	42	31	2	3
48	18	30	34	2	3
44	19	25	20	2	3
35	12	23	22	2	3
43	11	32	35	2	3
39	22	17	36	1	3
34	25	9	23	1	3
13	16	-3	21	1	3
73	13	60	17	1	3
25	11	14	27	1	3
27	13	14	15	1	3
			10	1	3

This is necessarily an observational study rather than a controlled experiment. There is no way that the researchers could have assigned children at random to parents in or out of lead-related occupations. Furthermore, the Exposed subjects were all chosen from the small group of children whose parents worked at one particular plant. They were not chosen from the larger population of children everywhere who might be exposed to lead as a result of their parents' working conditions.

If lead levels are unusually high in the Exposed group, it might be argued that the lead in their blood came from some source other than their parents' place of work: from lead solder in water pipes at home, from lead-paint dust at school, from air pollution, and so on. For this reason a properly chosen control group of children is crucial to the credibility of the study.

In principle, the children in the Control group should be subject to all of the same possible lead contaminants as those in the Exposed group, except for lead brought home from work by parents. In practice, the designers of this study chose to use two criteria in forming pairs: neighborhood and age. Neighborhood seems a reasonable choice because general environmental conditions, types of housing, and so on could vary greatly for children living in different neighborhoods. Controlling for age seems reasonable because lead poisoning is largely cumulative, so levels of lead might be higher in older children. Thus for each child in the Exposed group, researchers sought a paired child of the same age and living in the same neighborhood.

Summarizing Data (3). We begin by looking at dot plots of the data for the Exposed and Control groups:

We can see that over half of the children in the Exposed group have more lead in their blood than do any of the children in the Control group. This graphical comparison is not the most effective one we could make because it ignores the pairing of Exposed and Control children. Even so, it presents clear evidence that, on average, the Exposed children have more lead in their blood than do the Controls.

Notice that the lead levels of the Exposed group are much more diverse than those of the Control group. This suggests that some children in the Exposed group are getting a lot more lead, presumably from their working parents, than are others in this group. Perhaps some parents at the battery factory do not work in areas where they come into direct contact with lead. Perhaps some parents wear protective clothing that is left at work or they shower before they leave work. For this study, information on the exposure and hygiene of parents was collected by the investigators. Such factors were found to contribute to the diversity of the lead levels observed among the Exposed children.

Some toxicologists believe that *any* amount of lead may be detrimental to children, but all agree that the highest levels among the exposed children in our study are dangerously high. Specifically, it is generally agreed that children with lead levels above 40 micrograms per deciliter need medical treatment. Children above 60 on this scale should be immediately *hospitalized* for treatment (Miller and Kean, 1987). A quick glance at the dot plot shows that we are looking at some serious cases of lead poisoning in the Exposed group.

By plotting differences, we get an even sharper picture. For each matched pair of children the variable Diff shows how much more lead the exposed child has than his or her Control neighbor of the same age.

```
                                 .              .   .         .   .
                        .  . :  .: :...    :.:.   : :    ..      .       .               .
        -+---------+---------+---------+---------+---------+-----Diff
        -15        0        15        30        45        60
```

If we consider a hypothetical population of pairs of children, the difference measures the increased lead levels that may result from exposure via a parent working at the battery factory.

If parents who work at the battery factory were not bringing lead home with them, we would expect about half of these values to be positive and half to be negative. The lead values in the blood would vary, but in such a way that the Exposed child would have only a 50-50 chance of having the higher value. Thus, we would expect the dot plot to be centered near 0.

In contrast, look at the dot plot of the actual data. Almost every child in the Exposed group has a higher lead value than does the corresponding Control child. As a result, most of the differences are positive. The average of the differences is the *balance point* of the dot plot, located somewhat above 15. (In some respects we can read the dot plot quite precisely. In one pair out of 33, both children have the same value, to the nearest whole number as reported. In only four pairs does the Control child have the higher level of lead.)

The dot plot of the differences displays strong evidence that the children in the Exposed group have more lead than their Control counterparts. In the next section we will perform some formal statistical tests to check whether this effect is statistically significant, but we already suspect from this striking graph what the conclusion must be.

In this section we have looked directly at the *pairs* of children around which the study was built. It may take a bit more thought to deal with differences than to look at the separate variables Exposed and Control as we did in the previous section. But looking at pairs is best. If the effect had turned out to be weaker and if we had not thought to look at pairs, then we might have missed seeing the effect.

a. Obtain mean, median, standard deviation, and so on, for each of the three variables in LEADKIDS. The file LEADKIDS is found on the CD that came with your book.

1) Compare the median of the Exposed children with the maximum of the Controls. What statement in the discussion does this confirm?

2) Compare the difference between the individual means of the Exposed and Control groups with the mean of the differences. On average how much higher are the lead values for Exposed children?

b. In contrast to part (a), notice that the difference between the individual medians of the Exposed and Control groups is *not* the same as the median for Diff. Why not? Which figure based on medians would you use if you were trying to give the most accurate view of the increase in lead exposure due to a parent working at the battery factory?

6.41 Analyzing Data, Interpreting the Analyses, and Communicating Results (4). A paired *t* test for the difference data in Exercise 6.40 is shown here.

```
Paired T for Exposed - Control

                N       Mean      StDev     SE Mean
Exposed         33      31.85     14.41      2.51
Control         33      15.88      4.54      0.79
Difference      33      15.97     15.86      2.76

95% CI for mean difference: (10.34, 21.59)
T-Test of mean difference = 0 (vs not = 0):
    T-Value = 5.78  P-Value = 0.000
```

The *p*-value in the output reads .000, which means that it is smaller than .0005 (1 chance in 2,000). Thus, it is extremely unlikely that we would see data as extreme as those actually collected unless workers at the battery factory were contaminating their children. We reject the null hypothesis and conclude that the difference between the lead levels of children in the Exposed and Control groups is large enough to be statistically significant.

The next question is whether the difference between the two groups is large enough to be of practical importance. This is a judgment for people who know about lead poisoning to make, not for statisticians. The best estimate of the true (population) mean difference is 15.97, or about 16. On average, children of workers in the battery plant have about 16 μg/dl more lead than their peers whose parents do not work in a lead-related industry. Almost any toxicologist would deem this increase to be dangerous and unacceptable. (The mean of the Control group is also about 16. On average, the effect of having a parent who works in the battery factory is to double the lead level. Doubling the lead level brings the average value for Exposed children to about 32, which is getting close to the level where medical treatment is required. Also remember that some toxicologists believe that any amount of lead is harmful to the neurological development of children.)

a. Should the *t* test we did have been one-sided? In practice, we must make the decision to do a one-sided test *before* the data are collected. We might argue that having a parent working at the battery factory could not *decrease* a child's exposure to lead.

    1) Write the null hypothesis and its one-sided alternative in both words and symbols. Perform the test. How is its *p*-value related to the *p*-value for the two-sided test?

    2) It might be tempting to argue that children of workers at a lead-using factory could not have generally lower levels of lead than children in the rest of the population. But can you imagine a scenario in which the mean levels would really be lower for Exposed children?

b. We used a *t* test to confirm our impression that Exposed children have more lead in their blood than their Control counterparts. Although there is no clear reason to prefer nonparametric tests for these data, verify that they yield the same conclusion as the *t* test does.

CHAPTER 7

Inferences about Population Variances

7.1 Introduction

When people think of statistical inference, they usually think of inferences concerning population means. However, the population parameter that answers an experimenter's practical questions will vary from one situation to another. In many situations, the variability of a population's values is as important as the population mean. In the case of problems involving product improvement, product quality is defined as a product having mean value at the target value with low variability about the mean. For example, the producer of a drug product is certainly concerned with controlling the mean potency of tablets, but he or she must also worry about the variation in potency from one tablet to another. Excessive potency or an underdose could be very harmful to a patient. Hence, the manufacturer would like to produce tablets with the desired mean potency and with as little variation in potency (as measured by σ or σ^2) as possible. Another example is from the area of investment strategies. Investors search for a portfolio of stocks, bonds, real estate, and other investments having low risk. A measure used by investors to determine the uncertainty inherent in a particular portfolio is the variance in the value of the investments over a set period. At times, a portfolio with a high average value and a large standard deviation will have a value that is much lower than the average value. Investors thus need to examine the variability in the value of a portfolio along with its average value when determining its degree of risk.

7.2 Estimation and Tests for a Population Variance

The sample variance

$$s^2 = \frac{\Sigma (y - \bar{y})^2}{n - 1}$$

can be used for inferences concerning a population variance σ^2. For a random sample of n measurements drawn from a population with mean μ and variance σ^2,

unbiased estimator s^2 is an **unbiased estimator** of σ^2. If the population distribution is normal, then

295

FIGURE 7.1

Densities of the chi-square
(df = 5, 15, 30) distribution

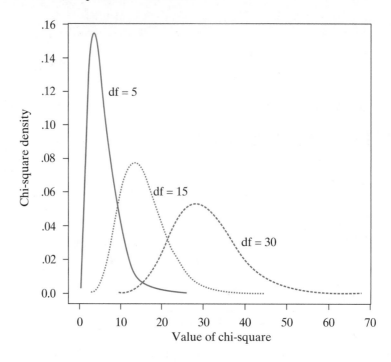

chi-square distribution
with df = *n* − 1

the sampling distribution of s^2 can be specified as follows. From repeated samples of size n from a normal population whose variance is σ^2, calculate the statistic $(n-1)s^2/\sigma^2$ and plot the histogram for these values. The shape of the histogram is similar to those depicted in Figure 7.1, because it can be shown that the statistic $(n-1)s^2/\sigma^2$ follows a **chi-square distribution with df = *n* − 1.** The mathematical formula for the chi-square (χ^2, where χ is the Greek letter chi) probability distribution is very complex so we will not display it. However, some of the properties of the distribution are as follows:

1. The chi-square distribution is positively skewed with values between 0 and ∞ (see Figure 7.1).
2. There are many chi-square distributions and they are labeled by the parameter degrees of freedom (df). Three such chi-square distributions are shown in Figure 7.1 with df = 5, 15, and 30.
3. The mean and variance of the chi-square distribution are given by μ = df and σ^2 = (2)df. For example, if the chi-square distribution has df = 30, then the mean and variance of that distribution are μ = 30 and σ^2 = 60.

Upper-tail values of the chi-square distribution can be found in Table 8 in the Appendix. Entries in the table are values of χ^2 that have an area a to the right under the curve. The degrees of freedom are specified in the left column of the table, and values of a are listed across the top of the table. Thus, for df = 14, the value of chi-square with an area a = .025 to its right under the curve is 26.12 (see Figure 7.2). To determine the value of chi-square with an area .025 to its left under the curve, we compute a = 1 − .025 and obtain 5.629 from Table 8 in the Appendix. Combining these two values, we have that the area under the curve between 5.629 and 26.12 is 1 − .025 − .025 = .95. (See Figure 7.2.) We can use this information to form a confidence interval for σ^2. Because the chi-square distribution is not symmetrical, the

FIGURE 7.2

Critical values of the chi-square distribution with df = 14

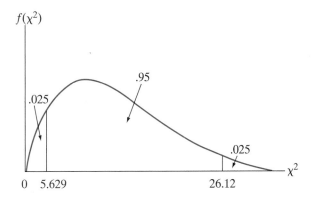

FIGURE 7.3

Upper-tail and lower-tail values of chi-square

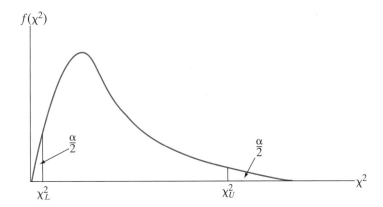

confidence intervals based on this distribution do not have the usual form, estimate ± error, as we saw for μ and $\mu_1 - \mu_2$. The $100(1 - \alpha)\%$ confidence interval for σ^2 is obtained by dividing the estimator of σ^2, s^2, by the lower and upper $\alpha/2$ percentiles, χ_L^2 and χ_U^2, as described here.

General Confidence Interval for σ^2 (or σ) with Confidence Coefficient $(1 - \alpha)$

$$\frac{(n-1)s^2}{\chi_U^2} < \sigma^2 < \frac{(n-1)s^2}{\chi_L^2}$$

where χ_U^2 is the upper-tail value of chi-square for df $= n - 1$ with area $\alpha/2$ to its right, and χ_L^2 is the lower-tail value with area $\alpha/2$ to its left (see Figure 7.3). We can determine χ_U^2 and χ_L^2 for a specific value of df by obtaining the critical value in Table 8 in the Appendix corresponding to $a = \alpha/2$ and $a = 1 - \alpha/2$, respectively. (*Note:* The confidence interval for σ is found by taking square roots throughout.)

EXAMPLE 7.1

The machine that fills 500-gram coffee containers for a large food processor is monitored by the quality control department. Ideally, the amount of coffee in a container should vary only slightly about the nominal 500-gram value. If the variation is large, then a large proportion of the containers will be either underfilled, thus cheating the customer, or overfilled, thus resulting in economic loss to the company. The machine was designed so that the weights of the 500-gram containers will have a normal distribution with mean value of 506.6 grams and a standard deviation of 4 grams. This produces a population of containers in which at most 5% of the containers weigh less than 500 grams. To maintain a population in which at most 5% of the containers are underweight, a random sample of 30 containers is selected every hour. These data are then used to determine whether the mean and standard deviation are maintained at their nominal values. The weights from one of the hourly samples are given here.

501.4	498.0	498.6	499.2	495.2	501.4	509.5	494.9	498.6	497.6
505.5	505.1	499.8	502.4	497.0	504.3	499.7	497.9	496.5	498.9
504.9	503.2	503.0	502.6	496.8	498.2	500.1	497.9	502.2	503.2

Estimate the mean and standard deviation in the weights of coffee containers filled during the hour, in which the random sample of 30 containers was selected using a 99% confidence interval.

Solution For these data, we find

$$\bar{y} = 500.453 \quad \text{and} \quad s = 3.433$$

To use our method for constructing a confidence interval for μ and σ, we must first check whether the weights are a random sample from a normal population. Figure 7.4 is a normal probability plot of the 30 weights. The 30 values fall near the straight line. Thus, the normality condition appears to be satisfied. The confidence coefficient for this example is $1 - \alpha = .99$. The upper-tail chi-square value can be obtained from Table 8 in the Appendix for df $= n - 1 = 29$ and $a = \alpha/2 = .005$. Similarly, the lower-tail chi-square value is obtained from Table 8, with $a = 1 - \alpha/2 = .995$. Thus,

$$\chi_L^2 = 13.12 \quad \text{and} \quad \chi_U^2 = 52.34$$

The 99% confidence interval for σ is then

$$\sqrt{\frac{29(3.433)^2}{52.34}} < \sigma < \sqrt{\frac{29(3.433)^2}{13.12}}$$

FIGURE 7.4

Normal probability plot of container weights

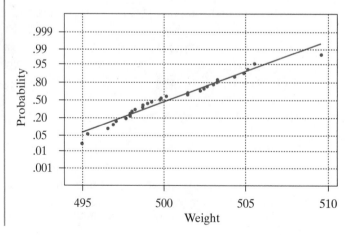

or

$$2.56 < \sigma < 5.10$$

Thus, we are 99% confident that the standard deviation in the weights of coffee cans lies between 2.56 and 5.10 grams. The designed value for σ, 4 grams, falls within our confidence interval. Using our results from Chapter 5, a 99% confidence interval for μ is

$$500.453 \pm 2.756 \frac{3.433}{\sqrt{30}} \qquad 500.453 \pm 1.73$$

or

$$498.7 < \mu < 502.2$$

Thus, it appears the machine is underfilling the containers, because 506.6 grams does not fall within the confidence limits.

In addition to estimating a population variance, we can construct a statistical test of the null hypothesis that σ^2 equals a specified value, σ_0^2. This test procedure is summarized here.

Statistical Test for σ^2 (or σ)

H_0: **1.** $\sigma^2 \leq \sigma_0^2$ H_a: **1.** $\sigma^2 > \sigma_0^2$
     2. $\sigma^2 \geq \sigma_0^2$ **2.** $\sigma^2 < \sigma_0^2$
     3. $\sigma^2 = \sigma_0^2$ **3.** $\sigma^2 \neq \sigma_0^2$

T.S.: $\chi^2 = \dfrac{(n-1)s^2}{\sigma_0^2}$

R.R: For a specified value of α,
  1. Reject H_0 if χ^2 is greater than χ_U^2, the upper-tail value for $a = \alpha$ and df $= n - 1$.
  2. Reject H_0 if χ^2 is less than χ_L^2, the lower-tail value for $a = 1 - \alpha$ and df $= n - 1$.
  3. Reject H_0 if χ^2 is greater than χ_U^2, based on $a = \alpha/2$ and df $= n - 1$, or less than χ_L^2, based on $a = 1 - \alpha/2$ and df $= n - 1$.

Check assumptions and draw conclusions.

EXAMPLE 7.2

A manufacturer of a specific pesticide useful in the control of household bugs claims that its product retains most of its potency for a period of at least 6 months. More specifically, it claims that the drop in potency from 0 to 6 months will vary in the interval from 0 to 8%. To test the manufacturer's claim, a consumer group obtained a random sample of 20 containers of pesticide from the manufacturer. Each can was tested for potency and then stored for a period of 6 months at room temperature. After the storage period, each can was again tested for potency. The percentage drop in potency was recorded for each can and is given here.

| 0.5 | 3.5 | 4.4 | 6.0 | 6.6 | 5.4 | 7.9 | 4.6 | 5.4 | 5.7 |
| 2.5 | 1.1 | 5.9 | 2.7 | 2.3 | 1.4 | 1.8 | 5.8 | 0.2 | 7.1 |

Use these data to determine whether there is sufficient evidence to indicate that the population of potency drops has more variability than claimed by the manufacturer. Use $\alpha = .05$.

Solution   The manufacturer claims that the population of potency reductions has a range of 8%. Dividing the range by 4, we obtain an approximate population standard deviation of $\sigma = 2\%$ (or $\sigma^2 = 4$).

The appropriate null and alternative hypotheses are

H_0: $\sigma^2 \le 4$ (i.e., we assume the manufacturer's claim is correct)

H_a: $\sigma^2 > 4$ (i.e., there is more variability than claimed by the manufacturer)

To use our inference techniques for variances, we must first check the normality of the data. From Figure 7.5 we observe that the plotted points fall nearly on the straight line. Thus, the normality condition appears to be satisfied. From the 20 data values, we compute $s^2 = 5.45$. The test statistic and rejection region are as follows:

T.S.: $\chi^2 = \dfrac{(n-1)s^2}{\sigma_0^2} = \dfrac{19(5.45)}{4} = 25.88$

R.R.: For $\alpha = .05$, we will reject H_0 if the computed value of chi-square is greater than 30.14, obtained from Table 8 in the Appendix for $a = .05$ and df $= 19$.

Checking assumptions and drawing conclusions: Because the computed value of chi-square, 25.88, is less than the critical value, 30.14, there is insufficient evidence to reject the manufacturer's claim, based on $\alpha = .05$. The p-value $= P(\chi_{19}^2 > 25.88) = .14$ can be found using a computer program. Using Table 8 in the Appendix, we can only conclude that p-value $> .10$, because the p-value $= P(\chi_{19}^2 > 25.88) > P(\chi_{19}^2 > 27.20) = .10$. The sample size is relatively small and the p-value is only moderately large, so the consumer group is not prepared to accept $H_0: \sigma^2 \le 4$. Rather, it would be wise to do additional testing with a larger sample size before reaching a definite conclusion.

FIGURE 7.5

Normal probability plot for potency data

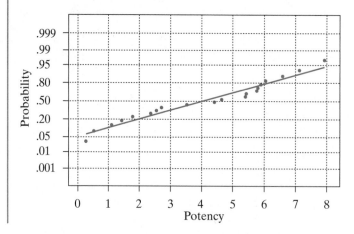

The inference methods about σ are based on the condition that the random sample is selected from a population having a normal distribution similar to the requirements for using t distribution–based inference procedures. However, when sample sizes are moderate to large ($n \geq 30$), the t distribution–based procedures can be used to make inferences about μ even when the normality condition does not hold because for moderate to large sample sizes the Central Limit Theorem provides that the sampling distribution of the sample mean is approximately normal. Unfortunately, the same type of result does not hold for the chi-square-based procedures for making inferences about σ; that is, if the population distribution is distinctly nonnormal, then these procedures for σ are not appropriate even if the sample size is large. Population nonnormality, in the form of skewness or heavy tails, can have serious effects on the nominal significance and confidence probabilities for σ. If a boxplot or normal probability plot of the sample data shows substantial skewness or a substantial number of outliers, the chi-square-based inference procedures should not be applied. There are some alternative approaches that involve computationally elaborate inference procedures. One such procedure is the bootstrap. Bootstrapping is a technique that provides a simple and practical way to estimate the uncertainty in sample statistics such as the sample variance. We can use bootstrap techniques to estimate the sampling distribution of sample variance. The estimated sampling distribution is then manipulated to produce confidence intervals for σ and rejection regions for tests of hypotheses about σ. Information about bootstrapping can be found in the books by Efron and Tibshirani (*An Introduction to the Bootstrap*, Chapman and Hall, New York, 1993) and by Manly (*Randomization, Bootstrap and Monte Carlo Methods in Biology*, Chapman and Hall, New York, 1998).

EXERCISES   **Basic Techniques**

7.1 Suppose that y has a χ^2 distribution with 27 df.
   a. Find $P(y > 46.96)$.   **b.** Find $P(y > 18.81)$.
   c. Find $P(y < 12.88)$.   **d.** What is $P(12.88 < y < 46.96)$?

7.2 For a χ^2 distribution with 11 df,
   a. Find $\chi^2_{.025}$.   **b.** Find $\chi^2_{.975}$.

Applications

Engin.   **7.3** A packaging line fills nominal 32-ounce tomato juice jars with a quantity of juice having a normal distribution with a mean of 32.30 ounces. The process should have a standard deviation smaller than .15 ounces per jar. (A larger standard deviation leads to too many underfilled and overfilled jars.) A random sample of 50 jars is taken every hour to evaluate the process. The data from one such sample are summarized here.

Normal probability plot of juice data

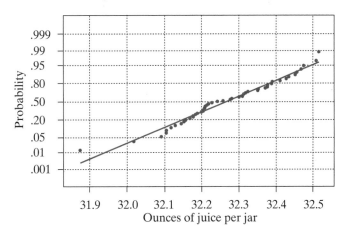

```
Descriptive Statistics for Juice Data

Variable        N       Mean   Median   TrMean   StDev   SE Mean
Juice Jars      50     32.267  32.248   32.270   0.135    0.019


Variable     Minimum   Maximum       Q1       Q3
Juice Jars    31.874    32.515   32.177   32.376
```

a. If the process yields jars having a normal distribution with a mean of 32.30 ounces and a standard deviation of .15 ounces, what proportion of the jars filled on the packaging line will be underfilled?

b. Does the normal probability plot suggest any violation of the conditions necessary to use the chi-square procedures for generating a confidence interval and a test of hypotheses about σ?

c. Construct a 95% confidence interval on the process standard deviation.

d. Do the data indicate that the process standard deviation is greater than .15? Use $\alpha = .05$.

e. Place bounds on the *p*-value of the test.

Engin.   **7.4** A leading researcher in the study of interstate highway accidents proposes that a major cause of many collisions on the interstates is not the speed of the vehicles but rather the *difference* in speeds of the vehicles. When some vehicles are traveling slowly while other vehicles are traveling at speeds greatly in excess of the speed limit, the faster-moving vehicles may have to change lanes quickly, which can increase the chance of an accident. Thus, when there is a large variation in the speeds of the vehicles in a given location on the interstate, there may be a larger number of accidents than when the traffic is moving at a more uniform speed. The reseacher believes that when the standard deviation in speed of vehicles exceeds 10 mph, the rate of accidents is greatly increased. During a 1-hour period of time, a random sample of 100 vehicles is selected from a section of an interstate known to have a high rate of accidents, and their speeds are recorded using a radar gun. The data are summarized here and in a boxplot.

```
Descriptive Statistics for Vehicle Speeds

Variable           N      Mean   Median   TrMean   StDev   SE Mean
Speed (mph)      100     64.48    64.20    64.46   11.35      1.13


Variable     Minimum   Maximum       Q1       Q3
Speed (mph)    37.85     96.51   57.42    71.05
```

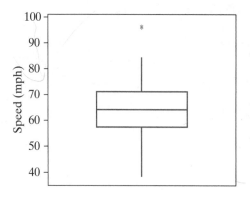

a. Does the boxplot suggest any violation of the conditions necessary to use the chi-square procedures for generating a confidence interval and a test of hypotheses about σ?

b. Estimate the standard deviation in the speeds of the vehicles on the interstate using a 95% confidence interval.

c. Do the data indicate at the 5% level that the standard deviation in vehicle speeds exceeds 10 mph?

Edu. **7.5** A large public school system was evaluating its elementary school reading program. In particular, educators were interested in the performance of students on a standardized reading test given to all third graders in the state. The mean score on the test was compared to the state average to determine the school system's rating. Also, the educators were concerned with the variation in scores. If the mean scores were at an acceptable level but the variation was high, this would indicate that a large proportion of the students still needed remedial reading programs. Also, a large variation in scores might indicate a need for programs for those students at the gifted level. Without accelerated reading programs, these students lose interest during reading classes. To obtain information about students early in the school year (the statewide test is given during the last month of the school year), a random sample of 150 third-grade students was given the exam used in the previous year. The possible scores on the reading test range from 0 to 100. The data are summarized here.

```
Descriptive Statistics for Reading Scores

Variable         N      Mean   Median   TrMean   StDev   SE Mean
Reading        150    70.571   71.226   70.514   9.537    0.779

Variable   Minimum    Maximum      Q1       Q3
Reading     44.509     94.570   65.085   76.144
```

a. Does the plot of the data suggest any violation of the conditions necessary to use the chi-square procedures for generating a confidence interval and a test of hypotheses about σ?

b. Estimate the variation in reading scores using a 99% confidence interval.

c. Do the data indicate that the variation in reading scores is greater than 90, the variation for all students taking the exam the previous year?

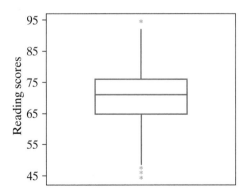

7.6 Place bounds on the p-value of the test in Exercise 7.5.

Engin. **7.7** Baseballs vary somewhat in their rebounding coefficient. A baseball that has a large rebound coefficient will travel further when the same force is applied to it than a ball with a smaller coefficient. To achieve a game in which each batter has an equal opportunity to hit a home run, the balls should have nearly the same rebound coefficient. A standard test has been developed to measure the rebound coefficient of baseballs. A purchaser of large quantities of baseballs requires that the mean coefficient value be 85 units and the standard deviation be less than 2 units.

A random sample of 81 baseballs is selected from a large batch of balls and tested. The data are summarized here.

Descriptive Statistics for Rebound Coefficient Data

Variable	N	Mean	Median	TrMean	StDev	SE Mean
Rebound	81	85.296	85.387	85.285	1.771	0.197

Variable	Minimum	Maximum	Q1	Q3
Rebound	80.934	89.687	84.174	86.352

a. Does the plot indicate any violation of the conditions underlying the use of the chi-square procedures for constructing confidence intervals or testing hypotheses about σ?

b. Is there sufficient evidence that the standard deviation in rebound coefficient for the batch of balls is less than 2?

c. Estimate the standard deviation of the rebound coefficients using a 95% confidence interval.

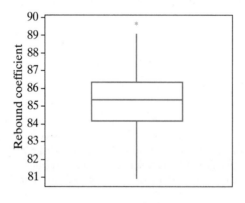

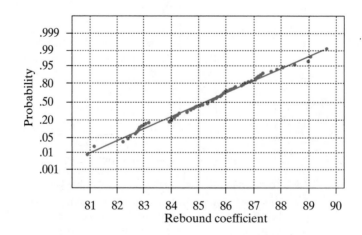

7.3 Estimation and Tests for Comparing Two Population Variances

We are often concerned about comparing the standard deviations of the two procedures. In many situations in which we are comparing two processes or two suppliers of a product, we need to compare the standard deviations of the populations associated with process measurements. Another major application of a test for the equality of two population variances is for **evaluating** the validity of the **equal variance condition** (that is, $\sigma_1^2 = \sigma_2^2$) for a two-sample t test. The test developed in this section requires that the two population distributions both have normal distributions. We are interested in comparing the variance of population 1, σ_1^2, to the variance of population 2, σ_2^2.

When random samples of sizes n_1 and n_2 have been independently drawn from two normally distributed populations, the ratio

$$\frac{s_1^2/\sigma_1^2}{s_2^2/\sigma_2^2} = \frac{s_1^2/s_2^2}{\sigma_1^2/\sigma_2^2}$$

possesses a probability distribution in repeated sampling referred to as an **F distribution.** The formula for the probability distribution is omitted here, but we will specify its properties.

Properties of the F Distribution

1. Unlike t or z but like χ^2, F can assume only positive values.
2. The F distribution, unlike the normal distribution or the t distribution but like the χ^2 distribution, is nonsymmetrical. (See Figure 7.6.)
3. There are many F distributions, and each one has a different shape. We specify a particular one by designating the degrees of freedom associated with s_1^2 and s_2^2. We denote these quantities by df_1 and df_2, respectively. (See Figure 7.6.)
4. Tail values for the F distribution are tabulated and appear in Table 9 in the Appendix.

Table 9 in the Appendix records upper-tail values of F corresponding to areas $a = .25, .10, .05, .025, .01, .005,$ and $.001$. The degrees of freedom for s_1^2, designated by df_1, are indicated across the top of the table; df_2, the degrees of freedom for s_2^2, appear in the first column to the left. Values of a are given in the next column. Thus, for $df_1 = 5$ and $df_2 = 10$, the critical values of F corresponding to $a = .25, .10, .05, .025, .01, .005,$ and $.001$ are, respectively, 1.59, 2.52, 3.33, 4.24, 5.64, 6.78, and 10.48. It follows that only 5% of the measurements from an F distribution with $df_1 = 5$ and $df_2 = 10$ will exceed 3.33 in repeated sampling. (See Figure 7.7.) Similarly, for $df_1 = 24$ and $df_2 = 10$, the critical values of F corresponding to tail areas of $a = .01$ and $.001$ are, respectively, 4.33 and 7.64.

A statistical test comparing σ_1^2 and σ_2^2 uses the test statistic s_1^2/s_2^2. When $\sigma_1^2 = \sigma_2^2$, $\sigma_1^2/\sigma_2^2 = 1$ and s_1^2/s_2^2 follows an F distribution with $df_1 = n_1 - 1$ and $df_2 = n_2 - 1$. Table 9 in the Appendix provides the upper-tail values of the F distribution. The lower-tail values are obtained from the following relationship. Let F_{a,df_1,df_2} be the

FIGURE 7.6

Densities of two
F distributions

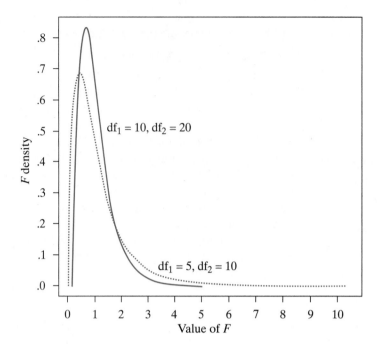

FIGURE 7.7

Critical value for
the *F* distributions
($df_1 = 5$, $df_2 = 10$)

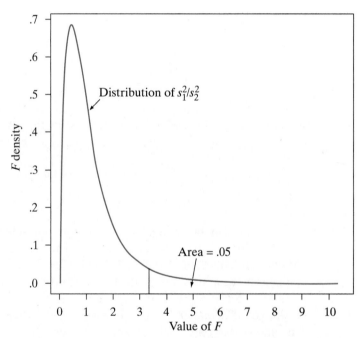

upper α percentile and $F_{1-\alpha,df_1,df_2}$ be the lower α percentile of an *F* distribution with df_1 and df_2. Then,

$$F_{1-\alpha,df_1,df_2} = \frac{1}{F_{\alpha,df_2,df_1}}$$

Note that the degrees of freedom have been reversed for the *F* percentile on the right-hand side of the equation.

EXAMPLE 7.3

Determine the lower .025 percentile for an F distribution with $df_1 = 4$ and $df_2 = 9$.

Solution   From Table 9 in the Appendix, the upper .025 percentile for the F distribution with $df_1 = 9$ and $df_2 = 4$ is 8.90. Thus,

$$F_{.975,4,9} = \frac{1}{F_{.025,9,4}} \quad \text{or} \quad F_{.975,4,9} = \frac{1}{8.90} = 0.11$$

For a one-tailed alternative hypothesis, the populations are designated 1 and 2 so that H_a is of the form $\sigma_1^2 > \sigma_2^2$. Then the rejection region is located in the upper-tail of the F distribution.

We summarize the test procedure next.

A Statistical Test Comparing σ_1^2 and σ_2^2

H_0:   **1.** $\sigma_1^2 \leq \sigma_2^2$ H_a: **1.** $\sigma_1^2 > \sigma_2^2$
          2. $\sigma_1^2 = \sigma_2^2$ **2.** $\sigma_1^2 \neq \sigma_2^2$

T.S.: $F = s_1^2/s_2^2$

R.R.:   For a specified value of α and with $df_1 = n_1 - 1$, $df_2 = n_2 - 1$,
   1. Reject H_0 if $F \geq F_{\alpha,df_1,df_2}$.
   2. Reject H_0 if $F \leq F_{1-\alpha/2,df_1,df_2}$ or if $F \geq F_{\alpha/2,df_1,df_2}$.

Check assumptions and draw conclusions.

EXAMPLE 7.4

Previously, we discussed an experiment in which company officials were concerned about the length of time a particular drug product retained its potency. A random sample of 10 bottles was obtained from the production line and each bottle was analyzed to determine its potency. A second sample of 10 bottles was obtained and stored in a regulated environment for 1 year. Potency readings were obtained on these bottles at the end of the year. The sample data were then used to place a confidence interval on $\mu_1 - \mu_2$, the difference in mean potencies for the two time periods.

We mentioned in Chapter 6 that in order to use the pooled t test in a statistical test for $\mu_1 - \mu_2$, we require that the samples be drawn from normal populations with possibly different means *but* with a common variance. Use the sample data summarized next to test the equality of the population variances. Use $\alpha = .05$. Sample 1 data are the readings taken immediately after production and sample 2 data are the readings taken 1 year after production. Draw conclusions.

Sample 1: $\bar{y}_1 = 10.37$, $s_1^2 = 0.105$, $n_1 = 10$

Sample 2: $\bar{y}_2 = 9.83$, $s_2^2 = 0.058$, $n_2 = 10$

Solution   The four parts of the statistical test of H_0: $\sigma_1^2 = \sigma_2^2$ follow.

H_0: $\sigma_1^2 = \sigma_2^2$

H_a: $\sigma_1^2 \neq \sigma_2^2$

T.S.: $F = \dfrac{s_1^2}{s_2^2} = \dfrac{0.105}{0.058} = 1.81$

Prior to setting the rejection region, we must first determine whether the two samples appear to be from normally distributed populations. After determining that this condition has been satisfied, we then determine the following rejection region.

R.R.: For a two-tailed test with $\alpha = .05$, we will reject H_0 if $F \geq F_{.025,9,9} = 4.03$ or if $F \leq F_{.975,9,9} = 1/F_{.025,9,9} = 1/4.03 = 0.25$.

Checking assumptions and drawing conclusions: Because 1.81 does not fall in the rejection region, we do not reject H_0: $\sigma_1^2 = \sigma_2^2$. The assumption of equality of variances appears to hold for the t methods used with these data.

We can now formulate a confidence interval for the ratio σ_1^2/σ_2^2.

General Confidence Interval for σ_1^2/σ_2^2 with Confidence Coefficient $(1 - \alpha)$

$$\frac{s_1^2}{s_2^2} F_L \leq \frac{\sigma_1^2}{\sigma_2^2} \leq \frac{s_1^2}{s_2^2} F_U$$

where $F_U = F_{\alpha/2,df_2,df_1}$ and $F_L = F_{1-\alpha/2,df_2,df_1} = 1/F_{\alpha/2,df_1,df_2}$, with $df_1 = n_1 - 1$ and $df_2 = n_2 - 1$. (*Note:* A confidence interval for σ_1/σ_2 is found by taking the square root of the endpoints of the confidence interval for σ_1^2/σ_2^2.)

It should be noted that although our estimation procedure for σ_1^2/σ_2^2 is appropriate for any confidence coefficient $(1 - \alpha)$, Appendix Table 9 allows us to construct confidence intervals for σ_1^2/σ_2^2 with the more commonly used confidence coefficients, such as .90, .95, .98, and .99. For more detailed tables of the F distribution, see Pearson and Hartley (1966).

EXAMPLE 7.5

The life length of an electrical component was studied under two operating voltages, 110 and 220. Ten different components were randomly assigned to operate at 110 volts and 16 different components were randomly assigned to operate at 220 volts. The times to failure (in hundreds of hours) for the 26 components were obtained and yielded the following summary statistics and normal probability plots.

Normal probability plot for life length under 110 volts

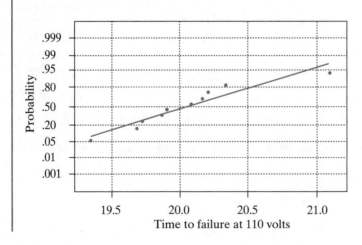

Normal probability plot for life length under 220 volts

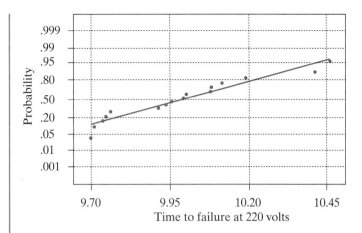

Voltage	Sample Size	Mean	Standard Deviation
110	10	20.04	.474
220	16	9.99	.233

The researchers wanted to estimate the relative size of the variation in life length under 110 and 220 volts. Use the data to construct a 90% confidence interval for σ_1/σ_2, the ratio of the standard deviations in life lengths for the components under the two operating voltages.

Solution Before constructing the confidence interval, it is necessary to check whether the two populations of life lengths were both normally distributed. From the normal probability plots, it appears that both samples of life lengths are from normal distributions. Next, we need to find the upper and lower $\alpha/2 = .10/2 = .05$ percentiles for the F distribution with $df_1 = 10 - 1 = 9$ and $df_2 = 16 - 1 = 15$. From Table 9 in the Appendix, we find

$$F_U = F_{.05,15,9} = 3.01 \quad \text{and} \quad F_L = F_{.95,15,9} = 1/F_{.05,9,15} = 1/2.59 = .386$$

Substituting into the confidence interval formula, we have a 90% confidence interval for σ_1^2/σ_2^2

$$\frac{(.474)^2}{(.233)^2}.386 \leq \frac{\sigma_1^2}{\sigma_2^2} \leq \frac{(.474)^2}{(.233)^2}3.01$$

$$1.5975 \leq \frac{\sigma_1^2}{\sigma_2^2} \leq 12.4569$$

It follows that our 90% confidence interval for σ_1/σ_2 is given by

$$\sqrt{1.5975} \leq \frac{\sigma_1}{\sigma_2} \leq \sqrt{12.4569} \quad \text{or} \quad 1.26 \leq \frac{\sigma_1}{\sigma_2} \leq 3.53$$

Thus, we are 90% confident that σ_1 is between 1.26 and 3.53 times as large as σ_2.

EXERCISES   **Basic Techniques**

7.8 Random samples of $n_1 = 8$ and $n_2 = 10$ observations were selected from populations 1 and 2, respectively. The corresponding sample variances were $s_1^2 = 7.4$ and $s_2^2 = 12.7$. Do the data provide sufficient evidence to indicate a difference between σ_1^2 and σ_2^2? Test by using $\alpha = .10$. What assumptions have you made?

7.9 An experiment was conducted to determine whether there was sufficient evidence to indicate that data variation within one population, say population A, exceeded the variation within a second population, population B. Random samples of $n_A = n_B = 8$ measurements were selected from the two populations and the sample variances were calculated to be

$$s_A^2 = 2.87 \qquad s_B^2 = .91$$

Do the data provide sufficient evidence to indicate that σ_A^2 is larger than σ_B^2? Test by using $\alpha = .05$.

Applications

Engin.   **7.10** A soft-drink firm is evaluating an investment in a new type of canning machine. The company has already determined that it will be able to fill more cans per day for the same cost if the new machines are installed. However, it must determine the variability of fills using the new machines and wants the variability from the new machines to be equal to or smaller than that currently obtained using the old machines. A study is designed in which random samples of 61 cans are selected from the output of both types of machines and the amount of fill (in ounces) is determined. The data are summarized in the following table and boxplots.

Summary Data for Canning Experiment			
Machine Type	Sample Size	Mean	Standard Deviation
Old	61	12.284	.231
New	61	12.197	.162

Boxplots of old machine and new machine (means are indicated by solid circles)

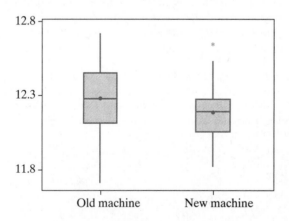

a. Estimate the standard deviations in fill for types of machines using 95% confidence intervals.
b. Do these data present sufficient evidence to indicate that the new type of machine has less variability of fills than the old machine?
c. Do the necessary conditions for conducting the inference procedures in parts (a) and (b) appear to be satisfied? Justify your answer.

Edu. **7.11** The Scholastic Aptitude Test (SAT) is an exam taken by most high scho
of their college admission requirements. A proposal has been made to alter t
the students take the exam on a computer. The exam questions will be selec
in the following fashion. For a given section of questions, if the student answe
tions posed correctly, then the following questions become increasingly difficult. If the student
provides incorrect answers for the initial questions asked in a given section, then the level of
difficulty of following questions does not increase. The final score on the exams will be standardized to take into account the overall difficulty of the questions on each exam. The testing agency
wants to compare the scores obtained using the new method of administering the exam to the
scores using the current method. A group of 182 high school students is randomly selected to participate in the study with 91 students randomly assigned to each of the two methods of administering the exam. The data are summarized in the following table and boxplots for the math portion
of the exam.

Summary Data for SAT Math Exams			
Testing Method	**Sample Size**	**Mean**	**Standard Deviation**
Computer	91	484.45	53.77
Conventional	91	487.38	36.94

Boxplots of conventional and computer methods (means are indicated by solid circles)

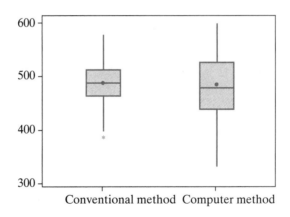

Conventional method Computer method

Evaluate the two methods of administering the SAT exam. Provide tests of hypotheses and confidence intervals. Are the means and standard deviations of scores for the two methods equivalent?
Justify your answer using $\alpha = .05$.

7.4 Tests for Comparing $t > 2$ Population Variances

In the previous section, we discussed a method for comparing variances from two
normally distributed populations based on taking independent random samples
from the populations. In many situations, we will need to compare more than two
populations. For example, we may want to compare the variability in the level of
nutrients of five different suppliers of a feed supplement or the variability in scores
of the students using SAT preparatory materials from the three major publishers
of those materials. Thus, we need to develop a statistical test that will allow us to
compare $t > 2$ population variances. We will consider two procedures. The first

procedure, Hartley's test, is very simple to apply but has the restriction that the population distributions must be normally distributed and the sample sizes equal. The second procedure, Levene's test, is more complex in its computations but does not restrict the population distributions or the sample sizes. Levene's test can be obtained from many of the statistical software packages. For example, SAS and Minitab both use Levene's test for comparing population variances.

Hartley F_{max} test
H. O. Hartley (1950) developed the **Hartley F_{max} test** for evaluating the hypotheses

$$H_0: \sigma_1^2 = \sigma_2^2 = \cdots = \sigma_t^2 \quad \text{vs.} \quad H_a: \sigma_i^2 \text{s not all equal}$$

The Hartley F_{max} requires that we have independent random samples of the same size n from t normally distributed populations. With the exception that we require $n_1 = n_2 = \cdots = n_t = n$, the Hartley test is a logical extension of the F test from the previous section for testing $t = 2$ variances. With s_i^2 denoting the sample variance computed from the ith sample, let $s_{min}^2 =$ the smallest of the s_i^2s and $s_{max}^2 =$ the largest of the s_i^2s. The Hartley F_{max} test statistic is

$$F_{max} = \frac{s_{max}^2}{s_{min}^2}$$

The test procedure is summarized here.

Hartley's F_{max} Test for Homogeneity of Population Variances

H_0: $\sigma_1^2 = \sigma_2^2 = \cdots = \sigma_t^2$ (homogeneity of variances)

H_a: Population variances are not all equal

T.S.: $F_{max} = \dfrac{s_{max}^2}{s_{min}^2}$

R.R.: For a specified value of α, reject H_0 if F_{max} exceeds the tabulated F values (Appendix Table 10) for α, t, and $df_2 = n - 1$, where n is the common sample size for the t random samples.

Check assumptions and draw conclusions.

We will illustrate the application of the Hartley test with the following example.

EXAMPLE 7.6

Wludyka and Nelson [*Technometrics* (1997) 39:274–285] describe the following experiment. In the manufacture of soft contact lenses, a monomer is injected into a plastic frame, the monomer is subjected to ultraviolet light and heated (the time, temperature, and light intensity are varied), the frame is removed, and the lens is hydrated. It is thought that temperature can be manipulated to target the power (the strength of the lens), so interest is in comparing the variability in power. The data are coded deviations from target power using monomers from three different suppliers. We wish to test $H_0: \sigma_1^2 = \sigma_2^2 = \sigma_3^2$.

Deviations from Target Power for Three Supplier

Supplier	Sample 1	2	3	4	5	6	7	8	9	n	s_i^2
1	191.9	189.1	190.9	183.8	185.5	190.9	192.8	188.4	189.0	9	8.69
2	178.2	174.1	170.3	171.6	171.7	174.7	176.0	176.6	172.8	9	6.89
3	218.6	208.4	187.1	199.5	202.0	211.1	197.6	204.4	206.8	9	80.22

Solution   Before conducting the Hartley test, we must check the normality condition. The data are evaluated for normality using a boxplot given in Figure 7.8. All three data sets appear to be from normally distributed populations. Thus, we will apply the Hartley F_{max} test to the data sets. From Appendix Table 10, with $\alpha = .05$, $t = 3$, and $df_2 = 9 - 1 = 8$, we have $F_{max,.05} = 6.00$. Thus, our rejection region will be

R.R.:   Reject H_0 if $F_{max} \geq F_{max,05} = 6.00$

$$s_{min}^2 = \min(8.69, 6.89, 80.22) = 6.89$$

and

$$s_{max}^2 = \max(8.69, 6.89, 80.22) = 80.22$$

Thus,

$$F_{max} = \frac{s_{max}^2}{s_{min}^2} = \frac{80.22}{6.89} = 11.64 > 6.00$$

Thus, we reject H_0 and conclude that the variances are not all equal.

FIGURE 7.8
Boxplot of deviations from target power for three suppliers

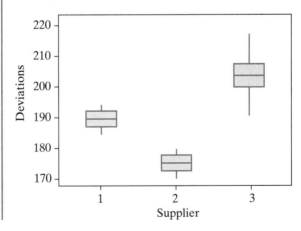

If the sample sizes are not all equal, we can take $n = n_{max}$, where n_{max} is the largest sample size. F_{max} no longer has an exact level α. In fact, the test is liberal in the sense that the probability of Type I error is slightly more than the nominal value α. Thus, the test is more likely to falsely reject H_0 than the test having all n_is equal when sampling from normal populations with the variances all equal.

The Hartley F_{max} test is quite sensitive to departures from normality. Thus, if the population distributions we are sampling from have a somewhat nonnormal distribution but the variances are equal, the F_{max} will reject H_0 and declare the variances to be unequal. The test is detecting the nonnormality of the population distributions, not the unequal variances. Thus, when the population distributions are nonnormal, the F_{max} is not recommended as a test of homogeneity of variances.

An alternative approach that does not require the populations to have normal distributions is the Levene test. However, the Levene test involves considerably more calculation than the Hartley test. Also, when the populations have a normal distribution, the Hartley test is more powerful than the Levene test. Conover, Johnson, and Johnson [*Technometrics* (1981) 23:351–361] conducted a simulation study of a variety of tests of homogeneity of variance, including the Hartley and Levene tests. They demonstrated the inflated α levels of the Hartley test when the populations have highly skewed distributions and recommended the Levene test as one of several alternative procedures.

The Levene test involves replacing the jth observation from sample i, y_{ij}, with the random variable $z_{ij} = |y_{ij} - \tilde{y}_i|$, where $\tilde{y}_i$ is the sample median of the ith sample. We then compute the Levene test statistic on the z_{ij}s.

Levene's Test for Homogeneity of Population Variances

H_0: $\sigma_1^2 = \sigma_2^2 = \cdots = \sigma_t^2$ (homogeneity of variances)

H_a: Population variances are not all equal

T.S.: $L = \dfrac{\sum_{i=1}^{t} n_i(\bar{z}_{i.} - \bar{z}_{..})^2/(t - 1)}{\sum_{i=1}^{t} \sum_{j=1}^{n_i} (z_{ij} - \bar{z}_{i.})^2/(N - t)}$

R.R.: For a specified value of α, reject H_0 if $L \geq F_{\alpha,df_1,df_2}$, where $df_1 = t - 1$, $df_2 = N - t$, $N = \sum_{i=1}^{t} n_i$, and F_{α,df_1,df_2} is the upper α percentile from the F distribution, Appendix Table 9.

Check assumptions and draw conclusions.

We will illustrate the computations for the Levene test in the following example. However, in most cases, we recommend using a computer software package such as SAS or Minitab for conducting the test.

EXAMPLE 7.7

Three different additives that are marketed for increasing the miles per gallon (mpg) for automobiles were evaluated by a consumer testing agency. Past studies have shown an average increase of 8% in mpg for economy automobiles after using the product for 250 miles. The testing agency wants to evaluate the variability in the increase in mileage over a variety of brands of cars within the economy class. The agency randomly selected 30 economy cars of similar age, number of miles on their odometer, and overall condition of the power train to be used in the study. It then randomly assigned 10 cars to each additive. The percentage increase in mpg obtained by each car was recorded for a 250-mile test drive. The testing agency wanted to evaluate whether there was a difference between the three additives with respect to their variability in the increase in mpg. The data are given here along with the intermediate calculations needed to compute the Levene's test statistic.

Solution Using the plots in Figures 7.9(a)–(d), we can observe that the samples from additive 1 and additive 2 do not appear to be samples from normally distributed

FIGURE 7.9   (a) Boxplots of additive 1, additive 2, and additive 3 (means are indicated by solid circles); (b–d) Normal probability plots for additives 1, 2, and 3

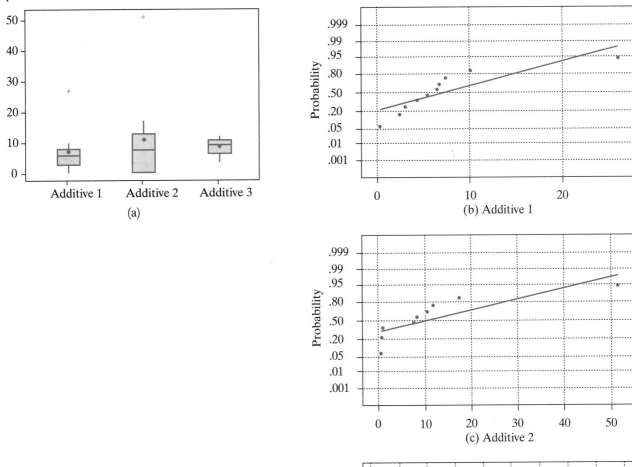

populations. Hence, we should not use Hartley's F_{max} test for evaluating differences in the variances in this example. The information in Table 7.1 will assist us in calculating the value of the Levene test statistic. The medians of the percentage increase in mileage, y_{ij}s, for the three additives are 5.80, 7.55, and 9.15. We then calculate the absolute deviations of the data values about their respective medians

Additive	y_{1j}	$\bar{y}_1$	$z_{1j} = \|y_{1j} - 5.80\|$	$\bar{z}_{1.}$	$(z_{1j} - 4.07)^2$	$(z_{ij} - 5.06)^2$
1	4.2	5.80	1.60	4.07	6.1009	11.9716
1	2.9		2.90		1.3689	4.6656
1	0.2		5.60		2.3409	0.2916
1	25.7		19.90		250.5889	220.2256
1	6.3		0.50		12.7449	20.7936
1	7.2		1.40		7.1289	13.3956
1	2.3		3.50		0.3249	2.4336
1	9.9		4.10		0.0009	0.9216
1	5.3		0.50		12.7449	20.7936
1	6.5		0.70		11.3569	19.0096

Additive	y_{2j}	$\bar{y}_2$	$z_{2j} = \|y_{2j} - 7.55\|$	$\bar{z}_{2.}$	$(z_{1j} - 8.88)^2$	$(z_{2j} - 5.06)^2$
2	0.2	7.55	7.35	8.88	2.3409	5.2441
2	11.3		3.75		26.3169	1.7161
2	0.3		7.25		2.6569	4.7961
2	17.1		9.55		0.4489	20.1601
2	51.0		43.45		1,195.0849	1,473.7921
2	10.1		2.55		40.0689	6.3001
2	0.3		7.25		2.6569	4.7961
2	0.6		6.95		3.7249	3.5721
2	7.9		0.35		72.7609	22.1841
2	7.2		0.35		72.7609	22.1841

Additive	y_{3j}	$\bar{y}_3$	$z_{3j} = \|y_{3j} - 9.15\|$	$\bar{z}_{3.}$	$(z_{1j} - 2.33)^2$	$(z_{3j} - 5.06)^2$
3	7.2	9.15	1.95	2.23	0.0784	9.6721
3	6.4		2.75		0.2704	5.3361
3	9.9		0.75		2.1904	18.5761
3	3.5		5.65		11.6964	0.3481
3	10.6		1.45		0.6084	13.0321
3	10.8		1.65		0.3364	11.6281
3	10.6		1.45		0.6084	13.0321
3	8.4		0.75		2.1904	18.5761
3	6.0		3.15		0.8464	3.6481
3	11.9		2.75		0.2704	5.3361
Total				5.06	1,742.6	1,978.4

—namely, $z_{1j} = |y_{1j} - 5.80|$, $z_{2j} = |y_{2j} - 7.55|$, and $z_{3j} = |y_{3j} - 9.15|$ for $j = 1, \ldots,$ 10. These values are given in column 4 of the table. Next, we calculate the three means of these values, $\bar{z}_{1.} = 4.07$, $\bar{z}_{2.} = 8.88$, and $\bar{z}_{3.} = 2.23$. Then, we calculate the squared deviations of the z_{ij}s about their respective means, $(z_{ij} - \bar{z}_{i.})^2$; that is, $(z_{1j} - 4.07)^2$, $(z_{2j} - 8.88)^2$, and $(z_{3j} - 2.23)^2$. These values are contained in column 6 of the table. Then we calculate the squared deviations of the z_{ij}s about the overall mean, $\bar{z}_{..} = 5.06$—that is, $(z_{ij} - \bar{z}_{..})^2 = (z_{ij} - 5.06)^2$. The last column in the table contains these values. The final step is to sum columns 6 and 7, yielding

$$T_1 = \sum_{i=1}^{3} \sum_{j=1}^{n_i} (z_{ij} - \bar{z}_{i.})^2 = 1742.6 \quad \text{and} \quad T_2 = \sum_{i=1}^{3} \sum_{j=1}^{n_i} (z_{ij} - \bar{z}_{..})^2 = 1{,}978.4$$

The value of Levene's test statistics, in an alternative form, is given by

$$L = \frac{(T_2 - T_1)/(t - 1)}{T_1/(N - t)} = \frac{(1{,}978.4 - 1{,}742.6)/(3 - 1)}{1{,}742.6/(30 - 3)} = 1.827$$

The rejection region for Levene's test is to reject H_0 if $L \geq F_{\alpha, t-1, N-t} = F_{.05, 3-1, 30-3} = 3.35$. Because $L = 1.827$, we fail to reject H_0: $\sigma_1^2 = \sigma_2^2 = \sigma_3^2$ and conclude that there is insufficient evidence of a difference in the population variances of the percentage increase in mpg for the three additives.

EXERCISES   **Applications**

7.12 In Example 7.7 we stated that the Hartley test was not appropriate because there was evidence that two of the population distributions were nonnormal. The Levene test was then applied to the data and it was determined that the data did not support a difference in the population variances at an $\alpha = .05$ level. The data yielded the following summary statistics:

Additive	Sample Size	Mean	Median	Standard Deviation
1	10	7.05	5.80	7.11
2	10	10.60	7.55	15.33
3	10	8.53	9.15	2.69

    a. Using the plots in Example 7.7, justify that the population distributions are not normal.
    b. Use the Hartley test to test for differences in the population variances.
    c. Are the results of the Hartley test consistent with those of the Levene test?
    d. Which test is more appropriate for this data set? Justify your answer.
    e. Which additive appears to be a better product? Justify your answer.

Edu.   **7.13** Many school districts are attempting to both reduce costs and motivate students by using computers as instructional aides. A study was designed to evaluate the use of computers in the classroom. A group of students enrolled in an alternative school were randomly assigned to one of four methods for teaching adding and multiplying fractions. The four methods were lectures only (L), lectures with remedial textbook assistance (L/R), lectures with computer assistance (L/C), and computer instruction only (C). After a 15-week instructional period, an exam was given. The students had taken an exam at the beginning of the 15-week period and the difference in the scores of the two exams is given in the following table. The school administrator wants to determine which method yields the largest increase in test scores and provides the most consistent gains in scores.

Method	Student 1	2	3	4	5	6	7	8	9	10
L	7	2	2	6	16	11	9	0	4	2
L/R	5	2	3	11	16	11	3			
L/C	9	12	2	17	12	20	20	31	21	
C	17	19	26	1	47	27	−8	10	20	

Boxplots of L, L/R, L/C, and C (means are indicated by solid circles)

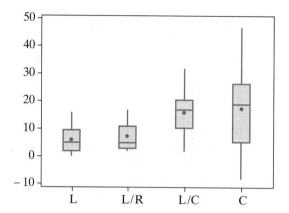

Which method of instruction appears to be the most successful? Provide all relevant tests, confidence intervals, and plots to justify your conclusion.

7.5 Encounters with Real Data: Evaluation of Methods for Detecting *E. coli*

Defining the Problem (1)

The outbreaks of bacterial disease in recent years due to the consumption of contaminated meat products have created a demand for new rapid pathogen-detecting methods that can be used in a meat surveillance program. The paper "Repeatability of the petrifilm HEC test and agreement with a hydrophobic grid membrane filtration method for the enumeration of *Escherichia coli* 0157:H7 on beef carcasses" [*Journal of Food Protection* (1998) 61:402–408] describes a formal comparison between a new microbial method for the detection of *E. coli,* the petrifilm HEC test, and an elaborate laboratory-based procedure, hydrophobic grid membrane filtration (HGMF). The HEC test is easier to inoculate, more compact to incubate, and safer to handle than conventional procedures. However, researchers had to compare the performances of the HEC test and the HGMF procedure to determine whether HEC was a viable method for detecting *E. coli.*

Collecting Data (2)

The developers of the HEC method sought answers to the following questions:

1. What parameters associated with the HEC and HGMF readings must be compared?
2. How many observations are necessary for a valid comparison of HEC and HGMF?
3. What type of experimental design will produce the most efficient comparison of HEC and HGMF?
4. What are the valid statistical procedures for making the comparisons?
5. What types of information should be included in a final report to document the evaluation of HEC and HGMF?

FIGURE 7.10
Hypothetical distribution of *E. coli* concentrations from HEC and HGMF

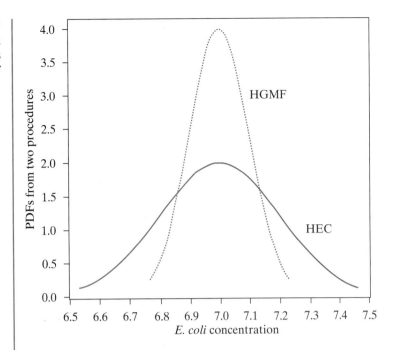

What aspects of the *E. coli* counts should be of interest to the researchers? A comparison of only the mean concentration would indicate whether or not the two procedures were in agreement with respect to the average readings over a large number of determinations. However, we would not know whether HEC was more variable in its determination of *E. coli* than HGMF. For example, consider the two distributions in Figure 7.10. Suppose the distributions represent the population of *E. coli* concentration determinations from HEC and HGMF for a situation in which the true *E. coli* concentration is 7 $\log_{10}$ CFU/ml. The distributions indicate that the HEC evaluation of a given meat sample may yield a reading very different from the true *E. coli* concentration, whereas the individual readings from HGMF are more likely to be near the true concentration. In this type of situation, it is crucial to compare both the means and standard deviations of the two procedures. In fact, we need to examine other aspects of the relationship between HEC and HGMF determinations to evaluate the comparability of the two procedures. We will examine these ideas in Chapters 11 and 12.

The experiment was designed to have two phases. Phase one of the study applied both procedures to pure cultures of *E. coli* representing 10^7 CFU/ml of strain E318N. Based on the specified degree of precision in estimating the *E. coli* level, it was determined that the HEC and HGMF procedures would be applied to 24 pure cultures each. Thus, there were two independent samples of size 24 each. The determinations yielded the *E. coli* concentrations in transformed metric (in $\log_{10}$ CFU/ml) given in Table 7.2. (The values in Table 7.2 were simulated using the summary statistics given in the paper. These data are also in the HECTEST file on the CD that came with your book.)

Phase two of the study applied both procedures to artificially contaminated beef. Portions of beef trim were obtained from three Holstein cows that had tested negative for *E. coli*. Eighteen portions of beef trim were obtained from the cows and then contaminated with *E. coli*. The HEC and HGMF procedures were applied

Sample	**HGMF**	**HEC**		**Sample**	**HGMF**	**HEC**		**Sample**	**HGMF**	**HEC**
1	6.65	6.67		9	6.89	7.08		17	7.07	7.25
2	6.62	6.75		10	6.90	7.09		18	7.09	7.28
3	6.68	6.83		11	6.92	7.09		19	7.11	7.34
4	6.71	6.87		12	6.93	7.11		20	7.12	7.37
5	6.77	6.95		13	6.94	7.11		21	7.16	7.39
6	6.79	6.98		14	7.03	7.14		22	7.28	7.45
7	6.79	7.03		15	7.05	7.14		23	7.29	7.58
8	6.81	7.05		16	7.06	7.23		24	7.30	7.54

TABLE 7.2

E. coli readings ($\log_{10}$ CFU/ml) from HGMF and HEC procedures

to a portion of each of the 18 samples. The two procedures yielded *E. coli* concentrations in transformed metric ($\log_{10}$ CFU/ml). The data in this case were 18 paired samples. The researchers were interested in determining a model to relate the two procedures' determinations of *E. coli* concentrations. We will only consider phase one in this chapter; phase two will be discussed in Chapter 11. The researchers next prepared the data for a statistical analysis following the steps described in Section 2.5.

Summarizing Data (3)

The researchers were interested in determining whether the two procedures yielded equivalent measures of *E. coli* concentrations. The boxplots of the experimental data are given in Figure 7.11. The two procedures appear to be very similar with respect to the width of box and length of whiskers, but HEC has a larger median than HGMF. The sample summary statistics are given here.

```
Descriptive Statistics

Variable        N      Mean    Median    TrMean     StDev    SE Mean
HEC            24    7.1346    7.1100    7.1373    0.2291     0.0468
HGMF           24    6.9529    6.9350    6.9550    0.2096     0.0428


Variable   Minimum   Maximum       Q1        Q3
HEC         6.6700    7.5400    6.9925    7.3250
HGMF        6.5600    7.3000    6.7900    7.1050
```

FIGURE 7.11

Boxplots of HEC and HGMF (means are indicated by solid circles)

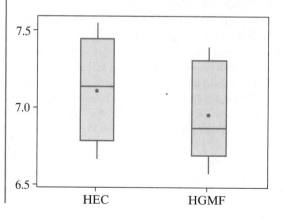

From the summary statistics, we note that HEC yields a larger mean concentration than HGMF. Also, the variability in concentration readings for HEC appears to be slightly greater than that for HGMF. Our initial conclusion might be that the two procedures yield different distributions of readings for their determination of *E. coli* concentrations. However, we need to determine whether the differences in their sample means and standard deviations infer a difference in the corresponding population values. We will summarize our findings after developing the appropriate procedures for comparing population variances.

Inferential problems about population variances are similar to the problems addressed in making inferences about the population mean. We must construct point estimators, confidence intervals, and the test statistics from the randomly sampled data to make inferences about the variability in the population values. We then can state our degree of certainty that observed differences in the sample data convey differences in the population parameters.

Analyzing Data, Interpreting the Analyses, and Communicating Results (4)

Because the objective of the study was to evaluate the HEC procedure for its performance in detecting *E. coli*, it is necessary to evaluate its repeatability and its agreement with an accepted method for *E. coli*—namely, the HGMF procedure. Thus, we need to compare both the level and variability in the two methods to determine *E. coli* concentrations; that is, we need to test hypotheses about both the means and standard deviations of HEC and HGMF *E. coli* concentrations. Recall that we had 24 independent observations from the HEC and HGMF procedures on pure cultures of *E. coli* having a specified level of 7 $\log_{10}$ CFU/ml. Prior to constructing confidence intervals or testing hypotheses, we must first check whether the data represent random samples from normally distributed populations. From the boxplots displayed in Figure 7.11 and the normal probability plots in Figure 7.12, the data from both procedures appear to follow a normal distribution.

We next test the hypotheses

$$H_0: \sigma_1^2 = \sigma_2^2 \quad \text{vs.} \quad H_a: \sigma_1^2 \neq \sigma_2^2$$

where we designate HEC as population 1 and HGMF as population 2. The summary statistics are given here.

Procedure	Sample Size	Mean	Standard Deviation
HEC	24	7.1346	.2291
HGMF	24	6.9529	.2096

For a two-tailed test with $\alpha = .05$, we will reject H_0 if

$$F_0 = s_1^2/s_2^2 \leq F_{.975,23,23} = 1/F_{.025,23,23} = 1/2.31 = .43 \quad \text{or} \quad F \geq F_{.025,23,23} = 2.31$$

Because $F_0 = (.2291)^2/(.2096)^2 = 1.19$ is not less than .43 or greater than 2.31, we fail to reject H_0. Using a computer software program, we determine that the *p*-value of the test statistic is 0.672. Thus, we can conclude that HEC appears to have a similar degree of variability as HGMF in its determination of *E. coli* concentration. To

FIGURE 7.12

Normal probability plots for HGMF and HEC

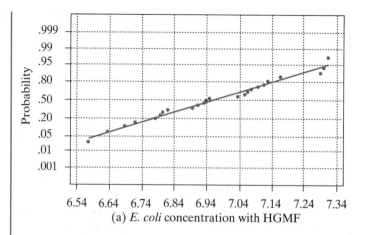

(a) *E. coli* concentration with HGMF

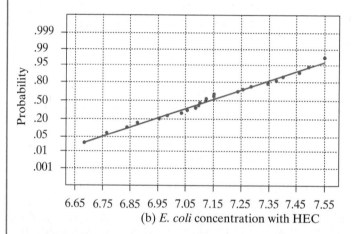

(b) *E. coli* concentration with HEC

obtain estimates of the variability in the HEC and HGMF readings, 95% confidence intervals on their standard deviations are given by (0.17, .23) for σ_{HEC} and (.16, .21) for σ_{HGMF}.

Both the HEC and HGMF *E. coli* concentration readings appear to be independent random samples from normal populations with a common standard deviation, so we can use a pooled *t* test to evaluate

$$H_0: \mu_1 = \mu_2 \quad \text{vs.} \quad H_a: \mu_1 \neq \mu_2$$

For a two-tailed test with $\alpha = .05$, we will reject H_0 if

$$|t| = \frac{|\bar{y}_1 - \bar{y}_2|}{s_p\sqrt{\dfrac{1}{n_1} + \dfrac{1}{n_2}}} \geq t_{.025,46} = 2.01$$

Because $t = 2.87$ is greater than 2.01, we reject H_0. The *p*-value is .0062. Thus, there is significant evidence that the average HEC *E. coli* concentration readings differ from the average HGMF readings. To estimate the average readings, 95% confidence intervals are given by (7.04, 7.23) for μ_{HEC} and (6.86, 7.04) for μ_{HGMF}. The HEC readings are on the average somewhat higher than the HGMF readings.

These findings then prepare us for the second phase of the study (Chapter 11), in which HEC and HGMF will be applied to the same sample of meats in a field study similar to what would be encountered in a meat quality monitoring setting. The two procedures have similar levels of variability, but HEC produced *E. coli* concentration readings higher than those of HGMF. Thus, the goal of phase two is to calibrate the HEC readings to the HGMF readings.

In order to communicate the results, we need to write a report summarizing our findings concerning phase one of the study. The report should include

1. Statement of objective for study
2. Description of study design and data collection procedures
3. Numerical and graphical summaries of data sets
4. Description of all inference methodologies:

   - *t* and *F* tests
   - *t*-based confidence intervals on means
   - Chi-square-based confidence intervals on standard deviations
   - Verification that all necessary conditions for using inference techniques were satisfied

5. Discussion of results and conclusions
6. Interpretation of findings relative to previous studies
7. Recommendations for future studies
8. Listing of data sets

7.6 Staying Focused

In this chapter, we discussed procedures for making inferences concerning population variances or, equivalently, population standard deviations. Estimation and statistical tests concerning σ make use of the chi-square distribution with $\mathrm{df} = n - 1$. Inferences concerning the ratio of two population variances or standard deviations use the F distribution with $\mathrm{df}_1 = n_1 - 1$ and $\mathrm{df}_2 = n_1 - 2$. Finally, when we developed tests concerning differences in $t > 2$ population variances, we used the Hartley or Levene test statistic.

The need for inferences concerning one or more population variances can be traced to our discussion of numerical descriptive measures of a population in Chapter 3. To describe or make inferences about a population of measurements, we cannot always rely on the mean, a measure of central tendency. Many times in evaluating or comparing the performance of individuals on a psychological test, the consistency of manufactured products emerging from a production line, or the yields of a particular variety of corn, we gain important information by studying the population variance.

Key Formulas

1. $100(1-\alpha)\%$ confidence interval for σ^2 (or σ)

$$\frac{(n-1)s^2}{\chi_U^2} < \sigma^2 < \frac{(n-1)s^2}{\chi_L^2} \quad \text{or} \quad \sqrt{\frac{(n-1)s^2}{\chi_U^2}} < \sigma < \sqrt{\frac{(n-1)s^2}{\chi_L^2}}$$

2. Statistical test for σ^2 (σ_0^2 specified)

$$\text{T.S.:} \quad \chi^2 = \frac{(n-1)s^2}{\sigma_0^2}$$

3. Statistical test for σ_1^2/σ_2^2

$$\text{T.S.:} \quad F = \frac{s_1^2}{s_2^2}$$

4. $100(1 - \alpha)\%$ confidence interval for σ_1^2/σ_2^2 (or σ_1/σ_2)

$$\frac{s_1^2}{s_2^2}F_L < \frac{\sigma_1^2}{\sigma_2^2} < \frac{s_1^2}{s_2^2}F_U$$

where

$$F_L = \frac{1}{F_{\alpha/2,\mathrm{df}_1,\mathrm{df}_2}} \quad \text{and} \quad F_U = F_{\alpha/2,\mathrm{df}_2,\mathrm{df}_1}$$

or

$$\sqrt{\frac{s_1^2}{s_2^2}F_L} < \frac{\sigma_1}{\sigma_2} < \sqrt{\frac{s_1^2}{s_2^2}F_U}$$

5. Statistical test for H_0: $\sigma_1^2 = \sigma_2^2 = \cdots = \sigma_t^2$
   a. When population distributions are normally distributed, the Hartley test should be used.

$$\text{T.S.:} \quad F_{\max} = \frac{s_{\max}^2}{s_{\min}^2}$$

   b. When population distributions are nonnormally distributed, the Levene test should be used.

$$\text{T.S.:} \quad L = \frac{\sum_{i=1}^t n_i(\bar{z}_{i.} - \bar{z}_{..})^2/(t-1)}{\sum_{i=1}^t \sum_{j=1}^{n_i} (z_{ij} - \bar{z}_{i.})^2/(N-t)}$$

where $z_{ij} = |y_{ij} - \tilde{y}_{i.}|$, $\tilde{y}_{i.} = \text{median}\,(y_{i1}, \ldots, y_{in_i})$, $\bar{z}_{i.} = \text{mean}\,(z_{i1}, \ldots, z_{in_i})$, and $\bar{z}_{..} = \text{mean}\,(z_{11}, \ldots, z_{tn_i})$.

Supplementary Exercises

Bus. **7.14** A consumer-protection magazine was interested in comparing tires purchased from two different companies that each claimed their tires would last 40,000 miles. A random sample of 10 tires of each brand was obtained and tested under simulated road conditions. The number of miles until the tread thickness reached a specified depth was recorded for all tires. The data are given next (in thousands of miles).

Brand	I	38.9	39.7	42.3	39.5	39.6	35.6	36.0	39.2	37.6	39.5
Brand	II	44.6	46.9	48.7	41.5	37.5	33.1	43.4	36.5	32.5	42.0

  a. Plot the data and compare the distributions of longevity for the two brands.
  b. Construct 95% confidence intervals on the means and standard deviations for the number of miles until tread wearout occurred for both brands.
  c. Does there appear to be a difference in wear characteristics for the two brands? Justify your statement with appropriate plots of the data, tests of hypotheses, and confidence intervals.

Med.   **7.15** A pharmaceutical company manufactures a particular brand of antihistamine tablets. In the quality control division, certain tests are routinely performed to determine whether the product being manufactured meets specific performance criteria prior to release of the product onto the market. In particular, the company requires that the potencies of the tablets lie in the range of 90–110% of the labeled drug amount.

  a. If the company is manufacturing 25-mg tablets, within what limits must tablet potencies lie?
  b. A random sample of 30 tablets is obtained from a recent batch of antihistamine tablets. The data for the potencies of the tablets are given next. Is the assumption of normality warranted for inferences about the population variances?
  c. Translate the company's 90–110% specifications on the range of the product potency into a statistical test concerning the population variance for potencies. Draw conclusions based on $\alpha = .05$.

24.1	27.2	26.7	23.6	26.4	25.2
25.8	27.3	23.2	26.9	27.1	26.7
22.7	26.9	24.8	24.0	23.4	25.0
24.5	26.1	25.9	25.4	22.9	24.9
26.4	25.4	23.3	23.0	24.3	23.8

Bus.   **7.16** The risk of an investment is measured in terms of the variance in the return that could be observed. Random samples of 10 yearly returns were obtained from two different portfolios. The data are given next (in thousands of dollars).

Portfolio 1	130	135	135	131	129	135	126	136	127	132
Portfolio 2	154	144	147	150	155	153	149	139	140	141

  a. Does portfolio 2 appear to have a higher risk than portfolio 1?
  b. Give a p-value for your test and place a confidence interval on the ratio of the standard deviations of the two portfolios.
  c. Provide a justification that the required conditions have been met for the inference procedures used in parts (a) and (b).

7.17 Refer to Exercise 7.16. Are there any differences in the average returns for the two portfolios? Indicate the method you used in arriving at a conclusion, and explain why you used it.

Med.   **7.18** Sales from weight-reducing agents marketed in the United States represent sizable amounts of income for many of the companies that manufacture these products. Psychological as well as physical effects often contribute to how well a person responds to the recommended therapy. Consider a comparison of two weight-reducing agents, A and B. In particular, consider the length of time people remain on the therapy. A total of 26 overweight males, matched as closely as possible physically, were randomly divided into two groups. Those in group 1 received preparation A and those assigned to group 2 received preparation B. The data are given here (in days).

Preparation A	42	47	12	17	26	27	28	26	34	19	20	27	34
Preparation B	35	38	35	36	37	35	29	37	31	31	30	33	44

Compare the lengths of times that people remain on the two therapies. Make sure to include all relevant plots, tests, confidence intervals, and a written conclusion concerning the two therapies.

7.19   Refer to Exercise 7.18. How would your inference procedures change if preparation A was an old product that had been on the market a number of years and preparation B was a new product, and we wanted to determine whether people would continue to use B a longer time in comparison to preparation A?

CHAPTER 8

The Completely Randomized Design

8.1   Introduction

In Chapter 6, we presented methods for comparing two population means, based on independent random samples. Very often the two-sample problem is a simplification of what we encounter in practical situations. For example, suppose we wish to compare the mean hourly wage for nonunion farm laborers from three different ethnic groups (African American, Anglo-American, and Hispanic) employed by a large produce company. Independent random samples of farm laborers would be selected from each of the three ethnic groups (populations). Then, using the information from the three sample means, we would try to make an inference about the corresponding population mean hourly wages. Most likely, the sample means would differ, but this does not necessarily imply a difference among the population means for the three ethnic groups. How do you decide whether the differences among the sample means are large enough to imply that the corresponding population means are different? We will answer this question using a statistical testing procedure called an *analysis of variance*.

The experimental design corresponding to this situation was discussed in Section 2.5 and was called a completely randomized design. We are interested in testing the equality of t (t is an integer ≥ 2) population means from normal populations with a common variance σ^2. Random samples of sizes $n_1, n_2, \ldots, n_t$ are drawn from the populations. In the terminology of the design of experiments, we assume that there are $n_1 + n_2 + \cdots + n_t$ homogeneous *experimental units* (people or objects on which a measurement is made). The treatments are randomly allocated to the experimental units in such a way that n_1 units receive treatment 1, n_2 receive treatment 2, and so on. The objective of the experiment is to make inferences about the corresponding treatment (population) means.

8.2 The Model for Observations in a Completely Randomized Design

Before we discuss how to analyze data from the experimental design, we need some notation.

Notation Needed for a Completely Randomized Design

y_{ij}: The jth sample observation selected from population i. For example, y_{23} denotes the third sample observation drawn from population 2.

n_i: The number of sample observations selected from population i.

N: The total sample size; $N = \Sigma\, n_i$.

$\bar{y}_{i.}$: The average of the n_i sample observations drawn from population i, $\bar{y}_{i.} = \Sigma_j\, y_{ij}/n_i$.

$\bar{y}_{..}$: The average of all sample observations; $\Sigma_i \Sigma_j\, y_{ij}/N$.

In this section, we will consider a model for the completely randomized design (sometimes referred to as a one-way classification). This model will demonstrate the types of settings for which testing procedures for a completely randomized design are appropriate. We can think of a model as a mathematical description of a physical setting. A model also enables us to computer-simulate the data that the physical process generates.

We will impose the following conditions concerning the sample measurements and the populations from which they are drawn:

1. The samples are independent random samples. Results from one sample in no way affect the measurements observed in another sample.
2. Each sample is selected from a normal population.
3. The mean and variance for population i are, respectively, μ_i and $\sigma^2\ (i = 1, 2, \ldots, t)$.

Figure 8.1 depicts a setting in which these three conditions are satisfied. The population distributions are normal with the same standard deviation. Note that populations III and IV have the same mean, which differs from the means of populations I and II. To summarize, we assume that the t populations are independently normally distributed with different means but a common variance σ^2.

model We can now formulate a **model** (equation) that encompasses these three conditions. Recall that we previously let y_{ij} denote the jth sample observation from population i.

$$y_{ij} = \mu + \alpha_i + \varepsilon_{ij}$$

This model states that y_{ij}, the jth sample measurement selected from population i, **terms** is the sum of three **terms.** The term μ denotes the overall mean across all t popula-

FIGURE 8.1

Distributions of four
populations that satisfy
our conditions

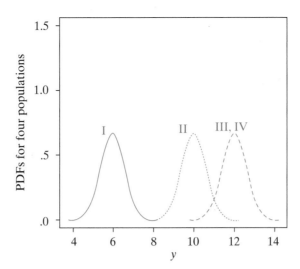

tions—that is, the mean of the population consisting of the observations from all t populations. The term α_i denotes the effect of population i on the overall variation in the observations. The terms μ and α_i are unknown constants, which will be estimated from the data obtained during the study or experiment. The term ε_{ij} represents the random deviation of y_{ij} about the ith population mean, μ_i. The ε_{ij}s are often referred to as *error terms*. The expression *error* is not to be interpreted as a mistake made in the experiment. Instead, the ε_{ij}s in the model represent the random variation of the y_{ij}s about their mean μ_i. The term *error* simply refers to the fact that the observations from the t populations differ by more than just their means. We assume that ε_{ij}s are independently normally distributed with a mean of 0 and a standard deviation of σ_ε. The independence condition can be interpreted as follows: The ε_{ij}s are independent if the size of the deviation of the y_{ij} observation from μ_i in no way affects the size of the deviation associated with any other observation.

Because y_{ij} is an observation from the ith population, it has mean μ_i. However, because the ε_{ij}s are distributed with mean 0, the mean or expected value of y_{ij}, denoted by $E(y_{ij})$, is

$$\mu_i = E(y_{ij}) = E(\mu + \alpha_i + \varepsilon_{ij}) = \mu + \alpha_i + E(\varepsilon_{ij}) = \mu + \alpha_i$$

that is, y_{ij} is a randomly selected observation from a population having mean $\mu_i = \mu + \alpha_i$. The effect α_i thus represents the deviation of the ith population mean μ_i from the overall mean μ. Thus, the α_is may assume a positive, zero, or negative value. Hence, the mean for population i can be greater than, equal to, or less than μ, the overall mean. The variance for each of the t populations can be shown to be σ_ε^2. Finally, because the ε_{ij}s are normally distributed, each of the t populations is normal. A summary of the assumptions for a completely randomized design is shown in Table 8.1.

The null hypothesis for a completely randomized design is that $\mu_1 = \mu_2 = \cdots = \mu_t$. Using our model, this is equivalent to the null hypothesis

$$H_0: \quad \alpha_1 = \alpha_2 = \cdots = \alpha_t = 0$$

Population	Population Mean	Population Variance	Sample Measurements
1	$\mu + \alpha_1$	σ_ε^2	$y_{11}, y_{12}, \ldots, y_{1n_1}$
2	$\mu + \alpha_2$	σ_ε^2	$y_{21}, y_{22}, \ldots, y_{2n_2}$
$\vdots$	$\vdots$	$\vdots$	$\vdots$
t	$\mu + \alpha_t$	σ_ε^2	$y_{t1}, y_{t2}, \ldots, y_{tn_t}$

If H_0 is true, then all populations have the same unknown mean μ. Indeed, many textbooks use the latter null hypothesis in a completely randomized design. The corresponding alternative hypothesis is

H_a: At least one of the α_is differs from 0.

In this section, we have presented a brief description of the model associated with the analysis of variance for a completely randomized design. Although some authors bypass an examination of the model, we believe it is a necessary part of a discussion of analysis of variance.

We have imposed several conditions on the populations from which the data are selected or, equivalently, on the experiments in which the data are generated, so we need to verify that these conditions are satisfied prior to making inferences about the population means. In Chapter 7, we discussed how to test the equality of variances condition using Hartley's $F_{\max}$ test or Levene's test. The normality condition is not as critical as the equal variance assumption when we have large sample sizes unless the populations are severely skewed or have very heavy tails. When we have small sample sizes, the normality condition and the equal variance condition become more critical. This situation presents a problem because there generally will not be enough observations from the individual population to test validly whether the normality or equal variance condition is satisfied. In the next section, we will discuss a technique that can at least partially overcome this problem. Also, some alternatives will be presented in later sections of this chapter that can be used when the populations have unequal variances or have nonnormal distributions. As we discussed in Chapter 6, the most critical of the three conditions is that the data values be independent. This condition can be met by carefully conducting the studies or experiments so as to not obtain data values that are dependent. In studies involving randomly selecting data from the t populations, we need to take care that the samples are truly random and that the samples from one population are not dependent on the values obtained from another population. In experiments in which t treatments are randomly assigned to experimental units, we need to make sure that the treatments are truly **randomly assigned.** Also, the experiments must be conducted so the experimental units do not interact with one another in a manner that could affect their responses.

8.3 Analyzing Data from a Completely Randomized Design: An Analysis of Variance

Refer back to the situation discussed in Section 8.1 where we wish to compare the mean hourly wage for nonunion farm laborers from three different ethnic groups. Data from a random sample of the hourly wages of five farm laborers are shown in

TABLE 8.2

A comparison of three sample means (small amount of within-sample variation)

	Sample from Population	
1	**2**	**3**
5.90	5.51	5.01
5.92	5.50	5.00
5.91	5.50	4.99
5.89	5.49	4.98
5.88	5.50	5.02
$\bar{y}_1 = 5.90$	$\bar{y}_2 = 5.50$	$\bar{y}_3 = 5.00$

TABLE 8.3

A comparison of three sample means (large amount of within-sample variation)

	Sample from Population	
1	**2**	**3**
5.90	6.31	4.52
4.42	3.54	6.93
7.51	4.73	4.48
7.89	7.20	5.55
3.78	5.72	3.52
$\bar{y}_1 = 5.90$	$\bar{y}_2 = 5.50$	$\bar{y}_3 = 5.00$

Table 8.2 (in dollars). Do these data present sufficient evidence to indicate differences among the three population means? A brief visual inspection of the data indicates very little variation within a sample, whereas the variability among the sample means is much larger. Because the variability among the sample means is large *in comparison to the* **within-sample variation,** we might conclude intuitively that the corresponding population means are different.

Table 8.3 illustrates a situation in which the sample means are the same as given in Table 8.2, but the variability within a sample is much larger and the **between-sample variation** is small relative to the within-sample variability. We are less likely to conclude that the corresponding population means differ based on these data.

The variations in the two sets of data, Tables 8.2 and 8.3, are shown graphically in Figure 8.2. The strong evidence indicating a difference in population means for the data of Table 8.2 is apparent in Figure 8.2(a). The lack of evidence indicating a difference in population means for the data of Table 8.3 is indicated by the overlapping of data points for the samples in Figure 8.2(b).

The preceding discussion, with the aid of Figure 8.2, indicates what we mean by an **analysis of variance.** All differences in sample means are judged statistically significant (or not) by comparing them to the variation within samples. The details of the testing procedure will be presented next.

Consider the data for a completely randomized design as arranged in Table 8.4. As discussed in the previous section, the model for a completely randomized design with t treatments and n_i observations per treatment can be written in the form

$$y_{ij} = \mu + \alpha_i + \varepsilon_{ij}$$

FIGURE 8.2

Dot diagrams for the data of Table 8.2 and Table 8.3: ○, measurement from sample 1; ●, measurement from sample 2; □, measurement from sample 3

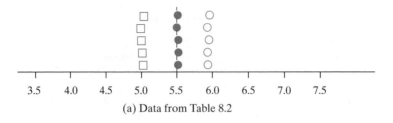

(a) Data from Table 8.2

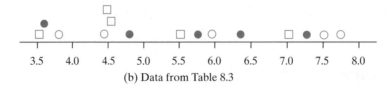

(b) Data from Table 8.3

TABLE 8.4

A completely randomized design

Treatment					Mean
1	y_{11}	y_{12}	$\cdots$	y_{1n_1}	$\bar{y}_{1.}$
2	y_{21}	y_{22}	$\cdots$	y_{2n_2}	$\bar{y}_{2.}$
$\vdots$	$\vdots$	$\vdots$	$\vdots$	$\vdots$	$\vdots$
t	y_{t1}	y_{t2}	$\cdots$	y_{tn_t}	$\bar{y}_{t.}$

where the terms of the model are defined as follows:

y_{ij}: Observation on jth experimental unit receiving treatment i.

μ: Overall treatment mean, an unknown constant.

α_i: An effect due to treatment i, an unknown constant.

ε_{ij}: A random error associated with the response from the jth experimental unit receiving treatment i. We require that the ε_{ij}s have a normal distribution with mean 0 and common variance σ_ε^2. In addition, the errors must be independent.

The conditions given here for our model can be shown to imply that the jth recorded response from the ith treatment y_{ij} is normally distributed with mean $\mu + \alpha_i$ and variance σ_ε^2. The treatment means differ by an amount α_i, the treatment effect. Thus, a test of

$$H_0: \mu_1 = \mu_2 = \cdots = \mu_t \quad \text{vs.} \quad H_a: \text{Not all } \mu_i\text{'s are equal}$$

is equivalent to testing

$$H_0: \alpha_1 = \alpha_2 = \cdots = \alpha_t = 0 \quad \text{vs.} \quad H_a: \text{Not all } \alpha_i\text{'s are 0}$$

total sum of squares   Our test statistic is developed using the idea of a partition of the **total sum of squares** (TSS) of the measurements about their mean $\bar{y}_{..} = \Sigma_{ij}\, y_{ij}/N$, which is defined as

$$\text{TSS} = \sum_{ij} (y_{ij} - \bar{y}_{..})^2$$

The total sum of squares is partitioned into two separate sources of variability: one due to variability among treatments and one due to the variability among the y_{ij}s within each treatment. The second source of variability is called *error* because it accounts for the variability that is not explained by treatment differences. The **partition of TSS** can be shown to take the following form:

partition of TSS

$$\sum_{ij} (y_{ij} - \bar{y}_{..})^2 = \sum_i n_i(\bar{y}_{i.} - \bar{y}_{..})^2 + \sum_{ij} (y_{ij} - \bar{y}_{i.})^2$$

When the number of replications is the same for all treatments—that is, $n_1 = n_2 = \cdots = n_t = n$—the partition becomes

$$\sum_{ij} (y_{ij} - \bar{y}_{..})^2 = n \sum_i (\bar{y}_{i.} - \bar{y}_{..})^2 + \sum_{ij} (y_{ij} - \bar{y}_{i.})^2$$

between-treatment
sum of squares

The first term on the right side of the equal sign measures the variability of the treatment means $\bar{y}_{i.}$ about the overall mean $\bar{y}_{..}$. Thus, it is called the **between-treatment sum of squares** (SST) and is a measure of the variability in the y'_{ij}s due to differences between the treatment means, μ_is. It is given by

$$\text{SST} = n \sum_i (\bar{y}_{i.} - \bar{y}_{..})^2$$

sum of squares for error

The second quantity is referred to as the **sum of squares for error** (SSE) and it represents the variability in the y'_{ij}s not explained by differences in the treatment means. This variability represents the differences in the experimental units prior to applying the treatments and the differences in the conditions that each experimental unit is exposed to during the experiment. It is given by

$$\text{SSE} = \sum_{ij} (y_{ij} - \bar{y}_{i.})^2$$

The quantities

$$\text{MST} = \frac{\text{SST}}{t-1} \quad \text{and} \quad \text{MSE} = \frac{\text{SSE}}{N-t}$$

are referred to as mean square treatment and mean square error, respectively. They are mean squares because they both are averages of squared deviations. We summarize this information in a table called an analysis of variance (AOV) table, as represented in Table 8.5.

unbiased estimates

When $H_0: \alpha_1 = \alpha_2 = \cdots = \alpha_t = 0$ is true, both MST and MSE are **unbiased estimates** of σ_ε^2, the variance of the experimental error. That is, when H_0 is true, both MST and MSE have a mean value in repeated sampling, called the **expected mean squares,** equal to σ_ε^2. We express these terms as

expected mean squares

$$E(\text{MST}) = \sigma_\varepsilon^2 \quad \text{and} \quad E(\text{MSE}) = \sigma_\varepsilon^2$$

TABLE 8.5
Analysis of variance table
for a completely
randomized design

Source	SS	df	MS	F
Treatments	SST	$t-1$	MST = SST/$(t-1)$	MST/MSE
Error	SSE	$N-t$	MSE = SSE/$(N-t)$	
Total	TSS	$N-1$		

Thus, we would expect the quantity $F = \text{MST/MSE}$ to be near 1 when H_0 is true. When H_a is true and there is a difference in the treatment means, the mean of MSE is still an unbiased estimate of σ_ε^2,

$$E(\text{MSE}) = \sigma_\varepsilon^2$$

However, MST is no longer unbiased for σ_ε^2. In fact, the expected mean square for treatments can be shown to be

$$E(\text{MST}) = \sigma_\varepsilon^2 + n\theta_T$$

where $\theta_T = \Sigma \, \alpha_i^2/(t - 1)$. When H_a is true, some of the α_i's are not zero and θ_T is positive. Thus, MST will tend to overestimate σ_ε^2. Hence, under H_a, the ratio $F = \text{MST/MSE}$ will tend to be greater than 1, and we will reject H_0 in the upper tail of the distribution of F. In particular, for selected values of the probability of Type I error α, we will reject $H_0: \alpha_1 = \alpha_2 = \cdots = \alpha_t = 0$ if the computed value of F exceeds $F_{\alpha, t-1, N-t}$, the critical value of F found in Table 9 in the Appendix with $a = \alpha$, $\text{df}_1 = t - 1$, $\text{df}_2 = N - t$. Note that df_1 and df_2 correspond to the degrees of freedom for MST and MSE, respectively, in the AOV table.

EXAMPLE 8.1

An important factor in road safety on rural roads is the use of reflective paint to mark the lanes on highways. This provides lane references for drivers on roads with little or no evening lighting. A problem with the currently used paint is that it does not maintain its reflectivity over long periods of time. A researcher will be conducting a study to compare three new paints (P_2, P_3, P_4) to the currently used paint (P_1). The paints will be applied to sections of highway 6 feet in length. The response variable will be the percentage decrease in reflectivity of the markings 6 months after application. There are 16 sections of highway, and each type of paint is randomly applied to four sections of highway. The reflective coating is applied to the highway, and 6 months later the decrease in reflectivity is computed at each section. The resulting measurements are given in Table 8.6. It appears that paint P_4 is able to maintain its reflectivity longer than the other three paints, because it has the smallest decrease in reflectivity. We will now attempt to confirm this observation by testing the hypotheses

$$H_0: \quad \mu_1 = \mu_2 = \mu_3 = \mu_4 \qquad H_a: \quad \text{Not all } \mu_i\text{'s are equal.}$$

We will construct the AOV table by computing the sum of squares using the formulas given previously:

TABLE 8.6

Reflectivity measurements

Section	1	2	3	4	Mean
Paint P_1	28	35	27	21	27.75
P_2	21	36	25	18	25
P_3	26	38	27	17	27
P_4	16	25	22	18	20.25

$$\bar{y}_{..} = \sum_{ij} y_{ij}/N = 400/16 = 25$$

$$\text{TSS} = \sum_{ij} (y_{ij} - \bar{y}_{..})^2$$

$$= (28 - 25)^2 + (35 - 25)^2 + \cdots + (22 - 25)^2 + (18 - 25)^2 = 692$$

$$\text{SST} = n\sum_{i} (\bar{y}_{i.} - \bar{y}_{..})^2$$

$$= 4[(27.75 - 25)^2 + (25 - 25)^2 + (27 - 25)^2 + (20.25 - 25)^2]$$

$$= 136.5$$

$$\text{SSE} = \text{TSS} - \text{SST} = 692 - 136.5 = 555.5$$

We can now complete the AOV table as follows:

Source	SS	df	MS	F	p-value
Treatments	136.5	3	45.5	.98	.4346
Error	555.5	12	46.292		
Total	692	15			

Because p-value $= .4346 > .05 = \alpha$, we fail to reject H_0. There is not significant evidence of a difference in the mean decrease in reflectivity for the four types of paints.

EXAMPLE 8.2

A clinical psychologist wished to compare three methods for reducing hostility levels in university students and used a certain test (HLT) to measure the degree of hostility. A high score on the test indicated great hostility. The psychologist used 24 students who obtained high and nearly equal scores in the experiment. Eight were selected at random from among the 24 problem cases and were treated with method 1. Seven of the remaining 16 students were selected at random and treated with method 2. The remaining nine students were treated with method 3. All treatments were continued for a one-semester period. Each student was given the HLT test at the end of the semester, with the results shown in Table 8.7. Use these data to perform an analysis of variance to determine whether there are differences among mean scores for the three methods. Use $\alpha = .05$.

TABLE 8.7
HLT test scores

Method	Test Scores									Mean	Sample Size
1	96	79	91	85	83	91	82	87		86.75	8
2	77	76	74	73	78	71	80			75.57	7
3	66	73	69	66	77	73	71	70	74	71.00	9
Total											24

Solution The null and alternative hypotheses are

$$H_0: \quad \mu_1 = \mu_2 = \mu_3$$

$$H_a: \quad \text{At least one of the population means differs from the rest.}$$

We can construct the AOV table by partitioning the TSS into SST (the between-treatment sum of squares) and SSE (the error sum of squares). For these data,

$$\bar{y}_{..} = \sum_{ij} y_{ij}/N = (96 + 79 + \cdots + 74)/24$$

$$= 1{,}862/24 = 77.58$$

$$\text{TSS} = \sum_{ij} (y_{ij} - \bar{y}_{..})^2$$

$$= (96 - 77.58)^2 + (79 - 77.58)^2 + \cdots + (74 - 77.58)^2$$

$$= 1477.80$$

and

$$\text{SST} = \sum_i n_i(\bar{y}_i - \bar{y}_{..})^2$$

$$= 8(86.75 - 77.58)^2 + 7(75.57 - 77.58)^2 + 9(71 - 77.58)^2$$

$$= 1090.66$$

Hence,

$$\text{SSE} = \text{TSS} - \text{SST}$$

$$= 1477.80 - 1090.66 = 387.14$$

The AOV table for these data is given in Table 8.8.

The critical value of F is obtained from Table 9 in the Appendix for $\alpha = .05$, $df_1 = 2$, and $df_2 = 21$; this value is 3.47. Because the computed value of F is 29.57, which exceeds the critical value 3.47, we reject the null hypothesis of equality of the mean scores for the three methods of treatment. We can only place an upper bound on the p-value because the largest value in Table 9 for $df_1 = 2$ and $df_2 = 21$ is 9.77, which corresponds to a probability of .001. Thus, with $p < .001$ there is a very strong rejection of the null hypothesis. From the three sample means, we observe that the mean for method 1 is considerably larger than the means for methods 2 and 3. The researcher needs to determine whether all three population means differ or the means for methods 2 and 3 are equal. Also, we may want to place confidence intervals on the three method means and on their differences; this would provide

TABLE 8.8
AOV table for Example 8.2

Source	SS	df	MS	F
Treatments	1090.66	2	545.33	29.58
Error	387.14	21	18.43	
Total	1477.80	23		

the researcher with information concerning the degree of differences in the three methods. In the next chapter, we will develop techniques to construct these types of inferences. The computer output shown here is consistent with the results we obtained, except for rounding errors in our hand calculations.

```
General Linear Models Procedure

Class Level Information

Class   Levels  Values
METHOD       3  1 2 3

Number of observations in data set = 24

Dependent Variable: SCORE

Source              DF   Sum of Squares  F Value  Pr > F
Model                2      1090.61904762   29.57  0.0001
Error               21       387.21428571
Corrected Total     23      1477.833333333
```

Before we leave the AOV for a completely randomized design, we want to show another equivalent way to display the partitioning of the total sum of squares based on the model

$$y_{ij} = \mu + \alpha_i + \varepsilon_{ij}$$

The parameters in this model have sample estimates $\hat{\mu} = \bar{y}_{..}$ and $\hat{\alpha}_i = \bar{y}_{i.} - \bar{y}_{..}$; hence, the partitioning of TSS can be written, as before, as

$$\text{TSS} = \sum_{ij}(y_{ij} - \bar{y}_{..})^2 = \sum_i n_i(\bar{y}_{i.} - \bar{y}_{..})^2 + \sum_{ij}(y_{ij} - \bar{y}_{i.})^2$$

$$= \text{SST} + \text{SSE}$$

Using the sample estimates, we have

$$\text{SST} = \sum_i n_i(\bar{y}_{i.} - \bar{y}_{..})^2 = \sum_i n_i(\hat{\alpha}_i)^2$$

and

$$\text{SSE} = \sum_{ij}(y_{ij} - \bar{y}_{i.})^2 = \sum_{ij}(y_{ij} - \hat{\mu} - \hat{\alpha}_i)^2 = \sum_{ij}(e_{ij})^2 = \sum_i(n_i - 1)s_i^2$$

where s_i^2 is the sample variance for data from the ith sample.

EXERCISES   **Applications**

Ag.   **8.1** A large laboratory has four types of devices used to determine the pH of soil samples. The laboratory wants to determine whether there are differences in the average readings given by these devices. The lab uses 24 soil samples having known pH in the study and randomly assigns six of the samples to each device. The soil samples are tested and the response recorded is the difference between the pH reading of the device and the known pH of the soil. These values, along with summary statistics, are given in the following table.

Device	Sample 1	2	3	4	5	6	Sample Size	Mean	Standard Deviation
A	−.307	−.294	.079	.019	−.136	−.324	6	−.1605	.1767
B	−.176	.125	−.013	.082	.091	.459	6	.0947	.2091
C	.137	−.063	.240	−.050	.318	.154	6	.1227	.1532
D	−.042	.690	.201	.166	.219	.407	6	.2735	.2492

a. Based on your intuition, is there evidence to indicate any difference among the mean differences in pH readings for the four devices?
b. Run an analysis of variance to confirm or reject your conclusion of part (a). Use $\alpha = .05$.
c. Compute the p-value of the F test in part (b).
d. What conditions must be satisfied for your analysis in parts (b) and (c) to be valid?
e. Suppose the 24 soil samples have widely different pH values. What problems may occur by simply randomly assigning the soil samples to the different devices?

Bus. **8.2** A cigarette manufacturer has advertised that it has developed a new brand of cigarette, LowTar, that has a lower average tar content than the major brands. To evaluate this claim, a consumer testing agency randomly selected 100 cigarettes from each of the four leading brands of cigarettes and 100 from the new brand. The tar content (in milligrams) of the cigarettes gave the following results:

Brand	$\bar{y}_i$	s_i	n_i
LowTar	9.64	.291	100
A	10.22	.478	100
B	10.77	.372	100
C	11.57	.352	100
D	13.59	.469	100

A boxplot of the data used to produce the table is given here.

Boxplots of tar content by brand for Exercise 8.2 (means are indicated by solid circles)

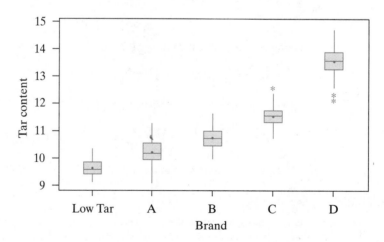

a. Based on the information contained in the boxplot, does the LowTar brand appear to have a lower average tar content than the other brands?
b. Using the computer output shown here, is there a significant ($\alpha = .01$) difference in the average tar content of the five brands of cigarettes?

c. What is the *p*-value of the test statistic in (b)?

d. What are the practical consequences of making a Type I error with respect to your test in (b)?

```
One-Way Analysis of Variance for Exercise 8.2

Analysis of Variance for Tar Cont
Source      DF        SS         MS        F       P
Brand        4    941.193    235.298   1478.39  0.000
Error      495     78.784      0.159
Total      499   1019.976
                                   Individual 95% CIs for Mean
                                   Based on Pooled StDev
Level       N      Mean      StDev   -+---------+---------+---------+-----
1         100     9.644      0.291   *)
2         100    10.221      0.478        *)
3         100    10.775      0.372               (*
4         100    11.570      0.352                      *)
5         100    13.592      0.469                                       *)
                                    -+---------+---------+---------+-----
Pooled StDev =    0.399               9.6      10.8      12.0      13.2
```

8.4 Checking on the AOV Conditions

The assumption of equal population variances and the assumption of normality of the populations have been made in several places in the text, such as for the *t* test when comparing two population means and now for the analysis of variance *F* test in a completely randomized design.

Let us consider first an experiment in which we wish to compare *t* population means based on independent random samples from each of the populations. Recall that we assume we are dealing with normal populations with a common variance σ_ε^2 and possibly different means. We could verify the assumption of equality of the population variances using Hartley's test or Levene's test in Chapter 7.

Several comments should be made here. Most practitioners do not routinely run Hartley's test. One reason is that the test is extremely sensitive to departures from normality. Thus, in checking one assumption (constant variance), the practitioner would have to be very careful about departures from another analysis of variance assumption (normality of the populations). Fortunately, as we mentioned in Chapter 6, the assumption of homogeneity (equality) of population variances is less critical when the sample sizes are nearly equal, where the variances can be markedly different and the *p*-values for an analysis of variance will still be only mildly distorted. Thus, we recommend that Hartley's test or Levene's test be used only for the more extreme cases. In these extreme situations where homogeneity of the population variances is a problem, a transformation of the data may help to stabilize the variances (see Ott and Longnecker, 2001, Chap. 8). Then inferences can be made from an analysis of variance.

The normality of the population distributions can be checked using normal probability plots or boxplots, as we discussed in Chapters 5 and 6, when the sample sizes are relatively large. However, in many experiments, the sample sizes may be as small as 5 to 10 observations from each population. In this case, the plots will not be a very reliable indication of whether the population distributions are normal. By taking into consideration the model we introduced in the previous section, the evaluation of the normal condition will be evaluated using a **residuals analysis.**

residuals analysis

From the model, we have $y_{ij} = \mu + \alpha_i + \varepsilon_{ij} = \mu_i + \varepsilon_{ij}$. Thus, we can write $\varepsilon_{ij} = y_{ij} - \mu_i$. Then if the condition of equal variances is valid, the ε_{ij}s are a random sample from a normal population. However, μ_i is an unknown constant; if we estimate μ_i with $\bar{y}_{i.}$ and let

$$e_{ij} = y_{ij} - \bar{y}_{i.}$$

then we can use the e_{ij}s to evaluate the normality assumption. Even when the individual n_is are small, we will have N residuals, which will provide a sufficient number of values to evaluate the normality condition. We can plot the e_{ij}s in a boxplot or a normality plot to evaluate whether the data appear to have been generated from normal populations.

EXAMPLE 8.3

Because many HMOs either do not cover mental health costs or provide only minimal coverage, ministers and priests often need to provide counseling to people suffering from mental illness. An interdenominational organization wanted to determine whether the clerics from different religions have different levels of awareness with respect to the causes of mental illness. Three random samples were drawn, one containing 10 Methodist ministers, a second containing 10 Catholic priests, and a third containing 10 Pentecostal ministers. Each of the 30 clerics was then examined, using a standard written test, to measure his or her knowledge about causes of mental illness. The test scores are listed in Table 8.9. Does there appear to be a significant difference in the mean test scores for the three religions?

Solution Prior to conducting an AOV test of the three means, we need to evaluate whether the conditions required for AOV are satisfied. Figure 8.3 is a boxplot of the mental illness scores by religion. There is an indication that the data may be some-

TABLE 8.9
Scores for clerics' knowledge of mental illness

Cleric	Methodist	Catholic	Pentecostal
1	62	62	37
2	60	62	31
3	60	24	15
4	25	24	15
5	24	22	14
6	23	20	14
7	20	19	14
8	13	10	5
9	12	8	3
10	6	8	2
$\bar{y}_{i.}$	30.50	25.90	15.00
s_i	21.66	20.01	11.33
n_i	10	10	10
Median $(\tilde{y}_i)$	23.5	21	14

FIGURE 8.3

Boxplots of score by religion
(means are indicated
by solid circles)

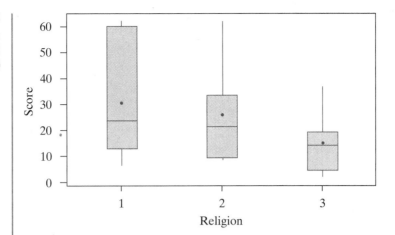

TABLE 8.10

Residuals e_{ij} for clerics'
knowledge of mental illness

Cleric	Methodist	Catholic	Pentecostal
1	31.5	36.1	22.0
2	29.5	36.1	16.0
3	29.5	−1.9	0.0
4	−5.5	−1.9	0.0
5	−6.5	−3.9	−1.0
6	−7.5	−5.9	−1.0
7	−10.5	−6.9	−1.0
8	−17.5	−15.9	−10.0
9	−18.5	−17.9	−12.0
10	−24.5	−17.9	−13.0

what skewed to the right. Thus, we will evaluate the normality condition. We need to obtain the residuals $e_{ij} = y_{ij} - \overline{y}_{i.}$. For example, $e_{11} = y_{11} - \overline{y}_{1.} = 62 - 30.50 = 31.50$. The remaining e_{ij}s are given in Table 8.10.

The residuals are then plotted in Figures 8.4 and 8.5. The boxplot in Figure 8.5 displays three outliers out of 30 residuals. It is very unlikely that 10% of the data

FIGURE 8.4

Normal probability plot
for residuals

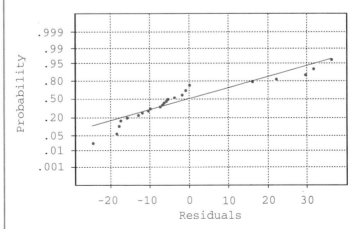

```
Average: -0.0000000        Anderson-Darling Normality Test
StDev: 17.5984                  A-Squared: 1.714
N: 30                           P-Value: 0.000
```

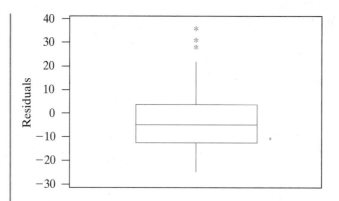

FIGURE 8.5
Boxplot of residuals

values are outliers if the residuals are in fact a random sample from a normal distribution. This is confirmed in the normal probability plot displayed in Figure 8.4, which shows a lack of concentration of the residuals about the straight line. Furthermore, the test of normality has a p-value less than .001, which indicates a strong departure from normality. Thus, we conclude that the data have nonnormal characteristics. In Section 8.5, we will provide an alternative to the F test from the AOV table, which is appropriate for this situation.

Because the data may be nonnormal, it is inappropriate to test for equal variances using Hartley's F_{max} test. Thus, we will use Levene's test. An examination of the formula for Levene's test reveals that once we make the conversion of the data from y_{ij} to $z_{ij} = |y_{ij} - \tilde{y}_i|$, where $\tilde{y}_i$ is the sample median of the ith data set, Levene's test is equivalent to the F test from AOV applied to the z_{ij}s. Thus, we can simply use the formulas from AOV to compute Levene's test. The z_{ij}s are given in Table 8.11 using the medians from Table 8.9.

Using the sample means given in the table, we compute the overall mean of the 30 data values:

$$\bar{z}_{..} = \sum_{i=1}^{3} n_i \bar{z}_{i.}/N = [10(15.70) + 10(12.90) + 10(7.40)]/30 = 360/30 = 12$$

TABLE 8.11
Transformed data set,
$z_{ij} = |y_{ij} - \tilde{y}_i|$

Cleric	Methodist	Catholic	Pentecostal
1	38.5	41	23
2	36.5	41	17
3	36.5	3	1
4	1.5	3	1
5	0.5	1	0
6	0.5	1	0
7	3.5	2	0
8	10.5	11	9
9	11.5	13	11
10	17.5	13	12
$\bar{z}_{i.}$	15.70	12.90	7.40
s_i	15.80	15.57	8.29

Using this value along with the means and standard deviations in Table 8.11, we can compute the sum of squares as follows:

$$\text{SST} = \sum_{i=1}^{3} n_i(\bar{z}_{i.} - \bar{z}_{..})^2 = 10(15.70 - 12)^2 + 10(12.90 - 12)^2 + 10(7.40 - 12)^2$$

$$= 356.6$$

and

$$\text{SSE} = \sum_{i=1}^{3} (n_i - 1)s_i^2 = (10 - 1)(15.80)^2 + (10 - 1)(15.57)^2$$

$$+ (10 - 1)(8.29)^2 = 5,047.10$$

The mean squares are MST = SST/$(t - 1)$ = 356.6/$(3 - 1)$ = 178.3 and MSE = SSE/$(N - t)$ = 5,047.10/$(30 - 3)$ = 186.9. Finally, we can next obtain the value of the Levene's test statistic from L = MST/MSE = 178.3/186.9 = .95. The critical value of L, using α = .05, is obtained from the F tables with df$_1$ = 2 and df$_2$ = 27. This value is 3.35, and thus we fail to reject the null hypothesis that the standard deviations are equal. The p-value is greater than .25 because the smallest value in the F table with df$_1$ = 2 and df$_2$ = 27 is 1.46, which corresponds to a probability of .25. Thus, we have a high degree of confidence that the three populations have the same variance.

In Section 8.5, we will present the Kruskal–Wallis test, which can be used when the populations are nonnormal but have identical distributions under the null hypothesis. This test requires, as a minimum, that the populations have the same variance. Thus, the Kruskal–Wallis test is not appropriate for the situation in which the populations have very different variances.

8.5 A Nonparametric Alternative: The Kruskal–Wallis Test

The concept of a rank sum test can be extended to a comparison of more than two populations. In particular, suppose that n_1 observations are drawn at random from population 1, n_2 from population 2, . . . , and n_k from population k. We may wish to test the hypothesis that the k samples were drawn from identical distributions. The following test procedure, sometimes called the Kruskal–Wallis test, is then appropriate.

Extension of the Rank Sum Test for More Than Two Populations

H_0: The k distributions are identical.

H_a: Not all the distributions are the same.

T.S.: $H = \dfrac{12}{N(N + 1)} \sum_i \dfrac{T_i^2}{n_i} - 3(N + 1)$

where n_i is the number of observations from sample i $(i = 1, 2, \ldots, k)$, N is the combined (total) sample size (that is, $N = \sum_i n_i$) and T_i denotes the sum of the ranks for the measurements in sample i after the combined sample measurements have been ranked.

R.R.: For a specified value of α, reject H_0 if H exceeds the critical value of χ^2 for $a = \alpha$ and df $= k - 1$.

Note: When there are a large number of ties in the ranks of the sample measurements, use

H'

$$H' = \frac{H}{1 - [\Sigma_j (t_j^3 - t_j)/(N^3 - N)]}$$

where t_i is the number of observations in the jth group of tied ranks.

Figure 8.6 displays population distributions under the alternative hypotheses of the Kruskal–Wallis test.

FIGURE 8.6

Four skewed population distributions identical in shape but shifted

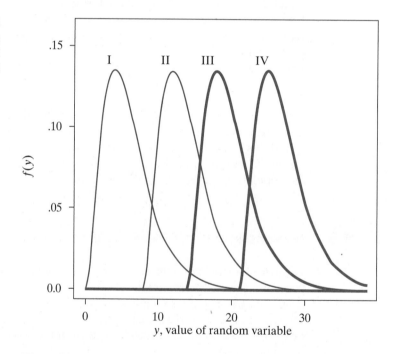

EXAMPLE 8.4

Refer to Example 8.3, where we determined that the clerics' test scores are not normally distributed but can be assumed to have the same variance. We will apply the Kruskal–Wallis test to the data set displayed in Table 8.9.

Use the data to determine whether the three groups of clerics differ with respect to their knowledge about the causes of mental illness. Use $\alpha = .05$.

Solution The null and research hypotheses for this example can be stated as follows:

H_0: There is no difference among the three groups with respect to knowledge about the causes of mental illness (i.e., the samples of scores were drawn from identical populations).

H_a: At least one of the three groups of clerics differs from the others with respect to knowledge about causes of mental illness.

Before computing H, we must first jointly rank the 30 test scores from lowest to highest. From Table 8.9, we see that 2 is the lowest test score, so we assign this cleric the rank of 1. Similarly, we give the scores 3, 4, and 6 the ranks 2, 3, and 4, respectively. Two clerics have a test score of 8, and because these two scores occupy the ranks 5 and 6, we assign each one a rank of 5.5—the average of the ranks 5 and 6. In a similar fashion, we can assign the remaining ranks to the test scores. Table 8.12 lists the 30 test scores and associated ranks (in parentheses).

TABLE 8.12

Scores for clerics' knowledge of mental illness in Example 8.3

Cleric	Methodist	Catholic	Pentecostal
1	62 (29)	62 (29)	37 (25)
2	60 (26.5)	62 (29)	31 (24)
3	60 (26.5)	24 (21)	15 (13.5)
4	25 (23)	24 (21)	15 (13.5)
5	24 (21)	22 (18)	14 (11)
6	23 (19)	20 (16.5)	14 (11)
7	20 (16.5)	19 (15)	14 (11)
8	13 (9)	10 (7)	5 (3)
9	12 (8)	8 (5.5)	3 (2)
10	6 (4)	8 (5.5)	2 (1)
Sum of Ranks	182.5	167.5	115

Note from Table 8.12 that the sums of the ranks for the three groups of clerics are 182.5, 167.5, and 115. Hence, the computed value of H is

$$H = \frac{12}{30(30+1)} \left(\frac{(182.5)^2}{10} + \frac{(167.5)^2}{10} + \frac{(115)^2}{10} \right) - 3(30+1)$$

$$= \frac{12}{930}(3{,}330.625 + 2{,}805.625 + 1{,}322.5) - 93 = 3.24$$

Because there are groups of tied ranks, we will use H' and compare its value to H. To do this we form the g groups composed of identical ranks, shown in the accompanying table.

Rank	Group	t_i	Rank	Group	t_i
1	1	1	15	11	1
2	2	1	16.5, 16.5	12	2
3	3	1	18	13	1
4	4	1	19	14	1
5.5, 5.5	5	2	21, 21, 21	15	3
7	6	1	23	16	1
8	7	1	24	17	1
9	8	1	25	18	1
11, 11, 11	9	3	26.5, 26.5	19	2
13.5, 13.5	10	2	29, 29, 29	20	3

From this information, we calculate the quantity

$$\sum_i \frac{(t_i^3 - t_i)}{N^3 - N}$$

$$= \frac{(2^3 - 2) + (3^3 - 3) + (2^3 - 2) + (2^3 - 2) + (3^3 - 3) + (2^3 - 2) + (3^3 - 3)}{30^3 - 30}$$

$$= .0036$$

Substituting this value into the formula for H', we have

$$H' = \frac{H}{1 - .0036} = \frac{3.24}{.9964} = 3.25$$

Thus, even with more than half of the measurements involved in ties, H' and H are nearly the same value. The critical value of the chi-square with $\alpha = .05$ and $df = k - 1 = 2$ can be found using Table 8 in the Appendix. This value is 5.991; we fail to reject the null hypothesis and conclude that there is no significant difference in the test scores of the three groups of clerics. It is interesting to note that the p-value for the Kruskal–Wallis test is .198, whereas the p-value from the AOV F test applied to the original test scores was .168. Thus, even though the data did not have a normal distribution, the F test from AOV is robust against departures from normality. Only when the data are extremely skewed or very heavy tailed do the Kruskal–Wallis test and the F test from AOV differ.

EXERCISES   Applications

Hort.   **8.3**   A team of researchers wants to compare the yields (in pounds) of five different varieties (A, B, C, D, E) of 4-year-old orange trees in one orchard. They obtain a random sample of seven trees of each variety from the orchard. The yields for these trees are presented here.

A	B	C	D	E
13	27	40	17	36
19	31	44	28	32
39	36	41	41	34
38	29	37	45	29
22	45	36	15	25
25	32	38	13	31
10	44	35	20	30

a. Using tests and plots of the data, determine whether the conditions for using the AOV are satisfied.
b. Conduct an AOV test of the null hypothesis that the five varieties have the same mean yield. Use $\alpha = .01$.
c. Use the Kruskal–Wallis test to test the null hypothesis that the five varieties have the same yield distributions. Use $\alpha = .01$.
d. Are the conclusions you reached in parts (b) and (c) consistent?

8.4   How do the research hypotheses tested by the AOV test and the Kruskal–Wallis test differ?

Engin.   **8.5**   In the manufacture of soft contact lenses, the actual strength (power) of the lens needs to be very close to the target value for the lenses to properly fit the customer's needs. In the paper, "An ANOM-type test for variances from normal populations" [*Technometrics* (1997) 39:274–283], a comparison of several suppliers is made relative to the consistency of the power of the lenses. The following table contains the deviations from the target power of lenses produced using materials from three different suppliers:

	Lens								
Supplier	**1**	**2**	**3**	**4**	**5**	**6**	**7**	**8**	**9**
A	189.9	191.9	190.9	183.8	185.5	190.9	192.8	188.4	189.0
B	156.6	158.4	157.7	154.1	152.3	161.5	158.1	150.9	156.9
C	218.6	208.4	187.1	199.5	202.0	211.1	197.6	204.4	206.8

a. Is there a significant difference in the distributions of deviations for the three suppliers? Use $\alpha = .01$.

b. Using the appropriate tests and plots given here, assess whether the data meet the necessary conditions to use an AOV to determine whether there is a significant difference in the mean deviations for the three suppliers.

c. Conduct an AOV with $\alpha = .05$ and compare your results with the conclusions from (a).

d. Suppose that a difference in mean deviation of 20 units has commercial consequences for the manufacture of the lenses. Does there appear to be a *practical* difference in the three suppliers?

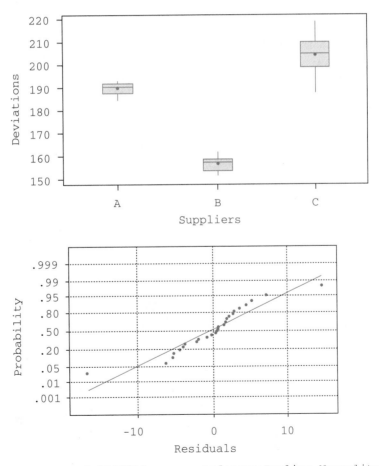

Average: -0.0000000
StDev: 5.53971
N: 27

Anderson-Darling Normality Test
A-Squared: 0.718
P-Value: 0.054

8.6 Multiple Comparisons

Earlier in this chapter, we introduced a procedure for testing the equality of t population means. We used the test statistic $F = \text{MST}/\text{MSE}$ to determine whether the between-treatment variability was large relative to the within-treatment variability. If the computed value of F for the sample data exceeded the critical value obtained from Table 9 in the Appendix, we rejected the null hypothesis $H_0: \mu_1 = \mu_2 = \cdots = \mu_t$ in favor of the alternative hypothesis

H_a: At least one of the t population means differs from the rest.

Multiple-comparison procedures

Although rejection of the null hypothesis does give us some information concerning the population means, we do not know which means differ from one another. For example, does μ_1 differ from μ_2 or μ_3? Does μ_3 differ from the average of μ_2, μ_4, and μ_5? Is there an increasing trend in the treatment means $\mu_1, \ldots, \mu_t$? **Multiple-comparison procedures** and contrasts have been developed to answer questions such as these. Although many multiple-comparison procedures have been proposed, we will focus on just a few of the more commonly used methods. After studying these few procedures, you should be able to evaluate the results of most published material using multiple comparisons or to suggest an appropriate multiple-comparison procedure in an experimental situation.

data dredging

data snooping

A word of caution: It is tempting to analyze only those comparisons that appear to be interesting after seeing the sample data. This practice has sometimes been called **data dredging** or **data snooping,** and the confidence coefficient for a single comparison does not reflect the after-the-fact nature of the comparison. For example, we know from previous work that the interval estimate for the difference between two population means using the formula

$$(\bar{y}_1 - \bar{y}_2) \pm t_{\alpha/2} s_p \sqrt{\frac{1}{n_1} + \frac{1}{n_2}}$$

has a confidence coefficient of $1 - \alpha$. Suppose we had run an analysis of variance to test the hypothesis

$$H_0: \quad \mu_1 = \mu_2 = \mu_3 = \mu_4 = \mu_5 = \mu_6$$

for six populations, but decided to compute a confidence interval for μ_1 and μ_2 only after we saw that the largest sample mean was $\bar{y}_1$ and the smallest was $\bar{y}_2$. In this situation, the confidence coefficient would not be $1 - \alpha$ as originally thought; that value applies only to a preplanned comparison, one planned before looking at the sample data.

One way to allow for data snooping after observing the sample data is to use a multiple-comparison procedure that has a confidence coefficient to cover all comparisons that could be done after observing the sample data. Two of these procedures are discussed in this chapter.

exploratory hypothesis generation

confirmation

The other possibility is to use data-snooping comparisons as a basis for generating hypotheses that must be confirmed in future experiments or studies. Here, the data-snooping comparisons serve an **exploratory,** or **hypothesis-generating,** role, and inferences would not be made based on the data snoop. Further experimentation would be done to **confirm** (or not) the hypothesis generated in the data snoop.

In many experiments, the researcher will want to compare all pairs of treatments or compare all treatments to a control. In these situations, there are many methods for testing these types of contrasts among the treatment means. A major difference among these multiple-comparison procedures is the type of error rate that each procedure controls. We will discuss two of these procedures, but before we do that we will examine two different types of Type I error rates.

Suppose we have a completely randomized design with t treatments and that we want to make m $(t > m)$ pairwise comparisons among the population means. Also suppose that each of the individual pairwise comparisons is done using a t test with Type I error α_I, called the **individual Type I error rate.**

We need to also consider the probability of falsely rejecting at least one of the m null hypotheses, called the **experimentwise Type I error rate** and denoted by α_E. The value of α_E takes into account that we are conducting m tests, each having an α_I chance of making a Type I error. Now, if the m tests are independent, then when all m null hypotheses are true the probability of falsely rejecting at least one of the m null hypotheses can be shown to be $\alpha_E = 1 - (1 - \alpha_I)^m$. Table 8.13 contains values of α_E for various values of m and α_I. We can observe from Table 8.13 that as the number of tests m increases for a given value of α_I, the probability of falsely rejecting H_0 on at least one of the m tests α_E becomes quite large. For example, if an experimenter wanted to compare $t = 20$ population means by using $m = 10$ independent, pairwise comparisons, the probability of falsely rejecting H_0 on at least one of the t tests could be as high as .401 when each individual test was performed with $\alpha_I = .05$.

In any practical problem, the tests will not be independent. Thus, the relationship between α_E and α_I is not generally as described in Table 8.13. It is difficult to obtain an expression equivalent to $\alpha_E = 1 - (1 - \alpha_I)^m$ for comparisons made with tests that are not independent. However, it can be shown that for most of the types of comparisons we will be making among the population means, the following upper bound exists for the experimentwise error rate:

$$\alpha_E \leq m\alpha_I$$

Thus, we know the largest possible value for α_E when we set the value of α_I for each of the individual tests. Suppose, for example, that we wish the experimentwise error rate for $m = 8$ comparisons among $t = 20$ population means to be at most .05. What value of α_I must we use on the 8 tests to achieve an overall error rate of

α_I, individual Type I error rate

α_E, experimentwise Type I error rate

TABLE 8.13

A comparison of the experimentwise error rate α_E for m independent pairwise comparisons among the t population means

m, Number of Comparisons	α_I Probability of a Type I Error on an Individual Test		
	.10	.05	.01
1	.100	.050	.010
2	.190	.097	.020
3	.271	.143	.030
4	.344	.185	.039
5	.410	.226	.049
⋮	⋮	⋮	⋮
10	.651	.401	.096

$\alpha_E = .05$? We can use the previous upper bound to determine that if we select

$$\alpha_I = \alpha_{E/m} = .05/8 = .00625$$

then we will have $\alpha_E \leq .05$. The only problem is that this procedure may be very conservative with respect to the experimentwise error rate, and hence an inflated probability of Type II error may result.

We will now discuss two different procedures for making pairwise comparisons of the treatment means in a completely randomized design.

Fisher's Least Significant Difference

Recall that we are interested in determining which population means differ after we have rejected the hypothesis of equality of t population means in an analysis of variance. R. A. Fisher (1949) developed a procedure for making pairwise comparisons among a set of t population means. The procedure is called Fisher's least significant difference (LSD).

The α-level of Fisher's LSD is valid for a given comparison only if the LSD is used for independent comparisons or for preplanned comparisons. However, because many people find Fisher's LSD easy to compute and hence use it for making all possible pairwise comparisons (particularly those that look "interesting" following the completion of the experiment), researchers recommend applying Fisher's LSD only after the F test for treatments has been shown to be significant. This revised approach is sometimes referred to as **Fisher's protected LSD.** Simulation studies [Carmer and Swanson (1973)] suggest that the error rate for the protected LSD is controlled on an experimentwise basis at a level approximately equal to the α-level for the F test.

Fisher's protected LSD

We will illustrate Fisher's protected procedure, but will continue to call it Fisher's LSD. This procedure is summarized here for a completely randomized design.

Fisher's Least Significant Difference Procedure

1. Perform an analysis of variance to test $H_0: \mu_1 = \mu_2 = \cdots = \mu_t$ against the alternative hypothesis that at least one of the means differs from the rest.
2. If there is insufficient evidence to reject H_0 using $F = \text{MST/MSE}$, proceed no further.
3. If H_0 is rejected, define the **least significant difference** (LSD) to be the observed difference between two sample means necessary to declare the corresponding population means different.
4. For a specified value of α, the least significant difference for comparing μ_i to μ_j is

least significant difference (LSD)

$$\text{LSD}_{ij} = t_{\alpha/2}\sqrt{\text{MSE}\left(\frac{1}{n_i} + \frac{1}{n_j}\right)}$$

where n_i and n_j are the respective sample sizes from population i and j and t is the critical t value (Table 3 of the Appendix) for $a = \alpha/2$ and df denoting the degrees of freedom for MSE. Note that for $n_1 = n_2 = \cdots = n_t = n$,

$$\text{LSD}_{ij} \equiv \text{LSD} = t_{\alpha/2}\sqrt{\frac{2\text{MSE}}{n}}$$

5. Then compare all pairs of sample means. If $|\bar{y}_{i.} - \bar{y}_{j.}| \geq \text{LSD}_{ij}$, declare the corresponding population means μ_i and μ_j different.
6. For each pairwise comparison of population means, the probability of a Type I error is fixed at a specified value of α.

Note: The LSD procedure is analogous to a two-sample t test for any two population means μ_i and μ_j, except that we use MSE, the pooled estimator of the population variance σ_ε^2 from all t samples rather than the pooled sample variance from samples i and j. Also, the degrees of freedom for the t value is df $= N - t$ from the analysis of variance, rather than $n_i + n_j - 2$.

EXAMPLE 8.5

Various agents are used to control weeds in crops. Of particular concern is the overuse of chemical agents. Although effective in controlling weeds, these agents may also drain into the underground water system and cause health problems. Thus, several new biological weed agents have been proposed to eliminate the contamination problem present in chemical agents. Researchers conducted a study of biological agents to assess their effectiveness in comparison to the chemical weed agents. The study consisted of a control (no agent), two biological agents (Bio1 and Bio2), and two chemical agents (Chm1 and Chm2). Thirty 1-acre plots of land were planted with hay. Six plots were randomly assigned to receive one of the five treatments. The hay was harvested and the total yield in tons per acre was recorded. The data are given here and the analysis of variance is shown in Table 8.14.

Agent	1	2	3	4	5
Type	None	Bio1	Bio2	Chm1	Chm2
$\bar{y}_{i.}$	1.175	1.293	1.328	1.415	1.500
s_i	.1204	.1269	.1196	.1249	.1265
n_i	6	6	6	6	6

Compare the population means using Fisher's LSD.

steps for LSD procedure

Solution We can solve this problem by following the five **steps** listed for the **LSD procedure.**

Step 1. We use the AOV table in Table 8.14. The F test of H_0: $\mu_1 = \mu_2 = \cdots = \mu_5$ is based on

$$F = \frac{\text{MST}}{\text{MSE}} = 5.96$$

TABLE 8.14
AOV table for the data of Example 8.5

Source	df	SS	MS	F	p-value
Treatment	4	.3648	.0912	5.96	.0016
Error	25	.3825	.0153		
Totals	29	.7472			

For $\alpha = .05$ with $df_1 = 4$ and $df_2 = 25$, we reject H_0 if F exceeds 2.76 (see Table 9 in the Appendix).

Steps 2, 3. Because 5.96 is greater than 2.76, we reject H_0 and conclude that at least one of the population means differs from the rest ($p = .0016$).

Step 4. The least significant difference for comparing two means based on samples of size 6 is then

$$\text{LSD} = t_{\alpha/2}\sqrt{\frac{2\text{MSE}}{6}} = 2.060\sqrt{\frac{2(.0153)}{6}} = .1471$$

Note that the appropriate t value (2.060) was obtained from Table 3 with $\alpha/2 = .025$ and $df = 25$.

Step 5. When we have equal sample sizes, it is convenient to use the following procedures rather than make all pairwise comparisons among the sample means, because the same LSD is to be used for all comparisons.

a. We rank the sample means from lowest to highest.

Agent	1	2	3	4	5
$\bar{y}_{i.}$	1.175	1.293	1.328	1.415	1.500

b. We compute the sample difference

$$\bar{y}_{\text{largest}} - \bar{y}_{\text{smallest}}$$

If this difference is greater than the LSD, we declare the corresponding population means significantly different from one another. Next we compute the sample difference

$$\bar{y}_{\text{2nd largest}} - \bar{y}_{\text{smallest}}$$

and compare the result to the LSD. We continue to make comparisons with $\bar{y}_{\text{smallest}}$:

$$\bar{y}_{\text{3rd largest}} - \bar{y}_{\text{smallest}}$$

and so on, until we find either that all sample differences involving $\bar{y}_{\text{smallest}}$ exceed the LSD (and hence the corresponding population means are different) or that a sample difference involving $\bar{y}_{\text{smallest}}$ is less than the LSD. In the latter case, we stop and make no further comparisons with $\bar{y}_{\text{smallest}}$. For our data, comparisons with $\bar{y}_{\text{smallest}}$, $\bar{y}_{1.}$ give the following results:

Comparison	Conclusion
$\bar{y}_{\text{largest}} - \bar{y}_{\text{smallest}} = \bar{y}_{5.} - \bar{y}_{1.} = .325$	>LSD; proceed
$\bar{y}_{\text{2nd largest}} - \bar{y}_{\text{smallest}} = \bar{y}_{4.} - \bar{y}_{1.} = .240$	>LSD; proceed
$\bar{y}_{\text{3rd largest}} - \bar{y}_{\text{smallest}} = \bar{y}_{3.} - \bar{y}_{1.} = .153$	>LSD; proceed
$\bar{y}_{\text{4th largest}} - \bar{y}_{\text{smallest}} = \bar{y}_{2.} - \bar{y}_{1.} = .118$	<LSD; stop

To summarize our results, we make the following diagram:

Agent   <u>1   2</u>   3   4   5

Those populations joined by the underline have means that are not significantly different from $\bar{y}_{1.}$. Note that agents 3, 4, and 5 have sample differences with agent 1 that exceed LSD and hence are not underlined.

c. We now make similar comparisons with $\bar{y}_{2nd\ smallest}$, $\bar{y}_{2.}$. In this case, we use the procedure of part (b).

Comparison	Conclusion
$\bar{y}_{5.} - \bar{y}_{2.} = .207$	>LSD; proceed
$\bar{y}_{4.} - \bar{y}_{2.} = .122$	<LSD; stop

Agent   1   <u>2   3   4</u>   5

d. Continue the comparisons with $\bar{y}_{3rd\ smallest}$, $\bar{y}_{3.}$ in this example.

Comparison	Conclusion
$\bar{y}_{5.} - \bar{y}_{3.} = .172$	>LSD; proceed
$\bar{y}_{4.} - \bar{y}_{3.} = .087$	<LSD; stop

Agent   1   2   <u>3   4</u>   5

e. Continue the comparisons with $\bar{y}_{4th\ smallest}$, $\bar{y}_{4.}$ in this example.

Comparison	Conclusion
$\bar{y}_{5.} - \bar{y}_{4.} = .085$	<LSD; stop

Agent   1   2   3   <u>4   5</u>

f. We can summarize steps (a) through (e) as follows:

Agent   <u>1   2</u>   3   4   5

Those populations not underlined by a common line are declared to have means that are significantly different according to the LSD criterion. Note that we can eliminate the third line from the top of part (f) because it is contained in the second line from the top. The revised summary of significant and nonsignificant results is

Agent   <u>1   2</u>   3   4   5

In conclusion, we have μ_1, μ_2, and μ_3 significantly less than μ_5. Also, μ_3 and μ_4 are significantly greater than μ_1.

Although the LSD procedure described in Example 8.5 may seem quite laborious, its application is quite simple. First, we run an analysis of variance. If we reject the null hypothesis of equality of the population means, we compute the LSD for all pairs of sample means. When the sample sizes are the same, this difference is a single number for all pairs. We can use the stepwise procedure described in steps 5(a) through 5(f) of Example 8.5. We need not write down all those steps, only the summary lines. The final summary, as given in step 5(f), gives a handy visual display of the pairwise comparisons using Fisher's LSD.

Several remarks should be made concerning the LSD method for pairwise comparisons. First, there is the possibility that the overall F test in our analysis of variance is significant but that no pairwise differences are significant using the LSD procedure. This apparent anomaly can occur because the null hypothesis $H_0: \mu_1 = \mu_2 = \cdots = \mu_t$ for the F test is equivalent to the hypothesis that all possible comparisons (paired or otherwise) among the population means are zero. For a given set of data, the comparisons that are significant might not be of the form $\mu_i - \mu_j$, the form we are using in our paired comparisons.

Fisher's confidence interval

Second, **Fisher's** LSD procedure can also be used to form a **confidence interval** for $\mu_i - \mu_j$. A $100(1 - \alpha)\%$ confidence interval has the form

$$(\bar{y}_{i.} - \bar{y}_{j.}) \pm \text{LSD}_{ij}$$

LSD for equal sample sizes

Third, when all **sample sizes** are the **same,** the **LSD** for all pairs is

$$t_{\alpha/2}\sqrt{\frac{2\text{MSE}}{n}}$$

Tukey's W Procedure

We are aware of the major drawback of a multiple-comparison procedure with a controlled per-comparison error rate. Even when $\mu_1 = \mu_2 = \cdots = \mu_t$, unless α, the per-comparison error rate (such as with Fisher's unprotected LSD), is quite small, there is a high probability of declaring at least one pair of means significantly different when running multiple comparisons. To avoid this, other multiple-comparison procedures have been developed that control different error rates.

Studentized range distribution

Tukey (1953) proposed a procedure that makes use of the **Studentized range distribution.** When more than two sample means are being compared, to test the largest and smallest sample means we could use the test statistic

$$\frac{\bar{y}_{\text{largest}} - \bar{y}_{\text{smallest}}}{s_p\sqrt{1/n}}$$

where n is the number of observations in each sample and s_p is a pooled estimate of the common population standard deviation σ. This test statistic is very similar to that for comparing two means, but it does not possess a t distribution. One reason it does not is that we have waited to determine which two sample means (and hence population means) we would compare until we observed the largest and smallest sample means. This procedure is quite different from that of specifying $H_0: \mu_1 - \mu_2 = 0$, observing $\bar{y}_{1.}$ and $\bar{y}_{2.}$, and forming a t statistic.

The quantity

$$\frac{\bar{y}_{\text{largest}} - \bar{y}_{\text{smallest}}}{s_p\sqrt{1/n}}$$

follows a Studentized range distribution. We will not discuss the properties of this distribution, but will illustrate its use in Tukey's multiple-comparison procedure for a completely randomized design.

Tukey's W Procedure	**1.** Rank the t sample means.
	2. Two populations means μ_i and μ_j are declared different if
W	$$\|\bar{y}_{i.} - \bar{y}_{j.}\| \geq W$$
	where
$q_\alpha(t, v)$	$$W = q_\alpha(t, v)\sqrt{\frac{MSE}{n}}$$
	MSE is the mean square error based on v degrees of freedom, $q_\alpha(t, v)$ is
upper-tail critical value of the Studentized range	the **upper-tail critical value of the Studentized range** for comparing t different populations, and n is the number of observations in each sample. A discussion follows showing how to obtain values of $q_\alpha(t, v)$ from Table 11 in the Appendix.
experimentwise error rate	**3.** The error rate that is controlled is an **experimentwise error rate.** Thus, the probability of observing an experiment with one or more pairwise comparisons falsely declared to be significant is specified at α.

We can obtain values of $q_\alpha(t, v)$ from Table 11 in the Appendix. Values of v are listed along the left column of the table with values of t across the top row. Upper-tail values for the Studentized range are then presented for $a = .05$ and .01. For example, in comparing 10 population means based on 9 degrees of freedom for MSE, the .05 upper-tail critical value of the Studentized range is $q_{.05}(10, 9) = 5.74$.

EXAMPLE 8.6

Refer to the data in Example 8.5. Use Tukey's W procedure with $\alpha = .05$ to make pairwise comparisons among the five population means.

Solution Step 1 is to rank the sample means from smallest to largest, to produce

Agent	1	2	3	4	5
$\bar{y}_{i.}$	1.175	1.293	1.328	1.415	1.500

For the experiment described in Example 8.5, we have

$t = 5$ (we are making pairwise comparisons among five means)
$v = 25$ (MSE had degrees of freedom equal to df_{Error} in the AOV)
$\alpha = .05$ (we specified α_E, the experimentwise error rate at .05)
$n = 6$ (there were six plots randomly assigned to each of the agents)

We find in Table 11 of the Appendix that

$$q_\alpha(t, v) = q_{.05}(5, 25) \approx 4.158$$

The absolute value of each difference in the sample means $|\bar{y}_{i.} - \bar{y}_{j.}|$ must then be compared to

$$W = q_\alpha(t, v)\sqrt{\frac{MSE}{n}} = 4.158\sqrt{\frac{.0153}{6}} = .2100$$

By substituting W for LSD, we can use the same stepwise procedure for comparing sample means that we used in step 5 of the solution to Example 8.5. Having ranked the sample means from low to high, we compare against $\bar{y}_{smallest}$, which is $\bar{y}_1$, as follows:

Comparison	Conclusion
$\bar{y}_{largest} - \bar{y}_{smallest} = \bar{y}_{5.} - \bar{y}_{1.} = .325$	$>W$; proceed
$\bar{y}_{2nd\ largest} - \bar{y}_{smallest} = \bar{y}_{4.} - \bar{y}_{1.} = .240$	$>W$; proceed
$\bar{y}_{3rd\ largest} - \bar{y}_{smallest} = \bar{y}_{3.} - \bar{y}_{1.} = .153$	$<W$; stop

To summarize our results we make the following diagram:

Agent $\underline{1\quad 2\quad 3}$ 4   5

Comparison with $\bar{y}_{2nd\ smallest}$, which is $\bar{y}_{2.}$, yields

Comparison	Conclusion
$\bar{y}_{5.} - \bar{y}_{2.} = .207$	$<W$; stop

Agent   1 $\underline{2\quad 3\quad 4\quad 5}$

Similarly, comparisons of $\bar{y}_{5.}$ with $\bar{y}_{3.}$ and $\bar{y}_{4.}$ yield

Agent   1   2 $\underline{3\quad 4\quad 5}$

Combining our results, we obtain

Agent $\underline{1\quad 2\quad 3}$ 4   5

which simplifies to

Agent $\underline{1\quad 2\quad 3}$ 4   5

All populations not underlined by a common line have population means that are significantly different from one another; that is, μ_4 and μ_5 are significantly larger than μ_1. No other pairs of means are significantly different.

By examining the multiple-comparison summaries using the least significant difference (Example 8.5) and Tukey's W procedure (Example 8.6), we see that Tukey's procedure is more conservative (declares fewer significant differences) than the LSD procedure. For example, when applying Tukey's procedure to the data in

Example 8.5 we found that μ_3 is no longer significantly larger than μ_1. Similarly, μ_2 and μ_3 are no longer significantly less than μ_5. The explanation for this is that, although both procedures have an experimentwise error rate, we have shown the per-comparison error rate of the protected LSD method to be larger than that for Tukey's W procedure.

A limitation of Tukey's procedure is the requirement that all the sample means be based on the same number of data values. There are no available tables for the case in which the sample sizes are unequal. If the sample sizes are only somewhat different, then Miller (1981) recommends replacing the value of n in the formula for W with the harmonic mean of the n_is:

$$ n = \frac{t}{\dfrac{1}{n_1} + \dfrac{1}{n_2} + \cdots + \dfrac{1}{n_t}} $$

For large departures from equal sample sizes, the experimenter should consider using the LSD procedure.

Tukey's confidence interval

Tukey's procedure can also be used to construct **confidence intervals** for comparing two means. However, unlike the confidence intervals that we can form from Fisher's LSD, Tukey's procedure enables us to construct simultaneous confidence intervals for all pairs of treatment differences. For a specified α level from which we compute W, the overall probability is $1 - \alpha$ that all differences $\mu_i - \mu_j$ will be included in an interval of the form

$$ (\bar{y}_{i.} - \bar{y}_{j.}) \pm W $$

That is, the probability is $1 - \alpha$ that all the intervals $(\bar{y}_{i.} - \bar{y}_{j.}) \pm W$ include the corresponding population differences $\mu_i - \mu_j$.

Exercises

Hort.

8.6 A horticulturist was investigating the phosphorus content of tree leaves from three different varieties of apple trees (1, 2, and 3). Random samples of five leaves from each of the three varieties were analyzed for phosphorus content. The data are given here as well as the AOV table.

Variety	Phosphorus Content					Sample Sizes	Means
1	.35	.40	.58	.50	.47	5	0.460
2	.65	.70	.90	.84	.79	5	0.776
3	.60	.80	.75	.73	.66	5	0.708
Total						15	0.648

Source	Sum of Squares	Degrees of Freedom	Mean Square	F Test
Treatment	.277	2	.277/2 = .138	.138/.008 = 17.25
Error	.0978	12	.0978/12 = .008	
Totals	.3748	14		

Run Tukey's procedure to make all pairwise comparisons based on $\alpha = .05$.

Med. **8.7** Researchers conducted an experiment to compare the effectiveness of four new weight-reducing agents to that of an existing agent. The researchers randomly divided a random sample of 50 males into five equal groups, with preparation A1 assigned to the first group, A2 to the second group, and so on. They then gave a prestudy physical to each person in the experiment and told him how many pounds overweight he was. A comparison of the mean number of pounds overweight for the groups showed no significant differences. The researchers then began the study program, and each group took the prescribed preparation for a fixed period of time. The weight losses recorded at the end of the study period are given here.

A1	12.4	10.7	11.9	11.0	12.4	12.3	13.0	12.5	11.2	13.1
A2	9.1	11.5	11.3	9.7	13.2	10.7	10.6	11.3	11.1	11.7
A3	8.5	11.6	10.2	10.9	9.0	9.6	9.9	11.3	10.5	11.2
A4	12.7	13.2	11.8	11.9	12.2	11.2	13.7	11.8	12.2	11.7
S	8.7	9.3	8.2	8.3	9.0	9.4	9.2	12.2	8.5	9.9

The standard agent is labeled S, and the four new agents are labeled A_1, A_2, A_3, and A_4. The data and a computer printout of an analysis are given here.

```
General Linear Models Procedure
Class Level Information

Class      Levels   Values

AGENT        5     1 2 3 4 S

Number of observations in data set = 50

Dependent Variable: L   WEIGHTLOSS

SOURCE                DF      Sum of Squares      F Value      Pr > F
Model                  4         61.61800000       15.68       0.0001
Error                 45         44.20700000
Corrected Total       49        105.82500000

               R-Square              C.V.               L Mean
               0.582263           9.035093           10.9700000

Source                DF      Type III SS      F Value      Pr > F
AGENT                  4       61.61800000       15.68       0.0001
-------------------------------------------------------------------
Level of           -------------L----------------
A            N      Mean                SD

1           10    12.0500000         0.82898867
2           10    11.0200000         1.12130876
3           10    10.2700000         1.02637442
4           10    12.2400000         0.75601293
S           10     9.2700000         1.15859110
-------------------------------------------------------------------
```

```
FISHER'S LSD  for variable: WEIGHTLOSS

Alpha= 0.05  df= 45  MSE= 0.982378
Critical Value of T= 2.01
Least Significant Difference= 0.8928

Means with the same letter are not significantly different.

T Grouping            Mean       N  A
                 A   12.2400     10  4
                 A   12.0500     10  1
                 B   11.0200     10  2
                 B   10.2700     10  3
                 C    9.2700     10  S
-----------------------------------------------------------------
Tukey's Studentized Range (HSD) Test for variable: L

Alpha= 0.05  df= 45  MSE= 0.982378
Critical Value of Studentized Range= 4.018
Minimum Significant Difference= 1.2595

Means with the same letter are not significantly different.

Tukey Grouping           Mean       N  A

                  A    12.2400     10  4
                  A    12.0500     10  1
              B   A    11.0200     10  2
              B   C    10.2700     10  3
                  C     9.2700     10  S
-----------------------------------------------------------------
Univariate Procedure

Variable=RESIDUAL

                Moments

N                50  Sum Wgts        50
Mean              0  Sum              0
Std Dev    0.949833  Variance  0.902184
Skewness   0.523252  Kurtosis  0.995801
Test of Normality:   P-value   0.6737

Variable=RESIDUAL

   Stem Leaf              #  Boxplot
     2 9                  1    0
     2 2                  1    0
     1 5                  1    |
     1 00003              5    |
     0 556679             6  +-----+
     0 0112233444        10  |  +  |
    -0 444433321100      12  +-----+
```

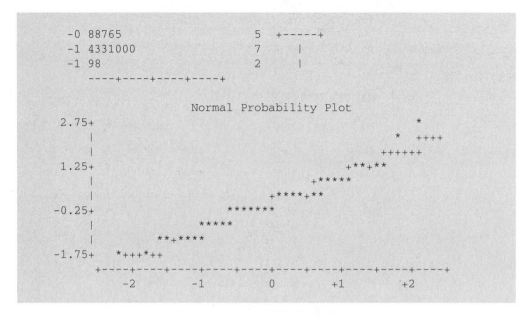

```
     -0 88765                         5    +-----+
     -1 4331000                       7    |
     -1 98                            2    |
          ----+----+----+----+
```

Run an analysis of variance to determine whether there are any significant differences among the five weight-reducing agents. Use $\alpha = .05$. Do any of the AOV assumptions appear to be violated? What conclusions do you reach concerning the mean weight loss achieved using the five different agents?

8.8 Refer to Exercise 8.7. Using the computer output included there, determine the significantly different pairs of means using the following procedures.
  a. Fisher's LSD, $\alpha = .05$
  b. Tukey's W, $\alpha = .05$

8.7 Encounters with Real Data: Effect of Timing of the Treatment of Port-Wine Stains with Lasers

Defining the Problem (1)

Port-wine stains are congenital vascular malformations that occur in an estimated three children per 1,000 births. The stigma of a disfiguring birthmark may have a substantial effect on a child's social and psychosocial adjustment. In 1985, the flash-pumped, pulsed-dye laser was advocated for the treatment of port-wine stains in children. Treatment with this type of laser was hypothesized to be more effective in children than in adults because the skin in children is thinner and the size of the port-wine stain is smaller; fewer treatments would therefore be necessary to achieve optimal clearance. These are all arguments for initiating treatment at an early age.

In a prospective study described in the paper "Effect of the timing of treatment of port-wine stains with the flash-lamp-pumped pulsed-dye laser" [*The New England Journal of Medicine* (1998) 338:1028–1033], the researchers investigated whether treatment at a young age yielded better results than treatment at an older age.

Collecting Data (2)

The researchers considered the following issues relative to the most effective treatment:

1. What objective measurements should be used to assess the effectiveness of the treatment in reducing the visibility of the port-wine stains?
2. How many different age groups should be considered for evaluating the treatment?
3. What type of experimental design would produce the most efficient comparison of the different treatments?
4. What are the valid statistical procedures for making the comparisons?
5. What types of information should be included in a final report to document for which age groups the laser treatment was most effective?

One hundred patients, 31 years of age or younger, with a previously untreated port-wine stain were selected for inclusion in the study. During the first consultation, the extent and location of the port-wine stain was recorded. Four age groups of 25 patients each were determined for evaluating whether the laser treatment was more effective for younger patients. Enrollment in an age group ended as soon as 25 consecutive patients had entered the group. A series of treatments was required to achieve optimal clearance of the stain. Before the first treatment, color slides were taken of each patient by a professional photographer in a studio under standardized conditions. Color of the skin was measured using a chromometer. The reproducibility of the color measurements was analyzed by measuring the same location twice in a single session before treatment. For each patient, subsequent color measurements were made at the same location. Treatment was discontinued if either the port-wine stain had disappeared or the three previous treatments had not resulted in any further lightening of the stain. The outcome measure of each patient was the reduction in the difference in color between the skin with the port-wine stain and the contralateral healthy skin.

Eleven of the 100 patients were not included in the final analysis due to a variety of circumstances that occurred during the study period. A variety of baseline characteristics were recorded for the 89 patients: sex, surface area and location of the port-wine stain, and any other medical conditions that might have implications for the effectiveness of the treatment. Also included were treatment characteristics such as average number of visits, level of radiation exposure, number of laser pulses per visit, and the occurrence of headaches after treatment. For all variables, there were no significant differences between the four age groups with respect to these characteristics.

The two variables of main interest to the researchers were the difference in color between port-wine stain and contralateral healthy skin before treatment and the improvement in this difference in color after a series of treatments. The before-treatment differences in color are presented in Figure 8.7. The boxplots demonstrate that there were no sizable differences in the color differences among the four groups. This is important, because if the groups differed prior to treatment, then the effect of age group on the effectiveness of the treatment may have been masked by preexisting differences. (The values in Table 8.15 were simulated using the summary statistics given in the paper. These data can also be found in the `PORTWINE` file on the CD that came with your book.)

FIGURE 8.7

Boxplots of stain color by age group (means are indicated by solid circles)

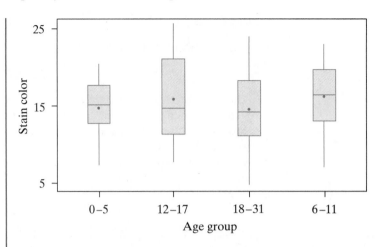

TABLE 8.15

Improvement in color of port-wine stains by age group

Patient	0–5 Years	6–11 Years	12–17 Years	18–31 Years
1	9.6938	13.4081	10.9110	1.4352
2	7.0027	8.2520	10.3844	10.7740
3	10.3249	12.0098	6.4080	8.4292
4	2.7491	7.4514	13.5611	4.4898
5	.5637	6.9131	3.4523	13.6303
6	8.0739	5.6594	9.5427	4.1640
7	.1440	8.7352	10.4976	5.4684
8	8.4572	.2510	4.6775	4.8650
9	2.0162	8.9991	24.7156	3.0733
10	6.1097	6.6154	4.8656	12.3574
11	9.9310	6.8661	.5023	7.9067
12	9.3404	5.5808	7.3156	9.8787
13	1.1779	6.6772	10.7833	2.3238
14	1.3520	8.2279	9.7764	6.7331
15	.3795	.1883	3.6031	14.0360
16	6.9325	1.9060	9.5543	.6678
17	1.2866	7.7309	5.3193	2.7218
18	8.3438	7.9143	3.0053	2.3195
19	9.2469	1.8724	11.0496	1.6824
20	.7416	12.5082	2.8697	1.8150
21	1.1072	6.2382	.1082	5.9665
22		11.2425		.5041
23		6.8404		5.4484
24		11.2774		

Next, the researchers prepared the data for a statistical analysis following the steps described in Section 2.5. The researchers needed to verify that the stain colors were properly recorded and that all computer files were consistent with the field data.

FIGURE 8.8

Boxplots of improvement
by age group (means are
indicated by solid circles)

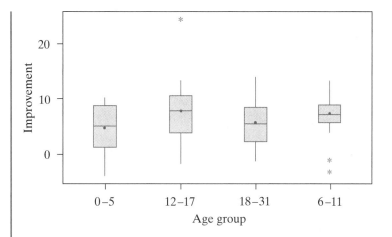

Summarizing Data (3)

The summary statistics are given in the following table along with boxplots for the four age groups (Figure 8.8). The 12–17 years group showed the greatest improvement, but the 6–11 years group had only a slightly smaller improvement. The other two groups had values at least 2 units less than the 12–17 years group. However, from the boxplots we can observe that the four groups do not appear to have that great a difference in improvement. We will develop an analysis of variance to confirm whether a statistically significant difference exists among the four age groups.

```
Descriptive Statistics for Port-Wine Stain Case Study

Variable            N      Mean   Median   TrMean   StDev   SE Mean
0-5   Years        21     4.999    6.110    4.974   3.916    0.855
6-11  Years        24     7.224    7.182    7.262   3.564   ·0.727
12-17 Years        21     7.757    7.316    7.270   5.456    1.191
18-31 Years        23     5.682    4.865    5.531   4.147    0.865


Variable       Minimum  Maximum       Q1       Q3
0-5   Years      0.144   10.325    1.143    8.852
6-11  Years      0.188   13.408    5.804    8.933
12-17 Years      0.108   24.716    3.528   10.640
18-31 Years      0.504   14.036    2.320    8.429
```

Analyzing Data, Interpreting the Analyses, and Communicating Results (4)

The objective of the study was to evaluate whether the treatment of port-wine stains was more effective for younger children than for older ones. Before we run an AOV, we must check on the assumptions. We observed in Figure 8.8 that the boxplots were nearly of the same width with no outliers and whiskers of the same length. The means and medians were of a similar size for each of the four age groups. Thus, the assumptions of AOV appear to be satisfied. To confirm this observation, we com-

FIGURE 8.9

Normal probability plot
of the residuals for the
improvement data

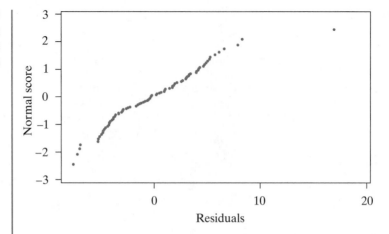

puted the residuals and plotted them in a normal probability plot (see Figure 8.9).
From this plot we can observe that, with the exception of one data value, the points
fall nearly on a straight line. Thus, there is a strong confirmation that the four populations of improvements in skin color have normal distributions.

Next, we can check on the equal variance assumption by using Hartley's test
or Levene's test. For Hartley's test, we obtain

$$F_{\max} = \frac{(5.46)^2}{(3.564)^2} = 2.35$$

The critical value of $F_{\max}$ for $\alpha = .05$, $t = 4$, and $df_2 = 20$ is 3.29. This test is only approximate because the sample sizes are unequal. However, the sample sizes are very
nearly the same: 21, 21, 23, and 24. Because $F_{\max}$ is not greater than 3.29, we conclude that there is not significant evidence that the four population variances differ.
Levene's test yields a value of $L = 1.050$ with a p-value of .375 and thus agrees with
the findings from Hartley's test. We feel comfortable that the normality and equal
variance conditions of the AOV procedure are satisfied.

The condition of independence of the data would be checked by discussing
with the researchers the manner in which the study was conducted. The sequenc-
ing of treatment and the evaluation of the color of the stains should have been performed such that the determination of improvement in color of one patient would
not in any way affect the determination of improvement in color of any other patient. The problems that may arise in this type of experiment, which can cause dependencies in the data, would be due to equipment problems, technician biases, any
relationships between patients, and other similar factors.

The research hypothesis is that the mean improvement in stain color after
treatment is different for the four age groups:

H_0: $\mu_1 = \mu_2 = \mu_3 = \mu_4$

H_a: At least one of the means differs from the rest.

The computer output for the AOV table is given here:

```
One-Way Analysis of Variance for Improvement in Stain Color

Source     DF      SS      MS      F      P
Age Group  3     108.0    36.0   1.95   0.128
Error      85   1572.5    18.5
Total      88   1680.5
                                Individual 95% CIs for Mean
                                Based on Pooled StDev
Level      N     Mean    StDev   -----+---------+---------+---------+-
0-5        21    4.999   3.916   (--------*--------)
06-11      24    7.224   3.564              (--------*--------)
12-17      21    7.757   5.456               (---------*--------)
18-31      23    5.682   4.147      (--------*--------)
                                -----+---------+---------+---------+-
Pooled StDev = 4.301              4.0       6.0       8.0      10.0
```

From the output, the p-value for the F test is .128. Thus, there is not a significant difference in the mean improvement for the four groups. We can also compute 95% confidence intervals for the mean improvements. The four intervals are provided in the computer output. They are computed using the pooled standard deviation, $\hat{\sigma} = \sqrt{\text{MSE}} = \sqrt{18.5} = 4.30$ with df = 85. Thus, the intervals are of the form

$$\bar{y}_{i.} \pm \frac{t_{.025,85}\hat{\sigma}}{\sqrt{n_i}} = \bar{y}_{i.} \pm \frac{(1.99)(4.30)}{\sqrt{n_i}}$$

The four intervals are presented here:

Age Group	$\bar{y}_{i.}$	95% Confidence Interval
0–5	4.999	(3.13, 6.87)
6–11	7.224	(5.48, 8.97)
12–17	7.757	(5.89, 9.62)
18–31	5.682	(3.90, 7.47)

From the confidence intervals, we can observe the overall effect in the estimation of the mean improvement in stain color for the four groups. The youngest group has the smallest improvement but its upper bound is greater than the lower bound for the age group having the greatest improvement. Note that these intervals did not come from one of the multiple comparison procedures of Section 8.6; therefore the confidence intervals are not simultaneous confidence intervals, and we cannot attribute a level of certainty to our conclusions. For these data, however, we can safely conclude that all pairs of treatment means are not significantly different, because the AOV F test failed to reject the null hypothesis.

The researchers did not confirm the hypothesis that treatment of port-wine stains at an early age is more effective than treatment at a later age. The researchers did conclude that their results had implications for the timing of therapy in children. Although facial port-wine stains can be treated effectively and safely early in life, treatment at a later age leads to similar results. Therefore, the age at which therapy is initiated should be based on a careful weighing of the anticipated benefit and the discomfort of treatment.

Formal documentation of this study and communication of the results require that we write a report summarizing our findings of this prospective study of the treatment of port-wine stains. It should also be pointed out that communicating the results of studies with negative results can be just as important as communicating positive findings. Unfortunately, journals usually only accept reports with positive findings and hence the published literature has a "positive" bias. The report should include

1. Statement of objective for study
2. Description of study design and data collection procedures
3. Discussion of why the results from 11 of the 100 patients were not included in the data analysis
4. Numerical and graphical summaries of data sets
5. Description of all inference methodologies:

   - AOV table and F test
   - t-based confidence intervals on means
   - Verification that all necessary conditions for using inference techniques were satisfied

6. Discussion of results and conclusions
7. Interpretation of findings relative to previous studies
8. Recommendations for future studies
9. Listing of data sets

8.8   Staying Focused

This chapter introduces an important method for analyzing data (an analysis of variance) which extends the results of Chapter 6 to include a comparison among t population means from a completely randomized design. The notation we introduced and the model we presented provide the basis for discussing the analysis of variance for data from other experimental designs (Chapter 9).

For a completely randomized design, an independent random sample is drawn from each of the t population (treatments). Using the sample data, we partitioned the total sum of squares (TSS) into two parts: SST, which measures the between-treatment variability, and SSE, which measure the within-treatment variability.

The decision to accept or reject the null hypothesis of equality of the t population means depends on the computed value of $F = \text{MST}/\text{MSE}$. Under H_0, both MST and MSE estimate σ_ε^2, the variance common to all t populations. Under the alternative hypothesis, MST estimates $\sigma_\varepsilon^2 + n\theta_T$, where θ_T is a positive quantity, whereas MSE still estimates σ_ε^2. Thus, large values of F indicate a rejection of H_0. Critical values for F are obtained from Table 9 in the Appendix for $\text{df}_1 = t - 1$ and $\text{df}_2 = N - t$. This test procedure, called an analysis of variance, is usually summarized in an analysis of variance (AOV) table.

Suppose we reject H_0 and conclude that at least one of the means differs from the rest; which ones differ from the others? Section 8.6 attacks this problem through procedures based on multiple comparisons.

In this chapter, we also discussed the assumptions underlying an analysis of variance for a completely randomized design. Independent random samples are absolutely necessary. The assumption of normality is least critical because we are dealing with means and the Central Limit Theorem holds for reasonable sample sizes. The equal variance assumption is critical only when the sample sizes are markedly different; this is a good argument for equal (or nearly equal) sample sizes. A test for equality of variances makes use of the F_{max} statistic, s_{max}^2/s_{min}^2, or Levene's test.

The topics in this chapter are certainly not covered in exhaustive detail. However, the material is sufficient for training the beginning researcher to be aware of the assumptions underlying his or her project and to consider running an alternative analysis (such as using a nonparametric statistical method, the Kruskal–Wallis test) when appropriate.

Key Formulas

1. Analysis of variance for a completely randomized design

$$\text{SST} = \sum_i n_i(\bar{y}_{i.} - \bar{y}_{..})^2 = \sum_i n_i(\hat{\alpha}_i)^2$$

$$\text{TSS} = \sum_{i,j} (y_{ij} - \bar{y}_{..})^2$$

$$\text{SSE} = \sum_{i,j} (y_{ij} - \bar{y}_{i.})^2 = \sum_i (n_i - 1)s_i^2 = \text{TSS} - \text{SST}$$

2. Model for a completely randomized design

$$y_{ij} = \mu + \alpha_i + \varepsilon_{ij}$$

where $\mu_i = \mu + \alpha_i$.

3. Conditions imposed on model:
   a. The t populations have normal distributions.
   b. $\sigma_1^2 = \cdots = \sigma_t^2 = \sigma_\varepsilon^2$
   c. Data consist of t independent random samples.

4. Check whether conditions are satisfied:
   a. Normality: Plots of residuals, $e_{ij} = y_{ij} - \bar{y}_{i.}$
   b. Homogeneity of variance: Hartley's test or Levene's test
   c. Independence: Careful review of how experiment or study was conducted

5. Kruskal–Wallis test (when population distributions are very non-normal)

   H_0: The k population distributions are identical.

   H_a: The k population distributions are shifted from one another.

$$\text{T.S.} = \frac{12}{N(N+1)} \sum_{i=1}^{k} \frac{T_i^2}{n_i} - 3(N+1)$$

6. Fisher's LSD procedure

$$LSD_{ij} = t_{\alpha/2}\sqrt{MSE\left(\frac{1}{n_i} + \frac{1}{n_j}\right)}$$

7. Tukey's W procedure

$$W = q_\alpha(t, v)\sqrt{\frac{MSE}{n}}$$

Supplementary Exercises

Hort.

8.9 Researchers from the Department of Fruit Crops at a university compared four different preservatives to be used in freezing strawberries. The researchers prepared the yield from a strawberry patch for freezing and randomly divided it into four equal groups. Within each group they treated the strawberries with the appropriate preservative and packaged them into eight small plastic bags for freezing at 0°C. The bags in group I served as a control group, while those in groups II, III, and IV were assigned one of three newly developed preservatives. After all 32 bags of strawberries were prepared, they were stored at 0°C for a period of 6 months. At the end of this time, the contents of each bag were allowed to thaw and then rated on a scale of 1 to 10 points for discoloration. (Note that a low score indicates little discoloration.) The ratings are given here.

Group I	10	8	7.5	8	9.5	9	7.5	7
Group II	6	7.5	8	7	6.5	6	5	5.5
Group III	3	5.5	4	4.5	3	3.5	4	4.5
Group IV	2	1	2.5	3	4	3.5	2	2

a. Use the following plots of the residuals and a test of the homogeneity of variances to assess whether the conditions needed to use AOV techniques are satisfied with this data set.
b. Test whether there is a difference in the mean ratings using $\alpha = .05$.
c. Place 95% confidence intervals on the mean ratings for each of the groups.
d. Confirm your results with the computer output given here.

```
One-Way Analysis of Variance for Exercise 8.9

Analysis of Variance for Ratings
Source      DF         SS         MS         F          P
Group        3    159.187     53.062      55.67      0.000
Error       28     26.687      0.953
Total       31    185.875
                                    Individual 95% CIs for Mean
                                    Based on Pooled StDev
Group       N        Mean      StDev   --+---------+---------+---------+----
I           8      8.3125     1.0670                                  (---*--)
II          8      6.4375     1.0155                         (--*---)
III         8      4.0000     0.8452              (---*---)
IV          8      2.5000     0.9636    (---*--)
                                       --+---------+---------+---------+----
Pooled StDev =     0.9763              2.0       4.0       6.0       8.0
```

Boxplots of ratings by group
for Exercise 8.9 (means are
indicated by solid circles)

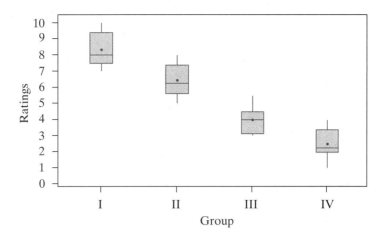

Normal probability plot of
residuals for Exercise 8.9

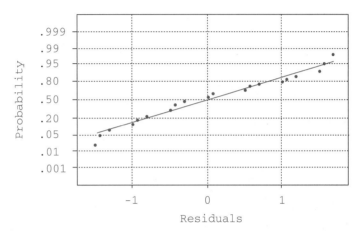

```
Average: 0                    Anderson-Darling Normality Test
StDev: 0.927840                   A-Squared: 0.503
N: 32                            P-Value: 0.191
```

8.10 Refer to Exercise 8.9. Compare the treatment means using Tukey's procedure. Do the conclusions change?

8.11 Refer to Exercise 8.9. In many situations in which the response is a rating rather than an actual measurement, it is recommended that the Kruskal–Wallis test be used.

    a. Apply the Kruskal–Wallis test to determine whether there is a shift in the distribution of ratings for the four groups.

    b. Is the conclusion reached using the Kruskal–Wallis test consistent with the conclusion reached in Exercise 8.9 using AOV?

Edu.   **8.12** Doing homework is a nightly routine for most school-age children. The article "Family involvement with middle-grades homework: effects of differential prompting" [*Journal of Experimental Education* (1997) 66:31–48] examines the question of whether parents' involvement with their children's homework is associated with improved academic performance. Seventy-four sixth graders and their families participated in the study. Researchers assigned the students, similar in student academic ability and background, to one of three mathematics classes taught by the same teacher and randomly assigned the classes to one of the three treatment groups.

    Group I, student/family prompt: Students were prompted to seek assistance from a family member, and the family was encouraged to provide assistance to students.

Group II, student prompt: Students were prompted to seek assistance from a family member, but there was no specific encouragement of family members to provide assistance to students.

Group III, no prompts: Students were not prompted to seek assistance from a family member nor were family members encouraged to provide assistance to students.

Thus, one class was assigned to each of the three treatment groups. The researchers gave the students a posttest, with the results given here.

Treatment Group	Number of Students	Mean Posttest Score
Student/family prompt	22	68%
Student prompt	22	66%
No prompt	25	67%

The researchers concluded that higher levels of family involvement were not associated with higher student achievement in this study.

   a. What is the population of interest in this study?
   b. Based on the data collected, to what population can the results of this study be inferred?
   c. What is the effective sample for each of the treatment groups; that is, how many experimental units were randomly assigned to each of the treatment groups?
   d. What criticisms do you have about the design of this study?
   e. Suggest an improved design for addressing the research hypothesis that family involvement improves student performance in mathematics classes.

Gov. **8.13** In a 1994 Senate subcommittee hearing, an executive of a major tobacco company testified that the accusation that nicotine was added to cigarettes was false. Tobacco company scientists stated that the amount of nicotine in cigarettes was completely determined by the size of tobacco leaf, with smaller leaves having greater nicotine content. Thus, the variation in nicotine content in cigarettes occurred due to a variation in the size of the tobacco leaves and was not due to any additives placed in the cigarettes by the company. Furthermore, the company argued that the size of the leaves varied depending on the weather conditions during the growing season, over which they had no control. To study whether smaller tobacco leaves had a higher nicotine content, a consumer health organization conducted the following experiment. The major factors controlling leaf size are temperature and the amount of water received by the plants during the growing season. The experimenters created four types of growing conditions for tobacco plants. Condition A was average temperature and rainfall amounts. Condition B was lower than average temperature and rainfall conditions. Condition C was higher temperatures with lower rainfall. Finally, Condition D was higher than normal temperatures and rainfall. The scientists then planted 10 tobacco plants under each of the four conditions in a greenhouse where temperature and amount of moisture were carefully controlled. After growing the plants, the scientists recorded the leaf size and nicotine content, which are given here.

Plant	A Leaf Size	B Leaf Size	C Leaf Size	D Leaf Size
1	27.7619	4.2460	15.5070	33.0101
2	27.8523	14.1577	5.0473	44.9680
3	21.3495	7.0279	18.3020	34.2074
4	31.9616	7.0698	16.0436	28.9766
5	19.4623	0.8091	10.2601	42.9229
6	12.2804	13.9385	19.0571	36.6827
7	21.0508	11.0130	17.1826	32.7229
8	19.5074	10.9680	16.6510	34.5668
9	26.2808	6.9112	18.8472	28.7695
10	26.1466	9.6041	12.4234	36.6952

Plant	A Nicotine	B Nicotine	C Nicotine	D Nicotine
1	10.0655	8.5977	6.7865	9.9553
2	9.4712	8.1299	10.9249	5.8495
3	9.1246	11.3401	11.3878	10.3005
4	11.3652	9.3470	9.7022	9.7140
5	11.3976	9.3049	8.0371	10.7543
6	11.2936	10.0193	10.7187	8.0262
7	10.6805	9.5843	11.2352	13.1326
8	8.1280	6.4603	7.7079	11.8559
9	10.5066	8.2589	7.5653	11.3345
10	10.6579	5.0106	9.0922	10.4763

a. Perform an analysis of variance to test whether there is a significant difference in the average leaf size under the four growing conditions. Use $\alpha = .05$.

b. What conclusions can you reach concerning the effect of growing conditions on the average leaf size?

c. Perform an analysis of variance to test whether there is a significant difference in the average nicotine content under the four growing conditions. Use $\alpha = .05$.

d. What conclusions can you reach concerning the effect of growing conditions on the average nicotine content?

e. Based on the conclusions you reached in (b) and (d), do you think the testimony of the tobacco companies' scientists is supported by this experiment? Justify your conclusions.

8.14   Using the plots given here, do the nicotine content data in Exercise 8.13 suggest violations of the AOV conditions? If you determine that the conditions are not met, perform an alternative analysis and compare your results to those of Exercise 8.13.

Boxplots of leaf size by group for Exercise 8.14 (means are indicated by solid circles)

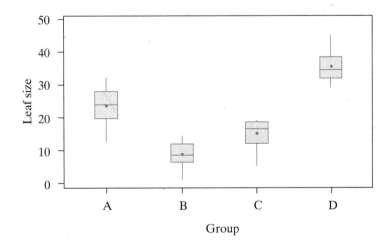

Probability plot of residuals
leaf size for Exercise 8.14

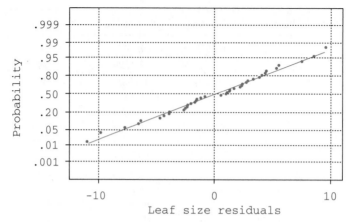

Average: 0.0000000   Anderson-Darling Normality Test
StDev: 4.75535   A-Squared: 0.205
N: 40   P-Value: 0.864

Boxplots of nicotine by group
for Exercise 8.14 (means are
indicated by solid circles)

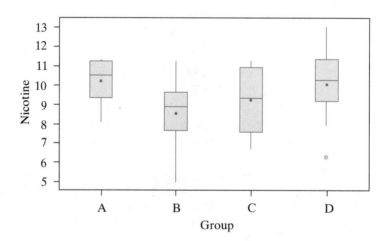

Probability plot of residuals
nicotine content for
Exercise 8.14

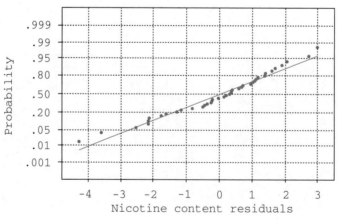

Average: 0.0000000   Anderson-Darling Normality Test
StDev: 1.62647   A-Squared: 0.443
N: 40   P-Value: 0.273

Engin. **8.15** A small corporation makes insulation shields for electrical wires using three different types of machines. The corporation wants to evaluate the variation in the inside diameter dimension of the shields produced by the machines. A quality engineer at the corporation randomly selects shields produced by each of the machines and records the inside diameters of each shield (in millimeters). She wants to determine whether the means and standard deviations of the three machines differ.

Shield	Machine A	Machine B	Machine C
1	18.1	8.7	29.7
2	2.4	56.8	18.7
3	2.7	4.4	16.5
4	7.5	8.3	63.7
5	11.0	5.8	18.9
6			107.2
7			19.7
8			93.4
9			21.6
10			17.8

a. Conduct a test for the homogeneity of the population variances. Use $\alpha = .05$.
b. Would it be appropriate to proceed with an analysis of variance based on the results of this test? Explain.
c. If the variances of the diameters are different, suggest a transformation that may alleviate their differences and then conduct an analysis of variance to determine whether the mean diameters differ. Use $\alpha = .05$.
d. Compare the results of your analysis in (c) to the computer output given here, which was an analysis of variance on the original diameters.
e. How could the engineer have designed her experiment differently if she had known that the variance of machine B and machine C were so much larger than that of machine A?

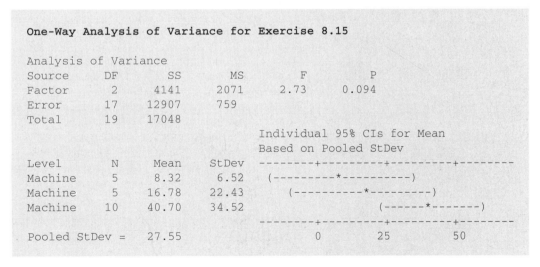

```
One-Way Analysis of Variance for Exercise 8.15

Analysis of Variance
Source      DF       SS        MS        F        P
Factor       2      4141      2071      2.73     0.094
Error       17     12907       759
Total       19     17048

                                  Individual 95% CIs for Mean
                                  Based on Pooled StDev
Level       N      Mean     StDev    --------+---------+---------+---------
Machine      5      8.32      6.52    (----------*----------)
Machine      5     16.78     22.43     (----------*---------)
Machine     10     40.70     34.52                    (------*-------)
                                      --------+---------+---------+---------
Pooled StDev =    27.55                       0        25        50
```

8.16 Refer to Exercise 8.15. The Kruskal–Wallis test is not as highly affected as the AOV test when the variances are unequal. Demonstrate this result by applying the Kruskal–Wallis test to both the original and transformed data and comparing the conclusions reached in this analysis for the data of Exercise 8.15.

8.17 Defining the Problem (1). Newborn mammals of some species are able to walk within minutes or hours of birth. Human infants are born with reflexes for certain movements that are necessary for walking, but they usually begin to walk somewhere between 10 and 18 months of age. The walking-related reflexes they had at birth largely disappear by the time they are 8 weeks old. Then it seems that similar movements have to be learned by trial and error before children can begin to walk on their own.

Zelazo, Zelazo, and Kolb (1972) performed an experiment to see whether appropriate exercises, started soon after birth, would help infants to retain the reflexes with which they were born and thus enable them to start walking unassisted at an earlier age than usual.

Collecting Data (2). Subjects for this study were 24 white male infants from middle-class families, enlisted with the aid of local physicians. These subjects were divided at random into four groups.

Active Exercise Group. Parents of infants in this group were taught how to administer exercises specifically targeted to reinforce walking-related reflexes. Parents conducted four 3-minute sessions each day, from age 1 week to age 8 weeks. Throughout this period, the investigators visited the infants' homes weekly to make tests of the relevant reflexes.

Passive Exercise Group. Infants in this group were treated exactly the same as those in the Active exercise group, *except* that the exercises administered to them were not designed to elicit any of the walking-related reflexes. A significant difference in the onset of walking between the Active and Passive groups would show that "maintaining" the reflexes can play a role in learning to walk.

Test-Only Group. The researchers did not specify exercises of any kind for these infants. But the infants were visited and tested in weeks 1 through 8 in the same way as those in the Active and Passive groups.

Control Group. These infants received no specified exercises and they were visited and tested only *twice:* at the first and eighth weeks. The existence of this group recognized that the weekly testing alone might have some effect.

All Groups. After 8 weeks the parents of each child were told the purpose of the experiment, given some advice on how they might encourage his motor development during the first year, and asked to report the age at which he first walked unassisted.

This is a *designed experiment* (completely randomized design) because it was possible to assign the children at random to carefully conceived treatment groups. There are no obvious ethical difficulties in making random assignments. It is difficult to imagine how the children could be damaged in any way by brief exercise sessions of the types conducted by their parents. The infants who did not receive the active exercises are at no special disadvantage: they should learn to walk about as early or late as if they had not been involved in the experiment.

Strictly speaking, any conclusions drawn from this experiment would apply only to white male infants who come from middle-class families and whose physicians would feel free to refer them for such a study. However, it is not obvious how or whether the mechanism of learning to walk could depend on gender, race, or social class.

The researchers obtained data of two kinds: (1) scores on walking-related reflexes collected for three of the four groups during the first 8 weeks and (2) the age at which each infant learned to walk. (The age at first walking is missing for one of the infants in the Control group.) For now, we'll focus on the age (in months) at which each infant learned to walk, shown here.

```
Complete listing of the dataset WALKKIDS

ActiveEx    PassivEx    TestOnly    Control
    9.00       11.00       11.50      13.25
    9.50       10.00       12.00      11.50
    9.75       10.00        9.00      12.00
   10.00       11.75       11.50      13.50
   13.00       10.50       13.25      11.50
    9.50       15.00       13.00
```

Summarizing Data (3).

 a. Compute the mean, median, and standard deviations for each sample.
 b. Use boxplots to display the sample data.
 c. What do you observe from parts (a) and (b)?

Analyzing Data, Interpreting the Analyses, and Communicating Results (4).

 d. An analysis of variance table for this completely randomized design is shown here. What conclusions can you draw?

```
Analysis of Variance for Months
Source     DF        SS        MS        F         P
Group       3      14.78      4.93      2.14      0.129
Error      19      43.69      2.30
Total      22      58.47

                                     Individual 95% CIs For Mean
                                     Based on Pooled StDev
Level       N       Mean      StDev   --+---------+---------+---------+----
1           6      10.125     1.447   (--------*-------)
2           6      11.375     1.896            (--------*-------)
3           6      11.708     1.520               (--------*--------)
4           5      12.350     0.962                  (--------*--------)
                                     --+---------+---------+---------+----
Pooled StDev =     1.516            9.0      10.5      12.0      13.5
```

 e. Do Tukey's procedure and Fisher's LSD for these data. Do your conclusions change?
 f. Recall that there were actually two kinds of data collected for this study. You have already seen the data on age at first unassisted walking. During the eight weekly visits (for all groups except the Control), data were also collected on reflexes that might be related to walking. For one of these reflexes, called the "walking" reflex, the number of reflex responses seen in a 1-minute session was recorded for each child each week. Weekly averages for the children in each group were reported in graphical form, which we have digitized (approximately) in the following table. (Unfortunately, no information was given about the variability of the five or six measurements that produced each of these means. Other reflexes, called "straightening" and "placing," were measured, but the results were less persuasive and not reported in detail.)

Average number of walking reflexes observed during weekly 1-minute tests

Group	Wk 1	Wk 2	Wk 3	Wk 4	Wk 5	Wk 6	Wk 7	Wk 8
Active Exercise	9	12	11	19	19	19	25	30
Passive Exercise	9	1	1	6	6	7	6	4
Test Only	6	6	8	4	3	3	5	4
Control	9			Not tested				1

Make an appropriate graphical display of these data (perhaps by hand). Do you think that the Active exercise group is different from the others?

 g. Write a formal report giving your view of the evidence that the Active exercises help infants maintain certain reflexes and thus enable them to walk alone at an earlier age than if these reflexes were lost. Depending on your background and on which of the previous questions you have answered, you may wish to discuss graphical displays, results of two-sample tests, uses of nonparametric tests, the power of AOV tests, the data on the walking reflexes in (f), possible relevant information in the data not elicited in any of the procedures suggested so far in this unit, and so on. Follow your instructor's guidance as to the length and format of this report.

8.18 **Defining the Problem (1).** There are approximately 50 million people in the United States who report having a physical handicap. Furthermore, it is estimated that the unemployment rate of noninstitutionalized handicapped people between the ages of 18 and 64 is nearly double the unemployment rate of people with no impairment. Thus, it appears that people with disabilities have a more difficult time obtaining employment. One of the problems confronting people having a handicap may be a bias by employers during the employment interview.

The paper "Interviewers' decisions related to applicant handicap type and rater empathy" [*Human Performance* (1990) 3:157–171] describes a study that examines these issues. The purpose of the study was to investigate whether different types of physical handicaps produce different empathy by raters and to examine whether interviewers' evaluations are affected by the type of handicap of the person being interviewed.

Collecting Data (2). The researchers videotaped five simulated employment interviews. To minimize bias across videotapes, the same male actors (job applicant and interviewer) and the same interview script, consisting of nine questions, were used in all five videotapes. The script was directed toward average qualifications of the applicant because this type of applicant is the most likely to be susceptible to interview biases. The videotapes differed with respect to type of applicant disability, but all were depicted as permanent disabilities. The five conditions were labeled wheelchair, Canadian crutches, hard of hearing, leg amputee, and nonhandicapped (control).

Each participant in the study was asked to rate the applicant's qualifications for a computer sales position based on the questions asked during the videotaped interview. Prior to viewing the videotape, each participant completed the Hogan Empathy Scale. The researchers decided to have each participant view only one of the five videotapes. Based on the variability in scores of raters in previous studies, the researchers decided they would require 14 raters for each videotape in order to obtain a precise estimate of the mean rating for each of the five handicap conditions. Seventy undergraduate students were selected to participate in the study. For each of the five videotapes, 14 students were randomly assigned to view the videotapes. After viewing each videotape, the participant rated the applicant on two scales: an 11-item scale assessing the rater's liking of the applicant and a 10-item scale that assessed the rater's evaluation of the applicant's job qualifications. For each scale, the average of the individual items form an overall assessment of the applicant. The researchers used these two variables to determine whether different types of physical handicaps are reacted to differently by raters and to determine the effect of rater empathy on evaluations of handicapped applicants.

These are some of the questions that the researchers were interested in:

1. Is there a difference in the average empathy scores of the 70 raters?
2. Do the raters' average qualification scores differ across the five handicap conditions?
3. Which pairs of handicap conditions produced different average qualification scores?
4. Is the average rating for the control group (no handicap) greater than the average ratings for the groups for all types of handicapped applicants?
5. Is the average qualification rating for the hard-of-hearing applicant group different from the average ratings for the groups whose applicants had a mobility handicap?
6. Is the average qualification rating for the crutches applicant group different from the average ratings of groups whose applicants were either amputees or in wheelchairs?
7. Is the average rating for the amputee applicant group different from the average rating for the wheelchair applicant group?

Summarizing Data (3). The researchers conducted the experiments and obtained the following data from the 70 raters of the applicants. The data in the following table are a summary of the empathy values.

Empathy values across the five handicap conditions

Condition	Control (None)	Hard of Hearing	Canadian Crutches	One-Leg Amputee	Wheelchair
Mean	21.43	22.71	20.43	20.86	19.86
Standard Deviation	3.032	3.268	3.589	3.035	3.348

The data in the next table are the applicant qualification scores of the 70 raters for the five handicap conditions along with their summary statistics. (These data were simulated using the summary statistics of the ratings given in the paper. These data are in the HANDICAP file on the CD that came with your book.)

Ratings of applicant qualification across the five handicap conditions

Control	Hard of Hearing	Amputee	Crutches	Wheelchair
6.1	2.1	4.1	6.7	3.0
4.6	4.8	6.1	6.7	3.9
7.7	3.7	5.9	6.5	7.9
4.2	3.5	5.0	4.6	3.0
6.1	2.2	6.1	7.2	3.5
2.9	3.4	5.7	2.9	8.1
4.6	5.5	1.1	5.2	6.4
5.4	5.2	4.0	3.5	6.4
4.1	6.8	4.7	5.2	5.8
6.4	0.4	3.0	6.6	4.6
4.0	5.8	6.6	6.9	5.8
7.2	4.5	3.2	6.1	5.5
2.4	7.0	4.5	5.9	5.0
2.9	1.8	2.1	8.8	6.2

```
Descriptive Statistics for Case Study

Variable              N          Mean      Median      TrMean       StDev      SE Mean
Control              14         4.900       4.600       4.875       1.638        0.438
Hard of Hearing      14         4.050       4.100       4.108       1.961        0.524
Amputee              14         4.436       4.600       4.533       1.637        0.437
Crutches             14         5.914       6.300       5.925       1.537        0.411
Wheelchair           14         5.364       5.650       5.333       1.633        0.436

Variable         Minimum     Maximum          Q1          Q3
Control            2.400       7.700       3.725       6.175
Hard of Hearing    0.400       7.000       2.175       5.575
Amputee            1.100       6.600       3.150       5.950
Crutches           2.900       8.800       5.050       6.750
Wheelchair         3.000       8.100       3.800       6.400
```

a. Plot the qualification scores in boxplots and discuss what you see based on these plots and the descriptive statistics shown here.

Analyzing Data, Interpreting the Analyses, and Communicating Results (4). The objective of the study was to investigate whether an interviewer's evaluation of a job applicant is affected by the physical handicap of the person being interviewed. Prior to testing hypotheses and making comparisons among the five treatments, we need to verify that the conditions under which the tests and multiple comparison procedures are valid have been satisfied in this study.

b. Based on your findings in part (a), does it appear the AOV assumptions hold?

c. A probability plot of the residuals is shown here. Does it confirm what you found in part (b)? Explain.

Boxplots of L, L/R, L/C, and C (means are indicated by solid circles)

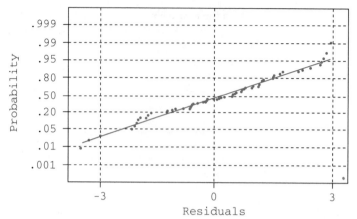

Average: -0.0000000
StDev: 1.63767
N: 70

Anderson-Darling Normality Test
A-Squared: 0.384
P-Value: 0.387

d. Check the equal variance assumption using Levene's test.

We check the condition of the independence of the data by discussing with the researchers the manner in which the study was conducted. It is important to make sure that the conditions in the room in which the interview tape was viewed remained constant throughout the study, so as not to introduce any distractions that could affect the raters' evaluations. Also, the initial check that the empathy scores were evenly distributed over the five groups of raters assures us that a difference in empathy levels did not exist in the five groups of raters prior to their evaluation of the applicants' qualifications. The research hypothesis is that the mean qualification ratings, μ_is, differ over the five handicap conditions:

H_0: $\mu_1 = \mu_2 = \mu_3 = \mu_4 = \mu_5$

H_a: At least one of the means differs from the rest.

The computer output for the AOV table is given here. The following notation is used in the output: Control (C), Hard of Hearing (H), Amputee (A), Crutches (R), and Wheelchair (W). Use the output to draw conclusions.

```
General Linear Models Procedure

Dependent Variable: Rating

Source              DF    Sum of Squares     F value     Pr > F

Model                4       30.47800000       2.68       0.0394
Error               65      185.05642857
Corrected Total     69      215.53442857
-----------------------------------------------------------------

Tukey's Studentized Range (HSD) Test for Variable: Rating

NOTE: This test controls the type I experimentwise error rate, but
      generally has a higher type II error rate than REGWQ.
```

```
Alpha= 0.05  df= 65  MSE= 2.847022
Critical Value of Studentized Range= 3.968
Minimum Significant Difference= 1.7894

Means with the same letter are not significantly different.

Tukey Grouping              Mean      N    HC

                    A       5.9143    14   R
             B      A       5.3643    14   W
             B      A       4.9000    14   C
             B      A       4.4357    14   A
             B              4.0500    14   H
-----------------------------------------------------------------
```

CHAPTER 9

More Complicated Experimental Designs

9.1   Introduction

In Chapter 2, we introduced the concepts involved in designing an experiment. These concepts are fundamental to the scientific process, in which hypotheses are formulated, experiments (studies) are planned, data are collected and analyzed, and conclusions are reached, which in turn leads to the formulation of new hypotheses. To obtain logical conclusions from the experiments (studies), it is mandatory that the hypotheses be precisely and clearly stated and that experiments be carefully designed, appropriately conducted, and properly analyzed. The analysis of a designed experiment requires the development of a model of the physical setting and a clear statement of the conditions under which this model is appropriate. Finally, a scientific report of the results of the experiment should contain graphical representations of the data, a verification of model conditions, a summary of the statistical analysis, and conclusions concerning the research hypotheses.

Chapter 8 introduced the nomenclature for the models and analyses for experimental designs. These were illustrated for a completely randomized design. We now turn our attention to other experimental designs. Sections 9.2 and 9.3 deal with extensions of the completely randomized design, in which the focus remains the same—namely, treatment mean comparisons—but where other nuisance variables must be controlled. For each of these designs, we will consider the arrangement of treatments, the advantages and disadvantages of the design, a model, and an analysis of variance for data from such a design. Section 9.4 introduces factorial experiments that focus on the evaluation of the effects of two or more independent variables (factors) on a response rather than on comparisons of treatment means as in the completely randomized design, randomized complete block design, and Latin square design. Particular attention is given to measuring the effects of each factor alone or in combination with the other factors. Not all designs focus on either comparison of treatment means or examination of the effects of factors on a response. In Section 9.5, we discuss designs that combine the attributes of the block designs of Sections 9.2 and 9.3 with those of factorial experiments in Section 9.4. The remaining sections of the chapter deal with estimation and comparisons of the treatment means for the different experimental designs and procedures to check the validity of model conditions.

9.2 Randomized Complete Block Design

Refer to the completely randomized design of Example 8.1, where a researcher studied the use of reflective paints to mark the lanes on highways. Each of the four types of paint—P_2, P_3, P_4, and P_1 (the currently used paint)—was randomly applied to four sections of the highway. The researcher is somewhat concerned about the results of the study described in Example 8.1 because he was certain that at least one of the paints would show some improvement over the currently used paint. He examines the road conditions and amount of traffic flow on the 16 sections used in the study and finds that the roadways had a very low traffic volume during the study period. He decides to redesign the study to improve the generalization of the results and will include four different locations having different amounts of traffic volumes in the new study.

We will now modify the reflective paint study to incorporate four different locations into the design. The researcher identifies four sections of roadway of length 6 feet at each of the four locations. If we randomly assigned the four paints to the 16 sections, we might end up with a randomization scheme such as the one listed in Table 9.1.

Even though we still have four observations for each treatment in this design, any differences that we may observe among the reflectivity of the road markings for the four types of paints may be due entirely to the differences in the road conditions and traffic volumes among the four locations. Because the factors location and type of paint are **confounded**, we cannot determine whether any observed differences in the decrease in reflectivity of the road markings are due to differences in the locations of the markings or due to differences in the type of paint used in creating the markings. This example illustrates a situation in which the 16 road markings are affected by an extraneous source of variability: the location of road marking. If the four locations present different environmental conditions or different traffic vol-

confounded

TABLE 9.1

Random assignment of the four paints to the 16 sections

Location			
1	**2**	**3**	**4**
P_1	P_2	P_3	P_4
P_1	P_2	P_3	P_4
P_1	P_2	P_3	P_4
P_1	P_2	P_3	P_4

TABLE 9.2

Randomized complete block assignment of the four paints to the 16 sections

Location			
1	2	3	4
P_2	P_2	P_1	P_1
P_1	P_4	P_3	P_2
P_3	P_1	P_4	P_4
P_4	P_3	P_2	P_3

umes, the 16 experimental units will not be a homogeneous set of units on which we can base an evaluation of the effects of the four treatments, the four types of paint.

The completely randomized design just described is not appropriate for this experimental setting. We need to use a randomized complete block design in order to take into account the differences that exist in the experimental units prior to assigning the treatments. One such randomization is listed in Table 9.2. Note that each location contains four sections of roadway, one section treated with each of the four paints. Hence, the variability in the reflectivity of paints due to differences in roadway conditions at the four locations can now be addressed and controlled. This will allow pairwise comparisons among the four paints that use the sample means to be free of the variability among locations. For example, if we ran the test

$$H_0: \quad \mu_{P_1} - \mu_{P_2} = 0$$

$$H_a: \quad \mu_{P_1} - \mu_{P_2} \neq 0$$

and rejected H_0, the differences between μ_{P_1} and μ_{P_2} would be due to a difference between the reflectivity properties of the two paints and not due to a difference among the locations because both paint P_1 and P_2 were applied to a section of roadway at each of the four locations.

The random assignment of the treatments to the experimental units is conducted separately within each block, the location of the roadways in this example. The four sections within a given location would tend to be more alike with respect to environmental conditions and traffic volume than sections of roadway in two different locations. Thus, we are in essence conducting four independent completely randomized designs, one for each of the four locations. By using the randomized complete block design, we have effectively filtered out the variability among the locations, enabling us to make more precise comparisons among the treatment means $\mu_{P_1}, \mu_{P_2}, \mu_{P_3},$ and μ_{P_4}.

In general, we can use a randomized complete block design to compare t treatment means when an extraneous source of variability (blocks) is present. If there are b different blocks, we randomly assign each of the t treatments to an experimental unit in each block in order to filter out the block-to-block variability. In our example, we had $t = 4$ treatments (types of paint) and $b = 4$ blocks (locations).

We can formally define a randomized complete block design as follows.

DEFINITION 9.1

A **randomized complete block design** is an experimental design for comparing t treatments in b blocks. The blocks consist of t homogeneous experimental units. Treatments are randomly assigned to experimental units within a block, with each treatment appearing exactly once in every block.

The randomized complete block design has certain advantages and disadvantages, as shown here.

Advantages and Disadvantages of the Randomized Complete Block Design

Advantages

1. The design is useful for comparing t treatment means in the presence of a single extraneous source of variability.
2. The statistical analysis is simple.
3. The design is easy to construct.
4. The design can be used to accommodate any number of treatments in any number of blocks.

Disadvantages

1. Because the experimental units within a block must be homogeneous, the design is best suited for a relatively small number of treatments.
2. The design controls for only one extraneous source of variability (due to blocks). Additional extraneous sources of variability tend to increase the error term, making it more difficult to detect treatment differences.
3. The effect of each treatment on the response must be approximately the same from block to block.

Consider the data for a randomized complete block design as arranged in Table 9.3. Note that although these data look similar to the data presentation for a completely randomized design (see Table 8.4), there is a difference in the way treatments were assigned to the experimental units.

TABLE 9.3
Data for a randomized complete block design

Treatment	Block				Mean
	1	**2**	$\cdots$	**b**	
1	y_{11}	y_{12}	$\cdots$	y_{1b}	$\bar{y}_{1.}$
2	y_{21}	y_{22}	$\cdots$	y_{2b}	$\bar{y}_{2.}$
$\vdots$	$\vdots$	$\vdots$	$\vdots$	$\vdots$	$\vdots$
t	y_{t1}	y_{t2}	$\cdots$	y_{tb}	$\bar{y}_{t.}$
Mean	$\bar{y}_{.1}$	$\bar{y}_{.2}$	$\cdots$	$\bar{y}_{.b}$	$\bar{y}_{..}$

The model for an observation in a randomized complete block design can be written in the form

$$y_{ij} = \mu + \alpha_i + \beta_j + \varepsilon_{ij}$$

where the terms of the model are defined as follows:

y_{ij}: Observation on experimental unit in jth block receiving treatment i.

μ: Overall mean, an unknown constant.

α_i: An effect due to treatment i, an unknown constant.

β_j: An effect due to block j, an unknown constant.

ε_{ij}: A random error associated with the response from an experimental unit in block j receiving treatment i. We require that the ε_{ij}s have a normal distribution with mean 0 and common variance σ_ε^2. In addition, the errors must be independent.

The conditions given for our model can be shown to imply that the recorded response from the ith treatment in the jth block, y_{ij}, is normally distributed with mean

$$E(y_{ij}) = \mu + \alpha_i + \beta_j$$

and variance σ_ε^2. Table 9.4 gives the population means (expected values) for the data of Table 9.3.

Several comments should be made concerning the table of expected values. First, any pair of observations that receive the same treatment (appear in the same row of Table 9.4) have population means that differ only by their block effects (β_j's). For example, the expected values associated with y_{11} and y_{12} (two observations receiving treatment 1) are

$$E(y_{11}) = \mu + \alpha_1 + \beta_1 \qquad E(y_{12}) = \mu + \alpha_1 + \beta_2$$

Thus, the difference in their means is

$$E(y_{11}) - E(y_{12}) = (\mu + \alpha_1 + \beta_1) - (\mu + \alpha_1 + \beta_2) = \beta_1 - \beta_2$$

which accounts for the fact that y_{11} was recorded in block 1 and y_{12} was recorded in block 2 but both were responses from experimental units receiving treatment 1. Thus, there is no treatment effect, but a possible block effect may be present. Second, two observations appearing in the same block (in the same column of Table 9.4) have means that differ by a treatment effect only. For example, y_{11} and y_{21} both appear in block 1. The difference in their means, from Table 9.4, is

$$E(y_{11}) - E(y_{21}) = (\mu + \alpha_1 + \beta_1) - (\mu + \alpha_2 + \beta_1) = \alpha_1 - \alpha_2$$

which accounts for the fact that the experimental units received different treatments but were observed in the same block. Hence, there is a possible treatment effect but no block effect. Finally, when two experimental units receive different treatments and are observed in different blocks, their expected values differ by effects due to both treatment differences and block differences. Thus, observations y_{11} and y_{22} have expectations that differ by

$$E(y_{11}) - E(y_{22}) = (\mu + \alpha_1 + \beta_1) - (\mu + \alpha_2 + \beta_2) = (\alpha_1 - \alpha_2) + (\beta_1 - \beta_2)$$

filtering

Using the information we have learned concerning the model for a randomized block design, we can illustrate the concept of **filtering** and show how the ran-

TABLE 9.4

Expected values for the y_{ij}s in a randomized complete block design

Treatment	Block 1	2	$\cdots$	b
1	$E(y_{11}) = \mu + \alpha_1 + \beta_1$	$E(y_{12}) = \mu + \alpha_1 + \beta_2$	$\cdots$	$E(y_{1b}) = \mu + \alpha_1 + \beta_b$
2	$E(y_{21}) = \mu + \alpha_2 + \beta_1$	$E(y_{22}) = \mu + \alpha_2 + \beta_2$	$\cdots$	$E(y_{2b}) = \mu + \alpha_2 + \beta_b$
$\vdots$	$\vdots$	$\vdots$	$\vdots$	$\vdots$
t	$E(y_{t1}) = \mu + \alpha_t + \beta_1$	$E(y_{t2}) = \mu + \alpha_t + \beta_2$	$\cdots$	$E(y_{tb}) = \mu + \alpha_t + \beta_b$

TABLE 9.5

Randomized complete block
design with $t = 3$ treatments
and $b = 3$ blocks

Block	Treatment		
1	1	2	3
2	1	3	2
3	3	1	2

domized block design filters out the variability due to blocks. Consider a randomized block design with $t = 3$ treatments (1, 2, and 3) laid out in $b = 3$ blocks as shown in Table 9.5. The model for this randomized block design is

$$y_{ij} = \mu + \alpha_i + \beta_j + \varepsilon_{ij} \quad (i = 1, 2, 3; j = 1, 2, 3)$$

Suppose we wish to estimate the difference in mean response for treatments 2 and 1—namely, $\alpha_2 - \alpha_1$. The difference in sample means, $\bar{y}_{2.} - \bar{y}_{1.}$, would represent a point estimate of $\alpha_2 - \alpha_1$. By substituting into our model, we have

$$\bar{y}_{1.} = \frac{1}{3} \sum_j y_{1j}$$

$$= \frac{1}{3}[(\mu + \alpha_1 + \beta_1 + \varepsilon_{11}) + (\mu + \alpha_1 + \beta_2 + \varepsilon_{12}) + (\mu + \alpha_1 + \beta_3 + \varepsilon_{13})]$$

$$= \mu + \alpha_1 + \bar{\beta} + \bar{\varepsilon}_1$$

where $\bar{\beta}$ represents the mean of the three block effects β_1, β_2, and β_3 and $\bar{\varepsilon}_1$ represents the mean of the three random errors ε_{11}, ε_{12}, and ε_{13}. Similarly, it is easy to show that

$$\bar{y}_{2.} = \mu + \alpha_2 + \bar{\beta} + \bar{\varepsilon}_2$$

and hence

$$\bar{y}_{2.} - \bar{y}_{1.} = (\alpha_2 - \alpha_1) + (\bar{\varepsilon}_2 - \bar{\varepsilon}_1)$$

Note how the block effects cancel, leaving the quantity $(\bar{\varepsilon}_2 - \bar{\varepsilon}_1)$ as the error of estimation using $\bar{y}_{2.} - \bar{y}_{1.}$ to estimate $\alpha_2 - \alpha_1$.

If a completely randomized design had been employed instead of a randomized block design, treatments would have been assigned to experimental units at random and it is quite unlikely that each treatment would have appeared in each block. When the same treatment appears more than once in a block and we calculate an estimate of $\alpha_2 - \alpha_1$ using $\bar{y}_{2.} - \bar{y}_{1.}$, all block effects do not cancel out as they did previously. Then the error of estimation includes not only $\bar{\varepsilon}_2 - \bar{\varepsilon}_1$ but also the block effects that do not cancel; that is,

$$\bar{y}_{2.} - \bar{y}_{1.} = \alpha_2 - \alpha_1 + [(\bar{\varepsilon}_2 - \bar{\varepsilon}_1) + (\text{block effects that do not cancel})]$$

Hence, the randomized block design filters out variability due to blocks by decreasing the error of estimation for a comparison of treatment means.

A plot of the expected values, μ_{ij} in Figure 9.1, demonstrates that the size of the difference between the means of observations receiving the same treatment but in different blocks (say, j and j') is the same for all treatments. That is,

$$\mu_{ij} - \mu_{ij'} = \beta_j - \beta_{j'} \quad \text{for all } i = 1, \ldots, t$$

FIGURE 9.1

Treatment means in a
randomized block design

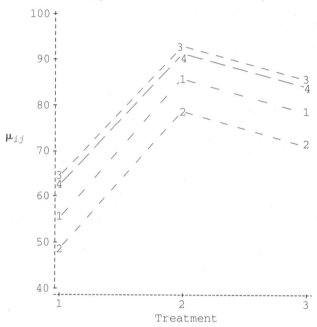

Plot of Treatment Mean by Treatment
(symbol is value of block)

A consequence of this condition is that the lines connecting the means having the
same treatment form a set of parallel lines.

The main goal in using the randomized complete block design was to examine differences in the t treatment means $\mu_{1.}, \mu_{2.}, \ldots, \mu_{t.}$, where $\mu_{i.}$ is the mean response of treatment i. The null hypothesis is *no difference among treatment means* versus the research hypothesis *treatment means differ*. That is,

$$H_0: \quad \mu_{1.} = \mu_{2.} = \cdots = \mu_{t.} \qquad H_a: \quad \text{At least one } \mu_{i.} \text{ differs from the rest.}$$

This set of hypotheses is equivalent to testing

$$H_0: \quad \alpha_1 = \alpha_2 = \cdots = \alpha_t = 0 \qquad H_a: \quad \text{At least one } \alpha_i \text{ is different from 0.}$$

The two sets of hypotheses are equivalent because, as we observed in Table 9.4, when comparing the mean response of two treatments (say, i and i') observed in the same block, the difference in their mean response is

$$\mu_{i.} - \mu_{i'.} = \alpha_i - \alpha_{i'}$$

Thus, under H_0, we are assuming that treatments have the same mean response within a given block. Our test statistic will be obtained by examining the model for a randomized block design and partitioning the total sum of squares to include terms for treatment effects, block effects, and random error effects. Using Table 9.3, we can introduce notation that is needed in the partitioning of the total sum of squares. This notation is presented here.

y_{ij}: Observation for treatment i in block j

t: Number of treatments

b: Number of blocks

$\bar{y}_{i.}$: Sample mean for treatment i, $\bar{y}_{i.} = \dfrac{1}{b}\Sigma_{j=1}^{b}\, y_{ij}$

$\bar{y}_{.j}$: Sample mean for block j, $\bar{y}_{.j} = \dfrac{1}{t}\Sigma_{i=1}^{t}\, y_{ij}$

$\bar{y}_{..}$: Overall sample mean, $\bar{y}_{..} = \dfrac{1}{bt}\Sigma_{ij}\, y_{ij}$

total sum of squares

The **total sum of squares** (TSS) of the measurements about their mean $\bar{y}_{..}$ is defined as before:

$$\text{TSS} = \sum_{ij}(y_{ij} - \bar{y}_{..})^2$$

This sum of squares will be partitioned into three separate sources of variability: one due to the variability among treatments, one due to the variability among blocks, and one due to the variability from all sources not accounted for by either treatment differences or block differences. We call this source of variability **error.** The **partition of TSS** follows from an examination of the randomized complete block model:

error

partition of TSS

$$y_{ij} = \mu + \alpha_i + \beta_j + \varepsilon_{ij}$$

The parameters in the model have sample estimates:

$$\hat{\mu} = \bar{y}_{..} \qquad \hat{\alpha}_i = \bar{y}_{i.} - \bar{y}_{..} \qquad \text{and} \qquad \hat{\beta}_j = \bar{y}_{.j} - \bar{y}_{..}$$

It can be shown algebraically that TSS takes the following form:

$$\sum_{ij}(y_{ij} - \bar{y}_{..})^2 = b\sum_{i}(\bar{y}_{i.} - \bar{y}_{..})^2 + t\sum_{j}(\bar{y}_{.j} - \bar{y}_{..})^2 + \sum_{ij}(y_{ij} - \bar{y}_{i.} - \bar{y}_{.j} + \bar{y}_{..})^2$$

We will interpret the terms in the partition using the parameter estimates. The first quantity on the right-hand side of the equal sign measures the variability of the treatment means $\bar{y}_{i.}$ from the overall mean $\bar{y}_{..}$. Thus,

$$\text{SST} = b\sum_{i}(\bar{y}_{i.} - \bar{y}_{..})^2 = b\sum_{i}(\hat{\alpha}_i)^2$$

between-treatment sum of squares

called the **between-treatment sum of squares,** is a measure of the variability in the y'_{ij}s due to differences in the treatment means. Similarly, the second quantity,

$$\text{SSB} = t\sum_{j}(\bar{y}_{.j} - \bar{y}_{..})^2 = t\sum_{j}(\hat{\beta}_j)^2$$

between-block sum of squares

sum of squares for error

measures the variability between the block means $\bar{y}_{.j}$ and the overall mean. It is called the **between-block sum of squares.** The third source of variability, referred to as the **sum of squares for error,** SSE, represents the variability in the $\bar{y}'_{ij}$s not accounted for by the block and treatment differences. There are several forms for this term:

Source	SS	df	MS	F
Treatments	SST	$t - 1$	$\text{MST} = \text{SST}/(t - 1)$	MST/MSE
Blocks	SSB	$b - 1$	$\text{MSB} = \text{SSB}/(b - 1)$	MSB/MSE
Error	SSE	$(b - 1)(t - 1)$	$\text{MSE} = \text{SSE}/(b - 1)(t - 1)$	
Total	TSS	$bt - 1$		

$$\text{SSE} = \sum_{ij} (e_{ij})^2 = \sum_{ij} (y_{ij} - \hat{\mu} - \hat{\alpha}_i - \hat{\beta}_j)^2 = \text{TSS} - \text{SST} - \text{SSB}$$

where $e_{ij} = y_{ij} - \hat{\mu} - \hat{\alpha}_i - \hat{\beta}_j$ are the residuals used to check model conditions. We can summarize our calculations in an AOV table as given in Table 9.6.

The test statistic for testing

$$H_0: \quad \alpha_1 = \alpha_2 = \cdots = \alpha_t = 0 \qquad H_a: \quad \text{At least one } \alpha_i \text{ is different from 0}$$

is the ratio

$$F = \frac{\text{MST}}{\text{MSE}}$$

unbiased estimates

expected mean squares

When $H_0: \alpha_1 = \alpha_2 = \cdots = \alpha_t = 0$ is true, both MST and MSE are **unbiased estimates** of σ_ε^2, the variance of the experimental error. That is, when H_0 is true, both MST and MSE have a mean value in repeated sampling, called the **expected mean squares,** equal to σ_ε^2. We express these terms as

$$E(\text{MST}) = \sigma_\varepsilon^2 \qquad E(\text{MSE}) = \sigma_\varepsilon^2$$

We thus expect $F = \text{MST}/\text{MSE}$ to have a value near 1.

When H_a is true, the expected value of MSE is still σ_ε^2. However, MST is no longer unbiased for σ_ε^2. In fact, the expected mean square for treatments can be shown to be

$$E(\text{MST}) = \sigma_\varepsilon^2 + b\theta_T \quad \text{where } \theta_T = \frac{1}{t - 1} \sum_i \alpha_i^2$$

Thus, a large difference in the treatment means will result in a large value for θ_T. The expected value of MST will then be larger than the expected value of MSE and we will expect $F = \text{MST}/\text{MSE}$ to be larger than 1. Thus, our test statistic F rejects H_0 when we observe a value of F larger than a value in the upper tail of the F distribution.

The preceding discussion leads to the following decision rule for a specified probability of a Type I error:

Reject $H_0: \quad \alpha_1 = \alpha_2 = \cdots = \alpha_t = 0 \quad$ when $F = \text{MST}/\text{MSE}$ exceeds $F_{a,\text{df}_1,\text{df}_2}$

where $F_{a,\text{df}_1,\text{df}_2}$ is from the F tables in Appendix Table 9 with $a =$ specified value of probability Type I error, $\text{df}_1 = \text{df}_{\text{MST}} = t - 1$, and $\text{df}_2 = \text{df}_{\text{MSE}} = (b - 1)(t - 1)$. Alternatively, we can compute the p-value for the observed value of the test statistic F_{obs} by computing

$$p\text{-value} = P(F_{\text{df}_1,\text{df}_2} > F_{\text{obs}})$$

where the F distribution with $\text{df}_1 = t - 1$ and $\text{df}_2 = (b - 1)(t - 1)$ is used to compute the probability. We would then compare the p-value to a selected value for the probability of Type I error, with small p-values supporting the research hypothesis and large p-values failing to reject H_0.

The block effects are generally assessed only to determine whether or not the blocking was efficient in reducing the variability in the experimental units. Thus, hypotheses about the block effects are not tested. However, we might still ask whether blocking has increased our precision for comparing treatment means in a given experiment. Let MSE_{RCB} and MSE_{CR} denote the mean square errors for a randomized complete block design and a completely randomized design, respectively. One measure of precision for the two designs is the variance of the estimate of the ith treatment mean, $\hat{\mu}_i = \bar{y}_{i.}\,(i = 1, 2, \ldots, t)$. For a randomized complete block design, the estimated variance of $\bar{y}_{i.}$ is $\text{MSE}_{\text{RCB}}/b$. For a completely randomized design, the estimated variance of $\bar{y}_{i.}$ is MSE_{CR}/r, where r is the number of observations (replications) of each treatment required to satisfy the relationship

$$\frac{\text{MSE}_{\text{CR}}}{r} = \frac{\text{MSE}_{\text{RCB}}}{b} \quad \text{or} \quad \frac{\text{MSE}_{\text{CR}}}{\text{MSE}_{\text{RCB}}} = \frac{r}{b}$$

relative efficiency
RE(RCB, CR)

The quantity r/b is called the **relative efficiency** of the randomized complete block design compared to a completely randomized design **RE(RCB, CR).** The larger the value of MSE_{CR} compared to MSE_{RCB}, the larger r must be to obtain the same level of precision for estimating a treatment mean in a completely randomized design as obtained using the randomized complete block design. Thus, if the blocking is effective, we expect the variability in the experimental units to be smaller in the randomized complete block design than in a completely randomized design. The ratio $\text{MSE}_{\text{CR}}/\text{MSE}_{\text{RCB}}$ should be large, which results in r being much larger than b. Thus, the amount of data needed to obtain the same level of precision in estimating μ_i is larger in the completely randomized design than in the randomized complete block design. When the blocking is not effective, then the ratio $\text{MSE}_{\text{CR}}/\text{MSE}_{\text{RCB}}$ is nearly 1 and r and b are equal.

In practice, evaluating the efficiency of the randomized complete block design relative to a completely randomized design cannot be accomplished because the completely randomized design was not conducted. However, we can use the mean squares from the randomized complete block design, MSB and MSE, to obtain the relative efficiency RE(RCB, CR) by using the formula

$$\text{RE(RCB, CR)} = \frac{\text{MSE}_{\text{CR}}}{\text{MSE}_{\text{RCB}}} = \frac{(b-1)\text{MSB} + b(t-1)\text{MSE}}{(bt-1)\text{MSE}}$$

When RE(RCB, CR) is much larger than 1, then r is greater than b and we can conclude that the blocking was efficient because many more observations would be required in a completely randomized design than would be required in the randomized complete block design.

EXAMPLE 9.1

A researcher conducted an experiment to compare the effects of three different insecticides on a variety of string beans. To obtain a sufficient amount of data, it was necessary to use four different plots of land. Because the plots had somewhat different soil fertility, drainage characteristics, and sheltering from winds, the researcher decided to conduct a randomized complete block design with the plots serving as the blocks. Each plot was subdivided into three rows. A suitable distance was maintained between rows within a plot so that the insecticides could be confined to a particular row. Each row was planted with 100 seeds and then main-

TABLE 9.7

Number of seedlings by insecticide and plot for Example 9.1

Insecticide	Plot 1	2	3	4	Insecticide Mean
1	56	48	66	62	58
2	83	78	94	93	87
3	80	72	83	85	80
Plot Mean	73	66	81	80	75

tained under the insecticide assigned to the row. The insecticides were randomly assigned to the rows within a plot so that each insecticide appeared in one row of each of the four plots. The response y_{ij} of interest was the number of seedlings that emerged per row. The data and means are given in Table 9.7.

    a. Write an appropriate statistical model for this experimental situation.
    b. Run an analysis of variance to compare the effectiveness of the three insecticides. Use $\alpha = .05$.
    c. Summarize your results in an AOV table.
    d. Compute the relative efficiency of the randomized block design relative to a completely randomized design.

Solution We recognize this experimental design as a randomized complete block design with $b = 4$ blocks (plots) and $t = 3$ treatments (insecticides) per block. The appropriate statistical model is

$$y_{ij} = \mu + \alpha_i + \beta_j + \varepsilon_{ij} \qquad i = 1, 2, 3 \qquad j = 1, 2, 3, 4$$

From the information in Table 9.7, we can estimate the treatment means $\mu_{i.}$ by $\hat{\mu}_{1.} = \bar{y}_{i.}$, which yields

$$\hat{\mu}_{i.} = 58 \qquad \hat{\mu}_{2.} = 87 \qquad \hat{\mu}_{3.} = 80$$

It appears that the rows treated with insecticide 1 yielded many fewer plants than the other two insecticides. We will next estimate the model parameters and construct the AOV table. Recall that $\hat{\mu} = \bar{y}_{..}, \hat{\alpha}_i = \bar{y}_{i.} - \bar{y}_{..}$, and $\hat{\beta}_j = \bar{y}_{.j} - \bar{y}_{..}$. Thus, with $\hat{\mu} = \bar{y}_{..} = 75$, we obtain

Insecticide Effects

$\hat{\alpha}_1 = 58 - 75 = -17$
$\hat{\alpha}_2 = 87 - 75 = 12$
$\hat{\alpha}_3 = 80 - 75 = 5$

Plot Effects

$\hat{\beta}_1 = 73 - 75 = -2$
$\hat{\beta}_2 = 66 - 75 = -9$
$\hat{\beta}_3 = 81 - 75 = 6$
$\hat{\beta}_4 = 80 - 75 = 5$

Substituting into the formulas for the sum of squares, we have

$$\text{TSS} = \sum_{ij} (y_{ij} - \bar{y}_{..})^2 = (56 - 75)^2 + (48 - 75)^2 + \cdots + (85 - 75)^2 = 2{,}296$$

$$\text{SST} = b \sum_i (\hat{\alpha}_i)^2 = 4[(-17)^2 + (12)^2 + (5)^2] = 1{,}832$$

$$\text{SSB} = t \sum_j (\hat{\beta}_j)^2 = 3[(-2)^2 + (-9)^2 + (6)^2 + (5)^2] = 438$$

TABLE 9.8

AOV table for the data of Example 9.1

Source	SS	df	MS	F	p-value
Treatments	1,832	2	916	211.38	.0001
Blocks	438	3	146	33.69	.0004
Error	26	6	4.3333		
Total	2,296	11			

By subtraction, we have

$$\text{SSE} = \text{TSS} - \text{SST} - \text{SSB} = 2{,}296 - 1{,}832 - 438 = 26$$

The analysis of variance table in Table 9.8 summarizes our results. Note that the mean square for a source in the AOV table is computed by dividing the sum of squares for that source by its degrees of freedom.

The F test for differences in the treatment means—namely,

$$H_0: \alpha_1 = \alpha_2 = \cdots = \alpha_t = 0 \quad \text{vs.} \quad H_a: \text{at least one } \alpha_i \text{ is different from } 0$$

makes use of the F statistic MST/MSE. Because the computed value of F, 211.38, is greater than the tabulated F-value, 5.14, based on $df_1 = 2$, $df_2 = 6$, and $a = .05$, we reject H_0 and conclude that there are significant (p-value $< .0001$) differences in the mean number of seedlings among the three insecticides.

We will next assess whether the blocking was effective in increasing the precision of the analysis relative to a completely randomized design. From the AOV table, we have MSB = 146 and MSE = 4.3333. Hence, the relative efficiency of this randomized block design relative to a completely randomized design is

$$\text{RE(RCB, CR)} = \frac{(b - 1)\text{MSB} + b(t - 1)\text{MSE}}{(bt - 1)\text{MSE}}$$

$$= \frac{(4 - 1)(146) + 4(3 - 1)(4.3333)}{[(4)(3) - 1](4.3333)} = 9.92$$

That is, approximately 10 times as many observations of each treatment would be required in a completely randomized design to obtain the same precision for estimating the treatment means as with this randomized complete block design. The plots were considerably different in their physical characteristics and hence it was crucial that blocking be used in this experiment.

The results in Example 9.1 are valid only if we can be assured that the conditions placed on the model are consistent with the observed data. Thus, we use the residuals $e_{ij} = y_{ij} - \hat{\mu} - \hat{\alpha}_i - \hat{\beta}_j$ to assess whether the conditions of normality, equal variance, and independence appear to be satisfied for the observed data. The following example includes the computer output for such an analysis.

EXAMPLE 9.2

The computer output for the experiment described in Example 9.1 is displayed here. Compare the results to those obtained using the definition of the sum of squares and assess whether the model conditions appear to be valid.

```
Dependent Variable: NUMBER OF SEEDLINGS


                                    Sum of           Mean
Source                    DF       Squares         Square   F Value   Pr > F
Model                      5     2270.0000       454.0000    104.77   0.0001
Error                      6       26.0000         4.3333
Corrected Total           11     2296.0000


Source                    DF     Type I SS    Mean Square   F Value   Pr > F
INSECTICIDES               2     1832.0000       916.0000    211.38   0.0001
PLOTS                      3      438.000        146.0000     33.69   0.0004

RESIDUAL ANALYSIS

Variable=RESIDUALS


                                  Moments

        N                        12   Sum Wgts              12
        Mean                      0   Sum                    0
        Std Dev            1.537412   Variance        2.363636
        Skewness           -0.54037   Kurtosis        -0.25385
                                                            Test of Normality
        W:Normal           0.942499   Pr<W              0.4938


     Stem Leaf                            #           Boxplot
        2 00                              2              |
        1 000                             3           +-----+
        0 000                             3           *--+--*
       -0                                              |    |
       -1 00                              2           +-----+
       -2 0                               1              |
       -3 0                               1              |
          ----+----+----+----+

Variable=RESIDUALS


                 Normal Probability Plot

      2.5+                                   *    ++*+++
         |                                 * * +*+++++
         |                           * * +*+++++
     -0.5+                        * +*++++
         |                    ++*++++
         |              +++*++
     -3.5++++++
          +----+----+----+----+----+----+----+----+----+----+
            -2        -1         0        +1        +2
```

Solution Note that our hand calculations yielded the same values as are given in the computer output. Generally, there will be some rounding errors in our hand calculations, which can lead to values that will differ from these given in the computer output. It is strongly recommended that a computer software program be used in the analysis of variance calculations because of the potential for rounding errors.

FIGURE 9.2

Residuals versus treatment means from Example 9.1

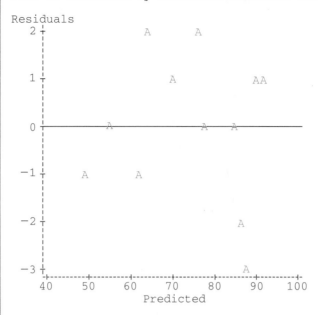

```
Plot of Residuals by Predicted Treatment Mean

Residuals
    2 +               A        A



    1 +                 A           AA



    0 +          A           A    A
      ┼─────────────────────────────────────────

   -1 +    A      A



   -2 +                        A



   -3 +                          A
      +---------+---------+---------+---------+---------+---------+
      40        50        60        70        80        90       100
                              Predicted
```

In assessing whether the model conditions have been met, we first note that in re-gard to the normality condition the test of H_0: residuals have normal distribution, the p-value from the Shapiro–Wilks test is p-value = .4938. Thus, we do not reject H_0 and the normality condition appears to be satisfied. Also, the stem-and-leaf plot, boxplot, and normal probability plot are consistent with the condition that the re-siduals have a normal distribution. Figure 9.2 is a plot of the residuals versus the es-timated treatment means. From this plot it appears that the variability in the resid-uals is somewhat constant across the treatments.

EXERCISES

Edu.

9.1 A researcher wanted to determine whether attending a Head Start program improves the academic performance of first graders from a low-income community. The researcher obtained a random sample of six children who attended a Head Start program and six who did not. There were large differences in the type of support the students received at home from their parents. Thus, after assessing the home environment of the twelve students, the researcher paired the students based on the similarities in their home environment. After completing the first grade, the students were given an overall aptitude examination. The results are shown here.

| | **Exam Score** | |
Pair	**Attended Head Start**	**Did Not Attend Head Start**
1	58	47
2	73	67
3	85	69
4	76	62
5	88	77
6	90	77

a. Do the students who attended a Head Start program appear to have higher mean aptitude scores than the students who did not attend such a program? Use $\alpha = .05$.

b. Give the efficiency of the randomized complete block design relative to a completely randomized design. Interpret your findings.

9.2 Refer to Exercise 9.1. Analyze these same data using the paired t test. Compare your results to the results from Exercise 9.1. (The F test for testing treatment differences in a randomized complete block design when there are only two treatments is equivalent to the paired t test of Chapter 6. This can be shown by noting that $t^2 = F$, where F is the F test for treatment differences.)

Engin. **9.3** An experiment compares four different mixtures of the components oxidizer, binder, and fuel used in the manufacturing of rocket propellant. The four mixtures under test, corresponding to settings of the mixture proportions for oxide, are shown here.

Mixture	Oxidizer	Binder	Fuel
1	.4	.4	.2
2	.4	.2	.4
3	.6	.2	.2
4	.5	.3	.2

To compare the four mixtures, five different samples of propellant are prepared from each mixture and readied for testing. Each of five investigators is randomly assigned one sample of each of the four mixtures and asked to measure the propellant thrust. These data are summarized next.

	Investigator				
Mixture	1	2	3	4	5
1	2,340	2,355	2,362	2,350	2,348
2	2,658	2,650	2,665	2,640	2,653
3	2,449	2,458	2,432	2,437	2,445
4	2,403	2,410	2,418	2,397	2,405

a. Identify the blocks and treatments for this experimental design.
b. Indicate the method of randomization.
c. Why is this design preferable to a completely randomized design?

9.4 Refer to Exercise 9.3.
a. Write a model for this experimental setting.
b. Estimate the parameters in the model.
c. Use the computer output shown here to conduct an analysis of variance. Use $\alpha = .05$.
d. What conclusions can you draw concerning the best mixture from the four tested? (*Note:* The higher the response value, the better is the rocket propellant thrust.)
e. Compute the relative efficiency of the randomized block design relative to a completely randomized design. Interpret this value. Were the blocks effective in reducing the variability in experimental units? Explain.

```
General Linear Models Procedure For Data in Exercise 9.3

Dependent Variable: THRUST

                                Sum of         Mean
Source                DF        Squares       Square  F Value   Pr > F
Model                  7      261713.45     37387.64   542.96   0.0001
Error                 12         826.30        68.86
Corrected Total       19      262539.75

                R-Square              C.V.     Root MSE          Y Mean
                0.996853          0.336807       8.2981          2463.8

Source                DF      Type I SS  Mean Square  F Value   Pr > F
M                      3      261260.95     87086.98  1264.73   0.0001
I                      4         452.50       113.12     1.64   0.2273
```

Psy. **9.5** An industrial psychologist working for a large corporation designs a study to evaluate the effect of background music on the typing efficiency of secretaries. The psychologist selects a random sample of seven secretaries from the secretarial pool. Each subject is exposed to three types of background music: no music, classical music, and hard rock music. The subject is given a standard typing test that combines an assessment of speed with a penalty for typing errors. The particular order of the three experiments is randomized for each of the seven subjects. The results are given here with a high score indicating a superior performance. This is a special type of randomized complete block design in which a single experimental unit serves as a block and receives all treatments.

Type of Music	Subject						
	1	2	3	4	5	6	7
No Music	20	17	24	20	22	25	18
Hard Rock	20	18	23	18	21	22	19
Classical	24	20	27	22	24	28	16

a. Write a statistical model for this experiment and estimate the parameters in your model.
b. Are there differences in the mean typing efficiency for the three types of music? Use $\alpha = .05$.
c. Does the additive model for a randomized complete block design appear to be appropriate? (*Hint:* Plot the data as was done in Figure 9.1.)
d. Compute the relative efficiency of the randomized block design relative to a completely randomized design. Interpret this value. Were the blocks effective in reducing the variability in experimental units? Explain.

9.6 Refer to Exercise 9.5. The computer output for the data in Exercise 9.5 follows. Compare your results with the results given here. Do the model conditions appear to be satisfied?

```
General Linear Models Procedure for Exercise 9.5

Dependent Variable: TYPING EFFICIENCY


                            Sum of          Mean
Source              DF      Squares         Square      F Value   Pr > F
Model                8      180.28571       22.53571       9.53   0.0004
Error               12       28.38095        2.36508
Corrected Total     20      208.66667


              R-Square          C.V.       Root MSE            Y Mean
              0.863989      7.208819         1.5379            21.333

Source              DF      Type I SS   Mean Square      F Value   Pr > F
M                    2      30.95238      15.47619          6.54   0.0120
S                    6     149.33333      24.88889         10.52   0.0003

RESIDUAL ANALYSIS:

Variable=RESIDUAL


                              Moments

         N                21  Sum Wgts           21
         Mean              0  Sum                 0
         Std Dev    1.191238  Variance     1.419048
         Skewness   -0.77527  Kurtosis     2.587721

         W:Normal   0.936418  Pr< W          0.1813

     Stem Leaf                       #          Boxplot
       2 5                           1             0
       1 03                          2             |
       0 001355789                   9          +--+--+
      -0 9985211                      7          +-----+
      -1 8                           1             |
      -2
      -3 3                           1             0
         ----+----+----+----+


                            Normal Probability Plot
      2.5+                                  ++*++++++
         |                              +++*+*++
         |                       ****+*+*+* *
     -0.5+             *  * *+***+++
         |        ++*+++++
         |++++++++++
     -3.5+       *

        +----+----+----+----+----+----+----+----+----+----+
            -2        -1         0        +1        +2
```

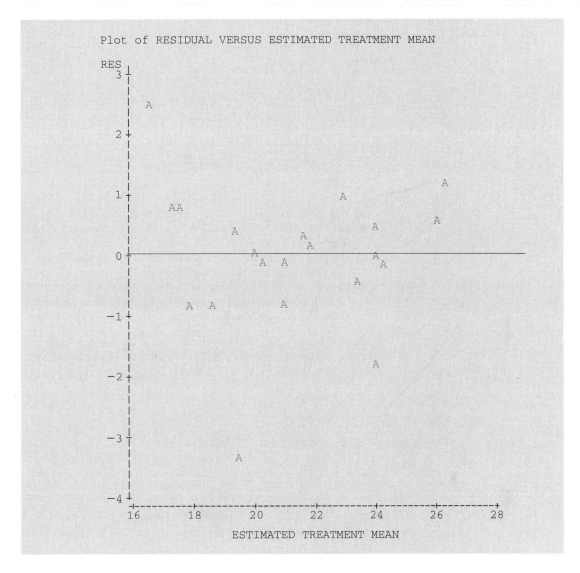

Plot of RESIDUAL VERSUS ESTIMATED TREATMENT MEAN

Psy. **9.7** A quality control engineer is considering implementing a workshop to instruct workers on the principles of total quality management (TQM). The program would be quite expensive to implement across the whole corporation; hence, the engineer has designed a study to evaluate which of four types of workshops would be most effective. The response variable will be the increase in productivity of the worker after participating in the workshop. Because the effectiveness of the workshop may depend on the worker's preconceived attitude concerning TQM, the workers are given an examination to determine their attitude prior to taking the workshop. Their attitudes are classified into five groups. There are four workers in each group, and the type of workshop is randomly assigned to the workers within each group. The increases in productivity are given here.

| Type of Workshop | Attitude | | | | | Mean |
	1	2	3	4	5	
A	33	38	39	42	62	42.8
B	35	37	43	47	71	46.6
C	40	42	45	52	74	50.6
D	54	50	55	62	84	61.0
Mean	40.5	41.75	45.5	50.75	72.75	50.25

a. Write a statistical model for this experiment and estimate the parameters in your model.
b. Are there differences in the mean increase in productivity for the four types of workshops? Use $\alpha = .05$.
c. Does the additive model for a randomized complete block design appear to be appropriate? (*Hint:* Plot the data as in Figure 9.1.)
d. Compute the relative efficiency of the randomized block design relative to a completely randomized design. Interpret this value. Were the blocks effective in reducing the variability in experimental units? Explain.

9.8 Refer to Exercise 9.7. The computer output for the data in Exercise 9.7 follows. Compare your results with the results given here. Do the model conditions appear to be satisfied?

```
General Linear Models Procedure for Exercise 9.7

Dependent Variable: INCREASE IN PRODUCTIVITY

                            Sum of            Mean
Source            DF       Squares          Square    F Value   Pr > F
Model              7     3708.0500        529.7214    114.12    0.0001
Error             12       55.7000          4.6417
Corrected Total   19     3763.7500

                R-Square            C.V.        Root MSE           Y Mean
                0.985201        4.287468          2.1545           50.250

Source            DF      Type I SS    Mean Square   F Value   Pr > F
W                  3       922.5500       307.5167     66.25    0.0001
A                  4      2785.5000       696.3750    150.03    0.0001

RESIDUAL ANALYSIS:

                              Moments
        N                  20    Sum Wgts            20
        Mean                0    Sum                  0
        Std Dev       1.712185   Variance      2.931579
        Skewness      0.207266   Kurtosis      0.149301

        W:Normal      0.985267   Pr<W            0.9725
```

```
Stem Leaf                          #        Boxplot
   3 7                             1           |
   2 8                             1           |
   1 029                           3           |
   0 5599                          4        +--+--+
  -0 88110                         5        *-----*
  -1 8321                          4        +-----+
  -2 5                             1           |
  -3 3                             1           |
       ----+----+----+----+
```

Normal Probability Plot

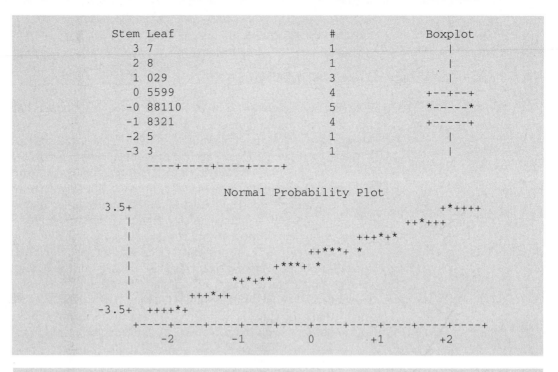

Plot of RESIDUALS VERSUS ESTIMATED TREATMENT MEAN

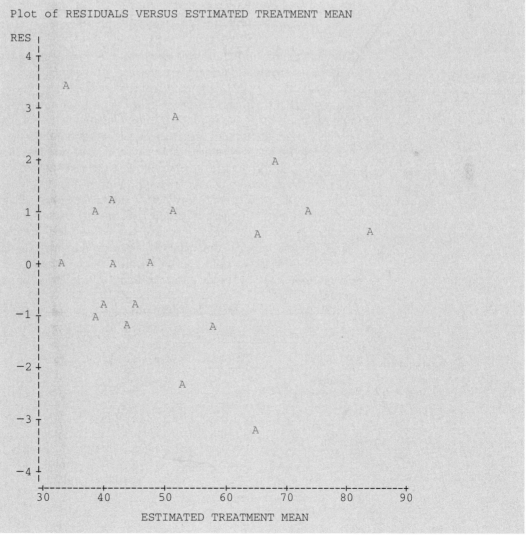

9.3 Latin Square Design

The randomized complete block design is used when there is one factor of interest and the experimenter wants to control a single source of extraneous variation. When there are two possible sources of extraneous variation, a **Latin square design** is the appropriate design for the experiment. Consider the following example.

A large law firm is studying which of four spreadsheets would be most appropriate for their secretarial pool. All four spreadsheets are from nationally known software companies and hence are acceptable to the company with respect to quality of their output. The final choice will thus be determined by which of the four is the easiest to learn. The software consultant for the firm notes that the time to learn various subroutines in a spreadsheet depends on the individual secretary and the type of problem being analyzed. Because the secretaries involved in the study will be unavailable for normal work, the law firm decides to use only four secretaries from the pool and restrict the time that they can be away from their regular assignments. Thus the consultant decides to have each secretary complete a task involving one of four types of problems. The factors to be considered in the study are

> Spreadsheet: A, B, C, D
> Secretary: 1, 2, 3, 4
> Problem: I (accounting), II (data tables), III (summary statistics), IV (graphics)

The factors secretary and problem type are extraneous sources of variation that must be taken into account but are not of central importance to the consultant. The response variable will be the length of time required for the secretary to complete the assigned task. Each secretary will be assigned four problems. The consultant at first considers using the randomized complete block design displayed in Table 9.9. In this design, the type of spreadsheet used is randomly assigned to the problems separately for each secretary. Suppose the type of problem has a strong influence on the amount of time needed to complete the task. In particular, problem type I is by far the most time-consuming of the four problems, whereas problem type IV is the least time-consuming. This design would then produce a strong negative bias for spreadsheet A because it was applied three times using type I problems and a positive bias to spreadsheet C because it was used three times on type IV problems. Thus, if it is found that spreadsheet C produces the shortest mean completion time for the four tasks, we could not be certain whether spreadsheet C was the better program or whether the results were due to three of its four tasks being type IV problems.

TABLE 9.9
A randomized complete block design for the spreadsheet study

Problem	Secretary			
	1	2	3	4
I	A	A	C	A
II	B	D	A	D
III	D	B	D	B
IV	C	C	B	C

TABLE 9.10
A Latin square design for the spreadsheet study

		Secretary		
Problem	**1**	**2**	**3**	**4**
I	A	B	C	D
II	B	C	D	A
III	C	D	A	B
IV	D	A	B	C

This example illustrates a situation in which the experimental units (problem) are affected by two sources of extraneous variation, the type of problem and the secretary solving the problem. We can modify the randomized complete block design to filter out the first source of variability, the variability among problem types, in addition to filtering out the second source, variability among secretaries. To do this, we restrict our randomization to ensure that each treatment appears in each row (problem type) and in each column (secretary). One such randomization is shown in Table 9.10. Note that the spreadsheets have been assigned to problem types and to secretaries so that each spreadsheet is applied once to each of the problem types and once to each of the secretaries. Hence, pairwise comparisons among spreadsheets that involve the sample means have been adjusted for the variability among problem types and secretaries.

Latin square design
This experimental design is called a **Latin square design.** In general, a Latin square design can be used to compare t treatment means in the presence of two extraneous sources of variability, which we block off into t rows and t columns. The t treatments are then randomly assigned to the rows and columns so that each treatment appears in every row and every column of the design (see Table 9.10).

The advantages and disadvantages of the Latin square design are listed here.

Advantages and Disadvantages of the Latin Square Design

Advantages

1. The design is particularly appropriate for comparing t treatment means in the presence of two sources of extraneous variation, each measured at t levels.
2. The analysis is quite simple.

Disadvantages

1. Although a Latin square can be constructed for any value of t, it is best suited for comparing t treatments when $5 \le t \le 10$.
2. Any additional extraneous sources of variability tend to inflate the error term, making it more difficult to detect differences among the treatment means.
3. The effect of each treatment on the response must be approximately the same across rows and columns.

The definition of a Latin square design is given here.

DEFINITION 9.2

A $t \times t$ **Latin square design** contains t rows and t columns. The t treatments are randomly assigned to experimental units within the rows and columns so that each treatment appears in every row and in every column.

The model for a response in a Latin square design can be written in the form

$$y_{ijk} = \mu + \alpha_k + \beta_i + \gamma_j + \varepsilon_{ijk}$$

where the terms of the model are defined as follows:

y_{ijk}: Observation on experimental unit in the ith row and jth column receiving treatment k.

μ: Overall mean, an unknown constant.

α_k: An effect due to treatment k, an unknown constant.

β_i: An effect due to row i, an unknown constant.

γ_j: An effect due to column j, an unknown constant.

ε_{ijk}: A random error associated with the response from an experimental unit in row i, column j, and treatment k. We require that the ε_{ijk}s have a normal distribution with mean 0 and common variance σ_ε^2. In addition, the errors must be independent.

The conditions given for our model can be shown to imply that the recorded response in the ith row and ith column, y_{ijk}, is normally distributed with mean

$$E(y_{ijk}) = \mu + \alpha_k + \beta_i + \gamma_j$$

and variance σ_ε^2.

filtering   We can use the model to illustrate how a Latin square design **filters** out extraneous variability due to row and column sources of variability. To illustrate, we will consider a Latin square design with $t = 4$ treatments (I, II, III, IV) and two sources of extraneous variability, each with $t = 4$ levels. This design is displayed in Table 9.11. If we wish to estimate $\alpha_3 - \alpha_1$, the difference in the mean response for treatments III and I, using the difference in sample means $\bar{y}_{..3} - \bar{y}_{..1}$, we can substitute into our model to obtain expressions for $\bar{y}_{..3}$ and $\bar{y}_{..1}$, carefully noting in which rows and columns the treatments appear. With y_{ijk} denoting the observation in row i, column j, and with treatment k, we have, from Table 9.11,

TABLE 9.11

A 4×4 Latin square design

	Column			
Row	**1**	**2**	**3**	**4**
1	I	II	III	IV
2	II	III	IV	I
3	III	IV	I	II
4	IV	I	II	III

$$\bar{y}_{..1} = \frac{1}{4}(y_{111} + y_{421} + y_{331} + y_{241})$$

$$= \mu + \alpha_1 + \frac{1}{4}(\beta_1 + \beta_2 + \beta_3 + \beta_4) + \frac{1}{4}(\gamma_1 + \gamma_2 + \gamma_3 + \gamma_4) + \bar{\varepsilon}_{..1}$$

where $\bar{\varepsilon}_{..1}$ is the mean of the random errors for the four observations on treatment I. Similarly,

$$\bar{y}_{..3} = \frac{1}{4}(y_{313} + y_{223} + y_{133} + y_{443})$$

$$= \mu + \alpha_3 + \frac{1}{4}(\beta_1 + \beta_2 + \beta_3 + \beta_4) + \frac{1}{4}(\gamma_1 + \gamma_2 + \gamma_3 + \gamma_4) + \bar{\varepsilon}_{..3}$$

Then the sample difference is

$$\bar{y}_{..3} - \bar{y}_{..1} = \alpha_3 - \alpha_1 + (\bar{\varepsilon}_{..3} - \bar{\varepsilon}_{..1})$$

and the error of estimation for $\alpha_3 - \alpha_1$ is $(\bar{\varepsilon}_{..3} - \bar{\varepsilon}_{..1})$

If a randomized block design had been used with blocks representing rows, treatments would be randomized within the rows only. It is quite possible for the same treatment to appear more than once in the same column. Then the sample difference would be

$$\bar{y}_{..3} - \bar{y}_{..1} = \alpha_3 - \alpha_1 + [(\bar{\varepsilon}_{..3} - \bar{\varepsilon}_{..1}) + (\text{column effects that do not cancel})]$$

Thus, the error of estimation would be inflated by the column effects that do not cancel out. Following the same reasoning, if a completely randomized design was used when a Latin square design was appropriate, the error of estimation would be inflated by both row and column effects that do not cancel out.

test for treatment effects   We can **test specific hypotheses concerning the parameters in our model.** In particular, we may wish to test the hypothesis of no difference among the t treatment means. This hypothesis can be stated in the form

$$H_0: \quad \alpha_1 = \alpha_2 = \cdots = \alpha_t = 0 \text{ (i.e., the } t \text{ treatment means are identical)}$$

The alternative hypothesis is

$$H_a: \quad \text{At least one of the } \alpha_i\text{s is not equal to zero (i.e., at least one treatment mean is different from the others).}$$

Our test statistic will be obtained by examining the model for a Latin square design and partitioning the total sum of squares to include terms for treatment effects, row effects, column effects, and random error effects.

total sum of squares   The **total sum of squares** of the measurements about their mean $\bar{y}_{...}$ is defined as before:

$$\text{TSS} = \sum_{ij} (y_{ijk} - \bar{y}_{...})^2$$

This sum of squares will be partitioned into four separate sources of variability: one due to the variability among treatments, one due to the variability among rows, one due to the variability among columns, and one due to the variability from all sources

error

partition of TSS

not accounted for by either treatment differences or block differences. We call this source of variability **error.** The **partition of TSS** follows from an examination of the Latin square model:

$$y_{ijk} = \mu + \alpha_k + \beta_i + \gamma_j + \varepsilon_{ijk}$$

The parameters in the model have sample estimates:

$$\hat{\mu} = \bar{y}_{...} \qquad \hat{\alpha}_k = \bar{y}_{..k} - \bar{y}_{...} \qquad \hat{\beta}_i = \bar{y}_{i..} - \bar{y}_{...} \qquad \hat{\gamma}_j = \bar{y}_{.j.} - \bar{y}_{...}$$

$$\text{TSS} = t \sum_k (\bar{y}_{..k} - \bar{y}_{...})^2 + t \sum_i (\bar{y}_{i..} - \bar{y}_{...})^2 + t \sum_j (\bar{y}_{.j.} - \bar{y}_{...})^2 + \text{SSE}$$

We will interpret the terms in the partition using the parameter estimates. The first quantity on the right-hand side of the equal sign measures the variability of the treatment means $\bar{y}_{..k}$ from the overall mean $\bar{y}_{...}$. Thus,

$$\text{SST} = t \sum_k (\bar{y}_{..k} - \bar{y}_{...})^2 = t \sum_k (\hat{\alpha}_k)^2$$

between-treatment sum of squares

called the **between-treatment sum of squares,** is a measure of the variability in the y_{ijk}s due to differences in the treatment means. The second quantity,

$$\text{SSR} = t \sum_i (\bar{y}_{i..} - \bar{y}_{...})^2 = t \sum_i (\hat{\beta}_i)^2$$

between-rows sum of squares

between-columns sum of squares

measures the variability between the row means $\bar{y}_{i..}$ and the overall mean. It is called the **between-rows sum of squares.** The third source of variability, referred to as the **between-columns sum of squares,** measures the variability between the column means $\bar{y}_{.j.}$ and the overall mean. It is given by

$$\text{SSC} = t \sum_j (\bar{y}_{.j.} - \bar{y}_{...})^2 = t \sum_j (\hat{\gamma}_j)^2$$

sum of squares for error

The final source of variability, designated as the **sum of squares for error,** SSE, represents the variability in the y_{ijk}s not accounted for by the row, column, and treatment differences. It is given by

$$\text{SSE} = \text{TSS} - \text{SST} - \text{SSR} - \text{SSC}$$

We can summarize our calculations in an AOV table as given in Table 9.12. The test statistic for testing

$$H_0: \quad \alpha_1 = \alpha_2 = \cdots = \alpha_t = 0 \qquad H_a: \quad \text{At least one } \alpha_k \text{ is different from } 0$$

is the ratio

$$F = \frac{\text{MST}}{\text{MSE}}$$

TABLE 9.12

Analysis of variance table for a $t \times t$ Latin square design

Source	SS	df	MS	F
Treatments	SST	$t - 1$	$\text{MST} = \text{SST}/(t - 1)$	MST/MSE
Rows	SSR	$t - 1$	$\text{MSR} = \text{SSR}/(t - 1)$	MSR/MSE
Columns	SSC	$t - 1$	$\text{MSC} = \text{SSC}/(t - 1)$	MSC/MSE
Error	SSE	$(t - 1)(t - 2)$	$\text{MSE} = \text{SSE}/(t - 1)(t - 2)$	
Total	TSS	$t^2 - 1$		

For our model,

$$E(\text{MSE}) = \sigma_\varepsilon^2 \quad \text{and} \quad E(\text{MST}) = \sigma_\varepsilon^2 + t\theta_T$$

where $\theta_T = \sum \alpha_k^2/(t-1)$. When H_0 is true, α_k equals 0 for all $k = 1, \ldots, t$, and hence $\theta_T = 0$. Thus, when H_0 is true we expect MST/MSE to be close to 1. However, under the research hypothesis, H_a, θ_T will be positive because at least one α_k is not 0. Thus, a large difference in the treatment means will result in a large value for θ_T. The expected value of MST will then be larger than the expected value of MSE and we expect $F = \text{MST/MSE}$ to be larger than 1. Thus, our test statistic F rejects H_0 when we observe a value of F larger than a value in the upper tail of the F distribution.

This discussion leads to the following decision rule for a specified probability of a Type I error:

Reject H_0: $\alpha_1 = \alpha_2 = \cdots = \alpha_t = 0$ when $F = \text{MST/MSE}$ exceeds $F_{a,\text{df}_1,\text{df}_2}$

where $F_{a,\text{df}_1,\text{df}_2}$ is from the F tables of Appendix Table 9 with $a =$ specified value of probability Type I error, $\text{df}_1 = \text{df}_{\text{MST}} = t - 1$, and $\text{df}_2 = \text{df}_{\text{MSE}} = (t-1)(t-2)$. Alternatively, we can compute the p-value for the observed value of the test statistic F_{obs} by computing

$$p\text{-value} = P(F_{\text{df}_1,\text{df}_2} > F_{\text{obs}})$$

where the F-distribution with $\text{df}_1 = t - 1$ and $\text{df}_2 = (t-1)(t-2)$ is used to compute the probability. We can then compare the p-value to a selected value for the probability of Type I error, with small p-values supporting the research hypothesis and large values of the p-value failing to reject H_0.

The row and column effects are generally assessed only to determine whether accounting for the two extraneous sources of variability was efficient in reducing the variability in the experimental units. Thus, hypotheses about the row and column effects are not generally tested. As with the randomized block design, we can compare the efficiency to that of the completely randomized design. We want to determine whether accounting for the row and column sources of variability has increased our precision for comparing treatment means in a given experiment. Let MSE_{LS} and MSE_{CR} denote the mean square errors for a Latin square design and a completely randomized design, respectively. The **relative efficiency** of the Latin square design compared to a completely randomized design is denoted **RE(LS, CR).** We can use the mean squares from the Latin square design, MSR, MSC, and MSE, to obtain the relative efficiency RE(LS, CR) by using the formula

$$\text{RE(LS, CR)} = \frac{\text{MSE}_{\text{CR}}}{\text{MSE}_{\text{LS}}} = \frac{\text{MSR} + \text{MSC} + (t-1)\text{MSE}}{(t+1)\text{MSE}}$$

margin notes: **relative efficiency**   **RE(LS, CR)**

When RE(LS, CR) is much larger than 1, we conclude that accounting for the row and/or column sources of variability was efficient, because many more observations would be required in a completely randomized design than would be required in Latin square design to obtain the same degree of precision in estimating the treatment means.

EXAMPLE 9.3

The law firm conducted a study of which spreadsheet to implement, and the data are shown in Table 9.13. Use these data to answer the following questions.

Problem	Secretary				Row Mean	Spreadsheet Mean
	1	**2**	**3**	**4**		
I	A(.3)	B(1.8)	C(.7)	D(1.2)	1.0	A: .45
II	B(1.4)	C(1.4)	D(1.1)	A(.5)	1.1	B: 1.5
III	C(.5)	D(1.5)	A(.5)	B(1.1)	.9	C: 1.05
IV	D(1.0)	A(.5)	B(1.7)	C(1.6)	1.2	D: 1.2
Column Mean	.8	1.3	1.0	1.1	1.05	

a. Write an appropriate statistical model for this experimental situation.
b. Run an analysis of variance to compare the mean time to completion for the spreadsheets. Use $\alpha = .05$.
c. Summarize your results in an AOV table.
d. Compute the relative efficiency of the Latin square design relative to a completely randomized design.

Solution We recognize this experimental design as a Latin square design with $t = 4$ rows (problems), $t = 4$ columns (secretaries), and $t = 4$ treatments (spreadsheets). The appropriate statistical model is

$$y_{ijk} = \mu + \alpha_k + \beta_i + \gamma_j + \varepsilon_{ijk} \qquad i, j, k = 1, 2, 3, 4$$

From the information in Table 9.13, we can estimate the treatment means μ_k by $\hat{\mu}_k = \bar{y}_{..k}$ yielding

$$\hat{\mu}_1 = .45 \qquad \hat{\mu}_2 = 1.5 \qquad \hat{\mu}_3 = 1.05 \qquad \hat{\mu}_4 = 1.2$$

It appears that spreadsheet A has a somewhat shorter mean time to completion than the other three spreadsheets. We will next estimate the model parameters and construct the AOV table. With $\hat{\mu} = \bar{y}_{...} = 1.05$, we obtain

Spreadsheet Effects

$\hat{\alpha}_1 = .45 - 1.05 = -.6$
$\hat{\alpha}_2 = 1.5 - 1.05 = .45$
$\hat{\alpha}_3 = 1.05 - 1.05 = 0$
$\hat{\alpha}_4 = 1.2 - 1.05 = .15$

Problem Effects

$\hat{\beta}_1 = 1 - 1.05 = -.05$
$\hat{\beta}_2 = 1.1 - 1.05 = .05$
$\hat{\beta}_3 = .9 - 1.05 = -.15$
$\hat{\beta}_4 = 1.2 - 1.05 = .15$

Secretary Effects

$\hat{\gamma}_1 = .8 - 1.05 = -.25$
$\hat{\gamma}_2 = 1.3 - 1.05 = .25$
$\hat{\gamma}_3 = 1 - 1.05 = -.05$
$\hat{\gamma}_4 = 1.1 - 1.05 = .05$

Substituting into the formulas for the sum of squares, we have

$$\text{TSS} = \sum_{ij} (y_{ijk} - \bar{y}_{...})^2 = (.3 - 1.05)^2 + (1.8 - 1.05)^2 + \cdots + (1.6 - 1.05)^2$$

$$= 3.66$$

$$\text{SST} = t \sum_{k} (\hat{\alpha}_k)^2 = 4[(-.6)^2 + (.45)^2 + (0)^2 + (.15)^2] = 2.34$$

$$\text{SSR} = t \sum_{i} (\hat{\beta}_i)^2 = 4[(-.05)^2 + (.05)^2 + (-.15)^2 + (.15)^2] = .2$$

$$\text{SSC} = t \sum_{j} (\hat{\gamma}_j)^2 = 4[(-.25)^2 + (.25)^2 + (-.05)^2 + (.05)^2] = .52$$

TABLE 9.14

AOV table for the spreadsheet
study in Example 9.4

Source	SS	df	MS	F	p-value
Spreadsheets	2.34	3	.78	7.8	.0171
Problem	.2	3	.067	.67	.6025
Secretary	.52	3	.173	1.73	.2592
Error	.6	6	.1		
Total	3.66	15			

By subtraction, we have

$$SSE = TSS - SST - SSB = 3.66 - 2.34 - .2 - .52 = .6$$

The analysis of variance table in Table 9.14 summarizes our results. Note that the mean squares for a source in the AOV table is computed by dividing the sum of squares for that source by its degrees of freedom.

The F test for differences in the treatment means—namely,

$$H_0: \alpha_1 = \alpha_2 = \cdots = \alpha_t = 0 \quad \text{vs.} \quad H_a: \text{At least one } \alpha_k \text{ is different from } 0$$

makes use of the F statistic MST/MSE. Because the computed value of F, 7.8, is greater than the tabulated F-value, based on $df_1 = 3$, $df_2 = 6$, and $a = .05$, we reject H_0 and conclude that there are significant (p-value $= .0171$) differences in the mean completion times of the four spreadsheets. It appears that spreadsheet A has a considerably shorter mean completion time than the other three spreadsheets.

Next we will assess whether taking into account the two extraneous sources of variation was effective in increasing the precision of the analysis relative to a completely randomized design. From the AOV table, we have MSR $= .067$, MSC $= .173$, and MSE $= .1$. Hence, the relative efficiency of this Latin square design relative to a completely randomized design is

$$RE(LS, CR) = \frac{MSR + MSC + (t-1)MSE}{(t+1)MSE}$$

$$= \frac{.067 + .173 + (4-1)(.1)}{(4+1)(.1)} = 1.08$$

That is, approximately 8% more observations per each treatment would be required in a completely randomized design to obtain the same precision for estimating the treatment means as with this Latin square design. The Latin square did not improve the precision of estimation that much because there was very little difference in the completion times for the different types of problems. There was a somewhat larger difference in the mean completion times for the secretaries, but these differences did not contribute a large variability to the overall study results.

The results in Example 9.3 are valid only if we can be assured the conditions placed on the model are consistent with the observed data. Thus, we use the residuals $e_{ijk} = y_{ijk} - \hat{\mu} - \hat{\alpha}_k - \hat{\beta}_i - \hat{\gamma}_j$ to assess whether the conditions of normality, equal variance, and independence appear to be satisfied for the observed data. The following example will display the computer output for such an analysis.

EXAMPLE 9.4

The computer output for the experiment described in Example 9.3 is displayed here. Compare the results to those obtained using the definition of the sum of squares and assess whether the model conditions appear to be valid.

Solution   Note that our hand calculations yielded nearly the same values as are given in the computer output. The differences are due to rounding errors. There can be large rounding errors in our hand calculations, which can lead to results that will differ from the values given in the computer output. It is strongly recommended that a computer software program be used in the analysis of variance calculations because of the potential for rounding errors. In assessing whether the model conditions have been met, we first note that in regard to the normality condition the test of H_0: residuals have normal distribution, the p-value from the Shapiro–Wilks test is p-value = .9098. Thus, we do not reject H_0 and the normality condition appears to be satisfied. Also, the stem-and-leaf plot, boxplot, and normal probability plot are also consistent with the condition that the residuals have a normal distribution. Figure 9.3 is a plot of the residuals versus the estimated treatment means. From this plot it appears that the variability in the residuals is somewhat constant across the treatments.

FIGURE 9.3

Residuals versus treatment means from Example 9.4

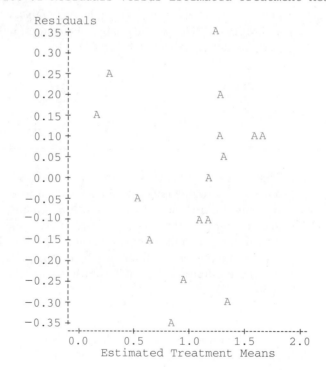

General Linear Models Procedure for Example 9.4

Dependent Variable: TIME TO COMPLETION

Source	DF	Sum of Squares	Mean Square	F value	Pr > F
Model	9	3.0600000	0.3400000	3.40	0.0751
Error	6	0.6000000	0.1000000		
Corrected Total	15	3.6600000			

	R-Square	C.V.	Root MSE		Y Mean
	0.836066	30.11693	0.3162		1.0500

Source	DF	Type I SS	Mean Square	F Value	Pr > F
R	3	0.2000000	0.0666667	0.67	0.6025
C	3	0.5200000	0.1733333	1.73	0.2592
T	3	2.3400000	0.7800000	7.80	0.0171

RESIDUAL ANALYSIS:

Variable-RESIDUAL

Moments

N	16	Sum Wgts	16
Mean	0	Sum	0
Std Dev	0.2	Variance	0.04
Skewness	-0.17143	Kurtosis	-0.6544
W:Normal	0.97711	Pr<W	0.9098

Variable=RESIDUALS

```
    Stem Leaf                      #           Boxplot
      3 5                          1              |
      2 05                         2              |
      1 0005                       4           +-----+
      0 05                         2           *--+--*
     -0 5                          1           |     |
     -1 500                        3           +-----+
     -2 5                          1              |
     -3 50                         2              |
        ----+----+----+----+
    Multiply Stem.Leaf by 10**-1
```

Normal Probability Plot

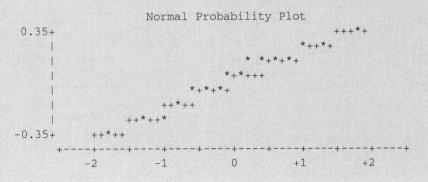

EXERCISES   Applications

Ag.   **9.9** An experiment compared two different fertilizer placements (broadcast, band) and two different rates of fertilizer flow on watermelon yields. Recent research has shown that broadcast application (scattering over the outer area) of fertilizer is superior to bands of fertilizer applied near
the seed for watermelon yields. For this experiment the investigators wished to compare two nitrogen–phosphorus–potassium (broadcast and band) fertilizers applied at a rate of 160–70–135
pounds per acre and two brands of micronutrients (A and B). These four combinations were to be
studied in a Latin square field plot.

   The treatments were randomly assigned according to a Latin square design conducted over
a large farm plot, which was divided into rows and columns. A watermelon plant dry weight was
obtained for each row–column combination 30 days after the emergence of the plants. The data
are shown next.

		Column						
Row	**1**		**2**		**3**		**4**	
1	1	1.75	3	1.43	4	1.28	2	1.66
2	2	1.70	1	1.78	3	1.40	4	1.31
3	4	1.35	2	1.73	1	1.69	3	1.41
4	3	1.45	4	1.36	2	1.65	1	1.73

Treatment 1 broadcast, A   Treatment 3 band, A
Treatment 2 broadcast, B   Treatment 4 band, B

   a. Write an appropriate statistical model for this experiment.
   b. Use the data to run an analysis of variance. Give the p-value for each test and draw
conclusions.

Engin.   **9.10** A petroleum company was interested in comparing the miles per gallon achieved by four
different gasoline blends (A, B, C, and D). Because there can be considerable variability due to
differences in driving characteristics and car models, these two extraneous sources of variability
were included as blocking variables in the study. The researcher selected four different brands of
cars and four different drivers. The drivers and brands of cars were assigned to blends in the manner displayed in the following table. The mileage (in mpg) obtained over each test run was recorded
as follows.

	Car Model			
Driver	**1**	**2**	**3**	**4**
1	A(15.5)	B(33.8)	C(13.7)	D(29.2)
2	B(16.3)	C(26.4)	D(19.1)	A(22.5)
3	C(10.5)	D(31.5)	A(17.5)	B(30.1)
4	D(14.0)	A(34.5)	B(19.7)	C(21.6)

   a. Write a model for this experimental setting.
   b. Estimate the parameters in the model.
   c. Conduct an analysis of variance. Use $\alpha = .05$.
   d. What conclusions can you draw concerning the best gasoline blend?

e. Compute the relative efficiency of the Latin square design relative to a completely randomized design. Interpret this value. Were the blocking variables effective in reducing the variability in experimental units? Explain.

f. If future studies were to be conducted, would you recommend using both car model and driver as blocking variables? Explain.

9.11 Refer to the following computer output for the data of Exercise 9.10.
a. Compare the results to those of Exercise 9.10.
b. Do the model conditions appear to be satisfied for this set of data? Explain.

OBS	R	C	T	Y
1	1	1	1	15.5
2	1	2	2	33.8
3	1	3	3	13.7
4	1	4	4	29.2
5	2	1	2	16.3
6	2	2	3	26.4
7	2	3	4	19.1
8	2	4	1	22.5
9	3	1	3	10.5
10	3	2	4	31.5
11	3	3	1	17.5
12	3	4	2	30.1
13	4	1	4	14.0
14	4	2	1	34.5
15	4	3	2	19.7
16	4	4	3	21.6

General Linear Models Procedure
Dependent Variable: MILES PER GALLON

Source	DF	Sum of Squares	Mean Square	F Value	Pr > F
Model	9	869.97563	96.66396	22.42	0.0006
Error	6	25.86375	4.31062		
Corrected Total	15	895.83937			

R-Square	C.V.	Root MSE	Y Mean
0.971129	9.333878	2.0762	22.244

Source	DF	Type I SS	Mean Square	F Value	Pr > F
R	3	8.33187	2.77729	0.64	0.6143
C	3	755.37188	251.79063	58.41	0.0001
T	3	106.27188	35.42396	8.22	0.0151

RESIDUALS ANALYSIS:
Variable = RESIDUALS

Moments

N	16	Sum Wgts	16
Mean	0	Sum	0
Std Dev	1.313107	Variance	1.72425
Skewness	0.043408	Kurtosis	-0.41
W:Normal	0.985867	Pr<W	0.9840

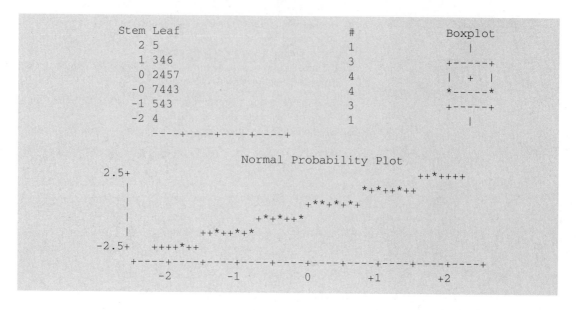

```
    Stem Leaf                              #        Boxplot
       2  5                                1           |
       1  346                              3        +-----+
       0  2457                             4        |  +  |
      -0  7443                             4        *-----*
      -1  543                              3        +-----+
      -2  4                                1           |
          ----+----+----+----+
```

```
                  Normal Probability Plot
     2.5+                                              ++*++++
        |                                         *+*++*++
        |                                     +**+*+*+
        |                               +*+*++*
        |                         ++*++*+*
    -2.5+    ++++*++
         +----+----+----+----+----+----+----+----+----+----+
              -2        -1         0        +1        +2
```

PLOT OF RESIDUALS VERSUS ESTIMATED TREATMENT MEANS FOR EXERCISE 9.10

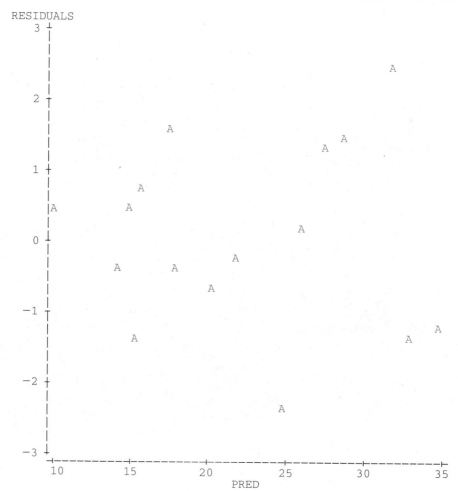

9.4 Factorial Treatment Structure in a Completely Randomized Design

For the three designs discussed in detail—the completely randomized design (Chapter 8), the randomized complete block design (Section 9.2), and the Latin square (Section 9.3)—we were concerned with the randomization technique and controlling extraneous sources of variation through the use of blocking variables. The completely randomized design was used to compare t treatments using homogeneous experimental units. Thus, there were no constraints on the randomization of treatments to experimental units. In the randomized complete block design and Latin square design, the experimental units are not homogeneous and must be grouped into sets (blocks) of homogeneous experimental units prior to randomly assigning the treatments. This results in a constraint on the randomization that is not present in the completely randomized design. The randomized complete block design involves identifying a single characteristic (blocking variable) of the experimental units, whereas the Latin square design allows the experimenter to use two different characteristics of the experimental units in creating the sets of homogeneous experimental units.

In this section, we will discuss the treatment design; that is, how treatments are constructed from several factors rather than just being t levels of a single factor. These types of experiments involve examining the effect of two or more independent variables on a response variable y.

The model for an observation in a completely randomized design with a two-factor factorial treatment structure and $n > 1$ replications can be written in the form

$$y_{ijk} = \mu + \alpha_i + \beta_j + \alpha\beta_{ij} + \varepsilon_{ijk}$$

where the terms of the model are defined as follows:

y_{ijk}: The response from the kth experimental unit receiving the ith level of factor A and the jth level of factor B.

μ: Overall mean, an unknown constant.

α_i: An effect due to the ith level of factor A, an unknown constant.

β_j: An effect due to the jth level of factor B, an unknown constant.

$\alpha\beta_{ij}$: An interaction effect of the ith level of factor A with the jth level of factor B, an unknown constant.

ε_{ijk}: A random error associated with the response from the kth experimental unit receiving the ith level of factor A combined with the jth level of factor B. We require that the ε_{ij}s have a normal distribution with mean 0 and common variance σ_ε^2. In addition, the errors must be independent.

The conditions given above for our model can be shown to imply that the recorded response from the kth experimental unit receiving the ith level of factor A combined with the jth level of factor B is normally distributed with mean

$$E(y_{ijk}) = \mu + \alpha_i + \beta_j + \alpha\beta_{ij}$$

and variance σ_ε^2.

	Factor B	
Factor A	**Level 1**	**Level 2**
Level 1	$\mu + \alpha_1 + \beta_1$	$\mu + \alpha_1 + \beta_2$
Level 2	$\mu + \alpha_2 + \beta_1$	$\mu + \alpha_2 + \beta_2$

To illustrate this model, consider the model for a two-factor factorial experiment with no interaction:

$$y_{ijk} = \mu + \alpha_i + \beta_j + \varepsilon_{ijk}$$

Expected values for a 2×2 factorial experiment are shown in Table 9.15.

This model assumes that the difference in population means (expected values) for any two levels of factor A is the same no matter what level of B we are considering. The same property holds when comparing two levels of factor B. For example, the difference in mean response for levels 1 and 2 of factor A is the same value, $\alpha_1 - \alpha_2$, no matter what level of factor B we are considering. Thus, a test for no differences among the two levels of factor A is of the form $H_0: \alpha_1 - \alpha_2 = 0$. Similarly, the difference between levels of factor B is $\beta_1 - \beta_2$ for either level of factor A, and a test of no difference between the factor B means is $H_0: \beta_1 - \beta_2 = 0$. This phenomenon was also noted for the randomized block design.

If the assumption of additivity of terms in the model does not hold, then we need a model that employs terms to account for **interaction.**

interaction

The expected values for a 2×2 factorial experiment with n observations per cell are presented in Table 9.16.

As can be seen from Table 9.16, the difference in mean response for levels 1 and 2 of factor A on level 1 of factor B is

$$(\alpha_1 - \alpha_2) + (\alpha\beta_{11} - \alpha\beta_{21})$$

but for level 2 of factor B this difference is

$$(\alpha_1 - \alpha_2) + (\alpha\beta_{12} - \alpha\beta_{22})$$

Because the difference in mean response for levels 1 and 2 of factor A is *not* the same for different levels of factor B, the model is no longer additive, and we say that the two factors interact.

DEFINITION 9.3

Two factors A and B are said to **interact** if the difference in mean responses for two levels of one factor is not constant across levels of the second factor.

	Factor B	
Factor A	**Level 1**	**Level 2**
Level 1	$\mu + \alpha_1 + \beta_1 + \alpha\beta_{11}$	$\mu + \alpha_1 + \beta_2 + \alpha\beta_{12}$
Level 2	$\mu + \alpha_2 + \beta_1 + \alpha\beta_{21}$	$\mu + \alpha_2 + \beta_2 + \alpha\beta_{22}$

In measuring the octane rating of gasoline, interaction can occur when two components of the blend are combined to form a gasoline mixture. The octane properties of the blended mixture may be quite different than would be expected by examining each component of the mixture. Interaction in this situation could have either a positive or negative effect on the performance of the blend, in which case the components are said to either potentiate or antagonize one another.

profile plot   We can amplify the notion of an interaction with the **profile plots** shown in Figure 9.4. As we see from Figure 9.4(a), when no interaction is present, the difference in the mean response between levels 1 and 2 of factor B (as indicated by the braces) is the same for both levels of factor A. However, for the other two illustrations in Figure 9.4, we see that the difference between the levels of factor B changes from level 1 to level 2 of factor A. For these cases, we have an interaction between the two factors.

Note that an interaction is not restricted to two factors. With three factors A, B, and C, we might have an interaction between factors A and B, A and C, and B and C, and the two-factor interactions would have interpretations that follow immediately from Definition 9.4. Thus, the presence of an AC interaction indicates that the difference in mean response for levels of factor A varies across levels of factor C. A three-way interaction among factors A, B, and C might indicate that the difference in mean response for levels of C changes across combinations of levels for factors A and B.

FIGURE 9.4

Illustrations of the absence and presence of interaction in a 2 × 2 factorial experiment: (a) factors A and B do not interact; (b) factors A and B interact; (c) factors A and B interact

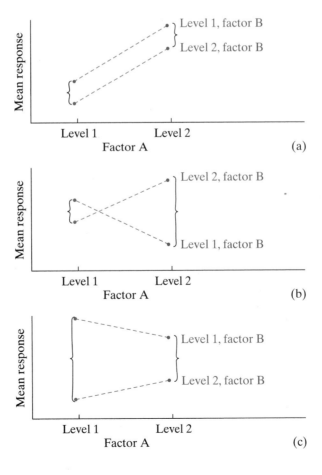

The analysis of variance for a factorial experiment with an interaction between the factors requires that we have $n > 1$ observations on each of the treatments (factor–level combinations). We will construct the analysis of variance table for a completely randomized two-factor experiment with a levels of factor A, b levels of factor B, and n observations on each of the ab treatments. Before partitioning the total sum of squares into its components we need the notation defined here.

y_{ijk}: Observation on the kth experimental unit receiving the ith level of factor A and jth level of factor B.

$\bar{y}_{i..}$: Sample mean for observations at the ith level of factor A,

$$\bar{y}_{i..} = \frac{1}{bn}\Sigma_{jk}\, y_{ijk}$$

$\bar{y}_{.j.}$: Sample mean for observations at the jth level of factor B,

$$\bar{y}_{.j.} = \frac{1}{an}\Sigma_{ik}\, y_{ijk}.$$

$\bar{y}_{ij.}$: Sample for observations at the ith level of factor A and the jth level of factor B, $\bar{y}_{ij.} = \frac{1}{n}\Sigma_k\, y_{ijk}.$

$\bar{y}_{...}$: Overall sample mean, $\bar{y}_{...} = \frac{1}{abn}\Sigma_{ijk}\, y_{ijk}.$

total sum of squares   The **total sum of squares** of the measurements about their mean $\bar{y}_{...}$ is defined as before:

$$\mathrm{TSS} = \sum_{ijk} (y_{ijk} - \bar{y}_{...})^2$$

This sum of squares will be partitioned into four sources of variability: two due to the main effects of factors A and B, one due to the interaction between factors A and B, and one due to the variability from all sources not accounted for by the main **error** effects and interaction. We call this source of variability **error**. The **partition of TSS** **partition of TSS** follows from an examination of the model:

$$y_{ijk} = \mu + \alpha_i + \beta_j + \alpha\beta_{ij} + \varepsilon_{ijk}$$

The parameters in the model have sample estimates:

$$\hat{\mu} = \bar{y}_{...} \qquad \hat{\alpha}_i = \bar{y}_{i..} - \bar{y}_{...} \qquad \hat{\beta}_j = \bar{y}_{.j.} - \bar{y}_{...} \qquad \widehat{\alpha\beta}_{ij} = (\bar{y}_{ij.} - \bar{y}_{...}) - \hat{\alpha}_i - \hat{\beta}_j$$
$$= \bar{y}_{ij.} - \bar{y}_{i..} - \bar{y}_{.j.} + \bar{y}_{...}$$

It can be shown algebraically that TSS takes the following form:

$$\sum_{ijk} (y_{ijk} - \bar{y}_{...})^2 = bn \sum_i (\bar{y}_{i..} - \bar{y}_{...})^2 + an \sum_j (\bar{y}_{.j.} - \bar{y}_{...})^2$$
$$+ n \sum_{ij} (\bar{y}_{ij.} - \bar{y}_{i..} - \bar{y}_{.j.} + \bar{y}_{...})^2 + \sum_{ijk} (y_{ijk} - \bar{y}_{ij.})^2$$

main effect of factor A   We will interpret the terms in the partition using the parameter estimates. The first quantity on the right-hand side of the equal sign measures the **main effect of factor A** and can be written as

TABLE 9.17

AOV table for a completely randomized two-factor factorial experiment

Source	SS	df	MS	F
Main Effect				
A	SSA	$a - 1$	MSA = SSA$/(a - 1)$	MSA/MSE
B	SSB	$b - 1$	MSB = SSB$/(b - 1)$	MSB/MSE
Interaction				
AB	SSAB	$(a - 1)(b - 1)$	MSAB = SSAB$/(a - 1)(b - 1)$	MSAB/MSE
Error	SSE	$ab(n - 1)$	MSE = SSE$/ab(n - 1)$	
Total	TSS	$abn - 1$		

$$SSA = bn \sum_i (\bar{y}_{i..} - \bar{y}_{...})^2 = bn \sum_i (\hat{\alpha}_i)^2$$

main effect of factor B

Similarly, the second quantity on the right-hand side of the equal sign measures the **main effect of factor B** and can be written as

$$SSB = an \sum_j (\bar{y}_{.j.} - \bar{y}_{...})^2 = an \sum_j (\hat{\beta}_j)^2$$

interaction effect of factors A and B

The third quantity measures the **interaction effect of factors A and B** and can be written as

$$SSAB = n \sum_{ij} (\bar{y}_{ij.} - \bar{y}_{i..} - \bar{y}_{.j.} + \bar{y}_{...})^2 = n \sum_{ij} (\widehat{\alpha\beta}_{ij})^2$$

sum of squares for error

The final term is the **sum of squares for error** (SSE) and represents the variability in the y'_{ijk}s not accounted for by the main effects and interaction effects. There are several forms for this term. Defining the residuals from the model as before, we have $e_{ijk} = y_{ijk} - \hat{\mu} - \hat{\alpha}_i - \hat{\beta}_j - \widehat{\alpha\beta}_{ij} = y_{ijk} - \bar{y}_{ij.}$. Therefore,

$$SSE = \sum_{ijk} (y_{ijk} - \bar{y}_{ij.})^2 = \sum_{ijk} (e_{ijk})^2$$

Alternatively, SSE = TSS − SST − SSB − SSAB. We summarize the partition of the sum of squares in the AOV table as given in Table 9.17. From the AOV table we observe that, if we have only one observation on each treatment, $n = 1$, then there are 0 degrees of freedom for error. Thus, if factors A and B interact and $n = 1$, there are no valid tests for interactions or main effects. However, if the factors do not interact, the interaction term can be used as the error term and we replace SSE with SSAB. However, it would be an exceedingly rare situation to run experiments with $n = 1$ because in most cases the researcher would not know prior to running the experiment whether or not factors A and B interact. Hence, in order to have valid tests for main effects and interactions, we need $n > 1$.

EXAMPLE 9.5

An experiment was conducted to determine the effects of four different pesticides on the yield of fruit from three different varieties (B_1, B_2, B_3) of a citrus tree. Eight trees from each variety were randomly selected from an orchard. The four pesticides were then randomly assigned to two trees of each variety and applications were made according to recommended levels. Yields of fruit (in bushels per tree) were obtained after the test period. The data appear in Table 9.18.

TABLE 9.18

Data for the 3 × 4 factorial experiment of fruit tree yield, $n = 2$ observations per treatment

		Pesticide, A			
Variety, B	**1**	**2**	**3**	**4**	
1	49	50	43	53	
	39	55	38	48	
2	55	67	53	85	
	41	58	42	73	
3	66	85	69	85	
	68	92	62	99	

a. Write an appropriate model for this experiment.

b. Set up an analysis of variance table and conduct the appropriate F tests of main effects and interactions using $\alpha = .05$.

c. Construct a plot of the treatment means, called a profile plot.

Solution The experiment described is a completely randomized 3 × 4 factorial experiment with factor A, pesticides, having $a = 4$ levels and factor B, variety, having $b = 3$ levels. There are $n = 2$ replications of the 12 factor–level combinations of the two factors.

a. The model for a 3 × 4 factorial experiment with interaction between the two factors is

$$y_{ijk} = \mu + \alpha_i + \beta_j + \alpha\beta_{ij} + \varepsilon_{ijk} \quad \text{for } i = 1, 2, 3, 4; j = 1, 2, 3; k = 1, 2$$

where μ is the overall mean yield per tree, α_i is the effect of the ith level of the pesticide, β_j is the effect of the jth variety of citrus tree, and $\alpha\beta_{ij}$ is the interaction effect of the ith level of pesticide with the jth variety of citrus tree.

b. In most experiments we strongly recommend using a computer software program to obtain the AOV table, but to illustrate the calculations we will construct the AOV for this example using the definitions of the individual sum of squares. First, we estimate the parameters in the model. To accomplish this we use the treatment means given in Table 9.19. Next, we obtain the parameter estimates:

Main Effects

$\hat{\alpha}_1 = \bar{y}_{1..} - \bar{y}_{...} = 53 - 61.46 = -8.46$

$\hat{\alpha}_2 = \bar{y}_{2..} - \bar{y}_{...} = 67.83 - 61.46 = 6.37$

$\hat{\alpha}_3 = \bar{y}_{3..} - \bar{y}_{...} = 51.17 - 61.46 = -10.29$

$\hat{\alpha}_4 = \bar{y}_{4..} - \bar{y}_{...} = 73.83 - 61.46 = 12.37$

Factor B

$\hat{\beta}_1 = \bar{y}_{.1.} - \bar{y}_{...} = 46.875 - 61.46 = -14.585$

$\hat{\beta}_2 = \bar{y}_{.2.} - \bar{y}_{...} = 59.25 - 61.46 = -2.21$

$\hat{\beta}_3 = \bar{y}_{.3.} - \bar{y}_{...} = 78.25 - 61.46 = 16.79$

TABLE 9.19

Sample means for factor–level combinations (treatments) of A and B

	Pesticide, A				
Variety, B	**1**	**2**	**3**	**4**	**Variety Means**
1	44	52.5	40.5	50.5	46.875
2	48	62.5	47.5	79	59.25
3	67	88.5	65.5	92	78.25
Pesticide Means	53	67.83	51.17	73.83	61.46

Interaction Effects

$$\widehat{\alpha\beta}_{11} = \bar{y}_{11.} - \bar{y}_{1..} - \bar{y}_{.1.} + \bar{y}_{...} = 44 - 53 - 46.875 + 61.46 = 5.585$$
$$\widehat{\alpha\beta}_{12} = \bar{y}_{12.} - \bar{y}_{1..} - \bar{y}_{.2.} + \bar{y}_{...} = 48 - 53 - 59.25 + 61.46 = -2.79$$
$$\widehat{\alpha\beta}_{13} = \bar{y}_{13.} - \bar{y}_{1..} - \bar{y}_{.3.} + \bar{y}_{...} = 67 - 53 - 78.25 + 61.46 = -2.79$$
$$\widehat{\alpha\beta}_{21} = \bar{y}_{21.} - \bar{y}_{2..} - \bar{y}_{.1.} + \bar{y}_{...} = 52.5 - 67.83 - 46.875 + 61.46 = -.745$$
$$\widehat{\alpha\beta}_{22} = \bar{y}_{22.} - \bar{y}_{2..} - \bar{y}_{.2.} + \bar{y}_{...} = 62.5 - 67.83 - 59.25 + 61.46 = -3.12$$
$$\widehat{\alpha\beta}_{23} = \bar{y}_{23.} - \bar{y}_{2..} - \bar{y}_{.3.} + \bar{y}_{...} = 88.5 - 67.83 - 78.25 + 61.46 = 3.88$$
$$\widehat{\alpha\beta}_{31} = \bar{y}_{31.} - \bar{y}_{3..} - \bar{y}_{.1.} + \bar{y}_{...} = 40.5 - 51.17 - 46.875 + 61.46 = 3.915$$
$$\widehat{\alpha\beta}_{32} = \bar{y}_{32.} - \bar{y}_{3..} - \bar{y}_{.2.} + \bar{y}_{...} = 47.5 - 51.17 - 59.25 + 61.46 = -1.46$$
$$\widehat{\alpha\beta}_{33} = \bar{y}_{33.} - \bar{y}_{3..} - \bar{y}_{.3.} + \bar{y}_{...} = 65.5 - 51.17 - 78.25 + 61.46 = -2.46$$
$$\widehat{\alpha\beta}_{41} = \bar{y}_{41.} - \bar{y}_{4..} - \bar{y}_{.1.} + \bar{y}_{...} = 50.5 - 73.83 - 46.875 + 61.46 = -8.745$$
$$\widehat{\alpha\beta}_{42} = \bar{y}_{42.} - \bar{y}_{4..} - \bar{y}_{.2.} + \bar{y}_{...} = 79 - 73.83 - 59.25 + 61.46 = 7.38$$
$$\widehat{\alpha\beta}_{43} = \bar{y}_{43.} - \bar{y}_{4..} - \bar{y}_{.3.} + \bar{y}_{...} = 92 - 73.83 - 78.25 + 61.46 = 1.38$$

We next calculate the total sum of squares. Note that there is rounding in the hand calculations, which results in the values for the sum of squares that are slightly different from the values obtained using a computer program.

$$TSS = \sum_{ijk} (y_{ijk} - \bar{y}_{...})^2 = (49 - 61.46)^2 + (50 - 61.46)^2 + \cdots + (99 - 61.46)^2$$
$$= 7{,}187.96$$

The main effect sum of squares is

$$SSA = bn \sum_i \hat{\alpha}_i^2 = (3)(2)[(-8.46)^2 + (6.37)^2 + (-10.29)^2 + (12.37)^2]$$
$$= 2{,}226.29$$

$$SSB = an \sum_i \hat{\beta}_j^2 = (4)(2)[(-14.585)^2 + (-2.21)^2 + (16.79)^2]$$
$$= 3{,}996.08$$

The interaction sum of squares is

$$SSAB = n \sum_i (\widehat{\alpha\beta}_{ij})^2 = (2)[(5.585)^2 + (-2.79)^2 + \cdots + (1.38)^2] = 456.92$$

The sum of squares error is obtained as

$$SSE = TSS - SSA - SSB - SSAB$$
$$= 7{,}187.96 - 2{,}226.29 - 3{,}996.08 - 456.92 = 508.67$$

The analysis of variance table for this completely randomized 3×4 factorial experiment with $n = 2$ replications per treatment is given in Table 9.20.

The first test of significance *must* be to test for an interaction between factors A and B, because if the interaction is significant then the main effects *may have no interpretation.* The F statistic is

$$F = \frac{MSAB}{MSE} = \frac{76.15}{42.39} = 1.80$$

TABLE 9.20

AOV table for fruit yield
experiment of Example 9.6

Source	SS	df	MS	F
Pesticide, A	2,226.29	3	742.10	17.51
Variety, B	3,996.08	2	1,998.04	47.13
Interaction, AB	456.92	6	76.15	1.80
Error	508.67	12	42.39	
Total	7,187.96	23		

FIGURE 9.5

Profile plot for fruit yield
experiment of Example 9.5

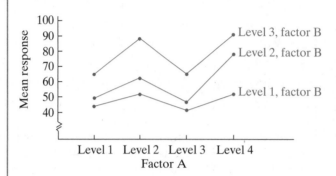

The computed value of F does not exceed the tabulated value of 3.00 for $a = .05$, $df_1 = 6$, $df_2 = 12$ in the F tables. Hence, we have insufficient evidence to indicate an interaction between pesticide levels and variety of trees levels. We can observe this lack of interaction by constructing a profile plot. Figure 9.5 contains a plot of the sample treatment means for this experiment.

From the profile plot we can observe that the differences in mean yields among the three varieties of citrus trees remain nearly constant across the four pesticide levels. That is, the three lines for the three varieties are nearly parallel lines and hence the interaction between the levels of variety and pesticide is not significant. Because the interaction is not significant, we can next test the main effects of the two factors. These tests separately examine the differences among the levels of variety and the levels of pecticides. For pesticides, the F statistic is

$$F = \frac{MSA}{MSE} = \frac{742.10}{42.39} = 17.51$$

The computed value of F does exceed the tabulated value of 3.49 for $a = .05$, $df_1 = 3$, and $df_2 = 12$ in the F tables. Hence, we have sufficient evidence to indicate a significant difference in the mean yields among the four pesticide levels. For varieties, the F statistic is

$$F = \frac{MSB}{MSE} = \frac{1,998.04}{42.39} = 47.13$$

The computed value of F does exceed the tabulated value of 3.89 for $a = .05$, $df_1 = 2$, and $df_2 = 12$ in the F tables. Hence, we have sufficient evidence to indicate a significant difference in the mean yields among the three varieties of citrus trees.

In Section 9.6, we will discuss how to explore which pairs of levels differ for both factors A and B.

FIGURE 9.6

Profile plot in which significant interactions are present, but interactions are orderly

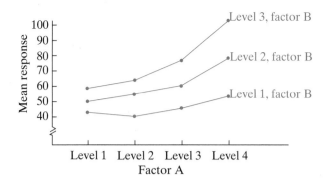

FIGURE 9.6

Profile plot in which significant interactions are present, but interactions are orderly

significant interaction

The results of an *F* test for main effects for factors A or B must be interpreted very carefully in the presence of a **significant interaction.** The first thing we do is to construct a profile plot using the sample treatment means, $\bar{y}_{ij}$. If the profile plot for Example 9.5 had appeared as shown in Figure 9.6, there would have been an indication of an interaction between factors A and B (pesticides and varieties). Provided that the MSE was not too large relative to MSAB, the *F* test for interaction would undoubtedly have been significant.

Would *F* tests for main effects have been appropriate for the profile plot in Figure 9.6? The answer is yes, because there is an *orderly* interaction; the *order* of the means for levels of factor B is always the same even though the *magnitude* of the differences between levels of factor B may change from level to level of factor A. Clearly, the profile plot in Figure 9.6 shows that the level 3 mean of factor B is always larger than the means for levels 1 and 2. Similarly, the level 2 mean for factor B is always larger than the mean for level 1 for factor B, no matter which level of factor A that we examine. However, we must be very careful in the conclusions we obtain from such a situation. If we find a significant difference in the levels of factor B, with mean response at level 3 larger than levels 1 and 2 of factor B across all levels of factor A, we may be led to conclude that level 3 of factor B produces significantly larger mean values than the other two levels of factor B. However, note that at level 1 of factor A, there is very little difference in the mean responses of the three levels of factor B. Thus, if we were to use level 1 of factor A, the three levels of factor B would produce equivalent mean responses. Thus, our conclusions about the differences in the mean responses among the levels of factor B are not consistent across the levels of factor A and may contradict the test for main effects of factor B at certain levels of factor A.

When the interaction is orderly, a test on main effects can be meaningful; however, not all interactions are orderly. The profile plot in Figure 9.7 shows a situation

FIGURE 9.7

Profile plot in which significant interactions are present, and interactions are disorderly

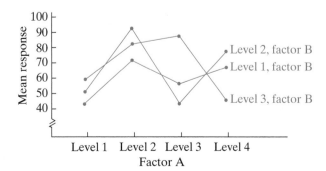

FIGURE 9.7

Profile plot in which significant interactions are present, and interactions are disorderly in which a test of main effects in the presence of a significant interaction might be misleading. A *disorderly* interaction, such as in Figure 9.7, can obscure the main effects. It is not that the tests are statistically incorrect; it is that they may lead to a misinterpretation of the results of the experiment. At level 1 of factor A, there is very little difference in the mean responses of the three levels of factor B. At level 3 of factor A, level 3 of factor B produces a much larger response than level 2 of factor B. In contradiction to this result, at level 4 of factor A, level 2 of factor B produces a much larger mean response than level 3 of factor B. Thus, when the two factors have significant interactions, conclusions about the differences in the mean responses among the levels of factor B must be made separately at *each level* of factor A. That is, a single conclusion about the levels of factor B does not hold for all levels of factor A.

When our experiment involves three factors, the calculations become considerably more complex. However, interpretations about main effects and interactions are similar to the interpretations when we have only two factors. With three factors A, B, and C, we might have an interaction between factors A and B, A and C, and B and C. The interpretations for these two-way interactions follow immediately from Definition 9.3. Thus, the presence of an AC interaction indicates that the differences in mean responses among the levels of factor A vary across the levels of factor C. The same care must be taken in making interpretations among main effects, as we discussed previously. A three-way interaction between factors A, B, and C might indicate that the differences in mean responses for levels of factor C change across combinations of levels for factors A and B. A second interpretation of a three-way interaction is that the pattern in the interactions between factors A and B changes across the levels of factor C. Thus, if a three-way interaction were present and we plotted a separate profile plot for the two-way interaction between factors A and B at each level of factor C, we would see decidedly different patterns in several of the profile plots.

The model for an observation in a completely randomized design with a three-factor factorial treatment structure and $n > 1$ replications can be written in the form

$$y_{ijkm} = \mu + \alpha_i + \beta_j + \gamma_k + \alpha\beta_{ij} + \alpha\gamma_{ik} + \beta\gamma_{jk} + \alpha\beta\gamma_{ijk} + \varepsilon_{ijkm}$$

where the terms of the model are defined as follows:

y_{ijkm}: The response from the mth experimental unit receiving the ith level of factor A, the jth level of factor B, and the kth level of factor C.

μ: Overall mean, an unknown constant.

α_i: An effect due to the ith level of factor A, an unknown constant.

β_j: An effect due to the jth level of factor B, an unknown constant.

γ_k: An effect due to the kth level of factor C, an unknown constant.

$\alpha\beta_{ij}$: A two-way interaction effect of the ith level of factor A with the jth level of factor B, an unknown constant.

$\alpha\gamma_{ik}$: A two-way interaction effect of the ith level of factor A with the kth level of factor C, an unknown constant.

$\beta\gamma_{jk}$: A two-way interaction effect of the jth level of factor B with the kth level of factor C, an unknown constant.

$\alpha\beta\gamma_{ijk}$: A three-way interaction effect of the ith level of factor A, the jth level of factor B, and the kth level of factor C, an unknown constant.

ε_{ijkm}: A random error associated with the response from the mth experimental unit receiving the ith level of factor A combined with the jth level of factor B and the kth level of factor C. We require that the εs have a normal distribution with mean 0 and common variance σ_ε^2. In addition, the errors must be independent.

The conditions given here for our model can be shown to imply that the recorded response from the mth experimental unit receiving the ith level of factor A combined with the jth level of factor B and the kth level of factor C is normally distributed with mean

$$E(y_{ijkm}) = \mu + \alpha_i + \beta_j + \gamma_k + \alpha\beta_{ij} + \alpha\gamma_{ik} + \beta\gamma_{jk} + \alpha\beta\gamma_{ijk}$$

and variance σ_ε^2.

The following notation will be helpful in obtaining the partition of the total sum of squares into its components for main effects, interactions, and error.

y_{ijkm}: Observation on the mth experimental unit receiving the ith level of factor A, jth level of factor B, and kth level of factor C.

$\bar{y}_{i...}$: Sample mean for observations at the ith level of factor A,

$$\bar{y}_{i...} = \frac{1}{bcn}\Sigma_{jkm}\, y_{ijkm}$$

$\bar{y}_{.j..}$: Sample mean for observations at the jth level of factor B,

$$\bar{y}_{.j..} = \frac{1}{acn}\Sigma_{ikm}\, y_{ijkm}$$

$\bar{y}_{..k.}$: Sample mean for observations at the kth level of factor C,

$$\bar{y}_{..k.} = \frac{1}{abn}\Sigma_{ijm}\, y_{ijkm}$$

$\bar{y}_{ij..}$: Sample mean for observations at the ith level of factor A and jth level of factor B,

$$\bar{y}_{ij..} = \frac{1}{cn}\Sigma_{km}\, y_{ijkm}$$

$\bar{y}_{i.k.}$: Sample mean for observations at the ith level of factor A and kth level of factor C

$$\bar{y}_{i.k.} = \frac{1}{bn}\Sigma_{jm}\, y_{ijkm}$$

$\bar{y}_{.jk.}$: Sample mean for observations at the jth level of factor B and kth level of factor C,

$$\bar{y}_{.jk.} = \frac{1}{an}\Sigma_{im}\, y_{ijkm}$$

$\bar{y}_{ijk.}$: Sample mean for observations at the ith level of factor A, jth level of factor B, and kth level of factor C,

$$\bar{y}_{ijk.} = \frac{1}{n}\Sigma_m\, y_{ijkm}$$

$\bar{y}_{....}$: Overall sample mean,

$$\bar{y}_{....} = \frac{1}{abcn}\Sigma_{ijkm}\, y_{ijkm}$$

The parameters in the model have sample estimates:

$$\hat{\mu} = \bar{y}_{....} \qquad \hat{\alpha}_i = \bar{y}_{i...} - \bar{y}_{....} \qquad \hat{\beta}_j = \bar{y}_{.j..} - \bar{y}_{....} \qquad \hat{\gamma}_k = \bar{y}_{..k.} - \bar{y}_{....}$$

$$\widehat{\alpha\beta}_{ij} = (\bar{y}_{ij..} - \bar{y}_{....}) - \hat{\alpha}_i - \hat{\beta}_j = \bar{y}_{ij..} - \bar{y}_{i...} - \bar{y}_{.j..} + \bar{y}_{....}$$

$$\widehat{\alpha\gamma}_{ik} = (\bar{y}_{i.k.} - \bar{y}_{....}) - \hat{\alpha}_i - \hat{\gamma}_k = \bar{y}_{i.k.} - \bar{y}_{i...} - \bar{y}_{..k.} + \bar{y}_{....}$$

$$\widehat{\beta\gamma}_{jk} = (\bar{y}_{.jk.} - \bar{y}_{....}) - \hat{\beta}_j - \hat{\gamma}_k = \bar{y}_{.jk.} - \bar{y}_{.j..} - \bar{y}_{..k.} + \bar{y}_{....}$$

$$\widehat{\alpha\beta\gamma}_{ijk} = (\bar{y}_{ijk.} - \bar{y}_{....}) - \hat{\alpha}_i - \hat{\beta}_j - \hat{\gamma}_k - \widehat{\alpha\beta}_{ij} - \widehat{\alpha\gamma}_{ik} - \widehat{\beta\gamma}_{jk}$$

$$= \bar{y}_{ijk.} - \bar{y}_{ij..} - \bar{y}_{i.k.} - \bar{y}_{.jk.} + \bar{y}_{i...} + \bar{y}_{.j..} + \bar{y}_{..k.} - \bar{y}_{....}$$

The residuals from the fitted model then become

$$e_{ijkm} = y_{ijkm} - \hat{\alpha}_i - \hat{\beta}_j - \hat{\gamma}_k - \widehat{\alpha\beta}_{ij} - \widehat{\alpha\gamma}_{ik} - \widehat{\beta\gamma}_{jk} - \widehat{\alpha\beta\gamma}_{ijk} = y_{ijkm} - \bar{y}_{ijk.}$$

Using these expressions, we can partition the total sum of squares for a three-factor factorial experiment with a levels of factor A, b levels of factor B, c levels of factor C, and n observations per factor–level combination (treatments) into sums of squares for main effects (variability between levels of a single factor), two-way interactions, a three-way interaction, and sum of squares for error.

main effects
The sums of squares for **main effects** are

$$\text{SSA} = bcn \sum_i (\hat{\alpha}_i)^2 = bcn \sum_i (\bar{y}_{i...} - \bar{y}_{....})^2$$

$$\text{SSB} = acn \sum_j (\hat{\beta}_j)^2 = acn \sum_j (\bar{y}_{.j..} - \bar{y}_{....})^2$$

$$\text{SSC} = abn \sum_k (\hat{\gamma}_k)^2 = abn \sum_k (\bar{y}_{..k.} - \bar{y}_{....})^2$$

two-way interactions
The sums of squares for **two-way interactions** are

$$\text{SSAB} = cn \sum_{ij} (\widehat{\alpha\beta}_{ij})^2 = cn \sum_{ij} (\bar{y}_{ij..} - \bar{y}_{....})^2 - \text{SSA} - \text{SSB}$$

$$\text{SSAC} = bn \sum_{ik} (\widehat{\alpha\gamma}_{ik})^2 = bn \sum_{ik} (\bar{y}_{i.k.} - \bar{y}_{....})^2 - \text{SSA} - \text{SSC}$$

$$\text{SSBC} = an \sum_{jk} (\widehat{\beta\gamma}_{jk})^2 = an \sum_{jk} (\bar{y}_{.jk.} - \bar{y}_{....})^2 - \text{SSB} - \text{SSC}$$

three-way interaction   The sum of squares for the **three-way interaction** is

$$\text{SSABC} = n \sum_{ijk} (\widehat{\alpha\beta\gamma}_{ijk})^2$$

$$= n \sum_{ijk} (\bar{y}_{ijk.} - \bar{y}_{....})^2 - \text{SSAB} - \text{SSAC} - \text{SSBC} - \text{SSA} - \text{SSB} - \text{SSC}$$

sum of squares for error   The **sum of squares for error** is given by

$$\text{SSE} = \sum_{ijkm} (e_{ijkm})^2$$

$$= \sum_{ijkm} (y_{ijkm} - \bar{y}_{ijk.})^2$$

$$= \text{TSS} - \text{SSA} - \text{SSB} - \text{SSC} - \text{SSAB} - \text{SSAC} - \text{SSBC} - \text{SSABC}$$

where $\text{TSS} = \sum_{ijkm} (y_{ijkm} - \bar{y}_{....})^2$.

The AOV table for a completely randomized three-factor factorial experiment with a levels of factor A, b levels of factor B, c levels of factor C, and n observations per each of the abc treatments (factor–level combinations) is given in Table 9.21.

From the AOV table, we observe that if we have only one observation on each treatment, $n = 1$, then there are 0 degrees of freedom for error. Thus, if the interaction terms are in the model and $n = 1$, there are no valid tests for interactions or main effects. However, if some of the interactions are known to be 0; then, these interaction terms can be combined to serve as the error term in order to test the remaining terms in the model. However, it would be a rare situation to run experiments with $n = 1$ because in most cases the researcher would not know prior to running the experiment which of the interactions would be 0. Hence, in order to have valid tests for main effects and interactions, we need $n > 1$.

The analysis of a three-factor experiment is somewhat complicated by the fact that, if the three-way interaction is significant, then we must handle the two-way interactions and main effects differently than when the three-way is not significant. The following diagram (Figure 9.8) from *Analysis of Messy Data,* by G. Milliken and D. Johnson (1984), provides a general method for analyzing three-factor experiments.

TABLE 9.21
AOV table for a completely randomized three-factor factorial experiment

Source	SS	df	MS	F
Main Effects				
A	SSA	$a - 1$	MSA = SSA/$(a - 1)$	MSA/MSE
B	SSB	$b - 1$	MSB = SSB/$(b - 1)$	MSB/MSE
C	SSC	$c - 1$	MSC = SSC/$(c - 1)$	MSC/MSE
Interactions				
AB	SSAB	$(a - 1)(b - 1)$	MSAB = SSAB/$(a - 1)(b - 1)$	MSAB/MSE
AC	SSAC	$(a - 1)(c - 1)$	MSAB = SSAC/$(a - 1)(c - 1)$	MSAC/MSE
BC	SSBC	$(b - 1)(c - 1)$	MSAB = SSBC/$(b - 1)(c - 1)$	MSBC/MSE
ABC	SSABC	$(a - 1)(b - 1)(c - 1)$	MSABC = SSABC/$(a - 1)(b - 1)(c - 1)$	MSABC/MSE
Error	SSE	$abc(n - 1)$	MSE = SSE/$abc(n - 1)$	
Total	TSS	$abcn - 1$		

FIGURE 9.8

Method for analyzing three-factor experiment

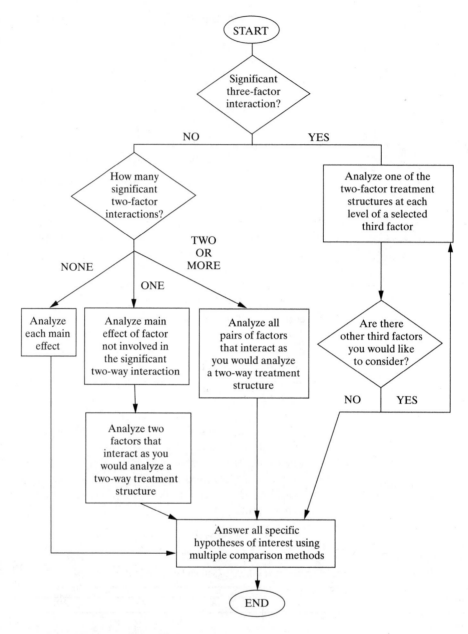

We will illustrate the analysis of variance of a three-factor experiment in Section 9.7.

EXERCISES   **Applications**

Psy.   **9.12** A large advertising firm specializes in creating television commercials for children's products. The firm wants to design a study to investigate factors that may affect the length of time a commercial is able to hold a child's attention. A preliminary study determines that two factors that may be important are the age of the child and the type of product being advertised. The firm wants to determine whether there were large differences in the mean length of time that the commercial is able to hold the child's attention depending on these two factors. If there proves to be a difference, the firm will then attempt to determine new types of commercials depending on the product and targeted age group. Three age groups are used:

A_1: 5–6 years $\qquad A_2$: 7–8 years $\qquad A_3$: 9–10 years

The types of products selected are

P_1: breakfast cereals $\qquad P_2$: video games

A group of 20 children is recruited in each age group and 10 are randomly assigned to watch a 60-second commercial for each of the two products. Researchers record their attention span during the viewing of the commercial. The data are given here.

Child	A_1–P_1	A_2–P_1	A_3–P_1	A_1–P_2	A_2–P_2	A_3–P_2
1	19	19	37	39	30	51
2	36	35	6	18	47	52
3	40	22	28	32	6	43
4	30	28	4	22	27	48
5	4	1	32	16	44	39
6	10	27	16	2	26	33
7	30	27	8	36	33	56
8	5	16	41	43	48	43
9	34	3	29	7	23	40
10	21	18	18	16	21	51
Mean	22.9	19.6	21.9	23.1	30.5	45.6

Mean by age group: $\quad A_1 \quad A_2 \quad A_3 \qquad$ Mean by product type: $\quad P_1 \quad P_2$
$\qquad\qquad\qquad\qquad$ 23.00 25.05 33.75 $\qquad\qquad\qquad\qquad\qquad$ 21.47 30.07

    a. Identify the design.
    b. Write a model for this situation, identifying all the terms in the model.
    c. Estimate the parameters in the model.
    d. Compute the sum of squares for the data and summarize the information in an AOV table.

9.13 Refer to Exercise 9.12.
    a. Draw a profile plot for the two factors, age and product type.
    b. Perform appropriate F tests and draw conclusions from these tests concerning the effect of age and product type on the mean attention span of the children.

9.14 The Minitab output for the data in Exercise 9.12 is shown here.
    a. Compare the AOV from Minitab to the one you obtained in Exercise 9.13.
    b. Use the residual plots to determine whether any of the conditions required for the validity of the F tests have been violated.

```
Two-Way Analysis of Variance
Analysis of Variance for Time
Source          DF        SS         MS        F        P
Age              2      1303.0      651.5     4.43    0.017
Product          1      2018.4     2018.4    13.72    0.001
Interaction      2      1384.3      692.1     4.70    0.013
Error           54      7944.0      147.1
Total           59     12649.7
```

Normal Probability Plot of Residuals

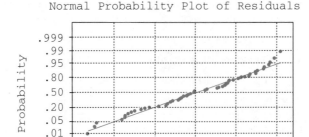

Residuals versus the Fitted Values
(response is Time)

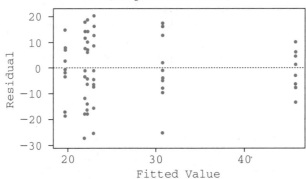

9.5   Factorial Treatment Structure in a Randomized Complete Block Design

In the previous section, we defined a factorial experiment to be an experiment in which the response y is observed at all factor–level combinations of the independent variables. The factor–level combinations of the independent variables (treatments) were randomly assigned to the experimental units; hence, we were employing a completely randomized design to investigate the effects of the factors on the response.

Sometimes the objectives of a study are such that we wish to investigate the effects of certain factors on a response while blocking out certain other extraneous sources of variability. Such situations require a block design with treatments from a factorial experiment. We will draw on our knowledge of block designs (randomized block designs and Latin square designs) to effectively block out the extraneous sources of variability in order to focus on the effects of the factors on the response of interest. This can be illustrated with the following example.

EXAMPLE 9.6

A nutritionist wants to study the percentage of protein content in bread made from three new types of flours and baked at three different temperatures. She would like to bake three loaves of bread from each of the nine flour–temperature combinations for a total of 27 loaves from which the percentage of protein would be determined. However, she is only able to bake nine loaves on any given day. Propose an appropriate experimental design.

Solution Because nine loaves can be baked on a given day, it would be possible to run a complete replication of the 3×3 factorial experiment on three different days to obtain the desired number of observations. The design is shown here.

	Day 1 Temperature			Day 2 Temperature			Day 3 Temperature		
Flour Type	1	2	3	1	2	3	1	2	3
A	y	y	y	y	y	y	y	y	y
B	y	y	y	y	y	y	y	y	y
C	y	y	y	y	y	y	y	y	y

Note that this design is really a randomized block design, where the blocks are days and the treatments are the nine factor–level combinations of the 3×3 factorial experiment. So, with the randomized block design, we are able to block or filter out the variability due to the nuisance variable, days, while comparing the treatments. Because the treatments are factor–level combinations from a factorial experiment, we can examine the effects of the two factors (A and B) on the response while filtering out the day-to-day variability.

The analysis of variance for this design follows from our discussions in Sections 9.2 and 9.4.

EXAMPLE 9.7

Construct an analysis of variance table identifying the sources of variability and the degrees of freedom for the 3×3 factorial experiment laid off in a randomized complete block design with $b = 3$ discussed in Example 9.6.

Solution The analysis of variance table for a randomized complete block design with $t = 9$ and $b = 3$ is shown here:

Source	SS	df
Treatments	SST	8
Blocks	SSB	2
Error	SSE	16
Total	TSS	26

Because the treatments of this randomized block are the nine factor–level combinations of a 3×3 factorial experiment, we can subdivide the sum of squares treatment (SST) into the sources of variability for a 3×3 factorial experiment from Section 9.4. The revised AOV table is shown here.

Source	SS	df
Treatments	SST	8
A	SSA	2
B	SSB	2
AB	SSAB	4
Blocks	SSB	2
Error	SSE	16
Total	TSS	26

So, rather than running an overall test to compare the treatment means using $F = \text{MST}/\text{MSE}$, we could conduct the analysis of variance for a factorial experiment to examine the interaction and main effects. These F tests would use the appropriate numerator mean squares (MSAB, MSA, and MSB) and MSE from this analysis.

EXERCISES   **Basic Techniques**

9.15   Diagram a design that has a 3×5 factorial experiment laid off in a randomized complete block design with $b = 3$ blocks. Give the complete analysis of variance table (sources, SSs, and dfs).

9.16   Diagram a design that has a $2 \times 4 \times 3$ factorial experiment laid off in a randomized complete block design with $b = 2$ blocks. Give the complete analysis of variance for this experimental design.

9.6 Estimation of Treatment Differences and Comparisons of Treatment Means

We have emphasized the analysis of variance associated with a randomized complete block design, a Latin square design, and factorial experiments. However, there are times when we might be more interested in estimating the difference in mean response for two treatments (different levels of the same factor or different combinations of levels). For example, an environmental engineer might be more interested in estimating the difference in the mean dissolved oxygen content for a lake before and after rehabilitative work than in testing to see whether there is a difference. Thus, the engineer is asking the question, "What is the difference in mean dissolved oxygen content?" instead of the question, "Is there a difference between the mean content before and after the cleanup project?"

Fisher's LSD procedure can be used to evaluate the difference in treatment means for a randomized complete block design, a Latin square design, and k-factor factorial experiments with various designs. Let $\bar{y}_i$ denote the mean response for treatment i, $\bar{y}_{i'}$ denote the mean response for treatment i', and n_t denote the number of observations in each treatment. A $100(1 - \alpha)\%$ confidence interval on $\mu_i - \mu_{i'}$, the difference in mean response for the two treatments, is defined as shown here.

100(1 − α)% Confidence Interval for the Difference in Treatment Means

$$(\bar{y}_i - \bar{y}_{i'}) \pm t_{\alpha/2} s_\varepsilon \sqrt{\frac{2}{n}}$$

where s_ε is the square root of MSE in the AOV table and $t_{\alpha/2}$ can be obtained from Table 3 in the Appendix for $a = \alpha/2$ and the degrees of freedom for MSE.

EXAMPLE 9.8

A company was interested in comparing three different display panels for use by air traffic controllers. Each display panel was to be examined under five different simulated emergency conditions. Thirty highly trained air traffic controllers with similar work experience were enlisted for the study. A random assignment of controllers to display panel–emergency conditions was made, with two controllers assigned to each factor–level combination. The time (in seconds) required to stabilize the emergency situation was recorded for each controller at a panel–emergency condition. These data appear in Table 9.22.

   a. Construct a profile plot.
   b. Run an analysis of variance that includes a test for interaction.

Solution

   a. The sample means are given in Table 9.23 and then displayed in a profile plot given in Figure 9.9. From the profile plot we observe that the difference in mean reaction time for controllers on any pair of different display

TABLE 9.22
Display panel data
(time in seconds)

Display Panel, B	Emergency Condition, A				
	1	2	3	4	5
1	18	31	22	39	15
	16	35	27	36	12
2	13	33	24	35	10
	15	30	21	38	16
3	24	42	40	52	28
	28	46	37	57	24

TABLE 9.23
Mean reaction times for
display panels–emergency
conditions study

Display Panel, B	Emergency Condition, A					Means $\bar{y}_{\cdot j\cdot}$
	1	2	3	4	5	
1	17	33	24.5	37.5	13.5	25.1
2	14	31.5	22.5	36.5	13	23.5
3	26	44	38.5	54.5	26	37.8
Means $\bar{y}_{i\cdot\cdot}$	19.0	36.2	28.5	42.8	17.5	$\bar{y}_{\cdot\cdot\cdot} = 28.8$

FIGURE 9.9
Profile plot of panel means for
each emergency condition

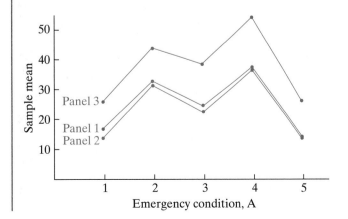

TABLE 9.24
AOV table for display panel–emergency condition study

```
General Linear Models Procedure

Dependent Variable: y, Stabilization Time

                                   Sum of          Mean
Source                    DF       Squares        Square    F Value   Pr > F
Model                     14     4122.8000      294.4857      41.67   0.0001
Error                     15      106.0000        7.0667
Corrected Total           29     4228.8000

               R-Square              C.V.      Root MSE                Y Mean
               0.974934          9.230279        2.6583                28.800

Source                    DF     Type I SS   Mean Square   F Value   Pr > F
D                          2     1227.8000      613.9000      86.87   0.0001
E                          4     2850.1333      712.5333     100.83   0.0001
D*E                        8       44.8667        5.6083       0.79   0.6167
```

panels remains relatively constant across all five emergency conditions. Panel 1 and panel 2 yield essentially the same mean reaction times across the five emergency conditions, whereas panel 3 produces mean reaction times that are consistently higher than the mean times for the other two panels. We will next confirm these observations using tests of hypotheses that take into account the variability of the reaction times about the observed mean times.

b. The computer output for the analysis of variance table is given in Table 9.24. The first test of hypothesis is for an interaction between the two factors, emergency condition and type of display panel. Because the computed value of F, 0.79, is less than the critical value of F, 2.64, for $a = .05$, $df_1 = 8$, and $df_2 = 15$, we have insufficient evidence (p-value = .6167) to indicate an interaction between emergency conditions and type of display panel. This confirms our observations from the profile plot. Because the interaction was not significant, we will next test for a main effect due to type of display panel. The computed value of F, 86.87, is greater than the critical value of F, 3.68, for $a = .05$, $df_1 = 2$, and $df_2 = 15$, so we have sufficient evidence (p-value < .0001) to indicate a significant difference in mean reaction time across the three types of display panels.

EXAMPLE 9.9

Refer to Example 9.8. The researchers were very interested in the size of the differences in mean reaction time among the three types of panels. Estimate these differences using 95% confidence intervals.

Solution Because there is not a significant interaction between type of display panel and type of emergency condition, the size of the differences in mean reaction times between the types of display panels should be relatively the same for all five types of emergency conditions. Thus, we can examine the main effect means for the three display panels, averaging over the five emergency conditions: $\hat{\mu}_j = \bar{y}_{.j.}$, for $j = 1, 2, 3$. From Table 9.23 we have

$$\bar{y}_{.1.} = 25.1 \qquad \bar{y}_{.2.} = 23.5 \qquad \bar{y}_{.3.} = 37.8$$

The t-value for $a = .025$ and df = 15 is 2.131; the estimate of σ_ε is

$$s_\varepsilon = \sqrt{\text{MSE}} = \sqrt{7.0667} = 2.66$$

The formula for a 95% confidence interval on the difference between the mean reaction times of two display panels, $\mu_{.j.} - \mu_{.j'.}$, is given by

$$\bar{y}_{.j.} - \bar{y}_{.j'.} \pm t_{\alpha/2} s_\varepsilon \sqrt{\frac{2}{n_t}}$$

For panels 2 and 3, we have $n_t = 10$ observations per panel, thus we have

$$37.8 - 23.5 \pm (2.131)(2.66)\sqrt{\frac{2}{10}}$$

$$14.3 \pm 2.54$$

that is, 11.76 to 16.84. Thus, we are 95% confident that the difference in the mean reaction times between display panel 2 and display panel 3 is between 11.76 and 16.84 seconds. Similarly, we can calculate confidence intervals on the differences between panels 1 and 3 and between panels 1 and 2.

After determining that there was a significant main effect using the F test, we proceed with two further inference procedures. First, we place confidence intervals on the difference between any pair of factor–level means, $\mu_{i..} - \mu_{i'..}$ for factor A or $\mu_{.j.} - \mu_{.j'.}$ for factor B, using the procedure illustrated in Example 9.9. This estimates the effect sizes for these two factors. Next, we want to determine which levels of the factors differ from the rest for each of the factors.

multiple comparison procedures   As discussed in Chapter 8, we apply one of the **multiple comparison procedures** in order to control the experimentwise error rate for comparing the several pairs of factor levels. There are $a(a-1)/2$ pairs for factor A and $b(b-1)/2$ pairs for factor B. The choice of which procedure to use once again depends on the experiment, as discussed in Chapter 8. Both of the procedures discussed in Chapter 8— Fisher's LSD and Tukey's procedure—can be performed for a randomized complete block design or a Latin square design. In the formulas given in Chapter 8 for these procedures, the degrees for MSE are obtained from the AOV table and the sample size n refers to the number of observations per mean value in the comparison—that is, the number of data values averaged to obtain $\bar{y}_{i..}$, for example.

EXAMPLE 9.10

Refer to Example 9.8 and the data in Tables 9.22 and 9.23. Use Tukey's W procedure to locate significant differences among display panels.

Solution   For Tukey's W procedure we use the formula presented in Chapter 8:

$$W = q_\alpha(t, v)\sqrt{\frac{\text{MSE}}{n}}$$

where MSE is from the AOV table, based on $v = 15$ degrees of freedom, and $q_\alpha(t, v)$ is the upper-tail critical value of the Studentized range (with $a = \alpha$) for comparing t different population means. The value of $q_\alpha(t, v)$ from Table 11 in the Appendix for comparing the three display panel means, each of which has 10 observations per sample mean, is

$$q_{.05}(3, 15) = 3.67$$

For 10 observations per mean, the value of W is

$$W = q_\alpha(t, v)\sqrt{\frac{\text{MSE}}{n}} = 3.67\sqrt{\frac{7.07}{10}} = 3.09$$

The display panel means are, from Table 9.23,

$$\bar{y}_{.1.} = 25.1 \qquad \bar{y}_{.2.} = 23.5 \qquad \bar{y}_{.3.} = 37.8$$

First we rank the sample means from lowest to highest:

Display panel	2	1	3
Means	23.5	25.1	37.8

If two means that differ (in absolute value) by more than $W = 3.09$, we declare them to be significantly different from each other. The results of our multiple comparison procedure are summarized here:

Display panel   2   1   3

Thus, display panels 1 and 2 both have a mean reaction time lower than panel 3, but we are unable to detect any difference between panels 1 and 2.

9.7   Encounters with Real Data: Texture of Low-Fat Bologna

Defining the Problem (1)

Dietary health concerns and consumer demand for low-fat products have prompted meat companies to develop a variety of low-fat meat products. Numerous ingredients have been evaluated as fat replacements with the goal of maintaining product yields and minimizing formulation costs while retaining acceptable palatability. The paper "Utilization of soy protein isolate and konjac blends in a low-fat bologna (model system)" [*Meat Science* (1999) 53:45–57] describes an experiment that examines several of these issues. The researchers determined that lowering the cost of production without affecting the quality of the low-fat meat product required the substitution of a portion of the meat block with non-meat ingredients such as soy protein isolates (SPI). Previous experiments have demonstrated SPI's effect on the characteristics of comminuted meats, but studies evaluating SPI's effect in low-fat meat applications are limited. Konjac flour has been incorporated into processed meat products to improve gelling properties and water-holding capacity while reducing fat content. Thus, when replacing meat with SPI, it is necessary to incorporate Konjac flour into the product to maintain the high-fat characteristics of the product.

Collecting Data (2), Summarizing Data (3)

The three factors identified for study were type of Konjac blend, amount of Konjac blend, and percentage of SPI substitution in the meat product. There were many other possible factors of interest, including cooking time, temperature, type of meat product, and length of curing. However, the researchers selected the commonly used levels of these factors in a commercial preparation of bologna and narrowed the study to the three most important factors. This resulted in an experiment having 12 treatments as displayed in Table 9.25.

The objective of this study was to evaluate various types of Konjac blends as a partial lean-meat replacement, and to characterize their effects in a very low-fat bologna model system. Two types of Konjac blends (KSS = Konjac flour/ starch and KNC = Konjac flour/carrageenan/starch), at levels .5 and 1%, and three meat protein replacement levels with SPI (1.1, 2.2, and 4.4%, DWB) were selected for evaluation.

The experiment was conducted as a completely randomized design with a $2 \times 2 \times 3$ three-factor factorial treatment structure and three replications of the 12 treatments. There were a number of response variables measured on the 36 runs of the experiment, but we will discuss the results for the texture of the final product as measured by an Instron universal testing machine. The mean responses are given in Table 9.26.

The researchers next prepared the data for a statistical analysis following the steps described in Section 2.5. The researchers needed to verify that the texture readings were properly recorded and that all computer files were consistent with the field data. (The values in Table 9.26 were simulated using the summary statistics given in the paper.)

TABLE 9.25

Treatment design for low-fat bologna study

Treatment	Level of Blend (%)	Konjac Blend	SPI (%)
1	.5	KSS	1.1
2	.5	KSS	2.2
3	.5	KSS	4.4
4	.5	KNC	1.1
5	.5	KNC	2.2
6	.5	KNC	4.4
7	1.0	KSS	1.1
8	1.0	KSS	2.2
9	1.0	KSS	4.4
10	1.0	KNC	1.1
11	1.0	KNC	2.2
12	1.0	KNC	4.4

TABLE 9.26

Mean values for meat texture in low-fat bologna study

Konjac Level (%)	Konjac Blend	SPI (%)	Texture Readings	Mean Texture
.5	KSS	1.1	107.3, 110.1, 112.6	110.0
.5	KSS	2.2	97.9, 100.1, 102.0	100.0
.5	KSS	4.4	86.8, 88.1, 89.1	88.0
.5	KNC	1.1	108.1, 110.1, 111.8	110.0
.5	KNC	2.2	108.6, 110.2, 111.2	110.0
.5	KNC	4.4	95.0, 95.4, 95.5	95.3
1.0	KSS	1.1	97.3, 99.1, 100.6	99.0
1.0	KSS	2.2	92.8, 94.6, 96.7	94.7
1.0	KSS	4.4	86.8, 88.1, 89.1	88.0
1.0	KNC	1.1	94.1, 96.1, 97.8	96.0
1.0	KNC	2.2	95.7, 97.6, 99.8	97.7
1.0	KNC	4.4	90.2, 92.1, 93.7	92.0

Analyzing Data, Interpreting the Analyses, and Communicating Results (4)

The researchers were interested in evaluating the relationship between mean texture of low-fat bologna as the percentage of SPI was increased and in comparing this relationship for the two types of Konjac blend at the set two levels. Because the number of calculations needed to obtain the sum of squares in a three-factor experiment are numerous and consequently may lead to significant rounding error, we will use a software program to obtain the results shown in Table 9.27.

The notation in the AOV table is as follows: L refers to Konjac level, B refers to type of Konjac blend, and P refers to the level of SPI. Because the three-way interaction in an AOV model was not significant (L*B*P, $p = .3106$), we next examine the two-way interactions. The three sets of two-way interactions had levels of significance (L*B, $p = .0008$), (L*P, $p < .0001$), and (B*P, $p < .0001$). Thus, all three were highly significant. To examine the types of relationships that may exist between the three factors, we need to obtain the sample means, $\bar{y}_{ij..}$, $\bar{y}_{i.k.}$, and $\bar{y}_{.jk.}$. These values are given in Table 9.28. The means in the table are then plotted in Figure 9.10 to yield the profile plots for the two-way interactions of level of Konjac with type of Konjac, level of Konjac with level of SPI, and type of Konjac with level of SPI.

From Figure 9.10, we can observe that there are considerable differences in the mean texture of the meat product depending on the type of Konjac, level of Konjac, and level of SPI in the meat product. When the level of Konjac is 1%, there is very little difference in the mean texture of the meat; however, at the .5% level, KNC Konjac produced a product with a higher mean texture than did the KSS blend of Konjac. When considering the effect of level of SPI on the mean texture of the bologna, we can observe that at a level of 1.1% SPI there was a sizeable difference between using .5% Konjac and 1% Konjac. As the level of SPI increased, the size of the difference decreased markedly. Furthermore, at a 1.1% level of SPI there was essentially no difference between the two blends of Konjac, but as the level of SPI increased, the KNC blend produced a meat product having a higher texture than the KSS blend. These observations about the relationships between

TABLE 9.27
AOV table for data in case study, a three-factor factorial experiment

```
General Linear Models Procedure

Dependent Variable: Texture of Meat:

                                Sum of          Mean
Source              DF         Squares         Square    F Value    Pr > F
Model               11     2080.28750      189.11705      62.40    0.0001
Error               24       72.74000        3.03083
Corrected Total     35     2153.02750

              R-Square            C.V.       Root MSE               Y Mean
              0.966215        1.769387        1.74093              98.3917

Source              DF       Type I SS    Mean Square    F Value    Pr > F
Main Effects:
L                    1      526.70250      526.70250     173.78    0.0001
B                    1      113.42250      113.42250      37.42    0.0001
P                    2     1090.11500      545.05750     179.84    0.0001
Interactions:
L*B                  1       44.22250       44.22250      14.59    0.0008
L*P                  2      182.53500       91.26750      30.11    0.0001
B*P                  2      115.84500       57.92250      19.11    0.0001
L*B*P                2        7.44500        3.72250       1.23    0.3106
```

TABLE 9.28
Table of means for low-fat bologna study

Level (%)	Blend	SPI (%)	Two-Way Means
.5	KSS	*	99.3
.5	KNC	*	105.1
1.0	KSS	*	93.9
1.0	KNC	*	95.2
.5	*	1.1	110.0
.5	*	2.2	105.0
.5	*	4.4	91.7
1.0	*	1.1	97.5
1.0	*	2.2	96.2
1.0	*	4.4	90.0
*	KSS	1.1	104.5
*	KSS	2.2	97.4
*	KSS	4.4	88.0
*	KNC	1.1	103.0
*	KNC	2.2	103.9
*	KNC	4.4	93.7

FIGURE 9.10

Profile plots of the
two-way interactions

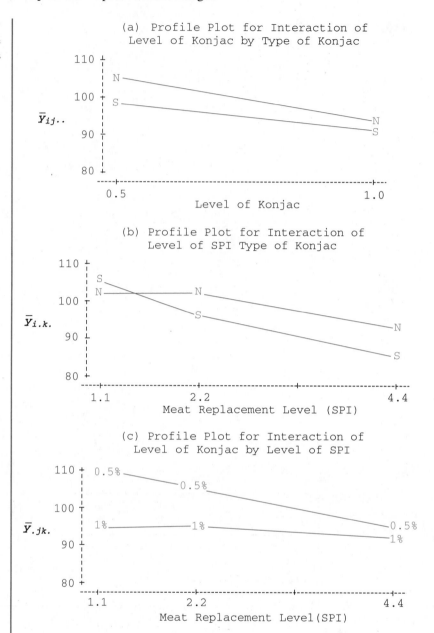

(a) Profile Plot for Interaction of
Level of Konjac by Type of Konjac

(b) Profile Plot for Interaction of
Level of SPI Type of Konjac

(c) Profile Plot for Interaction of
Level of Konjac by Level of SPI

the three factors and the mean texture of the meat product need to be confirmed
using multiple comparison procedures.

Figure 9.11 contains the residuals analysis for the texture data. We obtain the
residuals using the formula:

$$e_{ijkm} = y_{ijkm} - \hat{\alpha}_i - \hat{\beta}_j - \hat{\gamma}_k - \widehat{\alpha\beta}_{ij} - \widehat{\alpha\gamma}_{ik} - \widehat{\beta\gamma}_{jk} - \widehat{\alpha\beta\gamma}_{ijk} = y_{ijkm} - \overline{y}_{ijk.}$$

The summary statistics for the e_{ijkm}s reveal that the skewness is nearly 0 but the
kurtosis is nearly -1. An examination of the stem-and-leaf plot and boxplot reveals that the residuals are nearly symmetric but have a sharp peak near 0. The

FIGURE 9.11

Residuals analysis for
low-fat bologna study

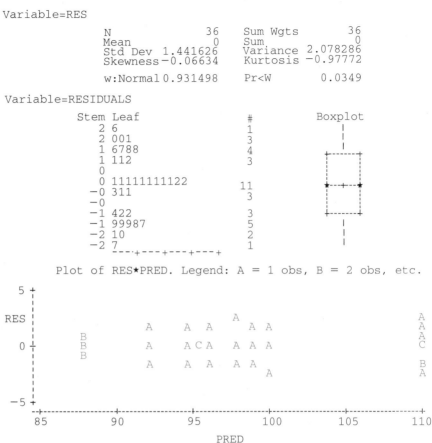

Shapiro–Wilk's test for normality has a *p*-value of .0349, which reflects the somewhat nonnormal nature of the residuals. However, because there are no outliers and very few residuals even near extreme in size, the normality assumption is nearly met. The plot of the residuals versus the estimated treatment means $\bar{y}_{ijk.}$ reveals a slight increase in variability as the mean texture readings increase. However, this increase is not large enough to overcome the natural robustness of the *F* test for small deviations from the model conditions. Thus, both the normality and equal variance conditions appear to be satisfied and we can conclude that the *F* tests in the AOV test are valid.

Because the three-way interaction, L*S*P, was not significant (*p*-value = .3106), we will examine the two-way interactions of interest to the researchers. They wanted to investigate the effect on mean texture of increasing the percentage of SPI in the meat product. Thus, we need to examine the differences in mean texture as a function of the percentage of SPI. Because there was a significant (*p*-value < .0001) interaction between SPI and level of Konjac and a significant (*p*-value < .0001) interaction between SPI and type of Konjac, we need to conduct four different mean separations of the levels of the percentage of SPI. The researchers were concerned about falsely declaring pairs different, so we will use Tukey's procedure.

First we compare the mean textures across the percentage of SPI separately for each of the two values of level of Konjac: 0.5 and 1.0%. The value of Tukey's W is given by

$$W = q_\alpha(t, \text{df}_{\text{error}})\sqrt{\frac{s_\varepsilon^2}{n_t}}$$

where $t = 3$, the number of levels of the percentage of SPI; $\text{df}_{\text{error}} = 24$, $s_\varepsilon^2 = 3.0308$ from the AOV table; and $n_t = 6$, the number of observations in each of the percentage of SPI means at each of the values of level of Konjac, because $\bar{y}_{i.k.}$ is based on six data values. Thus, from Table 11 in the Appendix we find $q_\alpha(t, \text{df}_{\text{error}}) = q_{.05}(3, 24) = 3.53$, which yields

$$W = 3.53\sqrt{\frac{3.0308}{6}} = 2.51$$

Thus, any pair of means $\bar{y}_{i.k.}$ and $\bar{y}_{i.k'.}$ that differ by more than 2.51 will be declared to be significantly different at the $\alpha = .05$ level. A summary of results is given in Table 9.29.

For the 0.5% level of Konjac, all three percentages of SPI yield significantly different mean textures, with the higher the level of the percentage of SPI, the lower the value for mean texture. For the 1.0% level of Konjac, the 1.1 and 2.2 percentages of SPI have nonsignificantly different mean textures, whereas the 4.4 percentage of SPI has a significantly lower mean texture compared to the 1.1 and 2.2 percentages. Thus, the relationship between the percentage of SPI and mean texture is different at the two levels of Konjac. Similarly, we obtain the following results (Table 9.30) for the relationship between mean texture and the percentage of SPI at the two types of Konjac. The values of all the quantities in W remain the same as before because the number of observations in each of the type of Konjac–percentage of SPI means, $\bar{y}_{ik}$, is $n_t = 6$. Thus, $W = 2.51$.

TABLE 9.29

Mean texture across levels of percentage of SPI at each level of Konjac

Level of Konjac (%)	SPI (%)		
	1.1	**2.2**	**4.4**
.5	110.0	105.0	9.17
	a	b	c
1.0	97.5	96.2	90.0
	a	a	b

TABLE 9.30

Mean texture across levels of percentage of SPI for each type of Konjac

Type of Konjac	SPI (%)		
	1.1	**2.2**	**4.4**
KSS	104.5	97.4	88.0
	a	b	c
KNC	103.0	103.9	93.7
	a	a	b

For the KSS Konjac, all three percentages of SPI yield significantly different mean textures. For KNC Konjac, the 1.1 and 2.2 percentages of SPI have nonsignificantly different mean textures, whereas the 4.4 percentage of SPI has a significantly lower mean texture in comparison to the 1.1 and 2.2 percentages. Thus, the relationship between percentage of SPI and mean texture is different for the two types of Konjac.

9.8   Staying Focused

In this chapter we expanded our discussion of the analysis of variance beyond the completely randomized design of Chapter 8 to include several different experimental designs and treatment structures. The designs considered were the randomized complete block design and the Latin square design. These designs illustrated how we can minimize the effect of undesirable variability from extraneous variables to obtain more precise comparisons among treatment means. The factorial treatment structure is useful in investigating the effect of one or more factors on an experimental response. Factorial treatments can be used in either a completely randomized, randomized complete block, or Latin square design. Thus, an experimenter may wish to examine the effects of two or more factors on a response while blocking out one or more extraneous sources of variability.

For each design discussed in this chapter, we presented a description of the design layout (including arrangement of treatments), potential advantages and disadvantages, a model, and the analysis of variance. Finally, we discussed how we could conduct multiple comparisons between treatment means for each of these designs.

Note that the designs presented in this chapter (and in Chapter 8) are only the most basic designs. Also, we have only dealt with the situation in which we have a *balanced design;* that is, a design in which each treatment (factor–level combination) is randomly assigned to the same number of experimental units. Thus, in a two-factor factorial experiment, we have a balanced design if each of the $t = ab$ treatments is observed on exactly the same number of experimental units. The formulas and analysis of balanced designs are somewhat simpler than those for unbalanced designs, which are beyond the scope of this book.

Key Formulas

1. One factor in a completely randomized design

Model: $y_{ij} = \mu + \alpha_i + \varepsilon_{ij}$

Sum of Squares:

Total $\text{TSS} = \sum_{ij} (y_{ij} - \overline{y}_{..})^2$

Treatment $\text{SST} = n \sum_i (\overline{y}_{i.} - \overline{y}_{..})^2 = n \sum_i (\hat{\alpha}_i)^2$

Error $\text{SSE} = \sum_{ij} (e_{ij})^2 = \sum_{ij} (y_{ij} - \hat{\mu} - \hat{\alpha}_i)^2 = \text{TSS} - \text{SST}$

2. One factor in a randomized complete block design

Model: $y_{ij} = \mu + \alpha_i + \beta_j + \varepsilon_{ij}$

Sum of Squares:

Total $\qquad$ TSS $= \Sigma_{ij} (y_{ij} - \bar{y}_{..})^2$

Treatment $\qquad$ SST $= b \Sigma_i (\bar{y}_{i.} - \bar{y}_{..})^2 = b \Sigma_i (\hat{\alpha}_i)^2$

Block $\qquad$ SSB $= t \Sigma_j (\bar{y}_{.j} - \bar{y}_{..})^2 = t \Sigma_j (\hat{\beta}_j)^2$

Error $\qquad$ SSE $= \Sigma_{ij} (e_{ij})^2 = \Sigma_{ij} (y_{ij} - \hat{\mu} - \hat{\alpha}_i - \hat{\beta}_j)^2 = $ TSS $-$ SST $-$ SSB

3. One factor in a Latin square design

Model: $y_{ijk} = \mu + \alpha_k + \beta_i + \gamma_j + \varepsilon_{ijk}$

Sum of Squares:

Total $\qquad$ TSS $= \Sigma_{ijk} (y_{ijk} - \bar{y}_{...})^2$

Treatment $\qquad$ SST $= t \Sigma_k (\bar{y}_{..k} - \bar{y}_{...})^2 = t \Sigma_k (\hat{\alpha}_k)^2$

Row $\qquad$ SSR $= t \Sigma_i (\bar{y}_{i..} - \bar{y}_{...})^2 = t \Sigma_i (\hat{\beta}_i)^2$

Column $\qquad$ SSC $= t \Sigma_j (\bar{y}_{.j.} - \bar{y}_{...})^2 = t \Sigma_j (\hat{\gamma}_j)^2$

Error $\qquad$ SSE $= $ TSS $-$ SST $-$ SSR $-$ SSC

4. Two-factor factorial treatment structure in a completely randomized design

Model: $y_{ijk} = \mu + \alpha_i + \beta_j + \alpha\beta_{ij} + \varepsilon_{ijk}$

Sum of Squares:

Total $\qquad$ TSS $= \Sigma_{ijk} (y_{ijk} - \bar{y}_{...})^2$

Factor A $\qquad$ SSA $= bn \Sigma_i (\bar{y}_{i..} - \bar{y}_{...})^2 = bn \Sigma_i (\hat{\alpha}_i)^2$

Factor B $\qquad$ SSB $= an \Sigma_j (\bar{y}_{.j.} - \bar{y}_{...})^2 = an \Sigma_j (\hat{\beta}_j)^2$

Interaction $\qquad$ SSAB $= n \Sigma_{ij} (\bar{y}_{ij.} - \bar{y}_{i..} - \bar{y}_{.j.} + \bar{y}_{...})^2 = n \Sigma_{ij} (\widehat{\alpha\beta}_{ij})^2$

Error $\qquad$ SSE $= \Sigma_{ijk} (y_{ijk} - \bar{y}_{ij.})^2 = \Sigma_{ijk} (e_{ijk})^2$

5. $100(1 - \alpha)\%$ confidence interval for difference in treatment means

$$(\bar{y}_i - \bar{y}_{i'}) \pm t_{\alpha/2, \mathrm{df}_{\mathrm{error}}} \, s_\varepsilon \sqrt{2/n}$$

Supplementary Exercises

Gov. $\quad$ **9.17** The city manager of a large midwestern city was negotiating with the three unions that represented the policemen, firemen, and building inspectors over the salaries for these groups of employees. The three unions claimed that the starting salaries were substantially different among the three groups, whereas in most cities there was not a significant difference in starting salaries among the three groups. To obtain information on starting salaries across the nation, the city manager decided to randomly select one city in each of eight geographical regions. The starting yearly salaries (in thousands of dollars) were obtained for each of the three groups in each of the eight regions. The data appear here.

Region	1	2	3	4	5	6	7	8	Mean
Policemen	32.3	33.2	30.8	30.5	30.1	30.2	28.4	27.9	30.42
Fireman	31.9	32.8	31.6	31.2	30.8	30.6	28.7	27.5	30.64
Inspectors	27.9	27.8	26.5	26.8	26.4	26.8	25.3	25.9	26.68
Region Mean	30.7	31.3	29.6	29.5	29.1	29.2	27.5	27.1	29.25

a. Write a model for this study, identifying all the terms in the model.
b. Using the analysis of variance from the Minitab computer output shown here, do the data suggest a difference in mean starting salary for the three groups of employees? Use $\alpha = .05$.
c. Give the level of significance for your test.
d. Which pairs of job types have significantly different starting salaries?

```
Two-Way Analysis of Variance for Salary
Source         DF        SS         MS         F          P
REGION          7      42.620      6.089     14.42      0.000
JOB             2      79.491     39.745     94.16      0.000
Error          14       5.909      0.422
Total          23     128.020
```

9.18 Refer to Exercise 9.17.
   a. Plot the data in a profile plot with factors job type and region. Does there appear to be an interaction between the two factors? If there were an interaction, could you test for it using the given data? If not, why not?
   b. Did the geographical region variable increase the efficiency of the design over conducting the study as a completely randomized design in which the city manager would have just randomly selected eight cities regardless of their location?
   c. Identify additional sources of variability that may need to be included in future studies.

Ag. **9.19** An experiment was set up to compare the effect of different soil pH values and calcium additives on the increase in trunk diameters for orange trees. Annual applications of elemental sulfur, gypsum, soda ash, and other ingredients were applied to provide pH value levels of 4, 5, 6, and 7. Three levels of a calcium supplement (100, 200, and 300 pounds per acre) were also applied. All factor–level combinations of these two variables were used in the experiment. At the end of a 2-year period, three diameters were examined at each factor–level combination. The data appear next.

pH Value	Calcium 100	200	300	pH Value	Calcium 100	200	300
	5.2	7.4	6.3		7.6	7.6	7.2
4.0	5.9	7.0	6.7	6.0	7.2	7.5	7.3
	6.3	7.6	6.1		7.4	7.8	7.0
	7.1	7.4	7.3		7.2	7.4	6.8
5.0	7.4	7.3	7.5	7.0	7.5	7.0	6.6
	7.5	7.1	7.2		7.2	6.9	6.4

      a. Construct a profile plot. What do the data suggest?
      b. Write an appropriate statistical model.
      c. Perform an analysis of variance and identify the experimental design. Use $\alpha = .05$.

9.20 Refer to Exercise 9.19.
      a. Use the computer output given here to test for interactions and main effects. Use $\alpha = .05$.
      b. What can you conclude about the effects of pH and calcium on the increase in the mean trunk diameters for orange trees?

```
General Linear Models Procedure for Exercise 9.20

Dependent Variable: Increase in Tree Diameter

                            Sum of           Mean
Source              DF      Squares          Square      F Value    Pr > F
Model               11      9.1830556        0.8348232    12.32     0.0001
Error               24      1.6266667        0.0677778
Corrected Total     35     10.8097222


          R-Square              C.V.          Root MSE              Y Mean
          0.849518          3.691335           0.2603              7.0528


Source              DF     Type III SS     Mean Square    F Value    Pr > F
PH                   3      4.4608333       1.4869444      21.94     0.0001
CA                   2      1.4672222       0.7336111      10.82     0.0004
PH*CA                6      3.2550000       0.5425000       8.00     0.0001
```

9.21 Refer to Exercises 9.19 and 9.20.
      a. Use Tukey's W procedure to determine the differences in mean increase in trunk diameters among the three calcium rates. Use $\alpha = .05$.
      b. Are your conclusions about the differences in mean increase in diameters among the three calcium rates the same for all four pH values?

```
          Level of                    --------------Y--------------
          PH          N               Mean                 SD
          4           9               6.50000000           0.75828754
          5           9               7.31111111           0.15365907
          6           9               7.40000000           0.25000000
          7           9               7.00000000           0.36400549


          Level of                    --------------Y--------------
          CA          N               Mean                 SD
          100         12              6.95833333           0.75252102
          200         12              7.33333333           0.28069179
          300         12              6.86666667           0.45193188
```

Level of PH	Level of CA	N	----------Y----------- Mean	SD
4	100	3	5.80000000	0.55677644
4	200	3	7.33333333	0.30550505
4	300	3	6.36666667	0.30550505
5	100	3	7.33333333	0.20816660
5	200	3	7.26666667	0.15275252
5	300	3	7.33333333	0.15275252
6	100	3	7.40000000	0.20000000
6	200	3	7.63333333	0.15275252
7	300	3	7.16666667	0.15275252
7	100	3	7.30000000	0.17320508
7	200	3	7.10000000	0.26457513
7	300	3	6.60000000	0.20000000

9.22 Refer to Exercise 9.19.

a. Use the residual analysis contained in the computer output given here to determine whether any of the conditions required to conduct an appropriate F test have been violated.

b. If any of the conditions have been violated, suggest ways to overcome these difficulties.

Variable-RESIDUALS

```
                  Moments
N                    36    Sum Wgts         36
Mean                  0    SUM               0
Std Dev        0.215583    Variance   0.046476
Skewness       -0.22276    Kurtosis   0.699897

W:Normal       0.986851    Pr<W         0.9562
```

Variable-RESIDUALS

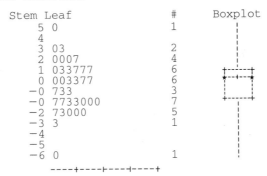

```
     Stem Leaf                 #      Boxplot
        5 0                    1         |
        4                                |
        3 03                   2         |
        2 0007                 4         |
        1 033777               6      +-----+
        0 003377               6      *--+--*
       -0 733                   3      +-----+
       -0 7733000               7         |
       -2 73000                 5         |
       -3 3                      1         |
       -4                                 |
       -5                                 |
       -6 0                     1         |
          ----+----+----+----+
```

Multiply Stem.Leaf by 10**-1

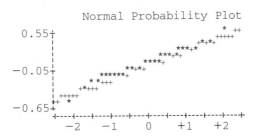

```
            Normal Probability Plot
                                    *  ++
    0.55+                         +++++
        |                    ***+*+
        |                  ***+*+
        |               ****+
        |              ****+
   -0.05+        **+*+
        |     ******+
        |    * *+++
        |   +*+++
        | ++++
        |++  *
   -0.65++
        +----+----+----+----+----+----+----+----+----+
            -2   -1    0   +1   +2
```

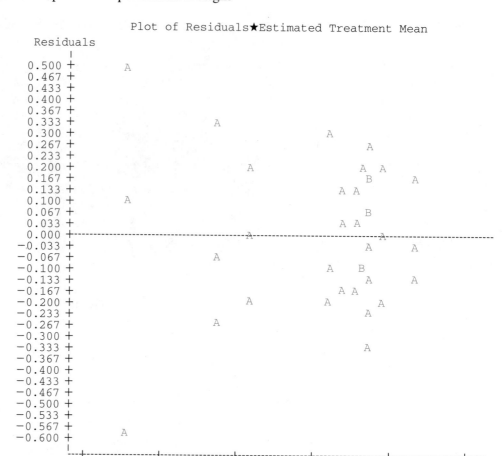

```
                        Plot of Residuals★Estimated Treatment Mean
     Residuals
        |
 0.500 +            A
 0.467 +
 0.433 +
 0.400 +
 0.367 +
 0.333 +                    A
 0.300 +                              A
 0.267 +                                   A
 0.233 +
 0.200 +                    A       A  A
 0.167 +                                B       A
 0.133 +                          A  A
 0.100 +      A
 0.067 +                                B
 0.033 +                          A  A
 0.000 +----------------------A----------------A-------------------------
-0.033 +                              A     A
-0.067 +            A
-0.100 +                          A  B
-0.133 +                              A     A
-0.167 +                          A  A
-0.200 +            A             A     A
-0.233 +                                A
-0.267 +            A
-0.300 +
-0.333 +                              A
-0.367 +
-0.400 +
-0.433 +
-0.467 +
-0.500 +
-0.533 +
-0.567 +
-0.600 +     A
        |
        ---+----------------+----------------+----------------+----------------+----------------+-----
          5.5              6.0              6.5              7.0              7.5              8.0

                              Estimated Treatment Mean
```

Env.   **9.23**   An experiment was conducted to investigate the heat loss for five different designs for commercial thermal panes. The researcher, in order to obtain results that would be applicable throughout most regions of the country, decided to evaluate the panes at five temperatures, 0°F, 20°F, 40°F, 60°F, and 80°F. A sample of 10 panes of each design was obtained. Two panes of each design were randomly assigned to each of the five exterior temperature settings. The interior temperature of the test was controlled at 70°F for all five exterior temperatures. The heat losses associated with the five pane designs are given here.

Exterior Temperature Setting (°F)	Pane Design				
	A	**B**	**C**	**D**	**E**
80	7.2, 7.8	7.1, 7.9	8.1, 8.8	8.3, 8.9	9.3, 9.8
60	8.1, 8.1	8.0, 8.9	8.2, 8.9	8.1, 8.8	9.2, 9.9
40	9.0, 9.9	9.2, 9.8	10.0, 10.8	10.2, 10.7	9.9, 9.0
20	9.2, 9.8	9.1, 9.9	10.1, 10.8	10.3, 10.9	9.3, 9.8
0	10.2, 10.8	10.1, 10.9	11.1, 11.8	11.3, 11.9	9.3, 9.9

a. Identify the experimental design and write an appropriate statistical model.
b. Is there a significant difference in the mean heat loss of the five pane designs? Use $\alpha = .05$. An AOV table for the data is given next.
c. Are the differences in the five designs consistent across the five temperatures? Use $\alpha = .05$ and a profile plot in reaching your conclusion.
d. Use Tukey's W procedure at an $\alpha = .05$ level to compare the mean heat loss for the five pane designs.

```
General Linear Models Procedure for Exercise 9.23

Dependent Variable: HEAT LOSS

Source                    DF    Sum of Squares    F Value     Pr > F

Model                     24      58.07280000      10.47       0.0001
Error                     25       5.78000000
Corrected Total           49      63.85280000


                   R-Square                 C.V.              Y Mean
                   0.909479              5.067797          9.48800000


Source                    DF    Type III SS       F Value     Pr > F
T                          4      39.77880000       43.01       0.0001
D                          4       7.32280000        7.92       0.0003
T*D                       16      10.97120000        2.97       0.0073


T     D    N        Mean        T      N       Mean        D      N       Mean
0     a    2     10.5000000                             a     10     9.01000000
0     b    2     10.5000000     0     10    10.7300000    b     10     9.09000000
0     c    2     11.4500000    20     10     9.9200000    c     10     9.86000000
0     d    2     11.6000000    40     10     9.8500000    d     10     9.94000000
0     e    2      9.5000000    60     10     8.6200000    e     10     9.54000000
20    a    2      9.5000000    80     10     8.3200000
20    b    2      9.5000000
20    c    2     10.4500000
20    d    2     10.6000000
```

Psy. **9.24** An experiment was conducted to examine the effects of different levels of reinforcement and different levels of isolation on children's ability to recall. A single analyst was to work with a random sample of 36 children selected from a relatively homogeneous group of fourth-grade students. Two levels of reinforcement (none and verbal) and three levels of isolation (20, 40, and 60 minutes) were to be used. Students were randomly assigned to the six treatment groups, with a total of six students being assigned to each group.

Each student was to spend a 30-minute session with the analyst. During this time, the student was to memorize a specific passage, with reinforcement provided as dictated by the group to which the student was assigned. Following the 30-minute session, the student was isolated for the time specified for his or her group and then tested for recall of the memorized passage. The data appear next.

Level of Reinforcement	Time of Isolation (minutes)					
	20		**40**		**60**	
None	26	19	30	36	6	10
	23	18	25	28	11	14
	28	25	27	24	17	19
Verbal	15	16	24	26	31	38
	24	22	29	27	29	34
	25	21	23	21	35	30

Use the computer output shown here to draw your conclusions.

```
General Linear Models Procedure for Exercise 9.24

Dependent Variable: TEST SCORE

                              Sum of          Mean
Source             DF         Squares         Square      F Value    Pr > F
Model               5       1410.8889        282.1778      17.88     0.0001
Error              30        473.3333         15.7778
Corrected Total    35       1884.2222

                R-Square           C.V.        Root MSE              Y Mean
                0.748791        16.70520         3.9721              23.778

Source             DF      Type III SS     Mean Square    F Value    Pr > F
REINFORCE           1       196.0000        196.0000       12.42     0.0014
TIME                2       156.2222         78.1111        4.95     0.0139
INTERACTION         2      1058.6667        529.3333       33.55     0.0001
```

Med. **9.25** Researchers were interested in the stability of a drug product stored for four storage times (1, 3, 6, and 9 months). The drug was manufactured with 30 mg/mL of active ingredient of a drug product, and the amount of active ingredient of the drug at the end of the storage period was to be determined. The drug was stored at a constant temperature of 30°C. Two laboratories were used in the study with three 2-mL vials of the drug randomly assigned to each of the four storage times. At the end of the storage time, the amount of the active ingredient was determined for each of the vials. A measure of the pH of the drug was also recorded for each vial. The data are given here.

Time (in months at 30°C)	Laboratory	mg/mL of Active Ingredient	pH	Time (in months at 30°C)	Laboratory	mg/mL of Active Ingredient	pH
1	1	30.03	3.61	1	2	30.12	3.87
1	1	30.10	3.60	1	2	30.10	3.80
1	1	30.14	3.57	1	2	30.02	3.84
3	1	30.10	3.50	3	2	29.90	3.70
3	1	30.18	3.45	3	2	29.95	3.80
3	1	30.23	3.48	3	2	29.85	3.75
6	1	30.03	3.56	6	2	29.75	3.90

Time (in months at 30°C)	Laboratory	mg/mL of Active Ingredient	pH	Time (in months at 30°C)	Laboratory	mg/mL of Active Ingredient	pH
6	1	30.03	3.74	6	2	29.85	3.90
6	1	29.96	3.81	6	2	29.80	3.90
9	1	29.81	3.60	9	2	29.75	3.77
9	1	29.79	3.55	9	2	29.85	3.74
9	1	29.82	3.59	9	2	29.80	3.76

a. Write a model relating the response measured on each vial to the factors length of storage time and laboratory.

b. Display an analysis of variance table for the model of part (a) without computing the necessary sum of squares.

9.26 Refer to Exercise 9.25. The computer output is shown for an analysis of variance for both dependent variables (i.e., y_1 = mg/mL of active ingredient and y_2 = pH). Draw conclusions about the stability of these 2-mL vials based on these analyses. Use $\alpha = .05$.

```
General Linear Models Procedure for Exercise 9.26

Dependent Variable: MG/ML

                           Sum of          Mean
Source              DF     Squares        Square      F Value      Pr > F
Model                7     0.4674000      0.0667714    27.30       0.0001
Error               16     0.0391333      0.0024458
Corrected Total     23     0.5065333

            R-Square          C.V.         Root MSE                  Y1 Mean
            0.922743        0.165090         0.0495                  29.957

Source              DF     Type III SS   Mean Square   F Value      Pr > F
TIME                 3     0.2937667     0.0979222     40.04        0.0001
LAB                  1     0.0912667     0.0912667     37.32        0.0001
TIME*LAB             3     0.0823667     0.0274556     11.23        0.0003

General Linear Models Procedure

Dependent Variable: pH

                           Sum of          Mean
Source              DF     Squares        Square      F Value      Pr > F
Model                7     0.4201625      0.0600232    21.47       0.0001
Error               16     0.0447333      0.0027958
Corrected Total     23     0.4648958

            R-Square          C.V.         Root MSE                  Y2 Mean
            0.903778        1.429232         0.0529                  3.6996

Source              DF     Type III SS   Mean Square   F Value      Pr > F
TIME                 3     0.1144458     0.0381486     13.64        0.0001
LAB                  1     0.2970375     0.2970375    106.24        0.0001
TIME*LAB             3     0.0086792     0.0028931      1.03        0.4038
```

9.27 Refer to Exercise 9.25. The same type of data on mg/mL and pH were generated at 40°C as were obtained at 30°C in Exercise 9.25. The data are shown here.

Time (in months at 40°C)	Laboratory	mg/mL of Active Ingredient	pH	Time (in months at 40°C)	Laboratory	mg/mL of Active Ingredient	pH
1	1	30.08	3.61	1	2	30.12	3.80
1	1	30.10	3.60	1	2	30.10	3.70
1	1	30.14	3.59	1	2	30.02	3.81
3	1	30.03	3.39	3	2	29.90	3.70
3	1	30.18	3.45	3	2	29.85	3.80
3	1	30.26	3.29	3	2	29.80	3.75
6	1	29.90	3.63	6	2	29.75	3.80
6	1	29.90	3.71	6	2	29.70	3.70
6	1	29.96	3.65	6	2	29.75	3.70
9	1	29.81	3.51	9	2	29.65	3.64
9	1	29.85	3.38	9	2	29.75	3.68
9	1	29.72	3.32	9	2	29.70	3.60

a. The computer output is shown here for the analysis of variance of part (a) for both response variables y_1 = mg/mL and y_2 = pH. What is the effect of the three factors on the mean active ingredient of the drug? Use $\alpha = .05$.

b. Using the computer output for the analysis of variance of part (a) for both response variables y_1 = mg/mL and y_2 = pH, what is the effect of the three factors on the mean pH of the drug? Use $\alpha = .05$ and profile plots in reaching your conclusions.

c. Are your conclusions about the effects of storage time and temperature on mean pH the same for both laboratories?

d. Are your conclusions about the effects of storage time and temperature on mean active ingredient of the drug the same for both laboratories?

```
General Linear Models Procedure for Exercise 9.27

Dependent Variable: MG/ML

                              Sum of          Mean
Source            DF         Squares        Square     F Value    Pr > F
Model             15       1.1475917     0.0765061      25.29     0.0001
Error             32       0.0968000     0.0030250
Corrected Total   47       1.2443917

                  R-Square          C.V.      Root MSE              Y1 Mean
                  0.922211      0.183719        0.0550               29.937

Source            DF     Type III SS   Mean Square    F Value    Pr > F
TIME               3       0.7360083     0.2453361      81.10     0.0001
LAB                1       0.2296333     0.2296333      75.91     0.0001
TIME*LAB           3       0.1443167     0.0481056      15.90     0.0001
TEMP               1       0.0184083     0.0184083       6.09     0.0192
TIME*TEMP          3       0.0120750     0.0040250       1.33     0.2817
LAB*TEMP           1       0.0027000     0.0027000       0.89     0.3519
TIME*LAB*TEMP      3       0.0044500     0.0014833       0.49     0.6914
```

```
General Linear Models Procedure

Dependent Variable: pH
                                   Sum of          Mean
Source                DF          Squares        Square     F Value    Pr > F
Model                 15        0.9848667     0.0656578       20.48    0.0001
Error                 32        0.1026000     0.0032063
Corrected Total       47        1.0874667

              R-Square              C.V.        Root MSE                 Y2 Mean
              0.905652          1.547802          0.0566                  3.6583

Source                DF     Type III SS    Mean Square    F Value    Pr > F
TIME                   3       0.2286000      0.0762000       23.77    0.0001
LAB                    1       0.5676750      0.5676750      177.05    0.0001
TIME*LAB               3       0.0555583      0.0185194        5.78    0.0028
TEMP                   1       0.0816750      0.0816750       25.47    0.0001
TIME*TEMP              3       0.0250917      0.0083639        2.61    0.0686
LAB*TEMP               1       0.0003000      0.0003000        0.09    0.7617
TIME*LAB*TEMP          3       0.0259667      0.0086556        2.70    0.0621
```

Bus.   **9.28** A manufacturer frequently sends small packages to a customer in another city via air freight, and, in many cases, it is important for a package to reach the customer as soon as possible. Three different firms offer air freight service, including pickup and delivery, on a 24-hour basis. The head of the manufacturer's shipping department would like to know whether the firms differ in speed of service and whether the time of day makes any difference. An experiment is designed to investigate these issues. Packages are sent at random times, and the air freight firm used for each package is also randomly chosen. The customer records the time that each package arrives, so that the time elapsed during shipment can be determined. These times are rounded to the nearest hour. The experimental results for a total of 54 packages are shown in the following table.

	Firm		
Time	**1**	**2**	**3**
Morning	8, 6, 6, 12, 7, 8	11, 11, 9, 10, 8, 11	7, 4, 6, 4, 9, 7
Afternoon	7, 10, 8, 11, 9, 11	10, 13, 10, 12, 11, 10	10, 8, 6, 5, 8, 6
Night	13, 11, 14, 11, 9, 12	12, 6, 9, 9, 10, 6	8, 11, 9, 9, 10, 12

a. What evidence is relevant for deciding whether the choice of best firm will be different at different times of the day? What conclusion can you draw using a 5% level of significance? Construct a graph that depicts the nature of any differences in firm as a function of the time of day.

b. Does any firm appear to be better than the other two firms? How could you compare the best firm and the second-best firm using a confidence interval?

```
General Linear Models Procedure for Exercise 9.28

Dependent Variable: SPEED

                             Sum of          Mean
Source                 DF    Squares        Square      F Value    Pr > F

Model                   8   154.37037     19.29630        6.06     0.0001
Error                  45   143.33333      3.18519
Corrected Total        53   297.70370

            R-Square         C.V.        Root MSE                  Y Mean
            0.518537      19.66822         1.7847                  9.0741

Source                 DF   Type III SS   Mean Square    F Value    Pr > F
TIME                    2    38.259259     19.129630        6.01     0.0049
FIRM                    2    50.037037     25.018519        7.85     0.0012
TIME*FIRM               4    66.074074     16.518519        5.19     0.0016

           MEAN SPEED FOR TREATMENTS AND EACH FACTOR

   TIME    FIRM    MEAN   |  TIME  MEAN      FIRM    MEAN
--------------------------|--------------------------------------
    A       1     9.33    |   A    9.167      1     9.611
    A       2    11.00    |   M    8.000      2     9.888
    A       3     7.16    |   N   10.056      3     7.722
    M       1     7.83    |
    M       2    10.00    |
    M       3     6.16    |
    N       1    11.66    |
    N       2     8.66    |
    N       3     9.83    |
```

Ag. **9.29** The yields of wheat (in pounds) are shown here for five farms. Five plots are selected based on their soil fertility at each farm with the most fertile plots designated as 1. The treatments (fertilizers) applied to each plot are shown in parentheses.

Farm	Fertility				
	1	**2**	**3**	**4**	**5**
1	(D) 10.3	(E) 8.6	(A) 6.7	(C) 7.6	(B) 5.8
2	(E) 8.8	(B) 6.7	(C) 6.7	(A) 4.8	(D) 6.0
3	(A) 6.3	(C) 8.3	(B) 6.8	(D) 8.0	(E) 8.8
4	(C) 8.9	(D) 7.4	(E) 8.2	(B) 6.2	(A) 4.4
5	(B) 7.3	(A) 4.4	(D) 7.7	(E) 6.8	(C) 6.7

a. Identify the designs.
b. Do an analysis of variance and draw conclusions concerning the five fertilizers. Use $\alpha = .01$.

9.30 Refer to Exercise 9.29. Run a multiple comparison procedure to make all pairwise comparisons of the treatment means. Identify which error rate was controlled.

```
General Linear Models Procedure for Exercises 9.29 and 9.30

Dependent Variable: YIELD
                                  Sum of           Mean
Source                  DF       Squares          Square    F Value    Pr > F
Model                   12     46.067200        3.838933       9.88    0.0002
Error                   12      4.663200        0.388600
Corrected Total         24     50.730400

                R-Square            C.V.        Root MSE              Y Mean
                0.908079        8.745481          0.6234              7.1280

Source                  DF    Type III SS    Mean Square    F Value    Pr > F

FARM                     4      6.522400       1.630600       4.20    0.0236
PLOT                     4     11.266400       2.816600       7.25    0.0033
FERT                     4     28.278400       7.069600      18.19    0.0001


     FARM     Mean   |   PLOT    Mean   |  FERTILIZER    Mean
    ------------------|-------------------|----------------------
      1      7.80   |    1      8.32   |    A          5.32
      2      6.60   |    2      7.08   |    B          6.56
      3      7.64   |    3      7.22   |    C          7.64
      4      7.02   |    4      6.68   |    D          7.88
      5      6.58   |    5      6.34   |    E          8.24
```

CHAPTER 10

Categorical Data

10.1   Introduction

Up to this point, we have been concerned primarily with sample data measured on a quantitative scale. However, we sometimes encounter situations in which levels of the variable of interest are identified by name or rank only and we are interested in the number of observations occurring at each level of the variable. Data obtained from these types of variables are called **categorical** or **count data.** For example, an item coming off an assembly line may be classified into one of three quality classes: acceptable, second, or reject. Similarly, a traffic study might require a count and classification of the type of transportation used by commuters along a major access road into a city. A pollution study might be concerned with the number of different alga species identified in samples from a lake and the number of times each species is identified. A consumer protection group might be interested in the results of a prescription fee survey to compare prices on a checklist of medications in different sections of a large city.

In this chapter, we will examine specific inferences that can be made from experiments involving categorical data.

10.2   Inferences about a Population Proportion π

In the binomial experiment discussed in Chapter 4, each trial results in one of two outcomes, which we labeled as either a success or a failure. We designated π as the probability of a success and $(1 - \pi)$ as the probability of a failure. Then the probability distribution for y, the number of successes in n identical trials, is

$$P(y) = \frac{n!}{y!(n - y)!}\pi^y(1 - \pi)^{n-y}$$

The point estimate of the binomial parameter π is one that we would choose intuitively. In a random sample of n from a population in which the proportion of elements classified as successes is π, the best estimate of the parameter π is the sample proportion of successes. Letting y denote the number of successes in the n sample trials, the sample proportion is

$$\hat{\pi} = \frac{y}{n}$$

We observed in Section 4.10 that y possesses a mound-shaped probability distribution that can be approximated by using a normal curve when

$$n \geq \frac{5}{\min(\pi, 1 - \pi)} \qquad \text{(or equivalently, } n\pi \geq 5 \text{ and } n(1 - \pi) \geq 5\text{)}$$

In a similar way, the distribution of $\hat{\pi} = y/n$ can be approximated by a normal distribution with a mean and a standard error as given here.

Mean and Standard Error of $\hat{\pi}$

$$\mu_{\hat{\pi}} = \pi$$

$$\sigma_{\hat{\pi}} = \sqrt{\frac{\pi(1 - \pi)}{n}}$$

The normal approximation to the distribution of $\hat{\pi}$ can be applied under the same condition as was required for approximating y by using a normal distribution. The traditional method for developing inference procedures for π are based on replacing π with $\hat{\pi}$ in the expression for $\sigma_{\hat{\pi}}$. However, in an article by Agresti and Coull (1998), it is shown that confidence intervals for π based on this method can be considerably inaccurate, even when n is large. They suggest the following adjustment to $\hat{\pi}$:

$$\hat{\pi} = \frac{y + 2}{n + 4}$$

This estimator was first suggested by Edwin Wilson in 1927. Hence, the authors name $\hat{\pi}$ the Wilson estimator of π.

A confidence interval can be obtained for π using $\hat{\pi}$ and the methods of Chapter 5 for μ by replacing $\bar{y}$ with $\hat{\pi}$ and $\sigma_{\bar{y}}$ with $\sigma_{\hat{\pi}}$. A general $100(1 - \alpha)\%$ confidence interval for the population proportion π is given here.

Confidence Interval for π, with Confidence Coefficient of $(1 - \alpha)$

$$\hat{\pi} \pm z_{\alpha/2}\hat{\sigma}_{\hat{\pi}}$$

where

$$\hat{\pi} = \frac{y + 2}{n + 4} \quad \text{and} \quad \hat{\sigma}_{\hat{\pi}} = \sqrt{\frac{\hat{\pi}(1 - \hat{\pi})}{n + 4}}$$

EXAMPLE 10.1

Researchers in the development of new treatments for cancer patients often evaluate the effectiveness of new therapies by reporting the proportion of patients who survive for a specified period of time after completion of the treatment. A new genetic treatment of 870 patients with a particular type of cancer resulted in 330 patients surviving at least 5 years after treatment. Estimate the proportion of all patients with the specified type of cancer who will survive at least 5 years after being administered this treatment. Use a 90% confidence interval.

Solution   For these data,

$$\hat{\pi} = \frac{330 + 2}{870 + 4} = .38$$

$$\hat{\sigma}_{\hat{\pi}} = \sqrt{\frac{(.38)(1 - .38)}{870 + 4}} = .016$$

The confidence coefficient for our example is .90. Recall from Chapter 5 that we can obtain $z_{\alpha/2}$ by looking up the z-value in Table 1 in the Appendix corresponding to an area of $(\alpha/2)$. For a confidence coefficient of .90, the z-value corresponding to an area of .05 is 1.645. Hence, the 90% confidence interval on the proportion of cancer patients who will survive at least 5 years after receiving the new genetic treatment is

$$.38 \pm 1.645(.016) \quad \text{or} \quad .38 \pm .026$$

The confidence interval for π is based on a normal approximation to a binomial, which is appropriate provided n is sufficiently large. The rule we have specified is that both $n\pi$ and $n(1 - \pi)$ should be at least 5, but because π is the unknown parameter, we'll require that $n\hat{\pi}$ and $n(1 - \hat{\pi})$ be at least 5. When the sample size is too small and violates this rule, the confidence interval usually will be too wide to be of any use. For example, with $n = 20$ and $\hat{\pi} = .2$, the rule is not satisfied because $n\hat{\pi} = 4$. The 95% confidence interval based on these data is $.025 < \pi < .375$, which is practically useless. Very few product managers would be willing to launch a new product if the expected increase in market share was between .025 and .375.

Another problem that arises in the estimation of π occurs when π is very close to zero or one. In these situations, the population proportion is often estimated to be 0 or 1, respectively, unless the sample size is extremely large. These estimates are not realistic because they suggest that either no successes or no failures exist in the population. Rather than estimating π using the formula $\hat{\pi}$ given previously, adjustments are provided to prevent the estimates from being so extreme. One of the proposed adjustments is to use

$$\hat{\pi}_{\text{Adj.}} = \frac{\frac{3}{8}}{(n + \frac{3}{4})} \quad \text{when } y = 0$$

and

$$\hat{\pi}_{\text{Adj.}} = \frac{(n + \frac{3}{8})}{(n + \frac{3}{4})} \quad \text{when } y = n$$

When computing the confidence interval for π in those situations where $y = 0$ or $y = n$, the confidence intervals using the normal approximation would not be valid. We can use the following confidence intervals, which are derived from using the binomial distribution.

100(1 − α)% Confidence Interval for π, when y = 0 or y = n

When $y = 0$, the confidence interval is $(0, 1 - (\alpha/2)^{1/n})$.
When $y = n$, the confidence interval is $((\alpha/2)^{1/n}, 1)$.

EXAMPLE 10.2

A new PC operating system is being developed. The designer claims the new system will be compatible with nearly all computer programs currently being run on Microsoft's Windows operating system. A sample of 50 programs are run and all 50 programs perform without error. Estimate π, the proportion of all Microsoft's Windows-compatible programs that will run without change on the new operating system. Compute a 95% confidence interval for π.

Solution If we use the standard estimator of π, we obtain

$$\hat{\pi} = \frac{50}{50} = 1.0$$

Thus, we would conclude that 100% of all programs that are Microsoft's Windows-compatible programs will run without alteration on the new operating system. Is this conclusion valid? Probably not, because we have only investigated a tiny fraction of all Microsoft's Windows-compatible programs. Thus, we will use the alternative estimators and confidence interval procedures. The point estimator is given by

$$\hat{\pi}_{Adj.} = \frac{(n + \frac{3}{8})}{(n + \frac{3}{4})} = \frac{(50 + \frac{3}{8})}{(50 + \frac{3}{4})} = .993$$

A 95% confidence interval for π is

$$((\alpha/2)^{1/n}, 1) = ((.05/2)^{1/50}, 1) = ((.025)^{.02}, 1) = (.929, 1.0)$$

We can now conclude that we are reasonably confident (95%) a high proportion (between 92.9% and 100%) of all programs that are Microsoft's Windows-compatible will run without alteration on the new operating system.

Keep in mind, however, that a sample size that is sufficiently large to satisfy the rule *does not* guarantee that the interval will be informative. It only judges the adequacy of the normal approximation to the binomial—the basis for the confidence level.

Sample size calculations for estimating π follow very closely the procedures we developed for inferences about μ. The required sample size for a $100(1 - \alpha)\%$ confidence interval for π of the form $\hat{\pi} \pm E$ (where E is specified) is found by solving the expression

$$z_{\alpha/2}\sigma_{\hat{\pi}} = E$$

for n. The result is shown here.

Sample Size Required for a 100(1 − α)% Confidence Interval for π of the Form $\hat{\pi} \pm E$

$$n = \frac{z_{\alpha/2}^2 \pi(1 - \pi)}{E^2}$$

Note: Because π is not known, either substitute an educated guess or use π = .5. Use of π = .5 will generate the largest possible sample size for the specified confidence interval width, 2E, and thus will give a conservative answer to the required sample size.

EXAMPLE 10.3

A large public opinion polling agency plans to conduct a national survey to determine the proportion of employed adults who fear losing their job within the next year. How many workers must the agency poll to estimate to within .02 using a 95% confidence interval?

Solution   By design, the agency wants the interval of the form $\hat{\pi} \pm .02$. The sample size necessary to achieve this accuracy is given by

$$n = \frac{z_{\alpha/2}^2 \pi(1 - \pi)}{E^2}$$

where $z_{\alpha/2} = 1.96$ and $E = .02$. If a previous survey has been run recently, we can use the sample proportion from that survey to substitute for π; otherwise, we can use π = .5. Using π = .5, the required sample size is

$$n = \frac{(1.96)^2(.5)(.5)}{(.02)^2} = 2,401$$

that is, 2,401 workers would have to be surveyed to estimate π to within .02.

A statistical test about a binomial parameter π is very similar to the large-sample test concerning a population mean presented in Chapter 5. These results are summarized next, with three different alternative hypotheses along with their corresponding rejection regions. Recall that only one alternative is chosen for a particular problem.

Summary of a Statistical Test for π, π_0 Is Specified

H_0:   **1.** $\pi \leq \pi_0$ H_a: **1.** $\pi > \pi_0$
      2. $\pi \geq \pi_0$ **2.** $\pi < \pi_0$
      3. $\pi = \pi_0$ **3.** $\pi \neq \pi_0$

T.S.: $z = \dfrac{\hat{\pi} - \pi_0}{\sigma_{\hat{\pi}}}$

R.R.:   For a probability α of a Type I error

    1. Reject H_0 if $z > z_\alpha$.
    2. Reject H_0 if $z < -z_\alpha$.
    3. Reject H_0 if $|z| > z_{\alpha/2}$.

Note: Under H_0,

$$\sigma_{\hat{\pi}} = \sqrt{\frac{\pi_0(1 - \pi_0)}{n}}$$

Also, n must satisfy both $n\pi_0 \geq 5$ and $n(1 - \pi_0) \geq 5$.

Check assumptions and draw conclusions.

EXAMPLE 10.4

Sports car owners in a town complain that the state vehicle inspection station judges their cars differently from family-style cars. Previous records indicate that 30% of all passenger cars fail the inspection on the first time through. In a random sample of 150 sports cars, 60 failed the inspection on the first time through. Is there sufficient evidence to indicate that the percentage of first failures for sports cars is higher than the percentage for all passenger cars? Use $\alpha = .05$.

Solution The appropriate statistical test is as follows:

$$H_0: \quad \pi \leq .30$$

$$H_a: \quad \pi > .30$$

$$\text{T.S.:} \quad z = \frac{\hat{\pi} - \pi_0}{\sigma_{\hat{\pi}}}$$

R.R.: For $\alpha = .05$, we will reject H_0 if $z > 1.645$.

Using the sample data,

$$\hat{\pi} = \frac{60}{150} = .4 \quad \text{and} \quad \sigma_{\hat{\pi}} = \sqrt{\frac{(.3)(.7)}{150}} = .037$$

Also,

$$n\pi_0 = 150(.3) = 45 \geq 5$$

and

$$n(1 - \pi_0) = 150(.7) = 105 \geq 5$$

The test statistic is then

$$z = \frac{.4 - .3}{.037} = 2.70$$

Because the observed value of z exceeds 1.645, we conclude that sports cars at the vehicle inspection station have a first-failure rate greater than .3 (p-value = .0035). However, we must be careful not to attribute this difference to a difference in standards for sports cars and family-style cars. Parallel testing of sports cars versus other cars would have to be conducted to eliminate other sources of variability that would perhaps account for the higher first-failure rate for sports cars.

Most computer packages do not include this test. However, by coding successes as 1s and failures as 0s, we can approximate this test. This trick works exactly if the package includes a z test. Specify σ as $\sqrt{(\pi_0)(1 - \pi_0)}$. If the package includes a one-sample t test, the result will be slightly different, but often close enough. For example, suppose we want to compare two versions of a product. We ask a random sample of 100 potential customers to choose between a new version of the product and the existing version. The new product will be adopted only if there is clear evidence that more than half of the customers prefer it. Thus we take the null hypothesis as $H_0: \pi \le .5$ and the research hypothesis as $H_a: \pi > .5$. If 68 of the 100 potential customers prefer the new product, we can run a z test and an approximate t test using Minitab. To do so, we input as data 68 1s and 32 0s in a column labeled "Yes_No." We obtained the following Minitab results; note that we specified σ as $\sqrt{(\pi_0)(1 - \pi_0)} = \sqrt{(.5)(1 - .5)} = .5$:

```
MTB > ZTest .5 .5 'Yes_No';
SUBC> Alternative 0.


Z-Test


Test of mu = 0.5000 vs mu not = 0.5000
The assumed sigma = 0.500


Variable      N      Mean      StDev     SE Mean        Z      P-Value
Yes_No      100    0.6800     0.4688      0.0500      3.60       0.0003


MTB > TTest 0.5 'Yes_No';
SUBC> Alternative 0.


T-Test of the Mean


Test of mu = 0.500 vs mu not = 0.5000


Variable      N      Mean      StDev     SE Mean        Z      P-Value
Yes_No      100    0.6800     0.4688      0.0469      3.84       0.0002
```

The results of the z-test procedure are exactly what we obtained by hand. The results of the t-test procedure are not quite the same, basically because that procedure used a sample standard deviation instead of $\sqrt{(\pi_0)(1 - \pi_0)}$. The conclusion is the same, however: The research hypothesis of a difference in preferences is strongly supported, with a very low p-value. Although the t test based on 1s and 0s is not exactly the same as the z test, usually the results are so similar that we needn't worry too much about the difference.

We said that the z test for π is approximate and works best if n is large and π_0 is not too near 0 or 1. A natural next question is: When can we use it? There are several rules to answer the question; none of them should be considered sacred. Our sense of the many studies that have been done is this: If either $n\pi_0$ or $n(1 - \pi_0)$ is less than about 2, treat the results of a z test very skeptically. If $n\pi_0$ and $n(1 - \pi_0)$ are at least 5, the z test should be reasonably accurate. For the same sample size,

sample-size requirement

tests based on extreme values of π_0 (for example, .001) are less accurate than tests for values of π_0, such as .05 or .10. For example, a test of $H_0: \pi = .0001$ with $n\pi_0 = 1.2$ is much more suspect than one for $H_0: \pi = .10$ with $n\pi_0 = 50$. If the issue becomes crucial, it's best to interpret the results skeptically or use Fisher's exact test (see Conover, 1998).

EXERCISES

Basic Techniques

10.1 Hypothetical sample results from a binomial experiment with $n = 150$ yielded $\hat{\pi} = .2$.
  a. Does this experiment satisfy the sample-test requirement for a confidence interval for π based on z? What sample sizes would be suspect, given the same sample proportion?
  b. Construct a 90% confidence interval for π.

10.2 Under what conditions can the formula $\hat{\pi} \pm z_{\alpha/2}\sigma_{\hat{\pi}}$ be used to express a confidence interval for π?

10.3 A random sample of 1,500 is drawn from a binomial population. If there are $y = 1,200$ successes,
  a. Construct a 95% confidence interval for π.
  b. Construct a 90% confidence interval for π.

Applications

Soc. **10.4** Experts have predicted that approximately 1 in 12 tractor-trailer units will be involved in an accident this year. One reason for this is that 1 in 3 tractor-trailer units has an imminently hazardous mechanical condition, probably related to the braking system on the vehicle. A survey of 50 tractor-trailer units passing through a weighing station confirmed that 19 had a potentially serious braking system problem.
  a. Do the binomial assumptions hold?
  b. Can a normal approximation to the binomial be applied here to get a confidence interval for π?
  c. Give a 95% confidence interval for π using these data. Is the interval informative? What could be done to decrease the width of the interval, assuming $\hat{\pi}$ remained the same?

Psy. **10.5** In a study of self-medication practices, a random sample of 1,230 adults completed a survey. Some of the medical conditions that were self-treated are shown here. Summarize the results of this part of the survey using a 95% confidence interval for each medical condition.

Medical Condition	Home Remedy	% Responding
Sore throat—not related to a cold	Salt water or baking soda mouthwash	30
Burns—other than sunburn	Cold water/butter	28
Overindulgence in alcohol	Homebrew	25
Overweight	Diet	22
Pain associated with injury	Hot or cold compress	21

Psy. **10.6** In the survey discussed in Exercise 10.5, 441 of the adults reported they had a cough or cold recently and 260 of the respondents said they had treated the condition with an over-the-counter (OTC) remedy. The data are summarized here.

Survey respondents reporting problem	441
Number of patients using any OTC remedy	260
Patients using specific classes of OTC remedies:	
Adult pain relievers	110
Adult cold caps/tabs	57
Cough remedies	44
Allergy/hay fever remedies	9
Liquid cold remedies	35
Sprays/inhalers	4
Children's pain reliever	22
Cough drops	13
Sore-throat lozenges/gum	9
Children's cold caps/tabs	13
Nose drops	9
Chest rubs/ointments	9
Anesthetic throat lozenges	4
Room vaporizers	4
Other product	4

    a. How might you organize and summarize these data? Would percentages help? Do the percentages add to 100%? Why or why not?

    b. Based on these data, which classes of OTC remedies could you summarize using a 95% confidence interval for π?

Med.   **10.7** Many individuals over the age of 40 develop an intolerance for milk and milk-based products. A dairy has developed a line of lactose-free products that are more tolerable to such individuals. To assess the potential market for these products, the dairy commissioned a market research study of individuals over 40 in its sales area. A random sample of 250 individuals showed that 86 of them suffer from milk intolerance. Based on the sample results, calculate a 90% confidence interval for the population proportion that suffers milk intolerance. Interpret this confidence interval.

Soc.   **10.8** Shortly before April 15 of the previous year, a team of sociologists conducted a survey to study their theory that tax cheaters tend to allay their guilt by holding certain beliefs. The team interviewed a total of 500 adults and asked them under what situations they think cheating on an income tax return is justified. The responses include:

     56% agree that "other people don't report all their income."
     50% agree that "the government is often careless with tax dollars."
     46% agree that "cheating can be overlooked if one is generally law abiding."

Assuming that the data are a simple random sample of the population of taxpayers (or tax-nonpayers), calculate 95% confidence intervals for the population proportion that agrees with each statement.

Pol. Sci.   **10.9** A national columnist recently reported the results of a survey on marriage and the family. Part of the column is paraphrased here.

The Ingredients of Marriage

    The Gallup people offered respondents a list of well-known ingredients. Here in the United States, such elements as faithfulness, mutual respect, and understanding ranked at the top. These were followed by enough money, same background, good housing, and agreement in politics. Seventy-five percent of the respondents voted for "a good sex life," 59% for children, 52% for common interests, 48% for "living away from in-laws," and 43% for "sharing household chores." (In West Germany, by contrast, only 52% voted for a good sex life and only 19% for sharing household chores.)

a. How could you display the results of this survey in a graph or table?

b. Would you use a confidence interval to convey more information about the true percentages expressing an opinion on the various ingredients of a good marriage? Why?

c. What qualms might you have about the way this survey has been reported?

Edu.   **10.10**   A substantial part of the U.S. population is "technologically illiterate," according to experts at a National Technological Literacy Conference organized by the National Science Foundation and Pennsylvania State University. At this conference, the results of a national survey of 2,000 adults showed that:

- 70% do not understand radiation.
- 40% think space-rocket launchings change the weather and that some unidentified flying objects are actually visitors from other planets.
- More than 80% do not understand how telephones work.
- 75% do not have a clear understanding of what computer software is.
- 72% do not understand the gross national product.

a. How might you display these data in a graph or table? Construct the display.

b. Many newspaper articles reporting the results of a survey give conclusions without sufficient details about the study for the reader to assess the data and reach a separate conclusion. What details are missing here but needed for you to reach your own conclusion?

Med.   **10.11**   The benign mucosal cyst is the most common lesion of a pair of sinuses in the upper jawbone. In a random sample of 800 males, 35 people were observed to have a benign mucosal cyst.

a. Would it be appropriate to use a normal approximation in conducting a statistical test of the null hypothesis H_0: $\pi \geq .096$ (the highest incidence in previous studies among males)? Explain.

b. Conduct a statistical test of the research hypothesis H_a: $\pi < .096$. Use $\alpha = .05$.

Pol. Sci.   **10.12**   National public opinion polls are based on interviews of as few as 1,500 people in a random sampling of public sentiment toward one or more issues. These interviews are commonly done in person, because mail returns are poor and telephone interviews tend to reach older people, thus biasing the results. Suppose that a random sample of 1,500 people determines the proportion of the adult public in agreement with recent energy conservation proposals.

a. If 560 indicate they favor the policies set forth by the current administration, estimate π, the proportion of adults holding a "favor" opinion. Use a 95% confidence interval. What is the half width of the confidence interval?

b. How many people must the survey include to have a 95% confidence interval with a half width of .01?

Env.   **10.13**   Researchers obtained a sample of 20 crayfish of all sizes from a large lake to estimate the proportion of crayfish that exhibit more than 9 (ppb) units of mercury. Of those sampled, eight exceeded 9 units. Use these data to estimate π, the proportion of all crayfish in the lake with a mercury level greater than 9, using a 95% confidence interval. Interpret the interval.

Engin.   **10.14**   A pharmaceutical firm has been investigating the possibility of having hospital personnel supplied with small disposable vials that can be used to perform many of the standard laboratory analyses. For a particular analysis, such as blood sugar, the technician would insert a measured amount of blood into an appropriate vial and observe its color when thoroughly mixed with the chemical already stored in the vial. By comparing the optical density of the combined fluid to a color-coded chart, the technician would have a reading on the blood sugar level of the patient. Obviously, the system must be tightly controlled to ensure that the vials are correctly sealed with the proper amount of chemical prior to shipment to the hospital laboratories. In a random sample of 100 vials from a batch of several thousand vials, all 100 vials had the correct amount of the chemical and were properly sealed.

   a. Estimate the proportion of defective vials in the batch.
   b. Construct a 99% confidence interval for the proportion of defective vials in the batch.
   c. Using your confidence interval from (b), test the pharmaceutical firm's claim that there are less than 1% defective vials per batch. What is the α level of your test?

10.3 Inferences about the Difference between Two Population Proportions, $\pi_1 - \pi_2$

Many practical problems involve the comparison of two binomial parameters. Social scientists may wish to compare the proportions of women who take advantage of prenatal health services for two communities representing different socioeconomic backgrounds. A director of marketing may wish to compare the public awareness of a new product recently launched and that of a competitor's product.

For comparisons of this type, we assume that independent random samples are drawn from two binomial populations with unknown parameters designated by π_1 and π_2. If y_1 successes are observed for the random sample of size n_1 from population 1 and y_2 successes are observed for the random sample of size n_2 from population 2, then the point estimates of π_1 and π_2 are the observed sample proportions $\hat{\pi}_1$ and $\hat{\pi}_2$, respectively.

$$\hat{\pi}_1 = \frac{y_1}{n_1} \quad \text{and} \quad \hat{\pi}_2 = \frac{y_2}{n_2}$$

This notation is summarized here.

Notation for Comparing Two Binomial Proportions

	Population	
	1	**2**
Population proportion	π_1	π_2
Sample size	n_1	n_2
Number of successes	y_1	y_2
Sample proportion	$\hat{\pi}_1 = \dfrac{y_1}{n_1}$	$\hat{\pi}_2 = \dfrac{y_2}{n_2}$

Inferences about two binomial proportions are usually phrased in terms of their difference $\pi_1 - \pi_2$, and we use the difference in sample proportions $\hat{\pi}_1 - \hat{\pi}_2$ as part of a confidence interval or statistical test. The sampling distribution for $\hat{\pi}_1 - \hat{\pi}_2$ can be approximated by a normal distribution with mean and standard error given by

$$\mu_{\hat{\pi}_1 - \hat{\pi}_2} = \pi_1 - \pi_2$$

and

$$\sigma_{\hat{\pi}_1 - \hat{\pi}_2} = \sqrt{\frac{\pi_1(1 - \pi_1)}{n_1} + \frac{\pi_2(1 - \pi_2)}{n_2}}$$

This approximation is appropriate if we apply the same requirements to both binomial populations that we applied in recommending a normal approximation to a binomial (see Chapter 4). Thus, the normal approximation to the distribution of $\hat{\pi}_1 - \hat{\pi}_2$ is appropriate if both $n\pi_i$ and $n(1 - \pi_i)$ are 5 or more for $i = 1, 2$. Because π_1 and π_2 are not known, the validity of the approximation is made by examining $n\hat{\pi}_i$ and $n(1 - \hat{\pi}_i)$ for $i = 1, 2$.

Confidence intervals and statistical tests about $\pi_1 - \pi_2$ are straightforward and follow the format we used for comparisons using $\mu_1 - \mu_2$. Interval estimation is summarized here; it takes the usual form, point estimate $\pm z$ (standard error).

100($1 - \alpha$)% Confidence Interval for $\pi_1 - \pi_2$

$$\hat{\pi}_1 - \hat{\pi}_2 \pm z_{\alpha/2}\sigma_{\hat{\pi}_1 - \hat{\pi}_2},$$

where

$$\sigma_{\hat{\pi}_1 - \hat{\pi}_2} = \sqrt{\frac{\pi_1(1 - \pi_1)}{n_1} + \frac{\pi_2(1 - \pi_2)}{n_2}}$$

Note: Substitute $\hat{\pi}_1$ and $\hat{\pi}_2$ for π_1 and π_2 in the formula for $\sigma_{\hat{\pi}_1 - \hat{\pi}_2}$. When the normal approximation is valid for $\hat{\pi}_1 - \hat{\pi}_2$, very little error will result from this substitution.

EXAMPLE 10.5

A company test-markets a new product in the Grand Rapids, Michigan, and Wichita, Kansas, metropolitan areas. The company's advertising in the Grand Rapids area is based almost entirely on television commercials. In Wichita, the company spends a roughly equal dollar amount on a balanced mix of television, radio, newspaper, and magazine ads. Two months after the ad campaign begins, the company conducts surveys to determine consumer awareness of the product.

	Grand Rapids	**Wichita**
Number interviewed	608	527
Number aware	392	413

Calculate a 95% confidence interval for the regional difference in the proportion of all consumers who are aware of the product.

Solution The sample awareness proportion is higher in Wichita, so let's make Wichita region 1.

$$\hat{\pi}_1 = 413/527 = .784 \qquad \hat{\pi}_2 = 392/608 = .645$$

The estimated standard error is

$$\sqrt{\frac{(.784)(.216)}{527} + \frac{(.645)(.355)}{608}} = .0264$$

Therefore, the 95% confidence interval is

$$(.784 - .645) - 1.96(.0264) \le \pi_1 - \pi_2 \le (.784 - .645) + 1.96(.0264)$$

or

$$.087 \le \pi_1 - \pi_2 \le .191$$

which indicates that somewhere between 8.7 and 19.1% more Wichita consumers than Grand Rapids consumers are aware of the product.

rule for sample sizes   This confidence interval method is based on the normal approximation to the binomial distribution. In Chapter 4, we indicated as a general rule that $n\hat{\pi}$ and $n(1 - \hat{\pi})$ should both be at least 5 to use this normal approximation. For this confidence interval to be used, the rule should hold for each sample. In practice, sample sizes that come even close to violating this rule aren't very useful because they lead to excessively wide confidence intervals. For instance, even though $n\hat{\pi}$ and $n(1 - \hat{\pi})$ are greater than 5 for both samples when $n_1 = 30$, $\hat{\pi}_1 = .20$ and $n_2 = 60$, $\hat{\pi}_2 = .10$, the 95% confidence interval is $-.06 \le \pi_1 - \pi_2 < .26$; π_1 could be anything from 6 percentage points lower than π_2 to 26 percentage points higher.

The reason for confidence intervals that seem very wide and unhelpful is that each measurement conveys very little information. In effect, each measurement conveys only one bit of data: a 1 for a success or a 0 for a failure. For example, surveys of the compensation of chief executive officers of companies often give a manager's age in years. If we replaced the actual age by a category such as "over 55 years old" versus "under 55," we definitely would have far less information. When there is little information per item, we need a large number of items to get an adequate total amount of information. Wherever possible, it is better to have a genuinely numerical measure of a result rather than mere categories. When numerical measurement isn't possible, relatively large sample sizes will be needed.

Hypothesis testing about the difference between two population proportions is based on the z statistic from a normal approximation. The typical null hypothesis is that there is no difference between the population proportions, although any specified value for $\pi_1 - \pi_2$ may be hypothesized. The procedure is very much like a t test of the difference of means, and is summarized next.

Statistical Test for the Difference between Two Population Proportions

H_0: **1.** $\pi_1 - \pi_2 \le 0$ H_a: **1.** $\pi_1 - \pi_2 > 0$
     2. $\pi_1 - \pi_2 \ge 0$ **2.** $\pi_1 - \pi_2 < 0$
     3. $\pi_1 - \pi_2 = 0$ **3.** $\pi_1 - \pi_2 \ne 0$

T.S.: $z = \dfrac{(\hat{\pi}_1 - \hat{\pi}_2)}{\sqrt{\dfrac{\hat{\pi}_1(1 - \hat{\pi}_1)}{n_1} + \dfrac{\hat{\pi}_2(1 - \hat{\pi}_2)}{n_2}}}$

R.R.: **1.** $z > z_\alpha$
     2. $z < -z_\alpha$
     3. $|z| > z_{\alpha/2}$

Check assumptions and draw conclusions.

Note: This test should be used only if $n_1\hat{\pi}_1$, $n_1(1 - \hat{\pi}_1)$, $n_2\hat{\pi}_2$, and $n_2(1 - \hat{\pi}_2)$ are all at least 5.

EXAMPLE 10.6

An educational researcher designs a study to compare the effectiveness of teaching English to non-English-speaking people by a computer software program and by the traditional classroom system. The researcher randomly assigns 125 students from a class of 300 to instruction using the computer. The remaining 175 students are instructed using the traditional method. At the end of a 6-month instructional period, all 300 students are given an examination with the results reported in the following table.

Exam Results	Computer Instruction	Traditional Instruction
Pass	94	113
Fail	31	62
Total	125	175

Does instruction using the computer software program appear to increase the proportion of students passing the examination in comparison to the pass rate using the traditional method of instruction? Use $\alpha = .05$.

Solution   Denote the proportion of all students passing the examination using the computer method of instruction and the traditional method of instruction by π_1 and π_2, respectively. We will test the hypotheses

$$H_0: \quad \pi_1 - \pi_2 \leq 0$$
$$H_a: \quad \pi_1 - \pi_2 > 0$$

We will reject H_0 if the test statistic z is greater than $z_{.05} = 1.645$. From the data we compute the estimates

$$\hat{\pi}_1 = \frac{94}{125} = .752 \quad \text{and} \quad \hat{\pi}_2 = \frac{113}{175} = .646$$

From these we compute the test statistic to be

$$z = \frac{\hat{\pi}_1 - \hat{\pi}_2}{\sqrt{\dfrac{\hat{\pi}_1(1 - \hat{\pi}_1)}{n_1} + \dfrac{\hat{\pi}_2(1 - \hat{\pi}_2)}{n_2}}} = \frac{.752 - .646}{\sqrt{\dfrac{.752(1 - .752)}{125} + \dfrac{.646(1 - .646)}{175}}} = 2.00$$

Because $z = 2.00$ is greater than 1.645, we reject H_0 and conclude that the observations support the hypothesis that the computer instruction has a higher pass rate than the traditional approach. The p-value of the observed data is given by p-value $= P(z \geq 2.00) = .0228$, using the standard normal tables. A 95% confidence interval on the effect size $\pi_1 - \pi_2$ is given by

$$.752 - .646 \pm 1.96\sqrt{\frac{.752(1 - .752)}{125} + \frac{.646(1 - .646)}{175}} \quad \text{or} \quad .106 \pm .104$$

We are 95% confident that the proportion passing the examination is between .2% and 21% higher for students using computer instruction than those using the traditional approach. For our conclusions to have a degree of validity, we need to check whether the sample sizes were large enough. Now, $n_1\hat{\pi}_1 = 94$, $n_1(1 - \hat{\pi}_1) = 31$, $n_2\hat{\pi}_2 = 113$, and $n_2(1 - \hat{\pi}_2) = 62$; thus all four quantities are greater than 5. Hence, the large sample criterion appears to be satisfied.

EXERCISES   Applications

Pol. Sci.   **10.15** A law student believes that the proportion of registered Republicans in favor of additional tax incentives is greater than the proportion of registered Democrats in favor of such incentives. The student acquires independent random samples of 200 Republicans and 200 Democrats and finds 109 Republicans and 86 Democrats in favor of additional tax incentives. Use these data to test $H_0: \pi_1 - \pi_2 \leq 0$ versus $H_a: \pi_1 - \pi_2 > 0$. Give the level of significance for your test.

Bus.   **10.16** A retail computer dealer is trying to decide between two methods for servicing customers' equipment. The first method emphasizes preventive maintenance; the second emphasizes quick response to problems. The dealer serves samples of customers by one of the two methods. After six months, it finds that 171 of 200 customers served by the first method are very satisfied with the service, as compared to 153 of 200 customers served by the second method. Execustat output, based on 1s and 0s for data, follows.

```
TWO SAMPLE ANALYSIS FOR SATISFIED BY METHOD

                                1                      2

Sample size      200               200
Mean                   0.855             0.765         diff. = 0.09
Variance               0.124598          0.180678      ratio = 0.689612

Std. deviation         0.352984          0.425063

95% confidence intervals
     mu1 - mu2: (0.0131926, 0.166807) assuming equal variances
     mu1 - mu2: (0.0131847, 0.166815) not assuming equal variances

HYPOTHESIS TEST-DIFFERENCE OF MEANS

Null hypothesis: difference of means = 0
Alternative: not equal
Equal variances assumed: no

Computed t statistic = 2.30362
            P value = 0.0218
```

Test the research hypothesis that the population proportions are different. Use $\alpha = .05$. State your conclusion carefully.

10.17 Locate a confidence interval for the difference of proportions in Exercise 10.16. Show that it reaches the same conclusion as the formal test about the research hypothesis.

Bus.   **10.18** The media selection manager for an advertising agency inserts the same advertisement for a client bank in two magazines, similarly placed in each. One month later, a market research study finds that 226 of 473 readers of the first magazine are aware of the banking services offered in the ad, as are 165 of 439 readers of the second magazine (readers of both magazines are excluded). The following Minitab output was based on the appropriate number of 1s and 0s as data:

```
MTB > TWOT 'AWARE?' 'MAGAZINE'

             Two
Sample T-Test and Confidence Interval

Twosample T for Aware?
Magazine     N     Mean    StDev    SE Mean
1           473   0.478    0.500      0.023
2           439   0.376    0.485      0.023

95% C.I. for mu 1 - mu 2: (0.038, 0.166)
T-Test mu 1 = mu 2 (vs not =): T = 3.13 P = 0.0018 DF = 908
```

a. Calculate by hand a 95% confidence interval for the difference of proportions of readers who are aware of the advertised services. Compare your answer to the interval given by Minitab.
b. Are the sample sizes adequate to use the normal approximation?
c. Does the confidence interval indicate that there is a statistically significant difference using $\alpha = .05$?

10.19   Using the output of Exercise 10.18, perform a formal test of the null hypothesis of equal populations. Use $\alpha = .05$.

Med.   **10.20**   There is a remedy for male pattern baldness—at least that's what millions of males hope since the FDA approved the compound minoxidil for this use. Minoxidil was investigated in a large, 27-center study in which patients were randomly assigned to receive topical minoxidil or an identical-appearing placebo. Ignoring the center-to-center variation, suppose the preliminary results were as follows:

	Sample Size	% with New Hair Growth
Minoxidil group	310	32
Placebo	309	20

a. Use these data to test H_0: $\pi_1 - \pi_2 = 0$ versus H_a: $\pi_1 - \pi_2 \neq 0$. Give the p-value for your test.
b. If you were working for the FDA, what additional information might you want to examine in this study?

Bio.   **10.21**   Is cocaine deadlier than heroin? A study reported in the *Journal of the American Medical Association* found that rats with unlimited access to cocaine had poorer health, had more behavior disturbances, and died at a higher rate than did a corresponding group of rats given unlimited access to heroin. The death rates after 30 days on the study were as follows:

	% Dead at 30 Days
Cocaine group	90
Heroin group	36

a. Suppose that 100 rats were used in each group. Conduct a test of H_0: $\pi_1 - \pi_2 \leq 0$ versus H_a: $\pi_1 - \pi_2 > 0$. Give the p-value for your test.
b. What implications are there for human use of the two drugs?

10.4   Inferences about Several Proportions: Chi-Square Goodness-of-Fit Test

We can extend the binomial sampling scheme of Chapter 4 to situations in which each trial results in one of k possible outcomes ($k > 2$). For example, a random sample of registered voters is classified according to political party (Republican, Democrat, Socialist, Green, Independent, etc.) or patients in a clinical trial are evaluated with respect to the degree of improvement in their medical condition (substantially improved, improved, no change, worse). This type of experiment or study is called a multinomial experiment with the characteristics listed here.

The Multinomial Experiment

1. The experiment consists of n identical trials.
2. Each trial results in one of k outcomes.
3. The probability that a single trial will result in outcome i is π_i for $i = 1, 2, \ldots, k$, and remains constant from trial to trial. (*Note:* $\sum \pi_i = 1$.)
4. The trials are independent.
5. We are interested in n_i, the number of trials resulting in outcome i. (*Note:* $\sum n_i = n$).

multinomial distribution

The probability distribution for the number of observations resulting in each of the k outcomes, called the **multinomial distribution,** is given by the formula

$$P(n_1, n_2, \ldots, n_k) = \frac{n!}{n_1! n_2! \cdots n_k!} \pi_1^{n_1} \pi_2^{n_2} \cdots \pi_k^{n_k}$$

Recall from Chapter 4, where we discussed the binomial probability distribution, that

$$n! = n(n - 1) \cdots 1$$

and

$$0! = 1$$

We can use the formula for the multinomial distribution to compute the probability of particular events.

EXAMPLE 10.7

Previous experience with the breeding of a particular herd of cattle suggests that the probability of obtaining one healthy calf from a mating is .83. Similarly, the probabilities of obtaining zero or two healthy calves are, respectively, .15 and .02. A farmer breeds three dams from the herd; find the probability of obtaining exactly three healthy calves.

Solution   Assuming the three dams are chosen at random, this experiment can be viewed as a multinomial experiment with $n = 3$ trials and $k = 3$ outcomes. These outcomes are listed with the corresponding probabilities.

Outcome	Number of Progeny	Probability, π_i
1	0	.15
2	1	.83
3	2	.02

Note that outcomes 1, 2, and 3 refer to the events that a dam produces zero, one, or two healthy calves, respectively. Similarly, n_1, n_2, and n_3 refer to the number of dams producing zero, one, or two healthy progeny, respectively. To obtain exactly three healthy progeny, we must observe one of the following possible events.

$$\text{A:} \begin{cases} \text{1 dam gives birth to no healthy progeny:} & n_1 = 1 \\ \text{1 dam gives birth to 1 healthy progeny:} & n_2 = 1 \\ \text{1 dam gives birth to 2 healthy progeny:} & n_3 = 1 \end{cases}$$

$$\text{B:}\quad \text{3 dams give birth to 1 healthy progeny:} \begin{cases} n_1 = 0 \\ n_2 = 3 \\ n_3 = 0 \end{cases}$$

For event A with $n = 3$ and $k = 3$,

$$P(n_1 = 1, n_2 = 1, n_3 = 1) = \frac{3!}{1!1!1!}(.15)^1(.83)^1(.02)^1 \cong .015$$

Similarly, for event B,

$$P(n_1 = 0, n_2 = 3, n_3 = 0) = \frac{3!}{0!3!0!}(.15)^0(.83)^3(.02)^0 = (.83)^3 \cong .572$$

Thus, the probability of obtaining exactly three healthy progeny from three dams is the sum of the probabilities for events A and B; namely, $.015 + .572 \approx .59$.

Our primary interest in the multinomial distribution is as a probability model underlying statistical tests about the probabilities $\pi_1, \pi_2, \ldots, \pi_k$. We will hypothesize specific values for the πs and then determine whether the sample data agree with the hypothesized values. One way to test such a hypothesis is to examine the observed number of trials resulting in each outcome and to compare this to the number we *expect* to result in each outcome. For instance, in our previous example, we gave the probabilities associated with zero, one, and two progeny as .15, .83, and .02. If we were to examine a sample of 100 mated dams, we would **expect to observe** 15 dams that produce no healthy progeny. Similarly, we would expect to observe 83 dams that produce one healthy calf and two dams that produce two healthy calves.

expected number of outcomes

DEFINITION 10.1

In a multinomial experiment in which each trial can result in one of k outcomes, the **expected number of outcomes** of type i in n trials is $n\pi_i$, where π_i is the probability that a single trial results in outcome i.

FIGURE 10.1

Chi-square probability
distribution for df = 4

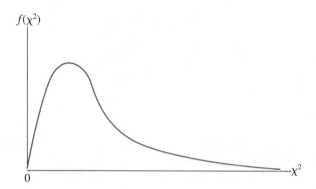

In 1900, Karl Pearson proposed the following test statistic to test the specified probabilities:

$$\chi^2 = \sum_i \left[\frac{(n_i - E_i)^2}{E_i} \right]$$

where n_i represents the number of trials resulting in outcome i and E_i represents the number of trials we expect to result in outcome i when the hypothesized probabilities represent the actual probabilities assigned to each outcome. Frequently, **cell probabilities** we will refer to the probabilities $\pi_1, \pi_2, \ldots, \pi_k$ as **cell probabilities,** one cell corresponding to each of the k outcomes. The observed numbers $n_1, n_2, \ldots, n_k$ corresponding **observed cell counts** to the k outcomes will be called **observed cell counts,** and the expected **expected cell counts** numbers $E_1, E_2, \ldots, E_k$ will be referred to as **expected cell counts.**

Suppose that we hypothesize values for the cell probabilities $\pi_1, \pi_2, \ldots, \pi_k$. We can then calculate the expected cell counts by using Definition 10.1 to examine how well the observed data fit, or agree, with what we expect to observe. Certainly, if the hypothesized π-values are correct, the observed cell counts n_i should not deviate greatly from the expected cell counts E_i, and the computed value of χ^2 should be small. Similarly, when one or more of the hypothesized cell probabilities are incorrect, the observed and expected cell counts will differ substantially, making χ^2 large.

chi-square distribution The distribution of the quantity χ^2 can be approximated by a **chi-square distribution** provided that the expected cell counts E_i are fairly large.

The chi-square goodness-of-fit test based on k specified cell probabilities will have $k - 1$ degrees of freedom. We will explain why we have $k - 1$ degrees of freedom at the end of this section. Upper-tail values of the test statistic

$$\chi^2 = \sum_i \left[\frac{(n_i - E_i)^2}{E_i} \right]$$

can be found in Table 8 in the Appendix. See Figure 10.1 for a chi-square distribution with df = 4.

We can now summarize the chi-square goodness-of-fit test concerning k specified cell probabilities.

**Chi-Square Goodness-
of-Fit Test**

H_0: $\pi_i = \pi_{i0}$ for categories $i = 1, \ldots, k$; π_{i0} are specified probabilities or proportions.

H_a: At least one of the cell probabilities differs from the hypothesized value.

T.S.: $\chi^2 = \sum \left[\dfrac{(n_i - E_i)^2}{E_i} \right]$, where n_i is the observed number in category i
and $E_i = n\pi_{i0}$ is the expected number under H_0.

R.R.:   Reject H_0 if χ^2 exceeds the tabulated critical value for $a = \alpha$ and df $= k - 1$.

Check assumptions and draw conclusions.

The approximation of the sampling distribution of the chi-square goodness-of-fit test statistic by a chi-square distribution improves as the sample size n becomes larger. The accuracy of the approximation depends on both the sample size n and the number of cells k. Cochran (1954) indicates that the approximation should be adequate if no E_i is less than 1 and no more than 20% of the E_is are less than 5. The values of n/k that provide adequate approximations for the chi-square goodness-of-fit test statistic tends to decrease as k increases. Agresti (1990) discusses situations in which the chi-squared approximation tends to be poor for studies having small observed cell counts even if the expected cell counts are moderately large. Agresti concludes that it is hopeless to determine a single rule concerning the appropriate sample size to cover all cases. However, we recommend applying Cochran's guidelines for determining whether the chi-square goodness-of-fit test statistic can be adequately approximated with a chi-square distribution. When some of the E_is are too small, there are several alternatives. Researchers combine levels of the categorical variable to increase the observed cell counts. However, combining categories should not be done unless there is a natural way to redefine the levels of the categorical variable that does not change the nature of the hypothesis to be tested. When it is not possible to obtain observed cell counts large enough to permit the chi-square approximation, Agresti (1990) discusses *exact* methods to test the hypotheses.

EXAMPLE 10.8

A laboratory is comparing a test drug to a standard drug preparation that is useful in the maintenance of patients suffering from high blood pressure. Over many clinical trials at many different locations, the standard therapy was administered to patients with comparable hypertension [as measured by the New York Heart Association (NYHA) Classification]. The lab then classified the responses to therapy for this large patient group into one of four response categories. Table 10.1 lists the categories and percentages of patients treated on the standard preparation who have been classified in each category.

The lab then conducted a clinical trial with a random sample of 200 patients with high blood pressure. All patients were required to be listed according to the

TABLE 10.1

Results of clinical trials using the standard preparation

Category	Percentage
Marked decrease in blood pressure	50
Moderate decrease in blood pressure	25
Slight decrease in blood pressure	10
Stationary or slight increase in blood pressure	15

TABLE 10.2

Sample data for example

Category	Observed Cell Counts
1	120
2	60
3	10
4	10

same hypertensive categories of the NYHA Classification as those studied under the standard preparation. Use the sample data in Table 10.2 to test the hypothesis that the cell probabilities associated with the test preparation are identical to those for the standard. Use $\alpha = .05$.

Solution This experiment possesses the characteristics of a multinomial experiment, with $n = 200$ and $k = 4$ outcomes.

Outcome 1: A person's blood pressure will decrease markedly after treatment with the test drug.

Outcome 2: A person's blood pressure will decrease moderately after treatment with the test drug.

Outcome 3: A person's blood pressure will decrease slightly after treatment with the test drug.

Outcome 4: A person's blood pressure will remain stationary or increase slightly after treatment with the test drug.

The null and alternative hypotheses are then

$$H_0: \quad \pi_1 = .50, \pi_2 = .25, \pi_3 = .10, \pi_4 = .15$$

and

$$H_a: \quad \text{At least one of the cell probabilities is different from the hypothesized value.}$$

Before computing the test statistic, we must determine the expected cell numbers. These data are given in Table 10.3.

Because all the expected cell numbers are relatively large, we may calculate the chi-square statistic and compare it to a tabulated value of the chi-square distribution.

TABLE 10.3

Observed and expected cell numbers for example

Category	Observed Cell Number, n_i	Expected Cell Number, E_i
1	120	$200(.50) = 100$
2	60	$200(.25) = 50$
3	10	$200(.10) = 20$
4	10	$200(.15) = 30$

$$\chi^2 = \sum_i \left[\frac{(n_i - E_i)^2}{E_i} \right]$$

$$= \frac{(120 - 100)^2}{100} + \frac{(60 - 50)^2}{50} + \frac{(10 - 20)^2}{20} + \frac{(10 - 30)^2}{30}$$

$$= 4 + 2 + 5 + 13.33 = 24.33$$

For the probability of a Type I error set at $\alpha = .05$, we look up the value of the chi-square statistic for $a = .05$ and df $= k - 1 = 3$. The critical value from Table 8 in the Appendix is 7.815.

R.R.: Reject H_0 if $\chi^2 > 7.815$.

Checking assumptions and drawing conclusions: The computed value of χ^2 is greater than 7.815, so we reject the null hypothesis and conclude that at least one of the cell probabilities differs from that specified under H_0. Practically, it appears that a much higher proportion of patients treated with the test preparation falls into the moderate and marked improvement categories. The p-value for this test is $p < .001$. (See Table 8 in the Appendix.)

The assumptions needed for running a chi-square goodness-of-fit test are those associated with a multinomial experiment, of which the key ones are independence of the trials and constant cell probabilities. Independence of the trials would be violated if, for example, several patients from the same family were included in the sample because hypertension has a strong hereditary component. The assumption of constant cell probabilities would be violated if the study were conducted over a period of time during which the standards of medical practice shifted, allowing for other "standard" therapies.

The test statistic for the chi-square goodness-of-fit test is the sum of k terms, which is the reason the degrees of freedom depend on k, the number of categories, rather than on n, the total sample size. However, there are only $k - 1$ degrees of freedom, rather than k, because the sum of the $n_i - E_i$ terms must be equal to $n - n = 0$; $k - 1$ of the observed minus expected differences are free to vary, but the last one (kth) is determined by the condition that the sum of the $n_i - E_i$ equals zero.

This goodness-of-fit test has been used extensively over the years to test various scientific theories. Unlike previous statistical tests, however, the hypothesis of interest is the null hypothesis, not the research (or alternative) hypothesis. Unfortunately, the logic behind running a statistical test does not hold. In the standard situation in which the research (alternative) hypothesis is the one of interest to the scientist, we formulate a suitable null hypothesis and gather data to reject H_0 in favor of H_a. Thus, we "prove" H_a by contradicting H_0.

We cannot do the same with the chi-square goodness-of-fit test. If a scientist has a set theory and wants to show that sample data conform to or fit that theory, she wants to accept H_0. From our previous work, there is the potential for committing a Type II error in accepting H_0. Here, as with other tests, the calculation of β probabilities is difficult. In general, for a goodness-of-fit test, the potential for committing a Type II error is high if n is small or if k, the number of categories, is large. Even if the expected cell counts E_i conform to our recommendations, the probability of a Type II error could be large. Therefore, the results of a chi-square goodness-of-fit test should be viewed suspiciously. Don't automatically accept the null hypothesis as fact given that H_0 was not rejected.

EXERCISES **Basic Techniques**

Bus. **10.22** Over the past 5 years, an insurance company has had a mix of 40% whole life policies, 20% universal life policies, 25% annual renewable-term (ART) policies, and 15% other types of policies. A change in this mix over the long haul could require a change in the commission structure, reserves, and possibly investments. A sample of 1,000 policies issued over the last few months gave the results shown here. Use these data to assess whether there has been a shift from the historical percentages. Give the p-value for your test. Which policies (if any) seem to be more popular?

Category	Observed Cell Number, n_i
Whole life	320
Universal life	280
ART	240
Other	160
Total	1,000

Soc. **10.23** A university and several industries developed a work-study program in the surrounding community. Students were to work with industrial sociologists during a 3-month internship. Equal numbers of students from the university worked in a chemical, a textile, and a pharmaceutical industry. Students completing the program were classified according to the industry in which they interned. Consider the following data as a random sample of the many students who could have completed the program. Test the null hypothesis that the probability that a finishing student interned in a pharmaceutical, chemical, or textile industry is 1/3. Use $\alpha = .01$ with n_i the number of students in group i finishing the program.

Group	n_i
Pharmaceutical	20
Chemical	13
Textile	30

Soc. **10.24** Researchers conducted an experiment to determine whether the proportion of mentally ill patients of each social class housed in a county facility agrees with the social class distribution of the county. The observed cell numbers for the 400 patients classified are given here.

Lower: 215   Upper-middle: 60
Lower-middle: 100   Upper: 25

Use these data to test the null hypothesis

$$\pi_1 = .25 \qquad \pi_3 = .20$$
$$\pi_2 = .48 \qquad \pi_4 = .07$$

where the πs are the hypothesized proportions of people in the respective social-class categories in the county. Use $\alpha = .05$ and draw conclusions.

Pol. Sci. **10.25** In previous presidential elections in a given locality, 50% of the registered voters were Republicans, 40% were Democrats, and 10% were registered as independents. Prior to the upcoming election, a random sample of 200 registered voters showed that 90 were registered as Republicans,

80 as Democrats, and 30 as independents. Test the research hypothesis that the distribution of registered voters is different from that in previous election years. Give the *p*-value for your test. Draw conclusions.

Med.   **10.26**   Previous experimentation with a drug product developed for the relief of depression was conducted with normal adults with no signs of depression. We will assume a large data bank is available from studies conducted with normals and, for all practical purposes, the data bank can represent the population of responses for normals. Each of the adults participating in one of these studies was asked to rate the drug as ineffective, mildly effective, or effective. The percentages of respondents in these categories were 60%, 30%, and 10%, respectively. In a new study of depressed adults, a random sample of 85 adults responded as follows:

> Ineffective: 30
> Mildly effective: 35
> Effective: 20

Is there evidence to indicate a different percentage distribution of responses for depressed adults than for nondepressed? Give the level of significance for your test and draw conclusions.

Bus.   **10.27**   A researcher obtained a sample of 125 securities analysts and asked each analyst to select four stocks on the New York Stock Exchange that were expected to outperform the Standard and Poor's Index over a 3-month period. One theory suggests that the securities analysts would be expected to do no better than chance and that the number of correct guesses from the four selected had a multinomial distribution as shown here.

Number correct	0	1	2	3	4
Multinomial probabilities (π_i)	.0625	.2500	.3750	.2500	.0625

If the number of correct guesses from the sample of 125 analysts had a frequency distribution as shown here, use these data to conduct a chi-square goodness-of-fit test. Use $\alpha = .05$. Draw conclusions.

Number correct	0	1	2	3	4
Frequency	3	23	51	39	9

10.5   Contingency Tables: Tests for Independence and Homogeneity

In Section 10.3, we showed a test for comparing two proportions. The data were simply counts of how many times we got a particular result in two samples. In this section, we extend that test. First, we present a single test statistic for testing whether several deviations of sample data from theoretical proportions could plausibly have occurred by chance.

When we first introduced probability ideas in Chapter 4, we started by using tables of frequencies (counts). At the time, we treated these counts as if they represented the whole population. In practice, we'll hardly ever know the complete population data; we'll usually have only a sample. When we have counts from a sample, **cross tabulations**   they're usually arranged in **cross tabulations** or **contingency tables.** In this section, **contingency tables**   we'll describe one particular test that is often used for such tables, a chi-square test of independence.

In Chapter 4, we introduced the idea of independence. In particular, we dis**dependence**   cussed the idea that **dependence** of variables means that one variable has some value for predicting the other. With sample data, there usually appears to be some degree of dependence. In this section, we develop a χ^2 test that assesses whether the perceived dependence in sample data may be a fluke—the result of random variability rather than real dependence.

First, the frequency data are to be arranged in a cross tabulation with r rows and c columns. The possible values of one variable determine the rows of the table, and the possible values of the other determine the columns. We denote the population proportion (or probability) falling in row i, column j as π_{ij}. The total proportion for row i is $\pi_{i.}$ and the total proportion for column j is $\pi_{.j}$. If the row and column variables are independent, then $\pi_{ij} = \pi_{i.}\pi_{.j}$. For instance, suppose that a personnel manager for a large firm wants to assess the popularity of three alternative flexible time-scheduling (flextime) plans among clerical workers in four different offices. The following indicates a set of proportions (π_{ij}) that exhibit independence. The proportion of all clerical workers who favor plan 2 and work in office 1 is $\pi_{21} = .03$, the proportion of all workers favoring plan 2 is $\pi_{2.} = .30$, and the proportion working in office 1 is $\pi_{.1} = .10$. Independence holds for that cell because $\pi_{21} = .03 = (\pi_{2.})(\pi_{.1}) = (.30)(.10)$. Independence also holds for all other cells.

Favored Plan	Office				Total
	1	**2**	**3**	**4**	**Total**
1	.05	.20	.15	.10	.50
2	.03	.12	.09	.06	.30
3	.02	.08	.06	.04	.20
Total	.10	.40	.30	.20	

The null hypothesis for this χ^2 test is independence. The research hypothesis specifies only that there is some form of dependence—that is, it is not true that $\pi_{ij} = \pi_{i.}\pi_{.j}$ in every cell of the table. The test statistic is once again the sum over all cells of

(observed value − expected values)²/expected value

The computation of expected values E_{ij} under the null hypothesis is different for the independence test than for the goodness-of-fit test. The null hypothesis of independence does not specify numerical values for the row probabilities $\pi_{i.}$ and column probabilities $\pi_{.j}$, so these probabilities must be estimated by the row and column relative frequencies. If $n_{i.}$ is the actual frequency in row i, estimate $\pi_{i.}$ by $\hat{\pi}_{i.} = n_{i.}/n$; similarly $\hat{\pi}_{.j} = n_{.j}/n$. Assuming the null hypothesis of independence is true, it follows that $\hat{\pi}_{ij} = \hat{\pi}_{i.}\hat{\pi}_{.j} = (n_{i.}/n)(n_{.j}/n)$.

DEFINITION 10.2   Under the hypothesis of independence, the **estimated expected value** in row i, column j is

$$\hat{E}_{ij} = n\hat{\pi}_{ij} = n\frac{(n_{i.})}{n}\frac{(n_{.j})}{n} = \frac{(n_{i.})(n_{.j})}{n}$$

the row total multiplied by the column total divided by the grand total.

EXAMPLE 10.9

Suppose that in the flexible time-scheduling illustration, a random sample of 216 workers yields the following frequencies:

	Office				
Favored Plan	**1**	**2**	**3**	**4**	**Total**
1	15	32	18	5	70
2	8	29	23	18	78
3	1	20	25	22	68
Total	24	81	66	45	216

Calculate a table of $\hat{E}_{ij}$ values.

Solution For row 1, column 1 the estimated expected number is

$$\hat{E}_{11} = \frac{(\text{row 1 total})(\text{column 1 total})}{\text{grand total}} = \frac{(70)(24)}{216} = 7.78$$

Similar calculations for all cells yield the following table.

	Office				
Plan	**1**	**2**	**3**	**4**	**Total**
1	7.78	26.25	21.39	14.58	70.00
2	8.67	29.25	23.83	16.25	78.00
3	7.56	25.50	20.78	14.17	68.01
Totals	24.01	81.00	66.00	45.00	216.01

Note that the row and column totals in the $\hat{E}_{ij}$ table equal (except for rounding error) the corresponding totals in the observed (n_{ij}) table.

χ^2 Test of Independence

H_0: The row and column variables are independent.

H_a: The row and column variables are dependent (associated).

T.S.: $\chi^2 = \sum_{i,j} (n_{ij} - \hat{E}_{ij})^2 / \hat{E}_{ij}$

R.R.: Reject H_0 if $\chi^2 > \chi_\alpha^2$, where χ_α^2 cuts off area α in a χ^2 distribution with $(r - 1)(c - 1)$ df; r = number of rows, c = number of columns.

Check assumptions and draw conclusions.

The test statistic is sometimes called the Pearson χ^2 statistic.

df for table

The **degrees of freedom** for the χ^2 test of independence relate to the number of cells in the two-way table that are free to vary while the marginal totals remain

TABLE 10.4
(a) One df in a 2 × 2 table; (b) two df in a 2 × 3 table

	Category B		Total		Category B			Total
Category A	*		16	**Category A**	*	*		51
			34					40
Totals	21	29	50	**Totals**	28	41	22	91
		(a)					(b)	

fixed. For example, in a 2 × 2 table (2 rows, 2 columns), only one cell entry is free to vary. Once that entry is fixed, we can determine the remaining cell entries by subtracting from the corresponding row or column total. In Table 10.4(a), we have indicated some (arbitrary) totals. The cell indicated by * could take any value (within the limits implied by the totals), but then all remaining cells would be determined by the totals. Similarly, with a 2 × 3 table (2 rows, 3 columns), two of the cell entries, as indicated by *, are free to vary. Once these entries are set, we determine the remaining cell entries by subtracting from the appropriate row or column total [see Table 10.4(b)]. In general, for a table with r rows and c columns, $(r - 1)(c - 1)$ of the cell entries are free to vary. This number represents the degrees of freedom for the χ^2 test of independence.

This χ^2 test of independence is also based on an approximation. The approximation should be adequate if no E_i is less than 1 and no more than 20% of the E_is are less than 5.

EXAMPLE 10.10

Carry out the χ^2 test of independence for the data of Example 10.9. First use $\alpha = .05$; then obtain a bound for the p-value.

Solution The null and alternative hypotheses are:

H_0: The popularity of the scheduling plan is independent of the office location.

H_a: The popularity of the scheduling plan depends on the office location.

The test statistic can be computed using the n_{ij} and E_{ij} from Example 10.9:

T.S.: $\chi^2 = \sum (n_{ij} - E_{ij})^2 / E_{ij}$

$$= (15 - 7.78)^2/7.78 + (32 - 26.25)^2/26.25$$

$$+ \cdots + (22 - 14.17)^2/14.17$$

$$= 6.70 + 1.26 + \cdots + 4.33 = 27.13$$

R.R.: For df $= (3 - 1)(4 - 1) = 6$ and $\alpha = .05$, the critical value from Table 8 in the Appendix is 12.59. Because $\chi^2 = 27.13$ exceeds 12.59, H_0 is rejected. In fact, because 27.13 is larger than the value in Table 8 for $a = .001$, the p-value is $< .001$.

Checking assumptions and drawing conclusions: Because each of the expected values exceeds 5, the χ^2 approximation should be good. There is strong evidence $(p < .001)$ that the popularity of the scheduling plan depends on the office location.

likelihood ratio statistic

There is an alternative χ^2 statistic called the **likelihood ratio statistic** that is often shown in computer outputs. It is defined as

$$\text{likelihood ratio } \chi^2 = \sum_{ij} n_{ij} \ln(n_{ij}/(n_{i.}n_{.j}))$$

where $n_{i.}$ is the total frequency in row i, $n_{.j}$ is the total in column j, and ln is the natural logarithm (base $e = 2.71828$). Its value should also be compared to the χ^2 distribution with the same $(r - 1)(c - 1)$ df. Although it isn't at all obvious, this form of the χ^2 independence test is approximately equal to the Pearson form. There is some reason to believe that the Pearson χ^2 yields a better approximation to table values, so we prefer to rely on it rather than on the likelihood ratio form.

The only function of a χ^2 test of independence is to determine whether apparent dependence in sample data may be a fluke, plausibly a result of random variation. Rejection of the null hypothesis indicates only that the apparent association is not reasonably attributable to chance. It does not indicate anything about the

strength of association

strength or type **of association.**

The same χ^2 test statistic applies to a slightly different sampling procedure. An implicit assumption of our discussion surrounding the χ^2 test of independence is that the data result from a single random sample from the whole population. Often, separate random samples are taken from the subpopulations defined by the column (or row) variable. In the flextime example (Example 10.9), the data might have resulted from separate samples (of respective sizes 24, 81, 66, and 45) from the four offices rather than from a single random sample of 216 workers.

In general, suppose the column categories represent c distinct subpopulations. Random samples of size $n_1, n_2, \ldots, n_c$ are selected from these subpopulations. The observations from each subpopulation are then classified into the r values of a categorical variable represented by the r rows in the contingency table. The research hypothesis is that there is a difference in the distribution of subpopulation units into the r levels of the categorical variable. The null hypothesis is that the set of r proportions for each subpopulation $(\pi_{1j}, \pi_{2j}, \ldots, \pi_{rj})$ is the same for all c subpopulations. Thus, the null hypothesis is given by

$$H_0: \quad (\pi_{11}, \pi_{21}, \ldots, \pi_{r1}) = (\pi_{12}, \pi_{22}, \ldots, \pi_{r2}) = \cdots = (\pi_{1c}, \pi_{2c}, \ldots, \pi_{rc})$$

test of homogeneity

The test is called a **test of homogeneity** of distributions. The mechanics of the test of homogeneity and the test of independence are identical. However, note that the sampling scheme and conclusions are different. With the test of independence, we randomly select n units from a single population and classify the units with respect to the values of two categorical variables. We then want to determine whether the two categorical variables are related to one another. In the test of homogeneity of proportions, we have c subpopulations from which we randomly select $n = n_1 + n_2 + \cdots + n_c$ units, which are classified according to the values of a single categorical variable. We want to determine whether the distribution of the subpopulation units to the values of the categorical variable is the same for all c subpopulations.

As we discussed in Section 10.4, the accuracy of the approximation of the sampling distribution of χ^2 by a chi-square distribution depends on both the sample size n and the number of cells k. Cochran (1954) indicates that the approximation should be adequate if no E_i is less than 1 and no more than 20% of the E_is are less than 5. Larntz (1978) and Koehler (1986) showed that χ^2 is valid with smaller sample sizes than the likelihood ratio test statistic. Agresti (1990) compares the nominal and actual α-levels for both test statistics for testing independence, for various sample sizes. The χ^2 test statistic appears to be adequate when n/k exceeds 1. Again, we recommend applying Cochran's guidelines for determining whether the chi-square test statistic can be adequately approximated with a chi-square distribution. When some of the E_{ij}s are too small, there are several alternatives. Researchers combine levels of the categorical variables to increase the observed cell counts. However, combining categories should not be done unless there is a natural way to redefine the levels of the categorical variables that does not change the nature of the hypothesis to be tested. When it is not possible to obtain observed cell counts large enough to permit the chi-square approximation, Agresti (1990) discusses *exact* methods to test the hypotheses. For example, the Fisher exact test is used when both categorical variables have only two levels.

EXAMPLE 10.11

Random samples of 200 individuals from major oil-producing and natural gas-producing states, 200 from coal states, and 400 from other states participate in a poll of attitudes toward five possible energy policies. Each respondent indicates the most preferred alternative from among the following:

1. Primarily emphasize conservation
2. Primarily emphasize domestic oil and gas exploration
3. Primarily emphasize investment in solar-related energy
4. Primarily emphasize nuclear energy development and safety
5. Primarily reduce environmental restrictions and emphasize coal-burning activities

The results are as follows:

Policy Choice	Oil/Gas States	Coal States	Other States	Total
1	50	59	161	270
2	88	20	40	148
3	56	52	188	296
4	4	3	5	12
5	2	66	6	74
Totals	200	200	400	800

Execustat output also carries out the calculations. The second entry in each cell is a percentage in the column.

```
                        Crosstabulation
                  OilGas       Coal       Other         Row
                                                        Total
        1             50         59         161          270
                     25.0       29.5        40.3        33.75

        2             88         20          40          148
                     44.0       10.0        10.0        18.50

        3             56         52         188          296
                     28.0       26.0        47.0        37.00

        4              4          3           5           12
                      2.0        1.5         1.3         1.50

        5              2         66           6           74
                      1.0       33.0         1.5         9.25

        Column       200        200         400          800
        Total       25.00      25.00       50.00       100.00

        Summary Statistics for Crosstabulation

            Chi-square        D.F.     P Value

              289.22            8      0.0000

    Warning: Some table cell counts < 5.
```

Conduct a χ^2 test of homogeneity of distributions for the three groups of states. Give the p-value for this test.

Solution A test that the corresponding population distributions are different makes use of the following table of expected values:

Policy Choice	Oil/Gas States	Coal States	Other States
1	67.5	67.5	135
2	37	37	74
3	74	74	148
4	3	3	6
5	18.5	18.5	37

We observe that the table of expected values has two E_{ij}s that are less than 5. However, our guideline for applying the chi-square approximation to the test statistic is met because only $2/15 = 13\%$ of the E_{ij}s are less than 5 and all the values are greater than 1. The test procedure is outlined here:

H_0: The column distributions are homogeneous.

H_a: The column distributions are not homogeneous.

T.S.: $\chi^2 = \sum (n_{ij} - \hat{E}_{ij})^2/\hat{E}_{ij}$

$= (50 - 67.5)^2/67.5 + (88 - 37)^2/37 + \cdots + (6 - 37)^2/37$

$= 289.22$

R.R.: Because the table value of χ^2 for df $= 8$ and $\alpha = .001$ is 26.12, p-value is $<.001$.

Checking assumptions and drawing conclusions: Even recognizing the limited accuracy of the χ^2 approximations, we can reject the hypothesis of homogeneity at some very small p-value. Percentage analysis, particularly of state type for a given policy choice, shows dramatic differences; for instance, 1% of those living in oil/gas states favor policy 5, compared to 33% of those in coal states who favor policy 5.

The χ^2 test described in this section has a limited but important purpose. This test only assesses whether the data indicate a statistically detectable (significant) relation among various categories. It does not measure how strong the apparent relation might be. A weak relation in a large data set may be detectable (significant); a strong relation in a small data set may be nonsignificant.

EXERCISES

H.R.

10.28 A personnel director for a large, research-oriented firm categorizes colleges and universities as most desirable, good, adequate, and undesirable for purposes of hiring their graduates. The director collects data on 156 recent graduates, and has each rated by a supervisor.

	Rating		
School	**Outstanding**	**Average**	**Poor**
Most desirable	21	25	2
Good	20	36	10
Adequate	4	14	7
Undesirable	3	8	6

Output from the Execustat computer package follows:

				Crosstabulation
		Outstanding	Average	Poor	Row
					Total
Most desirable		21	25	2	48
		43.8	52.1	4.2	30.77
Good		20	36	10	66
		30.3	54.5	15.2	42.31

```
Adequate                         4         14          7         25
                              16.0       56.0       28.0      16.03

Undesirable                      3          8          6         17
                              17.6       47.1       35.3      10.90

        Column                  48         83         25        156
        Total                30.77      53.21      16.03     100.00

                    Summary Statistics for Crosstabulation

        Chi-square                D.F.            P Value
             15.97                   6             0.0139

Warning: Some table cell counts < 5.
```

a. Locate the value of the χ^2 statistic.
b. Locate the p-value.
c. Can the director safely conclude that there is a relation between school type and rating?
d. Is there any problem in using the χ^2 approximation?

10.29 Refer to Exercise 10.28. Do the row percentages (the second entry in each cell of the output) reflect the existence of the relation we found in Exercise 10.28?

H.R. **10.30** A study of potential age discrimination considers promotions among middle managers in a large company. The data are as follows:

	Age				
	Under 30	**30–39**	**40–49**	**50 and Over**	**Total**
Promoted	9	29	32	10	80
Not promoted	41	41	48	40	170
Totals	50	70	80	50	

Minitab output follows:

```
MTB > Table 'promoted' 'agegroup';
SUBC>    Counts;
SUBC>    ColPercents;
SUBC>    ChiSquare.
 ROWS: promoted        COLUMNS: agegroup
             1           2          3          4        ALL

    1        9          29         32         10         80
          18.00       41.43      40.00      20.00      32.00
             9          29         32         10         80

    2       41          41         48         40        170
          82.00       58.57      60.00      80.00      68.00
            41          41         48         40        170
```

```
ALL          50              70            80           50          250
           100.00          100.00       100.00       100.00      100.00
             50              70            80           50          250


CHI-SQUARE =      `13.025     WITH D.F. = 3
   CELL CONTENTS -
                            COUNT
                            % OF COL
                            COUNT
```

a. Find the expected numbers under the hypothesis of independence.
b. Justify the indicated degrees of freedom.
c. Is there a statistically significant relation between age and promotions, using $\alpha = .05$?

10.31 Place bounds on the p-value in Exercise 10.30.

10.32 The data of Exercise 10.30 are combined as follows:

	Age		
	Up to 39	**40 and Over**	**Total**
Promoted	38	42	80
Not promoted	82	88	170
Total	120	130	

The Minitab results are as follows:

```
MTB > Table 'promoted' 'combined';
SUBC>    Counts;
SUBC>    ColPercents;
SUBC>    ChiSquare.
 ROWS:     promoted        COLUMNS: combined
                 1             2        ALL
     1          38            42         80
              31.67         32.31      32.00
                38            42         80

     2          82            88        170
              68.33         67.69      68.00
                82            88        170

ALL           120           130        250
            100.00        100.00     100.00
              120           130        250

CHI-SQUARE =           0.012    WITH D.F. = 1
```

a. Can the hypothesis of independence be rejected using a reasonable α?
b. What is the effect of combining age categories? Compare the answers to those for Exercise 10.30.

10.6 Odds and Odds Ratios

Another way to analyze count data on qualitative variables is to use the concept of odds. This approach is widely used in biomedical studies and could be useful in some market research contexts as well. The basic definition of odds is the ratio of the probability that an event happens to the probability that it does not happen.

DEFINITION 10.3

Odds of an event $A = \dfrac{P(A)}{1 - P(A)}$

If an event has probability 2/3 of happening, the odds are $\frac{2}{3}/\frac{1}{3} = 2$. Usually this is reported as "the odds of the event happening are 2 to 1." Odds are used in horse racing and other betting establishments. The horse racing odds are given as the odds against the horse winning. Therefore odds of 4 to 1 means that it is 4 times more likely the horse will lose (not win) than not. Based on the odds, a horse with 4 to 1 odds is a better bet than, say, a horse with 20 to 1 odds. What about a horse with 1 to 2 odds (or equivalently, .5 to 1) against winning? This horse is highly favored because it is twice as likely (2 to 1) that the horse will win as not.

In working with odds, just make certain what the event of interest is. Also it is easy to convert the odds of an event back to the probability of the event. For event A,

$$P(A) = \frac{\text{odds of event } A}{1 + \text{odds of event } A}$$

Thus, if the odds of a horse (not winning) are stated as 9 to 1, then the probability of the horse not winning is

$$\text{Probability (not winning)} = \frac{9}{1 + 9} = .9$$

Similarly, the probability of winning is .1.

Odds are a convenient way to see how the occurrence of a condition changes the probability of an event. Recall from Chapter 4 that the conditional probability of an event A given another event B is

$$P(A|B) = P(A \text{ and } B)/P(B)$$

The odds favoring an event A given another event B turn out after a little algebra to be

$$\frac{P(A|B)}{P(\text{not } A|B)} = \frac{P(A)}{P(\text{not } A)} \frac{P(B|A)}{P(B|\text{not } A)}$$

The initial odds are multiplied by the *likelihood ratio,* the ratio of the probability of the conditioning event given A to its probability given not A. If B is more likely to happen when A is true than when it is not, the occurrence of B makes the odds favoring A go up.

EXAMPLE 10.12

Consider both a population in which 1 of every 1,000 people carried the HIV virus and a test that yielded positive results for 95% of those who carry the virus and (false) positive results for 2% of those who do not carry it. If a randomly chosen person obtains a positive test result, should the odds that that person carries the HIV virus go up or go down? By how much?

Solution We certainly would think that a positive test result would increase the odds of carrying the virus. It would be a strange test indeed if a positive result decreased the chance of having the disease! Take the event A to be "carries HIV" and the event B to be "positive test result."

Before the test is made, the odds of a randomly chosen person carrying HIV are

$$\frac{.001}{.999} \approx .001$$

The occurrence of a positive test result causes the odds to change to

$$\frac{P(\text{HIV}|\text{positive})}{P(\text{not HIV}|\text{positive})} = \frac{P(\text{HIV})}{P(\text{not HIV})} \frac{P(\text{positive}|\text{HIV})}{P(\text{positive}|\text{not HIV})}$$

$$= \frac{.001}{.999} \frac{.95}{.02} = .0475$$

The odds of carrying HIV do go up given a positive test result, from about .001 (to 1) to about .0475 (to 1).

odds ratio   A closely related idea, widely used in biomedical studies, is the **odds ratio.** As the name indicates, it is the ratio of the odds of an event (for example, contracting a certain form of cancer) for one group (for example, men) to the odds of the same event for another group (for example, women). The odds ratio is usually defined using conditional probabilities but can be stated equally well in terms of joint probabilities.

DEFINITION 10.4

Odds Ratio of an Event for Two Groups

If A is any event with probabilities $P(A|\text{group 1})$ and $P(A|\text{group 2})$, the odds ratio is

$$\frac{P(A|\text{group 1})/[1 - P(A|\text{group 1})]}{P(A|\text{group 2})/[1 - P(A|\text{group 2})]}$$

$$= \frac{P(A \text{ and group 1})/[1 - P(A \text{ and group 1})]}{P(A \text{ and group 2})/[1 - P(A \text{ and group 2})]}$$

The odds ratio equals 1 if the event A is statistically independent of group.

For example, suppose we have the following table of frequencies of purchase of a special suspension package for two brands of sports sedans:

	Yes	No	Total
Brand 1	250	750	1,000
Brand 2	400	1600	2,000
Total	650	2350	3,000

We estimate the conditional probabilities of purchasing or not purchasing the package, given the brand, as

	Yes	No	Total
Brand 1	.250	.750	1.000
Brand 2	.200	.800	1.000

The odds ratio is (.250/.750)/(.200/.800) = 1.333, indicating that the odds of purchasing the package are 33.3% higher for purchasers of brand 1 than for purchasers of brand 2. It does not appear that purchasing the package is independent of brand (although we will have to allow for the fact that we have limited sample data). We could equally well have calculated the odds as

$$\frac{(250/3,000)(750/3,000)}{(400/3,000)/(1,600/3,000)} = \frac{250/750}{400/1,600} = 1.333$$

Inference about the odds ratio is usually done by way of the natural logarithm of the odds ratio. Recall that ln is the usual notation for the natural logarithm (base $e = 2.71828$) and that $\ln(1) = 0$. When the natural logarithm of the odds ratio is estimated from count data, it has approximately a normal distribution, with expected value the natural logarithm of the population odds ratio. Its standard error can be estimated by summing up 1/frequency for all four counts in a table, and then taking the square root. For the suspension package purchasing data, ln(odds ratio) = $\ln(1.333) = 0.2874$; the estimated standard error is

$$\sqrt{1/250 + 1/750 + 1/400 + 1/1,600} = 0.0920$$

With these results, we can compute a 95% confidence interval as

$$0.2874 - 1.96(0.0920) \leq \ln(\text{odds ratio}) \leq 0.2874 + 1.96(0.0920)$$

or

$$0.1071 \leq \ln(\text{odds ratio}) \leq 0.4677$$

We may convert this to a statement about the odds ratio by "unlogging"—exponentiating—the interval.

$$e^{0.1071} \leq \text{odds ratio} \leq e^{0.4677}$$

or

$$1.113 \leq \text{odds ratio} \leq 1.596$$

Because the interval does not include an odds ratio of 1.000 [or, equivalently, the interval for ln(odds ratio) does not include 0.000], we may conclude that there is a statistically detectable relation in the data.

TABLE 10.5

	Reinfarction?		
	Yes	**No**	
Compound 1	250 (5%)	4,750	$n_1 = 5{,}000$
Compound 2	100 (2%)	4,900	$n_2 = 5{,}000$

The odds ratio is a useful way to compare two population proportions π_1 and π_2 and may be more meaningful than their difference $(\pi_1 - \pi_2)$ when π_1 and π_2 are small. For example, suppose the rate of reinfarction for a sample of 5,000 coronary bypass patients treated with compound 1 is $\hat{\pi}_1 = .05$ and the corresponding rate for another sample of 5,000 coronary bypass patients treated with compound 2 is $\hat{\pi}_2 = .02$. Then their difference $\hat{\pi}_1 - \hat{\pi}_2 = .03$ may be less important and less informative than the odds ratio. See Table 10.5.

The reinfarction odds for compounds 1 and 2 are as follows:

$$\text{Compound 1 odds} = \frac{250/5{,}000}{4{,}750/5{,}000} = \frac{250}{4{,}750} = .053$$

$$\text{Compound 2 odds} = \frac{100/5{,}000}{4{,}900/5{,}000} = \frac{100}{4{,}900} = .020$$

The corresponding odds ratio is .053/.020 = 2.65. Note that although the difference in reinfarction rates is only .03, the odds of having a reinfarction after treatment with compound 1 are 2.65 as likely as a reinfarction following treatment with compound 2.

10.7 Encounters with Real Data: Smoking and Survival Rates

Defining the Problem (1), Collecting Data (2)

In the early 1970s a health survey was conducted in the Whickham district of England to study thyroid and heart disease. One-sixth of all registered voters were interviewed for this study. As is usual in such studies, extensive demographic and health-related information was collected for each subject.

This study was of such importance that 20 years later a follow-up was done. Almost all of the subjects from the original study were traced and additional information about their survival and health was collected.

The data set WHISMOKE contains some information on 1,314 women from the two Whickham studies: smoking habits, ages, and 20-year survival data. The columns of the data set and their category codes are shown below, and the data appear in the file WHISMOKE on the CD that came with your book.

```
c1      Survival      0 = Dead at the time of the follow-up,
                      1 = Alive

c2      Smoke         0 = Had never smoked at the time of the original study,
                      1 = Smoked at that time.
```

```
c3      AgeDca          Age category by 7 "decades" (at time of original study):
                        18 = 18-24,   25 = 25-34,   35 = 35-44,   45 = 45-54,
                        55 = 55-64,   65 = 65-74,   75 = 75 and above.

c4      AgeYMO          Ages combined into three categories:
                        1 = 18-34,   2 = 35-64,   3 = 65 and above.
```

The Whickham data are from David R. Appleton, Joyce M. French, and Mark P. J. Vanderpump, "Ignoring a covariate: An example of Simpson's Paradox" [*The American Statistician* (1996) 50(4):340–341]. The data, stratified according to the categories in `AgeDca`, are given there.

```
Portion of WHISMOKE Data Set

Subject   Survival    Smoke     AgeDca    AgeYMO
   1         1          0         35         2
   2         0          1         35         2
   3         0          1         65         3
   4         1          0         45         2
   5         1          0         35         2
   6         0          1         55         2
   7         1          1         35         2
   8         1          0         45         2
   9         1          1         55         2
  10         0          1         55         2
  11         1          0         18         1
  12         1          1         35         2
  13         1          0         35         2
  14         0          0         55         2
  15         1          1         45         2
   . . .
```

Each row of the data set corresponds to one of these 1,314 subjects. We do not know the actual order in which subject data were recorded during the first study; so the data set lists subjects in random order.

Omitted from WHISMOKE are the women who had smoked but stopped before the time of the original study or whose smoking status was not recorded then. The original study also included 1,285 men, but we look only at the data for 1,314 women.

Summarizing Data (3)

Here we begin to explore whether the Whickham data contain information about the effects of smoking on the health and survival of the women studied. The simplest summary that shows a connection between smoking and survival is displayed below. Both variables are binary: Survival (Dead or Alive) and Smoking Status (No or Yes). The result is a 2 × 2 contingency table.

```
Smoking Status and 20-Year Survival of 1,314 Women

                      Smoking Status
                      --------------
   Survival          No        Yes        Total
-------------------------------------------------
   Dead              230       139          369
   Alive             502       443          945
-------------------------------------------------
   Total             732       582         1314

Death Rate         0.314     0.239       0.281
```

Notice that the death rate is *lower* for the smokers than for the nonsmokers. If we interpret these death rates hastily, we might conclude that smoking is *beneficial.*

However, a large amount of other evidence collected over many years shows that smoking tends to damage health and shorten life spans. We should always be suspicious when the results of a single study appear to contradict common sense or the accumulated evidence of many other studies. We need to question whether the association we see here is real, and if so whether it says anything useful about the association between smoking and longevity for the women of Whickham.

Analyzing Data, Interpreting the Analyses, and Communicating Results (4)

The previous tabulation shows a striking result, namely that the observed death rate was lower for smokers than for nonsmokers. Here we will test whether the apparent association is statistically significant, using the chi-square test of independence. The results are shown next.

```
ROWS: Survival      COLUMNS: Smoke

              0         1        ALL

   0        230       139        369
         205.56    163.44     369.00
           1.70     -1.91        --

   1        502       443        945
         526.44    418.56     945.00
          -1.07      1.19        --

 ALL        732       582       1314
         732.00    582.00    1314.00
           --         --        --

CHI-SQUARE = 9.121,  D.F. = 1,  P-value = 0.003

  CELL CONTENTS --
                   COUNT
                   EXP FREQ
                   STD RES
```

It is now clear that the association we have observed between survival and smoking status in Whickham women is reliably present. A chi-square statistic of 9.121 or larger is very unlikely to be due to chance, and the low probability is quantified by the *p*-value. What we do not know is whether this association tells us anything meaningful about the health consequences of smoking. Practical interpretations of association can be slippery.

The false impression given by the contingency table lies in the unequal distribution of smokers among age groups. About 45% of the women in the *youngest* age group smoked at the time of the original study, but for these young smokers 20 more years may not have been long enough to experience life-threatening consequences of smoking. Relatively few of the women in the *oldest* group were smokers, yet it is hardly surprising that many of them had died 20 years later—of a variety of causes.

We now break the population into three age groups, as indicated by the variable `AgeYMO`. We construct two-way tables for each of the three strata, showing both observed cell counts and percentages of those alive and dead.

```
CONTROL: AgeYMO =  1
ROWS: Survival       COLUMNS: Smoke

                0         1       ALL

     0          6         5        11
             2.74      2.79      2.76

     1        213       174       387
            97.26     97.21     97.24

   ALL        219       179       398
           100.00    100.00    100.00


CONTROL: AgeYMO =  2
ROWS: Survival       COLUMNS: Smoke

                0         1       ALL

     0         59        92       151
            18.44     25.99     22.40

     1        261       262       523
            81.56     74.01     77.60

   ALL        320       354       674
           100.00    100.00    100.00


CONTROL: AgeYMO =  3
ROWS: Survival       COLUMNS: Smoke

                0         1       ALL

     0        165        42       207
            85.49     85.71     85.54
```

```
1        28          7        35
        14.51     14.29     14.46

ALL     193         49       242
        100.00    100.00    100.00


   CELL CONTENTS --
                     COUNT
                     % OF COL
```

The remarkable result here is that, in each of these three age groups, the death rate for smokers is *higher* than it is for nonsmokers, only slightly in the youngest group of 398 women (2.79 vs. 2.74%) and in the oldest group of 242 women (85.71 vs. 85.49%), but substantially in the middle group of 674 women (25.99 vs. 18.44%).

Three separate chi-square tests confirm that the slight differences for the oldest and youngest groups are not statistically significant, but that the large difference for the middle group is significant with a *p*-value of about 0.02. The following bar chart displays the nearly equal survival rates for smokers and nonsmokers in the youngest and oldest age groups of women and the significantly lower survival rate for smokers in the middle age group.

Smoking and 20-year survival rates (stratified by age)

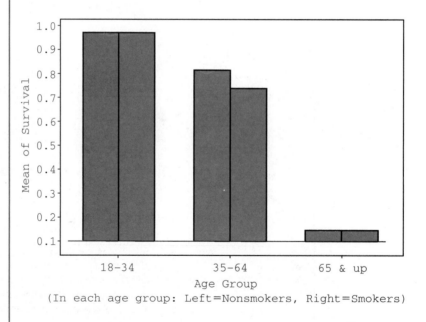

(In each age group: Left=Nonsmokers, Right=Smokers)

It seems worthwhile to comment and speculate a bit about these findings of higher death rates for smokers in all age groups:

- In the youngest age group, there have not been enough deaths for a differential death rate between smokers and nonsmokers to be significant. Moreover, even a much larger study might still fail to find an effect for this stratum because accumulated evidence from many studies indicates that, for most smokers, the fatal damage comes in later years.

- In the oldest age group, it might not have been possible to detect even a fairly large actual differential effect because the group is small (242) and contains very few smokers (49). It is also possible that a large proportion of smokers in this generation may already have died before the first study and thus may not be part of our data set. Recall that these subjects were 65 and older at the time of the first study—85 and older at the time of the second.
- In the middle age group, there is strong evidence that smoking is associated with shorter life. This is the largest group and it is about evenly divided between smokers and nonsmokers, so here we have the most power to detect a differential effect of smoking. Also this group, 35–64 years of age in the early 1970s, aging to 55–84 during the interim period, covers the span of years when other evidence indicates that smoking causes the most deaths. So it is not surprising that we have detected a significant association between shortening of life and smoking in this group of Whickham women.

Recall that when all 1,314 subjects were aggregated, the overall death rate was *lower* for smokers. This is an example of Simpson's Paradox: An apparent association between two categorical variables for aggregated data can be reversed in direction for each stratum when data are broken out according to a third categorical variable.

10.8 Staying Focused

In this chapter, we dealt with categorical data. Categorical data on a single variable arise in a number of situations. We first examined estimation and test procedures for a population proportion π and for two population proportions $(\pi_1 - \pi_2)$ based on independent samples. The extension of these procedures to comparing several population proportions (more than two) gave rise to the chi-square goodness-of-fit test.

Two-variable categorical data problems were discussed using the chi-square tests for independence and for homogeneity based on data displayed in an $r \times c$ contingency table. Finally, we discussed odds and odds ratios, which are especially useful in biomedical trials involving binomial proportions.

Key Formulas

1. Confidence interval for π

$$\hat{\pi} \pm z_{\alpha/2}\hat{\sigma}_{\hat{\pi}}$$

where

$$\hat{\pi} = \frac{y + 2}{n + 4}$$

and

$$\hat{\sigma}_{\hat{\pi}} = \sqrt{\frac{\hat{\pi}(1 - \hat{\pi})}{n + 4}}$$

2. Sample size required for a $100(1 - \alpha)\%$ confidence interval of the form $\hat{\pi} \pm E$

$$n = \frac{z_{\alpha/2}^2 \pi(1 - \pi)}{E^2}$$

(*Hint:* Use $\pi = .5$ if no estimate is available.)

3. Statistical test for π

$$\text{T.S.: } z = \frac{\hat{\pi} - \pi_0}{\sigma_{\hat{\pi}}}$$

where

$$\sigma_{\hat{\pi}} = \sqrt{\frac{\pi_0(1 - \pi_0)}{n}}$$

and

$$\hat{\pi} = \frac{y}{n}$$

4. Confidence interval for $\pi_1 - \pi_2$

$$\hat{\pi}_1 - \hat{\pi}_2 \pm z_{\alpha/2}\sigma_{\hat{\pi}_1 - \hat{\pi}_2}$$

where

$$\sigma_{\hat{\pi}_1 - \hat{\pi}_2} = \sqrt{\frac{\hat{\pi}_1(1 - \hat{\pi}_1)}{n_1} + \frac{\hat{\pi}_2(1 - \hat{\pi}_2)}{n_2}}$$

5. Statistical test for $\pi_1 - \pi_2$

$$\text{T.S.: } z = \frac{\hat{\pi}_1 - \hat{\pi}_2}{\sigma_{\hat{\pi}_1 - \hat{\pi}_2}}$$

where

$$\sigma_{\hat{\pi}_1 - \hat{\pi}_2} = \sqrt{\frac{\hat{\pi}_1(1 - \hat{\pi}_1)}{n_1} + \frac{\hat{\pi}_2(1 - \hat{\pi}_2)}{n_2}}$$

6. Multinomial distribution

$$P(n_1, n_2, \ldots, n_k) = \frac{n!}{n_1!n_2! \cdots n_k!}\pi_1^{n_1}\pi_2^{n_2} \cdots \pi_k^{n_k}$$

7. Chi-square goodness-of-fit test

$$\text{T.S.: } \chi^2 = \sum_i \left[\frac{(n_i - E_i)^2}{E_i}\right]$$

where

$$E_i = n\pi_{i0}$$

8. Chi-square test of independence

$$\chi^2 = \sum_{i,j} \left[\frac{(n_{ij} - E_{ij})^2}{E_{ij}}\right]$$

where

$$E_{ij} = \frac{(\text{row } i \text{ total})(\text{column } j \text{ total})}{n}$$

9. Odds of event $A = \dfrac{P(A)}{1 - P(A)}$

$\left(\text{in a binomial experiment, odds of a success} = \dfrac{\pi}{(1 - \pi)}\right)$

10. Odds ratio for binomial experiment, two groups

$$\frac{\text{odds for group 1}}{\text{odds for group 2}} = \frac{\pi_1/(1 - \pi_1)}{\pi_2/(1 - \pi_2)}$$

Supplementary Exercises

Soc.

10.33 A speaker who advises managers on how to avoid being unionized claims that only 25% of industrial workers favor union membership, 40% are indifferent, and 35% are opposed. In addition, the adviser claims that these opinions are independent of actual union membership. A random sample of 600 industrial workers yields the following data:

	Favor	Indifferent	Opposed	Total
Members	140	42	18	200
Nonmembers	70	198	132	400
Total	210	240	150	600

  a. What part of the data is relevant to the 25%, 40%, 35% claim?
  b. Test this hypothesis using $\alpha = .01$.

10.34 What can be said about the *p*-value in Exercise 10.33?

10.35 Test the hypothesis of independence in the data in Exercise 10.33. How conclusively is it rejected?

10.36 Calculate (for the data of Exercise 10.33) percentages of workers in favor of unionization, indifferent to it, and opposed to it; do this separately for members and for nonmembers. Do the percentages suggest there is a strong relation between membership and opinion?

Pol. Sci.

10.37 Three different television commercials are advertising an established product. The commercials are shown separately to theater panels of consumers; each consumer views only one of the possible commercials and then states an opinion of the product. Opinions range from 1 (very favorable) to 5 (very unfavorable). The data are as follows.

Commercial	Opinion					Total
	1	2	3	4	5	
A	32	87	91	46	44	300
B	53	141	76	20	10	300
C	41	93	67	36	63	300
Total	126	321	234	102	117	900

  a. Calculate expected frequencies under the null hypothesis of independence.
  b. How many degrees of freedom are available for testing this hypothesis?

c. Is there evidence that the opinion distributions are different for the various commercials? Use $\alpha = .01$.

10.38 State bounds on the p-value for Exercise 10.37.

Bus.   **10.39** A direct-mail retailer experimented with three different ways of incorporating order forms into its catalog. In type 1 catalogs, the form was at the end of the catalog; in type 2, it was in the middle; and in type 3, there were forms both in the middle and at the end. Each form was sent to a sample of 1,000 potential customers, none of whom had previously bought from the retailer. A code on each form allowed the retailer to determine which type it was; the number of orders received on each type of form was recorded. Excel was used to calculate expected frequencies and the χ^2 statistic. Excel's CHITEST function gave the p-value. The results are shown here.

	A	B	C	D	E	F	G	H
1	Observed			Received?			max. freq.	
2			No	Yes	Total			
3		1	944	56	1000		944	
4	Type form	2	961	39	1000		961	
5		3	915	85	1000		915	
6			2820	180	3000		2820	
7							lambda (column dep.)	
8	Expected						0	
9			No	Yes	Total			
10		1	940	60	1000			
11	Type form	2	940	60	1000			
12		3	940	60	1000			
13			2820	180	3000			
14								
15		chi sq.	19.18440					
16		p-value	0.00007					
17								

a. What does the null hypothesis of statistical independence indicate about the three types of order forms?
b. Can this null hypothesis be retained at normal α levels?

Bus.   **10.40** A programming firm had developed a more elaborate, more complex version of its spreadsheet program. A beta-test copy of the program was sent to a sample of users of the current program. From information supplied by the users, the firm rated the sophistication of each user; 1 indicated standard, basic applications of the program and 3 indicated the most complex applications. Each user indicated a preference between the current version and the test version, with 1 indicating a strong preference for the current version, 3 indicating no particular preference between the two versions, and 5 indicating a strong preference for the new version. The data were analyzed using JMP IN. Partial output is shown here.

```
SOPHIST By PREFER
Crosstabs
SOPHIST PREFER
Count    1         2         3         4         5         Row %
1        32        28        17        12        8         97
         32.99     28.87     17.53     12.37     8.25
2        10        24        16        6         4         60
         16.67     40.00     26.67     10.00     6.67
3        2         4         5         8         14        33
         6.06      12.12     15.15     24.24     42.42
         44        56        38        26        26        190


Tests
Source    DF      -LogLikelihood        RSquare (U)
Model     8       19.91046              0.1036
Error     180     172.23173
C Total   188     192.14219
Total Count       190


Test                     ChiSquare            Prob>ChiSq
Likelihood Ratio         39.821               < .0001
Pearson                  44.543               < .0001
```

a. Do the Row % entries suggest there is a relation between SOPHIST and PREFER? If the data showed no relation, what would be true of the Row % entries?

b. Does the (Pearson) chi-square computation indicate there is a statistically detectable (significant) relation at usual values of α?

Bio.   **10.41**   A carcinogenicity study was conducted to examine the tumor potential of a drug product scheduled for initial testing in humans. A total of 300 rats (150 males and 150 females) were studied for a 6-month period. At the beginning of the study, 100 rats (50 males, 50 females) were randomly assigned to the control group, 100 to the low-dose group, and the remaining 100 (50 males, 50 females) to the high-dose group. On each day of the 6-month period, the rats in the control group received an injection of an inert solution, whereas those in the drug groups received an injection of the solution plus drug. The sample data are shown in the accompanying table.

	Number of Tumors	
Rat Group	**One or More**	**None**
Control	10	90
Low dose	14	86
High dose	19	81

a. Give the percentage of rats with one or more tumors for each of the three treatment groups.

b. Conduct a test of whether there is a significant difference in the proportion of rats having one or more tumors for the three treatment groups with $\alpha = .05$.

c. Does there appear to be a drug-related problem regarding tumors for this drug product? That is, as the dose is increased, does there appear to be an increase in the proportion of rats with tumors?

10.42 SAS computer output for the data of Exercise 10.41 is shown here. Compare the results from the output with your results in Exercise 10.41.

```
RAT_GRP        N_TUMORS

Frequency    |
Expected     |
Cell Chi-Square|
Percent      |
Row Pct      |
Col Pct      |NONE     |ONE-MORE|   Total
---------------+---------+--------+
CONTROL      |      90 |     10 |    100
             | 85.667 | 14.333 |
             | 0.2192 | 1.3101 |
             |  30.00 |   3.33 |    33.33
             |  90.00 |  10.00 |
             |  35.02 |  23.26 |
---------------+---------+--------+
HIGHDOSE     |      81 |     19 |    100
             | 85.667 | 14.333 |
             | 0.2542 | 1.5194 |
             |  27.00 |   6.33 |    33.33
             |  81.00 |  19.00 |
             |  31.52 |  44.19 |
---------------+---------+--------+
LOWDOSE      |      86 |     14 |    100
             | 85.667 | 14.333 |
             | 0.0013 | 0.0078 |
             |  28.67 |   4.67 |    33.33
             |  86.00 |  14.00 |
             |  33.46 |  32.56 |
---------------+---------+--------+
Total              257       43 |    300
                 85.67    14.33 | 100.00

STATISTICS FOR TABLE OF RAT_GRP BY N_TUMORS

Statistic                        DF    Value    Prob
-------------------------------------------------------
Chi-Square                        2    3.312    0.191
Likelihood Ratio Chi-Square       2    3.327    0.189
Mantel-Haenszel Chi-Square        1    0.649    0.420
Phi Coefficient                        0.105
Contingency Coefficient                0.104
Cramer's V                             0.105

Sample Size = 300
```

Soc.   **10.43** A sociological study was conducted to determine whether there is a relationship between the length of time blue-collar workers remain in their first job and their level of education. From union membership records, a random sample of people was classified. The data are shown here.

Years on First Job	Years of Education			
	0–4.5	4.5–9	9–13.5	13.5
0–2.5	5	21	30	33
2.5–5	15	35	40	30
5–7.5	22	16	15	30
7.5	28	10	8	10

a. Use the computer output that follows to identify the expected cell numbers.
b. Test the research hypothesis that the variable "years on first job" is related to the variable "years of education."
c. Give the level of significance for the test.
d. Draw your conclusions using $\alpha = .05$.

```
YRS_JOB                 YRS_EDU

Frequency        |
Expected         |
Cell Chi-Square  |
Percent          |
Row Pct          |
Col Pct          |0-4.5    |13.5     |4.5-9    |9-13.5   | Total
-----------------+---------+---------+---------+-------+
0-2.5            |       5 |      33 |      21 |      30|     89
                 |  17.902 |  26.342 |  20.971 |  23.784|
                 |  9.2988 |  1.6829 | 394E-7  |  1.6243|
                 |    1.44 |    9.48 |    6.03 |    8.62|  25.57
                 |    5.62 |   37.08 |   23.60 |   33.71|
                 |    7.14 |   32.04 |   25.61 |   32.26|
-----------------+---------+---------+---------+-------+
2.5-5            |      15 |      30 |      35 |      40|    120
                 |  24.138 |  35.517 |  28.276 |  32.069|
                 |  3.4594 |   0.857 |   1.599 |  1.9614|
                 |    4.31 |    8.62 |   10.06 |   11.49|  34.48
                 |   12.50 |   25.00 |   29.17 |   33.33|
                 |   21.43 |   29.13 |   42.68 |   43.01|
-----------------+---------+---------+---------+-------+
5-7.5            |      22 |      30 |      16 |      15|     83
                 |  16.695 |  24.566 |  19.557 |  22.181|
                 |  1.6854 |   1.202 |  0.6471 |  2.3248|
                 |    6.32 |    8.62 |    4.60 |    4.31|  23.85
                 |   26.51 |   36.14 |   19.28 |   18.07|
                 |   31.43 |   29.13 |   19.51 |   16.13|
-----------------+---------+---------+---------+-------+
7.5              |      28 |      10 |      10 |       8|     56
                 |  11.264 |  16.575 |  13.195 |  14.966|
                 |  24.864 |   2.608 |  0.7738 |   3.242|
                 |    8.05 |    2.87 |    2.87 |    2.30|  16.09
                 |   50.00 |   17.86 |   17.86 |   14.29|
                 |   40.00 |    9.71 |   12.20 |    8.60|
-----------------+---------+---------+---------+-------+
Total                   70      103       82       93     348
                     20.11    29.60    23.56    26.72   100.00
```

```
STATISTICS FOR TABLE OF YRS_JOB BY YRS_EDU

Statistic                          DF      Value      Prob
------------------------------------------------------------
Chi-Square                          9      57.830     0.001
Likelihood Ratio Chi-Square         9      55.605     0.001
Mantel-Haenszel Chi-Square          1      31.376     0.001
Phi Coefficient                            0.408
Contingency Coefficient                    0.377
Cramer's V                                 0.235

Sample Size = 348
```

Psy.   **10.44**   Two researchers at Johns Hopkins University studied the use of drug products in the elderly. Patients in a recent study were asked the extent to which physicians counseled them with regard to their drug therapies. The researchers found the following:

- 25.4% of the patients said their physicians did not explain what the drug was supposed to do.
- 91.6% indicated they were not told how the drug might "bother" them.
- 47.1% indicated their physicians did not ask how the drug "helped" or "bothered" them after therapy was started.
- 87.7% indicated the drug was not changed after discussion on how the therapy was helping or bothering them.

    a. Assume that 500 patients were interviewed in this study. Summarize each of these results using a 95% confidence interval.

    b. Do you have any comments about the validity of any of these results?

10.45   **Defining the Problem (1).** The president of a motel chain is preparing to make a decision among four ownership groups competing for a franchise to open a new motel in a fast-growing edge city (a suburban area that has a large concentration of offices and shopping areas). The president of the chain has requested that a comparison of guest satisfaction be made for guests at motels currently operated by the four groups.

Collecting Data (2). All four groups have operated other motels in the geographic area, so the chain has requested addresses of former guests so that information about consumer satisfaction could be obtained by using a mailed customer-satisfaction questionnaire. Initially, an analysis of the most recent guest survey results obtained by the four groups from their customers was to form the basis for the analysis. However, it was determined that the type and phrasing of questions were considerably different for the four groups. This would make the analysis of the data impossible because the data would not be consistent from all four groups. A questionnaire was designed to determine customers' satisfaction with their stay at each of the four groups' motels. The two key areas of interest, as far as the president is concerned, are guests' ratings of building quality (room, restaurant, exercise facility, etc.) and service quality. Both are rated on a 1 to 5 scale, with 1 denoting poor, 3 denoting average, and 5 denoting excellent, in the guests' opinions. Based on the size of necessary power of tests of hypotheses and the precision of confidence interval estimators, it was determined that a sample of at least 100 guests from each of the four ownership groups was necessary. Because only about a 20% response rate could be anticipated from the mailed questionnaires, 500 recent guests of each group were randomly selected and sent a questionnaire.

Summarizing Data (3). Of course, the president would prefer to grant the franchise to the ownership group that had achieved the best ratings. The financial arrangements negotiated with the groups were very similar, so much of the decision hinged on the ratings. The president realized that the survey covered only a small fraction of the people who had stayed at the various groups' motels in the recent past. However, the survey was the most unbiased information available about how the chain's customers perceived the four contending groups.

The president is not sure whether to emphasize the building ratings or the service ratings. In addition, there is a question of whether to compare the entire range of ratings or to concentrate on the proportion of guests giving an above average (4 or 5) rating.

The data from the questionnaires were summarized into the following two tables, one for building quality ratings and one for service quality ratings. The president asks you to analyze the ratings and provide inferences concerning the relative standings of the four ownership groups. He is concerned about how a survey of such a small percentage of the guests can accurately estimate the quality ratings of *all* guests of the motels. The methods of this chapter will enable you to answer the questions posed by the president.

Frequencies of building ratings for case study

Ratings	Ownership Group				Totals
	G1	**G2**	**G3**	**G4**	
1	11	8	15	6	40
2	10	6	18	5	39
3	51	50	42	38	181
4	30	41	26	40	137
5	22	29	16	40	107
Totals	124	134	117	129	504

Frequencies of service ratings for case study

Ratings	Ownership Group				Totals
	G1	**G2**	**G3**	**G4**	
1	15	16	23	11	65
2	18	21	17	18	74
3	36	31	33	24	124
4	29	35	21	33	118
5	26	31	23	43	123
Totals	124	134	117	129	504

Analyzing Data, Interpreting the Analyses, and Communicating Results (4).

a. Analyze the data to answer the president's questions. Be sure to include confidence intervals, test of hypotheses, and any pertinent graphs. The computer output is given on the following pages.

b. Write a nontechnical explanation of what your analysis reveals.

ANALYSIS OF CASE STUDY

TABLE OF BUILDING RATINGS BY GROUP

BUILDING RATINGS   GROUP

Frequency | Expected | Cell Chi-Square| Percent | Row Pct | Col Pct |	G1	G2	G3	G4	Total
1	11 9.8413 0.1364 2.18 27.50 8.87	8 10.635 0.6528 1.59 20.00 5.97	15 9.2857 3.5165 2.98 37.50 12.82	6 10.238 1.7544 1.19 15.00 4.65	40 7.94
2	10 9.5952 0.0171 1.98 25.64 8.06	6 10.369 1.8409 1.19 15.38 4.48	18 9.0536 8.8406 3.57 46.15 15.38	5 9.9821 2.4866 9.99 12.82 3.88	39 7.74
3	51 44.532 0.9395 10.12 28.18 41.13	50 48.123 0.0732 9.92 27.62 37.31	42 42.018 759E-8 8.33 23.20 35.90	38 46.327 1.4969 7.54 20.99 29.46	181 35.91
4	30 33.706 0.4076 5.95 21.90 24.19	41 36.425 0.5747 8.13 29.93 30.60	26 31.804 1.059 5.16 18.98 22.22	40 35.065 0.6944 7.94 29.20 31.01	137 27.18
5	22 26.325 0.7107 4.37 20.56 17.74	29 28.448 0.0107 5.75 27.10 21.64	16 24.839 3.1455 3.17 14.95 13.68	40 27.387 5.809 7.94 37.38 31.01	107 21.23
Total	124 24.60	134 26.59	117 23.21	129 25.60	504 100.00

STATISTICS FOR TABLE OF BUILDING RATINGS BY GROUP

Statistic	DF	Value	Prob
Chi-Square	12	34.167	0.001
Likelihood Ratio Chi-Square	12	32.737	0.001
Mantel-Haenszel Chi-Square	1	4.139	0.042
Phi Coefficient		0.260	
Contingency Coefficient		0.252	
Cramer's V		0.150	

Sample Size = 504

ANALYSIS OF CASE STUDY

TABLE OF SERVICE RATINGS BY GROUP

```
SERVICE RATINGS                 GROUP
Frequency        |
Expected         |
Cell Chi-Square|
Percent          |
Row Pct          |
Col Pct          |G1       |G2       |G3        |G4       |    Total
---------------+-------+--------+---------+-------+
1                |      15|      16|       23|      11|       65
                 |  15.992|  17.282|   15.089|  16.637|
                 |  0.0615|  0.0951|   4.1473|  1.9099|
                 |    2.98|    3.17|     4.56|    2.18|    12.90
                 |   23.08|   24.62|    35.38|   16.92|
                 |   12.10|   11.94|    19.66|    8.53|
---------------+-------+--------+---------+-------+
2                |      18|      21|       17|      18|       74
                 |  18.206|  19.675|   17.179|   18.94|
                 |  0.0023|  0.0893|   0.0019|  0.0467|
                 |    3.57|    4.17|     3.37|    3.57|    14.68
                 |   24.32|   28.38|    22.97|   24.32|
                 |   14.52|   15.67|    14.53|   13.95|
---------------+-------+--------+---------+-------+
3                |      36|      31|       33|      24|      124
                 |  30.508|  32.968|   28.786|  31.738|
                 |  0.9887|  0.1175|    0.617|  1.8866|
                 |    7.14|    6.15|     6.55|    4.76|    24.60
                 |   29.03|   25.00|    26.61|   19.35|
                 |   29.03|   23.13|    28.21|   18.60|
---------------+-------+--------+---------+-------+
4                |      29|      35|       21|      33|      118
                 |  29.032|  31.373|   27.393|  30.202|
                 |  347E-7|  0.4193|   1.4919|  0.2591|
                 |    5.75|    6.94|     4.17|    6.55|    23.41
                 |   24.58|   29.66|    17.80|   27.97|
                 |   23.39|   26.12|    17.95|   25.58|
---------------+-------+--------+---------+-------+
5                |      26|      31|       23|      43|      123
                 |  30.262|  32.702|   28.554|  31.482|
                 |  0.6002|  0.0886|   1.0802|  4.2139|
                 |    5.16|    6.15|     4.56|    8.53|    24.40
                 |   21.14|   25.20|    18.70|   34.96|
                 |   20.97|   23.13|    19.66|   33.33|
---------------+-------+--------+---------+-------+
Total            |     124|     134|      117|     129|      504
                 |   24.60|   26.59|    23.21|   25.60|   100.00
```

SERVICE RATINGS

The FREQ Procedure

Statistics for Table of SERVICE RATINGS BY GROUP

Statistic	DF	Value	Prob
Chi-Square	12	18.1170	0.1122
Likelihood Ratio Chi-Square	12	17.6795	0.1258
Mantel-Haenszel Chi-Square	1	2.1672	0.1410
Phi Coefficient		0.1896	
Contingency Coefficient		0.1863	
Cramer's V		0.1095	
Sample Size = 504			

CHAPTER 11

Linear Regression and Correlation

11.1 Introduction

Predicting future values of a variable is a crucial management activity. Financial officers must predict future cash flows, production managers must predict needs for raw materials, and human resource managers must predict future personnel needs. Explanation of past variation is also important. Explaining the past variation in number of clients of a social service agency can help a manager understand demand for the agency's services. Finding the variables that explain deviations from an automobile component's specifications can help to improve the quality of that component. The basic idea of regression analysis is to use data on a *quantitative* independent variable to predict or explain variation in a *quantitative* dependent variable.

prediction vs. explanation

We can distinguish between **prediction** (reference to future values) and **explanation** (reference to current or past values). Because of the virtues of hindsight, explanation is easier than prediction. However, it is often clearer to use the term *prediction* to include both cases. Therefore, in this book, we sometimes blur the distinction between prediction and explanation.

unit of association

For prediction (or explanation) to make much sense, there must be some connection between the variable we're predicting (the dependent variable) and the variable we're using to make the prediction (the independent variable). No doubt, if you tried long enough, you could find 28 common stocks whose price changes over a year have been accurately predicted by the won–lost percentage of the 28 major league baseball teams on the Fourth of July. However, such a prediction is absurd because there is no connection between the two variables. Prediction requires a **unit of association;** there should be an entity that relates the two variables. With time-series data, the unit of association may simply be time. The variables may be measured at the same time period or, for genuine prediction, the independent variable may be measured at a time period before the dependent variable. For cross-sectional data, an economic or physical entity should connect the variables. If we are trying to predict the change in market share of various soft drinks, we should consider the promotional activity for those drinks, not the advertising for various brands of spaghetti sauce. The need for a unit of association seems obvious, but many predictions are made for situations in which no such unit is evident.

simple regression   In this chapter, we consider simple linear regression analysis, in which there is a single independent variable and the equation for predicting a dependent variable y is a linear function of a given independent variable x. Suppose, for example, that the director of a county highway department wants to predict the cost of a resurfacing contract that is up for bids. We can reasonably predict the costs to be a function of the road miles to be resurfaced. A reasonable first attempt is to use a linear production function. Let y = total cost of a project in millions of dollars, x = number of miles to be resurfaced, and $\hat{y}$ = the predicted cost, also in millions of dollars. A prediction equation $\hat{y} = 2.0 + 3.0x$ (for example) is a linear equation. The

intercept   constant term, 2.0 in this case, is the **intercept** term and is interpreted as the predicted value of y when $x = 0$. In the road resurfacing example, we may interpret the intercept as the fixed cost of beginning the project. The coefficient of x, 3.0 in this

slope   case, is the **slope** of the line, the predicted change in y when there is a one-unit change in x. In the road resurfacing example, if two projects differed by 1 mile in length, we predict that the longer project will cost 3 (million dollars) more than the shorter one. In general, we write the prediction equation as

$$\hat{y} = \hat{\beta}_0 + \hat{\beta}_1 x$$

where $\hat{\beta}_0$ is the intercept and $\hat{\beta}_1$ is the slope. See Figure 11.1.

The basic idea of simple linear regression is to use data to fit a prediction line that relates a dependent variable y and a single independent variable x. The first assumption in simple regression is that the relation is, in fact, linear. According to the

assumption of linearity   **assumption of linearity,** the slope of the equation does not change as x changes. In the road resurfacing example, we assume that there are no (substantial) economies or diseconomies from projects of longer mileage. There is little point in using simple linear regression unless the linearity assumption makes sense (at least roughly).

Linearity is not always a reasonable assumption, on its face. For example, if we tried to predict y = number of drivers that are aware of a car dealer's midsummer sale using x = number of repetitions of the dealer's radio commercial, the assumption of linearity means that the first broadcast of the commercial leads to

FIGURE 11.1
Linear prediction function

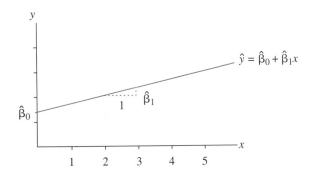

no greater an increase in aware drivers than the thousand and first. (You've heard commercials like that.) We strongly doubt that such an assumption is valid over a wide range of x values. It makes far more sense to us that the effect of repetition diminishes as the number of repetitions becomes larger, so a straight-line prediction wouldn't work well.

Assuming linearity, we would like to write y as a linear function of x: $y = \beta_0 + \beta_1 x$. However, according to such an equation, y is an exact linear function of x; no room is left for the inevitable errors (the deviation of actual y values from their predicted values). Therefore, corresponding to each y we introduce a **random error term** ε and assume the model

random error term

$$y = \beta_0 + \beta_1 x + \varepsilon$$

We assume the random variable y to be made up of a predictable part (a linear function of x) and an unpredictable part (the random error ε). The coefficients β_0 and β_1 are interpreted as the true, underlying intercept and slope. The error term ε includes the effects of all other factors, known or unknown. In the road resurfacing project, unpredictable factors such as strikes, weather conditions, and equipment breakdowns contribute to ε, as do factors such as hilliness or prerepair condition of the road—factors that might have been used in the prediction but were not. The combined effects of unpredictable and ignored factors yield the random error terms ε.

For example, one way to predict the gas mileage of various new cars (the dependent variable) based on their curb weight (the independent variable) is to assign each car to a different driver for, say, a 1-month period. What unpredictable and ignored factors might contribute to prediction error? Unpredictable (random) factors in this study include the driving habits and skills of the drivers, the type of driving done (city versus highway), and the number of stoplights encountered. Factors that would be ignored in a regression analysis of mileage and weight include engine size and type of transmission (manual versus automatic).

In regression studies, the values of the independent variable (the x_i values) are usually taken as predetermined constants, so the only source of randomness is the ε_i terms. Although most economic and business applications have fixed x_i values, this is not always the case. For example, suppose that x_i is the score of an applicant on an aptitude test and y_i is the productivity of the applicant. If the data are based on a random sample of applicants, x_i (as well as y_i) is a random variable. The question of fixed versus random in regard to x is not crucial for regression studies. If the x_is are random, we can simply regard all probability statements as conditional on the observed x_is.

When we assume that the x_is are constants, the only random portion of the model for $y_i = \beta_0 + \beta_1 x_i + \varepsilon_i$ is the random error term ε_i. We make the following formal assumptions.

Formal Assumptions of Regression Analysis

1. The relation is, in fact, linear, so that the errors all have expected value zero: $E(\varepsilon_i) = 0$ for all i.

2. The errors all have the same variance: $\mathrm{Var}(\varepsilon_i) = \sigma_\varepsilon^2$ for all i.

3. The errors are independent of one another.

4. The errors are all normally distributed; ε_i is normally distributed for all i.

FIGURE 11.2

Theoretical distribution of *y* in regression

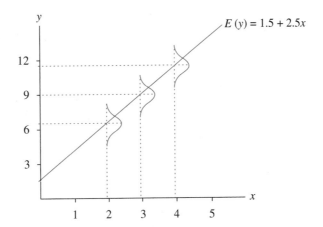

These assumptions are illustrated in Figure 11.2. The actual values of the dependent variable are distributed normally, with mean values falling on the regression line and the same standard deviation at all values of the independent variable. The only assumption not shown in the figure is independence from one measurement to another.

These are the formal assumptions, made in order to derive the significance tests and prediction methods that follow. We can begin to check these assumptions by looking at a **scatterplot** of the data. This is simply a plot of each (x, y) point, with the independent variable value on the horizontal axis and the dependent variable value on the vertical axis. Look to see whether the points basically fall around a straight line or whether there is a definite curve in the pattern. Also look to see whether there are any evident outliers falling far from the general pattern of the data. A scatterplot is shown in Figure 11.3(a).

scatterplot

smoothers

Recently, **smoothers** have been developed to sketch a curve through data without necessarily assuming any particular model. If such a smoother yields something close to a straight line, then linear regression is reasonable. One such method is called LOWESS (locally weighted scatterplot smoother). Roughly, a smoother takes a relatively narrow slice of data along the *x* axis, calculates a line that fits the data in that slice, moves the slice slightly along the *x* axis, recalculates the line, and

FIGURE 11.3   (a) Scatterplot and (b) LOWESS curve

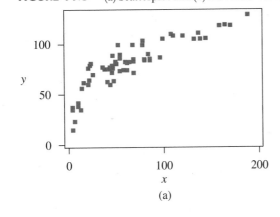

(a)

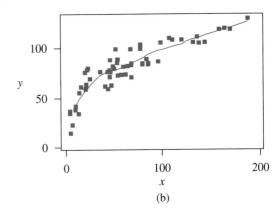

(b)

so on. Then all the little lines are connected in a smooth curve. The width of the slice is called the *bandwidth;* this may often be controlled in the computer program that does the smoothing. Figure 11.3(b) shows the scatterplot with a LOWESS curve through it. The scatterplot shows a curved relation; the LOWESS curve confirms that impression.

spline fit

Another type of scatterplot smoother is the **spline fit.** It can be understood as taking a narrow slice of data, fitting a curve (often a cubic equation) to the slice, moving to the next slice, fitting another curve, and so on. The curves are calculated in such a way as to form a connected, continuous curve.

Many economic relations are not linear. For example, any diminishing returns pattern will tend to yield a relation that increases, but at a decreasing rate. If the scatterplot does not appear linear, either by itself or when fitted with a LOWESS

transformation

curve, it can often be straightened out by a **transformation** of either the independent variable or the dependent variable. A good statistical computer package or a spreadsheet program will compute such functions as the square root of each value of a variable. The transformed variable should be thought of as simply another variable.

For example, a large city dispatches crews each spring to patch potholes in its streets. Records are kept of the number of crews dispatched each day and the number of potholes filled that day. A scatterplot of the number of potholes patched and the number of crews and the same scatterplot with a LOWESS curve through it are shown in Figure 11.4. The relation is not linear. Even without the LOWESS curve, the decreasing slope is obvious. That's not surprising; as the city sends out more crews, the crews will have less effective workers, will have to travel farther to find holes, and so on. All these reasons suggest that diminishing returns will occur.

We can try several transformations of the independent variable to find a more linear scatterplot. Three common transformations are the square root, natural logarithm, and inverse (1 divided by the variable). If we apply each of these transformations to the pothole repair data, we obtain the results shown in Figures 11.5(a)–(c), with LOWESS curves. The square root (a) and inverse transformations (c) don't really give us a straight line. The natural logarithm (b) works very well, however. Therefore, we use LnCrew as our independent variable.

Finding a good transformation often requires trial and error. Following are some suggestions to try for transformations. Note that there are *two* key questions to ask when examining a scatterplot. First, is the relation nonlinear? Second, is

FIGURE 11.4 Scatterplots for pothole data

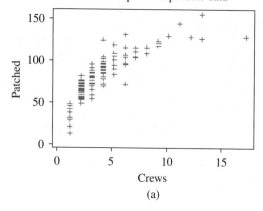

(a)

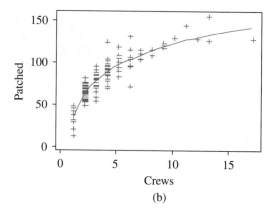

(b)

FIGURE 11.5

Scatterplots with transformed
predictors

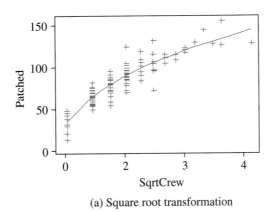

(a) Square root transformation

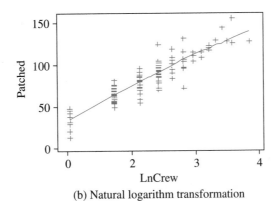

(b) Natural logarithm transformation

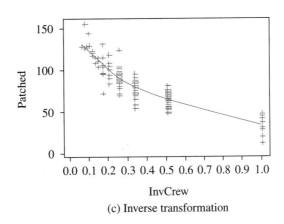

(c) Inverse transformation

there a pattern of increasing variability along the y (vertical) axis? If there is, the assumption of constant variance is questionable. These suggestions that follow don't cover all the possibilities but include the most common problems.

**Steps in Choosing
a Transformation**

1. If the plot indicates a relation that is increasing but at a decreasing rate and if variability around the curve is roughly constant, transform x using the square root, logarithm, or inverse transformation.

2. If the plot indicates a relation that is increasing at an increasing rate and if variability is roughly constant, try using both x and x^2 as predictors. Because this method uses two variables, the multiple regression methods of Chapter 12 are needed.

3. If the plot indicates a relation that increases to a maximum and then decreases and if variability around the curve is roughly constant, again try using both x and x^2 as predictors.

4. If the plot indicates a relation that is increasing at a decreasing rate and if variability around the curve increases as the predicted y value increases, try using y^2 as the dependent variable.

5. If the plot indicates a relation that is increasing at an increasing rate and if variability around the curve increases as the predicted y value increases, try using $\ln(y)$ as the dependent variable. It sometimes may also be helpful to use $\ln(x)$ as the independent variable. Note that a change in the natural logarithm corresponds quite closely to a percentage change in the original variable. Thus, the slope of a transformed variable can be interpreted quite well as a percentage change.

EXAMPLE 11.1

An airline has seen a very large increase in the number of free flights used by participants in its frequent flyer program. To try to predict the trend in these flights in the near future, the director of the program assembled data for the last 72 months. The dependent variable y is the number of thousands of free flights; the independent variable x is month number. A scatterplot with a LOWESS smoother, done using Minitab, is shown in Figure 11.6. What transformation is suggested?

Solution   The pattern shows flights increasing at an increasing rate. The LOWESS curve is definitely turning upward. In addition, variation (up and down) around the curve is increasing. The points around the high end of the curve (on the right, in this case) scatter much more than the ones around the low end of the curve. The increasing variability suggests transforming the y variable, and a natural logarithm (ln) transformation often works well. Minitab was used to compute the logarithms

FIGURE 11.6

Frequent flyer free flights by month

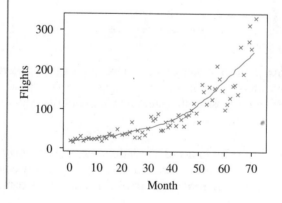

FIGURE 11.7

Result of logarithm transformation

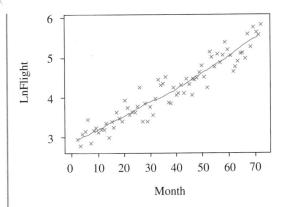

and replot the data, as shown in Figure 11.7. The pattern is much closer to a straight line, and the scatter around the line is much closer to constant.

We will have more to say about checking assumptions in Chapter 12. For a simple regression with a single predictor, careful checking of a scatterplot, ideally with a smooth curve fit through it, will help avoid serious blunders.

Once we have decided on any mathematical transformations, we must estimate the actual equation of the regression line. In practice, only sample data are available. The population intercept, slope, and error variance all have to be estimated from limited sample data. The assumptions we made in this section allow us to make inferences about the true parameter values from the sample data.

11.2   Estimating Model Parameters

The intercept β_0 and slope β_1 in the regression model

$$y = \beta_0 + \beta_1 x + \varepsilon$$

are population quantities. We must estimate these values from sample data. The error variance σ_ε^2 is another population parameter that must be estimated. The first regression problem is to obtain estimates of the slope, intercept, and variance; we discuss how to do so in this section.

The road resurfacing example in Section 11.1 is a convenient illustration. Suppose the following data for similar resurfacing projects in the recent past are available. Note that we do have a unit of association: The connection between a particular cost and mileage is that they're based on the same project.

Cost y_i (in millions of dollars):	6.0	14.0	10.0	14.0	26.0
Mileage x_i (in miles):	1.0	3.0	4.0	5.0	7.0

A first step in examining the relation between y and x is to plot the data as a scatterplot. Remember that each point in such a plot represents the (x, y) coordinates of one data entry, as in Figure 11.8. The plot makes it clear that there is an imperfect but generally increasing relation between x and y. A straight-line relation appears plausible; there is no evident transformation with such limited data.

The regression analysis problem is to find the best straight-line prediction. The most common criterion for "best" is based on squared prediction error. We find the

FIGURE 11.8

Scatterplot of cost versus mileage

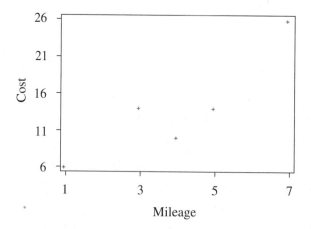

FIGURE 11.9

Prediction errors

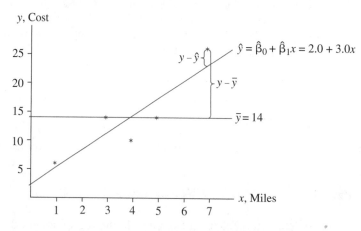

equation of the prediction line—that is, the slope $\hat{\beta}_1$ and intercept $\hat{\beta}_0$—that minimize the total squared prediction error. The method that accomplishes this goal is called

least-squares method   the **least-squares method** because it chooses $\hat{\beta}_0$ and $\hat{\beta}_1$ to minimize the quantity

$$\sum_i (y_i - \hat{y}_i)^2 = \sum_i [y_i - (\hat{\beta}_0 + \hat{\beta}_1 x_i)]^2$$

The prediction errors are shown on the plot of Figure 11.9 as vertical deviations from the line. The deviations are taken as vertical distances because we're trying to predict y values and errors should be taken in the y direction. For these data, the least-squares line can be shown to be $\hat{y} = 2.0 + 3.0x$; one of the deviations from it is indicated by the smaller brace. For comparison, the mean $\bar{y} = 14.0$ is also shown; deviation from the mean is indicated by the larger brace. The least-squares principle leads to some fairly long computations for the slope and intercept. Usually, these computations are done by computer.

DEFINITION 11.1   The **least-squares estimates of slope and intercept** are obtained as follows:

$$\hat{\beta}_1 = \frac{S_{xy}}{S_{xx}} \quad \text{and} \quad \hat{\beta}_0 = \bar{y} - \hat{\beta}_1 \bar{x}$$

where

$$S_{xy} = \sum_i (x_i - \bar{x})(y_i - \bar{y}) \quad \text{and} \quad S_{xx} = \sum_i (x_i - \bar{x})^2$$

Thus, S_{xy} is the sum of x deviations times y deviations and S_{xx} is the sum of x deviations squared.

For the road resurfacing data, $n = 5$ and

$$\sum x_i = 1.0 + \cdots + 7.0 = 20.0$$

so $\bar{x} = \dfrac{20.0}{5} = 4.0$

Similarly,

$$\sum y_i = 70.0 \quad \text{and} \quad \bar{y} = \frac{70.0}{5} = 14.0$$

Also,

$$\begin{aligned} S_{xx} &= \sum (x_i - \bar{x})^2 \\ &= (1.0 - 4.0)^2 + \cdots + (7.0 - 4.0)^2 \\ &= 20.00 \end{aligned}$$

and

$$\begin{aligned} S_{xy} &= \sum (x_i - \bar{x})(y_i - \bar{y}) \\ &= (1.0 - 4.0)(6.0 - 14.0) + \cdots + (7.0 - 4.0)(26.0 - 14.0) \\ &= 60.0 \end{aligned}$$

Thus,

$$\hat{\beta}_1 = \frac{60.0}{20.0} = 3.0 \quad \text{and} \quad \hat{\beta}_0 = 14.0 - (3.0)(4.0) = 2.0$$

From the value $\hat{\beta}_1 = 3$, we can conclude that the estimated average increase in cost for each additional mile is \$3,000,000.

EXAMPLE 11.2

Data from a sample of 10 pharmacies are used to examine the relation between prescription sales volume and the percentage of prescription ingredients purchased directly from the supplier. The sample data are shown here:

Pharmacy	Sales Volume, y (in thousands of dollars)	% of Ingredients Purchased Directly, x
1	25	10
2	55	18
3	50	25
4	75	40
5	110	50
6	138	63
7	90	42
8	60	30
9	10	5
10	100	55

a. Find the least-squares estimates for the regression line $\hat{y} = \hat{\beta}_0 + \hat{\beta}_1 x$.

b. Predict sales volume for a pharmacy that purchases 15% of its prescription ingredients directly from the supplier.

c. Plot the (x, y) data and the prediction equation $\hat{y} = \hat{\beta}_0 + \hat{\beta}_1 x$.

d. Interpret the value of $\hat{\beta}_1$ in the context of the problem.

Solution

a. The equation can be calculated by virtually any statistical computer package; for example, here is abbreviated Minitab output:

```
MTB > Regress 'Sales' on 1 variable 'Directly'

The regression equation is
Sales = 4.70 + 1.97 Directly

Predictor    Coef     Stdev    t-ratio      p
Constant     4.698    5.952      0.79    0.453
Directly     1.9705   0.1545    12.75    0.000
```

To see how the computer does the calculations, you can obtain the least-squares estimates from the following table:

y	x	$y - \bar{y}$	$x - \bar{x}$	$(x - \bar{x})(y - \bar{y})$	$(x - \bar{x})^2$
25	10	−46.3	−22.8	1,101.94	566.44
55	18	−16.3	−15.8	257.54	249.64
50	25	−21.3	−8.8	187.44	77.44
75	40	3.7	6.2	22.94	38.44
110	50	38.7	16.2	626.94	262.44
138	63	66.7	29.2	1,947.64	852.64
90	42	18.7	8.2	153.34	67.24
60	30	−11.3	−3.8	42.94	14.44
10	5	−61.3	−28.8	1,765.44	829.44
100	55	28.7	21.2	608.44	449.44
Totals 713	338	0	0	6,714.60	3,407.60
Means 71.3	33.8				

FIGURE 11.10

Sample data and least-squares prediction equation

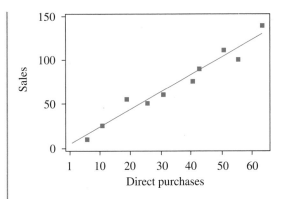

$$S_{xx} = \sum (x - \bar{x})^2 = 3{,}407.6$$

$$S_{xy} = \sum (x - \bar{x})(y - \bar{y}) = 6{,}714.6$$

Substituting into the formulas for $\hat{\beta}_0$ and $\hat{\beta}_1$,

$$\hat{\beta}_1 = \frac{S_{xy}}{S_{xx}} = \frac{6{,}714.6}{3{,}407.6} = 1.9704778 \qquad \text{rounded to } 1.97$$

$$\hat{\beta}_0 = \bar{y} - \hat{\beta}_1 \bar{x} = 71.3 - 1.9704778(33.8) = 4.6978519 \qquad \text{rounded to } 4.70$$

b. When $x = 15\%$, the predicted sales volume is $\hat{y} = 4.70 + 1.97(15) = 34.25$ (that is, \$34,250).

c. The (x, y) data and prediction equation are shown in Figure 11.10.

d. From $\hat{\beta}_1 = 1.97$, we conclude that if a pharmacy increases by 1% the percentage of ingredients purchased directly, then the estimated increase in average sales volume will be \$1,970.

EXAMPLE 11.3

Use the following Statistix output to identify the least-squares estimates for the road resurfacing data.

```
PREDICTOR
VARIABLES   COEFFICIENT  STD ERROR   STUDENT'S T  P

CONSTANT    2.00000 β₀    3.82970       0.52     0.6376
MILES       3.00000 β₁    0.85634       3.50     0.0394


R-SQUARED              0.8036  RESID. MEAN SQUARE (MSE)   14.6666
ADJUSTED R-SQUARED    0.7381  STANDARD DEVIATION          3.82970


SOURCE        DF    SS        MS        F        P

REGRESSION    1   180.000   180.000   12.27   0.0394
RESIDUAL      3   44.0000   14.6666
TOTAL         4   224.000
```

Solution   The intercept is shown in the COEFFICIENT column as $\hat{\beta}_0 = 2.00000$. The slope (coefficient of x = miles) is $\hat{\beta}_1 = 3.00000$.

high leverage point

The estimate of the regression slope can potentially be greatly affected by **high leverage points.** These are points that have very high or very low values of the independent variable—outliers in the x direction. They carry great weight in the estimate of the slope. A high leverage point that also happens to correspond to a y outlier is a **high influence point.** It will alter the slope and twist the line badly.

high influence point

A point has high influence if omitting it from the data will cause the regression line to change substantially. To have high influence, a point must first have high leverage and, in addition, must fall outside the pattern of the remaining points. Consider the two scatterplots in Figure 11.11. In plot (a), the point in the upper left corner is far to the left of the other points; it has a much lower x value and therefore has high leverage. If we draw a line through the other points, the line falls far below this point, so the point is an outlier in the y direction as well. Therefore, it also has high influence. Including this point would change the slope of the line greatly. In contrast, in plot (b), the y outlier point corresponds to an x value very near the mean and has low leverage. Including this point would pull the line upward, increasing the intercept, but it wouldn't increase or decrease the slope much at all. Therefore, it does not have great influence.

FIGURE 11.11

(a) High influence and (b) low influence points

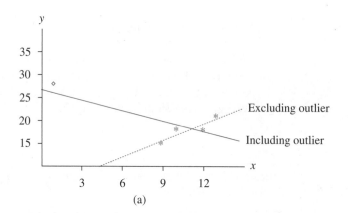

(a)

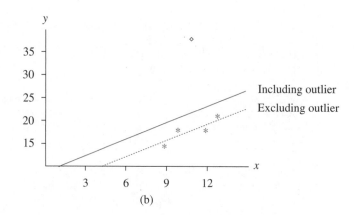

(b)

A high leverage point indicates only a *potential* distortion of the equation. Whether or not including the point will twist the equation depends on its influence (whether or not the point falls near a line through the remaining points). A point must have *both* high leverage and an outlying y value to qualify as a high influence point.

Mathematically, the effect of a point's leverage can be seen in the S_{xy} term that enters into the slope calculation. One of the many ways this term can be written is

$$S_{xy} = \sum (x_i - \bar{x}) y_i$$

We can think of this equation as a weighted sum of y values. The weights are large positive or negative numbers when the x value is far from its mean and has high leverage. The weight is almost 0 when x is very close to its mean and has low leverage.

diagnostic measures   Most computer programs that perform regression analyses will calculate one of several **diagnostic measures** of leverage and influence. We won't try to summarize all of these measures. We only note that very large values of any of these measures correspond to very high leverage or influence points. The distinction between high leverage (x outlier) and high influence (x outlier and y outlier) points is not universally agreed on yet. Check the program's documentation to see which definition is being used.

The standard error of the slope $\hat{\beta}_1$ is calculated by all statistical packages. Typically, it is shown in output in a column to the right of the coefficient column. Like any standard error, it indicates how accurately we can estimate the correct population or process value. The quality of estimation of $\hat{\beta}_1$ is influenced by two quantities: the error variance σ_ε^2 and the amount of variation in the independent variable S_{xx}:

$$\sigma_{\hat{\beta}_1} = \frac{\sigma_\varepsilon}{\sqrt{S_{xx}}}$$

The greater the variability σ_ε of the y value for a given value of x, the larger $\sigma_{\hat{\beta}_1}$ is. It makes sense that if there is high variability around the regression line, it is difficult to estimate that line. Also, the smaller the variation in x values (as measured by S_{xx}), the larger $\sigma_{\hat{\beta}_1}$ is. The slope is the predicted change in y per unit change in x; if x changes very little in the data, so that S_{xx} is small, it is difficult to estimate the rate of change in y accurately. If the price of a brand of diet soda has not changed for years, it is obviously hard to estimate the change in quantity demanded when price changes.

$$\sigma_{\hat{\beta}_0} = \sigma_\varepsilon \sqrt{\frac{1}{n} + \frac{\bar{x}^2}{S_{xx}}}$$

The standard error of the estimated intercept $\hat{\beta}_0$ is influenced by n, naturally, and also by the size of the square of the sample mean, $\bar{x}^2$, relative to S_{xx}. The intercept is the predicted y value when $x = 0$; if all the x_i are, for instance, large positive numbers, predicting y at $x = 0$ is a huge extrapolation from the actual data. Such extrapolation magnifies small errors, and the standard error of $\hat{\beta}_0$ is large. The ideal situation for estimating $\hat{\beta}_0$ is when $\bar{x} = 0$.

To this point, we have considered only the estimates of intercept and slope. We also have to estimate the true error variance σ_ε^2. We can think of this quantity residuals

as variance around the line, or as the mean squared prediction error. The estimate of σ_ε^2 is based on the **residuals** $y_i - \hat{y}_i$, which are the prediction errors in the sample. The estimate of σ_ε^2 based on the sample data is the sum of squared residuals divided by $n - 2$, the degrees of freedom. The estimated variance is often shown in computer output as MS(Error) or MS(Residual). Recall that MS stands for mean square and is always a sum of squares (SS) divided by the appropriate degrees of freedom:

$$s_\varepsilon^2 = \frac{\sum_i (y_i - \hat{y}_i)^2}{n - 2} = \frac{\text{SS(Residual)}}{n - 2}$$

In the computer output for Example 11.3, SS(Residual) is shown to be 44.0.

Just as we divide by $n - 1$ rather than by n in the ordinary sample variance s^2 (in Chapter 3), we divide by $n - 2$ in s_ε^2, the estimated variance around the line. The reduction from n to $n - 2$ occurs because in order to estimate the variability around the regression line, we must first estimate the two parameters β_0 and β_1 to obtain the estimated line. The effective sample size for estimating σ_ε^2 is thus $n - 2$. In our definition, s_ε^2 is undefined for $n = 2$, as it should be. Another argument is that dividing by $n - 2$ makes s_ε^2 an unbiased estimator of σ_ε^2. In the computer output in Example 11.3, $n - 2 = 5 - 2 = 3$ is shown as DF (degrees of freedom) for RESIDUAL and $s_\varepsilon^2 = 14.6666$ is shown as MS for RESIDUAL.

sample standard
deviation around the
regression line

residual standard
deviation

standard error of estimate

The square root s_ε of the sample variance is called the **sample standard deviation around the regression line,** the **standard error of estimate,** or the **residual standard deviation.** Because s_ε estimates σ_ε, the standard deviation of y_i, σ_ε estimates the standard deviation of the population of y values associated with a given value of the independent variable x. The Statistix output in Example 11.3 labels s_ε as STANDARD DEVIATION; it shows that s_ε, rounded off, is 3.830.

Like any other standard deviation, the residual standard deviation may be interpreted by the Empirical Rule. About 95% of the prediction errors will fall within ± 2 standard deviations of the mean error; the mean error is always 0 in the least-squares regression model. Therefore, a residual standard deviation of 3.830 means that about 95% of prediction errors will be less than $\pm 2(3.830) = 7.660$.

The estimates $\hat{\beta}_0$, $\hat{\beta}_1$, and s_ε are basic in regression analysis. They specify the regression line and the probable degree of error associated with y values for a given value of x. The next step is to use these sample estimates to make inferences about the true parameters.

EXAMPLE 11.4

The human resources director of a chain of fast-food restaurants studied the absentee rate of employees. Whenever employees called in sick or simply didn't appear, the restaurant manager had to find replacements in a hurry or else work shorthanded. The director had data on the number of absences per 100 employees per week (y) and the average number of months' experience at the restaurant (x) for 10 restaurants in the chain. The director expected that longer-term employees were more reliable and absent less often.

For the following data and Minitab output, do the following:

a. Examine the scatterplot and decide whether a straight line is a reasonable model.

b. Identify the least-squares estimates for β_0 and β_1 in the model $y = \beta_0 + \beta_1 x + \varepsilon$.

c. Predict y for $x = 19.5$.
d. Identify s_ε, the sample standard deviation about the regression line.
e. Interpret the value of $\hat{\beta}_1$.

y:	31.5	33.1	27.4	24.5	27.0	27.8	23.3	24.7	16.9	18.1
x:	18.1	20.0	20.8	21.5	22.0	22.4	22.9	24.0	25.4	27.3

```
MTB > Regress 'y' on 1 predictor 'x'.

The regression equation is
y = 64.7 - 1.75 x

Predictor        Coef      Stdev     t-ratio          p
Constant       64.672      6.762        9.56      0.000
x             -1.7487     0.2995       -5.84      0.000

s = 2.388      R-sq = 81.0%      R-sq(adj) = 78.6%

Analysis of Variance

SOURCE         DF          SS          MS          F          P
Regression      1      194.45      194.45      34.10      0.000
Error           8       45.61        5.70
Total           9      240.06
```

Solution

a. A scatterplot drawn by the Statistix package is shown in Figure 11.12; the data appear to fall approximately along a downward-sloping line. There is no reason to use a more complicated model.
b. The output shows the coefficients twice, with differing numbers of digits. The intercept (constant) is 64.672 and the slope (coefficient of x) is -1.7487. Note that the negative slope corresponds to a downward-sloping line.
c. The least-squares prediction value when $x = 19.5$ is

$$\hat{y} = 64.672 - 1.7487(19.5) = 30.57$$

FIGURE 11.12
Scatterplot of absences (y) versus average length of employment (x)

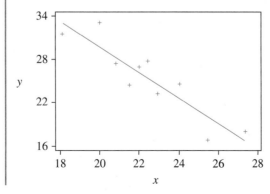

    d. The standard deviation around the line (the residual standard deviation) is shown as $s = 2.388$. Therefore, about 95% of the prediction errors should be less than $\pm 2(2.388) = 4.776$.

    e. From $\hat{\beta}_1 = -1.7487 \approx -1.75$, we conclude that for a 1-month increase in the average experience at a restaurant, there is an estimated decrease of 1.75 in the average number of absences per 100 employees per week.

EXERCISES   **Basic Techniques**

11.1 Plot the data shown here in a scatterplot and sketch a line through the points.

x	5	10	12	15	18	24
y	10	19	21	28	34	40

11.2 Refer to the data of Exercise 11.1. Find the least-squares prediction equation and compare it to the freehand regression line you sketched through the points.

11.3 A computer solution using SAS for the least-squares prediction equation to the data below is shown here.

x	10	18	25	40	50	63	42	30	5	55
y	25	55	50	75	110	138	90	60	10	100

```
SAS CODE:

option ls  = 70 ps = 55 nocenter nodate;
title 'EXERCISE 11.3';
data linreg;
    input X Y;
    CARDS;
10      25
18      55
25      50
40      75
50     110
63     138
42      90
30      60
 5      10
55     100
RUN;
PROC PLOT; PLOT Y*X='*';
PROC REG; MODEL Y = X;
OUTPUT OUT=NEW P=PRED R=RESID;
LABEL PRED='PREDICTED VALUE' RESID='RESIDUALS';
PROC PRINT; VAR Y X PRED RESID;
RUN;
```

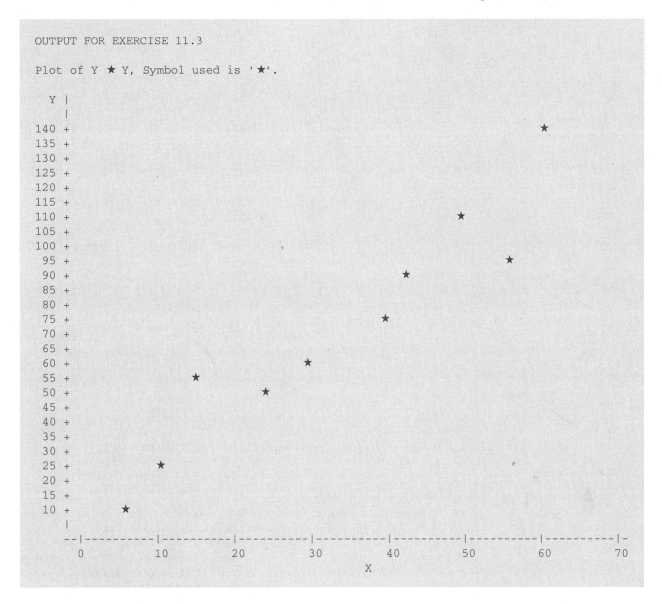

```
OUTPUT FOR EXERCISE 11.3

Plot of Y ★ Y, Symbol used is '★'.

  Y |
    |
140 +                                                              ★
135 +
130 +
125 +
120 +
115 +
110 +                                            ★
105 +
100 +
 95 +                                                  ★
 90 +                                        ★
 85 +
 80 +
 75 +                                  ★
 70 +
 65 +
 60 +                              ★
 55 +          ★
 50 +              ★
 45 +
 40 +
 35 +
 30 +
 25 +      ★
 20 +
 15 +
 10 +    ★
    |
 --|----------|----------|----------|----------|----------|----------|----------|-
   0         10         20         30         40         50         60         70
                                      X
```

```
Dependent Variable: Y

Analysis of Variance

                        Sum of          Mean
Source      DF         Squares        Square    F Value    Prob>F

Model        1    13230.96994    13230.96994    162.560    0.0001
Error        8       651.13006       81.39126
C Total      9    13882.10000
```

```
        Root MSE          9.02171      R-square      0.9531
        Dep Mean         71.30000      Adj R-sq      0.9472
        C.V.             12.65317

Parameter Estimates

                        Parameter       Standard     T for HO:
Variable   DF           Estimate          Error    Parameter = 0     Prob > |T|

INTERCEP    1           4.697852       5.95202071        0.789          0.4527
X           1           1.970478       0.15454842       12.750          0.0001

OBS      Y      X     PREDICTED     RESIDUALS
                        VALUES
  1      25     10       24.403         0.5974
  2      55     18       40.166        14.8335
  3      50     25       53.960        -3.9598
  4      75     40       83.517        -8.5170
  5     110     50      103.222         6.7783
  6     138     63      128.838         9.1620
  7      90     42       87.458         2.5421
  8      60     30       63.812        -3.8122
  9      10      5       14.550        -4.5502
 10     100     55      113.074       -13.0741
```

a. Determine the least-squares prediction equation from the output here and draw the regression line in the data plot.

b. Does the prediction equation seem to represent the data adequately?

c. Predict y for $x = 35$.

Applications

Ag.

11.4 A food processor conducted an experiment to examine the effect of different concentrations of pectin on the firmness of canned sweet potatoes, using three concentrations: 0, 1.5, and 3% pectin by weight. The processor packed six number 303 × 406 cans with sweet potatoes in a 25% (by weight) sugar solution. Two cans were randomly assigned to each of the pectin concentrations with the appropriate percentage of pectin added to the sugar syrup. The cans were then sealed and placed in a 25°C environment for 30 days. At the end of the storage time, the cans were opened and a firmness determination made for the contents of each can. These data appear here:

Pectin concentration	0%, 0%	1.5%, 1.5%	3.0%, 3.0%
Firmness reading	50.5, 46.8	62.3, 67.7	80.1, 79.2

a. Let x denote the pectin concentration of a can and y denote the firmness reading following the 30 days of storage at 25°C. Plot the sample data in a scatterplot.

b. Obtain least-squares estimates for the parameters in the model $y = \beta_0 + \beta_1 x + \varepsilon$.

11.5 Refer to Exercise 11.4. Predict the firmness for a can of sweet potatoes treated with a 1% concentration of pectin (by weight) after 30 days of storage at 25°C.

Bus.

11.6 A mail-order retailer spends considerable effort in picking orders—selecting the ordered items and assembling them for shipment. A small study took a sample of 100 orders. An experienced picker carried out the entire process. The time in minutes needed was recorded for each order. A scatterplot and spline fit, created using JMP, are shown. What sort of transformation is suggested by the plot?

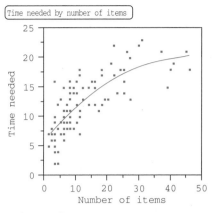

Time needed by number of items

Fitting ▶ — Smoothing spline fit, lambda = 10000

11.7 The order-picking time data in Exercise 11.6 were transformed by taking the square root of the number of items. A scatterplot of the result and the regression results follow.
   a. Does the transformed scatterplot appear reasonably linear?
   b. Write out the prediction equation based on the transformed data.

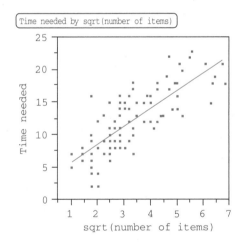

Time needed by sqrt(number of items)

Summary of Fit

RSquare	0.624567
RSquare Adj	0.620736
Root Mean Square Error	2.923232
Mean of Response	12.29
Observations (or Sum Wgts)	100

Analysis of Variance

Source	DF	Sum of Squares	Mean Square	F Ratio
Model	1	1393.1522	1393.15	163.0317
Error	98	837.4378	8.55	Prob>F
C Total	99	2230.5900		0.0000

Parameter Estimates

Term	Estimate	Std Error	t Ratio	Prob>|t|
Intercept	3.097869	0.776999	3.99	0.0001
sqrt (Number of items)	2.7633138	0.216418	12.77	0.0000

11.8 In the JMP output of Exercise 11.7, the residual standard deviation is called "Root Mean Square Error." Locate and interpret this number.

Bus. **11.9** As one part of a study of commercial bank branches, data are obtained on the number of independent businesses (x) located in sample zip code areas and the number of bank branches (y) located in these areas. The commercial centers of cities are excluded.

x:	92	116	124	210	216	267	306	378	415	502	615	703
y:	3	2	3	5	4	5	5	6	7	7	9	9

Output (StataQuest) for the analysis of the data is as follows:

```
. regress Branches Business

    Source |      SS          df         MS          Number of obs =      12
-----------+------------------------------           F( 1, 10)    = 172.60
     Model | 53.7996874       1     53.7996874       Prob > F     = 0.0000
  Residual | 3.11697922      10     .311697922       R-square     = 0.9452
-----------+------------------------------           Adj R-square = 0.9398
     Total   56.9166667      11     5.17424242       Root MSE     = .5583
```

```
-----------------------------------------------------------------------------
Branches |    Coef.     Std. Err.       t       P>|t|     [95% Conf. Interval]
---------+-------------------------------------------------------------------
Business | .0111049     .0008453     13.138    0.000     .0092216     .0129883
   _cons | 1.766846     .3211751      5.501    0.000     1.051223     2.482469
-----------------------------------------------------------------------------
```

a. Plot the data. Does a linear equation relating y to x appear plausible?
b. Locate the regression equation (with y as the dependent variable).
c. Interpret the value of $\hat{\beta}_1$ in the context of this problem.
d. Locate the sample residual standard deviation s_ε.

11.10 Does it appear that the variability of y increases with x in the data plot of Exercise 11.9? (This would violate the assumption of constant variance.)

Bus. **11.11** A realtor studied the relation between $x =$ yearly income (in thousands of dollars per year) of home purchasers and $y =$ sale price of the house (in thousands of dollars). The realtor gathered data from mortgage applications for 24 sales in the realtor's basic sales area in one season. Stata output was obtained, as shown after the data.

x:	25.0	28.5	29.2	30.0	31.0	31.5	31.9	32.0	33.0
y:	84.9	94.0	96.5	93.5	102.9	99.5	101.0	105.0	99.9

x:	33.5	34.0	35.9	36.0	39.0	39.0	40.5	40.9	42.5
y:	110.0	100.0	116.0	110.0	125.0	119.9	130.6	120.8	129.9

x:	44.0	45.0	50.0	54.6	65.0	70.0
y:	135.5	140.0	150.7	170.0	110.0	185.0

```
. regress Price Income

    Source |      SS          df         MS          Number of obs =      24
-----------+------------------------------           F( 1, 22)    = 45.20
     Model | 9432.58336       1     9432.58336       Prob > F     = 0.0000
  Residual | 4590.6746       22     208.667027       R-square     = 0.6726
-----------+------------------------------           Adj R-square = 0.6578
     Total | 14023.258       23     609.706868       Root MSE     = 14.445
```

```
-----------------------------------------------------------------------------
   Price |    Coef.     Std. Err.       t       P>|t|     [95% Conf. Interval]
---------+-------------------------------------------------------------------
  Income | 1.80264      .2681147      6.723    0.000     1.246604     2.358676
   _cons | 47.15048     10.93417      4.312    0.000     24.4744      69.82657
-----------------------------------------------------------------------------
```

```
.drop in 23
(1 observation deleted)

. regress Price Income

    Source |      SS          df        MS         Number of obs =      23
-----------+------------------------------------   F( 1, 21)     = 512.02
     Model | 13407.5437        1     13407.5437    Prob > F      = 0.0000
  Residual | 547.902031       21     26.185811     R-square      = 0.9606
-----------+------------------------------------   Adj R-square  = 0.9587
     Total | 13957.4457       22     634.429351    Root MSE      = 5.1172
------------------------------------------------------------------------------

     Price |   Coef.    Std. Err.     t      P>|t|    [95% Conf. Interval]
------------------------------------------------------------------------------
    Income | 2.461967   .108803    22.628    0.000    2.235699    2.688236
     _cons | 24.35755   4.286011    5.683    0.000    15.4443     33.27079
------------------------------------------------------------------------------
```

a. A scatterplot with a LOWESS smoother, drawn using Minitab, follows. Does the relation appear to be basically linear?

b. Are there any high leverage points? If so, which ones seem to have high influence?

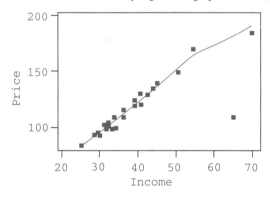

11.12 For Exercise 11.11,

  a. Locate the least-squares regression equation for the data.

  b. Interpret the slope coefficient. Is the intercept meaningful?

  c. Find the residual standard deviation.

11.13 The output of Exercise 11.11 also contains a regression line when we omit the point with $x = 65.0$ and $y = 110.0$. Does the slope change substantially? Why?

11.3 Inferences about Regression Parameters

The slope, intercept, and residual standard deviation in a simple regression model are all estimates based on limited data. As with all other statistical quantities, they are affected by random error. In this section, we consider how to allow for that random error. The concepts of hypothesis tests and confidence intervals that we have applied to means and proportions apply equally well to regression summary figures.

t test for β_1 The *t* distribution can be used to make significance tests and confidence intervals for the true slope and intercept. One natural null hypothesis is that the true slope β_1 equals 0. If this H_0 is true, a change in *x* yields no predicted change in *y*, and it follows that x has no value in predicting y. We know from the previous section that the sample slope $\hat{\beta}_1$ has the expected value β_1 and standard error

$$\sigma_{\hat{\beta}_1} = \sigma_{\varepsilon}\sqrt{\frac{1}{S_{xx}}}$$

In practice, σ_{ε} is not known and must be estimated by s_{ε}, the residual standard deviation. In almost all regression analysis computer outputs, the estimated standard error is shown next to the coefficient. A test of this null hypothesis is given by the t statistic

$$t = \frac{\hat{\beta}_1 - \beta_1}{\text{estimated standard error }(\hat{\beta}_1)} = \frac{\hat{\beta}_1 - \beta_1}{s_{\varepsilon}\sqrt{1/S_{xx}}}$$

The most common use of this statistic is shown in the following summary.

Summary of a Statistical Test for β_1

Hypotheses:

Case 1. $H_0: \beta_1 \leq 0$ vs. $H_a: \beta_1 > 0$
Case 2. $H_0: \beta_1 \geq 0$ vs. $H_a: \beta_1 < 0$
Case 3. $H_0: \beta_1 = 0$ vs. $H_a: \beta_1 \neq 0$

T.S.: $t = \dfrac{\hat{\beta}_1 - 0}{s_{\varepsilon}/\sqrt{S_{xx}}}$

R.R.: For df $= n - 2$ and Type I error α,
1. Reject H_0 if $t > t_{\alpha}$.
2. Reject H_0 if $t < -t_{\alpha}$.
3. Reject H_0 if $|t| > t_{\alpha/2}$.

Check assumptions and draw conclusions.

All regression analysis outputs show this t value.

In most computer outputs, this test is indicated after the standard error and labeled T TEST or T STATISTIC. Often, a p-value is also given, which eliminates the need to look up the t value in a table.

EXAMPLE 11.5

Use the computer output of Example 11.3 (reproduced here) to locate the value of the t statistic for testing $H_0: \beta_1 = 0$ in the road resurfacing example. Give the observed level of significance for the test.

PREDICTOR VARIABLES	COEFFICIENT	STD ERROR	STUDENT'S T	P
CONSTANT	2.00000	3.82970	0.52	0.6376
MILES	3.00000	0.85634	3.50	0.0394
R-SQUARED	0.8036	RESID. MEAN SQUARE (MSE)		14.6666
ADJUSTED R-SQUARED	0.7381	STANDARD DEVIATION		3.82970

```
SOURCE        DF      SS           MS          F        P

REGRESSION    1      180.000      180.000     12.27    0.0394
RESIDUAL      3      44.0000      14.6666
TOTAL         4      224.000
```

Solution It is clear from the output that the value of the test statistic in the column labeled STUDENT'S T is $t = 3.50$. The p-value for the two-tailed alternative $H_a: \beta_1 \neq 0$, labeled P, is .0394. Because this value is fairly small, we can reject the hypothesis that mileage has no effect on predicting cost.

EXAMPLE 11.6

The following data show mean ages of executives of 15 firms in the food industry and the previous year's percentage increase in earnings per share of the firms. Use the Systat output shown to test the hypothesis that executive age has no predictive value for change in earnings. Should a one-sided or two-sided alternative be used?

Mean age	x:	38.2	40.0	42.5	43.4	44.6	44.9	45.0	45.4
Change, earnings per share	y:	8.9	13.0	4.7	−2.4	12.5	18.4	6.6	13.5

	x:	46.0	47.3	47.3	48.0	49.1	50.5	51.6
	y:	8.5	15.3	18.9	6.0	10.4	15.9	17.1

```
DEP VAR: CHGEPS N: 15 MULTIPLE R: 0.383 SQUARED MULTIPLE R: 0.147
STANDARD ERROR OF ESTIMATE: 5.634

VARIABLE   COEFFICIENT   STD ERROR   STD COEF    T       P(2 TAIL)
CONSTANT     -16.991       18.866      0.000     0.901     0.384
MEANAGE        0.617        0.413      0.383     1.496     0.158

                    ANALYSIS OF VARIANCE

SOURCE       SUM-OF-SQUARES   DF    MEAN-SQUARE   F-RATIO     P
REGRESSION        71.055      1        71.055      2.239    0.158
RESIDUAL         412.602     13        31.739
```

Solution In the model $y = \beta_0 + \beta_1 x + \varepsilon$, the null hypothesis is $H_0: \beta_1 = 0$. The myth in American business is that younger managers tend to be more aggressive and harder driving, but it is also possible that the greater experience of the older executives leads to better decisions. Therefore, there is a good reason to choose a two-sided research hypothesis, $H_a: \beta_1 \neq 0$. The t statistic is shown in the output column marked T, reasonably enough. It shows $t = 1.496$, with a (two-sided) p-value of 0.158. There is not enough evidence to conclude that there is any relation between age and change in earnings for the year studied.

In passing, note that the interpretation of $\hat{\beta}_0$ is rather interesting in this example; it is the predicted change in earnings of a firm with the mean age of its managers equal to 0.

It is also possible to calculate a confidence interval for the true slope. This is an excellent way to communicate the likely degree of inaccuracy in the estimate of that slope. The confidence interval once again is simply the estimate plus or minus a t-table value times the standard error.

Confidence Interval for Slope β_1

$$\hat{\beta}_1 - t_{\alpha/2}s_\varepsilon\sqrt{\frac{1}{S_{xx}}} \leq \beta_1 \leq \hat{\beta}_1 + t_{\alpha/2}s_\varepsilon\sqrt{\frac{1}{S_{xx}}}$$

The required degrees of freedom for the table value $t_{\alpha/2}$ is $n - 2$, the error df.

EXAMPLE 11.7

Compute a 95% confidence interval for the slope β_1 using the output from Example 11.3.

Solution   In the output, $\hat{\beta}_1 = 3.000$ and the estimated standard error of $\hat{\beta}_1$ is shown as .856, rounded off. Because n is 5, there are $5 - 2 = 3$ df for error. The required table value for $\alpha/2 = .05/2 = .025$ is 3.182. The corresponding confidence interval for the true value of β_1 is then

$$3.00 \pm 3.182(.856) \quad \text{or} \quad .276 \text{ to } 5.724$$

The predicted cost per additional mile of resurfacing could be anywhere from $276,000 to $5,724,000. The enormous width of this interval results largely from the small sample size.

There is an alternative test, an F test, for the null hypothesis of no predictive value. It was designed to test the null hypothesis that *all* predictors have no value in predicting y. This test gives the same result as a two-sided t test of H_0: $\beta_1 = 0$ in simple linear regression; to say that all predictors have no value is to say that the (only) slope is 0. The F test is summarized next.

F Test for H_0; $\beta_1 = 0$

H_0: $\beta_1 = 0$

H_a: $\beta_1 \neq 0$

T.S.: $F = \dfrac{\text{SS(Regression)}/1}{\text{SS(Residual)}/(n-2)} = \dfrac{\text{MS(Regression)}}{\text{MS(Residual)}}$

R.R.:   With $df_1 = 1$ and $df_2 = n - 2$, reject H_0 if $F > F_\alpha$.

Check assumptions and draw conclusions.

SS(Regression) is the sum of squared deviations of predicted y values from the y mean. SS(Regression) $= \Sigma (\hat{y}_i - \bar{y})^2$. SS(Residual) is the sum of squared deviations of actual y values from predicted y values. SS(Residual) $= \Sigma (y_i - \hat{y}_i)^2$.

Virtually all computer packages calculate this F statistic. In the road resurfacing example, the output shows $F = 12.27$ with a p-value of .0394. Again, the hypothesis of no predictive value can be rejected. It is always true for simple linear regression problems that $F = t^2$; in the example, $12.27 = (3.50)^2$, to within rounding error. The F and two-sided t tests are equivalent in simple linear regression; they serve different purposes in multiple regression.

EXAMPLE 11.8

For the output of Example 11.4, reproduced here, use the F test for $H_0: \beta_1 = 0$. Show that $t^2 = F$.

```
The regression equation is
y = 64.7 - 1.75 x

Predictor         Coef       Stdev     t-ratio          p
Constant        64.672       6.762        9.56      0.000
x              -1.7487      0.2995       -5.84      0.000

s = 2.388          R-sq = 81.0%       R-sq(adj) = 78.6%


Analysis of Variance

SOURCE          DF          SS          MS          F          P
Regression       1      194.45      194.45      34.10      0.000
Error            8       45.61        5.70
Total            9      240.06
```

Solution   The F statistic is shown in the output as 34.10, with a p-value of 0.000 (indicating that the actual p-value is something smaller than 0.0005). Note that the t statistic is -5.84, and that $t^2 = (-5.84)^2 = 34.11$, equal to F, to within rounding error.

You should be able to work out comparable hypothesis testing and confidence interval formulas for the intercept β_0 using the estimated standard error of $\hat{\beta}_0$ as

$$\sigma_{\hat{\beta}_0} = s_\varepsilon \sqrt{\frac{1}{n} + \frac{\bar{x}^2}{S_{xx}}}$$

In practice, this parameter is of less interest than the slope. In particular, there is often no reason to hypothesize that the true intercept is zero (or any other particular value). Computer packages almost always test the null hypothesis of zero slope, but some don't bother with a test on the intercept term.

EXERCISES   **11.14**   Refer to the data of Exercise 11.6.
  a. Calculate a 90% confidence interval for β_1.
  b. What is the interpretation of $H_0: \beta_1 = 0$ in Exercise 11.6?
  c. What is the natural research hypothesis H_a for that problem?
  d. Do the data support H_a at $\alpha = .05$? Clearly state your assumptions.

11.15   Find the p-value of the test of $H_0: \beta_1 = 0$ for Exercise 11.14.

Bio.   **11.16**   The extent of disease transmission can be affected greatly by the viability of infectious organisms suspended in the air. Because of the infectious nature of the disease under study, the viability of these organisms must be studied in an airtight chamber. One way to do this is to disperse an aerosol cloud, prepared from a solution containing the organisms, into the chamber. The biological recovery at any particular time is the percentage of the total number of organisms suspended in the aerosol that are viable. The data in the accompanying table are the biological recovery percentages computed from 13 different aerosol clouds. For each of the clouds, recovery percentages were determined at different times.

   a. Plot the data.
   b. Because there is some curvature, use the log of the biological recovery to try to linearize the data.

Cloud	Time, x (in minutes)	Biological Recovery (%)
1	0	70.6
2	5	52.0
3	10	33.4
4	15	22.0
5	20	18.3
6	25	15.1
7	30	13.0
8	35	10.0
9	40	9.1
10	45	8.3
11	50	7.9
12	55	7.7
13	60	7.7

11.17   Refer to Exercise 11.16.
   a. Fit the linear regression model $y = \beta_0 + \beta_1 x + \varepsilon$, where y is the log of the biological recovery.
   b. Compute an estimate of σ_ε.
   c. Identify the standard errors of $\hat{\beta}_0$ and $\hat{\beta}_1$.

11.18   Refer to Exercise 11.16. Conduct a test of the null hypothesis that $\beta_1 = 0$. Use $\alpha = .05$.

Ag.   **11.19**   A researcher conducts an experiment to examine the relationship between the weight gain of chickens whose diets had been supplemented by different amounts of the amino acid lysine and the amount of lysine ingested. Because the percentage of lysine is known and we can monitor the amount of feed consumed, we can determine the amount of lysine eaten. A random sample of 12 2-week-old chickens was selected for the study. Each was caged separately and was allowed to eat at will from feed composed of a base supplemented with lysine. The sample data summarizing weight gains and amounts of lysine eaten over the test period are given here. (In the data, y represents weight gain in grams and x represents the amount of lysine ingested in grams.)
   a. Refer to the output. Does a linear model seem appropriate?
   b. From the output, obtain the estimated linear regression model $\hat{y} = \hat{\beta}_0 + \hat{\beta}_1 x$.

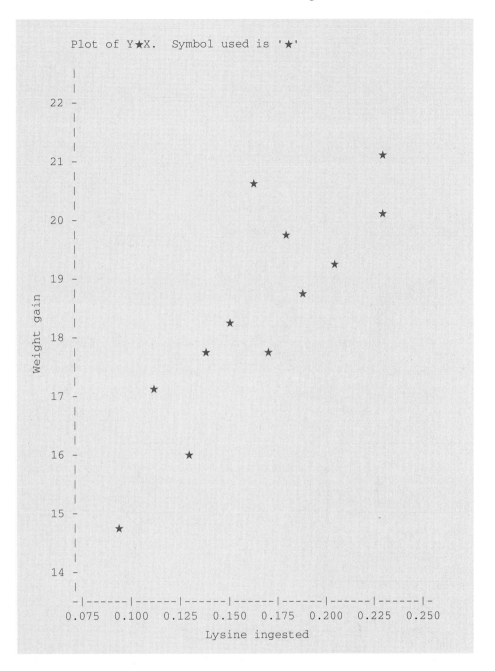

Chick	y	x	Chick	y	x
1	14.7	.09	7	17.2	.11
2	17.8	.14	8	18.7	.19
3	19.6	.18	9	20.2	.23
4	18.4	.15	10	16.0	.13
5	20.5	.16	11	17.8	.17
6	21.1	.23	12	19.4	.21

```
OUTPUT FOR EXERCISE 11.19

Dependent Variable: Y          WEIGHT GAIN

Analysis of Variance

                  Sum of        Mean
Source    DF     Squares       Square   F Value   Prob > F

Model      1    28.35785     28.35785    26.522    0.0004
Error     10    10.69215      1.06921
C Total   11    39.05000

       Root MSE     1.03403   R-square   0.7262
       Dep Mean    18.45000   Adj R-sq   0.6988
       C.V.         5.60449
```

```
Parameter Estimates

                 Parameter     Standard      T for H0:
Variable   DF    Estimate        Error    Parameter = 0   Prob > |T|

INTERCEP    1    12.508525    1.19168259        10.497       0.0001
X           1    35.827989    6.95693918 β̂₁      5.150       0.0004

                 Variable
Variable   DF    Label

INTERCEP    1    Intercept
X           1    LYSINE INGESTED

OBS     Y       X      PREDICTED    RESIDUALS
                       VALUES

 1    14.7    0.09     15.7330     -1.03304
 2    17.8    0.14     17.5244      0.27556
 3    19.6    0.18     18.9576      0.64244
 4    18.4    0.15     17.8827      0.51728
 5    20.5    0.16     18.2410      2.25900
 6    21.1    0.23     20.7490      0.35104
 7    17.2    0.11     16.4496      0.75040
 8    18.7    0.19     19.3158     -0.61584
 9    20.2    0.23     20.7490     -0.54896
10    16.0    0.13     17.1662     -1.16616
11    17.8    0.17     18.5993     -0.79928
12    19.4    0.21     20.0324     -0.63240
```

11.20 Refer to the output of Exercise 11.19.
  a. Estimate σ_ε^2.
  b. Identify the standard error of $\hat{\beta}_1$.

c. Conduct a statistical test of the research hypothesis that, for this diet preparation and length of study, there is a direct (positive) linear relationship between weight gain and the amount of lysine eaten.

Bus. **11.21** A firm that prints automobile bumper stickers investigates the relation between the total direct cost of a lot of stickers and the number produced in the printing run. The data are analyzed by the Execustat computer package. The relevant output is as follows:

```
            Simple Regression Analysis

Linear model: TotalCost = 99.777 + 5.19179*Runsize

               Table of Estimates

                        Standard     t        P
            Estimate     Error     Value    Value

Intercept    99.777     2.8273     35.29   0.0000
Slope        5.19179    0.0586455  88.53   0.0000


R-squared = 99.64%
Correlation coeff. = 0.998
Standard error of estimation = 12.2065
Durbin-Watson statistic = 2.67999

                    Analysis of Variance

                  Sum of                                            P
Source            Squares     D.F.    Mean Square   F-Ratio      Value

Model          1.16775e+006     1    1.16775e+006   7837.26    0.0000
Error          4171.98         28    148.999


Total (corr.)  1.17192e+006    29
```

a. Plot the data. Do you detect any difficulties with using a linear regression model? Can you see any blatant violations of assumptions? The raw data are as follows:

Runsize:	2.6	5.0	10.0	2.0	.8	4.0	2.5	.6	.8	1.0	2.0	
Total cost:	230	341	629	187	159	327	206	124	155	147	209	

Runsize:	3.0	.4	.5	5.0	20.0	5.0	2.0	1.0	1.5	.5	1.0	1.0
Total cost:	247	135	125	366	1146	339	208	150	179	128	155	143

Runsize:	.6	2.0	1.5	3.0	6.5	2.2	1.0
Total cost:	131	219	171	258	415	226	159

b. Write the estimated regression equation indicated in the output. Find the residual standard deviation.

c. Calculate a 95% confidence interval for the true slope. What are the interpretations of the intercept and slope in this problem?

11.22 Refer to the computer output of Exercise 11.21.

a. Locate the value of the t statistic for testing $H_0: \beta_1 = 0$.

b. Locate the p-value for this test. Is the p-value one-tailed or two-tailed? If necessary, calculate the p-value for the appropriate number of tails.

11.23 Refer to the computer output of Exercise 11.21.
  a. Locate the value of the F statistic and the associated p-value.
  b. How do the p-values for this F test and the t test in Exercise 11.22 compare? Why should this relation hold?

11.4 Predicting New y Values Using Regression

In all the regression analyses we have done so far, we have been summarizing and making inferences about relations in data that have already been observed. Thus, we have been predicting the past. One of the most important uses of regression is trying to forecast the future. In the road resurfacing example, the county highway director wants to predict the cost of a new contract that is up for bids. In a regression predicting quantity sold given price, a manager will want to predict the demand at a new price. In this section, we discuss how to make such regression forecasts and how to determine the plus or minus probable error factor.

There are two possible interpretations of a y prediction based on a given x. Suppose that the highway director substitutes $x = 6$ miles in the regression equation $\hat{y} = 2.0 + 3.0x$ and gets $\hat{y} = 20$. This can be interpreted as either

"The average cost $E(y)$ of *all* resurfacing contracts for 6 miles of road will be $20,000,000."

or

"The cost y of *this specific* resurfacing contract for 6 miles of road will be $20,000,000."

The best-guess prediction in either case is 20, but the plus or minus factor differs. It is easier to predict an average value $E(y)$ than an individual y value, so the plus or minus factor should be less for predicting an average. We discuss the plus or minus range for predicting an average first, with the understanding that this is an intermediate step toward solving the specific-value problem.

In the mean-value forecasting problem, suppose that the value of the predictor x is known. Because the previous values of x have been designated $x_1, \ldots, x_n$, let's call the new value x_{n+1}. Then $\hat{y}_{n+1} = \beta_0 + \beta_1 x_{n+1}$ is used to predict $E(y_{n+1})$. Because $\hat{\beta}_0$ and $\hat{\beta}_1$ are unbiased, $\hat{y}_{n+1}$ is an unbiased predictor of $E(y_{n+1})$. The standard error of $\hat{y}_{n+1}$ can be shown to be

$$\sigma_\varepsilon \sqrt{\frac{1}{n} + \frac{(x_{n+1} - \bar{x})^2}{S_{xx}}}$$

Here S_{xx} is the sum of squared deviations of the original n values of x_i; it can be calculated from most computer outputs as

$$\left(\frac{s_\varepsilon}{\text{standard error}(\hat{\beta}_1)}\right)^2$$

Again, t tables with $n - 2$ df (the error df) must be used. The usual approach to forming a confidence interval—namely, estimate plus or minus t (standard error)—yields a confidence interval for $E(y_{n+1})$. Some of the better statistical computer packages will calculate this confidence interval if a new x value is specified without specifying a corresponding y.

Confidence Interval for $E(y_{n+1})$

$$\hat{y}_{n+1} - t_{\alpha/2}s_{\varepsilon}\sqrt{\frac{1}{n} + \frac{(x_{n+1} - \bar{x})^2}{S_{xx}}} \leq E(y_{n+1})$$

$$\leq \hat{y}_{n+1} + t_{\alpha/2}s_{\varepsilon}\sqrt{\frac{1}{n} + \frac{(x_{n+1} - \bar{x})^2}{S_{xx}}}$$

where $t_{\alpha/2}$ cuts off area $\alpha/2$ in the right tail of the *t* distribution with $n - 2$ df.

For the resurfacing example, the computer output displayed here shows the estimated value of $E(y_{n+1})$ to be 20 when $x = 6$. The corresponding 95% confidence interval on $E(y_{n+1})$ is 12.29 to 27.71.

```
Resurfacing Data

PREDICTOR
VARIABLES      COEFFICIENT    STD ERROR    STUDENT'S T      P
CONSTANT          2.00000      3.82970         0.52       0.6376
MILES             3.00000      0.85634         3.50       0.0394


R-SQUARED                 0.8036    RESID. MEAN SQUARE (MSE)   14.6666
ADJUSTED R-SQUARED        0.7381    STANDARD DEVIATION          3.82970


SOURCE        DF      SS         MS        F       P


REGRESSION     1   180.000    180.000    12.27   0.0394
RESIDUAL       3   44.0000    14.6666
TOTAL          4   224.000


PREDICTED/FITTED VALUES OF COST

LOWER PREDICTED BOUND    5.5791    LOWER FITTED BOUND     12.291
PREDICTED VALUE         20.000     FITTED VALUE           20.000
UPPER PREDICTED BOUND   34.420     UPPER FITTED BOUND     27.708
SE (PREDICTED VALUE)     4.5313    SE (FITTED VALUE)       2.4221


PREDICTOR VALUES: MILES = 6.0000
```

The forecasting plus or minus term in the confidence interval for $E(y_{n+1})$ depends on the sample size *n* and the standard deviation around the regression line, as we might expect. It also depends on the squared distance of x_{n+1} from $\bar{x}$ (the mean of the previous x_i values) relative to S_{xx}. As x_{n+1} gets farther from $\bar{x}$, the term

$$\frac{(x_{n+1} - \bar{x})^2}{S_{xx}}$$

gets larger. When x_{n+1} is far away from the other *x* values, so that this term is large, the prediction is a considerable extrapolation from the data. Small errors in estimating the regression line are magnified by the extrapolation. The term **extrapolation penalty** $(x_{n+1} - \bar{x})^2/S_{xx}$ could be called an **extrapolation penalty** because it increases with the degree of extrapolation.

Extrapolation—predicting the results at independent variable values far from the data—is often tempting and always dangerous. Using it requires an assumption that the relation will continue to be linear, far beyond the data. By definition, you have no data to check this assumption. For example, a firm might find a negative correlation between the number of employees (ranging between 1,200 and 1,400) in a quarter and the profitability in that quarter; the fewer the employees, the greater the profit. It would be spectacularly risky to conclude from this fact that cutting the number of employees to 600 would vastly improve profitability. (Do you suppose we could have a negative number of employees?) Sooner or later, the declining number of employees must adversely affect the business so that profitability turns downward. The extrapolation penalty term actually understates the risk of extrapolation. It is based on the assumption of a linear relation, and that assumption gets very shaky for large extrapolations.

The confidence and prediction intervals also depend heavily on the assumption of constant variance. In some regression situations, the variability around the line increases as the predicted value increases, violating this assumption. In such a case, the confidence and prediction intervals will be too wide where there is relatively little variability and too narrow where there is relatively large variability. A scatterplot that shows a fan shape indicates nonconstant variance. In such a case, the confidence and prediction intervals are not very accurate.

EXAMPLE 11.9

For the data of Example 11.4, and the following Minitab output from that data, obtain a 95% confidence interval for $E(y_{n+1})$ based on an assumed x_{n+1} of 22.4. Compare the width of the interval to one based on an assumed x_{n+1} of 30.4.

```
MTB > regress 'y' on 1 variable 'x';
SUBC> predict at 22.4;
SUBC> predict at 30.4.

The regression equation is
y = 64.7 - 1.75 x

Predictor      Coef      Stdev    t-ratio        p
Constant     64.672      6.762       9.56    0.000
x           -1.7487     0.2995      -5.84    0.000

s = 2.388   R-sq = 81.0%    R-sq(adj) = 78.6%

Analysis of Variance

SOURCE        DF        SS        MS         F        p
Regression     1    194.45    194.45     34.10    0.000
Error          8     45.61      5.70
Total          9    240.06

    Fit    Stdev.Fit          95% C.I.                  95% P.I.
 25.500        0.755    ( 23.758,  27.242)       ( 19.723,  31.277)
 11.510        2.500    (  5.742,  17.278)       (  3.535,  19.485) XX

X denotes a row with X values away from the center
XX denotes a row with very extreme X values
```

Solution For $x_{n+1} = 22.4$, the first of the two Fit entries shows a predicted value equal to 25.5. The confidence interval is shown as 23.758 to 27.242. For $x_{n+1} = 30.4$, the predicted value is 11.51, with a confidence interval of 5.742 to 17.278. The second interval has a width of about 11.5, much larger than the first interval's width of about 3.5. The value $x_{n+1} = 30.4$ is far outside the range of x data; the extrapolation penalty makes the interval very wide.

Usually, the more relevant forecasting problem is that of predicting an individual y_{n+1} value rather than $E(y_{n+1})$. In most computer packages, the interval for predicting an individual value is called a **prediction interval.** The same best guess **prediction interval** $\hat{y}_{n+1}$ is used, but the forecasting plus or minus term is larger when predicting y_{n+1} than $E(y_{n+1})$. In fact, it can be shown that the plus or minus forecasting error using $\hat{y}_{n+1}$ to predict y_{n+1} is as follows.

Prediction Interval for y_{n+1}

$$\hat{y}_{n+1} - t_{\alpha/2}s_{\varepsilon}\sqrt{1 + \frac{1}{n} + \frac{(x_{n+1} - \bar{x})^2}{S_{xx}}} \le y_{n+1}$$

$$\le \hat{y}_{n+1} + t_{\alpha/2}s_{\varepsilon}\sqrt{1 + \frac{1}{n} + \frac{(x_{n+1} - \bar{x})^2}{S_{xx}}}$$

where $t_{\alpha/2}$ cuts off area $\alpha/2$ in the right tail of the t distribution with $n - 2$ df.

In the road resurfacing example, the corresponding 95% prediction limits for y_{n+1} when $x = 6$ are 5.58 to 34.42 (see output earlier in this section). The 95% intervals for $E(y_{n+1})$ and for y_{n+1} are plotted in Figure 11.13; the inner curves are for $E(y_{n+1})$ and the outer ones for y_{n+1}.

The only difference between prediction of a mean $E(y_{n+1})$ and prediction of an individual y_{n+1} is the term $+1$ in the standard error formula. The presence of this extra term indicates that predictions of individual values are less accurate than predictions of means. The extrapolation penalty term still applies, as does the warning that it understates the risk of extrapolation. If n is large and the extrapolation term is small, the $+1$ term dominates the square root factor in the prediction interval. In

FIGURE 11.13

Predicted versus observed values with 95% limits

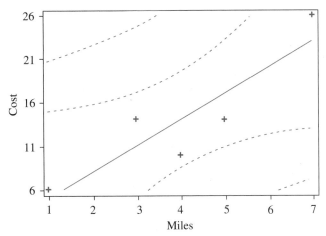

Cost = 2.0000 + 3.0000 * Miles   95% conf. and pred. intervals

such cases, the interval becomes approximately $\hat{y}_{n+1} - t_{\alpha/2}s_\varepsilon \leq y_{n+1} \leq \hat{y}_{n+1} + t_{\alpha/2}s_\varepsilon$. Thus, for large n, roughly 68% of the residuals (forecast errors) are less than $\pm 1 s_\varepsilon$ and 95% less than $\pm 2 s_\varepsilon$. There is not much point in devising rules for when to ignore the other terms in the square root factor. They are normally calculated in computer outputs and it does no harm to include them.

EXAMPLE 11.10

Using the output in Example 11.9 (reproduced here), find a 95% prediction interval for y_{n+1} with $x_{n+1} = 22.4$, and find the interval with $x_{n+1} = 30.4$. Compare these to widths estimated by the $\pm 2 s_\varepsilon$ rules just discussed.

```
MTB > regress 'y' on 1 variable 'x';
SUBC> predict at 22.4;
SUBC> predict at 30.4.

The regression equation is
y = 64.7 - 1.75 x

Predictor      Coef      Stdev    t-ratio        p
Constant     64.672      6.762       9.56    0.000
x           -1.7487     0.2995      -5.84    0.000

s = 2.388   R-sq = 81.0%    R-sq(adj) = 78.6%

Analysis of Variance

SOURCE         DF         SS         MS        F        p
Regression      1     194.45     194.45    34.10    0.000
Error           8      45.61       5.70
Total           9     240.06

   Fit    Stdev.Fit        95% C.I.              95% P.I.
25.500        0.755   ( 23.758, 27.242)    ( 19.723, 31.277)
11.510        2.500   (  5.742, 17.278)    (  3.535, 19.485) XX

X denotes a row with X values away from the center
XX denotes a row with very extreme X values
```

Solution   As in Example 11.9, $\hat{y}_{n+1} = 25.5$ if $x_{n+1} = 22.4$. The prediction interval is shown as

$$19.72 \leq y_{n+1} \leq 31.28$$

The $\pm 2 s_\varepsilon$ range is

$$25.5 - (2)(2.388) \leq y_{n+1} \leq 25.5 + (2)(2.388) \quad \text{or} \quad 20.72 \leq y_{n+1} \leq 30.28$$

The latter interval is a bit too narrow, mostly because the tabled t value with only 8 df is quite a bit larger than 2.

For $x_{n+1} = 30.4$, $\hat{y}_{n+1} = 11.51$, the 95% prediction interval is

$$3.54 \leq y_{n+1} \leq 19.48$$

The $\pm 2s_\varepsilon$ range is

$$11.5 - (2)(2.388) \leq y_{n+1} \leq 11.5 + (2)(2.388) \quad \text{or} \quad 6.72 \leq y_{n+1} \leq 16.28$$

The latter is much too narrow. Not only is the tabled t value larger than 2, but also the large extrapolation penalty is not reflected. The output labels this prediction XX and notes that the x value used is far from the data. Be warned.

EXERCISES   **Basic Techniques**

11.24   Refer to Exercise 11.16. For the least-squares equation

$$\hat{y} = \hat{\beta}_0 + \hat{\beta}_1 x$$

estimate the mean log biological recovery percentage at 30 minutes, using a 95% confidence interval.

11.25   Using the data of Exercise 11.16, construct a 95% prediction interval for the log biological recovery percentage at 30 minutes. Compare your result to the confidence interval on $E(y)$ of Exercise 11.24.

Applications

Engin.   **11.26** A chemist is interested in determining the weight loss y of a particular compound as a function of the amount of time the compound is exposed to the air. The data in the following table give the weight losses associated with $n = 12$ settings of the independent variable, exposure time.

Weight Loss and Exposure Time Data

Weight Loss, y (in pounds)	Exposure Time (in hours)
4.3	4
5.5	5
6.8	6
8.0	7
4.0	4
5.2	5
6.6	6
7.5	7
2.0	4
4.0	5
5.7	6
6.5	7

a. Find the least-squares prediction equation for the model

$$y = \beta_0 + \beta_1 x + \varepsilon$$

b. Test $H_0: \beta_1 \leq 0$; give the p-value for $H_a: \beta_1 > 0$; and draw conclusions.

11.27 Refer to Exercise 11.26 and the SAS computer output shown here.
a. Identify the 95% confidence bands for $E(y)$ when $4 \leq x \leq 7$.
b. Identify the 95% prediction bands for y, $4 \leq x \leq 7$.
c. Distinguish between the meaning of the confidence bands and prediction bands in parts (a) and (b).

```
Dependent Variable: Y        WEIGHT LOSS

Analysis of Variance

                Sum of      Mean
Source    DF    Squares    Square    F Value    Prob>F

Model      1   26.00417   26.00417    40.223    0.0001
Error     10    6.46500    0.64650
C Total   11   32.46917


    Root MSE    0.80405    R-square   0.8009
    Dep Mean    5.50833    Adj R-sq   0.7810
    C.V.       14.59701

Parameter Estimates

                Parameter    Standard    T for H0:
Variable   DF    Estimate      Error    Parameter=0    Prob > |T|

INTERCEP    1   -1.733333   1.16518239    -1.488        0.1677
X           1    1.316667   0.20760539     6.342        0.0001
```

X	Y	Predict Value	Std Err Predict	Lower95% Mean	Upper95% Mean	Lower95% Predict	Upper95% Predict	Residual
4	4.3	3.5333	0.388	2.6679	4.3987	1.5437	5.5229	0.7667
5	5.5	4.8500	0.254	4.2835	5.4165	2.9710	6.7290	0.6500
6	6.8	6.1667	0.254	5.6001	6.7332	4.2877	8.0456	0.6333
7	8.0	7.4833	0.388	6.6179	8.3487	5.4937	9.4729	0.5167
4	4.0	3.5333	0.388	2.6679	4.3987	1.5437	5.5229	0.4667
5	5.2	4.8500	0.254	4.2835	5.4165	2.9710	6.7290	0.3500
6	6.6	6.1667	0.254	5.6001	6.7332	4.2877	8.0456	0.4333
7	7.5	7.4833	0.388	6.6179	8.3487	5.4937	9.4729	0.0167
4	2.0	3.5333	0.388	2.6679	4.3987	1.5437	5.5229	-1.5333
5	4.0	4.8500	0.254	4.2835	5.4165	2.9710	6.7290	-0.8500
6	5.7	6.1667	0.254	5.6001	6.7332	4.2877	8.0456	-0.4667
7	6.5	7.4833	0.388	6.6179	8.3487	5.4937	9.4729	-0.9833

```
Sum of Residuals                    0
Sum of Squared Residuals       6.4650
Predicted Resid SS (Press)    10.0309
```

11.28 Another part of the output for Exercise 11.21 is shown here.

Table of Predicted Values

Row	Runsize	Predicted TotalCost	95.00% Prediction Limits Lower	Upper	95.00% Confidence Limits Lower	Upper
1	2	203.613	178.169	229.057	198.902	208.323

a. Predict the mean total direct cost for all bumper sticker orders with a print run of 2,000 stickers (that is, with Runsize = 2.0).

b. Locate a 95% confidence interval for this mean.

11.29 Does the prediction in Exercise 11.28 represent a major extrapolation?

11.30 Refer to Exercise 11.28.

a. Predict the total direct cost for a particular bumper sticker order with a print run of 2,000 stickers. Obtain a 95% prediction interval.

b. Would an actual total direct cost of $250 be surprising for this order?

11.5 Examining Lack of Fit in Linear Regression

In our study of linear regression, we have been concerned with how well a linear regression model $y = \beta_0 + \beta_1 x + \varepsilon$ fits, but only from an intuitive standpoint. We have examined a scatterplot of the data to see whether it looks linear and we have tested whether the slope differed from 0; however, we had no way of testing to see whether a higher-order model would be a more appropriate model for the relationship between y and x. This section will outline situations in which we can test for the validity of a linear regression model.

Pictures (or graphs) are always a good starting point for examining lack of fit. First, use a scatterplot of y versus x. Second, a plot of residuals $y_i - \hat{y}_i$ versus predicted values $\hat{y}_i$ may give an indication of the following problems:

1. Outliers or erroneous observations. In examining the residual plot, your eye will naturally be drawn to data points with unusually high (in absolute value) residuals.

2. Violation of the assumptions. For the model $y = \beta_0 + \beta_1 x + \varepsilon$, we have assumed a linear relation between y and the dependent variable x, and independent, normally distributed errors with a constant variance.

The residual plot for a model and data set that has none of these apparent problems looks much like the plot in Figure 11.14. Note from this plot that there are no extremely large residuals (and hence no apparent outliers) and there is no trend in the residuals to indicate that the linear model is inappropriate. When a higher-order model is more appropriate, a residual plot more like that shown in Figure 11.15 is observed.

A check of the constant variance assumption can be addressed in the y versus x scatterplot or with a plot of the residuals $(y_i - \hat{y}_i)$ versus x_i. For example, a pattern of residuals as shown in Figure 11.16 indicates homogeneous error variances across values of x; Figure 11.17 indicates that the error variances increase with increasing values of x.

FIGURE 11.14

Residual plot with no apparent pattern

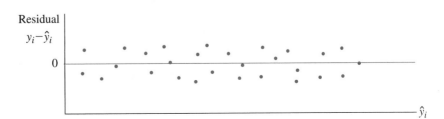

FIGURE 11.15

Residual plot showing the need for a higher-order model

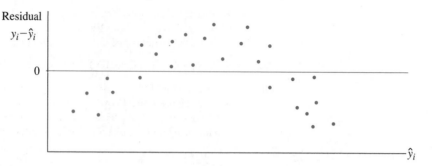

FIGURE 11.16

Residual plot showing homogeneous error variances

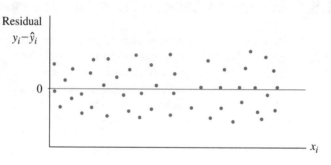

FIGURE 11.17

Residual plot showing error variances increasing with x

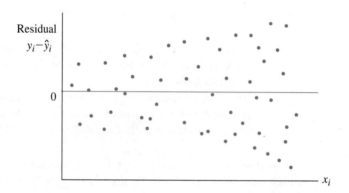

We illustrate some of the points we have learned so far about residuals by way of an example.

EXAMPLE 11.11

The manufacturer of a new brand of thermal panes examined the amount of heat loss by random assignment of three different panes to each of the three outdoor temperature settings being considered. For each trial, the window temperature was controlled at 68°F and 50% relative humidity.

Outdoor Temperature (°F)	Heat Loss
20	86, 80, 77
40	78, 84, 75
60	33, 38, 43

a. Plot the data.
b. Fit the linear regression model $y = \beta_0 + \beta_1 x + \varepsilon$ and test $H_0: \beta_1 = 0$ (give the p-value for your test).
c. Compute $\hat{y}_i$ and $y_i - \hat{y}_i$ for the nine observations. Plot $y_i - \hat{y}_i$ versus $\hat{y}_i$.
d. Does the constant variance assumption seem reasonable?

Solution The computer output shown here can be used to address the four parts of this example.

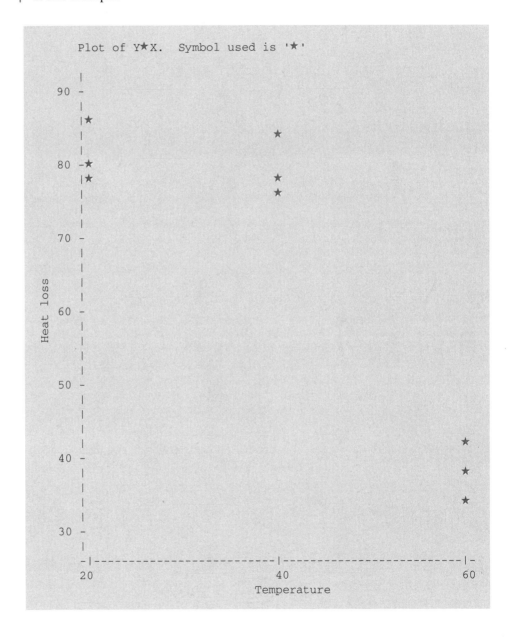

```
Dependent Variable:Y         HEAT LOSS

Analysis of Variance

                      Sum of         Mean
Source        DF      Squares       Square      F Value    Prob>F

Model          1    2773.50000   2773.50000     21.704     0.0023
Error          7     894.50000    127.78571
C Total        8    3668.00000

     Root SE      11.30423     R-square      0.7561
     Dep Mean     66.00000     Adj R-sq      0.7213
     C.V.         17.12763

Parameter Estimates

                 Parameter     Standard     T for H0:
Variable   DF     Estimate        Error    Parameter=0   Prob > |T|

INTERCEP    1   109.000000   9.96939762       10.933       0.0001
X           1    -1.075000   0.23074672       -4.659       0.0023



OBS     X     Y     PRED     RESID

 1     20    86    87.5      -1.5
 2     20    80    87.5      -7.5
 3     20    77    87.5     -10.5
 4     40    78    66.0      12.0
 5     40    84    66.0      18.0
 6     40    75    66.0       9.0
 7     60    33    44.5     -11.5
 8     60    38    44.5      -6.5
 9     60    43    44.5      -1.5
```

```
        Plot of RESID*PRED.   Symbol used is '*'

         |
         |
  18.5 +                          *
  17.5 +
  16.5 +
  15.5 +
  14.5 +
  13.5 +
  12.5 +                          *
  11.5 +
  10.5 +
   9.5 +                          *
   8.5 +
   7.5 +
   6.5 +
   5.5 +
   4.5 +
   3.5 +
   2.5 +
   1.5 +
   0.5 +
   0.0 +-------------------------------------------------------
  -0.5 +
  -1.5 +*                                                  *
  -2.5 +
  -3.5 +
  -4.5 +
  -5.5 +
  -6.5 +*                                                  *
  -7.5 +
  -8.5 +
  -9.5 +
 -10.5 +                                                   *
 -11.5 +*
       --|-----------------------------|-----------------------|
         44.5                         66.0                   87.5
                            Predicted value
```
(Residuals — vertical axis label)

a. The scatterplot of y versus x certainly shows a downward linear trend, and there may be evidence of curvature as well.

b. The linear regression model seems to fit the data well, and the test of H_0: $\beta_1 = 0$ is significant at the $p = .0023$ level. However, is this the best model for the data?

c. The plot of residuals $(y_i - \hat{y}_i)$ against the predicted values $\hat{y}_i$ is similar to Figure 11.15, suggesting that we may need additional terms in our model.

d. Because residuals associated with $x = 20$ (the first three), $x = 40$ (the second three), and $x = 60$ (the third three) are easily located, we really do not need a separate plot of residuals versus x to examine the constant variance assumption. It is clear from the original scatterplot and the residual plot shown that we do not have a problem.

How can we test for the apparent lack of fit of the linear regression model in Example 11.11? When there is more than one observation per level of the independent variable, we can conduct a test for lack of fit of the fitted model by partitioning SS(Residuals) into two parts, one **pure experimental error** and the other **lack of fit.** Let y_{ij} denote the response for the jth observation at the ith level of the independent variable. Then, if there are n_i observations at the ith level of the independent variable, the quantity

$$\sum_j (y_{ij} - \bar{y}_i)^2$$

provides a measure of what we will call pure experimental error. This sum of squares has $n_i - 1$ degrees of freedom.

Similarly, for each of the other levels of x, we can compute a sum of squares due to pure experimental error. The pooled sum of squares

$$SSP_{exp} = \sum_{ij} (y_{ij} - \bar{y}_i)^2$$

called the sum of squares for pure experimental error, has $\sum_i (n_i - 1)$ degrees of freedom. With SS_{Lack} representing the remaining portion of SSE, we have

$$SS(\text{Residuals}) = \underset{\substack{\text{due to pure}\\\text{experimental}\\\text{error}}}{SSP_{exp}} + \underset{\substack{\text{due to lack}\\\text{of fit}}}{SS_{Lack}}$$

If SS(Residuals) is based on $n - 2$ degrees of freedom in the linear regression model, then SS_{Lack} will have df $= n - 2 - \sum_i (n_i - 1)$.

Under the null hypothesis that our model is correct, we can form independent estimates of σ_{ε}^2, the model error variance, by dividing SSP_{exp} and SS_{Lack} by their respective degrees of freedom; these estimates are called **mean squares** and are denoted by MSP_{exp} and MS_{Lack}, respectively.

The test for lack of fit is summarized here.

A Test for Lack of Fit in Linear Regression

H_0: A linear regression model is appropriate.

H_a: A linear regression model is not appropriate.

T.S.: $F = \dfrac{MS_{Lack}}{MSP_{exp}}$,

where

$$MSP_{exp} = \frac{SSP_{exp}}{\sum (n_i - 1)} = \frac{\sum_{ij} (y_{ij} - \bar{y}_i)^2}{\sum_i (n_i - 1)}$$

and

$$MS_{Lack} = \frac{SS(Residuals) - SSP_{exp}}{n - 2 - \Sigma(n_i - 1)}$$

R.R.: For specified value of α, reject H_0 (the adequacy of the model) if the computed value of F exceeds the table value for $df_1 = n - 2 - \Sigma_i(n_i - 1)$ and $df_2 = \Sigma_i(n_i - 1)$.

Checking assumptions and drawing conclusions: If the F test is significant, this indicates that the linear regression model is inadequate. A nonsignificant result indicates that there is insufficient evidence to suggest that the linear regression model is inappropriate.

EXAMPLE 11.12

Refer to the data in Example 11.11. Conduct a test for lack of fit of the linear regression model.

Solution It is easy to show that the contributions to experimental error for the differential levels of x are as shown here.

Level of x	$\bar{y}_i$	Contribution to Pure Experimental Error $\Sigma_i(y_{ij} - \bar{y}_i)^2$	$n_i - 1$
20	81	42	2
40	79	42	2
60	38	50	2
Total		134	6

Summarizing these results, we have

$$SSP_{exp} = \sum_{ij}(y_{ij} - \bar{y}_i)^2 = 134$$

The output shown for Example 11.11 gives SS(Residual) = 894.5; hence, by subtraction,

$$SS_{Lack} = SS(Residual) - SSP_{exp} = 894.5 - 134 = 760.5$$

The sum of squares due to pure experimental error has $\Sigma_i(n_i - 1) = 6$ degrees of freedom; it therefore follows that with $n = 9$, SS_{Lack} has $n - 2 - \Sigma_i(n_i - 1) = 1$ degree of freedom. We find that

$$MSP_{exp} = \frac{SSP_{exp}}{6} = \frac{134}{6} = 22.33$$

and

$$MS_{Lack} = \frac{SS_{Lack}}{1} = 760.5$$

The F statistic for the test of lack of fit is

$$F = \frac{MS_{Lack}}{MSP_{exp}} = \frac{760.5}{22.33} = 34.06$$

Using $df_1 = 1$, $df_2 = 6$, and $\alpha = .05$, we will reject H_0 if $F \geq 5.99$.

Because the computed value of F exceeds 5.99, we reject H_0 and conclude that there is significant lack of fit for a linear regression model. The scatterplot shown in Example 11.11 confirms this nonlinearity.

To summarize: In situations for which there is more than one y-value at one or more levels of x, it is possible to conduct a formal test for lack of fit of the linear regression model. This test should precede any inferences made using the fitted linear regression line. If the test for lack of fit is significant, some higher-order polynomial in x may be more appropriate. A scatterplot of the data and a residual plot from the linear regression line should help in selecting the appropriate model. The selection of an appropriate model will be discussed in more detail along with multiple regression in Chapter 12.

If the F test for lack of fit is not significant, proceed with inferences based on the fitted linear regression line.

EXERCISES

Applications

Engin.

11.31 A manufacturer of laundry detergent was interested in testing a new product prior to market release. One area of concern was the relationship between the height of the detergent suds in a washing machine as a function of the amount of detergent added in the wash cycle. For a standard size washing machine tub filled to the full level, the manufacturer made random assignments of amounts of detergent and tested them on the washing machine. The data appear next.

Height, y	Amount, x
28.1, 27.6	6
32.3, 33.2	7
34.8, 35.0	8
38.2, 39.4	9
43.5, 46.8	10

a. Plot the data.
b. Fit a linear regression model.
c. Use a residual plot to investigate possible lack of fit.

11.32 Refer to Exercise 11.31.
a. Conduct a test for lack of fit of the linear regression model.
b. If the model is appropriate, give a 95% prediction band for y.

11.6 The Inverse Regression Problem (Calibration)

In experimental situations, we are often interested in estimating the value of the independent variable corresponding to a measured value of the dependent variable. This problem will be illustrated for the case in which the dependent variable y is linearly related to an independent variable x.

Consider the calibration of an instrument that measures the flow rate of a chemical process. Let x denote the actual flow rate and y denote a reading on the calibrating instrument. In the calibration experiment, the flow rate is controlled at n levels x_i, and the corresponding instrument readings y_i are observed. Suppose we assume a model of the form

$$y_i = \beta_0 + \beta_1 x_i + \varepsilon_i$$

where the ε_is are independent, identically distributed normal random variables with mean zero and variance σ_ε^2. Then, using the n data points (x_i, y_i), we can obtain the least-squares estimates $\hat{\beta}_0$ and $\hat{\beta}_1$. Sometime in the future the experimenter will be interested in estimating the flow rate x from a particular instrument reading y.

The most commonly used estimate is found by replacing $\hat{y}$ by y and solving the least-squares equation $\hat{y} = \hat{\beta}_0 + \hat{\beta}_1 x$ for x:

$$\hat{x} = \frac{y - \hat{\beta}_0}{\hat{\beta}_1}$$

Two different inverse prediction problems will be discussed here. The first is for predicting x corresponding to an *observed* value of y; the second is for predicting x corresponding to the mean of $m > 1$ values of y that were obtained independent of the regression data. The solution to the first inverse problem is shown here.

Case 1: Predicting x Based on an Observed y-Value

Predictor of x: $\quad \hat{x} = \dfrac{y - \hat{\beta}_0}{\hat{\beta}_1}$

$100(1 - \alpha)\%$ prediction limits for x:

$$\hat{x}_U = \bar{x} + \frac{1}{1 - c^2}[(\hat{x} - \bar{x}) + d]$$

$$\hat{x}_L = \bar{x} + \frac{1}{1 - c^2}[(\hat{x} - \bar{x}) - d]$$

where

$$d = \frac{t_{\alpha/2} s_\varepsilon}{\hat{\beta}_1}\sqrt{\frac{n+1}{n}(1 - c^2) + \frac{(\hat{x} - \bar{x})^2}{S_{xx}}} \qquad s_\varepsilon^2 = \frac{SSE}{n-2} \qquad c^2 = \frac{t_{\alpha/2}^2 s_\varepsilon^2}{\hat{\beta}_1^2 S_{xx}}$$

and $t_{\alpha/2}$ is based on df $= n - 2$.

Note that because

$$t = \frac{\hat{\beta}_1}{s_\varepsilon/\sqrt{S_{xx}}}$$

is the test statistic for $H_0: \beta_1 = 0$, $c = t_{\alpha/2}/t$. We will require that $|t| > t_{\alpha/2}$; that is, β_1 must be significantly different from zero. Then $c^2 < 1$ and $0 < (1 - c^2) < 1$. The greater the strength of the linear relationship between x and y, the larger the quantity $(1 - c^2)$, making the width of the prediction interval narrower. Note also that we will get a better prediction of x when $\hat{x}$ is closer to the center of the experimental region, as measured by $\bar{x}$. Combining a prediction at an end point of the experimental region with a weak linear relationship between x and y ($t \approx t_{\alpha/2}$ and $c^2 < 1$) can create extremely wide limits for the prediction of x.

EXAMPLE 11.13

An engineer is interested in calibrating a flow meter to be used on a liquid-soap production line. For the test, 10 different flow rates are fixed and the corresponding meter readings observed. The data are shown here. Use these data to place a 95% prediction interval on x, the actual flow rate corresponding to an instrument reading of 4.0.

Solution For these data, we find that $S_{xy} = 74.35$, $S_{xx} = 82.5$, and $S_{yy} = 67.065$. It follows that $\hat{\beta}_1 = 74.35/82.5 = .9012$, $\hat{\beta}_0 = \bar{y} - \hat{\beta}_1\bar{x} = 5.45 - (.9012)(5.5) = .4934$, and SS(Residual) $= S_{yy} - \hat{\beta}_1 S_{xy} = 67.065 - (.9012)(74.35) = .0608$. The estimate of σ_ε^2 is based on $n - 2 = 8$ degrees of freedom.

Data for the Calibration Problem

Flow Rate, x	Instrument Reading, y
1	1.4
2	2.3
3	3.1
4	4.2
5	5.1
6	5.8
7	6.8
8	7.6
9	8.7
10	9.5

$$s_\varepsilon^2 = \frac{\text{SS(Residual)}}{n - 2} = \frac{.0608}{8} = .0076$$

$$s_\varepsilon = .0872$$

For $\alpha = .05$, the t-value for df $= 8$ and $a = .025$ is 2.306.

$$c^2 = \frac{t_{\alpha/2}^2 s_\varepsilon^2}{\hat{\beta}_1^2 S_{xx}} = \frac{(2.306)^2(.0076)}{(.9012)^2(82.5)} = .0006$$

and $1 - c^2 = .9994$. Using $\hat{x} = (4.0 - .4934)/.9012 = 3.8910$, the upper and lower prediction limits for x when $y = 4.0$ are as follows:

$$\hat{x}_U = 5.5 + \frac{1}{.9994}\left[-1.6090 + \frac{2.306(.0872)}{.9012}\sqrt{\frac{11}{10}(.9994) + \frac{(-1.6090)^2}{82.5}}\right]$$

$$= 5.5 + \frac{1}{.9994}(-1.6090 + .2373) = 4.1274$$

$$\hat{x}_L = 5.5 + \frac{1}{.9994}(-1.6090 - .2373) = 3.6526$$

FIGURE 11.18
95% prediction interval for x when y = 4.0

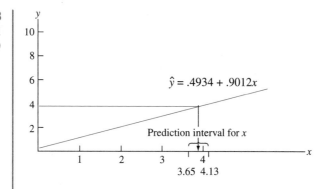

Thus, the 95% prediction limits for x are 3.65 to 4.13. These limits are shown in Figure 11.18.

The solution to the second inverse prediction problem is summarized next.

Case 2: Predicting x Based on m y-Values

Predicting the value of x corresponding to 100P% of the mean of m independent y values. For $0 \le P \le 1$,

$$\text{Predictor of } x : \hat{x} = \frac{P\bar{y}_m - \hat{\beta}_0}{\hat{\beta}_1}$$

$$\hat{x}_U = \bar{x} + \frac{1}{1 - c^2}[(\hat{x} - \bar{x}) + g]$$

$$\hat{x}_L = \bar{x} + \frac{1}{1 - c^2}[(\hat{x} - \bar{x}) - g]$$

where

$$g = \frac{t_{\alpha/2}}{\hat{\beta}_1}\sqrt{\left(s_{\bar{y}}^2 P^2 + \frac{s_\varepsilon^2}{n}\right)(1 - c^2) + \frac{(\hat{x} - \bar{x})^2 s_\varepsilon^2}{S_{xx}}}$$

and $\bar{y}_m$ and $s_{\bar{y}}$ are the mean and standard error, respectively, of m independent y-values.

EXERCISES

Applications

Ag.

11.33   A forester has become adept at estimating the volume (in cubic feet) of trees on a particular site prior to a timber sale. Because his operation has now expanded, he would like to train another person to assist in estimating the cubic-foot volume of trees. He decides to calibrate his assistant's estimations of actual tree volume. The forester selects a random sample of trees soon to be felled. For each tree, the assistant is to guess the cubic-foot volume y. The forester also obtains the actual cubic-foot volume x after the tree has been chopped down. From these data, the forester obtains the calibration curve for the model

$$y = \beta_0 + \beta_1 x + \varepsilon$$

In the near future he can then use the calibration curve to correct the assistant's estimates of tree volumes. The sample data are summarized here.

Tree	1	2	3	4	5	6	7	8	9	10
Estimated volume, y	12	14	8	12	17	16	14	14	15	17
Actual volume, x	13	14	9	15	19	20	16	15	17	18

Fit the calibration curve using the method of least squares. Do the data indicate that the slope is significantly greater than 0? Use $\alpha = .05$.

11.34 Refer to Exercise 11.33.
   a. Predict the actual tree volume for a tree the assistant estimates to have a cubic-foot volume of 13.
   b. Place a 95% prediction interval on x, the actual tree volume in part (a).

Med.   **11.35** A researcher obtains data from 24 patients to examine the relationship between dose (amount of drug) and cumulative urine volume (CUMVOL) for a drug product being studied as a diuretic. The data are shown here in the computer output. The initial fit of the data yielded a non-linear relationship between dose and CUMVOL. The researcher decided on the transformations natural logarithm of dose and arcsine of the square root of CUMVOL/100, labeled LOG (DOSE) and TRANS. CUMVOL on the output.
   a. Locate the linear regression equation. Identify the independent and dependent variables.
   b. Use the output to predict dose based on individual y values of 10, 14, and 19 cm^3. What are the corresponding 95% prediction limits for each of those cases?

```
OUTPUT FOR EXERCISE 11.35

OBS   DOSE   LOG (DOSE)   CUMVOL   TRANS. CUMVOL
  1    6.00    1.79176       7.1      0.26972
  2    6.00    1.79176      11.5      0.34598
  3    6.00    1.79176       8.4      0.29405
  4    6.00    1.79176       8.0      0.28676
  5    6.00    1.79176       9.4      0.31161
  6    6.00    1.79176      12.0      0.35374
  7    9.00    2.19722      13.2      0.37183
  8    9.00    2.19722      14.7      0.39348
  9    9.00    2.19722      12.7      0.36438
 10    9.00    2.19722      15.5      0.40465
 11    9.00    2.19722      18.4      0.44333
 12    9.00    2.19722      14.4      0.38923
 13   13.50    2.60269      12.1      0.35528
 14   13.50    2.60269      15.8      0.40878
 15   13.50    2.60269      13.8      0.38061
 16   13.50    2.60269      20.4      0.46863
 17   13.50    2.60269      22.7      0.49661
 18   13.50    2.60269      17.0      0.42499
 19   20.25    3.00815      19.8      0.46114
 20   20.25    3.00815      15.6      0.40603
 21   20.25    3.00815      25.3      0.52706
 22   20.25    3.00815      13.5      0.37624
 23   20.25    3.00815      24.8      0.52129
 24   20.25    3.00815      20.9      0.47481
 25   10.00    2.30259       .          .
 26   14.00    2.63906       .          .
 27   19.00    2.94444       .          .
```

OUTPUT FOR EXERCISE 11.35

Dependent Variable: Y   TRANSFORMED CUMVOL

Analysis of Variance

Source	DF	Sum of Squares	Mean Square	F Value	Prob>F
Model	1	0.06922	0.06922	32.750	0.0001
Error	22	0.04650	0.00211		
C Total	23	0.11572			

Root MSE	0.04597	R-square	0.5982	
Dep Mean	0.39709	Adj R-sq	0.5799	
C.V.	11.57773			

Parameter Estimates

Variable	DF	Parameter Estimate	Standard Error	T for H0: Parameter=0	Prob > |T|
INTERCEP	1	0.112770	0.05056109	2.230	0.0362
X	1	0.118470	0.02070143	5.723	0.0001

OBS	X	Y	PRED	L95PRED	U95PRED	L95MEAN	U95MEAN
1	1.79176	0.26972	0.32504	0.22429	0.42579	0.29247	0.35761
2	1.79176	0.34598	0.32504	0.22429	0.42579	0.29247	0.35761
3	1.79176	0.29405	0.32504	0.22429	0.42579	0.29247	0.35761
4	1.79176	0.28676	0.32504	0.22429	0.42579	0.29247	0.35761
5	1.79176	0.31161	0.32504	0.22429	0.42579	0.29247	0.35761
6	1.79176	0.35374	0.32504	0.22429	0.42579	0.29247	0.35761
7	2.19722	0.37183	0.37307	0.27537	0.47077	0.35175	0.39439
8	2.19722	0.39348	0.37307	0.27537	0.47077	0.35175	0.39439
9	2.19722	0.36438	0.37307	0.27537	0.47077	0.35175	0.39439
10	2.19722	0.40465	0.37307	0.27537	0.47077	0.35175	0.39439
11	2.19722	0.44333	0.37307	0.27537	0.47077	0.35175	0.39439
12	2.19722	0.38923	0.37307	0.27537	0.47077	0.35175	0.39439
13	2.60269	0.35528	0.42111	0.32341	0.51881	0.39979	0.44243
14	2.60269	0.40878	0.42111	0.32341	0.51881	0.39979	0.44243
15	2.60269	0.38061	0.42111	0.32341	0.51881	0.39979	0.44243
16	2.60269	0.46863	0.42111	0.32341	0.51881	0.39979	0.44243
17	2.60269	0.49661	0.42111	0.32341	0.51881	0.39979	0.44243
18	2.60269	0.42499	0.42111	0.32341	0.51881	0.39979	0.44243
19	3.00815	0.46114	0.46914	0.36839	0.56990	0.43658	0.50171
20	3.00815	0.40603	0.46914	0.36839	0.56990	0.43658	0.50171
21	3.00815	0.52706	0.46914	0.36839	0.56990	0.43658	0.50171
22	3.00815	0.37624	0.46914	0.36839	0.56990	0.43658	0.50171
23	3.00815	0.52129	0.46914	0.36839	0.56990	0.43658	0.50171
24	3.00815	0.47481	0.46914	0.36839	0.56990	0.43658	0.50171
25	2.30259	.	0.38556	0.28816	0.48296	0.36565	0.40546
26	2.63906	.	0.42542	0.32757	0.52327	0.40341	0.44742
27	2.94444	.	0.46160	0.36152	0.56168	0.43118	0.49201

Sum of Residuals	0
Sum of Squared Residuals	0.0465
Predicted Resid SS (Press)	0.0560

11.36   Refer to the output of Exercise 11.35. Suppose the investigator wanted to predict the dose of the diuretic that would produce a response equivalent to 50% (and 75%) of the response obtained from four patients treated with a known diuretic. The four values of CUMVOL are 10, 20, 30, and 12. Predict x and give appropriate limits for each of these situations.

11.7   Correlation

Once we have found the prediction line $\hat{y} = \hat{\beta}_0 + \hat{\beta}_1 x$, we need to measure how well it predicts actual values. One way to do so is to look at the size of the residual standard deviation in the context of the problem. About 95% of the prediction errors will be within $\pm 2 s_\varepsilon$. For example, suppose we are trying to predict the yield of a chemical process in which yields range from 0.50 to 0.94. If a regression model had a residual standard deviation of 0.01, we could predict most yields within ± 0.02—fairly accurate in context. However, if the residual standard deviation were 0.08, we could predict most yields within ± 0.16, which is not very impressive given that the yield range is only $0.94 - 0.50 = 0.44$. This approach, however requires that we know the context of the study well; an alternative, more general approach is based on the idea of correlation.

Suppose that we compare the squared prediction error for two prediction methods: one using the regression model, the other ignoring the model and always predicting the mean y value. In the road resurfacing example of previous sections, if we are given the mileage values x_i, we can use the prediction equation $\hat{y}_i = 2.0 + 3.0 x_i$ to predict costs. The deviations of actual values from predicted values, the residuals, measure prediction errors. These errors are summarized by the sum of squared residuals, $\text{SS(Residual)} = \sum (y_i - \hat{y}_i)^2$, which is 44 for these data. For comparison, if we were not given the x_i values, the best squared error predictor of y would be the mean value $\bar{y} = 14$, and the sum of squared prediction errors would, in this case, be $\sum_i (y_i - \bar{y}_i)^2 = \text{SS(Total)} = 224$. The proportionate reduction in error would be

$$\frac{\text{SS(Total)} - \text{SS(Residual)}}{\text{SS(Total)}} = \frac{224 - 44}{224} = .804$$

In words, use of the regression model reduces squared prediction error by 80.4%, which indicates a fairly strong relation between the mileage to be resurfaced and the cost of resurfacing.

correlation coefficient

This proportionate reduction in error is closely related to the **correlation coefficient** of x and y. A *correlation measures the strength of the linear relation between x and y*. The stronger the correlation, the better x predicts y.

Given n pairs of observations (x_i, y_i), we compute the sample correlation r as

$$r_{yx} = \frac{\sum (x_i - \bar{x})(y_i - \bar{y})}{\sqrt{S_{xx} S_{yy}}} = \frac{S_{xy}}{\sqrt{S_{xx} S_{yy}}}$$

where S_{xy} and S_{xx} are defined as before and

$$S_{yy} = \sum_i (y_i - \bar{y})^2 = \text{SS(Total)}$$

In the example,

$$r_{yx} = \frac{60}{\sqrt{(20)(224)}} = .896$$

FIGURE 11.19
Interpretation of r

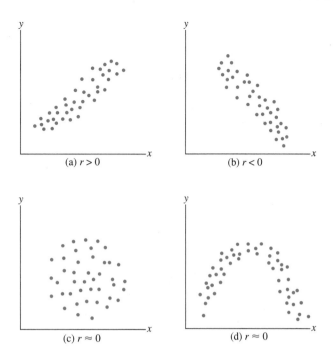

(a) $r > 0$ (b) $r < 0$

(c) $r \approx 0$ (d) $r \approx 0$

Generally, the correlation r_{yx} is a positive number if y tends to increase as x increases; r_{yx} is negative if y tends to decrease as x increases; and r_{yx} is zero if there is either no relation between changes in x and changes in y or there is a nonlinear relation such that patterns of increase and decrease in y (as x increases) cancel each other.

Figure 11.19 illustrates four possible situations for the values of r. In Figure 11.19(d), there is a strong relationship between y and x but $r = 0$. This is a result of symmetric positive and negative nearly linear relationships canceling one another. When $r = 0$, there is not a linear relationship between y and x. However, higher-order (nonlinear) relationships may exist. This situation illustrates the importance of plotting the data in a scatterplot. In Chapter 12, we will develop techniques for modeling nonlinear relationships between y and x.

EXAMPLE 11.14

Consider the following data:

y:	25	41	47	59	54	56	49	43	30
x:	10	20	20	30	30	30	40	40	50

a. Should the correlation be positive or negative?
b. Calculate the correlation.

Solution

a. Note that as x increases from 10 to 50, y first increases and then decreases. Therefore, the correlation should be small. The y values do not decrease quite back to where they started, so the correlation should be positive.
b. By hand calculation, the sample means are $\bar{x} = 30.0000$ and $\bar{y} = 44.8889$.

$$S_{xx} = (10 - 30.0000)^2 + \cdots + (50 - 30.0000)^2 = 1{,}200$$

$$S_{yy} = (25 - 44.8889)^2 + \cdots + (30 - 44.8889)^2 = 1{,}062.8889$$

$$S_{xy} = (10 - 30.0000)(25 - 44.8889) + \cdots + (50 - 30.0000)(30 - 44.8889)$$

$$= 140$$

$$r_{yx} = \frac{140}{\sqrt{(1{,}200)(1{,}062.8889)}} = .1240$$

The correlation is indeed a small positive number.

coefficient of determination

Correlation and regression predictability are closely related. The proportionate reduction in error for regression we defined earlier is called the **coefficient of determination.** The coefficient of determination is simply the square of the correlation coefficient,

$$r_{yx}^2 = \frac{SS(\text{Total}) - SS(\text{Residual})}{SS(\text{Total})}$$

which is the proportionate reduction in error. In the resurfacing example, $r_{yx} = .896$ and $r_{yx}^2 = .804$.

A correlation of zero indicates no predictive value in using the equation $\hat{y} = \hat{\beta}_0 + \hat{\beta}_1 x$; that is, we can predict y as well without knowing x as we can knowing x. A correlation of 1 or -1 indicates perfect predictability—a 100% reduction in error attributable to knowledge of x. A correlation coefficient should routinely be interpreted in terms of its squared value, the coefficient of determination. Thus, a correlation of $-.3$, say, indicates only a 9% reduction in squared prediction error. Many books and most computer programs use the equation

$$SS(\text{Total}) = SS(\text{Residual}) + SS(\text{Regression})$$

where

$$SS(\text{Regression}) = \sum_i (\hat{y}_i - \bar{y})^2$$

Because the equation can be expressed as $SS(\text{Residual}) = (1 - r_{yx}^2)SS(\text{Total})$, it follows that $SS(\text{Regression}) = r_{yx}^2 SS(\text{Total})$, which again says that regression on x explains a proportion r_{yx}^2 of the total squared error of y.

EXAMPLE 11.15

Find SS(Total), SS(Regression), and SS(Residual) for the data in Example 11.14.

Solution   SS(Total) = S_{yy}, which we computed to be 1,062.8889 in Example 11.14. We also found that $r_{yx} = .1240$, so $r_{yx}^2 = (.1240)^2 = .0154$. Using the fact that $SS(\text{Regression}) = r_{yx}^2 SS(\text{Total})$, we have SS(Regression) = (.0154)(1,062.8889) = 16.3685. Because SS(Residual) = SS(Total) − SS(Regression), SS(Residual) = 1,062.8889 − 16.3685 = 1,046.5204.

Note that SS(Regression) and r_{yx}^2 are very small. This suggests that x is not a good predictor of y. The reality, however, is that the relation between x and y is extremely nonlinear. A *linear* equation in x does not predict y very well, but a nonlinear equation would do far better.

FIGURE 11.20

Samples of size 1,000 from the
bivariate normal distribution

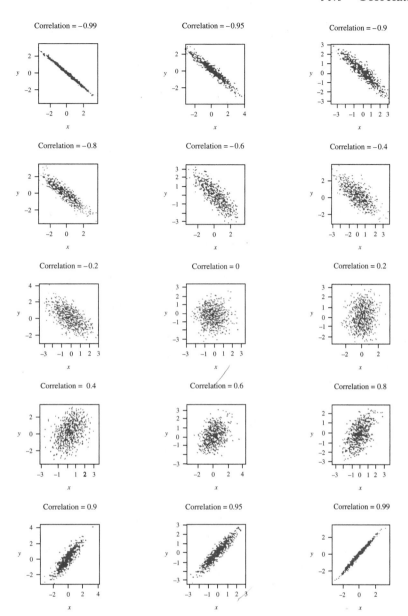

What values of r_{yx} indicate a strong relationship between y and x? Figure 11.20 displays 15 scatterplots obtained by randomly selecting 1,000 pairs (x_i, y_i) from 15 populations having bivariate normal distributions with correlations ranging from -0.99 to 0.99. We can observe that unless $|r_{yx}|$ is greater than 0.6 there is very little trend in the plot.

The sample correlation r_{yx} is the basis for estimation and significance testing of the population correlation ρ_{yx}. Statistical inferences are always based on assumptions. The assumptions of regression analysis—linear relation between x and y and constant variance around the regression line, in particular—are also **assumed** in **correlation inference.** In regression analysis, we regard the x values as predetermined constants. In correlation analysis, we regard the x values as randomly

**assumptions for
correlation inference**

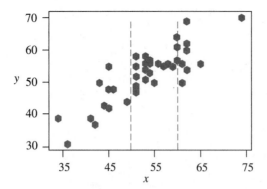

selected (and the regression inferences are conditional on the sampled *x* values). If the *x*s are not drawn randomly, it is possible that the correlation estimates are biased. In some texts, the additional assumption is made that the *x* values are drawn from a normal population. The inferences we make do not depend crucially on this normality assumption.

The most basic inference problem is potential bias in estimation of ρ_{yx}. A problem arises when the *x* values are predetermined, as often happens in regression analysis. The choice of *x* values can systematically increase or decrease the sample correlation. In general, a wide range of *x* values tends to increase the magnitude of the correlation coefficient and a small range to decrease it. This effect is shown in Figure 11.21. When all the points in this scatterplot are included, there is an obvious, strong correlation between *x* and *y*. Suppose, however, we consider only *x* values in the range between the dashed vertical lines. By eliminating the outside parts of the scatterplot, the sample correlation coefficient (and the coefficient of determination) are much smaller. Correlation coefficients can be affected by systematic choices of *x* values; the residual standard deviation is *not* affected systematically, although it may change randomly if part of the *x* range changes. Thus, it is a good idea to consider the residual standard deviation s_ε and the magnitude of the slope when you decide how well a linear regression line predicts *y*.

EXAMPLE 11.16

Suppose that a company has the following data on productivity *y* and aptitude test score *x* for 12 data-entry operators:

y:	41	39	47	51	43	40	57	46	50	59	61	52
x:	24	30	33	35	36	36	37	37	38	40	43	49

Is the correlation larger or smaller if we consider only the last six values?

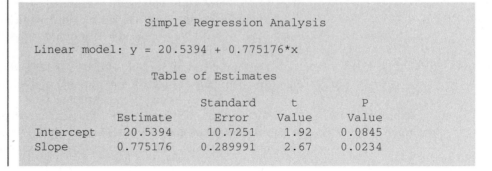

```
                    Simple Regression Analysis

    Linear model: y = 20.5394 + 0.775176*x

                     Table of Estimates

                             Standard       t          P
                  Estimate     Error     Value      Value
    Intercept     20.5394    10.7251      1.92     0.0845
    Slope        0.775176   0.289991      2.67     0.0234
```

```
R-squared = 41.68%
Correlation coeff. = 0.646
Standard error of estimation = 5.99236

File subset has been turned on, based on x>=37.

                  Simple Regression Analysis

Linear model: y = 44.7439 + 0.231707*x

                    Table of Estimates

                            Standard     t          P
                Estimate    Error      Value      Value
    Intercept   44.7439     24.8071     1.80       0.1456
    Slope       0.231707    0.606677    0.38       0.7219

R-squared = 3.52%
Correlation coeff. = 0.188
Standard error of estimation = 6.34357
```

Solution For all 12 observations, the output shows a correlation coefficient of .646; the residual standard deviation, labeled Standard error of estimation, is 5.992. For the six highest x scores, shown as the subset having x greater than or equal to 37, the correlation is .188 and the residual standard deviation is 6.344. In going from all 12 observations to the six observations with the highest x values, the correlation has decreased drastically, but the residual standard deviation has hardly changed at all.

Just as we can run a statistical test for β_i, we can do one for ρ_{yx}.

Summary of a
Statistical Test for ρ_{yx}

Hypotheses:

    Case 1. H_0: $\rho_{yx} \leq 0$ vs. H_0: $\rho_{yx} > 0$
    Case 2. H_0: $\rho_{yx} \geq 0$ vs. H_a: $\rho_{yx} < 0$
    Case 3. H_0: $\rho_{yx} = 0$ vs. H_a: $\rho_{yx} \neq 0$

T.S.: $t = r_{yx}\dfrac{\sqrt{n-2}}{\sqrt{1 - r_{yx}^2}}$

R.R.:   With $n - 2$ df and Type I error probability α,
    1. $t > t_\alpha$
    2. $t < -t_\alpha$
    3. $|t| > t_{\alpha/2}$

Check assumptions and draw conclusions.

We tested the hypothesis that the true slope is zero (in predicting resurfacing cost from mileage) in Example 11.5; the resulting t statistic was 3.50. For those data, we can calculate r_{yx} as .896421 and r_{yx}^2 as .803571. Hence, the correlation t statistic is

$$\frac{.896\sqrt{3}}{\sqrt{1-.803571}} = 3.50$$

An examination of the formulas for r and the slope $\hat{\beta}_1$ of the least-squares equation

$$\hat{y} = \hat{\beta}_0 + \hat{\beta}_1 x$$

yields the following relationship:

$$\hat{\beta}_1 = \frac{S_{xy}}{S_{xx}} = \frac{S_{xy}}{\sqrt{S_{xx}S_{yy}}}\sqrt{\frac{S_{yy}}{S_{xx}}} = r_{yx}\sqrt{\frac{S_{yy}}{S_{xx}}}$$

Thus, the t tests for a slope and for a correlation give identical results; it does not matter which form is used. It follows that the t test is valid for any choice of x values. The bias we mentioned previously does not affect the sign of the correlation.

EXAMPLE 11.17

Perform t tests for the null hypothesis of zero correlation and zero slope for the data of Example 11.16 (all observations). Use an appropriate one-sided alternative.

Solution   First, the appropriate H_a ought to be $\rho_{yx} > 0$ (and therefore $\beta_1 > 0$). It would be nice if an aptitude test had a positive correlation with the productivity score it was predicting! In Example 11.16, $n = 12$, $r_{yx} = .646$, and

$$t = \frac{.646\sqrt{12-2}}{\sqrt{1-(.646)^2}} = 2.68$$

Because this value falls between the tabled t values for df = 10, $\alpha = .025(2.228)$ and for df = 10, $\alpha = .01(2.764)$, the p-value lies between .010 and .025. Hence, H_0 may be rejected.

The t statistic for testing the slope β_1 is shown in the output of Example 11.16 as 2.67, which equals (to within rounding error) the correlation t statistic, 2.68.

The test for a correlation provides a neat illustration of the difference between statistical significance and statistical importance. Suppose that a psychologist has devised a skills test for production-line workers and tests it on a huge sample of 40,000. If the sample correlation between test score and actual productivity is .02, then

$$t = \frac{.02\sqrt{39,998}}{\sqrt{1-(.02)^2}} = 4.0$$

We would reject the null hypothesis at any reasonable α level, so the correlation is *statistically significant*. However, the test accounts for only $(.02)^2 = .0004$ of the squared error in skill scores, so it is *almost* worthless as a predictor. Remember, the rejection of the null hypothesis in a statistical test is the conclusion that the sample results cannot plausibly have occurred by chance when sampling from a population under which the null hypothesis was true. The test itself does not address the practical significance of the result. Clearly, for a sample size of 40,000, even a trivial sample correlation such as .02 is not likely to occur by mere luck of the draw. There is no practically meaningful relationship between these test scores and productivity scores in this example.

EXERCISES   **11.37** The output of Exercise 11.9 is reproduced here. Calculate the correlation coefficient r_{yx} from the R-square (r_{yx}^2) value. Should its sign be positive or negative?

```
. regress Branches Business
    Source |      SS         df         MS          Number of obs =      12
-----------+---------------------------------       F(1, 10)      = 172.60
     Model | 53.7996874      1    53.7996874        Prob > F      = 0.0000
  Residual | 3.11697922     10    .311697922        R-square      = 0.9452
-----------+---------------------------------       Adj R-square  = 0.9398
     Total | 56.9166667     11    5.17424242        Root MSE      = .5583
-----------------------------------------------------------------------------
  Branches |    Coef.    Std. Err.       t       P>|t|    [95% Conf. Interval]
-----------+-----------------------------------------------------------------
  Business | .0111049    .0008453     13.138     0.000    .0092216    .0129883
     _cons | 1.766846    .3211751      5.501     0.000    1.051223    2.482469
```

11.38 **a.** For the data in Exercise 11.37, test the hypothesis of no true correlation between x and y. Use a one-sided H_a and $\alpha = 0.05$.
b. Compare the result of this test to the t test of the slope found in the output.

Bus. **11.39** Suppose that an advertising campaign for a new product is conducted in 10 test cities. The intensity of the advertising x, measured as the number of exposures per evening of prime-time television, is varied across cities; the awareness percentage y is found by survey after the ad campaign:

x:	4.0	4.5	5.0	5.5	6.0	6.5	7.0	7.5	8.0	8.5
y:	10.1	10.3	10.4	21.7	36.7	51.5	67.0	68.5	68.2	69.3

```
MTB > Correlation 'Intensty' 'Aware'.
Correlation of Intensty and Aware = 0.956
```

a. Interpret the correlation coefficient r_{yx}.
b. Plot the data. Does the relation appear linear to you? Does it appear to be generally increasing?

Edu. **11.40** A survey of recent M.B.A. graduates of a business school obtained data on first-year salary and years of prior work experience. The following results were obtained using the Systat package.

CASE	EXPER	SALARY	CASE	EXPER	SALARY
1	8.000	53.900	12	10.000	53.500
2	5.000	52.500	13	2.000	38.300
3	5.000	49.000	14	2.000	37.200
4	11.000	65.100	15	5.000	51.300
5	4.000	51.600	16	13.000	64.700
6	3.000	52.700	17	1.000	45.300
7	3.000	44.500	18	5.000	47.000
8	3.000	40.100	19	1.000	43.800
9	0.000	41.100	20	5.000	47.400
10	13.000	66.900	21	5.000	40.200
11	14.000	37.900	22	7.000	52.800

23	4.000	40.700	38	2.000	50.600
24	3.000	47.300	39	4.000	41.800
25	3.000	43.700	40	1.000	44.400
26	7.000	61.800	41	5.000	46.600
27	7.000	51.700	42	1.000	43.900
28	9.000	56.200	43	4.000	45.000
29	6.000	48.900	44	1.000	37.900
30	6.000	51.900	45	2.000	44.600
31	4.000	36.100	46	7.000	46.900
32	6.000	53.500	47	5.000	47.600
33	5.000	50.400	48	1.000	43.200
34	1.000	38.700	49	1.000	41.600
35	13.000	60.100	50	0.000	39.200
36	1.000	38.900	51	1.000	41.700
37	6.000	48.400			

a. By scanning the numbers, can you sense there is a relation? In particular, does it appear that those with less experience have smaller salaries?

b. Can you notice any cases that seem to fall outside the pattern?

11.41 The data in Exercise 11.40 were plotted by Systat's influence plot. This plot is a scatterplot, with each point identified as to how much its removal would change the correlation. The larger the point, the more its removal would change the correlation. The plot is shown in the following figure. Does there appear to be an increasing pattern in the plot? Do any points clearly fall outside the basic pattern?

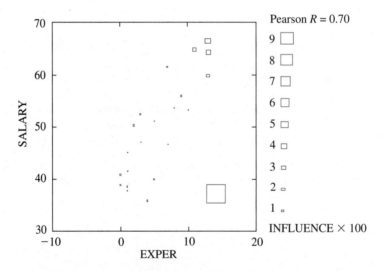

11.42 The 11th person in the data in Exercise 11.40 went to work for a family business in return for a low salary but a large equity in the firm. This case (the high influence point in the influence plot) was removed from the data and the results reanalyzed using Systat. A portion of the output follows.

```
DEP VAR: SALARY   N: 50   MULTIPLE R: 0.842 SQUARED MULTIPLE R: 0.709
ADJUSTED SQUARED MULTIPLE R:.703   STANDARD ERROR OF ESTIMATE: 4.071
VARIABLE        COEFFICIENT    STD ERROR     STD COEF    T    P(2 TAIL)
CONSTANT          39.188         0.971         0.000   40.353   0.000
EXPER              1.863         0.172         0.842   10.812   0.000
```

a. Should removing the high influence point in the plot increase or decrease the slope? Did it?
b. In which direction (larger or smaller) should the removal of this point change the residual standard deviation? Did it? How large was the change?
c. How should the removal of this point change the correlation? How large was this change?

11.8   Encounters with Real Data: Comparison of Two Methods for Detecting *E. coli*

Defining the Problem (1)

Section 7.5, Encounters with Real Data, described a new microbial method for the detection of *E. coli,* Petrifilm HEC test. The researcher wanted to evaluate the agreement of the results obtained using the HEC test with results obtained from an elaborate laboratory-based procedure, hydrophobic grid membrane filtration (HGMF). The HEC test is easier to inoculate, more compact to incubate, and safer to handle than conventional procedures. However, prior to using the HEC procedure it was necessary to compare the readings from the HEC test to readings from the HGMF procedure obtained on the same meat sample to determine whether the two procedures were yielding the same readings. If the readings differed but an equation could be obtained that could closely relate the HEC reading to the HGMF reading, then the researchers could calibrate the HEC readings to predict the readings that would have been obtained using the HGMF test procedure. If the HEC test results were unrelated to the HGMF test procedure results, then the HEC test could not be used in the field to detect *E. coli.*

Collecting Data (2)

In Chapter 7, we described phase one of the experiment. Phase two of the study was to apply both procedures to artificially contaminated beef. Portions of beef trim were obtained from three Holstein cows that had tested negatively for *E. coli.* Eighteen portions of beef trim were obtained from the cows and then contaminated with *E. coli.* The HEC and HGMF procedures were applied to a portion of each of the 18 samples. The two procedures yielded *E. coli* concentrations in transformed metric ($\log_{10}$ CFU/ml). The data in this case were 18 paired samples and are given here.

RUN	HEC	HGMF	RUN	HEC	HGMF
1	0.50	0.42	10	1.20	1.25
2	0.06	0.20	11	0.93	0.83
3	0.20	0.42	12	2.27	2.37
4	0.61	0.33	13	2.02	2.21
5	0.20	0.42	14	2.32	2.44
6	0.56	0.64	15	2.14	2.28
7	−0.82	−0.82	16	2.09	2.69
8	0.67	1.06	17	2.30	2.43
9	1.02	1.21	18	−0.10	1.07

Next the researchers prepared the data for a statistical analysis following the steps described in Section 2.5. They carefully reviewed the experimental procedures to make sure that each pair of meat samples was nearly identical so as not to introduce any differences in the HEC and HGMF readings that were not part of the differences in the two procedures. During such a review, procedural problems during run 18 were discovered and this pair of observations was excluded from the analysis.

Summarizing Data (3)

The researchers were interested in determining whether the two procedures yielded measures of *E. coli* concentrations that were strongly related. The scatterplot of the experimental data is given here.

A 45° line was placed in the scatterplot to display the relative agreement between the readings from the two procedures. If the plotted points fell on this line, then the two procedures would be in complete agreement in their determination of *E. coli* concentrations. The 17 points fall close to the line but have some variation about it. Thus, the researchers wanted to determine the degree of agreement and then obtain an equation that related the readings from the two procedures. If the readings from the two procedures could be accurately related using a regression equation, the researchers wanted to predict the reading of the HGMF procedure given the HEC reading on a meat sample. This would enable them to compare *E. coli* concentrations obtained from meat samples in the field using the HEC procedure to the readings obtained in the laboratory using the HGMF procedure.

Plot of HEC-method versus HGMF-method

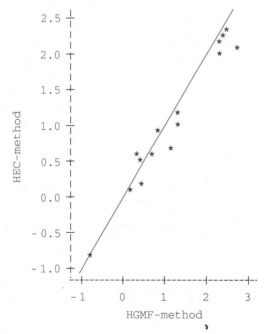

NOTE: 2 obs hidden.

Analyzing Data, Interpreting the Analyses, and Communicating Results (4)

The researchers were interested in assessing the degree to which the HEC and HGMF procedures agreed in determining the level of *E. coli* concentrations in meat samples. If there was a strong relationship between the two sets of readings, they then also wanted to obtain the inverse regression equations so they could predict HGMF readings from the HEC readings obtained in the field. We first obtain the regression relationship with HEC serving as the dependent variable y and HGMF as the independent variable x because the HGMF procedure has a known reliability in determining *E. coli* concentrations.

The computer output for analyzing the 17 pairs of *E. coli* concentrations is given here along with a plot of the residuals.

```
Dependent Variable: HEC                          HEC-METHOD

Analysis of Variance

                     Sum of          Mean
Source        DF     Squares         Square     F Value    Prob>F

Model          1     14.22159        14.22159   441.816    0.0001
Error         15      0.48283         0.03219
C Total       16     14.70442


     Root MSE        0.17941      R-square      0.9672
     Dep Mean        1.07471      Adj R-sq      0.9650
     C.V.           16.69413


                  Parameter      Standard     T for H0:
Variable   DF     Estimate        Error       Parameter=0     Prob > |T|

INTERCEP    1    -0.023039      0.06797755       -0.339         0.7394
HGMF        1     0.915685      0.04356377       21.019         0.0001
```

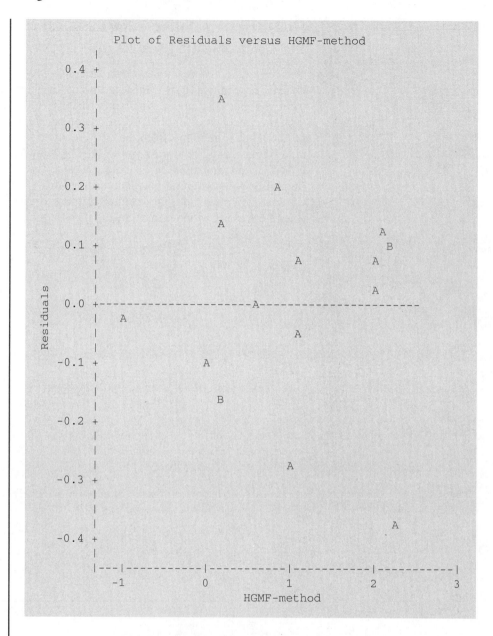

The R^2_{yx} value of .9672 indicates a strong linear relationship between HEC and HGMF concentrations. An examination of the residual plots does not indicate the necessity for higher-order terms in the model or heterogeneity in the variances. The least-squares equation relating HEC (y) to HGMF (x) concentrations is given here.

$$\hat{y} = -.023 + .9157x$$

Thus, we can assess whether there is an exact relationship between the two methods of determining *E. coli* concentrations by testing the hypotheses:

$$H_0: \beta_0 = 0, \beta_1 = 1 \quad \text{vs.} \quad H_{\hat{a}}: \beta_0 \neq 0 \quad \text{or} \quad \beta_1 \neq 1$$

If H_0 is accepted, then we have a strong indication that the relationship $\hat{y} = 0 + 1x$ is valid. That is, HEC and HGMF are yielding essentially the same values for

E. coli concentrations. From the output, we have a *p*-value = .7394 for testing $H_0: \beta_0 = 0$ and we can test $H_0: \beta_1 = 1$ using the test statistic:

$$t = \frac{\hat{\beta}_1 - 1}{\widehat{SE}(\hat{\beta}_1)} = \frac{.915685 - 1}{.04356377} = -1.935$$

The *p*-value of the test statistic is $\Pr(|t_{15}| \geq 1.935) = .0721$. To obtain an overall α-value of .05, we evaluate the hypotheses of $H_0: \beta_0 = 0$ and $H_0: \beta_1 = 1$ individually using $\alpha = .025$; that is, we reject either of the two hypotheses only if one of the *p*-values is less than .025. Because our *p*-values are .7394 and .0721, we fail to reject either null hypothesis and conclude that the data do not support the hypothesis that HEC and HGMF yield significantly different *E. coli* concentrations.

Because there are only 17 pairs of HEC and HGMF determinations, we construct the calibration curves to determine the degree of accuracy to which HEC (*y*) concentration readings predict HGMF (*x*) readings. Using the calibration equations, we obtain

$$\hat{x} = (y + .023)/.9157$$

with 95% prediction intervals

$$\hat{x}_L = 1.1988 + 1.0104(\hat{x} - 1.1988 - d)$$

$$\hat{x}_U = 1.1988 + 1.0104(\hat{x} - 1.1988 + d)$$

with $d = .4175\sqrt{1.0479 + (\hat{x} - 1.1988)^2/16.9612}$.

We next plot $\hat{x}_L$ and $\hat{x}_U$ for *y* ranging from -1 to $+2$ to obtain an indication of the range of values that would be obtained in predicting HGMF readings from observed HEC readings.

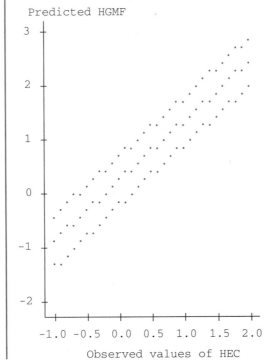

```
Plot of predicted HGMF for observed HEC
        with 95% prediction bounds
   Predicted HGMF
 3 +

 2 +

 1 +

 0 +

-1 +

-2 +
    +---+---+---+---+---+---+
   -1.0 -0.5  0.0  0.5  1.0  1.5  2.0
        Observed values of HEC
```

The width of the 95% prediction intervals is slightly less than one unit for most values of HEC. Thus, HEC determinations in the field of *E. coli* concentrations in the −1 to 2 range result in 95% prediction intervals for the corresponding HGMF determinations. This degree of accuracy is not acceptable. One way to reduce the width of the intervals is to conduct an expanded study involving considerably more observations than the 17 obtained in this study, provided the same general degree of relationship held between HEC and HGMF in the new study.

11.9 Staying Focused

This chapter introduces regression analysis and is devoted to simple regression, using only one independent variable to predict a dependent variable. The basic questions involve the nature of the relation (linear or curved), the amount of variability around the predicted value, whether that variability is constant over the range of prediction, how useful the independent variable is in predicting the dependent variable, and how much to allow for sampling error. The key concepts of the chapter include the following:

1. The data should be plotted in a scatterplot. A smoother such as LOWESS or a spline curve is useful in deciding whether a relation is nearly linear or is clearly curved. Curved relations can often be made nearly linear by transforming either the independent variable or the dependent variable or both.
2. The coefficients of a linear regression are estimated by least squares, which minimizes the sum of squared residuals (actual values minus predicted values). Because squared error is involved, this method is sensitive to outliers.
3. Observations that are extreme in the x (independent variable) direction have high leverage in fitting the line. If a high leverage point also falls well off the line, it has high influence, in that removing the observation substantially changes the fitted line. A high influence point should be omitted if it comes from a different population than the remainder. If it must be kept in the data, a method other than least squares should be used.
4. Variability around the line is measured by the standard deviation of the residuals. This standard deviation may be interpreted using the Empirical Rule. The standard deviation sometimes increases as the predicted value increases. In such a case, try transforming the dependent variable.
5. Hypothesis tests and confidence intervals for the slope of the line (and, less interestingly, the intercept) are based on the t distribution. If there is no relation, the slope is 0. The line is estimated most accurately if there is a wide range of variation in the x variable.
6. The fitted line may be used to forecast at a new x value, again using the t distribution. This forecasting is potentially inaccurate if the new x value is extrapolated far from the previous ones.
7. A standard method of measuring the strength of relation is the coefficient of determination, the square of the correlation. This measure is diminished by nonlinearity or by an artificially limited range of x variation.

One of the most important uses of statistics for managers is prediction. A manager may want to forecast the cost of a particular contracting job given the size of that job, to forecast the sales of a particular product given the current rate of growth of the gross national product, or to forecast the number of parts that will be produced given a certain size workforce. The statistical method most widely used in making predictions is *regression analysis.*

In the regression approach, past data on the relevant variables are used to develop and evaluate a prediction equation. The variable that is being predicted by this equation is the dependent variable. A variable that is being used to make the prediction is an independent variable. In this chapter, we discuss regression methods involving a single independent variable. In Chapter 12, we extend these methods to multiple regression, the case of several independent variables.

A number of tasks can be accomplished in a regression study:

1. The data can be used to obtain a prediction equation.
2. The data can be used to estimate the amount of variability or uncertainty around the equation.
3. The data can be used to identify unusual points far from the predicted value, which may represent unusual problems or opportunities.
4. Because the data are only a sample, inferences can be made about the true (population) values for the regression quantities.
5. The prediction equation can be used to predict a reasonable range of values for future values of the dependent variable.
6. The data can be used to estimate the degree of correlation between dependent and independent variables, a measure that indicates how strong the relation is.

Key Formulas

1. Least-squares estimates of slope and intercept

$$\hat{\beta}_1 = \frac{S_{xy}}{S_{xx}}$$

and

$$\hat{\beta}_0 = \bar{y} - \hat{\beta}_1 \bar{x}$$

where

$$S_{xy} = \sum_i (x_i - \bar{x})(y_i - \bar{y})$$

and

$$S_{xx} = \sum_i (x_i - \bar{x})^2$$

2. Estimate of σ_ε^2

$$s_\varepsilon^2 = \frac{\sum_i (y_i - \hat{y}_i)^2}{n - 2} = \frac{SS(\text{Residual})}{n - 2}$$

3. Statistical test for β_1

$$H_0: \quad \beta_1 = 0 \text{ (two-tailed)}$$

$$\text{T.S.:} \quad t = \frac{\hat{\beta}_1}{\dfrac{s_\varepsilon}{\sqrt{S_{xx}}}}$$

4. Confidence interval for β_1

$$\hat{\beta}_1 \pm t_{\alpha/2} s_\varepsilon \sqrt{\frac{1}{S_{xx}}}$$

5. F test for $H_0: \beta_1 = 0$ (two-tailed)

$$\text{T.S.:} \quad F = \frac{\text{MS(Regression)}}{\text{MS(Residual)}}$$

6. Confidence interval for $E(y_{n+1})$

$$\hat{y}_{n+1} \pm t_{\alpha/2} s_\varepsilon \sqrt{\frac{1}{n} + \frac{(x_{n+1} - \bar{x})^2}{S_{xx}}}$$

7. Prediction interval for y_{n+1}

$$\hat{y}_{n+1} \pm t_{\alpha/2} s_\varepsilon \sqrt{1 + \frac{1}{n} + \frac{(x_{n+1} - \bar{x})^2}{S_{xx}}}$$

8. Test for lack of fit in linear regression

$$\text{T.S.:} \quad F = \frac{\text{MS}_{\text{Lack}}}{\text{MSP}_{\text{exp}}}$$

where

$$\text{MSP}_{\text{exp}} = \frac{\text{SSP}_{\text{exp}}}{\sum_i (n_i - 1)} = \frac{\sum_{ij} (y_{ij} - \bar{y}_i)^2}{\sum_i (n_i - 1)}$$

and

$$\text{MS}_{\text{Lack}} = \frac{\text{SS(Residual)} - \text{SSP}_{\text{exp}}}{(n - 2) - \sum_i (n_i - 1)}$$

9. Prediction limits for x based on a single y value

$$\hat{x} = \frac{y - \hat{\beta}_0}{\hat{\beta}_1}$$

$$\hat{x}_U = \bar{x} + \frac{1}{1 - c^2}[(\hat{x} - \bar{x}) + d]$$

$$\hat{x}_L = \bar{x} + \frac{1}{1 - c^2}[(\hat{x} - \bar{x}) - d]$$

where

$$c^2 = \frac{t_{\alpha/2}^2 s_\varepsilon^2}{\hat{\beta}_1^2 S_{xx}}$$

and

$$d = \frac{t_{\alpha/2}s_\varepsilon}{\hat{\beta}_1}\sqrt{\frac{n+1}{n}(1-c^2) + \frac{(\hat{x}-\overline{x})^2}{S_{xx}}}$$

10. Prediction interval for x based on m y-values

$$\hat{x}_U = \overline{x} + \frac{1}{1-c^2}[(\hat{x}-\overline{x}) + g]$$

$$\hat{x}_L = \overline{x} + \frac{1}{1-c^2}[(\hat{x}-\overline{x}) - g]$$

where

$$\hat{x} = \frac{P\overline{y}_m - \hat{\beta}_0}{\hat{\beta}_1}$$

and

$$g = \frac{t_{\alpha/2}}{\hat{\beta}_1}\sqrt{\left(s_{\overline{y}}^2 P^2 + \frac{s_\varepsilon^2}{n}\right)(1-c^2) + \frac{(\hat{x}-\overline{x})^2 s_\varepsilon^2}{S_{xx}}}$$

11. Correlation coefficient

$$r_{yx} = \frac{S_{xy}}{\sqrt{S_{xx}S_{yy}}} = \hat{\beta}_1\sqrt{\frac{S_{xx}}{S_{yy}}}$$

12. Coefficient of determination

$$r_{yx}^2 = \frac{SS(Total) - SS(Residual)}{SS(Total)}$$

13. Statistical test for ρ_{yx}

$$H_0: \quad \rho_{yx} = 0 \text{ (two-tailed)}$$

$$\text{T.S.:} \quad t = r_{yx}\frac{\sqrt{n-2}}{\sqrt{1-r_{yx}^2}}$$

Supplementary Exercises

Gov. **11.43** A government agency responsible for awarding contracts for much of its research work is under careful scrutiny by a number of private companies. One company examines the relationship between the amount of the contract (in tens of thousands of dollars) and the length of time between the submission of the contract proposal and contract approval:

Length (in months) y:	3	4	6	8	11	14	20
Size ($\times$ \$10,000) x:	1	5	10	50	100	500	1000

A plot of y versus x and Stata output follow.

```
.regress Length Size
  Source |       SS            df        MS            Number of obs =       7
---------+------------------------------              F(  1,     5) =   33.78
   Model | 191.389193          1    191.389193         Prob > F      =  0.0021
Residual | 28.3250928          5    5.66501856         R-square      =  0.8711
---------+------------------------------              Adj R-square  =  0.8453
   Total | 219.714286          6    36.6190476         Root MSE      =  2.3801
----------------------------------------------------------------------------

  Length |    Coef.    Std. Err.      t      P>|t|    [95% Conf. Interval]
----------------------------------------------------------------------------
    Size | .0148652    .0025575    5.812    0.002    .008291    .0214394
   _cons | 5.890659    1.086177    5.423    0.003   3.098553    8.682765
----------------------------------------------------------------------------
```

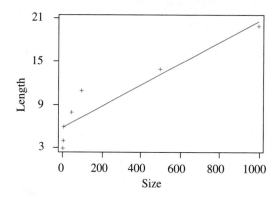

a. What is the least-squares line?

b. Conduct a test of the null hypothesis $H_0: \beta_1 \leq 0$. Give the p-value for your test, assuming $H_a: \beta_1 > 0$.

11.44   Refer to the data of Exercise 11.43. A plot of y versus the natural logarithm of x is shown and more Stata output is given here.

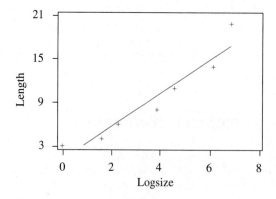

```
.regress Length lnSize
  Source |      SS          df         MS              Number of obs =        7
---------+-----------------------------------          F(  1,     5) =    49.20
   Model | 199.443893        1      199.443893          Prob > F      =   0.0009
Residual | 20.2703932        5      4.05407863          R-square      =   0.9077
---------+-----------------------------------          Adj R-square  =   0.8893
   Total | 219.714286        6      36.6190476          Root MSE      =   2.0135

----------------------------------------------------------------------------------
  Length |    Coef.    Std. Err.      t      P>|t|     [95% Conf. Interval]
----------------------------------------------------------------------------------
  lnSize | 2.307015    .3289169     7.014    0.000     1.461508     3.152523
   _cons | 1.007445    1.421494     0.709    0.510    -2.646622     4.661511
----------------------------------------------------------------------------------
```

a. What is the regression line using ln x as the independent variable?

b. Conduct a test of $H_0: \beta_1 \leq 0$, and give the level of significance for a one-sided alternative, $H_a: \beta_1 > 0$.

11.45 Use the results of Exercises 11.43 and 11.44 to determine which regression model provides the better fit. Give reasons for your choice.

11.46 Use the model you prefer for the data in Exercise 11.44 to predict the length of time in months before approval of a $750,000 contract. Give a rough estimate of a 95% prediction interval.

Env. **11.47** An airline studying fuel use by a certain type of aircraft obtains data on 100 flights. The air mileage x in hundreds of miles and the actual fuel use y in gallons are recorded. Statistix output follows and a plot is shown.

```
UNWEIGHTED LEAST SQUARES LINEAR REGRESSION OF GALLONS

PREDICTOR
VARIABLES      COEFFICIENT    STD ERROR      STUDENT'S T      P

CONSTANT          140.074       44.1293          3.17       0.0099
MILES             0.61896       0.04855         12.75       0.0000

R-SQUARED              0.9420    RESID. MEAN SQUARE (MSE)     1182.34
ADJUSTED R-SQUARED     0.9362    STANDARD DEVIATION           34.3852

SOURCE         DF        SS            MS           F         P

REGRESSION      1     1.921E+05    1.921E+05     162.48    0.0000
RESIDUAL       10      11823.4      1182.34
TOTAL          11     2.039E+05

PREDICTED/FITTED VALUES OF GALLONS

LOWER PREDICTED BOUND      678.33      LOWER FITTED BOUND       733.68
PREDICTED VALUE            759.03      FITTED VALUE             759.03
UPPER PREDICTED BOUND      839.73      UPPER FITTED BOUND       784.38
SE (PREDICTED VALUE)       36.218      SE (FITTED VALUE)        11.377

UNUSUALNESS (LEVERAGE)     0.1095
PERCENT COVERAGE           95.0
CORRESPONDING T            2.23

PREDICTOR VALUES: MILES = 1000.0
```

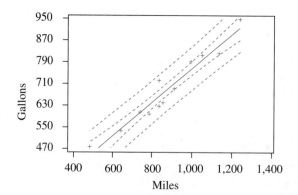

a. Locate the regression equation.
b. What are the sample correlation coefficient and coefficient of determination? Interpret these numbers.
c. Is there any point in testing $H_0: \beta_1 \leq 0$?

11.48 Refer to the data and output in Exercise 11.47.
a. Predict the mean fuel use of all 1,000-mile flights. Give a 95% confidence interval.
b. Predict the fuel use of a particular 1,000-mile flight. Would a use of 628 gallons be considered exceptionally low?

Soc.   **11.49** Suburban towns often spend a large fraction of their municipal budgets on public safety (police, fire, and ambulance) services. A taxpayers' group felt that very small towns were likely to spend large amounts per person because they have such small financial bases. The group obtained data on the per capita expenditure for public safety of 29 suburban towns in a metropolitan area, as well as the population of each town. The data were analyzed using the Minitab package. A regression model with dependent variable 'expendit' and independent variable 'townpopn' yields the following output.
a. If the taxpayers' group is correct, what sign should the slope of the regression model have?
b. Does the slope in the output confirm the opinion of the group?

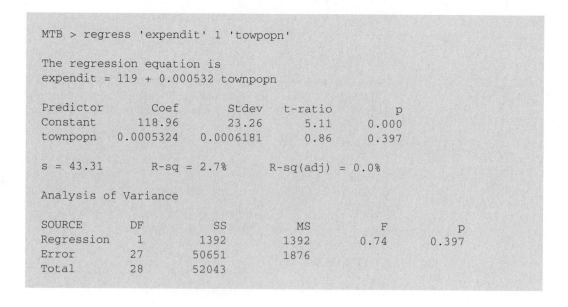

```
MTB > regress 'expendit' 1 'towpopn'

The regression equation is
expendit = 119 + 0.000532 townpopn

Predictor        Coef        Stdev     t-ratio          p
Constant       118.96        23.26        5.11      0.000
townpopn     0.0005324    0.0006181        0.86      0.397


s = 43.31         R-sq = 2.7%        R-sq(adj) = 0.0%

Analysis of Variance

SOURCE         DF           SS          MS          F         p
Regression      1         1392        1392       0.74     0.397
Error          27        50651        1876
Total          28        52043
```

```
Unusual Observations
Obs. townpopn    expendit        Fit  Stdev.Fit   Residual    St. Resid
   8    74151      334.00      158.43      25.32     175.57        5.00RX

R denotes an obs. with a large st. resid.
X denotes an obs. whose X value gives it large influence.
```

11.50 Minitab produced a scatterplot and LOWESS smoothing of the data in Exercise 11.49, shown here. Does this plot indicate that the regression line is misleading? Why?

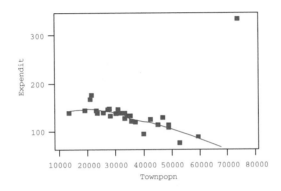

11.51 One town in the database of Exercise 11.49 is the home of an enormous regional shopping mall. A very large fraction of the town's expenditure on public safety is related to the mall; the mall management pays a yearly fee to the township that covers these expenditures. That town's data were removed from the database and the remaining data were reanalyzed by Minitab. A scatterplot is shown next.

    a. Explain why removing this one point from the data changed the regression line so substantially.

    b. Does the revised regression line appear to conform to the opinion of the taxpayers' group in Exercise 11.49?

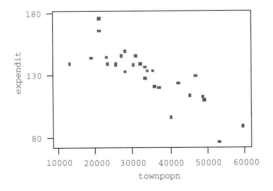

11.52 Regression output for the data in Exercise 11.49, excluding the one unusual town, is shown here. How has the slope changed from the one obtained previously?

```
MTB > regress 'expendit' 1 'townpopn'

The regression equation is
expendit = 184 - 0.00158 townpopn

Predictor         Coef        Stdev    t-ratio          p
Constant       184.240        7.481      24.63      0.000
townpopn    -0.0015766    0.0002099      -7.51      0.000

s = 12.14        R-sq = 68.5%    R-sq(adj) = 67.2%
```

```
Analysis of variance

SOURCE         DF          SS         MS        F        p
Regression      1      8322.7     8322.7    56.43    0.000
Error          26      3834.5      147.5
Total          27     12157.2

Unusual Observations
Obs. townpopn    expendit       Fit    Stdev.Fit    Residual    St.Resid
   5    40307       96.00    120.69         2.66      -24.69      -2.08R
   6    13457      139.00    163.02         4.87      -24.02      -2.16R
  13    59779       89.00     89.99         5.89       -0.99      -0.09 X
  22    21701      176.00    150.03         3.44       25.97       2.23R
  27    53322       76.00    100.17         4.67      -24.17      -2.16R

R denotes an obs. with a large st. resid.
X denotes an obs. whose X value gives it large influence.
```

Med.　**11.53** Researchers measured the specific activity of the enzyme sucrase extracted from portions of the intestines of 24 patients who underwent an intestinal bypass. After the sections were extracted, they were homogenized and analyzed for enzyme activity (Carter, 1981). Two different methods can be used to measure the activity of sucrase: the homogenate method and the pellet method. Data for the 24 patients are shown here for the two methods.

Patient	Homogenate Method, y	Pellet Method, x
Sucrase Activity as Measured by the Homogenate and Pellet Methods		
1	18.88	70.00
2	7.26	55.43
3	6.50	18.87
4	9.83	40.41
5	46.05	57.43
6	20.10	31.14
7	35.78	70.10
8	59.42	137.56

**Sucrase Activity as Measured by the Homogenate
and Pellet Methods**

Patient	Homogenate Method, y	Pellet Method, x
9	58.43	221.20
10	62.32	276.43
11	88.53	316.00
12	19.50	75.56
13	60.78	277.30
14	77.92	331.50
15	51.29	133.74
16	77.91	221.50
17	36.65	132.93
18	31.17	85.38
19	66.09	142.34
20	115.15	294.63
21	95.88	262.52
22	64.61	183.56
23	37.71	86.12
24	100.82	226.55

Relationship between Homogenate and Pellet

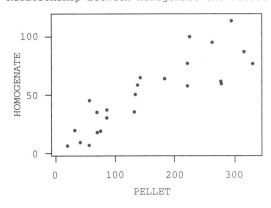

a. Examine the scatterplot of the data. Might a linear model adequately describe the relationship between the two methods?

```
Regression Analysis: HOMOGENATE versus PELLET

The regression equation is
HOMOGENATE = 10.3 + 0.267 PELLET

Predictor        Coef      SE Coef         T           P
Constant        10.335       5.995       1.72       0.099
PELLET         0.26694     0.03251       8.21       0.000

S = 15.62       R-Sq = 75.4%      R-Sq(adj) = 74.3%
```

```
Analysis of Variance

Source            DF        SS        MS        F        P
Regression         1     16440     16440    67.41    0.000
Residual Error    22      5366       244
Total             23     21806
```

Obs	PELLET	HOMOGENA	Fit	SE Fit	Residual	St Resid
1	70	18.88	29.02	4.24	-10.14	-0.67
2	55	7.26	25.13	4.57	-17.87	-1.20
3	19	6.50	15.37	5.49	-8.87	-0.61
4	40	9.83	21.12	4.93	-11.29	-0.76
5	57	46.05	25.67	4.52	20.38	1.36
6	31	20.10	18.65	5.17	1.45	0.10
7	70	35.78	29.05	4.24	6.73	0.45
8	138	59.42	47.06	3.24	12.36	0.81
9	221	58.43	69.38	3.83	-10.95	-0.72
10	276	62.32	84.13	5.04	-21.81	-1.48
11	316	88.53	94.69	6.10	-6.16	-0.43
12	76	19.50	30.50	4.13	-11.00	-0.73
13	277	60.78	84.36	5.07	-23.58	-1.60
14	332	77.92	98.83	6.53	-20.91	-1.47
15	134	51.29	46.04	3.27	5.25	0.34
16	222	77.91	69.46	3.83	8.45	0.56
17	133	36.65	45.82	3.28	-9.17	-0.60
18	85	31.17	33.13	3.93	-1.96	-0.13
19	142	66.09	48.33	3.22	17.76	1.16
20	295	115.15	88.98	5.52	26.17	1.79
21	263	95.88	80.41	4.70	15.47	1.04
22	184	64.61	59.33	3.31	5.28	0.35
23	86	37.71	33.32	3.92	4.39	0.29
24	227	100.82	70.81	3.92	30.01	1.99

Regression Line for Homogenate versus Pellet
HOMOGENATE = 10.3348 + 0.266940 PELLET
S = 15.6169   R-Sq = 75.4%   R-Sq(adj) = 74.3%

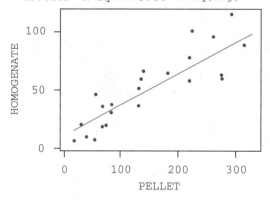

b. Examine the residual plot. Are there any potential problems uncovered by the plot?

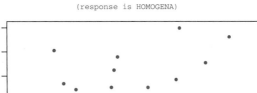

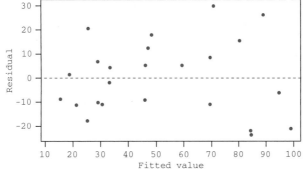

c. In general, the pellet method is more time-consuming than the homogenate method, yet it provides a more accurate measure of sucrase activity. How could you estimate the pellet reading based on a particular homogenate reading?

d. How would you develop a confidence (prediction) interval about your point estimate?

11.54   Defining the Problem (1). The correlation coefficient r measures the linear association between two variables x and y. A common mistake in interpreting r for two variables is to infer that a strong correlation indicates that one variable causes the other. In scientific studies, a strong correlation may be an important step in establishing causation, but in observational studies it is not uncommon for a third variable, which is not observed, to be the reason there is a strong correlation between x and y.

In the June 1986 issue of *Consumer Reports,* the results of tests on 54 brands of hot dogs were published in an effort to provide consumers with information on the contents of hot dogs.

Collecting Data (2). A subset of these data are shown here (with the brands omitted).

c1	Calories	Calories per hot dog
c2	Sodium	Milligrams of sodium per hot dog
c3	CalPerOz	Calories per ounce
c4	SodPerOz	Milligrams of sodium per ounce

Calories	Sodium	CalPerOz	SodPerOz
186	495	93.0	247.5
181	477	90.5	238.5
176	425	88.0	212.5
149	322	93.1	201.3
184	482	92.0	241.0
190	587	95.0	293.5
158	370	92.2	215.8
139	322	92.7	214.7
175	479	87.5	239.5
148	375	92.5	234.4
152	330	88.7	192.5

111	300	74.0	200.0
141	386	88.1	241.3
153	401	95.6	250.6
190	645	95.0	322.5
157	440	98.1	275.0
131	317	87.3	211.3
149	319	93.1	199.4
135	298	90.0	198.7
132	253	88.0	168.7
173	458	86.5	229.0
191	506	95.5	253.0
182	473	91.0	236.5
190	545	95.0	272.5
172	496	86.0	248.0
147	360	91.9	225.0
146	387	91.3	241.9
139	366	86.9	228.8
175	507	87.5	253.5
136	393	68.0	196.5
179	405	89.5	202.5
153	372	95.6	232.5
107	144	74.9	100.8
195	511	97.5	255.5
135	405	84.4	253.2
140	428	87.5	267.5
138	339	86.3	211.9
129	430	80.6	268.8
132	375	82.5	234.4
102	396	63.8	247.5
106	383	66.3	239.4
94	387	58.8	241.9
102	542	51.0	271.0
87	359	54.4	224.4
99	357	61.9	223.1
170	528	85.0	264.0
113	513	56.5	256.5
135	426	67.5	213.0
142	513	71.0	256.5
86	358	71.7	298.3
143	581	71.5	290.5
152	588	76.0	294.0
146	522	73.0	261.0
144	545	72.0	272.5

Summarizing Data (3). We see from the scatterplot that there is a noticeable association between the variables `Calories` and `Sodium`. This association is somewhat surprising because sodium in hot dogs comes mainly from salt and preservatives, which contain no calories. So it is not immediately clear why the two variables should be correlated.

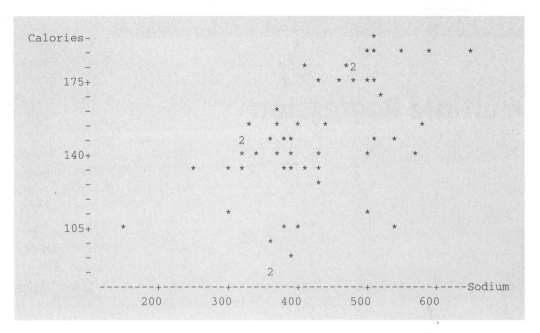

Note: In this character scatterplot the standard plotting symbol * is replaced by the appropriate number when several data points lie at the same position on the plot.

Analyzing Data, Interpreting the Analyses and Communicating Results (4). The correlation $r = 0.516$ is significantly different from 0. It is unlikely to be an artifact of the random variation in our sample.

```
Correlation of Calories and Sodium = 0.516, P-Value = 0.000
```

What is the explanation of this highly significant positive association? Do manufacturers of hot dogs load the high-calorie ones with extra salt and preservatives? As it turns out, they don't.

a. From the *Consumer Reports* article, it is clear that the hot dogs varied in size from 1.2 oz to 2.0 oz with 1.5, 1.6, and 2 oz being the most common sizes. Knowing this, it's not surprising that larger hot dogs, in general, have more calories and sodium than smaller hot dogs. To eliminate the effect of size, make a scatterplot and compute the correlation between CalPerOz and SodPerOz. Do you see any association?

b. You may be thinking that it was a stupid mistake for *Consumer Reports* to publish data that were not adjusted for weight. However, the writers reasoned that people eat hot dogs by the *hot dog,* not by the *ounce.* The purpose of the article was to provide basic consumer information about the contents of hot dogs, not to study correlation. Which do you think is the more helpful measure for the typical consumer: calories per hot dog or calories per ounce? You are entitled to your opinion, but defend your answer.

CHAPTER 12

Multiple Regression

12.1 Introduction

The simplest type of regression model relating the dependent variable y to a quantitative independent variable x is the one discussed in Chapter 11,

$$y = \beta_0 + \beta_1 x + \varepsilon$$

expected value of ε

$$E(\varepsilon) = 0$$

Under the assumption that the average value of ε (also called the **expected value of ε**) for a given value of x is $E(\varepsilon) = 0$, this model indicates that the expected value of y for a given value of x is described by the straight line

$$E(y) = \beta_0 + \beta_1 x$$

Not all data sets are adequately described by a model for which the expectation is a straight line. For example, consider the data of Table 12.1, which gives the yields (in bushels) for 14 equal-size plots planted in tomatoes for different levels of fertilization. It is evident from the scatterplot in Figure 12.1 that a linear equation will not adequately represent the relationship between yield and the amount of fertilizer applied to the plot. The reason for this is that, whereas a modest amount of fertilizer may enhance the crop yield, too much fertilizer can be destructive.

A model for this physical situation might be

$$y = \beta_0 + \beta_1 x + \beta_2 x^2 + \varepsilon$$

Again with the assumption that $E(\varepsilon) = 0$, the expected value of y for a given value of x is

$$E(y) = \beta_0 + \beta_1 x + \beta_2 x^2$$

One such line is plotted in Figure 12.1, superimposed on the data of Table 12.1.

TABLE 12.1

Yield of 14 equal-size plots of tomato plantings for different amounts of fertilizer

Plot	Yield, y (in bushels)	Amount of Fertilizer, x (in pounds per plot)
1	24	12
2	18	5
3	31	15
4	33	17
5	26	20
6	30	14
7	20	6
8	25	23
9	25	11
10	27	13
11	21	8
12	29	18
13	29	22
14	26	25

FIGURE 12.1

Scatterplot of the yield versus fertilizer data in Table 12.1

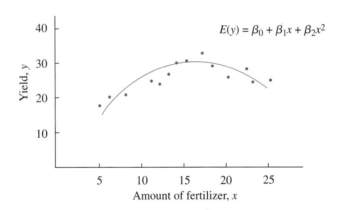

A general polynomial regression model relating a dependent variable y to a single quantitative independent variable x is given by

$$y = \beta_0 + \beta_1 x + \beta_2 x^2 + \cdots + \beta_p x^p + \varepsilon$$

with

$$E(y) = \beta_0 + \beta_1 x + \beta_2 x^2 + \cdots + \beta_p x^p$$

The choice of p and hence the choice of an appropriate regression model will depend on the experimental situation.

multiple regression model

The **multiple regression model** that relates a dependent variable y to a set of quantitative independent variables is a direct extension of a polynomial regression model in one independent variable. We write the multiple regression model as

$$y = \beta_0 + \beta_1 x_1 + \beta_2 x_2 + \cdots + \beta_k x_k + \varepsilon$$

Any of the independent variables may be powers of other independent variables; for example, x_2 might be x_1^2. In fact, there are many other possibilities; x_3 might be

cross-product term

a **cross-product term** equal to x_1x_2, x_4 might be log x_1, and so on. The only restriction is that no x is a perfect linear function of any other x.

first-order model

The simplest type of multiple regression equation is a **first-order model**, in which each of the independent variables appears, but there are no cross-product terms or terms in powers of the independent variables. For example, when three quantitative independent variables are involved, the first-order multiple regression model is

$$y = \beta_0 + \beta_1x_1 + \beta_2x_2 + \beta_3x_3 + \varepsilon$$

For these first-order models, we can attach some meaning to the βs. The parameter β_0 is the y-intercept, which represents the expected value of y when each x is zero. For cases in which it does not make sense to have each x be zero, β_0 (or its estimate) should be used only as part of the prediction equation and not given an interpretation by itself.

partial slopes

The other parameters $(\beta_1, \beta_2, \ldots, \beta_k)$ in the multiple regression equation are sometimes called **partial slopes.** In linear regression, the parameter β_1 is the slope of the regression line and it represents the expected change in y for a unit increase in x. In a first-order multiple regression model, β_1 represents the expected change in y for a unit increase in x_1 *when all other xs are held constant.* In general then, $\beta_j\,(j \neq 0)$ represents the expected change in y for a unit increase in x_j while holding all other xs constant. The usual assumptions for a multiple regression model are shown here.

Assumptions for Multiple Regression

1. The mathematical form of the relation is correct, so $E(\varepsilon_i) = 0$ for all i.
2. $\text{Var}(\varepsilon_i) = \sigma_\varepsilon^2$ for all i.
3. The ε_is are independent.
4. ε_i is normally distributed.

There is an additional assumption that is implied when we use a first-order multiple regression model. Because the expected change in y for a unit change in x_j is constant and does not depend on the value of any other x, we are in fact assuming that

additive effects

the effects of the independent variables are **additive.**

EXAMPLE 12.1

A brand manager for a new food product collected data on y = brand recognition (percent of potential consumers who can describe what the product is), x_1 = length in seconds of an introductory TV commercial, and x_2 = number of repetitions of the commercial over a 2-week period. What does the brand manager assume if a first-order model

$$\hat{y} = 0.31 + 0.042x_1 + 1.41x_2$$

is used to predict y?

Solution First, the manager assumes a straight-line, consistent rate of change. The manager assumes that a 1-second increase in length of the commercial will lead to a 0.042 percentage point increase in recognition, whether the increase is from, say, 10 to 11 seconds or from 59 to 60 seconds. Also, every additional repetition of the commercial is assumed to give a 1.41 percentage point increase in recognition, whether it is the second repetition or the twenty-second.

Second, there is a no-interaction assumption. The first-order model assumes that the effect of an additional repetition (that is, an increase in x_2) of a given length commercial (that is, holding x_1 constant) doesn't depend on *where* that length is held constant (at 10 seconds, 27 seconds, 60 seconds, or whatever).

When might the additional assumption of additivity be warranted? Figure 12.2(a) shows a scatterplot of y versus x_1; Figure 12.2(b) shows the same plot with an ID attached to the different levels of a second independent variable x_2 (where x_2 takes on the values of 1, 2, or 3). From Figure 12.2(a), we see that y is approximately linear in x_1. The parallel lines of Figure 12.2(b) corresponding to the three levels of the independent variable x_2 indicate that the expected change in y for a unit change in x_1 remains the same no matter which level of x_2 is used. These data suggest that the effects of x_1 and x_2 are additive; hence, a first-order model of the form $y = \beta_0 + \beta_1 x_1 + \beta_2 x_2 + \varepsilon$ is appropriate.

interaction   Figure 12.3 displays a situation in which **interaction** is present between the variables x_1 and x_2. Even though a scatterplot of y versus x_1 is as shown in Figure 12.2(a), the nonparallel lines of Figure 12.3 indicate that the expected change in y for a unit change in x_1 now depends on the level of x_2. When this occurs, the independent variables x_1 and x_2 are said to interact. A first-order model, which assumes additivity of the effects, is not appropriate here. At the very least, we would include a cross-product term $(x_1 x_2)$ in the model.

The simplest model allowing for interaction between x_1 and x_2 is

$$y = \beta_0 + \beta_1 x_1 + \beta_2 x_2 + \beta_3 x_1 x_2 + \varepsilon$$

Note that for a given value of x_2 (say, $x_2 = 2$), the expected value of y is

$$E(y) = \beta_0 + \beta_1 x_1 + \beta_2(2) + \beta_3 x_1(2)$$
$$= (\beta_0 + 2\beta_2) + (\beta_1 + 2\beta_3)x_1$$

FIGURE 12.2

(a) Scatterplot of y versus x_1;
(b) Scatterplot of y versus x_1, indicating additivity of effects for x_1 and x_2

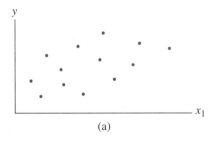
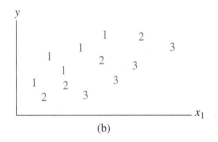

(a)   (b)

FIGURE 12.3

Scatterplot of y versus x_1, indicating nonadditivity (interaction) of effects between x_1 and x_2

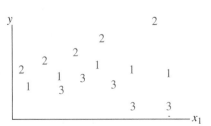

Here the intercept and slope are $(\beta_0 + 2\beta_2)$ and $(\beta_1 + 2\beta_3)$, respectively. The corresponding intercept and slope for $x_2 = 3$ can be shown to be $(\beta_0 + 3\beta_2)$ and $(\beta_1 + 3\beta_3)$. Clearly, the slopes of the two regression lines are not the same, and hence we have nonparallel lines.

Not all experiments can be modeled using a first-order multiple regression model. For these situations, in which a higher-order multiple regression model may be appropriate, it will be more difficult to assign a literal interpretation to the βs because of the presence of terms that contain cross-products or powers of the independent variables. Our focus will be on finding a multiple regression model that provides a good fit to the sample data, not on interpreting individual βs, except as they relate to the overall model.

The models that we have described briefly have been for regression problems for which the experimenter is interested in developing a model to relate a response to one or more *quantitative* independent variables. The problem of modeling an experimental situation is not restricted to the quantitative independent-variable case.

Consider the problem of writing a model for an experimental situation in which a response y is related to a set of *qualitative* independent variables or to both quantitative and qualitative independent variables. For the first situation (relating y to one or more qualitative independent variables), let us suppose that we want to compare the average number of lightning discharges per minute for a storm, as measured from two different tracking posts located 30 miles apart. If we let y denote the number of discharges recorded on an oscilloscope during a 1-minute period, we can write the following two models:

For tracking post 1: $y = \mu_1 + \varepsilon$

For tracking post 2: $y = \mu_2 + \varepsilon$

Thus, we assume that observations at tracking post 1 randomly bob about a population mean μ_1. Similarly, at tracking post 2, observations differ from a population mean μ_2 by a random amount ε. These two models are not new and could have been used to describe observations when comparing two population means in Chapter 6. What is new is that we can combine these two models into a single model of the form

$$y = \beta_0 + \beta_1 x_1 + \varepsilon$$

dummy variable where β_0 and β_1 are unknown parameters, ε is a random error term, and x_1 is a **dummy variable** with the following interpretation. We let

$x_1 = 1$ if an observation is obtained from tracking post 2

$x_1 = 0$ if an observation is obtained from tracking post 1

For observations obtained from tracking post 1, we substitute $x_1 = 0$ into our model to obtain

$$y = \beta_0 + \beta_1(0) + \varepsilon = \beta_0 + \varepsilon$$

Hence, $\beta_0 = \mu_1$, the population mean for observations from tracking post 1. Similarly, by substituting $x_1 = 1$ in our model, the equation for observations from tracking post 2 is

$$y = \beta_0 + \beta_1(1) + \varepsilon = \beta_0 + \beta_1 + \varepsilon$$

Because $\beta_0 = \mu_1$ and $\beta_0 + \beta_1$ must equal μ_2, we have $\beta_1 = \mu_2 - \mu_1$, the difference in means between observations from tracking posts 2 and 1.

This model, $y = \beta_0 + \beta_1 x_1 + \varepsilon$, which relates y to the qualitative independent variable tracking post, can be extended to a situation in which the qualitative variable has more than two levels. We do this by using more than one dummy variable. Consider a completely randomized design in which we're interested in four levels of a single qualitative variable. As indicated in Chapter 8, we call these levels treatments. We can write the model

$$y = \beta_0 + \beta_1 x_1 + \beta_2 x_2 + \beta_3 x_3 + \varepsilon$$

where

$$x_1 = 1 \quad \text{if treatment 2,} \qquad x_1 = 0 \quad \text{otherwise}$$

$$x_2 = 1 \quad \text{if treatment 3,} \qquad x_2 = 0 \quad \text{otherwise}$$

$$x_3 = 1 \quad \text{if treatment 4,} \qquad x_3 = 0 \quad \text{otherwise}$$

To interpret the βs in this equation, it is convenient to construct a table of the expected values. Because ε has expectation zero, the general expression for the expected value of y is

$$E(y) = \beta_0 + \beta_1 x_1 + \beta_2 x_2 + \beta_3 x_3$$

The expected value for observations on treatment 1 is found by substituting $x_1 = 0$, $x_2 = 0$, and $x_3 = 0$; after this substitution, we find $E(y) = \beta_0$. The expected value for observations on treatment 2 is found by substituting $x_1 = 1$, $x_2 = 0$, and $x_3 = 0$ into the $E(y)$ formula; this substitution yields $E(y) = \beta_0 + \beta_1$. Substitutions of $x_1 = 0$, $x_2 = 1$, $x_3 = 0$ and $x_1 = 0$, $x_2 = 0$, $x_3 = 1$ yield expected values for treatments 3 and 4, respectively. These expected values are summarized in Table 12.2.

TABLE 12.2
Expected values for an experiment with four treatments

Treatment			
1	**2**	**3**	**4**
$E(y) = \beta_0$	$E(y) = \beta_0 + \beta_1$	$E(y) = \beta_0 + \beta_2$	$E(y) = \beta_0 + \beta_3$

If we identify the mean of treatment 1 as μ_1, the mean of treatment 2 as μ_2, and so on, then from Table 12.2 we have

$$\mu_1 = \beta_0 \qquad \mu_2 = \beta_0 + \beta_1 \qquad \mu_3 = \beta_0 + \beta_2 \qquad \mu_4 = \beta_0 + \beta_3$$

Solving these equations for the βs, we have

$$\beta_0 = \mu_1 \qquad \beta_1 = \mu_2 - \mu_1 \qquad \beta_2 = \mu_3 - \mu_1 \qquad \beta_3 = \mu_4 - \mu_1$$

Any comparison among the treatment means can be phrased in terms of the βs. For example, the comparison $\mu_4 - \mu_3$ can be written as $\beta_3 - \beta_2$, and $\mu_3 - \mu_2$ can be written as $\beta_2 - \beta_1$.

EXAMPLE 12.2

Consider a hypothetical situation for an experiment with four treatments ($t = 4$) in which we know the means for the four treatments. If $\mu_1 = 7$, $\mu_2 = 9$, $\mu_3 = 6$, and $\mu_4 = 15$, determine values for β_0, β_1, β_2, and β_3 in the model

$$y = \beta_0 + \beta_1 x_1 + \beta_2 x_2 + \beta_3 x_3 + \varepsilon$$

where

$$x_1 = 1 \quad \text{if treatment 2} \qquad x_1 = 0 \quad \text{otherwise}$$

$$x_2 = 1 \quad \text{if treatment 3} \qquad x_2 = 0 \quad \text{otherwise}$$

$$x_3 = 1 \quad \text{if treatment 4} \qquad x_3 = 0 \quad \text{otherwise}$$

Solution Based on what we saw in Table 12.2, we know that

$$\beta_0 = \mu_1 \qquad \beta_1 = \mu_2 - \mu_1 \qquad \beta_2 = \mu_3 - \mu_1 \qquad \beta_3 = \mu_4 - \mu_1$$

Using the known values for μ_1, μ_2, μ_3, and μ_4, it follows that

$$\beta_0 = 7 \qquad \beta_1 = 9 - 7 = 2 \qquad \beta_2 = 6 - 7 = -1 \qquad \beta_3 = 15 - 7 = 8$$

EXAMPLE 12.3

Refer to Example 12.2. Express $\mu_3 - \mu_2$ and $\mu_3 - \mu_4$ in terms of the βs. Check your findings by substituting values for the βs.

Solution Using the relationship between the βs and the μs, we can see that

$$\beta_2 - \beta_1 = (\mu_3 - \mu_1) - (\mu_2 - \mu_1) = \mu_3 - \mu_2$$

and

$$\beta_2 - \beta_3 = (\mu_3 - \mu_1) - (\mu_4 - \mu_1) = \mu_3 - \mu_4$$

Substituting computed values for the βs, we have

$$\beta_2 - \beta_1 = -1 - (2) = -3$$

and

$$\beta_2 - \beta_3 = -1 - (8) = -9$$

These computed values are identical to the known differences for $\mu_3 - \mu_2$ and $\mu_3 - \mu_4$, respectively.

EXAMPLE 12.4

Use dummy variables to write the model for an experiment with t treatments. Identify the βs.

Solution We can write the model in the form

$$y = \beta_0 + \beta_1 x_1 + \beta_2 x_2 + \cdots + \beta_{t-1} x_{t-1} + \varepsilon$$

where

$$x_1 = 1 \quad \text{if treatment 2} \qquad x_1 = 0 \quad \text{otherwise}$$

$$x_2 = 1 \quad \text{if treatment 3} \qquad x_2 = 0 \quad \text{otherwise}$$

$$\vdots \qquad\qquad\qquad\qquad \vdots$$

$$x_{t-1} = 1 \quad \text{if treatment } t \qquad x_{t-1} = 0 \quad \text{otherwise}$$

The table of expected values is

	Treatment		
1	**2**	$\cdots$	t
$E(y) = \beta_0$	$E(y) = \beta_0 + \beta_1$	$\cdots$	$E(y) = \beta_0 + \beta_{t-1}$

from which we obtain

$$\beta_0 = \mu_1$$
$$\beta_1 = \mu_2 - \mu_1$$
$$\vdots$$
$$\beta_{t-1} = \mu_t - \mu_1$$

In the procedure just described, we have a response related to the qualitative variable treatments, and for t levels of the treatments we enter $(t - 1)$ βs into our model, using dummy variables. The model shown in this section using dummy variables is equivalent to the model for a completely randomized design in Chapter 8.

12.2 Estimating Multiple Regression Coefficients

The multiple regression model relates a response y to a set of quantitative independent variables. For a random sample of n measurements, we can write the ith observation as

$$y_i = \beta_0 + \beta_1 x_{i1} + \beta_2 x_{i2} + \cdots + \beta_k x_{ik} + \varepsilon_i \qquad (i = 1, 2, \ldots, n; n > k)$$

where $x_{i1}, x_{i2}, \ldots, x_{ik}$ are the settings of the quantitative independent variables corresponding to the observation y_i.

To find least-squares estimates for $\beta_0, \beta_1, \ldots,$ and β_k in a multiple regression model, we follow the same procedure that we did for a linear regression model in Chapter 11. We obtain a random sample of n observations; we find the least-squares prediction equation

$$\hat{y} = \hat{\beta}_0 + \hat{\beta}_1 x_1 + \cdots + \hat{\beta}_k x_k$$

by choosing $\hat{\beta}_0, \hat{\beta}_1, \ldots, \hat{\beta}_k$ to minimize SS(Residual) $= \sum_i (y_i - \hat{y}_i)^2$. However, although it was easy to write down the solutions to $\hat{\beta}_0$ and $\hat{\beta}_1$ for the linear regression model,

$$y = \beta_0 + \beta_1 x + \varepsilon$$

we must find the estimates for $\beta_0, \beta_1, \ldots, \beta_k$ by solving a set of simultaneous equations, called the *normal equations*, shown here.

	y_i	$\hat{\beta}_0$	$x_{i1}\hat{\beta}_1$	$\cdots$	$x_{ik}\hat{\beta}_k$
1	$\sum y_i =$	$n\hat{\beta}_0 +$	$\sum x_{i1}\hat{\beta}_1$	$+\cdots+$	$\sum x_{ik}\hat{\beta}_k$
x_{i1}	$\sum x_{i1}y_i =$	$\sum x_{i1}\hat{\beta}_0 +$	$\sum x_{i1}^2\hat{\beta}_1$	$+\cdots+$	$\sum x_{i1}x_{ik}\hat{\beta}_k$
$\vdots$	$\vdots$	$\vdots$	$\vdots$		$\vdots$
x_{ik}	$\sum x_{ik}y_i =$	$\sum x_{ik}\hat{\beta}_0 +$	$\sum x_{ik}x_{i1}\hat{\beta}_1$	$+\cdots+$	$\sum x_{ik}^2\hat{\beta}_k$

Note the pattern associated with these equations. By labeling the rows and columns as we have done, we can obtain any term in the normal equations by multiplying the row and column elements and summing. For example, the last term in the second equation is found by multiplying the row element (x_{i1}) by the column element $(x_{ik}\hat{\beta}_k)$ and summing; the resulting term is $\sum x_{i1}x_{ik}\hat{\beta}_k$. Because all terms in the normal equations can be formed in this way, it is fairly simple to write down the equations to be solved to obtain the least-squares estimates $\hat{\beta}_0, \hat{\beta}_1, \ldots, \hat{\beta}_k$. The solution to these equations is not necessarily trivial; that's why we'll enlist the help of various statistical software packages for their solution.

EXAMPLE 12.5

In Exercise 11.26, we presented data for the weight loss of a compound for the different amounts of time the compound was exposed to the air. Additional information was also available on the humidity of the environment during exposure. The complete data are presented in Table 12.3.

TABLE 12.3

Weight loss, exposure time, and relative humidity data

Weight Loss, y (pounds)	Exposure Time, x_1 (hours)	Relative Humidity, x_2
4.3	4	.20
5.5	5	.20
6.8	6	.20
8.0	7	.20
4.0	4	.30
5.2	5	.30
6.6	6	.30
7.5	7	.30
2.0	4	.40
4.0	5	.40
5.7	6	.40
6.5	7	.40

a. Set up the normal equations for this regression problem if the assumed model is

$$y = \beta_0 + \beta_1 x_1 + \beta_2 x_2 + \varepsilon$$

where x_1 is exposure time and x_2 is relative humidity.

b. Use the computer output shown here to determine the least-squares estimates of β_0, β_1, and β_2. Predict weight loss for 6.5 hours of exposure and a relative humidity of .35.

```
OUTPUT FOR EXAMPLE 12.5

OBS    WT_LOSS    TIME    HUMID

 1      4.3       4.0     0.20
 2      5.5       5.0     0.20
 3      6.8       6.0     0.20
 4      8.0       7.0     0.20
```

```
 5     4.0      4.0     0.30
 6     5.2      5.0     0.30
 7     6.6      6.0     0.30
 8     7.5      7.0     0.30
 9     2.0      4.0     0.40
10     4.0      5.0     0.40
11     5.7      6.0     0.40
12     6.5      7.0     0.40
13      .       6.5     0.35
```

Dependent Variable: WT_LOSS   WEIGHT LOSS

Analysis of Variance

Source	DF	Sum of Squares	Mean Square	F Value	Prob>F
Model	2	31.12417	15.56208	104.133	0.0001
Error	9	1.34500	0.14944		
C Total	11	32.46917			

Root MSE	0.38658	R-square	0.9586
Dep Mean	5.50833	Adj R-sq	0.9494
C.V.	7.01810		

Parameter Estimates

Variable	DF	Parameter Estimate	Standard Error	T for H0: Parameter=0	Prob > |T|
INTERCEP	1	0.666667	0.69423219	0.960	0.3620
TIME	1	1.316667	0.09981464	13.191	0.0001
HUMID	1	-8.000000	1.36676829	-5.853	0.0002

OBS	WT_LOSS	PRED	RESID	L95MEAN	U95MEAN
1	4.3	4.33333	-0.03333	3.80985	4.85682
2	5.5	5.65000	-0.15000	5.23519	6.06481
3	6.8	6.96667	-0.16667	6.55185	7.38148
4	8.0	8.28333	-0.28333	7.75985	8.80682
5	4.0	3.53333	0.46667	3.11091	3.95576
6	5.2	4.85000	0.35000	4.57346	5.12654
7	6.6	6.16667	0.43333	5.89012	6.44321
8	7.5	7.48333	0.01667	7.06091	7.90576
9	2.0	2.73333	-0.73333	2.20985	3.25682
10	4.0	4.05000	-0.05000	3.63519	4.46481
11	5.7	5.36667	0.33333	4.95185	5.78148
12	6.5	6.68333	-0.18333	6.15985	7.20682
13	.	6.42500	.	6.05269	6.79731

```
Sum of Residuals                      0
Sum of Squared Residuals        1.3450
Predicted Resid SS (Press)      2.6123
```

Solution

a. The three normal equations for this model are shown here.

	y_i	$\hat{\beta}_0$	$x_{i1}\hat{\beta}_1$	$x_{i2}\hat{\beta}_2$
1	$\sum y_i =$	$n\hat{\beta}_0 +$	$\sum x_{i1}\hat{\beta}_1 +$	$\sum x_{i2}\hat{\beta}_2$
x_{i1}	$\sum x_{i1}y_i = \sum x_{i1}\hat{\beta}_0 +$		$\sum x_{i1}^2\hat{\beta}_1 +$	$\sum x_{i1}x_{i2}\hat{\beta}_2$
x_{i2}	$\sum x_{i2}y_i = \sum x_{i2}\hat{\beta}_0 +$		$\sum x_{i2}x_{i1}\hat{\beta}_1 +$	$\sum x_{i2}^2\hat{\beta}_2$

For these data, we have

$$\sum y_i = 66.10 \qquad \sum x_{i1} = 66 \qquad \sum x_{i2} = 3.60$$

$$\sum x_{i1}y_i = 383.3 \qquad \sum x_{i2}y_i = 19.19 \qquad \sum x_{i1}x_{i2} = 19.8$$

$$\sum x_{i1}^2 = 378 \qquad \sum x_{i2}^2 = 1.16$$

Substituting these values into the normal equation yields the result shown here:

$$66.1 = 12\hat{\beta}_0 + 66\hat{\beta}_1 + 3.6\hat{\beta}_2$$

$$383.3 = 66\hat{\beta}_0 + 378\hat{\beta}_1 + 19.8\hat{\beta}_2$$

$$19.19 = 3.6\hat{\beta}_0 + 19.8\hat{\beta}_1 + 1.16\hat{\beta}_2$$

b. The normal equations in part (a) can be solved to determine $\hat{\beta}_0$, $\hat{\beta}_1$, and $\hat{\beta}_2$. The solution will agree with that shown here in the output. The least-squares prediction equation is

$$\hat{y} = 0.667 + 1.317x_1 - 8.000x_2$$

where x_1 is exposure time and x_2 is relative humidity. Substituting $x_1 = 6.5$ and $x_2 = .35$, we have

$$\hat{y} = 0.667 + 1.317(6.5) - 8.000(.35) = 6.428$$

This value agrees with the predicted value shown as observation 13 in the output, except for rounding errors.

There are many software programs that provide the calculations to obtain least-squares estimates for parameters in the general linear model (and hence for multiple regression). The output of such programs typically has a list of variable names, together with the estimated partial slopes, labeled COEFFICIENTS (or ESTIMATES or PARAMETERS). The intercept term $\hat{\beta}_0$ is usually called INTERCEPT (or CONSTANT); sometimes it is shown along with the slopes but with no variable name.

EXAMPLE 12.6

The data for three variables (shown here) are analyzed with the Excel spreadsheet program. Identify the estimates of the partial slopes and the intercept.

y:	25	34	28	40	36	42	44	53	49
x_1:	−10	−10	−10	0	0	0	10	10	10
x_2:	−5	0	5	−5	0	5	−5	0	5

	Coefficients	Standard Error	t Stat	P-value
Intercept	39.0	1.256	31.055	7.4E-08
X1	0.983	0.154	6.393	0.0007
X2	0.333	0.308	1.084	0.3202

Solution   The intercept value 39.0 is labeled as such. The estimated partial slopes .983 and .333 are associated with x_1 and x_2, respectively. Most programs label the coefficients similarly, in a column.

The coefficient of an independent variable x_j in a multiple regression equation does not, in general, equal the coefficient that applies to that variable in a simple linear regression. In multiple regression, the coefficient refers to the effect of changing that x_j variable while other independent variables stay constant. In simple linear regression, all other potential independent variables are ignored. If other independent variables are correlated with x_j (and therefore don't tend to stay constant while x_j changes), simple linear regression with x_j as the only independent variable captures not only the direct effect of changing x_j but also the indirect effect of the associated changes in other xs. In multiple regression, by holding the other xs constant, we eliminate that indirect effect.

EXAMPLE 12.7

Compare the coefficients of x_1 in the multiple regression model and in the simple (one-predictor) regression model shown in the following StataQuest output. Explain why the two coefficients differ.

```
. regress y  x1 x2

------------------------------------------------------------------------------
       y |   Coef.    Std. Err.      t      P>|t|      [95% Conf. Interval]
---------+--------------------------------------------------------------------
      x1 |      1     1.870829     0.535    0.646      -7.049526     9.049526
      x2 |      3     4.1833       0.717    0.548     -14.99929     20.99929
   _cons |     10     1.183216     8.452    0.014       4.909033    15.09097
------------------------------------------------------------------------------

. regress y x1

------------------------------------------------------------------------------
       y |   Coef.    Std. Err.      t      P>|t|      [95% Conf. Interval]
---------+--------------------------------------------------------------------
      x1 |    2.2     .7659417     2.872    0.064       -.2375683     4.637568
   _cons |     10     1.083205     9.232    0.003       6.552758    13.44724
------------------------------------------------------------------------------

. correlate  y x1 x2

             |        y        x1        x2
---------+---------------------------------
        y|   1.0000
       x1|   0.8563    1.0000
       x2|   0.8704    0.8944    1.0000
```

Solution   In the multiple regression model, the coefficient is shown as 1, but in the simple regression model, it's 2.2. The difference occurs because the two xs are correlated (correlation .8944 in the output). In the multiple regression model, we're thinking of varying x_1 while holding x_2 constant; in the simple regression model, we're thinking of varying x_1 and letting x_2 go wherever it goes.

residual standard deviation

In addition to estimating the intercept and partial slopes, it is important to estimate the **residual standard deviation** s_ε, sometimes called the *standard error of estimate*. The residuals are defined as before, as the difference between the observed value and the predicted value of y:

$$y_i - \hat{y}_i = y_i - (\hat{\beta}_0 + \hat{\beta}_1 x_{i1} + \hat{\beta}_2 x_{i2} + \cdots + \hat{\beta}_k x_{ik})$$

The sum of squared residuals, SS(Residual), also called SS(Error), is defined exactly as it sounds. Square the prediction errors and sum the squares:

$$\text{SS(Residual)} = \Sigma\,(y_i - \hat{y}_i)^2$$
$$= \Sigma\,[y_i - (\hat{\beta}_0 + \hat{\beta}_1 x_{i1} + \hat{\beta}_2 x_{i2} + \cdots + \hat{\beta}_i x_{ik})]^2$$

The df for this sum of squares is $n - (k + 1)$. One df is subtracted for the intercept and 1 df is subtracted for each of the k partial slopes. The mean square residual, MS(Residual), also called MS(Error), is the residual sum of squares divided by $n - (k + 1)$. Finally, the residual standard deviation s_ε is the square root of MS(Residual).

The residual standard deviation may be called *std dev, standard error of estimate,* or *root MSE*. If the output is not clear, you can take the square root of MS(Residual) by hand. As always, interpret the standard deviation by the Empirical Rule. About 95% of the prediction errors will be within ± 2 standard deviations of the mean (and the mean error is automatically zero):

$$s_\varepsilon = \sqrt{\text{MS(Residual)}} = \sqrt{\frac{\text{SS(Residual)}}{n - (k + 1)}}$$

EXAMPLE 12.8

Identify SS(Residual) and s_ε in the output shown here for the data of Example 12.6.

	A	B	C	D	E	F
1						
2	The regression equation is y=39.0 + .983 x1 + .333 x2					
3						
4	Predictor	Coef	Stdev	t-ratio	p	
5						
6	Constant	39.000	1.256	31.05	0.000	
7	x1	0.9833	0.1538	6.39	0.001	
8	x2	0.3333	0.3076	1.08	0.320	
9						
10	s=3.768	R-sq=87.5%	R-sq(adj)=83.3%			

	A	B	C	D	E	F
11						
12						
13	Analysis of Variance					
14						
15						
16	SOURCE	DF	SS	MS	F	P
17	Regression	2	596.83	298.420	21.02	0.002
18	Error	6	85.17	14.190		
19	Total	8	682.00			

Solution   In the section of the output labeled Analysis of Variance, SS(Residual) is shown as SS(Error) = 85.17, with 6 df. MS(Error) is 14.19. The residual standard deviation is indicated by $s = 3.768$. Note that $3.768 = \sqrt{14.19}$ to within rounding error.

The residual standard deviation is crucial in determining the probable error of a prediction using the regression equation. The precise standard error to be used in forecasting an individual y value is stated in Section 12.4. A rough approximation, ignoring extrapolation and df effects, is that the probable error is $\pm 2s_\varepsilon$. This approximation can be used as a rough indicator of the forecasting quality of a regression model.

EXAMPLE 12.9

The admissions office of a business school develops a regression model that uses aptitude test scores and class rank to predict the grade average (4.00 = straight A; 2.00 = C average, the minimum graduation average; 0.00 = straight F). The residual standard deviation is $s_\varepsilon = .46$. Does this value suggest a highly accurate prediction?

Solution   A measure of the probable error of prediction is $2s_\varepsilon = .92$. For example, if a predicted average is 2.80, then an individual's grade is roughly between $2.80 - .92 = 1.88$ (not good enough to graduate) and $2.80 + .92 = 3.72$ (good enough to graduate magna cum laude)! This is *not* an accurate forecast.

EXERCISES   **Applications**

Med.   **12.1** A pharmaceutical firm would like to obtain information on the relationship between the dose level and potency of a drug product. To do this, each of 15 test tubes is inoculated with a virus culture and incubated for 5 days at 30°C. Three test tubes are randomly assigned to each of the five different dose levels to be investigated (2, 4, 8, 16, and 32 mg). Each tube is injected with only one dose level and the response of interest (a measure of the protective strength of the product against the virus culture) is obtained. The data are given here.

Dose Level	Response
2	5, 7, 3
4	10, 12, 14
8	15, 17, 18
16	20, 21, 19
32	23, 24, 29

a. Plot the data.
b. Fit a linear regression model to these data.
c. What other regression model might be appropriate?
d. SAS computer output is shown for both a linear and quadratic regression equation. Which regression equation appears to fit the data better? Why?

```
OUTPUT FOR EXERCISE 12.1

OBS   DOSE   RESPONSE
  1     2        5
  2     2        7
  3     2        3
  4     4       10
  5     4       12
  6     4       14
  7     8       15
  8     8       17
  9     8       18
 10    16       20
 11    16       21
 12    16       19
 13    32       23
 14    32       24
 15    32       29
```

Dependent Variable: RESPONSE   PROTECTIVE STRENGTH

Analysis of Variance

Source	DF	Sum of Squares	Mean Square	F Value	Prob>F
Model	1	590.91613	590.91613	44.280	0.0001
Error	13	173.48387	13.34491		
C Total	14	764.40000			

Root MSE	3.65307	R-square	0.7730
Dep Mean	15.80000	Adj R-sq	0.7556
C.V.	23.12069		

Parameter Estimates

Variable	DF	Parameter Estimate	Standard Error	T for H0: Parameter=0	Prob > |T|
INTERCEP	1	8.666667	1.42786770	6.070	0.0001
DOSE	1	0.575269	0.08645016	6.654	0.0001

OBS	DOSE	RESPONSE	PRED	RESID
1	2	5	9.8172	-4.81720
2	2	7	9.8172	-2.81720
3	2	3	9.8172	-6.81720
4	4	10	10.9677	-0.96774
5	4	12	10.9677	1.03226
6	4	14	10.9677	3.03226

```
   7        8        15       13.2688    1.73118
   8        8        17       13.2688    3.73118
   9        8        18       13.2688    4.73118
  10       16        20       17.8710    2.12903
  11       16        21       17.8710    3.12903
  12       16        19       17.8710    1.12903
  13       32        23       27.0753   -4.07527
  14       32        24       27.0753   -3.07527
  15       32        29       27.0753    1.92473

Sum of Residuals                              0
Sum of Squared Residuals              173.4839
Predicted Resid SS (Press)            238.0013
```

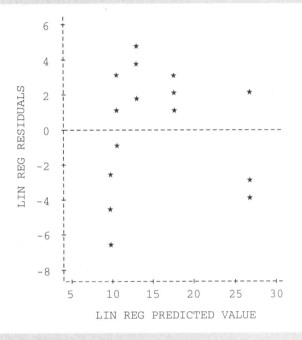

```
QUADRATIC REGRESSION ANALYSIS

Dependent Variable: RESPONSE   PROTECTIVE STRENGTH


Analysis of Variance

                        Sum of         Mean
Source        DF        Squares        Square      F Value    Prob>F

Model          2      673.82062      336.91031      44.634    0.0001
Error         12       90.57938        7.54828
C Total       14      764.40000


     Root MSE      2.74741    R-square     0.8815
     Dep Mean     15.80000    Adj R-sq     0.8618
     C.V.         17.38869
```

```
Parameter Estimates

                  Parameter      Standard      T for H0:
   Variable DF    Estimate         Error     Parameter=0    Prob > |T|

   INTERCEP   1   4.483660     1.65720388        2.706        0.0191
   DOSE       1   1.506325     0.28836373        5.224        0.0002
   DOSE2      1  -0.026987     0.00814314       -3.314        0.0062

   OBS    DOSE   RESPONSE    PREDICTED     RESIDUAL

     1      2        5         7.3884      -2.38836
     2      2        7         7.3884      -0.38836
     3      2        3         7.3884      -4.38836
     4      4       10        10.0772      -0.07717
     5      4       12        10.0772       1.92283
     6      4       14        10.0772       3.92283
     7      8       15        14.8071       0.19292
     8      8       17        14.8071       2.19292
     9      8       18        14.8071       3.19292
    10     16       20        21.6762      -1.67615
    11     16       21        21.6762      -0.67615
    12     16       19        21.6762      -2.67615
    13     32       23        25.0512      -2.05123
    14     32       24        25.0512      -1.05123
    15     32       29        25.0512       3.94877
```

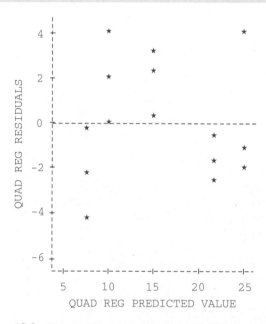

12.2   Refer to the data in Exercise 12.1. Often a logarithmic transformation can be used on the dose levels to linearize the response with respect to the independent variable.

a. Obtain the logarithms of the five dose levels.

b. Let x_i denote the log dose and fit the model

$$y = \beta_0 + \beta_1 x_1 + \varepsilon$$

A residual plot is shown here in the output.

c. Compare your results in part (b) to those for Exercise 12.1. Does the logarithmic transformation provide a better linear fit than that in Exercise 12.1?

```
          REGRESSION ANALYSIS USING NATURAL LOGARITHM OF DOSE

Dependent Variable: RESPONSE   PROTECTIVE STRENGTH

Analysis of Variance

                        Sum of          Mean
Source          DF     Squares         Square      F Value      Prob>F

Model            1    710.53333      710.53333     171.478      0.0001
Error           13     53.86667        4.14359
C Total         14    764.40000


        Root MSE        2.03558     R-square     0.9295
        Dep Mean       15.80000     Adj R-sq     0.9241
        C.V.           12.88342

Parameter Estimates

                    Parameter       Standard      T for H0:
Variable DF         Estimate          Error     Parameter=0     Prob > |T|

INTERCEP  1         1.200000       1.23260547         0.974         0.3480
LOG_DOSE  1         7.021116       0.53616972        13.095         0.0001
```

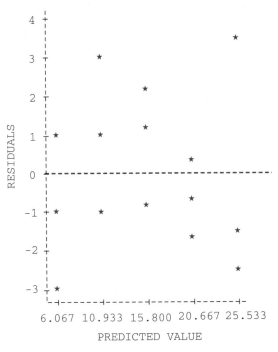

Engin. 12.3 A manufacturer tested the abrasive effect of a wear tester for experimental fabrics on a particular fabric while run at six different machine speeds. Forty-eight identical 5-inch-square pieces of fabric were cut, with eight squares randomly assigned to each of the six machine speeds,

100, 120, 140, 160, 180, and 200 revolutions per minute (rev/min). The order of assignment of the squares to the machine was random, with each square tested for a 3-minute period at the appropriate machine setting. The amount of wear was measured and recorded for each square. The data appear here.

Machine Speed (in rev/min)	Wear
100	23.0, 23.5, 24.4, 25.2, 25.6, 26.1, 24.8, 25.6
120	26.7, 26.1, 25.8, 26.3, 27.2, 27.9, 28.3, 27.4
140	28.0, 28.4, 27.0, 28.8, 29.8, 29.4, 28.7, 29.3
160	32.7, 32.1, 31.9, 33.0, 33.5, 33.7, 34.0, 32.5
180	43.1, 41.7, 42.4, 42.1, 43.5, 43.8, 44.2, 43.6
200	54.2, 43.7, 53.1, 53.8, 55.6, 55.9, 54.7, 54.5

a. Generate a graph of the data. (The variability within a speed is about the same for all speeds, so you can save time while still maintaining the trend by plotting the sample mean for each speed.)
b. What type of regression model appears appropriate?
c. Output for linear, quadratic, and cubic regression models is shown on the following pages. Which regression equation gives a better fit? Why?
d. Is there anything peculiar about the data? What might have happened?

```
            LINEAR REGRESSION ANALYSIS FOR WEAR TESTER DATA

Dependent Variable: FABRIC WEAR


Analysis of Variance

                      Sum of          Mean
Source       DF       Squares         Square      F Value     Prob>F

Model         1     4326.79207     4326.79207     291.474     0.0001
Error        46      682.84710       14.84450
C Total      47     5009.63917


    Root MSE        3.85286     R-square       0.8637
    Dep Mean       34.92917     Adj R-sq       0.8607
    C.V.           11.03048

Parameter Estimates

                   Parameter      Standard      T for H0:
Variable   DF      Estimate         Error     Parameter=0    Prob > |T|

INTERCEP    1      -6.765476     2.50470943      -2.701        0.0096
SPEED       1       0.277964     0.01628129      17.073        0.0001

                Variable
Variable   DF     Label
```

```
INTERCEP   1  Intercept
SPEED      1  MACHINE SPEED


           OBS    SPEED    WEAR      PRED1       RESID1

            1      100     23.0     21.0310      1.96905
            2      100     23.5     21.0310      2.46905
            3      100     24.4     21.0310      3.36905
            4      100     25.2     21.0310      4.16905
            5      100     25.6     21.0310      4.56905
            6      100     26.1     21.0310      5.06905
            7      100     24.8     21.0310      3.76905
            8      100     25.6     21.0310      4.56905
            9      120     26.7     26.5902      0.10976
           10      120     26.1     26.5902     -0.49024
           11      120     25.8     26.5902     -0.79024
           12      120     26.3     26.5902     -0.29024
           13      120     27.2     26.5902      0.60976
           14      120     27.9     26.5902      1.30976
           15      120     28.3     26.5902      1.70976
           16      120     27.4     26.5902      0.80976
           17      140     28.0     32.1495     -4.14952
           18      140     28.4     32.1495     -3.74952
           19      140     27.0     32.1495     -5.14952
           20      140     28.8     32.1495     -3.34952
           21      140     29.8     32.1495     -2.34952
           22      140     29.4     32.1495     -2.74952
           23      140     28.7     32.1495     -3.44952
           24      140     29.3     32.1495     -2.84952
           25      160     32.7     37.7088     -5.00881
           26      160     32.1     37.7088     -5.60881
           27      160     31.9     37.7088     -5.80881
           28      160     33.0     37.7088     -4.70881
           29      160     33.5     37.7088     -4.20881
           30      160     33.7     37.7088     -4.00881
           31      160     34.0     37.7088     -3.70881
           32      160     32.5     37.7088     -5.20881
           33      180     43.1     43.2681     -0.16810
           34      180     41.7     43.2681     -1.56810
           35      180     42.4     43.2681     -0.86810
           36      180     42.1     43.2681     -1.16810
           37      180     43.5     43.2681      0.23190
           38      180     43.8     43.2681      0.53190
           39      180     44.2     43.2681      0.93190
           40      180     43.6     43.2681      0.33190
           41      200     54.2     48.8274      5.37262
           42      200     43.7     48.8274     -5.12738
           43      200     53.1     48.8274      4.27262
           44      200     53.8     48.8274      4.97262
           45      200     55.6     48.8274      6.77262
           46      200     55.9     48.8274      7.07262
           47      200     54.7     48.8274      5.87262
           48      200     54.5     48.8274      5.67262
```

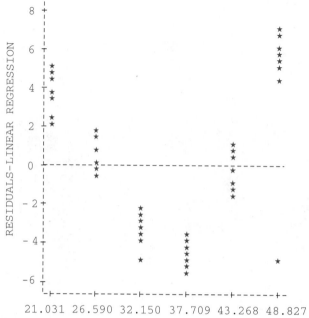

NOTE: 6 obs hidden

QUADRATIC REGRESSION ANALYSIS FOR WEAR TESTER DATA

Dependent Variable: FABRIC WEAR

Analysis of Variance

Source	DF	Sum of Squares	Mean Square	F Value	Prob>F
Model	2	4839.89302	2419.94651	641.532	0.0001
Error	45	169.74614	3.77214		
C Total	47	5009.63917			

Root MSE	1.94220	R-square	0.9661	
Dep Mean	34.92917	Adj R-sq	0.9646	
C.V.	5.56039			

Parameter Estimates

Variable	DF	Parameter Estimate	Standard Error	T for H0: Parameter=0	Prob > |T|
INTERCEP	1	63.139286	6.12529888	10.308	0.0001
SPEED	1	-0.705071	0.08468583	-8.326	0.0001
SPEED2	1	0.003277	0.00028096	11.663	0.0001

Variable	DF	Variable Label
INTERCEP	1	Intercept
SPEED	1	MACHINE SPEED
SPEED2	1	SPEED SQUARED

OBS	SPEED	WEAR	PRED2	RESID2
1	100	23.0	25.4000	-2.40000
2	100	23.5	25.4000	-1.90000
3	100	24.4	25.4000	-1.00000
4	100	25.2	25.4000	-0.20000
5	100	25.6	25.4000	0.20000
6	100	26.1	25.4000	0.70000
7	100	24.8	25.4000	-0.60000
8	100	25.6	25.4000	0.20000
9	120	26.7	25.7164	0.98357
10	120	26.1	25.7164	0.38357
11	120	25.8	25.7164	0.08357
12	120	26.3	25.7164	0.58357
13	120	27.2	25.7164	1.48357
14	120	27.9	25.7164	2.18357
15	120	28.3	25.7164	2.58357
16	120	27.4	25.7164	1.68357
17	140	28.0	28.6543	-0.65429
18	140	28.4	28.6543	-0.25429
19	140	27.0	28.6543	-1.65429
20	140	28.8	28.6543	0.14571
21	140	29.8	28.6543	1.14571
22	140	29.4	28.6543	0.74571
23	140	28.7	28.6543	0.04571
24	140	29.3	28.6543	0.64571
25	160	32.7	34.2136	-1.51357
26	160	32.1	34.2136	-2.11357
27	160	31.9	34.2136	-2.31357
28	160	33.0	34.2136	-1.21357
29	160	33.5	34.2136	-0.71357
30	160	33.7	34.2136	-0.51357
31	160	34.0	34.2136	-0.21357
32	160	32.5	34.2136	-1.71357
33	180	43.1	42.3943	0.70571
34	180	41.7	42.3943	-0.69429
35	180	42.4	42.3943	0.00571
36	180	42.1	42.3943	-0.29429
37	180	43.5	42.3943	1.10571
38	180	43.8	42.3943	1.40571
39	180	44.2	42.3943	1.80571
40	180	43.6	42.3943	1.20571
41	200	54.2	53.1964	1.00357
42	200	43.7	53.1964	-9.49643
43	200	53.1	53.1964	-0.09643
44	200	53.8	53.1964	0.60357

45	200	55.6	53.1964	2.40357
46	200	55.9	53.1964	2.70357
47	200	54.7	53.1964	1.50357
48	200	54.5	53.1964	1.30357

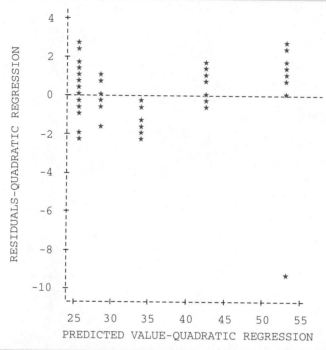

CUBIC REGRESSION ANALYSIS FOR WEAR TESTER DATA

Dependent Variable: FABRIC WEAR

Analysis of Variance

Source	DF	Sum of Squares	Mean Square	F Value	Prob>F
Model	3	4846.78202	1615.59401	436.494	0.0001
Error	44	162.85714	3.70130		
C Total	47	5009.63917			

Root MSE	1.92388	R-square	0.9675	
Dep Mean	34.92917	Adj R-sq	0.9653	
C.V.	5.50794			

Parameter Estimates

Variable	DF	Parameter Estimate	Standard Error	T for H0: Parameter=0	Prob > |T|
INTERCEP	1	18.872619	33.00952220	0.572	0.5704
SPEED	1	0.238477	0.69668199	0.342	0.7338
SPEED2	1	-0.003208	0.00476113	-0.674	0.5040
SPEED3	1	0.000014410	0.00001056	1.364	0.1794

Variable	DF	Variable Label
INTERCEP	1	Intercept
SPEED	1	MACHINE SPEED
SPEED2	1	SPEED SQUARED
SPEED3	1	SPEED CUBED

OBS	SPEED	WEAR	PRED3	RESID3
1	100	23.0	25.0542	-2.05417
2	100	23.5	25.0542	-1.55417
3	100	24.4	25.0542	-0.65417
4	100	25.2	25.0542	0.14583
5	100	25.6	25.0542	0.54583
6	100	26.1	25.0542	1.04583
7	100	24.8	25.0542	-0.25417
8	100	25.6	25.0542	0.54583
9	120	26.7	26.2006	0.49940
10	120	26.1	26.2006	-0.10060
11	120	25.8	26.2006	-0.40060
12	120	26.3	26.2006	0.09940
13	120	27.2	26.2006	0.99940
14	120	27.9	26.2006	1.69940
15	120	28.3	26.2006	2.09940
16	120	27.4	26.2006	1.19940
17	140	28.0	28.9310	-0.93095
18	140	28.4	28.9310	-0.53095
19	140	27.0	28.9310	-1.93095
20	140	28.8	28.9310	-0.13095
21	140	29.8	28.9310	0.86905
22	140	29.4	28.9310	0.46905
23	140	28.7	28.9310	-0.23095
24	140	29.3	28.9310	0.36905
25	160	32.7	33.9369	-1.23690
26	160	32.1	33.9369	-1.83690
27	160	31.9	33.9369	-2.03690
28	160	33.0	33.9369	-0.93690
29	160	33.5	33.9369	-0.43690
30	160	33.7	33.9369	-0.23690
31	160	34.0	33.9369	0.06310
32	160	32.5	33.9369	-1.43690
33	180	43.1	41.9101	1.18988
34	180	41.7	41.9101	-0.21012
35	180	42.4	41.9101	0.48988
36	180	42.1	41.9101	0.18988
37	180	43.5	41.9101	1.58988
38	180	43.8	41.9101	1.88988
39	180	44.2	41.9101	2.28988
40	180	43.6	41.9101	1.68988
41	200	54.2	53.5423	0.65774
42	200	43.7	53.5423	-9.84226
43	200	53.1	53.5423	-0.44226
44	200	53.8	53.5423	0.25774

45	200	55.6	53.5423	2.05774
46	200	55.9	53.5423	2.35774
47	200	54.7	53.5423	1.15774
48	200	54.5	53.5423	0.95774

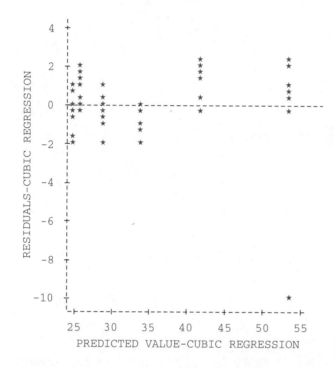

12.4 Refer to the data in Exercise 12.3. Suppose that another variable was controlled and that the first four squares at each speed were treated with a .2 concentration of protective coating, whereas the second four squares were treated with a .4 concentration of the same coating. x_1 denotes the machine speed and x_2 denotes the concentration of the protective coating. Fit these models using available statistical software. Which model seems to provide a better fit to the data? Why?

$$y = \beta_0 + \beta_1 x_1 + \beta_2 x_1^2 + \beta_3 x_2 + \varepsilon$$
$$y = \beta_0 + \beta_1 x_1 + \beta_2 x_1^2 + \beta_3 x_2 + \beta_4 x_1 x_2 + \beta_5 x_1^2 x_2 + \varepsilon$$

12.3 Inferences in Multiple Regression

We make inferences about any of the parameters in the general linear model (and hence in multiple regression) as we did for β_0 and β_1 in the linear regression model, $y = \beta_0 + \beta_1 x + \varepsilon$.

coefficient of determination

Before we do this, however, we must introduce the *coefficient of determination*, R^2. The **coefficient of determination,** R^2, is defined and interpreted very much like the r^2 value in Chapter 11. (The customary notation is R^2 for multiple regression and r^2 for simple linear regression.) As in Chapter 11, we define the coefficient of determination as the proportional reduction in the squared error of y, which we obtain by knowing the values of $x_1, \ldots, x_k$. For example, if we have the multiple regression model with three x values, and $R^2_{y \cdot x_1 x_2 x_3} = .736$, then we can account for 73.6% of the variability of the y values by variability in x_1, x_2, and x_3. Formally,

$$R^2_{y \cdot x_1 \cdots x_k} = \frac{\text{SS(Total)} - \text{SS(Residual)}}{\text{SS(Total)}}$$

where

$$\text{SS(Total)} = \Sigma \, (y_i - \bar{y})^2$$

EXAMPLE 12.10

Locate the value of $R^2_{y \cdot x_1 x_2}$ in the computer output of Example 12.8.

Solution We want R-sq = 87.5%, not the one that is adjusted (adj). Alternatively, SS(Total) = 682.00 and SS(Residual) = 85.17 are shown in the output, and we can compute $R^2_{y \cdot x_1 x_2} = (682.00 - 85.17)/682.00 = .875$.

There is no general relation between the multiple R^2 from a multiple regression equation and the individual coefficients of determination $r^2_{yx_1}, r^2_{yx_2}, \ldots, r^2_{yx_k}$ other than the multiple R^2 must be at least as big as any of the individual r^2 values. If all the independent variables are themselves perfectly uncorrelated with one another, then multiple R^2 is just the sum of the individual r^2 values. Equivalently, if all the xs are uncorrelated with one another, SS(Regression) for the all-predictors model is equal to the sum of SS(Regression) values for simple regressions using one x at a time. If the xs are correlated, it is much more difficult to break apart the overall predictive value of $x_1, x_2, \ldots, x_k$ as measured by $R^2_{y \cdot x_1 \cdots x}$ into separate pieces that can be attributable to x_1 alone, to x_2 alone, $\ldots$, and to x_k alone.

collinearity   When the independent variables are themselves correlated, **collinearity** (sometimes called *multicollinearity*) is present. In multiple regression, we are trying to separate out the predictive value of several predictors. When the predictors are highly correlated, this task is very difficult. For example, suppose that we try to explain variation in regional housing sales over time, using gross domestic product (GDP) and national disposable income (DI) as two of the predictors. DI has been almost exactly a fraction of GDP, so the correlation of these two predictors will be extremely high. Now, is variation in housing sales attributable more to variation in GDP or to variation in DI? Good luck taking those two apart! It is very likely that either predictor alone will explain variation in housing sales almost as well as both together.

Collinearity is usually present to some degree in a multiple regression study. It is a small problem for slightly correlated xs but a more severe one for highly correlated xs. Thus, if collinearity occurs in a regression study—and it usually does to some degree—it is not easy to break apart the overall $R^2_{y \cdot x_1 x_2 \cdots x_k}$ into separate components associated with each x variable. The correlated xs often account for overlapping pieces of the variability in y, so that often, but not inevitably,

$$R^2_{y \cdot x_1 x_2 \cdots x_k} < r^2_{yx1} + r^2_{yx2} + \cdots + r^2_{yxk}$$

sequential sums   Many statistical computer programs will report **sequential sums of squares.**
of squares   These SS are *incremental* contributions to SS(Regression) when the independent variables enter the regression model in the order you specify to the program. Sequential sums of squares depend heavily on the particular order in which the independent variables enter the model. Again, the trouble is collinearity. For example,

if all variables in a regression study are strongly and positively correlated (as often happens in economic data), whichever independent variable happens to be entered first typically accounts for most of the explainable variation in y and the remaining variables add little to the sequential SS. The explanatory power of any x given all the other xs (which is sometimes called the *unique predictive value* of that x) is small. When the data exhibit severe collinearity, separating out the predictive value of the various independent variables is very difficult indeed.

EXAMPLE 12.11

Interpret the sequential sums of squares in the following output for the data of Example 12.6. If x_2 and x_1 are used as predictors (in that order), do we obtain the same sequential sums of squares numbers?

```
MTB > Correlation 'y' 'x1' 'x2'.

                y        x1

x1          0.922
x2          0.156    0.000


MTB > Regress 'y' 2 'x1' 'x2'.

The regression equation is
y = 39.0 + 0.983 x1 + 0.333 x2

Predictor       Coef      Stdev     t-ratio        p
Constant      39.000      1.256       31.05    0.000
x1            0.9833     0.1538        6.39    0.001
x2            0.3333     0.3076        1.08    0.320


s = 3.768     R-sq = 87.5%    R-sq(adj) = 83.3%

Analysis of Variance

SOURCE          DF          SS          MS          F        p
Regression       2      596.83      298.42      21.02    0.002
Error            6       85.17       14.19
Total            8      682.00


SOURCE          DF      SEQ SS
x1               1      580.17
x2               1       16.67
```

Solution The SEQ SS column shows that x_1 by itself accounts for 580.17 of the total variation in y and that adding x_2 after x_1 accounts for another 16.67 of the y variation. This example is a rarity in that the predictors are completely uncorrelated; in this unusual case, the order of adding predictors does not matter.

```
MTB > Regress 'y' 2 'x2' 'x1'.

The regression equation is
y = 39.0 + 0.333 x2 + 0.983 x1

Predictor        Coef      Stdev    t-ratio        p
Constant       39.000      1.256      31.05    0.000
x2             0.3333     0.3076       1.08    0.320
x1             0.9833     0.1538       6.39    0.001

s = 3.768      R-sq = 87.5%    R-sq(adj) = 83.3%

Analysis of Variance

SOURCE         DF         SS         MS        F        p
Regression      2     596.83     298.42    21.02    0.002
Error           6      85.17      14.19
Total           8     682.00

SOURCE         DF     SEQ SS
x2              1      16.67
x1              1     580.17
```

The ideas of Section 12.2 involve point (best guess) estimation of the regression coefficients and the standard deviation s_ε. Because these estimates are based on sample data, they will be in error to some extent, and a manager should allow for that error in interpreting the model. We now present tests about the partial slope parameters in a multiple regression model.

First, we examine a test of an overall null hypothesis about the partial slopes $(\beta_1, \beta_2, \ldots, \beta_k)$ in the multiple regression model. According to this hypothesis, $H_0: \beta_1 = \beta_2 = \cdots = \beta_k = 0$; that is, none of the variables included in the multiple regression has any predictive value at all. This is the nullest of null hypotheses; it says that all those carefully chosen predictors are absolutely useless. The research hypothesis is a very general one—namely, H_a: At least one $\beta_j \neq 0$. This merely says that there is some predictive value somewhere in the set of predictors.

The test statistic is the F statistic of Chapter 11. To state the test, we first define the sum of squares attributable to the regression of y on the variables x_1, $x_2, \ldots, x_k$. We designate this sum of squares as SS(Regression); it is also called SS(Model) or the explained sum of squares. It is the sum of squared differences between predicted values and the mean y value.

DEFINITION 12.1

$$\text{SS(Regression)} = \sum (\hat{y} - \overline{y})^2$$
$$\text{SS(Total)} = \sum (y_i - \overline{y})^2$$
$$= \text{SS(Regression)} + \text{SS(Residual)}$$

Unlike SS(Total) and SS(Residual), we don't interpret SS(Regression) in terms of prediction error. Rather, it measures the extent to which the predictions $\hat{y}_i$ vary as the xs vary. If SS(Regression) = 0, the predicted y values ($\hat{y}$) are all the same. In such a case, information about the xs in the specified model is useless in predicting y. If SS(Regression) is large relative to SS(Residual), the indication is that there is real predictive value in the independent variables $x_1, x_2, \ldots, x_k$. We state the test statistic in terms of mean squares rather than sums of squares. As always, a mean square is a sum of squares divided by the appropriate df.

F Test of H_0:
$\beta_1 = \beta_2 = \cdots = \beta_k = 0$

H_0: $\beta_1 = \beta_2 = \cdots = \beta_k = 0$

H_a: At least one $\beta_j \neq 0$.

T.S.: $F = \dfrac{\text{SS(Regression)}/k}{\text{SS(Residual)}/[n - (k + 1)]} = \dfrac{\text{MS(Regression)}}{\text{MS(Residual)}}$

R.R.: With $df_1 = k$ and $df_2 = n - (k + 1)$, reject H_0 if $F > F_\alpha$.

Check assumptions and draw conclusions.

EXAMPLE 12.12

a. Locate SS(Regression) in the computer output of Example 12.11, reproduced here.
b. Locate the F statistic.
c. Can we safely conclude that the independent variables x_1 and x_2 together have at least some predictive power?

```
MTB > regress c1 on 2 vars c2 c3

The regression equation is
y = 39.0 + 0.983 x1 + 0.333 x2

Predictor      Coef       Stdev     t-ratio        p
Constant     39.000       1.256       31.05    0.000
x1           0.9833      0.1538        6.39    0.001
x2           0.3333      0.3076        1.08    0.320

s = 3.768    R-sq = 87.5%    R-sq(adj) = 83.3%

Analysis of Variance

SOURCE        DF          SS          MS         F        p
Regression     2      596.83      298.42     21.02    0.002
Error          6       85.17       14.19
Total          8      682.00
```

Solution

a. SS(Regression) is shown in the Analysis of Variance section of the output as 596.83.

b. The MS(Regression) and MS(Residual) values are also shown there. MS(Residual) is labeled as MS(Error), a common alternative name.

$$F = \frac{MS(\text{Regression})}{MS(\text{Residual})} = \frac{298.42}{14.19} = 21.02$$

c. For $df_1 = 2$, $df_2 = 6$, and $\alpha = .01$, the tabled F value is 10.92. Therefore, we have strong evidence (p-value well below .01, shown as .002) to reject the null hypothesis and conclude that the xs collectively have at least some predictive value.

This F test may also be stated in terms of R^2. Recall that $R^2_{y \cdot x_1 \cdots x_k}$ measures the reduction in squared error for y attributed to knowledge of all the x predictors. Because the regression of y on the xs accounts for a proportion $R^2_{y \cdot x_1 \cdots x_k}$ of the total squared error in y,

$$SS(\text{Regression}) = R^2_{y \cdot x_1 \cdots x_k} SS(\text{Total})$$

The remaining fraction, $1 - R^2$, is incorporated in the residual squared error:

$$SS(\text{Residual}) = (1 - R^2_{y \cdot x_1 \cdots x_k})SS(\text{Total})$$

The overall F test statistic can be rewritten as

F and R^2
$$F = \frac{MS(\text{Regression})}{MS(\text{Residual})} = \frac{R^2_{y \cdot x_1 \cdots x_k}/k}{(1 - R^2_{y \cdot x_1 \cdots x_k})/[n - (k + 1)]}$$

This statistic is to be compared with tabulated F values for $df_1 = k$ and $df_2 = n - (k + 1)$.

EXAMPLE 12.13

A large city bank studies the relation of average account size in each of its branches to per capita income in the corresponding zip code area, number of business accounts, and number of competitive bank branches. The data are analyzed by Statistix, as shown here.

```
CORRELATIONS (PEARSON)

          ACCTSIZE   BUSIN     COMPET
BUSIN     -0.6934
COMPET     0.8196   -0.6527
INCOME     0.4526    0.1492    0.5571


UNWEIGHTED LEAST SQUARES LINEAR REGRESSION OF ACCTSIZE


PREDICTOR
VARIABLES     COEFFICIENT    STD ERROR    STUDENT'S T      P       VIF

CONSTANT        0.15085      0.73776         0.20       0.8404
BUSIN          -0.00288      8.894E-04      -3.24       0.0048    5.2
COMPET         -0.00759      0.05810        -0.13       0.8975    7.4
INCOME          0.26528      0.10127         2.62       0.0179    4.3
```

```
R-SQUARED              0.7973      RESID. MEAN SQUARE (MSE)    0.03968
ADJUSTED R-SQUARED     0.7615      STANDARD DEVIATION          0.19920

SOURCE        DF      SS          MS         F        P

REGRESSION     3      2.65376     0.88458    22.29    0.0000
RESIDUAL      17      0.67461     0.03968
TOTAL         20      3.32838
```

a. Identify the multiple regression prediction equation.
b. Use the R^2 value shown to test $H_0: \beta_1 = \beta_2 = \beta_3 = 0$. (*Note: n* = 21.)

Solution

a. From the output, the multiple regression forecasting equation is

$$\hat{y} = 0.15085 - 0.00288x_1 - 0.00759x_2 + 0.26528x_3$$

b. The test procedure based on R^2 is

H_0: $\beta_1 = \beta_2 = \beta_3 = 0$

H_a: At least one β_j differs from zero.

T.S.: $F = \dfrac{R^2_{y \cdot x_1 x_2 x_3}/3}{(1 - R^2_{y \cdot x_1 x_2 x_3})/(21 - 4)} = \dfrac{.7973/3}{.2027/17} = 22.29$

R.R.: For $df_1 = 3$ and $df_2 = 17$, the critical .05 value of F is 3.20.

Because the computed F statistic, 22.29, is greater than 3.20, we reject H_0 and conclude that one or more of the x values has some predictive power. This also follows because the p-value, shown as .0000, is (much) less than .05. Note that the F value we compute is the same as that shown in the output.

Rejection of the null hypothesis of this F test is not an overwhelmingly impressive conclusion. This rejection merely indicates that there is good evidence of *some* degree of predictive value *somewhere* among the independent variables. It does not give any direct indication of how strong the relation is, nor any indication of which individual independent variables are useful. The next task, therefore, is to make inferences about the individual partial slopes.

To make these inferences, we need the estimated standard error of each partial slope. As always, the standard error for any estimate based on sample data indicates how accurate that estimate should be. These standard errors are computed and shown by most regression computer programs. They depend on three things: the residual standard deviation, the amount of variation in the predictor variable, and the degree of correlation between that predictor and the others. The expression that we present for the standard error is useful in considering the effect of collinearity (correlated independent variables), but it is *not* a particularly good way to do the computation. Let a computer program do the arithmetic.

DEFINITION 12.2

Estimated standard error of $\hat{\beta}_j$ in a multiple regression:

$$s_{\hat{\beta}_j} = s_\varepsilon \sqrt{\frac{1}{\sum (x_{ij} - \overline{x}_j)^2 (1 - R^2_{x_j \cdot x_1 \cdots x_{j-1} x_{j+1} \cdots x_k})}}$$

where $R^2_{x_j \cdot x_1 \cdots x_{j-1} x_{j+1} \cdots x_k}$ is the R^2 value obtained by letting x_j be the *dependent* variable in a multiple regression, with all other xs independent variables. Note that s_ε is the residual standard deviation for the multiple regression of y on $x_1, x_2, \ldots, x_k$.

effect of collinearity

As in simple regression, the larger the residual standard deviation, the larger the uncertainty in estimating coefficients. Also, the less variability there is in the predictor, the larger is the standard error of the coefficient. The most important use of the formula for estimated standard error is to illustrate the **effect of collinearity.** If the independent variable x_j is highly collinear with one or more other independent variables, $R^2_{x_j \cdot x_1 \cdots x_{j-1} x_{j+1} \cdots x_k}$ is by definition near 1 and $1 - R^2_{x_j \cdot x_1 \cdots x_{j-1} x_{j+1} \cdots x_k}$ is near zero. Division by a near-zero number yields a very large standard error. Thus, one important effect of severe collinearity is that it results in very large standard errors of partial slopes and therefore very inaccurate estimates of those slopes.

variance inflation factor

The term $1/(1 - R^2_{x_j \cdot x_1 \cdots x_{j-1} x_{j+1} \cdots x_k})$ is called the **variance inflation factor** (VIF). It measures how much the variance (square of the standard error) of a coefficient is increased because of collinearity. This factor is printed out by some computer packages and is helpful in assessing how serious the collinearity problem is. If the VIF is 1, there is no collinearity at all. If it is very large, such as 10 or more, collinearity is a serious problem.

A large standard error for any estimated partial slope indicates a large probable error for the estimate. The partial slope $\hat{\beta}_j$ of x_j estimates the effect of increasing x_j by one unit while all other xs remain constant. If x_j is highly collinear with other xs, when x_j increases, the other xs also vary rather than staying constant. Therefore, it is difficult to estimate β_j, and its probable error is large when x_j is severely collinear with other independent variables.

The standard error of each estimated partial slope $\hat{\beta}_j$ is used in a confidence interval and statistical test for $\hat{\beta}_j$. The confidence interval follows the familiar format of estimate plus or minus (table value) (estimated standard error). The table value is the t table with the error df, $n - (k + 1)$.

DEFINITION 12.3

The **confidence interval** for β_j is

$$\hat{\beta}_j - t_{\alpha/2} s_{\hat{\beta}_j} \le \beta_j \le \hat{\beta}_j + t_{\alpha/2} s_{\hat{\beta}_j}$$

where $t_{\alpha/2}$ cuts off area $\alpha/2$ in the tail of a t distribution with df $= n - (k + 1)$, the error df.

EXAMPLE 12.14

Calculate a 95% confidence interval for β_1 in the two-predictor model for the data in Example 12.7. The relevant output follows.

```
. regress y  x1 x2

---------------------------------------------------------------------------
       y |  Coef.   Std. Err.     t     P>|t|      [95% Conf. Interval]
---------+-----------------------------------------------------------------
      x1 |     1    1.870829    0.535   0.646     -7.049526     9.049526
      x2 |     3    4.1833      0.717   0.548    -14.99929     20.99929
   _cons |    10    1.183216    8.452   0.014      4.909033    15.09097
---------------------------------------------------------------------------
```

Solution $\hat{\beta}_1$ is 1.00 and the standard error is shown as 1.870829. The t value that cuts off an area of .025 in a t distribution with df $= n - (k + 1) = 5 - (2 + 1) = 2$ is 4.303. The confidence interval is $1.00 - 4.303 (1.870829) \le \beta_1 \le 1.00 + 4.303(1.870829)$, or $-7.050 \le \beta_1 \le 9.050$. The output shows this interval to more decimal places.

EXAMPLE 12.15

Locate the estimated partial slope for x_2 and its standard error in the output in Example 12.12. Calculate a 90% confidence interval for β_2.

```
MTB > Regress 'y' on 2 vars 'x1' 'x2'

The regression equation is
y = 39.0 + 0.983 x1 + 0.333 x2

Predictor       Coef      Stdev    t-ratio        p
Constant      39.000      1.256      31.05    0.000
x1            0.9833     0.1538       6.39    0.001
x2            0.3333     0.3076       1.08    0.320
```

Solution $\hat{\beta}_2$ is .3333 with standard error (labeled Stdev) .3076. The tabled t value is 1.943 [tail area .05, $9 - (2 + 1) = 6$ df]. The desired interval is $.3333 - 1.943(.3076) \le \beta_2 \le .3333 + 1.943 (.3076)$, or $-.2644 \le \beta_2 \le .9310$.

interpretation of H_0: $\beta_j = 0$ The usual null hypothesis for inference about β_j is H_0: $\beta_j = 0$. This hypothesis does not assert that x_j has no predictive value by itself. It asserts that it has no *additional* predictive value over and above that contributed by the other independent variables; that is, if all other xs had already been used in a regression model and then x_j was added last, the prediction would not improve. The test of H_0: $\beta_j = 0$ measures whether x_j has any additional (e.g., unique) predictive value. The t test of this H_0 is summarized next.

Summary for Testing β_j

H_0: **1.** $\beta_j \le 0$ H_a: **1.** $\beta_j > 0$
2. $\beta_j \ge 0$ **2.** $\beta_j < 0$
3. $\beta_j = 0$ **3.** $\beta_j \ne 0$

T.S.: $t = \hat{\beta}_j / s_{\hat{\beta}_j}$

R.R.: **1.** $t > t_\alpha$
2. $t < -t_\alpha$
3. $|t| > t_{\alpha/2}$

where t_α cuts off a right-tail area α in the t distribution with df $= n - (k + 1)$.

Check assumptions and draw conclusions.

This test statistic is shown by virtually all multiple regression programs.

EXAMPLE 12.16

a. Use the information given in Example 12.14 to test $H_0: \beta_1 = 0$ at $\alpha = .05$. Use a two-sided alternative.
b. Is the conclusion of the test compatible with the confidence interval?

Solution

a. The test statistic for $H_0: \beta = 0$ versus $H_a: \beta_1 \neq 0$ is $t = \hat{\beta}_1 / s_{\hat{\beta}_1} = 1.00/1.871 = .535$. Because the .025 point for the t distribution with $5 - (2 + 1) = 2$ df is 4.303, H_0 must be retained; x_1 has not been shown to have any additional predictive power in the presence of the other independent variable x_2.
b. The 95% confidence interval includes zero, which also indicates that $H_0: \beta_1 = 0$ must be retained at $\alpha = .05$, two-tailed.

EXAMPLE 12.17

Locate the t statistic for testing $H_0: \beta_2 \leq 0$ in the output of Example 12.15. Can $H_a: \beta_2 > 0$ be supported at any of the usual α levels?

Solution The t statistics are shown under the heading t-ratio. For x_2, the t statistic is 1.08. The t table value for 6 df and $\alpha = .10$ is 1.440, so H_0 cannot be rejected even at $\alpha = .10$. Alternatively, the p-value is $.320/2 = .160$, larger than $\alpha = .10$, so again H_0 cannot be rejected.

The multiple regression F and t tests that we discuss in this chapter test different null hypotheses. It sometimes happens that the F test results in the rejection of $H_0: \beta_1 = \beta_2 = \cdots = \beta_k = 0$, whereas no t test of $H_0: \beta_j = 0$ is significant. In such a case, we can conclude that there is predictive value in the equation as a whole, but we cannot identify the specific variables that have predictive value. Remember that each t test is testing the unique predictive value. Does this variable add predictive value, given all the other predictors? When two or more predictor variables are highly correlated among themselves, it often happens that no x_j can be shown to have significant, unique predictive value, even though the xs together have been shown to be useful. If we are trying to predict housing sales based on gross domestic product and disposable income, we probably cannot prove that GDP adds value given DI or that DI adds value given GDP.

EXERCISES

Bus.

12.5 A feeder airline transports passengers from small cities to a single larger hub airport. A regression study tried to predict the revenue generated by each of 22 small cities, based on the distance of each city (in miles) from the hub and on the population of the small cities. The correlations and scatterplots were obtained as shown.

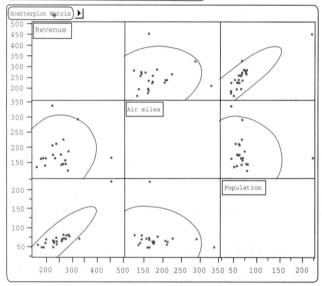

```
Correlations

Variable    Revenue   Air miles  Population
Revenue     1.0000    0.1133     0.8632
Air miles   0.1133    1.0000     -0.1502
Population  0.8632    -0.1502    1.0000
```

a. Are the variables severely correlated?
b. Might there be a problem with high leverage points?

12.6 In Exercise 12.5, we considered predicting revenue from each origin as a function of population and air miles to the hub airport. More JMP output is shown here. Is there a clear indication that the two independent variables together have at least some value in predicting revenue?

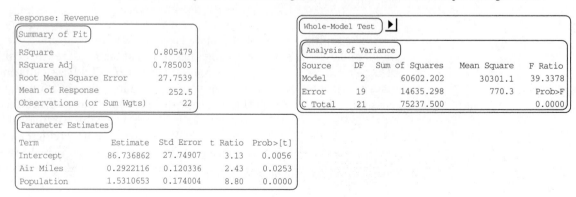

Response: Revenue

Summary of Fit

RSquare	0.805479
RSquare Adj	0.785003
Root Mean Square Error	27.7539
Mean of Response	252.5
Observations (or Sum Wgts)	22

Whole-Model Test

Analysis of Variance

Source	DF	Sum of Squares	Mean Square	F Ratio
Model	2	60602.202	30301.1	39.3378
Error	19	14635.298	770.3	Prob>F
C Total	21	75237.500		0.0000

Parameter Estimates

Term	Estimate	Std Error	t Ratio	Prob>|t|
Intercept	86.736862	27.74907	3.13	0.0056
Air Miles	0.2922116	0.120336	2.43	0.0253
Population	1.5310653	0.174004	8.80	0.0000

	## 12.4 Testing a Subset of Regression Coefficients

F test for several β_js

In the last section, we presented an *F* test for testing *all* the coefficients in a regression model and a *t* test for testing *one* coefficient. Another **F test** of the null hypothesis tests that *several* **of the true coefficients** are zero—that is, that several of the predictors have no value given the others. For example, if we try to predict the prevailing wage rate in various geographical areas for clerical workers based on the national minimum wage, national inflation rate, population density in the area, and median apartment rental price in the area, we might well want to test whether the variables related to area (density and apartment price) added anything, given the national variables.

A null hypothesis for this situation says that the true coefficients of density and apartment price are zero. According to this null hypothesis, these two independent variables together have no predictive value once minimum wage and inflation are included as predictors.

The idea is to compare the SS(Regression) or R^2 values when density and apartment price are excluded and when they are included in the prediction equation. When they are included, the R^2 is automatically at least as large as the R^2 when they are excluded because we can predict at least as well with more information as with less. Similarly, SS(Regression) will be larger for the complete model. The *F* test for this null hypothesis tests whether the gain is more than could be expected by chance alone. In general, let *k* be the total number of predictors, and let *g* be the number of predictors with coefficients not hypothesized to be zero ($g < k$). Then $k - g$ represents the number of predictors with coefficients that are hypothesized to be zero. The idea is to find SS(Regression) values using all predictors (the **complete model**) and using only the *g* predictors that do not appear in the null hypothesis (the **reduced model**). Once these have been computed, the test proceeds as outlined next. The notation is easier if we assume that the reduced model contains $\beta_1, \beta_2, \ldots, \beta_g$, so that the variables in the null hypothesis are listed last.

complete and reduced models

***F* Test of a Subset of Predictors**

H_0: $\beta_{g+1} = \beta_{g+2} = \cdots = \beta_k = 0$

H_a: H_0 is not true.

T.S.: $F = \dfrac{[\text{SS(Regression, complete)} - \text{SS(Regression, reduced)}]/(k - g)}{\text{SS(Residual, complete)}/[n - (k + 1)]}$

R.R.: $F > F_\alpha$, where F_α cuts off a right-tail of area α of the *F* distribution with $\text{df}_1 = (k - g)$ and $\text{df}_2 = [n - (k + 1)]$.

Check assumptions and draw conclusions.

EXAMPLE 12.18

A state fisheries commission wants to estimate the number of bass caught in a given lake during a season in order to restock the lake with the appropriate number of young fish. The commission could get a fairly accurate assessment of the seasonal catch by extensive netting sweeps of the lake before and after a season, but this technique is much too expensive to be done routinely. Therefore, the commission samples a number of lakes and records y, the seasonal catch (thousands of bass per square mile of lake area); x_1, the number of lakeshore residences per square mile of lake area; x_2, the size of the lake in square miles; $x_3 = 1$ if the lake has public access, 0 if not; and x_4, a structure index. (Structures are weed beds, sunken trees, drop-offs, and other living places for bass.) The data are shown here.

y	x_1	x_2	x_3	x_4
3.6	92.2	.21	0	81
.8	86.7	.30	0	26
2.5	80.2	.31	0	52
2.9	87.2	.40	0	64
1.4	64.9	.44	0	40
.9	90.1	.56	0	22
3.2	60.7	.78	0	80
2.7	50.9	1.21	0	60
2.2	86.1	.34	1	30
5.9	90.0	.40	1	90
3.3	80.4	.52	1	74
2.9	75.0	.66	1	50
3.6	70.0	.78	1	61
2.4	64.6	.91	1	40
.9	50.0	1.10	1	22
2.0	50.0	1.24	1	50
1.9	51.2	1.47	1	37
3.1	40.1	2.21	1	61
2.6	45.0	2.46	1	39
3.4	50.0	2.80	1	53

The commission is convinced that x_1 and x_2 are important variables in predicting y because they both reflect how intensively the lake has been fished. There is some question as to whether x_3 and x_4 are useful as additional predictor variables. Therefore, regression models (with all xs entering linearly) are run with and without x_3 and x_4. Relevant portions of the Minitab output follow.

```
MTB > regress 'catch' on 4 variables 'residenc' 'size' 'access'
'structur'
The regression equation is
catch = -1.94 + 0.0193 residenc + 0.332 size + 0.836 access
        + 0.0477 structur

Predictor        Coef       Stdev    t-ratio         p
Constant      -1.9378      0.9081      -2.13     0.050
residenc      0.01929     0.01018       1.90     0.077
size           0.3323      0.2458       1.35     0.196
access         0.8355      0.2250       3.71     0.002
structur     0.047714    0.005056       9.44     0.000

s = 0.4336      R-sq = 88.2%      R-sq(adj) = 85.0%
```

```
Analysis of Variance

SOURCE       DF        SS        MS        F        p
Regression    4    21.0474    5.2619    27.98    0.000
Error        15     2.8206    0.1880
Total        19    23.8680

SOURCE       DF     SEQ SS
residenc      1     0.2780
size          1     1.5667
access        1     2.4579
structur      1    16.7448

MTB > regress 'catch' on 2 vars 'residenc' 'size'

The regression equation is
catch = - 0.11 + 0.0310 residenc + 0.679 size

Predictor        Coef      Stdev    t-ratio        p
Constant       -0.107      2.336      -0.05    0.964
residenc      0.03102    0.02650       1.17    0.258
size          0.6794     0.6178        1.10    0.287

s = 1.138      R-sq = 7.7%      R-sq(adj) = 0.0%

Analysis of Variance

SOURCE       DF        SS        MS        F        p
Regression    2     1.845     0.922     0.71    0.505
Error        17    22.023     1.295
Total        19    23.868
```

a. Write the complete and reduced models.
b. Write the null hypothesis for testing that the omitted variables have no (incremental) predictive value.
c. Perform an F test for this null hypothesis.

Solution

a. The complete and reduced models are, respectively,

$$y_i = \beta_0 + \beta_1 x_{i1} + \beta_2 x_{i2} + \beta_3 x_{i3} + \beta_4 x_{i4} + \varepsilon_i$$

and

$$y_i = \beta_0 + \beta_1 x_{i1} + \beta_2 x_{i2} + \varepsilon_i$$

The corresponding multiple regression forecasting equations based on the sample data are

$$\text{Complete: } \hat{y} = -1.94 + .0193x_1 + .332x_2 + .836x_3 + .0477x_4$$

$$\text{Reduced: } \hat{y} = -.11 + .0310x_1 + .679x_2$$

b. The appropriate null hypothesis of no predictive power for x_3 and x_4 is $H_0\colon \beta_3 = \beta_4 = 0$.

c. The test statistic for the H_0 of part (b) makes use of SS(Regression, complete) = 21.0474, SS(Regression, reduced) = 1.845, SS(Residual, complete) = 2.8206, $k = 4$, $g = 2$, and $n = 20$:

$$\text{T.S.:} \quad F = \frac{[\text{SS(Regression, complete)} - \text{SS(Regression, reduced)}]/(4 - 2)}{\text{SS(Residual, complete)}/(20 - 5)}$$

$$= \frac{(21.0474 - 1.845)/2}{2.8206/15} = 51.059$$

The tabled value $F_{.01}$ for 2 and 15 df is 6.36. The value of the test statistic is much larger than the tabled value, so we have conclusive evidence that the Access and Structur variables add predictive value ($p < .001$).

EXERCISES | **12.7** | Two models based on the data in Example 12.13 were calculated, with the following results.

```
CORRELATIONS (PEARSON)

          ACCTSIZE  BUSIN     COMPET
BUSIN    -0.6934
COMPET    0.8196   -0.6527
INCOME    0.4526    0.1492    0.5571


CASES INCLUDED 21   MISSING CASES 0

(Model 1)
UNWEIGHTED LEAST SQUARES LINEAR REGRESSION OF ACCTSIZE


PREDICTOR
VARIABLES    COEFFICIENT   STD ERROR    STUDENT'S T      P        VIF

CONSTANT       0.15085      0.73776        0.20        0.8404
BUSIN         -0.00288     8.894E-04      -3.24        0.0048     5.2
COMPET        -0.00759      0.05810       -0.13        0.8975     7.4
INCOME         0.26528      0.10127        2.62        0.0179     4.3


R-SQUARED              0.7973    RESID. MEAN SQUARE (MSE)    0.03968
ADJUSTED R-SQUARED    0.7615      STANDARD DEVIATION           0.19920


SOURCE      DF      SS          MS        F        P

REGRESSION   3    2.65376    0.88458    22.29    0.0000
RESIDUAL    17    0.67461    0.03968
TOTAL       20    3.32838


(Model 2)
UNWEIGHTED LEAST SQUARES LINEAR REGRESSION OF ACCTSIZE
```

```
PREDICTOR
VARIABLES       COEFFICIENT      STD ERROR      STUDENT'S T        P

CONSTANT           0.12407        0.96768           0.13        0.8993
INCOME             0.20191        0.09125           2.21        0.0394

R-SQUARED                0.2049    , RESID. MEAN SQUARE (MSE)    0.13928
ADJUSTED R-SQUARED   0.1630          STANDARD DEVIATION           0.37321

SOURCE          DF       SS           MS          F           P

REGRESSION       1     0.68192      0.68192      4.90       0.0394
RESIDUAL        19     2.64645      0.13928
TOTAL           20     3.32838

CASES INCLUDED 21   MISSING CASES 0
```

a. Locate R^2 for the reduced model, with INCOME as the only predictor.
b. Locate R^2 for the complete model.
c. Compare the values in (a) and (b). Does INCOME provide an adequate fit?

12.8 Calculate the F statistic in Exercise 12.7, based on the sums of squares shown in the output. Interpret the results of the F test.

Soc.   **12.9**   An automobile financing company uses a rather complex credit rating system for car loans. The questionnaire requires substantial time to fill out, taking sales staff time and risking alienating the customer. The company decides to see whether three variables (Age, Monthly family income, and Debt payments as a fraction of income) will reproduce the credit score reasonably accurately. Data were obtained on a sample (with no evident biases) of 500 applications. The complicated rating score was calculated and served as the dependent variable in a multiple regression. Some results from JMP are shown.

a. How much of the variation in ratings is accounted for by the three predictors?
b. Use this number to verify the computation of the overall F statistic.
c. Does the F test clearly show that the three independent variables have predictive value for the rating score?

Response: Rating score

Summary of Fit

RSquare	0.979566
RSquare Adj	0.979443
Root Mean Square Error	2.023398
Mean of Response	65.044
Observations (or Sum Wgts)	500

Parameter Estimates

Term	Estimate	Std Error	t Ratio	Prob>|t|
Intercept	54.657197	0.634791	86.10	0.0000
Age	0.0056098	0.011586	0.48	0.6285
Monthly income	0.0100597	0.000157	64.13	0.0000
Debt fraction	-39.95239	0.883684	-45.21	0.0000

Effect Test

Source	Nparm	DF	Sum of Squares	F Ratio	Prob>F
Age	1	1	0.960	0.2344	0.6285
Monthly income	1	1	16835.195	4112.023	0.0000
Debt fraction	1	1	8368.627	2044.05	0.0000

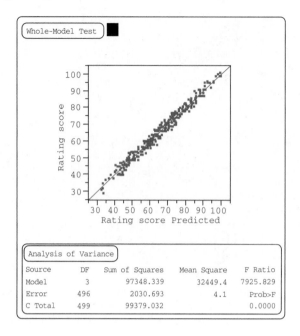

12.10 The credit rating data (Exercise 12.9) were reanalyzed, using only the monthly income variable as a predictor. JMP results are shown.

   a. By how much has the regression sum of squares been reduced by eliminating age and debt percentage as predictors?

   b. Do these variables add statistically significant (at normal α levels) predictive value, once income is given?

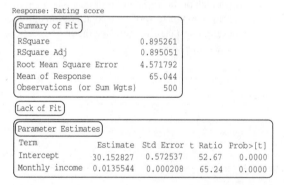

Engin. 12.11 A chemical firm tests the yield that results from the presence of varying amounts of two catalysts. Yields are measured for five different amounts of catalyst 1 paired with four different amounts of catalyst 2. A second-order model is fit to approximate the anticipated nonlinear relation. The variables are y = yield, x_1 = amount of catalyst 1, x_2 = amount of catalyst 2, $x_3 = x_1^2$, $x_4 = x_1 x_2$, and $x_5 = x_2^2$. Selected output from the regression analysis is shown here.

```
                    Multiple Regression Analysis

Dependent variable: Yield

                         Table of Estimates

                            Standard      t         P
               Estimate       Error     Value     Value

Constant       50.0195      4.3905      11.39     0.0000
Cat1            6.64357     2.01212      3.30     0.0052
Cat2            7.3145      2.73977      2.67     0.0183
Cat1Sq         -1.23143     0.301968    -4.08     0.0011
Cat1Cat2       -0.7724      0.319573    -2.42     0.0299
Cat2Sq         -1.1755      0.50529     -2.33     0.0355

R-squared = 86.24%
Adjusted R-squared = 81.33%
Standard error of estimation = 2.25973

                       Analysis of Variance

              Sum of                                         P
Source        Squares    D.F.    Mean Square   F-Ratio     Value
Model         448.193      5       89.6386      17.55      0.0000
Error          71.489     14        5.10636

Total (corr.)  519.682    19

                   Conditional Sums of Squares

              Sum of                                         P
Source        Squares    D.F.    Mean Square   F-Ratio     Value

Cat1          286.439      1       286.439      56.09      0.0000
Cat2           19.3688     1        19.3688      3.79      0.0718
Cat1Sq         84.9193     1        84.9193     16.63      0.0011
Cat1Cat2       29.8301     1        29.8301      5.84      0.0299
Cat2Sq         27.636      1        27.636       5.41      0.0355

Model         448.193      5

                    Multiple Regression Analysis

Dependent variable: Yield

                         Table of Estimates

                            Standard      t         P
               Estimate       Error     Value     Value

Constant       70.31        2.57001     27.36     0.0000
Cat1           -2.676       0.560822    -4.77     0.0002
Cat2           -0.8802      0.70939     -1.24     0.2315
```

```
R-squared = 58.85%
Adjusted R-squared = 54.00%
Standard error of estimation = 3.54695
```

| | Sum of | | Analysis of Variance | | P |
Source	Squares	D.F.	Mean Square	F-Ratio	Value
Model	305.808	2	152.904	12.15	0.0005
Error	213.874	17	12.5808		
Total (corr.)	519.682	19			

a. Write the estimated complete model.
b. Write the estimated reduced model.
c. Locate the R^2 values for the complete and reduced models.
d. Is there convincing evidence that the addition of the second-order terms improves the predictive ability of the model?

12.5 Comparing the Slopes of Several Regression Lines

This topic represents a special case of the general problem of constructing a multiple regression equation for both qualitative and quantitative independent variables. The best way to illustrate this particular problem is by way of an example.

EXAMPLE 12.19

An investigator was interested in comparing the responses of rats to different doses of two drug products (A and B). The study called for a sample of 60 rats of a particular strain to be randomly allocated into two equal groups. The first group of rats was to receive drug A, with 10 rats randomly assigned to each of three doses (5, 10, and 20 mg). Similarly, the 30 rats in group 2 were to receive drug B, with 10 rats randomly assigned to the 5-, 10-, and 20-mg doses. In the study, each rat received its assigned dose, and after a 30-minute observation period it was scored for signs of anxiety on a 0-to-30-point scale. Assume that a rat's anxiety score is a linear function of the dosage of the drug. Write a model relating a rat's scores to the two independent variables drug product and drug dose. Interpret the βs.

Solution For this experimental situation, we have one qualitative variable (drug product) and one quantitative variable (drug dose). Letting x_1 denote the drug dose, we have the model

$$y = \beta_0 + \beta_1 x_1 + \beta_2 x_2 + \beta_3 x_1 x_2 + \varepsilon$$

where

$x_1 = $ drug dose

$x_2 = 1$ if product B $x_2 = 0$ otherwise

The expected value for y in our model is

$$E(y) = \beta_0 + \beta_1 x_1 + \beta_2 x_2 + \beta_3 x_1 x_2$$

Substituting $x_2 = 0$ and $x_2 = 1$, respectively, for drugs A and B, we obtain the expected rat anxiety score for a given dose:

drug A: $\quad E(y) = \beta_0 + \beta_1 x_1$

drug B: $\quad E(y) = \beta_0 + \beta_1 x_1 + \beta_2 + \beta_3 x_1 = (\beta_0 + \beta_2) + (\beta_1 + \beta_3)x_1$

linear regression lines

These two expected values represent **linear regression lines.** The parameters in the model can be interpreted in terms of the slopes and intercepts associated with these regression lines. In particular,

y-intercept

$\quad \beta_0$: **y-intercept** for product A regression line

slope

$\quad \beta_1$: **slope** of product A regression line

$\quad \beta_2$: difference in y-intercepts of regression lines for products B and A

$\quad \beta_3$: difference in slopes of regression lines for products B and A

intersecting lines

parallel lines

Figure 12.4(a) indicates a situation in which $\beta_3 \neq 0$ (that is, there is an interaction between the two variables drug product and drug dose). Thus, the regression lines are not parallel. Figure 12.4(b) indicates a case in which $\beta_3 = 0$ (no interaction), which results in parallel regression lines.

FIGURE 12.4

Comparing two regression lines

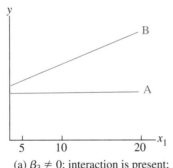 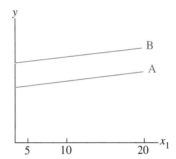

(a) $\beta_3 \neq 0$; interaction is present; intersecting lines $\qquad$ (b) $\beta_3 = 0$; interaction is not present; parallel lines

Indeed, many other experimental situations are possible, depending on the signs and magnitudes of the parameters $\beta_0, \beta_1, \beta_2,$ and β_3.

EXAMPLE 12.20

Sample data for the experiment discussed in Example 12.19 are listed in Table 12.4. The response of interest is an anxiety score obtained from trained investigators. Use these data to fit the general linear model

$$y = \beta_0 + \beta_1 x_1 + \beta_2 x_2 + \beta_3 x_1 x_2 + \varepsilon$$

Of particular interest to the experimenter is a comparison between the slopes of the regression lines. A difference in slopes would indicate that the drug products have different effects on the anxiety of the rats. Conduct a statistical test of the equality of the two slopes. Use $\alpha = .05$.

Solution Using the complete model

$$y = \beta_0 + \beta_1 x_1 + \beta_2 x_2 + \beta_3 x_1 x_2 + \varepsilon$$

TABLE 12.4

Rat anxiety scores

Drug Product	Drug Dose (mg)		
	5	**10**	**20**
A	15 16	18 16	20 17
	16 15	17 15	19 18
	18 16	18 19	21 21
	13 17	19 18	18 20
	19 15	20 16	19 17
	av = 16	av = 17.6	av = 19.0
B	16 15	19 18	24 23
	17 15	21 20	25 24
	18 18	22 21	23 22
	17 17	23 22	25 26
	15 16	20 19	25 24
	av = 16.4	av = 20.5	av = 24.1

we obtain a least-squares fit of

$$\hat{y} = 15.30 + .19x_1 - .70x_2 + .30x_1x_2$$

with SS(Regression, complete) = 442.10 and SS(Residual, complete) = 133.63. (See the computer output that follows.)

The reduced model corresponding to the null hypothesis H_0: $\beta_3 = 0$ (that is, the slopes are the same) is

$$y = \beta_0 + \beta_1x_1 + \beta_2x_2 + \varepsilon$$

```
REGRESSION ANALYSIS OF ANXIETY TREATMENTS-COMPLETE MODEL

Model: MODEL1
Dependent Variable: SCORE


Analysis of Variance

                    Sum of          Mean
Source      DF      Squares         Square       F Value      Prob>F

Model        3      442.10476       147.36825     61.758      0.0001
Error       56      133.62857         2.38622
C Total     59      575.73333


        Root MSE        1.54474      R-square     0.7679
        Dep Mean       18.93333      Adj R-sq     0.7555
        C.V.            8.15884
```

```
Parameter Estimates

                      Parameter      Standard     T for H0:
Variable    DF        Estimate          Error    Parameter=0    Prob > |T|

INTERCEP    1        15.300000     0.59827558        25.573        0.0001
DOSE        1         0.191429     0.04522538         4.233        0.0001
PRODUCT     1        -0.700000     0.84608944        -0.827        0.4116
PRD_DOSE    1         0.300000     0.06395835         4.691        0.0001

                      Variable
Variable    DF        Label

INTERCEP    1     Intercept
DOSE        1     DRUG DOSE LEVEL
PRODUCT     1     DRUG PRODUCT
PRD_DOSE    1     PRODUCT TIMES DOSE


REGRESSION ANALYSIS OF ANXIETY TREATMENTS-REDUCED MODEL

Model: MODEL1
Dependent Variable: SCORE


Analysis of Variance

                     Sum of         Mean
Source       DF      Squares        Square      F Value     Prob>F

Model         2     389.60476     194.80238      59.656      0.0001
Error        57     186.12857       3.26541
C Total      59     575.73333

     Root MSE        1.80705     R-square      0.6767
     Dep Mean       18.93333     Adj R-sq      0.6654
     C.V.            9.54425

Parameter Estimates

                      Parameter      Standard     T for H0:
Variable    DF        Estimate          Error    Parameter=0    Prob > |T|

INTERCEP    1        13.550000     0.54711020        24.766        0.0001
DOSE        1         0.341429     0.03740940         9.127        0.0001
PRODUCT     1         2.800000     0.46657715         6.001        0.0001

                      Variable
Variable    DF        Label

INTERCEP    1     Intercept
DOSE        1     DRUG DOSE LEVEL
PRODUCT     1     DRUG PRODUCT
```

for which we obtain

$$\hat{y} = 13.55 + .34x_1 - 2.80x_2$$

and SS(Regression, reduced) = 389.60. The reduction in the sum of squares for error attributed to $x_1 x_2$ is

$$\text{SS}_{\text{drop}} = \text{SS(Regression, complete)} - \text{SS(Regression, reduced)}$$
$$= 442.10 - 389.60 = 52.50$$

It follows that

$$F = \frac{[\text{SS(Regression, complete)} - \text{SS(Regression, reduced)}]/k - g}{\text{SS(Residual, complete)}/[n - (k+1)]}$$

$$= \frac{52.50/1}{133.63/56} = 22.00.$$

Because the observed value of F exceeds 4.00, the value for $\text{df}_1 = 1$, $\text{df}_2 = 56$ and $a = .05$ in Appendix Table 9, we reject H_0 and conclude that the slopes for the two groups are different. Note that we could have obtained the same result by testing $H_0: \beta_3 = 0$ using a t test. From the computer output, the t statistic is 4.69, which is significant at the .0001 level. For this type of test, the t statistic and F statistic are related; namely, $t^2 = F$ (here $4.69^2 \approx 22$).

The results presented here for comparing the slope of two regression lines can be readily extended to the comparison of three or more regression lines by including additional dummy variables and all possible interaction terms between the quantitative variable x_1 and the dummy variables. Thus, for example, in comparing the slopes of three regression lines, the model will contain the quantitative variable x_1, two dummy variables x_2 and x_3, and two interaction terms $x_1 x_2$ and $x_1 x_3$.

EXERCISES

Applications

Med.

12.12 An experimenter wished to compare the potencies of three different drug products. To do this, 12 test tubes were inoculated with a culture of the virus under study and incubated for 2 days at 35°C. Four dosage levels (.2, .4, .8, and 1.6 μg per tube) were to be used from each of the three drug products, with only one dose–drug product combination for each of the 12 test-tube cultures. One means of comparing the drug products is to examine their slopes (with respect to dose).

a. Write a general linear model relating the response y to the independent variables dose and drug product. Make the expected response a linear function of log dose (x_1). Identify the parameters in the model.

b. Change the model in part (a) to reflect that the three drug products have the same slope.

12.13 Refer to Exercise 12.12.

a. Use the following data to make a comparison among the three slopes. Fit a complete and a reduced model for your test. Use $\alpha = .05$.

Dose	Drug Product A	B	C
.2	2.0	1.8	1.3
.4	4.3	4.1	2.0
.8	6.5	4.9	2.8
1.6	8.9	5.7	3.4

b. Is there evidence to indicate the slopes are equal?
c. Suggest how you could test the null hypothesis that the intercepts are all equal to zero.

12.6 Logistic Regression

In many research studies, the response variable may be represented as one of two possible values. Thus, the response variable is a binary random variable taking on the values 0 and 1. For example, in a study of a suspected carcinogen, aflatoxin B_1, a number of levels of the compound were fed to test animals. After a period of time, the animals were sacrificed and the number of animals having liver tumors was recorded. The response variable is $y = 1$ if the animal has a tumor and $y = 0$ if the animal fails to have a tumor. Similarly, a bank wants to determine which customers are most likely to repay their loan. Thus, they want to record a number of independent variables that describe the customer's reliability and then determine whether these variables are related to the binary variable, $y = 1$ if the customer repays the loan and $y = 0$ if the customer fails to repay the loan. A model that relates a binary variable y to explanatory variables will be developed next.

When the response variable y is binary, the distribution of y depends on a single parameter, the probability $p = \Pr(y = 1)$. We want to relate p to a linear combination of the independent variables. The difficulty is that p varies between zero and one, whereas linear combinations of the explanatory variables can vary between $-\infty$ and $+\infty$. In Chapter 10, we introduced the transformation of probabilities into odds. As the probabilities vary between zero and one, the odds vary between zero and infinity. By taking the logarithm of the odds, we will have a transformed variable that will vary between $-\infty$ and $+\infty$ when the probabilities vary between zero and one. The model often used to study the association between a binary response and a set of explanatory variables is given by **logistic regression analysis.** In this model, the natural logarithm of the odds is related to the explanatory variables by a linear model. We will consider the situation where we have a single independent variable, but this model can be generalized to multiple independent variables. Let $p(x)$ be the probability that y equals 1 when the independent variable equals x. We model the log of the odds to a linear model in x, a **simple logistic regression model:**

logistic regression analysis

simple logistic regression model

$$\ln\left(\frac{p(x)}{1 - p(x)}\right) = \beta_0 + \beta_1 x$$

This transformation can be formulated directly in terms of $p(x)$ as

$$p(x) = \frac{e^{\beta_0 + \beta_1 x}}{1 + e^{\beta_0 + \beta_1 x}}$$

For example, the probability of a tumor being present in an animal exposed to x units of the aflatoxin B_1 is given by $p(x)$ as expressed by the preceding equation. The values of β_0 and β_1 can be estimated from the observed data using maximum likelihood estimation.

We can interpret the parameters β_0 and β_1 in the logistic regression model in terms of $p(x)$. The intercept parameter β_0 permits the estimation of the probability of the event associated with $y = 1$ when the independent variable $x = 0$. For example, the probability of a tumor being present when the animal is not exposed

FIGURE 12.5

Logistic regression functions

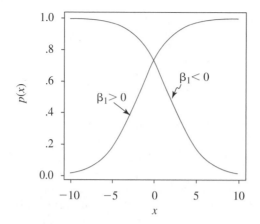

to aflatoxin B_1 corresponds to the probability of $y = 1$ when $x = 0$—that is, $p(0)$. The logistic regression model yields

$$p(0) = \frac{e^{\beta_0}}{1 + e^{\beta_0}}$$

The slope parameter β_1 measures the degree of association between the probability of the event occurring and the value of the independent variable x. When $\beta_1 = 0$, the probability of the event occurring is not associated with size of the value of x. In our example, the chance of an animal developing a liver tumor remains constant no matter the amount of aflatoxin B_1 the animal was exposed to. Figure 12.5 displays two simple logistic regression functions. If $\beta_1 > 0$, the probability of the event occurring increases as the value of the independent variable increases. If $\beta_1 < 0$, the probability of the event occurring decreases as the value of the independent variable increases.

In the situation where both β_0 and β_1 are zero, the event is as likely to occur as not to occur because

$$p(x) = \frac{e^0}{1 + e^0} = \frac{1}{1 + 1} = \frac{1}{2}$$

This indicates that the probability of the occurrence of the event given by $y = 1$ is not related to the independent variable x. Thus, the model is noninformative in determining the probability of the event's occurrence, hence an equal chance of the occurrence or nonoccurrence of the event no matter the value of the independent variable.

Whether we are using the simple logistic regression model or multiple logistic regression models, the computational techniques used to estimate the model parameters require the use of computer software. We will use an example to illustrate the use of logistic regression models.

EXAMPLE 12.21

A study reported by A. F. Smith (1967) recorded the level of an enzyme, creatinine kinase (CK), for patients who were suspected of having had a heart attack. The objective of the study was to assess whether measuring the amount of CK on admission to the hospital was a useful diagnostic indicator of whether patients admitted with a diagnosis of a heart attack had really had a heart attack. The enzyme CK was measured in 360 patients on admission to the hospital. After a period of time, a doctor reviewed the records of these patients to decide which of the 360 patients had actually had a heart attack. The data are given in the following table with the CK values given as the midpoint of the range of values in each of 13 classes of values.

CK Value	Number of Patients with Heart Attack	Number of Patients without Heart Attack
20	2	88
60	13	26
100	30	8
140	30	5
180	21	0
220	19	1
260	18	1
300	13	1
340	19	0
380	15	0
420	7	0
460	8	0
500	35	0

The computer output for obtaining the estimated logistic regression curve and 95% confidence intervals on the predicted probabilities of a heart attack are given here.

```
        LOGISTIC REGRESSION ANALYSIS EXAMPLE

The LOGISTIC Procedure

Data Set: WORK.LOGREG
Response Variable (Events): R
Response Variable (Trials): N
Number of Observations: 13
Link Function: Logit

     Response Profile

Ordered  Binary
  Value  Outcome      Count

     1   EVENT          230
     2   NO EVENT       130
```

```
Model Fitting Information and Testing Global Null Hypothesis BETA=0

                                  Intercept
                    ·Intercept       and
Criterion             Only       Covariates    Chi-Square for Covariates

AIC                  472.919       191.773              .
SC                   476.806       199.545              .
-2 LOG L             470.919       187.773      283.147 with 1 DF (p=0.0001)
Score                   .             .        159.142 with 1 DF (p=0.0001)


               Analysis of Maximum Likelihood Estimates


                Parameter  Standard    Wald      Pr >      Standardized
Variable DF     Estimate    Error   Chi-Square Chi-Square    Estimate

INTERCPT 1       -3.0284    0.3670    68.0948    0.0001
CK       1        0.0351    0.00408   73.9842    0.0001       3.100511


              LOGISTIC REGRESSION ANALYSIS EXAMPLE

        OBS     CK      PRED        LCL        UCL

         1      20    0.08897     0.05151    0.14937
         2      60    0.28453     0.21224    0.36988
         3     100    0.61824     0.51935    0.70821
         4     140    0.86833     0.78063    0.92436
         5     180    0.96410     0.91643    0.98502
         6     220    0.99094     0.97067    0.99724
         7     260    0.99776     0.99000    0.99950
         8     300    0.99945     0.99662    0.99991
         9     340    0.99986     0.99886    0.99998
        10     380    0.99997     0.99962    1.00000
        11     420    0.99999     0.99987    1.00000
        12     460    1.00000     0.99996    1.00000
        13     500    1.00000     0.99999    1.00000
```

a. Is CK level significantly related to the probability of a heart attack through the logistic regression model?

b. From the computer output, obtain the estimated coefficients β_0 and β_1.

c. Construct the estimated probability of a heart attack as a function of CK level. In particular, estimate this probability for a patient having a CK level of 140.

Solution

a. From the computer output, we obtain, p-value $= .0001$ for testing the hypothesis $H_0: \beta_1 = 0$ versus $H_0: \beta_1 \neq 0$ in the logistic regression model. Thus, CK is significantly related to the probability of a heart attack.

b. From the computer output, we obtain $\hat{\beta}_0 = -3.0284$ and $\hat{\beta}_1 = 0.0351$. Note that $\hat{\beta}_1$ is positive. This indicates that patients having higher levels of CK are associated with a larger probability that a heart attack had occurred.

c. The estimated probability of a heart attack as a function of CK level in the patient is given by

$$p(\widehat{CK}) = \frac{e^{-3.0284 + .0351 \cdot CK}}{1 + e^{-3.0284 + .0351 \cdot CK}}$$

We can use this formula to calculate the probability that a patient had a heart attack when the CK level in the patient was 140. This value is given by

$$p(\widehat{CK}) = \frac{e^{-3.0284 + .0351 \cdot 140}}{1 + e^{-3.0284 + .0351 \cdot 140}} = \frac{e^{1.886}}{1 + e^{1.886}} = .868$$

From the computer printout, we obtain 95% confidence intervals for this probability as .781 to .924. Thus, we are 95% confident that between 78.1 and 92.4% of patients with a CK level of 140 had a heart attack. The estimated probabilities of a heart attack along with 95% confidence intervals on these probabilities are plotted in Figure 12.6. We note that the estimated probability of a heart attack increases very rapidly with increasing CK levels in the patients. This would indicate that CK levels are a useful indicator of heart attack potential.

FIGURE 12.6

Estimated probability of heart attack with 95% confidence limits

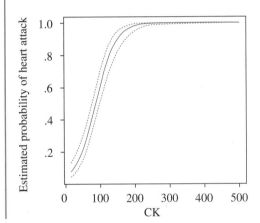

EXERCISES **Applications**

Engin. **12.14** A quality control engineer studied the relationship between years of experience of a system control engineer and the capacity of the engineer to complete within a given time a complex control design including the debugging of all computer programs and control devices. A group of 25 engineers having a wide difference in experience (measured in months of experience) were given the same control design project. The results of the study are given in the following table with $y = 1$ if the project was successfully completed in the allocated time and $y = 0$ if the project was not successfully completed.

Months of Experience	Project Success	Months of Experience	Project Success
2	0	15	1
4	0	16	1
5	0	17	0
6	0	19	1
7	0	20	1
8	1	22	0
8	1	23	1
9	0	24	1
10	0	27	1
10	0	30	0
11	1	31	1
12	1	32	1
13	0		

a. Use the computer output given here to determine whether experience is associated with the probability of completing the task.

b. Compute the probability of successfully completing the task for an engineer having 24 months of experience. Place a 95% confidence interval on your estimate.

```
SAS Code for Logistic Regression
option ls=70 ps=55 nocenter nodate;
data logreg;
input x y @@;
label x='MONTHS EXPERIENCE' y='SUCCESS INDICATOR';cards;
2 0 4 0 5 0 6 0 7 0 8 1 8 1 9 0 10 0 10 0 11 1 12 1
13 0 15 1 16 1 17 0 19 1 20 1 22 0 23 1 24 1 27 1 30 0 31 1 32 1
run;
proc print;
proc logistic descending;
model y=x;
output out=new p=pred lower=lcl upper=ucl;
proc sort; by x;
proc print; var x pred lcl ucl;
run;


---------------------------------------------------------------------


The LOGISTIC Procedure

Data Set: WORK.LOGREG
Response Variable: Y           SUCCESS INDICATOR
Response Levels: 2
Number of Observations: 25
Link Function: Logit
```

```
         Response Profile

Ordered
   Value       Y      Count

      1         1        13
      2         0        12

          Analysis of Maximum Likelihood Estimates

            Parameter Standard    Wald      Pr >     Standardized
Variable DF Estimate   Error   Chi-Square Chi-Square  Estimate

INTERCPT 1  -1.6842   0.9451    3.1759     0.0747         .
X        1   0.1194   0.0589    4.1091     0.0427      0.585706
```

The LOGISTIC Procedure

Association of Predicted Probabilities and Observed Responses

```
Concordant = 77.6%        Somers' D = 0.551
Discordant = 22.4%        Gamma     = 0.551
Tied       = 0.0%         Tau-a     = 0.287
(156 pairs)               c         = 0.776
```

The LOGISTIC Procedure

OBS	X	PRED	95% Lower Limit	95% Upper Limit
1	2	0.19070	0.04320	0.55155
2	4	0.23029	0.06487	0.56339
3	5	0.25213	0.07884	0.57042
4	6	0.27530	0.09518	0.57839
5	7	0.29974	0.11399	0.58749
6	8	0.32538	0.13526	0.59794
7	8	0.32538	0.13526	0.59794
8	9	0.35211	0.15884	0.61001
9	10	0.37980	0.18434	0.62397
10	10	0.37980	0.18434	0.62397
11	11	0.40830	0.21117	0.64011
12	12	0.43742	0.23858	0.65863
13	13	0.46698	0.26568	0.67964
14	15	0.52660	0.31574	0.72839
15	16	0.55623	0.33753	0.75512
16	17	0.58547	0.35684	0.78239
17	19	0.64199	0.38830	0.83514
18	20	0.66894	0.40092	0.85917
19	22	0.71954	0.42133	0.90040
20	23	0.74299	0.42962	0.91732
21	24	0.76512	0.43691	0.93186

22	27	0.82333	0.45436	0.96307
23	30	0.86958	0.46732	0.98065
24	31	0.88253	0.47097	0.98447
25	32	0.89435	0.47436	0.98756

12.15 An additive to interior house paint has been recently developed that may greatly increase the ability of the paint to resist staining. An investigation was conducted to determine whether the additive is safe when children were exposed to it. Various amounts of the additive were fed to test animals and the number of animals developing liver tumors was recorded. The data are given in the following table.

	Amount (ppm)					
	0	**10**	**25**	**50**	**100**	**200**
Number of Test Animals	30	20	20	30	30	30
Number of Animals with Tumors	0	2	2	7	25	30

a. Use the computer output given here to determine whether the amount of additive given to the test animals is associated with the probability of a tumor developing in the liver of the animal.

b. Compute the probability of a tumor developing in the liver of a test animal exposed to 100 ppm of the additive. Place a 95% confidence interval on your estimate.

```
SAS Code for Exercise

option ls=70 ps=55 nocenter nodate;
TITLE 'OUTPUT FOR EXERCISE';
data logreg;
input x R N @@;
label x='AMOUNT (PPM)' ;cards;
0 0 30 10 2 20 25 2 20 50 7 30 100 25 30 200 30 30
run;
proc print;
proc logistic descending;
model R/N=x;
output out=new p=pred lower=lcl upper=ucl;
proc sort; by x;
proc print; var x pred lcl ucl;
run;


-----------------------------------------------------------------

OUTPUT FOR EXERCISE

OBS     X     R     N

 1      0     0    30
 2     10     2    20
 3     25     2    20
 4     50     7    30
 5    100    25    30
 6    200    30    30
```

```
The LOGISTIC Procedure

        Response Profile

Ordered  Binary
  Value  Outcome      Count

      1  EVENT          66
      2  NO EVENT       94


              Analysis of Maximum Likelihood Estimates

                   Parameter Standard      Wald       Pr >    Standardized
        Variable DF Estimate   Error    Chi-Square Chi-Square   Estimate

        INTERCPT 1   -3.6429   0.5530     43.3998    0.0001        .
        X        1    0.0521   0.00824    39.9911    0.0001     2.044518

              Analysis of Maximum
              Likelihood Estimates

                    Odds    Variable
        Variable    Ratio    Label

        INTERCPT      .      Intercept
        X           1.053    AMOUNT (PPM)

                          95% Lower   95% Upper
        OBS     X    PRED    Limit       Limit

         1      0   0.02551  0.00878    0.07182
         2     10   0.04221  0.01681    0.10203
         3     25   0.08783  0.04308    0.17077
         4     50   0.26156  0.16907    0.38142
         5    100   0.82738  0.66925    0.91905
         6    200   0.99886  0.98818    0.99989
```

12.7 Encounters with Real Data: The Problem of Creating an Airline Schedule

Problem Definition (1)

Major airlines publish system schedules of their flights, in booklets or online. If we intend to fly from Atlanta to San Francisco, there is no need to try to predict how long the trip is scheduled to take; we can just look at a timetable or call the airline. But if our interest is to learn something about how airline schedules are put together, then regression techniques can be helpful.

Collecting Data (2)

The data set DELTAFLY contains the schedules of 100 selected nonstop domestic flights on Delta Air Lines (effective February 1994). The variables and data are shown here as well as in the file DELTAFLY on the CD that came with your book.

c1	From/To	Airport codes for cities of departure/arrival (text variable)
c2	FlightNo	Delta Air Lines flight number
c3	Miles	Distance between cities in miles
c4	TravTime	Elapsed travel time (in hours)
c5	TravDir	Travel Direction (coded)

	c1	c2	c3	c4	c5
Row No	From/To	FltNo	Miles	TravTime	TravDir
1	ABQ-DFW	568	569	1.56667	0
2	ABQ-DFW	346	569	1.70000	0
3	ANC-SEA	646	1448	3.16667	1
4	ATL-BHM	1291	134	0.75000	0
5	ATL-BOS	786	946	2.36667	1
6	ATL-BOS	624	946	2.56667	1
7	ATL-CLE	710	554	1.61667	0
8	ATL-DAY	615	432	1.41667	0
9	ATL-DEN	445	1208	3.38333	-2
10	ATL-DEN	499	1208	3.28333	-2
11	ATL-HNL	53	4501	9.58333	-2
12	ATL-TYS	429	152	0.90000	0
13	ATL-LAS	749	1746	3.96667	-2
14	ATL-LEX	918	303	1.13333	0
15	ATL-LAX	1565	1946	4.58333	-2
16	ATL-MEM	1677	332	1.20000	-2
17	ATL-BNA	784	214	1.00000	-1
18	ATL-MSY	375	425	1.41667	-1
19	ATL-MSY	771	425	1.36667	-1
20	ATL-JFK	1212	765	2.21667	1
21	ATL-JFK	1298	765	2.08333	1
22	ATL-OMA	1480	821	2.35000	-2
23	ATL-SNA	716	1919	4.78333	-2
24	ATL-PHX	345	1587	4.05000	-2
25	ATL-PIT	662	526	1.48333	0
26	ATL-PDX	51	2172	4.91667	-2
27	ATL-STL	674	484	1.71667	-2
28	ATL-SFO	217	2139	5.25000	-2
29	ATL-SFO	1792	2139	5.28333	-2
30	ATL-TLH	495	223	0.96667	0
31	ATL-DCA	1204	543	1.70000	1
32	ATL-DCA	978	543	1.65000	1
33	AUS-DFW	945	183	0.93333	0
34	BWI-CVG	1199	430	1.58333	-2
35	BGR-BOS	603	201	0.91667	0
36	BOS-ATL	1103	946	2.58333	-1
37	BOS-ATL	1085	946	2.66667	-1

38	BOS-ATL	651	946	2.71667	-1
39	BOS-DFW	629	1561	4.08333	-1
40	BOS-JFK	1837	182	1.03333	1
41	CVG-CLE	398	221	0.95000	0
42	DFW-ATL	16	731	2.00000	2
43	DFW-ATL	298	731	2.13333	2
44	DFW-ATL	1120	731	1.86667	2
45	DFW-DEN	475	645	1.96667	-1
46	DFW-MSY	490	447	1.25000	1
47	DFW-JFK	286	1395	3.16667	1
48	DFW-SLC	331	988	2.86667	-1
49	DFW-SAN	859	1171	3.11667	-2
50	DFW-SFO	273	1465	3.73333	-2
51	DEN-ATL	1108	1208	2.75000	2
52	DEN-SLC	1666	381	1.33333	-2
53	FAI-ANC	1172	261	0.91667	0
54	HNL-DFW	16	3784	7.00000	2
55	HNL-LAX	1578	2555	4.96667	2
56	HOU-ATL	1123	689	1.85000	2
57	JAX-ATL	962	270	1.15000	0
58	MCI-DFW	1058	459	1.50000	0
59	LAS-ATL	496	1746	3.78333	2
60	LAX-ATL	182	1946	3.93333	2
61	LAX-ATL	178	1946	4.08333	2
62	LAX-ATL	188	1946	3.75000	2
63	LAX-LAS	1605	236	1.00000	1
64	LAX-JFK	308	2477	5.33333	2
65	LAX-SLC	1609	590	1.71667	1
66	LAX-SFO	1517	337	1.25000	-1
67	MSP-ORD	1878	345	1.26667	0
68	BNA-ATL	586	214	0.96667	1
69	MSY-ATL	616	425	1.33333	1
70	MSY-DFW	825	447	1.56667	-1
71	JFK-DFW	285	1395	4.08333	-1
72	JFK-ATL	497	765	2.28333	-1
73	JFK-ATL	149	765	2.41667	-1
74	JFK-CVG	1997	592	2.16667	-2
75	JFK-SFO	1699	2588	6.25000	-2
76	OAK-SLC	1216	588	1.55000	2
77	SNA-SLC	1488	588	1.70000	1
78	PHL-ATL	1127	665	2.05000	-1
79	PHX-ATL	610	1587	3.40000	2
80	PHX-LAX	1230	370	1.33333	-2
81	PDX-ATL	52	2172	4.41667	2
82	RDU-ATL	1049	356	1.16667	1
83	RNO-DFW	938	1345	3.06667	1
84	SLC-ATL	160	1589	3.50000	2
85	SLC-SAN	678	626	1.88333	-1
86	SLC-SFO	1073	599	1.70000	-2
87	SAT-ATL	762	874	2.16667	2
88	SAN-ATL	276	1891	3.81667	2
89	SAN-DFW	618	1171	2.83333	2
90	SFO-ATL	764	2139	4.30000	2

```
  91     SFO-ATL     294    2139    4.33333      2
  92     SFO-DFW     340    1465    3.26667      2
  93     SFO-LAX     808     337    1.18333      1
  94     SFO-SLC    1567     599    1.61667      2
  95     SEA-ATL     292    2181    4.36667      1
  96     SEA-ANC    1717    1448    3.30000     -1
  97     SEA-LAX    1944     954    2.50000      0
  98     DCA-ATL     277     543    1.83333     -1
  99     DCA-BOS    1862     403    1.30000     -1
 100     DCA-MCO     557     759    2.33333      0
```

In 1994, Delta's flights varied greatly in length, from short local flights, mainly in the southeastern states, to transcontinental flights and ones serving Hawaii and Alaska. (No flights on the associated commuter services, collectively called Delta Connection, are included in the data set.)

Summarizing Data (3)

We begin by showing descriptive statistics and dot plots for the variables `Miles` and `TravTime`.

```
MTB > Describe 'Miles' 'TravTime'

                    N      MEAN    MEDIAN    TRMEAN     STDEV    SEMEAN
Miles             100    1023.6     745.0     951.1     798.7      79.9
TravTime          100     2.586     2.150     2.452     1.536     0.154

                  MIN       MAX        Q1        Q3
Miles           134.0    4501.0     430.5    1465.0
TravTime        0.750     9.583     1.417     3.475

MTB > DotPlot 'Miles' 'TravTime'


                  .  .
              .  :  :
             :.. :  :
            :::::.::   .          :
            :::::::::  : :  :   : :.
          .::::::::::: :.:::: : ::  .:          .        .
          +---------+---------+---------+---------+---------+-------Miles
          0      1000      2000      3000      4000      5000



              :
             :  :  ..
             :  :.::.
            :::::::::.::..   .:   ...        .
          .::::::::::::::::::::.:::::...: :       .       .            .
          -----+---------+---------+---------+---------+---------+-TravTime
             1.6       3.2       4.8       6.4       8.0       9.6
```

Flights for our data set were chosen to cover a wide variety of departure times, distances, and directions of flight. They are not a random sample. For example, it would not be appropriate to use 1,024 miles as an estimate of the mean length of the population of all Delta flights. For our sample, the dot plots show skewness to the right for both travel time and flight miles. The longest flight (4,500 miles, 9.6 hours) goes from Atlanta to Honolulu.

Not surprisingly, a scatterplot of the variables TravTime and Miles shows a very strong linear association: Flights over longer distances take more time.

Scatterplot of scheduled flight times and flight miles

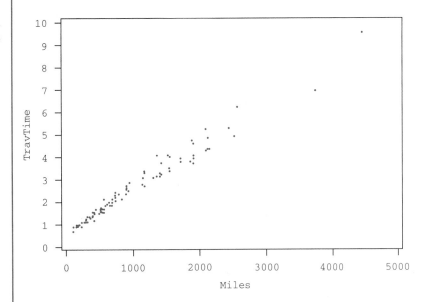

Analyzing Data, Interpreting the Analyses, and Communicating Results (4)

Because of this relationship, it seems worthwhile to find the regression line of $y =$ TravTime on $x =$ Miles. The strength of the linear association is quantified by the coefficient of determination $R^2 = 97\%$. The standard deviation $s_y = 1.5$ hr of the scheduled flight times is much larger than the standard deviation $s_\varepsilon = 0.27$ hr (denoted S in the printout that follows) of the vertical scatter about the regression line. The coefficient of determination is often interpreted as the percentage of the variance of scheduled travel times that is explained by the regression on the number of miles flown.

```
MTB > Regress 'TravTime' 1 'Miles';
SUBC>   SResiduals 'SRES1';

The regression equation is
TravTime = 0.646 + 0.00189 Miles

Predictor        Coef        StDev         T          P
Constant      0.64637      0.04339      14.90      0.000
Miles       0.00189446   0.00003349     56.58      0.000

S = 0.2661      R-Sq = 97.0%      R-Sq(adj) = 97.0%
```

```
Analysis of Variance

Source            DF         SS          MS         F         P
Regression         1      226.63      226.63    3200.81     0.000
Residual Error    98        6.94        0.07
Total             99      233.57

Unusual Observations
Obs     Miles    TravTime        Fit    StDev Fit     Residual     St Resid
 11      4501      9.5830      9.1733      0.1194       0.4097         1.72 X
 28      2139      5.2500      4.6986      0.0459       0.5514         2.10R
 29      2139      5.2830      4.6986      0.0459       0.5844         2.23R
 54      3784      7.0000      7.8150      0.0962      -0.8150        -3.28RX
 55      2555      4.9670      5.4867      0.0578      -0.5197        -2.00R
 62      1946      3.7500      4.3330      0.0408      -0.5830        -2.22R
 71      1395      4.0830      3.2891      0.0294       0.7939         3.00R
 75      2588      6.2500      5.5492      0.0588       0.7008         2.70R

R denotes an observation with a large standardized residual
X denotes an observation whose X value gives it large influence.
```

The regression equation `TravTime = .646 + .00189 Miles` has a practical interpretation. The constant term of about .65 hours (or 39 minutes) corresponds roughly to the scheduled time spent on the ground before takeoff and after landing. The coefficient of `Miles` indicates that each extra mile traveled takes about .00189 hours (or 6.8 seconds) of flying time. One mile in .00189 hours is equivalent to a ground speed of 1/.00189 = 529 mph. These numbers seem quite sensible.

The residual plot of these data, given here, shows that the residuals fan out; that is, the residuals become more variable as distance and time increase. This behavior in the residuals is a signal that something beyond what we have already discussed helps Delta schedule its flights.

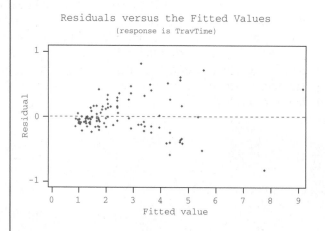

Residuals versus the Fitted Values
(response is TravTime)

Because major air currents around Earth flow from west to east, air travel time depends on the direction of travel. Eastbound flights usually have a tail wind that makes them go faster; westbound flights are usually slowed by head winds. Airlines have to take wind effects into account when planning their schedules.

Codes that provide crude indicators of flight direction are provided in the variable `TravDir`. If the direction of a flight is primarily westerly, against the prevailing winds, we assigned -2. Flights headed in an easterly direction were assigned the value $+2$. The value 0 was assigned to flights that travel mainly north or south. Flights traveling on a diagonal were coded $+1$ (SE and NE) or -1 (SW and NW). For example, we assigned the direction code $+1$ to a flight traveling northeast from MSY (New Orleans) to JFK (New York).

We begin our study of the effect of flight direction by making a labeled scatterplot of the standardized residuals against *flight miles*. The plotting symbol for each flight shows its direction. Aside from the symbols for directions, the overall effect is similar to that of the residual plot, with *fitted values* on the horizontal axis. (The plots look similar because fits are a linear function of flight miles, so the only distinction is a linear scale change on the horizontal axis.)

Standardized residuals of flight times (showing flight directions)

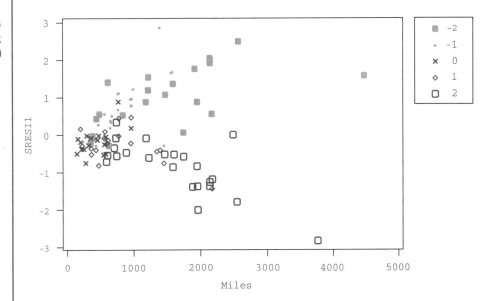

It is clear from the plotting symbols that the positive residuals come mainly from flights with a westerly component (large and small gray symbols) and the negative ones from those with an easterly component (large and small open symbols). None of the residuals for north–south flights are far from the 0 line. The information revealed by this plot of residuals suggests that we can improve our model for airline scheduling by taking flight directions into account.

We now have enough information about flight direction to make a more accurate model of airline scheduling. Because the effect of direction is magnified as distance traveled increases, we form the new variable `DirEffct` and use it in a multiple regression model.

```
MTB > Let 'DirEffct' = 'Miles'*'TravDir'
MTB > Regress TravTime 2 'Miles' 'DirEffct'

The regression equation is
TravTime = 0.647 + 0.00191 Miles -0.000089 DirEffct
```

```
Predictor         Coef        Stdev      t-ratio        p
Constant       0.64714      0.02621       24.69     0.000
Miles        0.00190543  0.00002025       94.12     0.000
Mi*Dir      -0.00008852  0.00000676      -13.10     0.000

s = 0.1607       R-sq = 98.9%   R-sq (adj) = 98.9%

Analysis of Variance

SOURCE          DF           SS          MS        F          p
Regression       2       231.08      115.54   4471.71     0.000
Error           97         2.51        0.03
Total           99       233.59

SOURCE          DF       SEQ SS
Miles            1       226.64
DirEfct          1         4.44

Unusual Observations
Obs. Miles   TravTime        Fit   Stdev.Fit   Residual   St.Resid
 11   4501     9.5833    10.0203      0.0969    -0.4370     -3.41RX
 13   1746     3.9667     4.2831      0.0332    -0.3165     -2.01R
 39   1561     4.0833     3.7597      0.0228     0.3236      2.03R
 54   3784     7.0000     7.1874      0.0753    -0.1874     -1.32 X
 64   2477     5.3333     4.9284      0.0459     0.4050      2.63R
 71   1395     4.0833     3.4287      0.0207     0.6546      4.11R
 75   2588     6.2500     6.0366      0.0514     0.2134      1.40 X

R denotes an obs. with a large st. resid.
X denotes an obs. whose X value gives it large influence.
```

The coefficient of determination is almost 99%, showing that by taking both flight distance and direction into account we have explained almost all of the variability in scheduled flight times.

Notice that the ground time of about 40 minutes, implied by the constant term, and the air speed of about 525 mph, implied by the coefficient of `Miles`, remain essentially unchanged from our previous regression model, which did not take account of direction. (These numbers might have changed considerably if we had not selected a good balance of east- and westbound flights, so that `DirEffct` has a small mean.) Also notice that the coefficient of the direction-effect term is negative, indicating that eastbound flights take less time.

Among the flights that fit this model worst are the longest flight (Atlanta to Honolulu, Observation 11), which spent much of its time in relatively uncluttered air space over Mexico and the Pacific, and a flight from JFK to Dallas-Fort Worth (Observation 71). The latter flight may have been a particularly difficult case for schedule planners at Delta. During the period when the schedule information was current, we checked the actual departure and arrival times of this flight for several days and found that it often departed more than 15 minutes late or arrived more than 15 minutes early.

12.8 Staying Focused

This chapter consolidates the material for expressing a response y as a function of one or more independent variables. Multiple regression models (where all the independent variables are quantitative) and models that incorporate information on qualitative variables were discussed. After presenting various models and the interpretation of βs in these models, we presented the normal equations used in obtaining the least-squares estimates $\hat{\beta}$.

A confidence interval and statistical test about an individual parameter β_j were developed using $\hat{\beta}_j$ and the standard error of $\hat{\beta}_j$. We also considered a statistical test about a set of βs, a confidence interval for $E(y)$ based on a set of xs, and a prediction interval for a given set of xs.

All of these inferences involve a fair to moderate amount of numerical calculation unless statistical software programs of packages are available. However, even these methods become unmanageable as the number of independent variables increases. Thus, the message should be very clear. Inferences about multiple regression models should be done using available computer software to facilitate the analysis and to minimize computational errors. Our job in these situations is to review and interpret the output.

Aside from a few exercises that will probe your understanding of the mechanics involved with these calculations, most of the exercises in the remainder of this chapter will make extensive use of computer output.

Here are some reminders about multiple regression concepts:

1. Regression coefficients in a first-order model (one not containing transformed values, such as squares of a variable or product terms) should be interpreted as partial slopes—the predicted change in a dependent variable when an independent variable is increased by one unit while other variables are held constant.

2. Correlations are important not only between an independent variable and the dependent variable, but also between independent variables. Collinearity—correlation between independent variables—implies that regression coefficients will change as variables are added to or deleted from a regression model.

3. The effectiveness of a regression model can be indicated not only by the R^2 value but also by the residual standard deviation. It's often helpful to use that standard deviation to see roughly how much of a plus or minus must be allowed around a prediction.

4. As always, the various statistical tests in a regression model only indicate how strong the evidence is that the apparent pattern is more than random. They don't directly indicate how good a predictive model is. In particular, a large overall F statistic may merely indicate a weak prediction in a large sample.

5. A t test in a multiple regression assesses whether that independent variable adds unique predictive value as a predictor in the model. It is quite possible that several variables may not add a statistically detectable amount of unique predicted value, yet deleting all of them from the model causes a serious drop in predictive value. This is especially true when there is severe collinearity.

6. The variance inflation factor (VIF) is a useful indicator of the overall impact of collinearity in estimating the coefficient of an independent variable. The higher the VIF number, the more serious is the impact of collinearity on the accuracy of a slope estimate.

7. Extrapolation in multiple regression can be subtle. A new set of x values may not be unreasonable when considered one by one, but the combination of values may be far outside the range of previous data.

Key Formulas

1. $R^2_{y \cdot x_1 \cdots x_k} = \dfrac{\text{SS(Total)} - \text{SS(Residual)}}{\text{SS(Total)}}$

where

$$\text{SS(Total)} = \Sigma \, (y_i - \bar{y})^2$$

and

$$\text{SS(Residual)} = \Sigma \, (y_i - \hat{y}_i)^2$$

2. $\text{SS(Regression)} = \Sigma \, (\hat{y}_i - \bar{y}_i)^2$

and

$$\text{SS(Total)} = \Sigma \, (y_i - \bar{y})^2 = \text{SS(Regression)} + \text{SS(Residual)}$$

3. F test for $H_0: \beta_1 = \beta_2 = \cdots = \beta_k = 0$

$$F = \dfrac{\text{SS(Regression)}/k}{\text{SS(Residual)}/[n - (k + 1)]}$$

4. $s_{\hat{\beta}_j} = s_\varepsilon \sqrt{\dfrac{1}{\Sigma_i \, (x_{ij} - \bar{x}_j)^2 (1 - R^2_{x_j \cdot x_1 \cdots x_{j-1} x_{j+1} \cdots x_k})}}$

and

$$s^2_\varepsilon = \dfrac{\text{MS(Residual)}}{n - (k + 1)}$$

5. Confidence interval for β_j

$$\hat{\beta}_j - t_{\alpha/2} s_{\hat{\beta}_j} \le \beta_j \le \hat{\beta}_j + t_{\alpha/2} s_{\hat{\beta}_j}$$

6. Statistical test for β_j

$$\text{T.S.:} \quad t = \dfrac{\hat{\beta}_j}{s_{\hat{\beta}_j}}$$

7. Testing a subset of predictors

$$H_0: \quad \beta_{g+1} = \beta_{g+2} = \cdots = \beta_k = 0$$

$$\text{T.S.:} \quad F = \dfrac{[\text{SS(Regression, complete)} - \text{SS(Regression, reduced)}]/(k - g)}{\text{SS(Residual, complete)}/[n - (k + 1)]}$$

Supplementary Exercises

12.16 Refer to the analyses of the data set DELTAFLY in Section 12.7. What happens if you do a multiple regression with the explanatory variables Miles and TravDir (i.e., you don't adjust the direction effect for distance)?

12.17 Refer to Exercise 12.16. Try using three explanatory variables Miles, TravDir, and DirEffct. What happens to TravDir in the presence of DirEffct? If you were scheduling a new 500-mile, west-bound flight, how much travel time would you allow for?

Bus.   **12.18** A study of demand for imported subcompact cars incorporates data from 12 metropolitan areas. The variables are as follows:

Demand:	Imported subcompact car sales as a percentage of total sales
Educ:	Average number of years of schooling completed by adults
Income:	Per capita income
Popn:	Area population
Famsize:	Average size of intact families

Minitab output is as follows.

```
MTB > Regress 'Demand' 4 'Educ' 'Income' 'Popn' 'Famsize'.

The regression equation is
Demand = - 1.3 + 5.55 Educ + 0.89 Income + 1.92 Popn - 11.4 Famsize

Predictor        Coef       Stdev     t-ratio         p
Constant        -1.32       57.98       -0.02     0.982
Educ            5.550       2.702        2.05     0.079
Income          0.885       1.308        0.68     0.520
Popn            1.925       1.371        1.40     0.203
Famsize       -11.389       6.669       -1.71     0.131

s = 2.686       R-sq = 96.2%    R-sq(adj) = 94.1%

Analysis of Variance

SOURCE        DF          SS          MS          F          p
Regression     4     1295.70      323.93      44.89      0.000
Error          7       50.51        7.22
Total         11     1346.22

SOURCE        DF      SEQ SS
Educ           1     1239.95
Income         1       32.85
Popn           1        1.86
Famsize        1       21.04

Unusual Observations
Obs.     Educ      Demand       Fit   Stdev.Fit   Residual   St.Resid
   9      9.3      13.100     9.760       2.149      3.340      2.07R

R denotes an obs. with a large st. resid.
```

a. Write the regression equation. Place the standard error of each coefficient below the coefficient, perhaps in parentheses.
b. Locate R^2 and the residual standard deviation.
c. The Unusual Observations entry in the output indicates that observation 9 had a value 2.07 standard deviations away from the predicted Fit value. Does this indicate that observation 9 is a very serious outlier?

12.19 Summarize the conclusions of the F test and the various t tests in the output of Exercise 12.18.

12.20 Another analysis of the data of Exercise 12.18 uses only Educ and Famsize to predict Demand. The output is as follows.

```
MTB > Regress 'Demand' 2 'Educ' 'Famsize'.

The regression equation is
Demand = - 19.2 + 7.79 Educ + 9.46 Famsize

Predictor        Coef      Stdev    t-ratio        p
Constant       -19.17      45.87      -0.42    0.686
Educ            7.793      2.490       3.13    0.012
Famsize        -9.464      5.207      -1.82    0.103

s = 2.939      R-sq = 94.2%    R-sq(adj) = 92.9%

Analysis of Variance

SOURCE        DF         SS        MS        F        p
Regression     2    1268.48    634.24    73.43    0.000
Error          9      77.73      8.64
Total         11    1346.22
```

a. Locate the R^2 value for this reduced model.
b. Test the null hypothesis that the true coefficients of Income and Popn are zero. Use $\alpha = .05$. What is the conclusion?

Engin.   **12.21** The manager of documentation for a computer software firm wants to forecast the time required to document moderate-size computer programs. Records are available for 26 programs. The variables are y = number of writer-days needed, x_1 = number of subprograms, x_2 = average number of lines per subprogram, $x_3 = x_1 x_2$, $x_4 = x_2^2$, and $x_5 = x_1 x_2^2$. A portion of the output from a regression analysis of the data is shown here.

```
                Multiple Regression Analysis

Dependent variable: Y

                    Table of Estimates

                               Standard         t         P
              Estimate          Error       Value     Value

Constant      -16.8198         11.631       -1.45    0.1636
X1             1.47019          0.365944     4.02    0.0007
X2             0.994778         0.611441     1.63    0.1194
X1X2          -0.0240071        0.0237565   -1.01    0.3243
X2Sq          -0.01031          0.007374    -1.40    0.1774
X1X2Sq        -0.000249574      0.000351779  0.71    0.4862
```

```
R-squared = 91.72%
Adjusted R-squared = 89.65%
Standard error of estimation = 3.39011
Durbin-Watson statistic = 2.12676
Mean absolute error = 2.4127
```

```
                    Analysis of Variance

                 Sum of                                                  P
Source           Squares      D.F.     Mean Square    F-Ratio      Value

Model            2546.03        5         509.205       44.31      0.0000
Error            229.857       20         11.4929

Total (corr.)    2775.88       25
```

a. Write the multiple regression model and locate the residual standard deviation.

b. Does x_3 have a significant, unique predictive value?

12.22 The model $y = \beta_0 + \beta_1 x_1 + \beta_2 x_2 + \varepsilon$ is fit to the data of Exercise 12.21. Selected output is shown here.

```
                Multiple Regression Analysis

Dependent variable: Y

                    Table of Estimates

                            Standard          t            P
                Estimate      Error        Value        Value

Constant        0.840085     3.43375        0.24       0.8089
X1              1.01583      0.0792925     12.81       0.0000
X2              0.0558262    0.0515066      1.08       0.2897
R-squared = 90.64%
Adjusted R-squared = 89.83%
Standard error of estimation = 3.36066
Durbin-Watson statistic = 2.2053
Mean absolute error = 2.57584
```

```
                    Analysis of Variance

                 Sum of                                                  P
Source           Squares      D.F.     Mean Square    F-Ratio      Value

Model            2516.12        2         1258.06      111.39      0.0000
Error            259.763       23         11.294

Total (corr.)    2775.88       25
```

a. Write the complete and reduced-form estimated models.

b. Is the improvement in R^2 obtained by adding x_3, x_4, and x_5 statistically significant at $\alpha = .05$? Approximately what is the p-value for this test?

Ag.   **12.23**   A producer of various feed additives for cattle conducts a study of the number of days of feedlot time required to bring beef cattle to market weight. Eighteen steers of essentially identical age and weight are purchased and brought to a feedlot. Each steer is fed a diet with a specific combination of protein content, antibiotic concentration, and percentage of feed supplement. The data are as follows:

STEER:	1	2	3	4	5	6	7	8	9
PROTEIN:	10	10	10	10	10	10	15	15	15
ANTIBIO:	1	1	1	2	2	2	1	1	1
SUPPLEM:	3	5	7	3	5	7	3	5	7
TIME:	88	82	81	82	83	75	80	80	75
STEER:	10	11	12	13	14	15	16	17	18
PROTEIN:	15	15	15	20	20	20	20	20	20
ANTIBIO:	2	2	2	1	1	1	2	2	2
SUPPLEM:	3	5	7	3	5	7	3	5	7
TIME:	77	76	72	79	74	75	74	70	69

Computer output from a Systat regression analysis follows.

```
CORRELATIONS (PEARSON)


            TIME      PROTEIN      ANTIBIO
PROTEIN   -0.7111
ANTIBIO   -0.4180    0.0000
SUPPLEM   -0.4693    0.0000      -0.0000


CASES INCLUDED 18 MISSING CASES 0

UNWEIGHTED LEAST SQUARES LINEAR REGRESSION OF TIME


PREDICTOR
VARIABLES     COEFFICIENT     STD ERROR     STUDENT'S T        P        VIF


CONSTANT        102.708        2.31037        44.46        0.0000
PROTEIN         -0.83333       0.09870        -8.44        0.0000      1.0
ANTIBIO         -4.00000       0.80589        -4.96        0.0002      1.0
SUPPLEM         -1.37500       0.24675        -5.57        0.0001      1.0


R-SQUARED              0.9007     RESID. MEAN SQUARE (MSE)     2.92261
ADJUSTED R-SQUARED     0.8794     STANDARD DEVIATION           1.70956


SOURCE        DF        SS          MS          F          P


REGRESSION     3      371.083     123.694     42.32     0.0000
RESIDUAL      14       40.9166      2.92261
TOTAL         17      412.000

PREDICTED/FITTED VALUES OF TIME


LOWER PREDICTED BOUND      73.566     LOWER FITTED BOUND     76.469
PREDICTED VALUE            77.333     FITTED VALUE           77.333
UPPER PREDICTED BOUND      81.100     UPPER FITTED BOUND     78.197
SE (PREDICTED VALUE)       1.7564     SE (FITTED VALUE)      0.4029
```

```
UNUSUALNESS (LEVERAGE)    0.0556
PERCENT COVERAGE          95.0
CORRESPONDING T           2.14

PREDICTOR VALUES: PROTEIN = 15.000, ANTIBIO = 1.5000, SUPPLEM = 5.0000
```

    a. Write the regression equation.
    b. Find the standard deviation.
    c. Find the R^2 value.
    d. How much of a collinearity problem is there with these data?

12.24 Refer to Exercise 12.23.
    a. Predict the feedlot time required for a steer fed 15% protein, 1.5% antibiotic concentration, and 5% supplement.
    b. Do these values of the independent variables represent a major extrapolation from the data?
    c. Give a 95% confidence interval for the mean time predicted in part (a).

12.25 The data of Exercise 12.23 are also analyzed by a regression model using only protein content as an independent variable, with the following output.

```
UNWEIGHTED LEAST SQUARES LINEAR REGRESSION OF TIME

PREDICTOR
VARIABLES     COEFFICIENT    STD ERROR    STUDENT'S T      P

CONSTANT       89.8333       3.20219       28.05        0.0000
PROTEIN        -0.83333      0.20598       -4.05        0.0009

R-SQUARED             0.5057    RESID. MEAN SQUARE (MSE)   12.7291
ADJUSTED R-SQUARED    0.4748    STANDARD DEVIATION         3.56779

SOURCE        DF      SS         MS        F        P

REGRESSION     1    208.333    208.333    16.37    0.0009
RESIDUAL      16    203.666    12.7291
TOTAL         17    412.000
```

    a. Write the regression equation.
    b. Find the R^2 value.
    c. Test the null hypothesis that the coefficients of ANTIBIO and SUPPLEM are zero at $\alpha = .05$.

Bus. **12.26** The market research manager of a catalog clothing supplier has begun an investigation of what factors determine the typical order size the supplier receives from customers. From the sales records stored on the company's computer, the manager obtained average order size data for 180 zip code areas. A part-time intern looked up the latest census information on per capita income, average years of formal education, and median price of an existing house in each of these zip code areas. (The intern couldn't find house price data for two zip codes, and entered 0 for those areas.) The manager also was curious whether climate had any bearing on order size, and included data on the average daily high temperature in winter and in summer.

    The market research manager has asked for your help in analyzing the data. The output provided is only intended as a first try. The manager would like to know whether there was any evidence that the temperature variables mattered much, and also which of the other variables seemed useful. There is some question about whether putting in 0 for the missing house price data was the right thing to do or whether that might distort the results. Please provide a basic, not too technical explanation of the results in this output and any other analyses you choose to perform.

```
MTB > name c1 'AvgOrder' c2 'Income' c3 'Educn' &
CONT> c4 'HousePr' c5 'WintTemp' c6 'SummTemp'
MTB > correlations of c1-c6

            AvgOrder    Income    Educn   HousePr  WintTemp
Income        0.205
Educn         0.171     0.913
HousePr       0.269     0.616    0.561
WintTemp     -0.134    -0.098    0.014    0.066
SummTemp     -0.068    -0.115    0.005    0.018    0.481


MTB > regress c1 on 5 variables in c2-c6

The regression equation is
AvgOrder = 36.2 + 0.078 Income - 0.019 Educn
   + 0.0605 HousePr - 0.223 WintTemp + 0.006 SummTemp

Predictor        Coef      Stdev    t-ratio        p
Constant        36.18      12.37       2.92    0.004
Income         0.0780     0.4190       0.19    0.853
Educn         -0.0189     0.5180      -0.04    0.971
HousePr       0.06049    0.02161       2.80    0.006
WintTemp      -0.2231     0.1259      -1.77    0.078
SummTemp       0.0063     0.1646       0.04    0.969

s = 4.747       R-sq = 9.6%      R-sq(adj) = 7.0%

Analysis of Variance

SOURCE         DF         SS        MS       F         p
Regression      5     417.63     83.53    3.71     0.003
Error         174    3920.31     22.53
Total         179    4337.94

SOURCE         DF    SEQ SS
Income          1    182.94
Educn           1      7.18
HousePr         1    142.63
WintTemp        1     84.84
SummTemp        1      0.03

Unusual Observations
Obs.   Income AvgOrder      Fit  Stdev.Fit  Residual   St.Resid
  25     17.1   23.570   36.555      0.632   -12.985     -2.76R
  78     11.9   24.990   34.950      0.793    -9.960     -2.13R
  83     13.4   36.750   29.136      2.610     7.614      1.92X
```

```
    87     14.3    45.970    35.918    0.463    10.052    2.13R
   111     11.1    21.720    33.570    0.802   -11.850   -2.53R
   113     10.4    43.500    33.469    0.817    10.031    2.15R
   143     16.1    20.350    27.915    3.000    -7.565   -2.06RX
   149     13.2    44.970    35.369    0.604     9.601    2.04R
   169     13.5    44.650    34.361    0.660    10.289    2.19R
   180     13.7    23.050    34.929    0.469   -11.879   -2.51R
 R denotes an obs. with a large st. resid.
 X denotes an obs. whose X value gives it large influence.
```

12.27 The accompanying table gives demographic data for 12 male patients with congestive heart failure enrolled in a study of an experimental compound.

Demographic Data for Patients with Heart Failure (NYHA Class III or IV)

| | | | | | Baseline | |
| | | | | | **Cardiac Index** | **Pulmonary Capillary Wedge Pressure** |
Patient	**Age (yrs)**	**Disease Duration**	**Height (cm)**	**Weight (kg)**	**(L/min/m²)**	**(mm Hg)**
01	67	5 yr	172.0	57.0	1.6	40
02	45	2 yr	170.0	67.0	2.4	25
03	59	8 yr	172.7	102.0	2.2	39
04	63	1 yr	175.3	74.9	1.7	39
05	55	1 yr	172.7	92.0	2.3	34
06	65	1 yr	178.0	90.0	1.6	36
07	62	2 yr	163.0	67.0	1.4	36
08	60	1 yr	182.5	72.0	2.2	17
09	72	2 yr	168.0	71.0	1.3	37
10	44	3 mo	163.0	68.0	2.4	28
11	63	5 yr	172.0	82.0	2.1	38
12	63	1 yr	163.0	64.0	1.1	36

a. Summarize these data using a boxplot for each variable.
b. Construct scatterplots to display (1) age by cardiac index (CI) and by pulmonary capillary wedge pressure (PCWP) and (2) disease duration by CI and by PCWP. Is there evidence of a correlation between age and CI or PCWP? What about correlation between duration of disease and CI or PCWP?

12.28 The data of Exercise 12.27 were used to fit several multiple regression models; y_1 = CI, y_2 = PCWP, x_1 = age, x_2 = disease duration.

    I. $y_1 = \beta_0 + \beta_1 x_1 + \beta_2 x_2 + \varepsilon$
    II. $y_1 = \beta_0 + \beta_1 x_1 + \beta_2 x_2 + \beta_3 x_1 x_2 + \varepsilon$
    III. $y_2 = \beta_0 + \beta_1 x_1 + \beta_2 x_2 + \varepsilon$
    IV. $y_2 = \beta_0 + \beta_1 x_1 + \beta_2 x_2 + \beta_3 x_1 x_2 + \varepsilon$

```
                    REGRESSION ANALYSIS, MODEL I

Dependent Variable: CI

Analysis of Variance

                          Sum of        Mean
Source          DF       Squares       Square      F Value    Prob>F
Model            2       1.56955      0.78478        9.298     0.0065
Error            9       0.75961      0.08440
C Total         11       2.32917


      Root MSE      0.29052     R-square      0.6739
      Dep Mean      1.85833     Adj R-sq      0.6014
      C.V.         15.63333

Parameter Estimates
                  Parameter      Standard     T for H0:
Variable  DF       Estimate         Error    Parameter=0    Prob > |T|
INTERCEP   1       4.475622    0.63976685         6.996        0.0001
AGE        1      -0.046203    0.01083529        -4.264        0.0021
DURATION   1       0.060395    0.03852829         1.568        0.1514


                    REGRESSION ANALYSIS, MODEL II

Dependent Variable: CI

Analysis of Variance

                          Sum of        Mean
Source          DF       Squares       Square      F Value    Prob>F
Model            3       1.57161      0.52387        5.532     0.0237
Error            8       0.75755      0.09469
C Total         11       2.32917


      Root MSE      0.30772     R-square      0.6748
      Dep Mean      1.85833     Adj R-sq      0.5528
      C.V.         16.55915

Parameter Estimates
                  Parameter      Standard     T for H0:
Variable  DF       Estimate         Error    Parameter=0    Prob > |T|
INTERCEP   1       4.599307    1.07814691         4.266        0.0027
AGE        1      -0.048340    0.01848097        -2.616        0.0309
DURATION   1      -0.022410    0.56287924        -0.040        0.9692
AGE_DUR    1       0.001376    0.00932590         0.147        0.8864


                    Variable
Variable  DF        Label

INTERCEP   1    Intercept
AGE        1
DURATION   1
AGE_DUR    1    AGE TIMES DURATION
```

```
                    REGRESSION ANALYSIS, MODEL III


Dependent Variable: PCWP

Analysis of Variance
                         Sum of        Mean
Source          DF       Squares       Square      F Value    Prob>F
Model            2      221.88101    110.94051      3.259      0.0862
Error            9      306.36899     34.04100
C Total         11      528.25000


     Root MSE      5.83447      R-square      0.4200
     Dep Mean     33.75000      Adj R-sq      0.2911
     C.V.         17.28731


Parameter Estimates
                    Parameter       Standard     T for H0:
Variable   DF       Estimate          Error      Parameter=0     Prob > |T|
INTERCEP    1       7.298786      12.84835977        0.568        0.5839
AGE         1       0.400475       0.21760372        1.840        0.0989
DURATION    1       1.021327       0.77375900        1.320        0.2194



                    REGRESSION ANALYSIS, MODEL IV

Dependent Variable: PCWP

Analysis of Variance
                         Sum of        Mean
Source          DF       Squares       Square      F Value    Prob>F
Model            3      228.56515     76.18838      2.034      0.1878
Error            8      299.68485     37.46061
C Total         11      528.25000


     Root MSE      6.12051      R-square      0.4327
     Dep Mean     33.75000      Adj R-sq      0.2199
     C.V.         18.13484


Parameter Estimates
                    Parameter       Standard     T for H0:
Variable   DF       Estimate          Error      Parameter=0     Prob > |T|
INTERCEP    1      14.344026      21.44389171        0.669        0.5224
AGE         1       0.278775       0.36757883        0.758        0.4700
DURATION    1      -3.695301      11.19543293       -0.330        0.7498
AGE_DUR     1       0.078352       0.18548824        0.422        0.6838


                    Variable
Variable   DF       Label
INTERCEP    1   Intercept
AGE         1
DURATION    1
AGE_DUR     1   AGE TIMES DURATION
```

a. Which model provides the best fit to the cardiac index data? To the pulmonary capillary wedge pressure data?

b. Do these analyses confirm what you concluded in Exercise 12.27? Explain.

12.29 Defining the Problem (1). Engineers for a manufacturer of power tools for home use were trying to design an electric drill that did not heat up under strenuous use. The three key design factors were insulation thickness, quality of the wire used in the motor, and size of the vents in the body of the drill.

Collecting Data (2). The engineers had learned a little about off-line quality control, so they designed an experiment that varied these design factors. They created 10 drills using each combination of the three design factors, split them into two lots, and tested the lots under two (supposedly equivalent) torture tests. The temperature of each drill was measured at the end of each test; for each lot, the mean temperature and the logarithm of the variance of temperatures were computed. The engineers wanted to minimize both the mean and the logarithm of the variance.

For this experiment, there are three key design factors:

IT is the insulating thickness of the drill (IT = 2, 3, 4, 5, or 6)
QW is the quality of wire used in the motor (QW = 6, 7, or 8)
VS is the size of the vent in the body of the drill (VS = 10, 11, or 12)

There are $5 \times 3 \times 3 = 45$ different combinations of these design factors. For each combination of factors, 10 drills were made and divided into two lots of five drills. Each drill was subjected to a torture test and the temperature recorded at the end of the test. Avtem represents the average temperature for the five drills of a lot for a given combination of the design factors. The 90 measurements are given here ($5 \times 3 \times 3 \times 2$ lots = 90). Also included in the data set are

logv = logarithm of the variance of temperatures for a given combination
of factors and lot

and I2, Q2, and V2, which are squared terms for the three design factors computed as

(design factor − mean design factor)^2

avtem	logv	IT	QW	VS	I2	Q2	V2	Lot
185	3.6	2	6	10	4	1	1	1
176	3.7	2	6	10	4	1	1	2
177	3.6	2	6	11	4	1	0	1
184	3.7	2	6	11	4	1	0	2
178	3.6	2	6	12	4	1	1	1
169	3.4	2	6	12	4	1	1	2
185	3.2	2	7	10	4	0	1	1
184	3.2	2	7	10	4	0	1	2
180	3.2	2	7	11	4	0	0	1
184	3.5	2	7	11	4	0	0	2
179	3.0	2	7	12	4	0	1	1
173	3.2	2	7	12	4	0	1	2
179	2.9	2	8	10	4	1	1	1
185	2.7	2	8	10	4	1	1	2
180	2.8	2	8	11	4	1	0	1
180	2.7	2	8	11	4	1	0	2
169	2.9	2	8	12	4	1	1	1
177	2.8	2	8	12	4	1	1	2
172	3.6	3	6	10	1	1	1	1
171	3.9	3	6	10	1	1	1	2
172	3.8	3	6	11	1	1	0	1

avtem	logv	IT	QW	VS	I2	Q2	V2	Lot
167	3.6	3	6	11	1	1	0	2
165	3.3	3	6	12	1	1	1	1
159	3.4	3	6	12	1	1	1	2
169	3.0	3	7	10	1	0	1	1
174	3.3	3	7	10	1	0	1	2
163	3.3	3	7	11	1	0	0	1
170	3.3	3	7	11	1	0	0	2
169	3.2	3	7	12	1	0	1	1
163	3.2	3	7	12	1	0	1	2
178	2.7	3	8	10	1	1	1	1
165	2.7	3	8	10	1	1	1	2
167	2.8	3	8	11	1	1	0	1
171	2.8	3	8	11	1	1	0	2
166	2.9	3	8	12	1	1	1	1
166	2.7	3	8	12	1	1	1	2
161	3.7	4	6	10	0	1	1	1
162	3.7	4	6	10	0	1	1	2
169	3.4	4	6	11	0	1	0	1
162	3.7	4	6	11	0	1	0	2
159	3.5	4	6	12	0	1	1	1
168	3.4	4	6	12	0	1	1	2
169	3.1	4	7	10	0	0	1	1
165	3.2	4	7	10	0	0	1	2
163	3.2	4	7	11	0	0	0	1
168	3.4	4	7	11	0	0	0	2
160	2.9	4	7	12	0	0	1	1
154	3.1	4	7	12	0	0	1	2
169	2.8	4	8	10	0	1	1	1
156	2.9	4	8	10	0	1	1	2
168	2.7	4	8	11	0	1	0	1
161	2.7	4	8	11	0	1	0	2
156	2.6	4	8	12	0	1	1	1
158	2.7	4	8	12	0	1	1	2
164	3.7	5	6	10	1	1	1	1
163	3.7	5	6	10	1	1	1	2
161	3.7	5	6	11	1	1	0	1
158	3.4	5	6	11	1	1	0	2
154	3.4	5	6	12	1	1	1	1
162	3.7	5	6	12	1	1	1	2
163	2.8	5	7	10	1	0	1	1
166	3.0	5	7	10	1	0	1	2
159	3.3	5	7	11	1	0	0	1
156	3.3	5	7	11	1	0	0	2
152	3.3	5	7	12	1	0	1	1
150	3.3	5	7	12	1	0	1	2
165	2.9	5	8	10	1	1	1	1
156	2.7	5	8	10	1	1	1	2
155	2.8	5	8	11	1	1	0	1

(continued)

avtem	logv	IT	QW	VS	I2	Q2	V2	Lot
155	3.2	5	8	11	1	1	0	2
149	2.6	5	8	12	1	1	1	1
152	2.9	5	8	12	1	1	1	2
165	3.4	6	6	10	4	1	1	1
160	3.7	6	6	10	4	1	1	2
157	3.7	6	6	11	4	1	0	1
149	3.7	6	6	11	4	1	0	2
149	3.8	6	6	12	4	1	1	1
145	3.7	6	6	12	4	1	1	2
154	3.4	6	7	10	4	0	1	1
153	3.2	6	7	10	4	0	1	2
150	3.0	6	7	11	4	0	0	1
156	3.1	6	7	11	4	0	0	2
146	3.2	6	7	12	4	0	1	1
153	3.3	6	7	12	4	0	1	2
161	2.8	6	8	10	4	1	1	1
160	2.9	6	8	10	4	1	1	2
156	2.9	6	8	11	4	1	0	1
150	2.7	6	8	11	4	1	0	2
149	2.9	6	8	12	4	1	1	1
151	2.8	6	8	12	4	1	1	2

These data are found in the file ELECTRIC on the CD for this book. The engineers want the data analyzed and they want to minimize both the mean and the logarithm of the variance for temperatures. They have asked you to try to figure out which of the design factors seem to affect the mean (and by how much), which design factors affect the variance, which squared terms seem to matter, and, finally, whether the lot number (corresponding to the type of test) is relevant.

a. The initial analysis of the data involves plotting the performance variables avtem and logv versus the design factors in order to obtain an idea of which design factors may be related to the performance variables. Also, these plots display whether the relationships between performance variables and design factors are linear or require a higher-order model.

1. Construct the six scatterplots of performance variables versus design factors; that is, plot avtem versus IT, QW, and VS. Then plot logv versus IT, QW, and VS.

2. Examine the three scatterplots involving avtem and describe the relationships, if any, between avtem and the three design factors.

3. Examine the three scatterplots involving logv and describe the relationships, if any, between logv and the three design factors.

b. After examining the scatterplots, we want to examine a number of models to determine which model provides the best overall fit to the avtem data without overfitting the model—that is, placing too many terms in the model. Fit the following models to the avtem data.

Model 1: $\text{avtem} = \beta_0 + \beta_1 \text{IT} + \beta_2 \text{QW} + \beta_3 \text{VS} + \varepsilon$

Model 2: $\text{avtem} = \beta_0 + \beta_1 \text{IT} + \beta_2 \text{QW} + \beta_3 \text{VS} + \beta_4 \text{I2} + \beta_5 \text{Q2} + \beta_6 \text{V2} + \varepsilon$

Model 3: $\text{avtem} = \beta_0 + \beta_1 \text{IT} + \beta_2 \text{QW} + \beta_3 \text{VS} + \beta_4 \text{IT} \cdot \text{QW} + \beta_5 \text{IT} \cdot \text{VS}$
$$+ \beta_6 \text{QW} \cdot \text{VS} + \varepsilon$$

Model 4: $\text{avtem} = \beta_0 + \beta_1 \text{IT} + \beta_2 \text{QW} + \beta_3 \text{VS} + \beta_4 \text{I2} + \beta_5 \text{Q2} + \beta_6 \text{V2}$
$$+ \beta_7 \text{IT} \cdot \text{QW} + \beta_8 \text{IT} \cdot \text{VS} + \beta_9 \text{QW} \cdot \text{VS} + \varepsilon$$

1. Based on the values of R^2 for the four models, which model would you select as providing the best fit to the data?
2. Test the hypothesis that model 2 is not significantly different from model 1. Use $\alpha = .05$.
3. Test the hypothesis that model 3 is not significantly different from model 1. Use $\alpha = .05$.
4. Test the hypothesis that model 4 is not significantly different from model 3. Use $\alpha = .05$.
5. Test the hypothesis that model 4 is not significantly different from model 2. Use $\alpha = .05$.
6. Using the scatterplots and the results of parts (b1)–(b5), choose which model you would recommend to the engineers. Explain your reasons for selecting the model.

c. After examining the scatterplots, we want to examine a number of models to determine which model provides the best overall fit to the logv data without overfitting the model—that is, without placing too many terms in the model. Fit the following models to the logv data.

$$\text{Model 1:} \quad \text{logv} = \beta_0 + \beta_1 \text{IT} + \beta_2 \text{QW} + \beta_3 \text{VS} + \varepsilon$$

$$\text{Model 2:} \quad \text{logv} = \beta_0 + \beta_1 \text{IT} + \beta_2 \text{QW} + \beta_3 \text{VS} + \beta_4 \text{I2} + \beta_5 \text{Q2} + \beta_6 \text{V2} + \varepsilon$$

$$\text{Model 3:} \quad \text{logv} = \beta_0 + \beta_1 \text{IT} + \beta_2 \text{QW} + \beta_3 \text{VS} + \beta_4 \text{IT} \cdot \text{QW} + \beta_5 \text{IT} \cdot \text{VS} + \beta_6 \text{QW} \cdot \text{VS} + \varepsilon$$

$$\text{Model 4:} \quad \text{logv} = \beta_0 + \beta_1 \text{IT} + \beta_2 \text{QW} + \beta_3 \text{VS} + \beta_4 \text{I2} + \beta_5 \text{Q2} + \beta_6 \text{V2} + \beta_7 \text{IT} \cdot \text{QW} + \beta_8 \text{IT} \cdot \text{VS} + \beta_9 \text{QW} \cdot \text{VS} + \varepsilon$$

1. Based on the values of R^2 for the four models, which model would you select as providing the best fit to the data?
2. Test the hypothesis that model 2 is not significantly different from model 1. Use $\alpha = .05$.
3. Test the hypothesis that model 3 is not significantly different from model 1. Use $\alpha = .05$.
4. Test the hypothesis that model 4 is not significantly different from model 3. Use $\alpha = .05$.
5. Test the hypothesis that model 4 is not significantly different from model 2. Use $\alpha = .05$.
6. Using the scatterplots and the results of parts (c1)–(c5), choose which model you would recommend to the engineers. Explain your reasons for selecting the model.

CHAPTER 13

Communicating and Documenting the Results of Analyses

13.1   Introduction

In Chapter 1, we introduced the subject of statistics as the science of learning from data. The four steps in learning from data, which closely parallel the steps in the scientific method—(1) defining the problem, (2) collecting data, (3) summarizing data, (4) analyzing data, interpreting the analyses, and communicating results—form the major outline for this book. We have kept these four steps in focus as we progressed through the book by showing these steps, however briefly, in our Encounters with Real Data.

In the previous chapters, we have discussed particular statistical methods, how those methods are applied to specific data sets, and how findings from statistical analyses in the form of computer output are interpreted. And, except for a brief, optional discussion on Section 2.7, we have not concentrated on the processing steps that the researcher follows between the time the data are received and the time they are available in computer-readable form for analysis, nor have we discussed in detail the form and content of the report that summarizes the results of a statistical analysis. In this chapter, we consider the data-processing steps and statistical report writing. This chapter is not a complete manual with all the tools required; rather, it is an overview—what a manager or researcher should know about these steps. The chapter reflects, as an example, the standard procedures in the pharmaceutical industry, which is highly regulated. Procedures differ somewhat in other industries and organizations. We begin this chapter with some general pitfalls and difficulties you might encounter in communicating effectively.

13.2   The Difficulty of Good Communication

We have spent time throughout the book making sense of data; the final step in this process is the communication of results. How might you communicate the results of a study or survey? The list of possibilities is almost endless, including all forms of verbal and written communication. There is quite a range of possibilities for verbal and written communication. For example, written communication within a company or organization can vary from an informal short note or memo to a formal project report (Figure 13.1).

Communicating the results of a statistical analysis in concise, unambiguous terms is difficult. In fact, descriptions of most things are difficult. For example, try to describe the person sitting next to you so precisely that a stranger could select the individual from a group of others having similar physical characteristics. It is not an easy task. Fingerprints, voiceprints, faceprints, and photographs—all pictorial descriptions—are some of the most precise methods of human identification. The description of a set of measurements is also a difficult task. However, like the description of a person, it can be accomplished more easily by using graphics or pictorial methods.

Cave drawings convey to us scattered bits of information about the life of prehistoric people. Similarly, vast quantities of knowledge about the ancient lives and cultures of the Babylonians, Egyptians, Greeks, and Romans are brought to life by means of drawings and sculpture. Art has been used to convey a picture of various lifestyles, history, and culture in all ages. Not surprisingly, the use of graphs and tables along with a written description can help to convey the meaning of a statistical analysis.

In reading the results of a statistical analysis and in communicating the results of our own analyses, we must be careful not to distort them because of the way we present the data and results. You have all heard the expression, "It is easy to lie with statistics." The idea is *not* new. The famous British statesman Disraeli is quoted as

FIGURE 13.1
Forms of written and
verbal communication

Verbal Communication	
Informal conversation	Formal presentation

Written Communication	
Internal (e.g., within company, university, etc.)	External
Memorandum	Letter (or letter to editor)
Formal project report	Scientific journal article

saying, "There are three kinds of lies: lies, damned lies, and statistics." Where do things go wrong?

First of all, the distortion of truth can occur only when we communicate. And because communication can be accomplished with graphs, pictures, sound, aroma, taste, words, numbers, or any other means devised to reach our senses, distortions can occur using any one or any combination of these methods of communication.

In this respect, statements that we make could be misleading to others because we might have omitted something in the explanation of the data-gathering stage or with the analyses done. For example, we might unintentionally fail to clearly explain the meaning of a numerical statement, or we might omit some background information that is necessary for a clear interpretation of the results. Even a correct statement may appear to be distorted if the reader lacks knowledge of elementary statistics. Thus a very clear expression of an inference using a 95% confidence interval is meaningless to a person who has not been exposed to the introductory concepts of statistics.

Now we will look at some potential hurdles to effective communication that we must carefully consider when we present the results of a statistical analysis—or when we try to interpret what someone else has presented.

13.3 Communication Hurdles: Graphical Distortions

Pictures can easily distort the truth. The marketing of many products, including soft drinks, beers, cosmetics, clothing, automobiles, and many more, involves the use of attractive, youthful models. The not-so-subtle impression we are left with is that (somehow) by using the product, we, too, will look like these models. Have you ever stepped back from one of these commercials and wondered how the commercial message relates to the quality and usefulness of the product? Have you thought about how you are being misled by a commercial? The use of sex appeal to sell products is very prominent, and we seem to accept this type of distortion. The beer ad article shown here, which appeared in *USA Today,* March 15, 2000, illustrates how we are manipulated through these types of ads.

Sex Appeal Slipping Back into Beer Ads

'Risqué' TV leads to more liberal promotions

By Michael McCarthy
USA TODAY

NEW YORK—The "babes" in socalled Beer & Babes advertising are back.

Nearly a decade ago national soul-searching about sexual harrassment in the wake of the Anita Hill controversy forced beer marketers to purge traditional and what were seen as sexist symbols, such as "The Swedish Bikini Team," from their ads.

Now, brewers including Miller Brewing, Anheuser-Busch and Heineken are injecting sex appeal back into TV ads.

"We're in a new decade, and tastes are changing," says marketing consultant Laura Ries.

As TV is more "risque," beer ads reflect a "more liberal environment," she says. "For a while, it was dogs, cats and penguins. Now, they're reverting back to sex sells. It's an attention grabber."

Others still see it as sexist sells. "We definitely see a trend," say Sonia Ossorio, chairwoman of the media watch committee of the National Organization for Women (NOW). And she warns that the new spots are "more insidious" because they take a humorous rather than leering approach.

Recent ads illustrating the trend:

▶**Miller.** The brewer will begin a campaign for Miller Lite during NCAA basketball tournament games starting this week. The ads resurrect their classic slogan, "It's Miller Time," and classic images. In one, three guys in a deserted bar on a rainy day pass the time rubbing their eyes. They completely miss two scantily clad and soaked beauties who duck in briefly to dry off.

In another, an unlucky buddy who draws the designated-driver straw ends up with the real prize. Three beautiful exchange students looking for a ride home. His pals look on enviously.

"They're done in a humorous way," says Miller's senior vice president of marketing Bob Mikulay. The commercials are "about the connection that guys make when they are hanging out over a beer."

Miller Lite could use a jump-start. Its 8% share of the U.S. market is down from 10% a decade ago, says Eric Shepard, executive editor of *Beer Marketer's Insights.*

▶**Heineken.** The brewer's "It's all about the Beer" campaign uses frank sexual imagery.

In one literally over-the-top TV spot, called "The Premature Pour," a guy spies a beautiful woman across the room in a bar. She slowly and seductively pours her Heineken into a tall glass. When he tries to do the same, he fumbles, pouring too fast and spilling foam all over the place.

In a spot called "The Wrong Bar Car," a guy on a train sits drinking the wrong beer in a bar car full of weirdos and losers also drinking mundane beer. Out the window he spies on the next track a beautiful blonde in a bar car full of partying Heineken drinkers. To his chagrin she waves as his train pulls out.

▶**Anheuser-Busch.** You don't have to be a Fellini fan to get the symbolism in its recent "Stranger on a Train" spot for Michelob Light.

A woman enters a train compartment and sits next to one of two guys sitting opposite each other. The train goes into a tunnel. When it emerges, she has switched seats and is cuddling the guy drinking a Michelob Light. He grins triumphantly at his rival.

Ries believes the new round of ads have brought "beercake" full circle. "The ads show equal opportunity sexism. The women are in charge, and the guys are the butt of the jokes," she says.

Cheryl Berman, chief creative officer for Leo Burnett, doesn't see this as a return to old-style beer ads. "The ads are more about male bonding than anything else," she says.

And she thinks this work is an improvement on some of the odd beer campaigns that fell flat in recent years, such as Miller's fictional ad man "Dick."

"There were a lot of awards-show ads," she says. They won really big at (industry) creative shows—but the public didn't buy into them."

Rick Boyko of Miller Lite ad agency Ogilvy & Mather says the campaign will show female friends enjoying "Miller Time," too. "It resonates with today's consumers. I have a 21-year-old daughter who says 'It's Miller Time.'"

But the poster boy for sexist beer advertising knows the pitfalls firsthand.

Ad agency executive Patrick Scullin lost his job and nearly his career after Stroh yanked his infamous "Swedish Bikini Team" ads for Old Milwaukee beer from the airwaves in 1991. Stroh beat a hasty retreat after a group of women employees sued, alleging the ad campaign helped create an atmosphere encouraging sexual harassment.

"My ZIP code became Siberia," says Scullin, now creative director at Ames Scullin O'Haire in Atlanta. "They were intended as a spoof. But they turned into a media circus."

Scullin doesn't see a big change in the new ads. "The women just wear tight jeans instead of bikinis," he says. "People don't like to admit it. But as long as there's biology, there will be sexism in beer advertising."

Television, Internet, and catalog pictures of products are frequently more attractive than the real thing, but we usually take this type of distortion for granted.

Statistical pictures are the histograms, frequency polygons, pie charts, and bar graphs of Chapter 3. These drawings or displays of numerical results are difficult to combine with sketches of lovely women or handsome men and hence are secure from the most common form of graphic distortion. However, other distortions are possible. One could shrink or stretch the axes, thus distorting the actual results. The idea behind these distortions is that shallow and steep slopes are commonly associated with small and large increases, respectively.

For example, suppose we want to examine the unemployment rate over a 12-month period. We might show the upward trend as shown in Figure 13.2. In this graph, the increase in the unemployment rate is apparent, but it does not appear to be that great. On the other hand, we could represent the same data in a much different light, as shown in Figure 13.3. In this graph, the vertical axis is stretched and does not include 0. Note the impression of a substantial rise that is indicated by the

FIGURE 13.2

12-month unemployment rate, 2000–2001

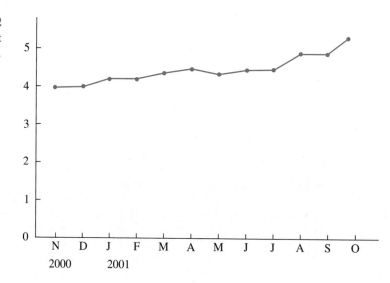

FIGURE 13.3

12-month unemployment rate, 2000–2001

Source: Department of Labor. This graph appeared in the December 7, 2001, edition of the *Kansas City Star.*

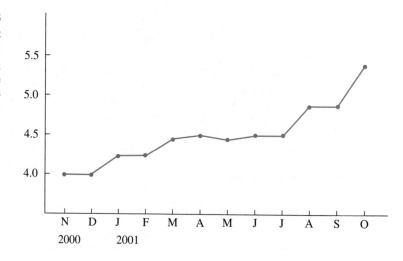

steeper slope. Another way to achieve the same effect—to decrease or increase a slope—is to stretch or shrink the horizontal axis.

When we present data in the form of bar graphs, histograms, frequency polygons, or other figures, we must be careful not to shrink or stretch axes because doing so will catch most readers off-guard. Increases or decreases in responses should be judged large or small depending on the arbitrary importance to the observer of the change, not on the slopes shown in graphic representations. In reality, most people look only at the slopes in the pictures.

13.4 Communication Hurdles: Biased Samples

One of the most common statistical distortions occurs because the experimenter unwittingly (or sometimes knowingly) samples the wrong population. That is, he or she draws the sample from a set of measurements that is not the proper population of interest.

For example, suppose that we want to assess the reaction of taxpayers to a proposed park and recreation center for children. A random sample of households is selected, and interviewers are sent to those households in the sample. Unfortunately, no one is at home in 40% of the sample households, so we randomly select and substitute other households in the city to make up the deficit. The resulting sample is selected from the wrong population, and the sample is therefore said to be biased.

The specified population of interest in the household survey is the collection of opinions that is obtained from the complete set of all households in the city. In contrast, the sample was drawn from a much smaller population or subset of this group—the set of opinions from householders who were at home when the sample was taken. It is possible that the fractions of householders favoring the park in these two populations are equal, and no damage was done by confining the sampling to those at home. However, it is much more likely that those at home had small children and that this group would yield a higher fraction in favor of the park than would the city as a whole. Thus, we have a biased sample because it is loaded in favor of families with small children. Perhaps a better way to see the difficulty is to note that we unwittingly selected the sample only from a special subset of the population of interest.

Biased samples frequently result from surveys that use mailed questionnaires. In a sense, the investigator lets the selection and number of the sampling units depend on the interests, available time, and various other personal characteristics of the individuals who receive the questionnaires. Extremely busy and energetic people may drop the questionnaires into the nearest wastebasket; you rarely hear from those low-energy folk who are uninterested or who are engrossed with other activities. Most often, the respondents are activists—those who are highly in favor, those who are opposed, or those who have something to gain from a certain outcome of the survey. In telephone surveys, just think how many people have caller ID or use their answering machine to screen calls, choosing to answer only selected calls; this can have a tremendous impact on the results of the survey.

Although numerous newscasters and analysts use election results as an expression of public opinion on major issues, it is a well-known fact that voting results represent a biased sample of public opinion. Those who vote represent much less than half of the eligible voters; they are individuals who desire to exercise their rights and responsibilities as citizens or are individuals who have been specially motivated to participate. The resultant subset of voters is not representative of the interests and opinions of all eligible voters in the country.

Sampling the wrong population also occurs when people attempt to extrapolate experimental results from one population to another. Numerous experimental results have been published about the effect of various products (e.g., saccharin) in inducing cancer in moles, rats, the breasts of beagles, and so forth. These results are often used to imply that humans have a high risk of developing cancer after frequent or extended exposure to the product. These inferences are not always justified because the experimental results were not obtained on humans. It is quite possible that humans are capable of resisting much higher doses than rats, or perhaps humans may be completely resistant for some reason. Drug induction of cancer in small mammals *does* indicate a need for concern and caution by humans, but it does not prove that the drug is definitely harmful to humans. Note that we are not criticizing experimentation in various species of animals—it is frequently the only way we can obtain any information about potential toxicity in human beings. We simply point out that the experimenter is knowingly sampling a population that is only similar (and quite likely not too similar) to the one of interest.

Engineers also test "rats" instead of "humans." The rats, in this context, are miniature models or pilot plants of a new engineering system. Experiments on the models occasionally yield results that differ substantially from the results of the larger, real systems. So again, we see a sampling from the wrong population, but it is the best the engineer can do because of the economics of the situation. Funds are not usually available to test a number of full-scale models prior to production.

Many other examples could be given of biased samples or of sampling from the wrong populations. The point is that when we communicate the results of a study or survey we should be clear about how the sample was drawn and whether it was *randomly* selected from the population of interest. How often are we told how a survey was conducted and what questions were asked when the results of "polls" are presented in the media? If this information is not given in the published results of a survey or experiment, the reader should take the inferences with a grain of salt.

13.5 Communication Hurdles: Sample Size

Distortions can occur when the sample size is not discussed. For example, suppose you read that a survey indicates that approximately 75% of a sample favor a new high-rise building complex. Further investigation might reveal that the investigator sampled only four people. When three out of the four favored the project, the investigator decided to stop the survey. Of course, we exaggerate with this example; but we could also have revealed inconclusive results based on a sample of 25, even though many buyers would consider this sample size to be large enough. As you well know, very large samples are required to achieve adequate information in sampling binomial populations.

Fortunately, many publications now provide more information about the sample size and how opinion surveys are conducted. Not too many years ago, it was rare to find how many people were sampled, much less how they were sampled. The situation is different now. In fact, sometimes the media have gone too far in an attempt to be completely open about how a survey was done. A case in point is the following article from the *Wall Street Journal*. How many of us understand much more than the number of people sampled and the approximate plus or minus (confidence interval)? It would take a person well trained in statistics and survey sampling to interpret what was done. Again, the moral of the story is simple: Try to communicate in unambiguous terms.

How Poll Was Conducted

The Wall Street Journal/NBC News poll was based on nationwide telephone interviews conducted last Friday through Monday with 4,159 adults age 18 or older. There were 2,630 likely voters.

The sample was drawn from a complete list of telephone exchanges, chosen so that each region of the country was represented in proportion to its population. Households were selected by a method that gave all telephone numbers, listed and unlisted, a proportionate chance of being included. The results of the survey were weighted to adjust for variations in the sample relating to education, age, race, gender, and religion.

Chances are 19 of 20 that if all adults in the United States had been surveyed using the same questionnaire, the findings would differ from these poll results by no more than two percentage points in either direction. The margin of error for subgroups may be larger.

13.6 Preparing Data for Statistical Analysis

We begin with a discussion of the steps involved in processing data from a study. In practice, these steps may consume 75% of the total effort from the receipt of the raw data to the presentation of results from the analysis. What are these steps, why are they so important, and why are they so time-consuming?

To answer these questions, let's list the major data-processing steps in the cycle, which begin with receipt of the data and end when the statistical analysis begins. Then we'll discuss each step separately.

Steps in Preparing Data for Analysis

1. Receiving the raw data source
2. Creating the database from the raw data source
3. Editing the database
4. Correcting and clarifying the raw data source
5. Finalizing the database
6. Creating data files from the database

raw data source

1. Receiving the raw data source. For each study that is to be summarized and analyzed, the data arrive in some form, which we'll refer to as the **raw data source.** For a clinical trial, the raw data source is usually case report forms, sheets of $8\frac{1}{2}'' \times 11''$ paper that have been used to record study data for each patient entered into the study. For other types of studies, the raw data source may be sheets of paper from a laboratory notebook, a magnetic tape (or any other form of machine-readable data), hand tabulations, and so on.

data trail

It is important to retain the raw data source because it is the beginning of the **data trail,** which leads from the raw data to the conclusions drawn from a study. Many consulting operations involved with the analysis and summarizing of many different studies keep a log that contains vital information related to the study and raw data source. General information contained in a study log is shown next.

Log for Study Data

1. Data received, and from whom
2. Study investigator
3. Statistician (and others) assigned
4. Brief description of study
5. Treatments (compounds, preparations, and so on) studied
6. Raw data source
7. Response(s) measured
8. Reference number for study
9. Estimated (actual) completion date
10. Other pertinent information

Later, when the study has been analyzed and results have been communicated, additional information can be added to the log on how the study results were communicated, where these results are recorded, what data files have been saved, and where these files are stored.

2. Creating the database from the raw data source. For most studies that are scheduled for a statistical analysis, a machine-readable database is created. The steps taken to create the database and the eventual form of the database vary from one operation to another, depending on the software systems to be used in the statistical analysis. However, we can give a few guidelines based on the form of the entry system.

When the data are to be *key-entered* at a terminal, the raw data are first checked for legibility. Any illegible numbers or letters or other problems should be brought to the attention of the study coordinator. Then a coding guide that assigns column numbers and variable names to the data is filled out. Certain codes for missing values (for example, those not available) are also defined here. Also, it is helpful to give a brief description of each variable. The data file keyed in at the terminal is referred to as the **machine-readable database.** A listing of the contents of the database should be obtained and checked carefully against the raw data source. Any errors should be corrected at the terminal and verified against an updated listing.

Sometimes data are received in machine-readable form. In these situations, the magnetic tape or disk file is considered to be the database. You must, however, have a coding guide to read the database. Using the coding guide, obtain a listing of the contents of the database and check it *carefully* to see that all numbers and characters look reasonable and that proper formats were used to create the file. Any problems that arise must be resolved before proceeding further.

Some data sets are so small that it is not necessary to create a machine-readable data file from the raw data source. Instead, calculations can be performed by hand or the data can be entered into an electronic calculator. In these situations, check any calculations to see that they make sense. Don't believe everything you see; redoing the calculations is not a bad idea.

3. Editing the database. The types of edits done and the completeness of the editing process really depend on the type of study and how concerned you are about the accuracy and completeness of the data prior to analysis. For example, in using SAS files it is wise to examine the minimum, maximum, and frequency distribution for each variable to make certain nothing looks unreasonable.

machine-readable database

Certain other checks should be made. Plot the data and look for problems.
logic checks Also, certain **logic checks** should be done, depending on the structure of the data.
For example, if data are recorded for patients during several different visits, then
the data recorded for visit 2 cannot be earlier than the data for visit 1; similarly, if a
patient is lost to follow-up after visit 2, we cannot have any data for that patient at
later visits.

For small data sets, we can do these data edits by hand, but for large data sets
the job may be too time-consuming and tedious. If machine editing is required, look
for a software system that allows the user to specify certain data edits. Even so, for
more complicated edits and logic checks, it may be necessary to have a customized
edit program written in order to machine-edit the data. This programming chore
can be a time-consuming step; plan for this well in advance of receipt of the data.

4. Correcting and clarifying the raw data source. Questions frequently arise
concerning the legibility or accuracy of the raw data during any one of the steps
from the receipt of the raw data to the communication of the results from the statistical analysis. We have found it helpful to keep a list of these problems or discrepancies in order to define the data trail for a study. If a correction (or clarification) is
required to the raw data source, this should be indicated on the form and the appropriate change made to the raw data source. If no correction is required, this
should be indicated on the form as well. Keep in mind that the machine-readable
database should be changed to reflect any changes made to the raw data source.

5. Finalizing the database. You may have been led to believe that all data for
a study arrive at one time. This, of course, is not always the case. For example,
with a marketing survey, different geographic locations may be surveyed at different times, and hence those responsible for data processing do not receive all
the data at once. All these subsets of data, however, must be processed through the
cycles required to create, edit, and correct the database. Eventually, the study is declared complete and the data are processed into the database. At this time, the
database should be reviewed again and final corrections made before beginning the
analysis. This is because, for large data sets, the analysis and summarizing chores
take considerable staff and computer time. It's better to agree on a final database
analysis than to have to repeat all analyses on a changed database at a later date.

6. Creating data files from the database. Generally, one or two sets of data
files are created from the machine-readable database. The first set, referred to as
original files **original files,** reflects the basic structure of the database. A listing of the files is
checked against the database listing to verify that the variables have been read with
correct formats and missing value codes have been retained. For some studies, the
original files are actually used for editing the database.

work files A second set of data files, called **work files,** may be created from the original
files. Work files are designed to facilitate the analysis. They may require restructuring of the original files, a selection of important variables, or the creation or addition of new variables by insertion, computation, or transformation. A listing of
the work files is checked against that of the original files to ensure proper restructuring and variable selection. Computed and transformed variables are checked by
hand calculations to verify the program code.

If your original and work files are SAS data sets, you should use the documentation features provided by SAS. At the time an SAS data set is created, a descriptive label for the data set of up to 40 characters should be assigned. The label
can be stored with the data set and imprinted wherever the contents procedure is
used to print the data set's contents. All variables can be given descriptive names,
up to 8 characters in length, which are meaningful to those involved in the project.

In addition, variable labels up to 40 characters in length can be used to provide additional information. Title statements can be included in the SAS code to identify the project and describe each job. For each file, a listing (proc print) and a dictionary (proc contents) can be retained.

For files created from the database using other software packages, use the labeling and documentation features available in the computer program.

Even if appropriate statistical methods are applied to data, the conclusions drawn from the study are only as good as the data on which they are based. So you be the judge. The amount of time spent on these data-processing chores before analysis really depends on the nature of the study, the quality of the raw data source, and how confident you want to be about the completeness and accuracy of the data.

13.7 Guidelines for a Statistical Analysis and Report

In this section, we briefly discuss a few guidelines for performing a statistical analysis and list some important elements of a statistical report used to communicate results. The statistical analysis of a large study can usually be broken down into three types of analyses: (1) preliminary analyses, (2) primary analyses, and (3) backup analyses.

preliminary analyses

The **preliminary analyses,** which are often descriptive or graphic, familiarize the statistician with the data and provide a foundation for all subsequent analyses. These analyses may include frequency distributions, histograms, descriptive statistics, an examination of comparability of the treatment groups, correlations, or univariate and bivariate plots.

primary analyses
backup analyses

Primary analyses address the objectives of the study and the analyses on which conclusions are drawn. **Backup analyses** include alternate methods for examining the data that confirm the results of the primary analyses; they may also include new statistical methods that are not as readily accepted as the more standard methods. Several guidelines for analyses follow.

Preliminary, Primary, and Backup Analyses

1. Perform the analyses with software that has been extensively tested.
2. Label the computer output to reflect which study is analyzed, what subjects (animals, patients, and so on) are used in the analysis, and a brief description of the analysis preferred. For example, TITLE statements in SAS are very helpful.
3. Use variable labels and value labels (for example, 0 = none, 1 = mild) on the output.
4. Provide a list of the data used in each analysis.
5. Check the output *carefully* for all analyses. Did the job run successfully? Are the sample sizes, means, and degrees of freedom correct? Other checks may be necessary as well.
6. Save all preliminary, primary, and backup analyses that provide the informational base from which study conclusions are drawn.

After the statistical analysis is completed, conclusions must be drawn and the results communicated to the intended audience. Sometimes it is necessary to communicate these results as a formal, written, statistical report. A general outline for a statistical report that we have found useful and informative follows.

General Outline for a Statistical Report

1. Summary
2. Introduction
3. Experimental design and study procedures
4. Descriptive statistics
5. Statistical methodology
6. Results and conclusions
7. Discussion
8. Data listings

13.8 Documentation and Storage of Results

The final part of this cycle of data processing, analysis, and summarizing concerns the documentation and storage of results. For formal statistical analyses that are subject to careful scrutiny by others, it is important to provide detailed documentation for all data processing and the statistical analyses so the data trail is clear and the database or work files readily accessible. Then the reviewer can follow what has been done, redo it, or extend the analyses. The elements of a documentation and storage file depend on the particular setting in which you work. The contents for a general documentation storage file are as follows.

Study Documentation and Storage File

1. Statistical report
2. Study description
3. Random code (used to assign subjects to treatment groups)
4. Important correspondence
5. File creation information
6. Preliminary, primary, and backup analyses
7. Raw data source
8. A data management sheet, which includes the study log, as well as information on the storage of the data files

The major thrust behind the documentation and storage file is that we want to provide a clear data and analysis trail for our own use, or for someone else's use, should there be a need to revisit the data. For any given situation, ask yourself whether such documentation is necessary and, if so, how detailed it must be. A good test of the completeness and understandability of your documentation is to ask a colleague, who is unfamiliar with your project but knowledgeable in your field, to try to reconstruct and even redo the primary analyses you did. If he or she can navigate through your documentation trail, you have done the job.

13.9 Staying Focused

In this chapter, we have discussed how to present the results of a statistical analysis of data and some of the problems with effectively communicating these results to the intended audience. The task is not easy. Some of the obstacles or hurdles standing in the way of effectively communicating the results of statistical analyses include graphical distortions, biased sampling, and omitting a discussion of the sample size and sampling technique. With some understanding of these obstacles, we can better critique and understand communications aimed at us and also do a better job communicating the results of our analyses to others.

The final topic in this chapter dealt with the documentation and storage of results. Having completed your analyses, drawn conclusions, and communicated these results to the intended audience, the temptation is to postpone or eliminate the documentation and storage of your results. However, it is worth your time to assess the potential for revisiting your analyses in the future and determine what steps should be taken to facilitate this process (for you or for others).

Finally, we should put our statistical analyses in the context of the practical problem(s) being addressed. The report of statistical analyses will not necessarily be the answer to an important question; it is only *part* of the answer. For example, we may demonstrate that a tablet delivers a drug more quickly than a capsule, but this is not the only consideration in the decision to market the capsule or tablet form. Such factors as cost, palatability, and stability will also be considered. Some of the relevant analyses addressing these considerations are not statistical.

Supplementary Exercise

Class Project

Each person in the class chooses a commercial he or she has heard or seen lately and critiques it for possible distortions, as well as making suggestions for improvement with regard to clarity of message or other communicative concern. (*Note:* We do *not* ask the class to improve it from a commercial standpoint.) The class then presents these findings.

APPENDIX

Statistical Tables

$Z_{.025} = 1.96, Z_{.05} = 1.645, Z_{.005} = 2.576, Z_{.1} = 1.282$

Shaded area $= Pr(Z \le -z)$

TABLE 1
Standard normal curve areas

z	0.00	0.01	0.02	0.03	0.04	0.05	0.06	0.07	0.08	0.09
−3.4	0.0003	0.0003	0.0003	0.0003	0.0003	0.0003	0.0003	0.0003	0.0003	0.0002
−3.3	0.0005	0.0005	0.0005	0.0004	0.0004	0.0004	0.0004	0.0004	0.0004	0.0003
−3.2	0.0007	0.0007	0.0006	0.0006	0.0006	0.0006	0.0006	0.0005	0.0005	0.0005
−3.1	0.0010	0.0009	0.0009	0.0009	0.0008	0.0008	0.0008	0.0008	0.0007	0.0007
−3.0	0.0013	0.0013	0.0013	0.0012	0.0012	0.0011	0.0011	0.0011	0.0010	0.0010
−2.9	0.0019	0.0018	0.0018	0.0017	0.0016	0.0016	0.0015	0.0015	0.0014	0.0014
−2.8	0.0026	0.0025	0.0024	0.0023	0.0023	0.0022	0.0021	0.0021	0.0020	0.0019
−2.7	0.0035	0.0034	0.0033	0.0032	0.0031	0.0030	0.0029	0.0028	0.0027	0.0026
−2.6	0.0047	0.0045	0.0044	0.0043	0.0041	0.0040	0.0039	0.0038	0.0037	0.0036
−2.5	0.0062	0.0060	0.0059	0.0057	0.0055	0.0054	0.0052	0.0051	0.0049	0.0048
−2.4	0.0082	0.0080	0.0078	0.0075	0.0073	0.0071	0.0069	0.0068	0.0066	0.0064
−2.3	0.0107	0.0104	0.0102	0.0099	0.0096	0.0094	0.0091	0.0089	0.0087	0.0084
−2.2	0.0139	0.0136	0.0132	0.0129	0.0125	0.0122	0.0119	0.0116	0.0113	0.0110
−2.1	0.0179	0.0174	0.0170	0.0166	0.0162	0.0158	0.0154	0.0150	0.0146	0.0143
−2.0	0.0228	0.0222	0.0217	0.0212	0.0207	0.0202	0.0197	0.0192	0.0188	0.0183
−1.9	0.0287	0.0281	0.0274	0.0268	0.0262	0.0256	0.0250	0.0244	0.0239	0.0233
−1.8	0.0359	0.0351	0.0344	0.0336	0.0329	0.0322	0.0314	0.0307	0.0301	0.0294
−1.7	0.0446	0.0436	0.0427	0.0418	0.0409	0.0401	0.0392	0.0384	0.0375	0.0367
−1.6	0.0548	0.0537	0.0526	0.0516	0.0505	0.0495	0.0485	0.0475	0.0465	0.0455
−1.5	0.0668	0.0655	0.0643	0.0630	0.0618	0.0606	0.0594	0.0582	0.0571	0.0559
−1.4	0.0808	0.0793	0.0778	0.0764	0.0749	0.0735	0.0721	0.0708	0.0694	0.0681
−1.3	0.0968	0.0951	0.0934	0.0918	0.0901	0.0885	0.0869	0.0853	0.0838	0.0823
−1.2	0.1151	0.1131	0.1112	0.1093	0.1075	0.1056	0.1038	0.1020	0.1003	0.0985
−1.1	0.1357	0.1335	0.1314	0.1292	0.1271	0.1251	0.1230	0.1210	0.1190	0.1170
−1.0	0.1587	0.1562	0.1539	0.1515	0.1492	0.1469	0.1446	0.1423	0.1401	0.1379
−0.9	0.1841	0.1814	0.1788	0.1762	0.1736	0.1711	0.1685	0.1660	0.1635	0.1611
−0.8	0.2119	0.2090	0.2061	0.2033	0.2005	0.1977	0.1949	0.1922	0.1894	0.1867
−0.7	0.2420	0.2389	0.2358	0.2327	0.2296	0.2266	0.2236	0.2206	0.2177	0.2148
−0.6	0.2743	0.2709	0.2676	0.2643	0.2611	0.2578	0.2546	0.2514	0.2483	0.2451
−0.5	0.3085	0.3050	0.3015	0.2981	0.2946	0.2912	0.2877	0.2843	0.2810	0.2776
−0.4	0.3446	0.3409	0.3372	0.3336	0.3300	0.3264	0.3228	0.3192	0.3156	0.3121
−0.3	0.3821	0.3783	0.3745	0.3707	0.3669	0.3632	0.3594	0.3557	0.3520	0.3483
−0.2	0.4207	0.4168	0.4129	0.4090	0.4052	0.4013	0.3974	0.3936	0.3897	0.3859
−0.1	0.4602	0.4562	0.4522	0.4483	0.4443	0.4404	0.4364	0.4325	0.4286	0.4247
−0.0	0.5000	0.4960	0.4920	0.4880	0.4840	0.4801	0.4761	0.4721	0.4681	0.4641

z	Area
−3.50	0.00023263
−4.00	0.00003167
−4.50	0.00000340
−5.00	0.00000029

H_0:
$\mu > \mu_0$
$\mu < \mu_0$
$\mu \ne \mu_0$

RR approach:
Reject H_0 if $z > z_\alpha$
Reject H_0 if $z < -z_\alpha$
Reject H_0 if $z > z_{\alpha/2}$

P-value approach:
Reject H_0 if $p < \alpha$
Reject H_0 if $p < \alpha$
Reject H_0 if $p < \alpha$

Source: Computed by M. Longnecker using Splus.

TABLE 1
Standard normal curve areas

z	0.00	0.01	0.02	0.03	0.04	0.05	0.06	0.07	0.08	0.09
0.0	0.5000	0.5040	0.5080	0.5120	0.5160	0.5199	0.5239	0.5279	0.5319	0.5359
0.1	0.5398	0.5438	0.5478	0.5517	0.5557	0.5596	0.5636	0.5675	0.5714	0.5753
0.2	0.5793	0.5832	0.5871	0.5910	0.5948	0.5987	0.6026	0.6064	0.6103	0.6141
0.3	0.6179	0.6217	0.6255	0.6293	0.6331	0.6368	0.6406	0.6443	0.6480	0.6517
0.4	0.6554	0.6591	0.6628	0.6664	0.6700	0.6736	0.6772	0.6808	0.6844	0.6879
0.5	0.6915	0.6950	0.6985	0.7019	0.7054	0.7088	0.7123	0.7157	0.7190	0.7224
0.6	0.7257	0.7291	0.7324	0.7357	0.7389	0.7422	0.7454	0.7486	0.7517	0.7549
0.7	0.7580	0.7611	0.7642	0.7673	0.7704	0.7734	0.7764	0.7794	0.7823	0.7852
0.8	0.7881	0.7910	0.7939	0.7967	0.7995	0.8023	0.8051	0.8078	0.8106	0.8133
0.9	0.8159	0.8186	0.8212	0.8238	0.8264	0.8289	0.8315	0.8340	0.8365	0.8389
1.0	0.8413	0.8438	0.8461	0.8485	0.8508	0.8531	0.8554	0.8577	0.8599	0.8621
1.1	0.8643	0.8665	0.8686	0.8708	0.8729	0.8749	0.8770	0.8790	0.8810	0.8830
1.2	0.8849	0.8869	0.8888	0.8907	0.8925	0.8944	0.8962	0.8980	0.8997	0.9015
1.3	0.9032	0.9049	0.9066	0.9082	0.9099	0.9115	0.9131	0.9147	0.9162	0.9177
1.4	0.9192	0.9207	0.9222	0.9236	0.9251	0.9265	0.9279	0.9292	0.9306	0.9319
1.5	0.9332	0.9345	0.9357	0.9370	0.9382	0.9394	0.9406	0.9418	0.9429	0.9441
1.6	0.9452	0.9463	0.9474	0.9484	0.9495	0.9505	0.9515	0.9525	0.9535	0.9545
1.7	0.9554	0.9564	0.9573	0.9582	0.9591	0.9599	0.9608	0.9616	0.9625	0.9633
1.8	0.9641	0.9649	0.9656	0.9664	0.9671	0.9678	0.9686	0.9693	0.9699	0.9706
1.9	0.9713	0.9719	0.9726	0.9732	0.9738	0.9744	0.9750	0.9756	0.9761	0.9767
2.0	0.9772	0.9778	0.9783	0.9788	0.9793	0.9798	0.9803	0.9808	0.9812	0.9817
2.1	0.9821	0.9826	0.9830	0.9834	0.9838	0.9842	0.9846	0.9850	0.9854	0.9857
2.2	0.9861	0.9864	0.9868	0.9871	0.9875	0.9878	0.9881	0.9884	0.9887	0.9890
2.3	0.9893	0.9896	0.9898	0.9901	0.9904	0.9906	0.9909	0.9911	0.9913	0.9916
2.4	0.9918	0.9920	0.9922	0.9925	0.9927	0.9929	0.9931	0.9932	0.9934	0.9936
2.5	0.9938	0.9940	0.9941	0.9943	0.9945	0.9946	0.9948	0.9949	0.9951	0.9952
2.6	0.9953	0.9955	0.9956	0.9957	0.9959	0.9960	0.9961	0.9962	0.9963	0.9964
2.7	0.9965	0.9966	0.9967	0.9968	0.9969	0.9970	0.9971	0.9972	0.9973	0.9974
2.8	0.9974	0.9975	0.9976	0.9977	0.9977	0.9978	0.9979	0.9979	0.9980	0.9981
2.9	0.9981	0.9982	0.9982	0.9983	0.9984	0.9984	0.9985	0.9985	0.9986	0.9986
3.0	0.9987	0.9987	0.9987	0.9988	0.9988	0.9989	0.9989	0.9989	0.9990	0.9990
3.1	0.9990	0.9991	0.9991	0.9991	0.9992	0.9992	0.9992	0.9992	0.9993	0.9993
3.2	0.9993	0.9993	0.9994	0.9994	0.9994	0.9994	0.9994	0.9995	0.9995	0.9995
3.3	0.9995	0.9995	0.9995	0.9996	0.9996	0.9996	0.9996	0.9996	0.9996	0.9997
3.4	0.9997	0.9997	0.9997	0.9997	0.9997	0.9997	0.9997	0.9997	0.9997	0.9998

z	Area
3.50	0.99976737
4.00	0.99996833
4.50	0.99999660
5.00	0.99999971

Handwritten annotations:

$$H_0: \quad \text{Reject:} \quad H_a:$$
$$\pi \le \pi_0 \qquad Z > Z_\alpha \qquad \pi_1 - \pi_2 \le 0$$
$$\pi \ge \pi_0 \qquad Z < -Z_\alpha \qquad \pi_1 - \pi_2 \ge 0$$
$$\pi = \pi_0 \qquad |Z| > Z_{\alpha/2} \qquad \pi_1 - \pi_2 = 0$$
$$0. > 5 \qquad\qquad 0.$$

$$S^2 = \frac{\sum_{i=1}^{n} Y_i^2 - \frac{(\sum Y_i)^2}{n}}{n-1}$$

$$Q_3 = (.75)(n) \Rightarrow \text{integer} \rightarrow .5$$
$$Q_1 = (.25)(n) \quad \text{decimal round up.}$$

$$IQR = Q_3 - Q_1$$
inner fence
$$Q_1 - 1.5(IQR)$$
$$Q_3 + 1.5(IQR)$$

goodness of fit
0. no $E_i < 1$ and at most 20% of $E_i < 5$
1. $H_0: \pi_i^2, \ldots$
   $H_a:$ at least one cell differs.
   $k - 1 = df.$
   χ^2

fail to reject

TABLE 2

Random numbers

Line/ Col.	(1)	(2)	(3)	(4)	(5)	(6)	(7)	(8)	(9)	(10)	(11)	(12)	(13)	(14)
1	10480	15011	01536	02011	81647	91646	69179	14194	62590	36207	20969	99570	91291	90700
2	22368	46573	25595	85393	30995	89198	27982	53402	93965	34095	52666	19174	39615	99505
3	24130	48360	22527	97265	76393	64809	15179	24830	49340	32081	30680	19655	63348	58629
4	42167	93093	06243	61680	07856	16376	39440	53537	71341	57004	00849	74917	97758	16379
5	37570	39975	81837	16656	06121	91782	60468	81305	49684	60672	14110	06927	01263	54613
6	77921	06907	11008	42751	27756	53498	18602	70659	90655	15053	21916	81825	44394	42880
7	99562	72905	56420	69994	98872	31016	71194	18738	44013	48840	63213	21069	10634	12952
8	96301	91977	05463	07972	18876	20922	94595	56869	69014	60045	18425	84903	42508	32307
9	89579	14342	63661	10281	17453	18103	57740	84378	25331	12566	58678	44947	05585	56941
10	85475	36857	53342	53988	53060	59533	38867	62300	08158	17983	16439	11458	18593	64952
11	28918	69578	88231	33276	70997	79936	56865	05859	90106	31595	01547	85590	91610	78188
12	63553	40961	48235	03427	49626	69445	18663	72695	52180	20847	12234	90511	33703	90322
13	09429	93969	52636	92737	88974	33488	36320	17617	30015	08272	84115	27156	30613	74952
14	10365	61129	87529	85689	48237	52267	67689	93394	01511	26358	85104	20285	29975	89868
15	07119	97336	71048	08178	77233	13916	47564	81056	97735	85977	29372	74461	28551	90707
16	51085	12765	51821	51259	77452	16308	60756	92144	49442	53900	70960	63990	75601	40719
17	02368	21382	52404	60268	89368	19885	55322	44819	01188	65255	64835	44919	05944	55157
18	01011	54092	33362	94904	31273	04146	18594	29852	71585	85030	51132	01915	92747	64951
19	52162	53916	46369	58586	23216	14513	83149	98736	23495	64350	94738	17752	35156	35749
20	07056	97628	33787	09998	42698	06691	76988	13602	51851	46104	88916	19509	25625	58104
21	48663	91245	85828	14346	09172	30168	90229	04734	59193	22178	30421	61666	99904	32812
22	54164	58492	22421	74103	47070	25306	76468	26384	58151	06646	21524	15227	96909	44592
23	32639	32363	05597	24200	13363	38005	94342	28728	35806	06912	17012	64161	18296	22851
24	29334	27001	87637	87308	58731	00256	45834	15398	46557	41135	10367	07684	36188	18510
25	02488	33062	28834	07351	19731	92420	60952	61280	50001	67658	32586	86679	50720	94953

Source: Abridged from William H. Beyer, ed., *Handbook of Tables for Probability and Statistics,* 2nd ed. © The Chemical Rubber Co., 1968. Used by permission of CRC Press, Inc.

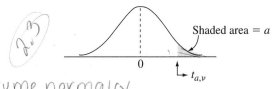

Shaded area = a

assume normalcy

$t_{a,v}$

TABLE 3

Percentage points of Student's t distribution

df = $n_1 + n_2 - 2$ pooled = t
2 populations; non-pooled = t'

$H_0: \mu_1 - \mu_2 \le D_0$
$\mu_1 - \mu_2 \ge D_0$
$\mu_1 - \mu_2 = D_0$

RR-approach:
$t \ge t_\alpha$
$t \le -t_\alpha$
$|t| \ge t_{\alpha/2}$

$S_p = \sqrt{\dfrac{(n_1-1)S_1^2 + (n_2-1)S_2^2}{n_1 + n_2 - 2}}$

df/a =	.40	.25	.10	.05	.025	.01	.005	.001	.0005
1	0.325	1.000	3.078	6.314	12.706	31.821	63.657	318.309	636.619
2	0.289	0.816	1.886	2.920	4.303	6.965	9.925	22.327	31.599
3	0.277	0.765	1.638	2.353	3.182	4.541	5.841	10.215	12.924
4	0.271	0.741	1.533	2.132	2.776	3.747	4.604	7.173	8.610
5	0.267	0.727	1.476	2.015	2.571	3.365	4.032	5.893	6.869
6	0.265	0.718	1.440	1.943	2.447	3.143	3.707	5.208	5.959
7	0.263	0.711	1.415	1.895	2.365	2.998	3.499	4.785	5.408
8	0.262	0.706	1.397	1.860	2.306	2.896	3.355	4.501	5.041
9	0.261	0.703	1.383	1.833	2.262	2.821	3.250	4.297	4.781
10	0.260	0.700	1.372	1.812	2.228	2.764	3.169	4.144	4.587
11	0.260	0.697	1.363	1.796	2.201	2.718	3.106	4.025	4.437
12	0.259	0.695	1.356	1.782	2.179	2.681	3.055	3.930	4.318
13	0.259	0.694	1.350	1.771	2.160	2.650	3.012	3.852	4.221
14	0.258	0.692	1.345	1.761	2.145	2.624	2.977	3.787	4.140
15	0.258	0.691	1.341	1.753	2.131	2.602	2.947	3.733	4.073
16	0.258	0.690	1.337	1.746	2.120	2.583	2.921	3.686	4.015
17	0.257	0.689	1.333	1.740	2.110	2.567	2.898	3.646	3.965
18	0.257	0.688	1.330	1.734	2.101	2.552	2.878	3.610	3.922
19	0.257	0.688	1.328	1.729	2.093	2.539	2.861	3.579	3.883
20	0.257	0.687	1.325	1.725	2.086	2.528	2.845	3.552	3.850
21	0.257	0.686	1.323	1.721	2.080	2.518	2.831	3.527	3.819
22	0.256	0.686	1.321	1.717	2.074	2.508	2.819	3.505	3.792
23	0.256	0.685	1.319	1.714	2.069	2.500	2.807	3.485	3.768
24	0.256	0.685	1.318	1.711	2.064	2.492	2.797	3.467	3.745
25	0.256	0.684	1.316	1.708	2.060	2.485	2.787	3.450	3.725
26	0.256	0.684	1.315	1.706	2.056	2.479	2.779	3.435	3.707
27	0.256	0.684	1.314	1.703	2.052	2.473	2.771	3.421	3.690
28	0.256	0.683	1.313	1.701	2.048	2.467	2.763	3.408	3.674
29	0.256	0.683	1.311	1.699	2.045	2.462	2.756	3.396	3.659
30	0.256	0.683	1.310	1.697	2.042	2.457	2.750	3.385	3.646
35	0.255	0.682	1.306	1.690	2.030	2.438	2.724	3.340	3.591
40	0.255	0.681	1.303	1.684	2.021	2.423	2.704	3.307	3.551
50	0.255	0.679	1.299	1.676	2.009	2.403	2.678	3.261	3.496
60	0.254	0.679	1.296	1.671	2.000	2.390	2.660	3.232	3.460
120	0.254	0.677	1.289	1.658	1.980	2.358	2.617	3.160	3.373
inf.	0.253	0.674	1.282	1.645	1.960	2.326	2.576	3.090	3.291

Source: Computed by M. Longnecker using Splus.   one population

H_0	RR-approach:	P-value:	df = n-1
$\mu \le \mu_0$	$t \ge t_\alpha$	$P(t \ge comp.t)$	
$\mu \ge \mu_0$	$t \le -t_\alpha$	$P(t \le comp\,t)$	
$\mu \ne \mu_0$	$|t| \ge t_{\alpha/2}$	$2P(t \ge |comp\,t|)$	

TABLE 4

Probability of Type II error
curves for $\alpha = .01$ (one-sided)

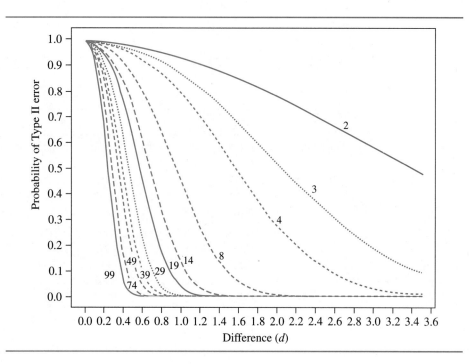

Source: Computed by M. Longnecker using SAS.

TABLE 4

Probability of Type II error
curves for $\alpha = .05$ (one-sided)

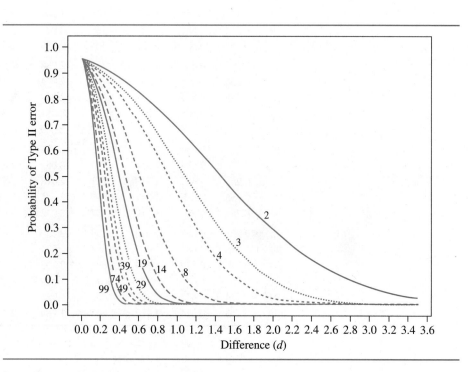

Source: Computed by M. Longnecker using SAS.

TABLE 4

Probability of Type II error
curves for $\alpha = .01$ (two-sided)

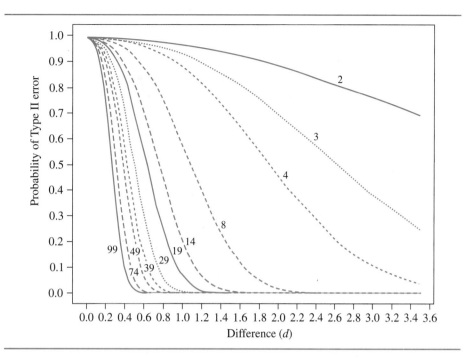

Source: Computed by M. Longnecker using SAS.

TABLE 4

Probability of Type II error
curves for $\alpha = .05$ (two-sided)

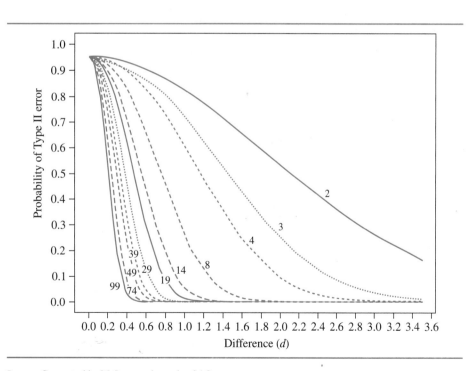

Source: Computed by M. Longnecker using SAS.

TABLE 5

Percentage points for confidence intervals on the median and the sign test: $C_{\alpha,n}$

$\alpha(2)$	.20	.10	.05	.02	.01	.005	.002	$\alpha(2)$	.20	.10	.05	.02	.01	.005	.002
$\alpha(1)$	.10	.05	.025	.01	.005	.0025	.001	$\alpha(1)$	.10	.05	.025	.01	.005	.0025	.001
n								n							
1	*	*	*	*	*	*	*	26	9	8	7	6	6	5	4
2	*	*	*	*	*	*	*	27	9	8	7	7	6	5	5
3	*	*	*	*	*	*	*	28	10	9	8	7	6	6	5
4	0	*	*	*	*	*	*	29	10	9	8	7	7	6	5
5	0	0	*	*	*	*	*	30	10	10	9	8	7	6	6
6	0	0	0	*	*	*	*	31	11	10	9	8	7	7	6
7	1	0	0	0	*	*	*	32	11	10	9	8	8	7	6
8	1	1	0	0	0	*	*	33	12	11	10	9	8	8	7
9	2	1	1	0	0	0	*	34	12	11	10	9	9	8	7
10	2	1	1	0	0	0	0	35	13	12	11	10	9	8	8
11	2	2	1	1	0	0	0	36	13	12	11	10	9	9	8
12	3	2	2	1	1	0	0	37	14	13	12	10	10	9	8
13	3	3	2	1	1	1	0	38	14	13	12	11	10	9	9
14	4	3	2	2	1	1	1	39	15	13	12	11	11	10	9
15	4	3	3	2	2	1	1	40	15	14	13	12	11	10	9
16	4	4	3	2	2	2	1	41	15	14	13	12	11	11	10
17	5	4	4	3	2	2	1	42	16	15	14	13	12	11	10
18	5	5	4	3	3	2	2	43	16	15	14	13	12	11	11
19	6	5	4	4	3	3	2	44	17	16	15	13	13	12	11
20	6	5	5	4	3	3	2	45	17	16	15	14	13	12	11
21	7	6	5	4	4	3	3	46	18	16	15	14	13	13	12
22	7	6	5	5	4	4	3	47	18	17	16	15	14	13	12
23	7	7	6	5	4	4	3	48	19	17	16	15	14	13	12
24	8	7	6	5	5	4	4	49	19	18	17	15	15	14	13
25	8	7	7	6	5	5	4	50	19	18	17	16	15	14	13

Note: An * means that no test or confidence interval of this level exists.

Source: Computed by M. Longnecker using Splus.

TABLE 6

Critical values of T_L and T_U for the Wilcoxon rank sum test: independent samples. Test statistic is rank sum associated with smaller sample (if equal sample sizes, either rank sum can be used)

a. $\alpha = .025$ one-tailed; $\alpha = .05$ two-tailed

n_2 \ n_1	3		4		5		6		7		8		9		10	
	T_L	T_U	T_L	T_U	T_L	T_U	T_L	T_U	T_L	T_U	T_L	T_U	T_L	T_U	T_L	T_U
3	5	16	6	18	6	21	7	23	7	26	8	28	8	31	9	33
4	6	18	11	25	12	28	12	32	13	35	14	38	15	41	16	44
5	6	21	12	28	18	37	19	41	20	45	21	49	22	53	24	56
6	7	23	12	32	19	41	26	52	28	56	29	61	31	65	32	70
7	7	26	13	35	20	45	28	56	37	68	39	73	41	78	43	83
8	8	28	14	38	21	49	29	61	39	73	49	87	51	93	54	98
9	8	31	15	41	22	53	31	65	41	78	51	93	63	108	66	114
10	9	33	16	44	24	56	32	70	43	83	54	98	66	114	79	131

b. $\alpha = .05$ one-tailed; $\alpha = .10$ two-tailed

n_2 \ n_1	3		4		5		6		7		8		9		10	
	T_L	T_U	T_L	T_U	T_L	T_U	T_L	T_U	T_L	T_U	T_L	T_U	T_L	T_U	T_L	T_U
3	6	15	7	17	7	20	8	22	9	24	9	27	10	29	11	31
4	7	17	12	24	13	27	14	30	15	33	16	36	17	39	18	42
5	7	20	13	27	19	36	20	40	22	43	24	46	25	50	26	54
6	8	22	14	30	20	40	28	50	30	54	32	58	33	63	35	67
7	9	24	15	33	22	43	30	54	39	66	41	71	43	76	46	80
8	9	27	16	36	24	46	32	58	41	71	52	84	54	90	57	95
9	10	29	17	39	25	50	33	63	43	76	54	90	66	105	69	111
10	11	31	18	42	26	54	35	67	46	80	57	95	69	111	83	127

Source: From F. Wilcoxon and R. A. Wilcox, *Some Rapid Approximate Statistical Procedures* (Pearl River, N.Y. Lederle Laboratories, 1964), pp. 20–23. Reproduced with the permission of American Cyanamid Company.

TABLE 7

Critical values for the Wilcoxon signed-rank test [$n = 5(1)54$]

One-Sided	Two-Sided	$n = 5$	$n = 6$	$n = 7$	$n = 8$	$n = 9$
$p = .1$	$p = .2$	2	3	5	8	10
$p = .05$	$p = .1$	0	2	3	5	8
$p = .025$	$p = .05$		0	2	3	5
$p = .01$	$p = .02$			0	1	3
$p = .005$	$p = .01$				0	1
$p = .0025$	$p = .005$					0
$p = .001$	$p = .002$					

One-Sided	Two-Sided	$n = 15$	$n = 16$	$n = 17$	$n = 18$	$n = 19$
$p = .1$	$p = .2$	36	42	48	55	62
$p = .05$	$p = .1$	30	35	41	47	53
$p = .025$	$p = .05$	25	29	34	40	46
$p = .01$	$p = .02$	19	23	27	32	37
$p = .005$	$p = .01$	15	19	23	27	32
$p = .0025$	$p = .005$	12	15	19	23	27
$p = .001$	$p = .002$	8	11	14	18	21

One-Sided	Two-Sided	$n = 25$	$n = 26$	$n = 27$	$n = 28$	$n = 29$
$p = .1$	$p = .2$	113	124	134	145	157
$p = .05$	$p = .1$	100	110	119	130	140
$p = .025$	$p = .05$	89	98	107	116	126
$p = .01$	$p = .02$	76	84	92	101	110
$p = .005$	$p = .01$	68	75	83	91	100
$p = .0025$	$p = .005$	60	67	74	82	90
$p = .001$	$p = .002$	51	58	64	71	79

One-Sided	Two-Sided	$n = 35$	$n = 36$	$n = 37$	$n = 38$	$n = 39$
$p = .1$	$p = .2$	235	250	265	281	297
$p = .05$	$p = .1$	213	227	241	256	271
$p = .025$	$p = .05$	195	208	221	235	249
$p = .01$	$p = .02$	173	185	198	211	224
$p = .005$	$p = .01$	159	171	182	194	207
$p = .0025$	$p = .005$	146	157	168	180	192
$p = .001$	$p = .002$	131	141	151	162	173

One-Sided	Two-Sided	$n = 45$	$n = 46$	$n = 47$	$n = 48$	$n = 49$
$p = .1$	$p = .2$	402	422	441	462	482
$p = .05$	$p = .1$	371	389	407	426	446
$p = .025$	$p = .05$	343	361	378	396	415
$p = .01$	$p = .02$	312	328	345	362	379
$p = .005$	$p = .01$	291	307	322	339	355
$p = .0025$	$p = .005$	272	287	302	318	334
$p = .001$	$p = .002$	249	263	277	292	307

Source: Computed by P. J. Hildebrand.

TABLE 7
(*continued*)

One-Sided	Two-Sided	n = 10	n = 11	n = 12	n = 13	n = 14
$p = .1$	$p = .2$	14	17	21	26	31
$p = .05$	$p = .1$	10	13	17	21	25
$p = .025$	$p = .05$	8	10	13	17	21
$p = .01$	$p = .02$	5	7	9	12	15
$p = .005$	$p = .01$	3	5	7	9	12
$p = .0025$	$p = .005$	1	3	5	7	9
$p = .001$	$p = .002$	0	1	2	4	6

One-Sided	Two-Sided	n = 20	n = 21	n = 22	n = 23	n = 24
$p = .1$	$p = .2$	69	77	86	94	104
$p = .05$	$p = .1$	60	67	75	83	91
$p = .025$	$p = .05$	52	58	65	73	81
$p = .01$	$p = .02$	43	49	55	62	69
$p = .005$	$p = .01$	37	42	48	54	61
$p = .0025$	$p = .005$	32	37	42	48	54
$p = .001$	$p = .002$	26	30	35	40	45

One-Sided	Two-Sided	n = 30	n = 31	n = 32	n = 33	n = 34
$p = .1$	$p = .2$	169	181	194	207	221
$p = .05$	$p = .1$	151	163	175	187	200
$p = .025$	$p = .05$	137	147	159	170	182
$p = .01$	$p = .02$	120	130	140	151	162
$p = .005$	$p = .01$	109	118	128	138	148
$p = .0025$	$p = .005$	98	107	116	126	136
$p = .001$	$p = .002$	86	94	103	112	121

One-Sided	Two-Sided	n = 40	n = 41	n = 42	n = 43	n = 44
$p = .1$	$p = .2$	313	330	348	365	384
$p = .05$	$p = .1$	286	302	319	336	353
$p = .025$	$p = .05$	264	279	294	310	327
$p = .01$	$p = .02$	238	252	266	281	296
$p = .005$	$p = .01$	220	233	247	261	276
$p = .0025$	$p = .005$	204	217	230	244	258
$p = .001$	$p = .002$	185	197	209	222	235

One-Sided	Two-Sided	n = 50	n = 51	n = 52	n = 53	n = 54
$p = .1$	$p = .2$	503	525	547	569	592
$p = .05$	$p = .1$	466	486	507	529	550
$p = .025$	$p = .05$	434	453	473	494	514
$p = .01$	$p = .02$	397	416	434	454	473
$p = .005$	$p = .01$	373	390	408	427	445
$p = .0025$	$p = .005$	350	367	384	402	420
$p = .001$	$p = .002$	323	339	355	372	389

$df = n-1$

left tail

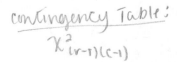

assume normal data
one-parameter:
$H_0: \sigma_0^2 \le \sigma^2$
$\sigma_0^2 \ge \sigma^2$
$\sigma_0^2 = \sigma^2$

Reject:
$\chi^2 > \chi_a^2$
$\chi^2 < \chi_{1-\alpha}^2$
$\chi^2 > \chi_{\alpha/2}^2$

a population:
$H_0: \sigma_1^2 \le \sigma_2^2$
$\sigma_1^2 = \sigma_2^2$

Reject:
$F \ge F_{\alpha, df_1, df_2}$
$F \ge F_{\alpha/2, df_1, df_2}$

assume normality

TABLE 8
Percentage points of the chi-square distribution

df/a =	.999	.995	.99	.975	.95	.90
1	.000002	.000039	.000157	.000982	.003932	.01579
2	.002001	.01003	.02010	.05064	.1026	.2107
3	.02430	.07172	.1148	.2158	.3518	.5844
4	.09080	.2070	.2971	.4844	.7107	1.064
5	.2102	.4117	.5543	.8312	1.145	1.610
6	.3811	.6757	.8721	1.237	1.635	2.204
7	.5985	.9893	1.239	1.690	2.167	2.833
8	.8571	1.344	1.646	2.180	2.733	3.490
9	1.152	1.735	2.088	2.700	3.325	4.168
10	1.479	2.156	2.558	3.247	3.940	4.865
11	1.834	2.603	3.053	3.816	4.575	5.578
12	2.214	3.074	3.571	4.404	5.226	6.304
13	2.617	3.565	4.107	5.009	5.892	7.042
14	3.041	4.075	4.660	5.629	6.571	7.790
15	3.483	4.601	5.229	6.262	7.261	8.547
16	3.942	5.142	5.812	6.908	7.962	9.312
17	4.416	5.697	6.408	7.564	8.672	10.09
18	4.905	6.265	7.015	8.231	9.390	10.86
19	5.407	6.844	7.633	8.907	10.12	11.65
20	5.921	7.434	8.260	9.591	10.85	12.44
21	6.447	8.034	8.897	10.28	11.59	13.24
22	6.983	8.643	9.542	10.98	12.34	14.04
23	7.529	9.260	10.20	11.69	13.09	14.85
24	8.085	9.886	10.86	12.40	13.85	15.66
25	8.649	10.52	11.52	13.12	14.61	16.47
26	9.222	11.16	12.20	13.84	15.38	17.29
27	9.803	11.81	12.88	14.57	16.15	18.11
28	10.39	12.46	13.56	15.31	16.93	18.94
29	10.99	13.12	14.26	16.06	17.71	19.77
30	11.59	13.79	14.95	16.79	18.49	20.60
40	17.92	20.71	22.16	24.43	26.51	29.05
50	24.67	27.99	29.71	32.36	34.76	37.69
60	31.74	35.53	37.48	40.48	43.19	46.46
70	39.04	43.28	45.44	48.76	51.74	55.33
80	46.52	51.17	53.54	57.15	60.39	64.28
90	54.16	59.20	61.75	65.65	69.13	73.29
100	61.92	67.33	70.06	74.22	77.93	82.36
120	77.76	83.85	86.92	91.57	95.70	100.62
240	177.95	187.32	191.99	198.98	205.14	212.39

contingency Table:
$\chi_{(r-1)(c-1)}^2$

TABLE 8
(continued)

Right tail

a = .10	.05	.025	.01	.005	.001	df
2.706	3.841	5.024	6.635	7.879	10.83	1
4.605	5.991	7.378	9.210	10.60	13.82	2
6.251	7.815	9.348	11.34	12.84	16.27	3
7.779	9.488	11.14	13.28	14.86	18.47	4
9.236	11.07	12.83	15.09	16.75	20.52	5
10.64	12.59	14.45	16.81	18.55	22.46	6
12.02	14.07	16.01	18.48	20.28	24.32	7
13.36	15.51	17.53	20.09	21.95	26.12	8
14.68	16.92	19.02	21.67	23.59	27.88	9
15.99	18.31	20.48	23.21	25.19	29.59	10
17.28	19.68	21.92	24.72	26.76	31.27	11
18.55	21.03	23.34	26.22	28.30	32.91	12
19.81	22.36	24.74	27.69	29.82	34.53	13
21.06	23.68	26.12	29.14	31.32	36.12	14
22.31	25.00	27.49	30.58	32.80	37.70	15
23.54	26.30	28.85	32.00	34.27	39.25	16
24.77	27.59	30.19	33.41	35.72	40.79	17
25.99	28.87	31.53	34.81	37.16	42.31	18
27.20	30.14	32.85	36.19	38.58	43.82	19
28.41	31.41	34.17	37.57	40.00	45.31	20
29.62	32.67	35.48	38.93	41.40	46.80	21
30.81	33.92	36.78	40.29	42.80	48.27	22
32.01	35.17	38.08	41.64	44.18	49.73	23
33.20	36.42	39.36	42.98	45.56	51.18	24
34.38	37.65	40.65	44.31	46.93	52.62	25
35.56	38.89	41.92	45.64	48.29	54.05	26
36.74	40.11	43.19	46.96	49.65	55.48	27
37.92	41.34	44.46	48.28	50.99	56.89	28
39.09	42.56	45.72	49.59	52.34	58.30	29
40.26	43.77	46.98	50.89	53.67	59.70	30
51.81	55.76	59.34	63.69	66.77	73.40	40
63.17	67.50	71.42	76.15	79.49	86.66	50
74.40	79.08	83.30	88.38	91.95	99.61	60
85.53	90.53	95.02	100.43	104.21	112.32	70
96.58	101.88	106.63	112.33	116.32	124.84	80
107.57	113.15	118.14	124.12	128.30	137.21	90
118.50	124.34	129.56	135.81	140.17	149.45	100
140.23	146.57	152.21	158.95	163.65	173.62	120
268.47	277.14	284.80	293.89	300.18	313.44	240

Source: Computed by P. J. Hildebrand.

Independence: $\chi^2_{(r-1)(c-1)}$
0, no $E_{ij} < 1$, at most 20% < 5
Ho: row and columns are independent
Ha: not independent

Homogeneity:
same as independence
but with more categories

Fischers:
@: >5 or nat

$m = \max(0, n_{1.} + n_{.1}, -n_{..})$
$m+ = \min(n_{1.}, n_{.1}) = n_{..}$

P-value
$Ha: \pi_1 > \pi_2 \Rightarrow \sum$ for N_{11} to $m+$
$\pi_1 < \pi_2 \Rightarrow \sum$ for N_{11} to $m-$
$\pi_1 \neq \pi_2 \Rightarrow N_{11}$ t all $< N_{11}$

F_a

TABLE 9

Percentage points of the F distribution (df_2 between 1 and 6)

df_2	a	df_1									
		1	2	3	4	5	6	7	8	9	10
1	.25	5.83	7.50	8.20	8.58	8.82	8.98	9.10	9.19	9.26	9.32
	.10	39.86	49.50	53.59	55.83	57.24	58.20	58.91	59.44	59.86	60.19
	.05	161.4	199.5	215.7	224.6	230.2	234.0	236.8	238.9	240.5	241.9
	.025	647.8	799.5	864.2	899.6	921.8	937.1	948.2	956.7	963.3	968.6
	.01	4052	5000	5403	5625	5764	5859	5928	5981	6022	6056
2	.25	2.57	3.00	3.15	3.23	3.28	3.31	3.34	3.35	3.37	3.38
	.10	8.53	9.00	9.16	9.24	9.29	9.33	9.35	9.37	9.38	9.39
	.05	18.51	19.00	19.16	19.25	19.30	19.33	19.35	19.37	19.38	19.40
	.025	38.51	39.00	39.17	39.25	39.30	39.33	39.36	39.37	39.39	39.40
	.01	98.50	99.00	99.17	99.25	99.30	99.33	99.36	99.37	99.39	99.40
	.005	198.5	199.0	199.2	199.2	199.3	199.3	199.4	199.4	199.4	199.4
	.001	998.5	999.0	999.2	999.2	999.3	999.3	999.4	999.4	999.4	999.4
3	.25	2.02	2.28	2.36	2.39	2.41	2.42	2.43	2.44	2.44	2.44
	.10	5.54	5.46	5.39	5.34	5.31	5.28	5.27	5.25	5.24	5.23
	.05	10.13	9.55	9.28	9.12	9.01	8.94	8.89	8.85	8.81	8.79
	.025	17.44	16.04	15.44	15.10	14.88	14.73	14.62	14.54	14.47	14.42
	.01	34.12	30.82	29.46	28.71	28.24	27.91	27.67	27.49	27.35	27.23
	.005	55.55	49.80	47.47	46.19	45.39	44.84	44.43	44.13	43.88	43.69
	.001	167.0	148.5	141.1	137.1	134.6	132.8	131.6	130.6	129.9	129.2
4	.25	1.81	2.00	2.05	2.06	2.07	2.08	2.08	2.08	2.08	2.08
	.10	4.54	4.32	4.19	4.11	4.05	4.01	3.98	3.95	3.94	3.92
	.05	7.71	6.94	6.59	6.39	6.26	6.16	6.09	6.04	6.00	5.96
	.025	12.22	10.65	9.98	9.60	9.36	9.20	9.07	8.98	8.90	8.84
	.01	21.20	18.00	16.69	15.98	15.52	15.21	14.98	14.80	14.66	14.55
	.005	31.33	26.28	24.26	23.15	22.46	21.97	21.62	21.35	21.14	20.97
	.001	74.14	61.25	56.18	53.44	51.71	50.53	49.66	49.00	48.47	48.05
5	.25	1.69	1.85	1.88	1.89	1.89	1.89	1.89	1.89	1.89	1.89
	.10	4.06	3.78	3.62	3.52	3.45	3.40	3.37	3.34	3.32	3.30
	.05	6.61	5.79	5.41	5.19	5.05	4.95	4.88	4.82	4.77	4.74
	.025	10.01	8.43	7.76	7.39	7.15	6.98	6.85	6.76	6.68	6.62
	.01	16.26	13.27	12.06	11.39	10.97	10.67	10.46	10.29	10.16	10.05
	.005	22.78	18.31	16.53	15.56	14.94	14.51	14.20	13.96	13.77	13.62
	.001	47.18	37.12	33.20	31.09	29.75	28.83	28.16	27.65	27.24	26.92
6	.25	1.62	1.76	1.78	1.79	1.79	1.78	1.78	1.78	1.77	1.77
	.10	3.78	3.46	3.29	3.18	3.11	3.05	3.01	2.98	2.96	2.94
	.05	5.99	5.14	4.76	4.53	4.39	4.28	4.21	4.15	4.10	4.06
	.025	8.81	7.26	6.60	6.23	5.99	5.82	5.70	5.60	5.52	5.46
	.01	13.75	10.92	9.78	9.15	8.75	8.47	8.26	8.10	7.98	7.87
	.005	18.63	14.54	12.92	12.03	11.46	11.07	10.79	10.57	10.39	10.25
	.001	35.51	27.00	23.70	21.92	20.80	20.03	19.46	19.03	18.69	18.41

$$F_V = F_{\alpha/2, df_1, df_2}$$
$$F_L = F_{\alpha/2, df_2, df_1} \Rightarrow F_L = \frac{1}{F_{\alpha/2, df_2, df_1}}$$
$$\left(\frac{S_1^2}{S_2^2 F_{\alpha/2, df_1, df_2}} \qquad \frac{S_1^2 F_{\alpha/2, df_2, df_1}}{S_2^2} \right)$$

TABLE 9
(*continued*)

12	15	20	24	30	40	60	120	240	inf.	a	df$_2$
				df$_1$							
9.41	9.49	9.58	9.63	9.67	9.71	9.76	9.80	9.83	9.85	.25	1
60.71	61.22	61.74	62.00	62.26	62.53	62.79	63.06	63.19	63.33	.10	
243.9	245.9	248.0	249.1	250.1	251.1	252.2	253.3	253.8	254.3	.05	
976.7	984.9	993.1	997.2	1001	1006	1010	1014	1016	1018	.025	
6106	6157	6209	6235	6261	6287	6313	6339	6353	6366	.01	
3.39	3.41	3.43	3.43	3.44	3.45	3.46	3.47	3.47	3.48	.25	2
9.41	9.42	9.44	9.45	9.46	9.47	9.47	9.48	9.49	9.49	.10	
19.41	19.43	19.45	19.45	19.46	19.47	19.48	19.49	19.49	19.50	.05	
39.41	39.43	39.45	39.46	39.46	39.47	39.48	39.49	39.49	39.50	.025	
99.42	99.43	99.45	99.46	99.47	99.47	99.48	99.49	99.50	99.50	.01	
199.4	199.4	199.4	199.5	199.5	199.5	199.5	199.5	199.5	199.5	.005	
999.4	999.4	999.4	999.5	999.5	999.5	999.5	999.5	999.5	999.5	.001	
2.45	2.46	2.46	2.46	2.47	2.47	2.47	2.47	2.47	2.47	.25	3
5.22	5.20	5.18	5.18	5.17	5.16	5.15	5.14	5.14	5.13	.10	
8.74	8.70	8.66	8.64	8.62	8.59	8.57	8.55	8.54	8.53	.05	
14.34	14.25	14.17	14.12	14.08	14.04	13.99	13.95	13.92	13.90	.025	
27.05	26.87	26.69	26.60	26.50	26.41	26.32	26.22	26.17	26.13	.01	
43.39	43.08	42.78	42.62	42.47	42.31	42.15	41.99	41.91	41.83	.005	
128.3	127.4	126.4	125.9	125.4	125.0	124.5	124.0	123.7	123.5	.001	
2.08	2.08	2.08	2.08	2.08	2.08	2.08	2.08	2.08	2.08	.25	4
3.90	3.87	3.84	3.83	3.82	3.80	3.79	3.78	3.77	3.76	.10	
5.91	5.86	5.80	5.77	5.75	5.72	5.69	5.66	5.64	5.63	.05	
8.75	8.66	8.56	8.51	8.46	8.41	8.36	8.31	8.28	8.26	.025	
14.37	14.20	14.02	13.93	13.84	13.75	13.65	13.56	13.51	13.46	.01	
20.70	20.44	20.17	20.03	19.89	19.75	19.61	19.47	19.40	19.32	.005	
47.41	46.76	46.10	45.77	45.43	45.09	44.75	44.40	44.23	44.05	.001	
1.89	1.89	1.88	1.88	1.88	1.88	1.87	1.87	1.87	1.87	.25	5
3.27	3.24	3.21	3.19	3.17	3.16	3.14	3.12	3.11	3.10	.10	
4.68	4.62	4.56	4.53	4.50	4.46	4.43	4.40	4.38	4.36	.05	
6.52	6.43	6.33	6.28	6.23	6.18	6.12	6.07	6.04	6.02	.025	
9.89	9.72	9.55	9.47	9.38	9.29	9.20	9.11	9.07	9.02	.01	
13.38	13.15	12.90	12.78	12.66	12.53	12.40	12.27	12.21	12.14	.005	
26.42	25.91	25.39	25.13	24.87	24.60	24.33	24.06	23.92	23.79	.001	
1.77	1.76	1.76	1.75	1.75	1.75	1.74	1.74	1.74	1.74	.25	6
2.90	2.87	2.84	2.82	2.80	2.78	2.76	2.74	2.73	2.72	.10	
4.00	3.94	3.87	3.84	3.81	3.77	3.74	3.70	3.69	3.67	.05	
5.37	5.27	5.17	5.12	5.07	5.01	4.96	4.90	4.88	4.85	.025	
7.72	7.56	7.40	7.31	7.23	7.14	7.06	6.97	6.92	6.88	.01	
10.03	9.81	9.59	9.47	9.36	9.24	9.12	9.00	8.94	8.88	.005	
17.99	17.56	17.12	16.90	16.67	16.44	16.21	15.98	15.86	15.75	.001	

If $\frac{S_1^2}{S_2^2} > a$, use non-pooled t

If $< a$, use pooled

$$t = \frac{\bar{y} - (\mu_1 - \mu_2)}{s/\sqrt{n}}$$

TABLE 9

Percentage points of the F distribution (df$_2$ between 7 and 12)

df$_2$	a	1	2	3	4	5	6	7	8	9	10
						df$_1$					
7	.25	1.57	1.70	1.72	1.72	1.71	1.71	1.70	1.70	1.69	1.69
	.10	3.59	3.26	3.07	2.96	2.88	2.83	2.78	2.75	2.72	2.70
	.05	5.59	4.74	4.35	4.12	3.97	3.87	3.79	3.73	3.68	3.64
	.025	8.07	6.54	5.89	5.52	5.29	5.12	4.99	4.90	4.82	4.76
	.01	12.25	9.55	8.45	7.85	7.46	7.19	6.99	6.84	6.72	6.62
	.005	16.24	12.40	10.88	10.05	9.52	9.16	8.89	8.68	8.51	8.38
	.001	29.25	21.69	18.77	17.20	16.21	15.52	15.02	14.63	14.33	14.08
8	.25	1.54	1.66	1.67	1.66	1.66	1.65	1.64	1.64	1.63	1.63
	.10	3.46	3.11	2.92	2.81	2.73	2.67	2.62	2.59	2.56	2.54
	.05	5.32	4.46	4.07	3.84	3.69	3.58	3.50	3.44	3.39	3.35
	.025	7.57	6.06	5.42	5.05	4.82	4.65	4.53	4.43	4.36	4.30
	.01	11.26	8.65	7.59	7.01	6.63	6.37	6.18	6.03	5.91	5.81
	.005	14.69	11.04	9.60	8.81	8.30	7.95	7.69	7.50	7.34	7.21
	.001	25.41	18.49	15.83	14.39	13.48	12.86	12.40	12.05	11.77	11.54
9	.25	1.51	1.62	1.63	1.63	1.62	1.61	1.60	1.60	1.59	1.59
	.10	3.36	3.01	2.81	2.69	2.61	2.55	2.51	2.47	2.44	2.42
	.05	5.12	4.26	3.86	3.63	3.48	3.37	3.29	3.23	3.18	3.14
	.025	7.21	5.71	5.08	4.72	4.48	4.32	4.20	4.10	4.03	3.96
	.01	10.56	8.02	6.99	6.42	6.06	5.80	5.61	5.47	5.35	5.26
	.005	13.61	10.11	8.72	7.96	7.47	7.13	6.88	6.69	6.54	6.42
	.001	22.86	16.39	13.90	12.56	11.71	11.13	10.70	10.37	10.11	9.89
10	.25	1.49	1.60	1.60	1.59	1.59	1.58	1.57	1.56	1.56	1.55
	.10	3.29	2.92	2.73	2.61	2.52	2.46	2.41	2.38	2.35	2.32
	.05	4.96	4.10	3.71	3.48	3.33	3.22	3.14	3.07	3.02	2.98
	.025	6.94	5.46	4.83	4.47	4.24	4.07	3.95	3.85	3.78	3.72
	.01	10.04	7.56	6.55	5.99	5.64	5.39	5.20	5.06	4.94	4.85
	.005	12.83	9.43	8.08	7.34	6.87	6.54	6.30	6.12	5.97	5.85
	.001	21.04	14.91	12.55	11.28	10.48	9.93	9.52	9.20	8.96	8.75
11	.25	1.47	1.58	1.58	1.57	1.56	1.55	1.54	1.53	1.53	1.52
	.10	3.23	2.86	2.66	2.54	2.45	2.39	2.34	2.30	2.27	2.25
	.05	4.84	3.98	3.59	3.36	3.20	3.09	3.01	2.95	2.90	2.85
	.025	6.72	5.26	4.63	4.28	4.04	3.88	3.76	3.66	3.59	3.53
	.01	9.65	7.21	6.22	5.67	5.32	5.07	4.89	4.74	4.63	4.54
	.005	12.23	8.91	7.60	6.88	6.42	6.10	5.86	5.68	5.54	5.42
	.001	19.69	13.81	11.56	10.35	9.58	9.05	8.66	8.35	8.12	7.92
12	.25	1.46	1.56	1.56	1.55	1.54	1.53	1.52	1.51	1.51	1.50
	.10	3.18	2.81	2.61	2.48	2.39	2.33	2.28	2.24	2.21	2.19
	.05	4.75	3.89	3.49	3.26	3.11	3.00	2.91	2.85	2.80	2.75
	.025	6.55	5.10	4.47	4.12	3.89	3.73	3.61	3.51	3.44	3.37
	.01	9.33	6.93	5.95	5.41	5.06	4.82	4.64	4.50	4.39	4.30
	.005	11.75	8.51	7.23	6.52	6.07	5.76	5.52	5.35	5.20	5.09
	.001	18.64	12.97	10.80	9.63	8.89	8.38	8.00	7.71	7.48	7.29

TABLE 9
(*continued*)

				df_1								
12	**15**	**20**	**24**	**30**	**40**	**60**	**120**	**240**	**inf.**	*a*	**df_2**
1.68	1.68	1.67	1.67	1.66	1.66	1.65	1.65	1.65	1.65	.25	**7**
2.67	2.63	2.59	2.58	2.56	2.54	2.51	2.49	2.48	2.47	.10	
3.57	3.51	3.44	3.41	3.38	3.34	3.30	3.27	3.25	3.23	.05	
4.67	4.57	4.47	4.41	4.36	4.31	4.25	4.20	4.17	4.14	.025	
6.47	6.31	6.16	6.07	5.99	5.91	5.82	5.74	5.69	5.65	.01	
8.18	7.97	7.75	7.64	7.53	7.42	7.31	7.19	7.13	7.08	.005	
13.71	13.32	12.93	12.73	12.53	12.33	12.12	11.91	11.80	11.70	.001	
1.62	1.62	1.61	1.60	1.60	1.59	1.59	1.58	1.58	1.58	.25	**8**
2.50	2.46	2.42	2.40	2.38	2.36	2.34	2.32	2.30	2.29	.10	
3.28	3.22	3.15	3.12	3.08	3.04	3.01	2.97	2.95	2.93	.05	
4.20	4.10	4.00	3.95	3.89	3.84	3.78	3.73	3.70	3.67	.025	
5.67	5.52	5.36	5.28	5.20	5.12	5.03	4.95	4.90	4.86	.01	
7.01	6.81	6.61	6.50	6.40	6.29	6.18	6.06	6.01	5.95	.005	
11.19	10.84	10.48	10.30	10.11	9.92	9.73	9.53	9.43	9.33	.001	
1.58	1.57	1.56	1.56	1.55	1.54	1.64	1.53	1.53	1.53	.25	**9**
2.38	2.34	2.30	2.28	2.25	2.23	2.21	2.18	2.17	2.16	.10	
3.07	3.01	2.94	2.90	2.86	2.83	2.79	2.75	2.73	2.71	.05	
3.87	3.77	3.67	3.61	3.56	3.51	3.45	3.39	3.36	3.33	.025	
5.11	4.96	4.81	4.73	4.65	4.57	4.48	4.40	4.35	4.31	.01	
6.23	6.03	5.83	5.73	5.62	5.52	5.41	5.30	5.24	5.19	.005	
9.57	9.24	8.90	8.72	8.55	8.37	8.19	8.00	7.91	7.81	.001	
1.54	1.53	1.52	1.52	1.51	1.51	1.50	1.49	1.49	1.48	.25	**10**
2.28	2.24	2.20	2.18	2.16	2.13	2.11	2.08	2.07	2.06	.10	
2.91	2.85	2.77	2.74	2.70	2.66	2.62	2.58	2.56	2.54	.05	
3.62	3.52	3.42	3.37	3.31	3.26	3.20	3.14	3.11	3.08	.025	
4.71	4.56	4.41	4.33	4.25	4.17	4.08	4.00	3.95	3.91	.01	
5.66	5.47	5.27	5.17	5.07	4.97	4.86	4.75	4.69	4.64	.005	
8.45	8.13	7.80	7.64	7.47	7.30	7.12	6.94	6.85	6.76	.001	
1.51	1.50	1.49	1.49	1.48	1.47	1.47	1.46	1.45	1.45	.25	**11**
2.21	2.17	2.12	2.10	2.08	2.05	2.03	2.00	1.99	1.97	.10	
2.79	2.72	2.65	2.61	2.57	2.53	2.49	2.45	2.43	2.40	.05	
3.43	3.33	3.23	3.17	3.12	3.06	3.00	2.94	2.91	2.88	.025	
4.40	4.25	4.10	4.02	3.94	3.86	3.78	3.69	3.65	3.60	.01	
5.24	5.05	4.86	4.76	4.65	4.55	4.45	4.34	4.28	4.23	.005	
7.63	7.32	7.01	6.85	6.68	6.52	6.35	6.18	6.09	6.00	.001	
1.49	1.48	1.47	1.46	1.45	1.45	1.44	1.43	1.43	1.42	.25	**12**
2.15	2.10	2.06	2.04	2.01	1.99	1.96	1.93	1.92	1.90	.10	
2.69	2.62	2.54	2.51	2.47	2.43	2.38	2.34	2.32	2.30	.05	
3.28	3.18	3.07	3.02	2.96	2.91	2.85	2.79	2.76	2.72	.025	
4.16	4.01	3.86	3.78	3.70	3.62	3.54	3.45	3.41	3.36	.01	
4.91	4.72	4.53	4.43	4.33	4.23	4.12	4.01	3.96	3.90	.005	
7.00	6.71	6.40	6.25	6.09	5.93	5.76	5.59	5.51	5.42	.001	

TABLE 9

Percentage points of the F distribution (df_2 between 13 and 18)

df_2	a	1	2	3	4	5	6	7	8	9	10
										df_1	
13	.25	1.45	1.55	1.55	1.53	1.52	1.51	1.50	1.49	1.49	1.48
	.10	3.14	2.76	2.56	2.43	2.35	2.28	2.23	2.20	2.16	2.14
	.05	4.67	3.81	3.41	3.18	3.03	2.92	2.83	2.77	2.71	2.67
	.025	6.41	4.97	4.35	4.00	3.77	3.60	3.48	3.39	3.31	3.25
	.01	9.07	6.70	5.74	5.21	4.86	4.62	4.44	4.30	4.19	4.10
	.005	11.37	8.19	6.93	6.23	5.79	5.48	5.25	5.08	4.94	4.82
	.001	17.82	12.31	10.21	9.07	8.35	7.86	7.49	7.21	6.98	6.80
14	25	1.44	1.53	1.53	1.52	1.51	1.50	1.49	1.48	1.47	1.46
	.10	3.10	2.73	2.52	2.39	2.31	2.24	2.19	2.15	2.12	2.10
	.05	4.60	3.74	3.34	3.11	2.96	2.85	2.76	2.70	2.65	2.60
	.025	6.30	4.86	4.24	3.89	3.66	3.50	3.38	3.29	3.21	3.15
	.01	8.86	6.51	5.56	5.04	4.69	4.46	4.28	4.14	4.03	3.94
	.005	11.06	7.92	6.68	6.00	5.56	5.26	5.03	4.86	4.72	4.60
	.001	17.14	11.78	9.73	8.62	7.92	7.44	7.08	6.80	6.58	6.40
15	.25	1.43	1.52	1.52	1.51	1.49	1.48	1.47	1.46	1.46	1.45
	.10	3.07	2.70	2.49	2.36	2.27	2.21	2.16	2.12	2.09	2.06
	.05	4.54	3.68	3.29	3.06	2.90	2.79	2.71	2.64	2.59	2.54
	.025	6.20	4.77	4.15	3.80	3.58	3.41	3.29	3.20	3.12	3.06
	.01	8.68	6.36	5.42	4.89	4.56	4.32	4.14	4.00	3.89	3.80
	.005	10.80	7.70	6.48	5.80	5.37	5.07	4.85	4.67	4.54	4.42
	.001	16.59	11.34	9.34	8.25	7.57	7.09	6.74	6.47	6.26	6.08
16	.25	1.42	1.51	1.51	1.50	1.48	1.47	1.46	1.45	1.44	1.44
	.10	3.05	2.67	2.46	2.33	2.24	2.18	2.13	2.09	2.06	2.03
	.05	4.49	3.63	3.24	3.01	2.85	2.74	2.66	2.59	2.54	2.49
	.025	6.12	4.69	4.08	3.73	3.50	3.34	3.22	3.12	3.05	2.99
	.01	8.53	6.23	5.29	4.77	4.44	4.20	4.03	3.89	3.78	3.69
	.005	10.58	7.51	6.30	5.64	5.21	4.91	4.69	4.52	4.38	4.27
	.001	16.12	10.97	9.01	7.94	7.27	6.80	6.46	6.19	5.98	5.81
17	.25	1.42	1.51	1.50	1.49	1.47	1.46	1.45	1.44	1.43	1.43
	.10	3.03	2.64	2.44	2.31	2.22	2.15	2.10	2.06	2.03	2.00
	.05	4.45	3.59	3.20	2.96	2.81	2.70	2.61	2.55	2.49	2.45
	.025	6.04	4.62	4.01	3.66	3.44	3.28	3.16	3.06	2.98	2.92
	.01	8.40	6.11	5.18	4.67	4.34	4.10	3.93	3.79	3.68	3.59
	.005	10.38	7.35	6.16	5.50	5.07	4.78	4.56	4.39	4.25	4.14
	.001	15.72	10.66	8.73	7.68	7.02	6.56	6.22	5.96	5.75	5.58
18	.25	1.41	1.50	1.49	1.48	1.46	1.45	1.44	1.43	1.42	1.42
	.10	3.01	2.62	2.42	2.29	2.20	2.13	2.08	2.04	2.00	1.98
	.05	4.41	3.55	3.16	2.93	2.77	2.66	2.58	2.51	2.46	2.41
	.025	5.98	4.56	3.95	3.61	3.38	3.22	3.10	3.01	2.93	2.87
	.01	8.29	6.01	5.09	4.58	4.25	4.01	3.84	3.71	3.60	3.51
	.005	10.22	7.21	6.03	5.37	4.96	4.66	4.44	4.28	4.14	4.03
	.001	15.38	10.39	8.49	7.46	6.81	6.35	6.02	5.76	5.56	5.39

TABLE 9
(*continued*)

12	15	20	24	30	40	60	120	240	inf.	*a*	df$_2$
					df$_1$						
1.47	1.46	1.45	1.44	1.43	1.42	1.42	1.41	1.40	1.40	.25	**13**
2.10	2.05	2.01	1.98	1.96	1.93	1.90	1.88	1.86	1.85	.10	
2.60	2.53	2.46	2.42	2.38	2.34	2.30	2.25	2.23	2.21	.05	
3.15	3.05	2.95	2.89	2.84	2.78	2.72	2.66	2.63	2.60	.025	
3.96	3.82	3.66	3.59	3.51	3.43	3.34	3.25	3.21	3.17	.01	
4.64	4.46	4.27	4.17	4.07	3.97	3.87	3.76	3.70	3.65	.005	
6.52	6.23	5.93	5.78	5.63	5.47	5.30	5.14	5.05	4.97	.001	
1.45	1.44	1.43	1.42	1.41	1.41	1.40	1.39	1.38	1.38	.25	**14**
2.05	2.01	1.96	1.94	1.91	1.89	1.86	1.83	1.81	1.80	.10	
2.53	2.46	2.39	2.35	2.31	2.27	2.22	2.18	2.15	2.13	.05	
3.05	2.95	2.84	2.79	2.73	2.67	2.61	2.55	2.52	2.49	.025	
3.80	3.66	3.51	3.43	3.35	3.27	3.18	3.09	3.05	3.00	.01	
4.43	4.25	4.06	3.96	3.86	3.76	3.66	3.55	3.49	3.44	.005	
6.13	5.85	5.56	5.41	5.25	5.10	4.94	4.77	4.69	4.60	.001	
1.44	1.43	1.41	1.41	1.40	1.39	1.38	1.37	1.36	1.36	.25	**15**
2.02	1.97	1.92	1.90	1.87	1.85	1.82	1.79	1.77	1.76	.10	
2.48	2.40	2.33	2.29	2.25	2.20	2.16	2.11	2.09	2.07	.05	
2.96	2.86	2.76	2.70	2.64	2.59	2.52	2.46	2.43	2.40	.025	
3.67	3.52	3.37	3.29	3.21	3.13	3.05	2.96	2.91	2.87	.01	
4.25	4.07	3.88	3.79	3.69	3.58	3.48	3.37	3.32	3.26	.005	
5.81	5.54	5.25	5.10	4.95	4.80	4.64	4.47	4.39	4.31	.001	
1.43	1.41	1.40	1.39	1.38	1.37	1.36	1.35	1.35	1.34	.25	**16**
1.99	1.94	1.89	1.87	1.84	1.81	1.78	1.75	1.73	1.72	.10	
2.42	2.35	2.28	2.24	2.19	2.15	2.11	2.06	2.03	2.01	.05	
2.89	2.79	2.68	2.63	2.57	2.51	2.45	2.38	2.35	2.32	.025	
3.55	3.41	3.26	3.18	3.10	3.02	2.93	2.84	2.80	2.75	.01	
4.10	3.92	3.73	3.64	3.54	3.44	3.33	3.22	3.17	3.11	.005	
5.55	5.27	4.99	4.85	4.70	4.54	4.39	4.23	4.14	4.06	.001	
1.41	1.40	1.39	1.38	1.37	1.36	1.35	1.34	1.33	1.33	.25	**17**
1.96	1.91	1.86	1.84	1.81	1.78	1.75	1.72	1.70	1.69	.10	
2.38	2.31	2.23	2.19	2.15	2.10	2.06	2.01	1.99	1.96	.05	
2.82	2.72	2.62	2.56	2.50	2.44	2.38	2.32	2.28	2.25	.025	
3.46	3.31	3.16	3.08	3.00	2.92	2.83	2.75	2.70	2.65	.01	
3.97	3.79	3.61	3.51	3.41	3.31	3.21	3.10	3.04	2.98	.005	
5.32	5.05	4.78	4.63	4.48	4.33	4.18	4.02	3.93	3.85	.001	
1.40	1.39	1.38	1.37	1.36	1.35	1.34	1.33	1.32	1.32	.25	**18**
1.93	1.89	1.84	1.81	1.78	1.75	1.72	1.69	1.67	1.66	.10	
2.34	2.27	2.19	2.15	2.11	2.06	2.02	1.97	1.94	1.92	.05	
2.77	2.67	2.56	2.50	2.44	2.38	2.32	2.26	2.22	2.19	.025	
3.37	3.23	3.08	3.00	2.92	2.84	2.75	2.66	2.61	2.57	.01	
3.86	3.68	3.50	3.40	3.30	3.20	3.10	2.99	2.93	2.87	.005	
5.13	4.87	4.59	4.45	4.30	4.15	4.00	3.84	3.75	3.67	.001	

TABLE 9

Percentage points of the F distribution (df_2 between 19 and 24)

df_2	a	1	2	3	4	5	6	7	8	9	10
							df_1				
19	.25	1.41	1.49	1.49	1.47	1.46	1.44	1.43	1.42	1.41	1.41
	.10	2.99	2.61	2.40	2.27	2.18	2.11	2.06	2.02	1.98	1.96
	.05	4.38	3.52	3.13	2.90	2.74	2.63	2.54	2.48	2.42	2.38
	.025	5.92	4.51	3.90	3.56	3.33	3.17	3.05	2.96	2.88	2.82
	.01	8.18	5.93	5.01	4.50	4.17	3.94	3.77	3.63	3.52	3.43
	.005	10.07	7.09	5.92	5.27	4.85	4.56	4.34	4.18	4.04	3.93
	.001	15.08	10.16	8.28	7.27	6.62	6.18	5.85	5.59	5.39	5.22
20	.25	1.40	1.49	1.48	1.47	1.45	1.44	1.43	1.42	1.41	1.40
	.10	2.97	2.59	2.38	2.25	2.16	2.09	2.04	2.00	1.96	1.94
	.05	4.35	3.49	3.10	2.87	2.71	2.60	2.51	2.45	2.39	2.35
	.025	5.87	4.46	3.86	3.51	3.29	3.13	3.01	2.91	2.84	2.77
	.01	8.10	5.85	4.94	4.43	4.10	3.87	3.70	3.56	3.46	3.37
	.005	9.94	6.99	5.82	5.17	4.76	4.47	4.26	4.09	3.96	3.85
	.001	14.82	9.95	8.10	7.10	6.46	6.02	5.69	5.44	5.24	5.08
21	.25	1.40	1.48	1.48	1.46	1.44	1.43	1.42	1.41	1.40	1.39
	.10	2.96	2.57	2.36	2.23	2.14	2.08	2.02	1.98	1.95	1.92
	.05	4.32	3.47	3.07	2.84	2.68	2.57	2.49	2.42	2.37	2.32
	.025	5.83	4.42	3.82	3.48	3.25	3.09	2.97	2.87	2.80	2.73
	.01	8.02	5.78	4.87	4.37	4.04	3.81	3.64	3.51	3.40	3.31
	.005	9.83	6.89	5.73	5.09	4.68	4.39	4.18	4.01	3.88	3.77
	.001	14.59	9.77	7.94	6.95	6.32	5.88	5.56	5.31	5.11	4.95
22	.25	1.40	1.48	1.47	1.45	1.44	1.42	1.41	1.40	1.39	1.39
	.10	2.95	2.56	2.35	2.22	2.13	2.06	2.01	1.97	1.93	1.90
	.05	4.30	3.44	3.05	2.82	2.66	2.55	2.46	2.40	2.34	2.30
	.025	5.79	4.38	3.78	3.44	3.22	3.05	2.93	2.84	2.76	2.70
	.01	7.95	5.72	4.82	4.31	3.99	3.76	3.59	3.45	3.35	3.26
	.005	9.73	6.81	5.65	5.02	4.61	4.32	4.11	3.94	3.81	3.70
	.001	14.38	9.61	7.80	6.81	6.19	5.76	5.44	5.19	4.99	4.83
23	.25	1.39	1.47	1.47	1.45	1.43	1.42	1.41	1.40	1.39	1.38
	.10	2.94	2.55	2.34	2.21	2.11	2.05	1.99	1.95	1.92	1.89
	.05	4.28	3.42	3.03	2.80	2.64	2.53	2.44	2.37	2.32	2.27
	.025	5.75	4.35	3.75	3.41	3.18	3.02	2.90	2.81	2.73	2.67
	.01	7.88	5.66	4.76	4.26	3.94	3.71	3.54	3.41	3.30	3.21
	.005	9.63	6.73	5.58	4.95	4.54	4.26	4.05	3.88	3.75	3.64
	.001	14.20	9.47	7.67	6.70	6.08	5.65	5.33	5.09	4.89	4.73
24	.25	1.39	1.47	1.46	1.44	1.43	1.41	1.40	1.39	1.38	1.38
	.10	2.93	2.54	2.33	2.19	2.10	2.04	1.98	1.94	1.91	1.88
	.05	4.26	3.40	3.01	2.78	2.62	2.51	2.42	2.36	2.30	2.25
	.025	5.72	4.32	3.72	3.38	3.15	2.99	2.87	2.78	2.70	2.64
	.01	7.82	5.61	4.72	4.22	3.90	3.67	3.50	3.36	3.26	3.17
	.005	9.55	6.66	5.52	4.89	4.49	4.20	3.99	3.83	3.69	3.59
	.001	14.03	9.34	7.55	6.59	5.98	5.55	5.23	4.99	4.80	4.64

TABLE 9

(*continued*)

				df$_1$								
12	**15**	**20**	**24**	**30**	**40**	**60**	**120**	**240**	**inf.**	*a*	**df$_2$**
1.40	1.38	1.37	1.36	1.35	1.34	1.33	1.32	1.31	1.30	.25	**19**
1.91	1.86	1.81	1.79	1.76	1.73	1.70	1.67	1.65	1.63	.10	
2.31	2.23	2.16	2.11	2.07	2.03	1.98	1.93	1.90	1.88	.05	
2.72	2.62	2.51	2.45	2.39	2.33	2.27	2.20	2.17	2.13	.025	
3.30	3.15	3.00	2.92	2.84	2.76	2.67	2.58	2.54	2.49	.01	
3.76	3.59	3.40	3.31	3.21	3.11	3.00	2.89	2.83	2.78	.005	
4.97	4.70	4.43	4.29	4.14	3.99	3.84	3.68	3.60	3.51	.001	
1.39	1.37	1.36	1.35	1.34	1.33	1.32	1.31	1.30	1.29	.25	**20**
1.89	1.84	1.79	1.77	1.74	1.71	1.68	1.64	1.63	1.61	.10	
2.28	2.20	2.12	2.08	2.04	1.99	1.95	1.90	1.87	1.84	.05	
2.68	2.57	2.46	2.41	2.35	2.29	2.22	2.16	2.12	2.09	.025	
3.23	3.09	2.94	2.86	2.78	2.69	2.61	2.52	2.47	2.42	.01	
3.68	3.50	3.32	3.22	3.12	3.02	2.92	2.81	2.75	2.69	.005	
4.82	4.56	4.29	4.15	4.00	3.86	3.70	3.54	3.46	3.38	.001	
1.38	1.37	1.35	1.34	1.33	1.32	1.31	1.30	1.29	1.28	.25	**21**
1.87	1.83	1.78	1.75	1.72	1.69	1.66	1.62	1.60	1.59	.10	
2.25	2.18	2.10	2.05	2.01	1.96	1.92	1.87	1.84	1.81	.05	
2.64	2.53	2.42	2.37	2.31	2.25	2.18	2.11	2.08	2.04	.025	
3.17	3.03	2.88	2.80	2.72	2.64	2.55	2.46	2.41	2.36	.01	
3.60	3.43	3.24	3.15	3.05	2.95	2.84	2.73	2.67	2.61	.005	
4.70	4.44	4.17	4.03	3.88	3.74	3.58	3.42	3.34	3.26	.001	
1.37	1.36	1.34	1.33	1.32	1.31	1.30	1.29	1.28	1.28	.25	**22**
1.86	1.81	1.76	1.73	1.70	1.67	1.64	1.60	1.59	1.57	.10	
2.23	2.15	2.07	2.03	1.98	1.94	1.89	1.84	1.81	1.78	.05	
2.60	2.50	2.39	2.33	2.27	2.21	2.14	2.08	2.04	2.00	.025	
3.12	2.98	2.83	2.75	2.67	2.58	2.50	2.40	2.35	2.31	.01	
3.54	3.36	3.18	3.08	2.98	2.88	2.77	2.66	2.60	2.55	.005	
4.58	4.33	4.06	3.92	3.78	3.63	3.48	3.32	3.23	3.15	.001	
1.37	1.35	1.34	1.33	1.32	1.31	1.30	1.28	1.28	1.27	.25	**23**
1.84	1.80	1.74	1.72	1.69	1.66	1.62	1.59	1.57	1.55	.10	
2.20	2.13	2.05	2.01	1.96	1.91	1.86	1.81	1.79	1.76	.05	
2.57	2.47	2.36	2.30	2.24	2.18	2.11	2.04	2.01	1.97	.025	
3.07	2.93	2.78	2.70	2.62	2.54	2.45	2.35	2.31	2.26	.01	
3.47	3.30	3.12	3.02	2.92	2.82	2.71	2.60	2.54	2.48	.005	
4.48	4.23	3.96	3.82	3.68	3.53	3.38	3.22	3.14	3.05	.001	
1.36	1.35	1.33	1.32	1.31	1.30	1.29	1.28	1.27	1.26	.25	**24**
1.83	1.78	1.73	1.70	1.67	1.64	1.61	1.57	1.55	1.53	.10	
2.18	2.11	2.03	1.98	1.94	1.89	1.84	1.79	1.76	1.73	.05	
2.54	2.44	2.33	2.27	2.21	2.15	2.08	2.01	1.97	1.94	.025	
3.03	2.89	2.74	2.66	2.58	2.49	2.40	2.31	2.26	2.21	.01	
3.42	3.25	3.06	2.97	2.87	2.77	2.66	2.55	2.49	2.43	.005	
4.39	4.14	3.87	3.74	3.59	3.45	3.29	3.14	3.05	2.97	.001	

TABLE 9

Percentage points of the F distribution (df$_2$ between 25 and 30)

		df$_1$									
df$_2$	a	1	2	3	4	5	6	7	8	9	10
25	.25	1.39	1.47	1.46	1.44	1.42	1.41	1.40	1.39	1.38	1.37
	.10	2.92	2.53	2.32	2.18	2.09	2.02	1.97	1.93	1.89	1.87
	.05	4.24	3.39	2.99	2.76	2.60	2.49	2.40	2.34	2.28	2.24
	.025	5.69	4.29	3.69	3.35	3.13	2.97	2.85	2.75	2.68	2.61
	.01	7.77	5.57	4.68	4.18	3.85	3.63	3.46	3.32	3.22	3.13
	.005	9.48	6.60	5.46	4.84	4.43	4.15	3.94	3.78	3.64	3.54
	.001	13.88	9.22	7.45	6.49	5.89	5.46	5.15	4.91	4.71	4.56
26	.25	1.38	1.46	1.45	1.44	1.42	1.41	1.39	1.38	1.37	1.37
	.10	2.91	2.52	2.31	2.17	2.08	2.01	1.96	1.92	1.88	1.86
	.05	4.23	3.37	2.98	2.74	2.59	2.47	2.39	2.32	2.27	2.22
	.025	5.66	4.27	3.67	3.33	3.10	2.94	2.82	2.73	2.65	2.59
	.01	7.72	5.53	4.64	4.14	3.82	3.59	3.42	3.29	3.18	3.09
	.005	9.41	6.54	5.41	4.79	4.38	4.10	3.89	3.73	3.60	3.49
	.001	13.74	9.12	7.36	6.41	5.80	5.38	5.07	4.83	4.64	4.48
27	.25	1.38	1.46	1.45	1.43	1.42	1.40	1.39	1.38	1.37	1.36
	.10	2.90	2.51	2.30	2.17	2.07	2.00	1.95	1.91	1.87	1.85
	.05	4.21	3.35	2.96	2.73	2.57	2.46	2.37	2.31	2.25	2.20
	.025	5.63	4.24	3.65	3.31	3.08	2.92	2.80	2.71	2.63	2.57
	.01	7.68	5.49	4.60	4.11	3.78	3.56	3.39	3.26	3.15	3.06
	.005	9.34	6.49	5.36	4.74	4.34	4.06	3.85	3.69	3.56	3.45
	.001	13.61	9.02	7.27	6.33	5.73	5.31	5.00	4.76	4.57	4.41
28	.25	1.38	1.46	1.45	1.43	1.41	1.40	1.39	1.38	1.37	1.36
	.10	2.89	2.50	2.29	2.16	2.06	2.00	1.94	1.90	1.87	1.84
	.05	4.20	3.34	2.95	2.71	2.56	2.45	2.36	2.29	2.24	2.19
	.025	5.61	4.22	3.63	3.29	3.06	2.90	2.78	2.69	2.61	2.55
	.01	7.64	5.45	4.57	4.07	3.75	3.53	3.36	3.23	3.12	3.03
	.005	9.28	6.44	5.32	4.70	4.30	4.02	3.81	3.65	3.52	3.41
	.001	13.50	8.93	7.19	6.25	5.66	5.24	4.93	4.69	4.50	4.35
29	.25	1.38	1.45	1.45	1.43	1.41	1.40	1.38	1.37	1.36	1.35
	.10	2.89	2.50	2.28	2.15	2.06	1.99	1.93	1.89	1.86	1.83
	.05	4.18	3.33	2.93	2.70	2.55	2.43	2.35	2.28	2.22	2.18
	.025	5.59	4.20	3.61	3.27	3.04	2.88	2.76	2.67	2.59	2.53
	.01	7.60	5.42	4.54	4.04	3.73	3.50	3.33	3.20	3.09	3.00
	.005	9.23	6.40	5.28	4.66	4.26	3.98	3.77	3.61	3.48	3.38
	.001	13.39	8.85	7.12	6.19	5.59	5.18	4.87	4.64	4.45	4.29
30	.25	1.38	1.45	1.44	1.42	1.41	1.39	1.38	1.37	1.36	1.35
	.10	2.88	2.49	2.28	2.14	2.05	1.98	1.93	1.88	1.85	1.82
	.05	4.17	3.32	2.92	2.69	2.53	2.42	2.33	2.27	2.21	2.16
	.025	5.57	4.18	3.59	3.25	3.03	2.87	2.75	2.65	2.57	2.51
	.01	7.56	5.39	4.51	4.02	3.70	3.47	3.30	3.17	3.07	2.98
	.005	9.18	6.35	5.24	4.62	4.23	3.95	3.74	3.58	3.45	3.34
	.001	13.29	8.77	7.05	6.12	5.53	5.12	4.82	4.58	4.39	4.24

TABLE 9
(*continued*)

12	15	20	24	30	40	60	120	240	inf.	a	df₂
1.36	1.34	1.33	1.32	1.31	1.29	1.28	1.27	1.26	1.25	.25	**25**
1.82	1.77	1.72	1.69	1.66	1.63	1.59	1.56	1.54	1.52	.10	
2.16	2.09	2.01	1.96	1.92	1.87	1.82	1.77	1.74	1.71	.05	
2.51	2.41	2.30	2.24	2.18	2.12	2.05	1.98	1.94	1.91	.025	
2.99	2.85	2.70	2.62	2.54	2.45	2.36	2.27	2.22	2.17	.01	
3.37	3.20	3.01	2.92	2.82	2.72	2.61	2.50	2.44	2.38	.005	
4.31	4.06	3.79	3.66	3.52	3.37	3.22	3.06	2.98	2.89	.001	
1.35	1.34	1.32	1.31	1.30	1.29	1.28	1.26	1.26	1.25	.25	**26**
1.81	1.76	1.71	1.68	1.65	1.61	1.58	1.54	1.52	1.50	.10	
2.15	2.07	1.99	1.95	1.90	1.85	1.80	1.75	1.72	1.69	.05	
2.49	2.39	2.28	2.22	2.16	2.09	2.03	1.95	1.92	1.88	.025	
2.96	2.81	2.66	2.58	2.50	2.42	2.33	2.23	2.18	2.13	.01	
3.33	3.15	2.97	2.87	2.77	2.67	2.56	2.45	2.39	2.33	.005	
4.24	3.99	3.72	3.59	3.44	3.30	3.15	2.99	2.90	2.82	.001	
1.35	1.33	1.32	1.31	1.30	1.28	1.27	1.26	1.25	1.24	.25	**27**
1.80	1.75	1.70	1.67	1.64	1.60	1.57	1.53	1.51	1.49	.10	
2.13	2.06	1.97	1.93	1.88	1.84	1.79	1.73	1.70	1.67	.05	
2.47	2.36	2.25	2.19	2.13	2.07	2.00	1.93	1.89	1.85	.025	
2.93	2.78	2.63	2.55	2.47	2.38	2.29	2.20	2.15	2.10	.01	
3.28	3.11	2.93	2.83	2.73	2.63	2.52	2.41	2.35	2.29	.005	
4.17	3.92	3.66	3.52	3.38	3.23	3.08	2.92	2.84	2.75	.001	
1.34	1.33	1.31	1.30	1.29	1.28	1.27	1.25	1.24	1.24	.25	**28**
1.79	1.74	1.69	1.66	1.63	1.59	1.56	1.52	1.50	1.48	.10	
2.12	2.04	1.96	1.91	1.87	1.82	1.77	1.71	1.68	1.65	.05	
2.45	2.34	2.23	2.17	2.11	2.05	1.98	1.91	1.87	1.83	.025	
2.90	2.75	2.60	2.52	2.44	2.35	2.26	2.17	2.12	2.06	.01	
3.25	3.07	2.89	2.79	2.69	2.59	2.48	2.37	2.31	2.25	.005	
4.11	3.86	3.60	3.46	3.32	3.18	3.02	2.86	2.78	2.69	.001	
1.34	1.32	1.31	1.30	1.29	1.27	1.26	1.25	1.24	1.23	.25	**29**
1.78	1.73	1.68	1.65	1.62	1.58	1.55	1.51	1.49	1.47	.10	
2.10	2.03	1.94	1.90	1.85	1.81	1.75	1.70	1.67	1.64	.05	
2.43	2.32	2.21	2.15	2.09	2.03	1.96	1.89	1.85	1.81	.025	
2.87	2.73	2.57	2.49	2.41	2.33	2.23	2.14	2.09	2.03	.01	
3.21	3.04	2.86	2.76	2.66	2.56	2.45	2.33	2.27	2.21	.005	
4.05	3.80	3.54	3.41	3.27	3.12	2.97	2.81	2.73	2.64	.001	
1.34	1.32	1.30	1.29	1.28	1.27	1.26	1.24	1.23	1.23	.25	**30**
1.77	1.72	1.67	1.64	1.61	1.57	1.54	1.50	1.48	1.46	.10	
2.09	2.01	1.93	1.89	1.84	1.79	1.74	1.68	1.65	1.62	.05	
2.41	2.31	2.20	2.14	2.07	2.01	1.94	1.87	1.83	1.79	.025	
2.84	2.70	2.55	2.47	2.39	2.30	2.21	2.11	2.06	2.01	.01	
3.18	3.01	2.82	2.73	2.63	2.52	2.42	2.30	2.24	2.18	.005	
4.00	3.75	3.49	3.36	3.22	3.07	2.92	2.76	2.68	2.59	.001	

df₁

TABLE 9

Percentage points of the F distribution (df_2 at least 40)

df_2	a	1	2	3	4	5	6	7	8	9	10
40	.25	1.36	1.44	1.42	1.40	1.39	1.37	1.36	1.35	1.34	1.33
	.10	2.84	2.44	2.23	2.09	2.00	1.93	1.87	1.83	1.79	1.76
	.05	4.08	3.23	2.84	2.61	2.45	2.34	2.25	2.18	2.12	2.08
	.025	5.42	4.05	3.46	3.13	2.90	2.74	2.62	2.53	2.45	2.39
	.01	7.31	5.18	4.31	3.83	3.51	3.29	3.12	2.99	2.89	2.80
	.005	8.83	6.07	4.98	4.37	3.99	3.71	3.51	3.35	3.22	3.12
	.001	12.61	8.25	6.59	5.70	5.13	4.73	4.44	4.21	4.02	3.87
60	.25	1.35	1.42	1.41	1.38	1.37	1.35	1.33	1.32	1.31	1.30
	.10	2.79	2.39	2.18	2.04	1.95	1.87	1.82	1.77	1.74	1.71
	.05	4.00	3.15	2.76	2.53	2.37	2.25	2.17	2.10	2.04	1.99
	.025	5.29	3.93	3.34	3.01	2.79	2.63	2.51	2.41	2.33	2.27
	.01	7.08	4.98	4.13	3.65	3.34	3.12	2.95	2.82	2.72	2.63
	.005	8.49	5.79	4.73	4.14	3.76	3.49	3.29	3.13	3.01	2.90
	.001	11.97	7.77	6.17	5.31	4.76	4.37	4.09	3.86	3.69	3.54
90	.25	1.34	1.41	1.39	1.37	1.35	1.33	1.32	1.31	1.30	1.29
	.10	2.76	2.36	2.15	2.01	1.91	1.84	1.78	1.74	1.70	1.67
	.05	3.95	3.10	2.71	2.47	2.32	2.20	2.11	2.04	1.99	1.94
	.025	5.20	3.84	3.26	2.93	2.71	2.55	2.43	2.34	2.26	2.19
	.01	6.93	4.85	4.01	3.53	3.23	3.01	2.84	2.72	2.61	2.52
	.005	8.28	5.62	4.57	3.99	3.62	3.35	3.15	3.00	2.87	2.77
	.001	11.57	7.47	5.91	5.06	4.53	4.15	3.87	3.65	3.48	3.34
120	.25	1.34	1.40	1.39	1.37	1.35	1.33	1.31	1.30	1.29	1.28
	.10	2.75	2.35	2.13	1.99	1.90	1.82	1.77	1.72	1.68	1.65
	.05	3.92	3.07	2.68	2.45	2.29	2.18	2.09	2.02	1.96	1.91
	.025	5.15	3.80	3.23	2.89	2.67	2.52	2.39	2.30	2.22	2.16
	.01	6.85	4.79	3.95	3.48	3.17	2.96	2.79	2.66	2.56	2.47
	.005	8.18	5.54	4.50	3.92	3.55	3.28	3.09	2.93	2.81	2.71
	.001	11.38	7.32	5.78	4.95	4.42	4.04	3.77	3.55	3.38	3.24
240	.25	1.33	1.39	1.38	1.36	1.34	1.32	1.30	1.29	1.27	1.27
	.10	2.73	2.32	2.10	1.97	1.87	1.80	1.74	1.70	1.65	1.63
	.05	3.88	3.03	2.64	2.41	2.25	2.14	2.04	1.98	1.92	1.87
	.025	5.09	3.75	3.17	2.84	2.62	2.46	2.34	2.25	2.17	2.10
	.01	6.74	4.69	3.86	3.40	3.09	2.88	2.71	2.59	2.48	2.40
	.005	8.03	5.42	4.38	3.82	3.45	3.19	2.99	2.84	2.71	2.61
	.001	11.10	7.11	5.60	4.78	4.25	3.89	3.62	3.41	3.24	3.09
inf.	.25	1.32	1.39	1.37	1.35	1.33	1.31	1.29	1.28	1.27	1.25
	.10	2.71	2.30	2.08	1.94	1.85	1.77	1.72	1.67	1.63	1.60
	.05	3.84	3.00	2.60	2.37	2.21	2.10	2.01	1.94	1.88	1.83
	.025	5.02	3.69	3.12	2.79	2.57	2.41	2.29	2.19	2.11	2.05
	.01	6.63	4.61	3.78	3.32	3.02	2.80	2.64	2.51	2.41	2.32
	.005	7.88	5.30	4.28	3.72	3.35	3.09	2.90	2.74	2.62	2.52
	.001	10.83	6.91	5.42	4.62	4.10	3.74	3.47	3.27	3.10	2.96

TABLE 9

(*continued*)

12	15	20	24	30	40	60	120	240	inf.	*a*	df$_2$
											df$_1$
1.31	1.30	1.28	1.26	1.25	1.24	1.22	1.21	1.20	1.19	.25	**40**
1.71	1.66	1.61	1.57	1.54	1.51	1.47	1.42	1.40	1.38	.10	
2.00	1.92	1.84	1.79	1.74	1.69	1.64	1.58	1.54	1.51	.05	
2.29	2.18	2.07	2.01	1.94	1.88	1.80	1.72	1.68	1.64	.025	
2.66	2.52	2.37	2.29	2.20	2.11	2.02	1.92	1.86	1.80	.01	
2.95	2.78	2.60	2.50	2.40	2.30	2.18	2.06	2.00	1.93	.005	
3.64	3.40	3.14	3.01	2.87	2.73	2.57	2.41	2.32	2.23	.001	
1.29	1.27	1.25	1.24	1.22	1.21	1.19	1.17	1.16	1.15	.25	**60**
1.66	1.60	1.54	1.51	1.48	1.44	1.40	1.35	1.32	1.29	.10	
1.92	1.84	1.75	1.70	1.65	1.59	1.53	1.47	1.43	1.39	.05	
2.17	2.06	1.94	1.88	1.82	1.74	1.67	1.58	1.53	1.48	.025	
2.50	2.35	2.20	2.12	2.03	1.94	1.84	1.73	1.67	1.60	.01	
2.74	2.57	2.39	2.29	2.19	2.08	1.96	1.83	1.76	1.69	.005	
3.32	3.08	2.83	2.69	2.55	2.41	2.25	2.08	1.99	1.89	.001	
1.27	1.25	1.23	1.22	1.20	1.19	1.17	1.15	1.13	1.12	.25	**90**
1.62	1.56	1.50	1.47	1.43	1.39	1.35	1.29	1.26	1.23	.10	
1.86	1.78	1.69	1.64	1.59	1.53	1.46	1.39	1.35	1.30	.05	
2.09	1.98	1.86	1.80	1.73	1.66	1.58	1.48	1.43	1.37	.025	
2.39	2.24	2.09	2.00	1.92	1.82	1.72	1.60	1.53	1.46	.01	
2.61	2.44	2.25	2.15	2.05	1.94	1.82	1.68	1.61	1.52	.005	
3.11	2.88	2.63	2.50	2.36	2.21	2.05	1.87	1.77	1.66	.001	
1.26	1.24	1.22	1.21	1.19	1.18	1.16	1.13	1.12	1.10	.25	**120**
1.60	1.55	1.48	1.45	1.41	1.37	1.32	1.26	1.23	1.19	10	
1.83	1.75	1.66	1.61	1.55	1.50	1.43	1.35	1.31	1.25	.05	
2.05	1.94	1.82	1.76	1.69	1.61	1.53	1.43	1.38	1.31	.025	
2.34	2.19	2.03	1.95	1.86	1.76	1.66	1.53	1.46	1.38	.01	
2.54	2.37	2.19	2.09	1.98	1.87	1.75	1.61	1.52	1.43	.005	
3.02	2.78	2.53	2.40	2.26	2.11	1.95	1.77	1.66	1.54	.001	
1.25	1.23	1.21	1.19	1.18	1.16	1.14	1.11	1.09	1.07	.25	**240**
1.57	1.52	1.45	1.42	1.38	1.33	1.28	1.22	1.18	1.13	10	
1.79	1.71	1.61	1.56	1.51	1.44	1.37	1.29	1.24	1.17	.05	
2.00	1.89	1.77	1.70	1.63	1.55	1.46	1.35	1.29	1.21	.025	
2.26	2.11	1.96	1.87	1.78	1.68	1.57	1.43	1.35	1.25	.01	
2.45	2.28	2.09	1.99	1.89	1.77	1.64	1.49	1.40	1.28	.005	
2.88	2.65	2.40	2.26	2.12	1.97	1.80	1.61	1.49	1.35	.001	
1.24	1.22	1.19	1.18	1.16	1.14	1.12	1.08	1.06	1.00	.25	**inf.**
1.55	1.49	1.42	1.38	1.34	1.30	1.24	1.17	1.12	1.00	10	
1.75	1.67	1.57	1.52	1.46	1.39	1.32	1.22	1.15	1.00	.05	
1.94	1.83	1.71	1.64	1.57	1.48	1.39	1.27	1.19	1.00	.025	
2.18	2.04	1.88	1.79	1.70	1.59	1.47	1.32	1.22	1.00	.01	
2.36	2.19	2.00	1.90	1.79	1.67	1.53	1.36	1.25	1.00	.005	
2.74	2.51	2.27	2.13	1.99	1.84	1.66	1.45	1.31	1.00	.001	

Source: Computed by P. J. Hildebrand.

TABLE 10

Percentage points of $F_{max} = s^2_{max}/s^2_{min}$

Upper 5% Points

df_2 \ t	2	3	4	5	6	7	8	9	10	11	12
2	39.0	87.5	142	202	266	333	403	475	550	626	704
3	15.4	27.8	39.2	50.7	62.0	72.9	83.5	93.9	104	114	124
4	9.60	15.5	20.6	25.2	29.5	33.6	37.5	41.1	44.6	48.0	51.4
5	7.15	10.8	13.7	16.3	18.7	20.8	22.9	24.7	26.5	28.2	29.9
6	5.82	8.38	10.4	12.1	13.7	15.0	16.3	17.5	18.6	19.7	20.7
7	4.99	6.94	8.44	9.70	10.8	11.8	12.7	13.5	14.3	15.1	15.8
8	4.43	6.00	7.18	8.12	9.03	9.78	10.5	11.1	11.7	12.2	12.7
9	4.03	5.34	6.31	7.11	7.80	8.41	8.95	9.45	9.91	10.3	10.7
10	3.72	4.85	5.67	6.34	6.92	7.42	7.87	8.28	8.66	9.01	9.34
12	3.28	4.16	4.79	5.30	5.72	6.09	6.42	6.72	7.00	7.25	7.48
15	2.86	3.54	4.01	4.37	4.68	4.95	5.19	5.40	5.59	5.77	5.93
20	2.46	2.95	3.29	3.54	3.76	3.94	4.10	4.24	4.37	4.49	4.59
30	2.07	2.40	2.61	2.78	2.91	3.02	3.12	3.21	3.29	3.36	3.39
60	1.67	1.85	1.96	2.04	2.11	2.17	2.22	2.26	2.30	2.33	2.36
∞	1.00	1.00	1.00	1.00	1.00	1.00	1.00	1.00	1.00	1.00	1.00

Upper 1% Points

df_2 \ t	2	3	4	5	6	7	8	9	10	11	12
2	199	448	729	1036	1362	1705	2063	2432	2813	3204	3605
3	47.5	85	120	151	184	21(6)	24(9)	28(1)	31(0)	33(7)	36(1)
4	23.2	37	49	59	69	79	89	97	106	113	120
5	14.9	22	28	33	38	42	46	50	54	57	60
6	11.1	15.5	19.1	22	25	27	30	32	34	36	37
7	8.89	12.1	14.5	16.5	18.4	20	22	23	24	26	27
8	7.50	9.9	11.7	13.2	14.5	15.8	16.6	17.9	18.9	19.8	21
9	6.54	8.5	9.9	11.1	12.1	13.1	13.9	14.7	15.3	16.0	16.6
10	5.85	7.4	8.6	9.6	10.4	11.1	11.8	12.4	12.9	13.4	13.9
12	4.91	6.1	6.9	7.6	8.2	8.7	9.1	9.5	9.9	10.2	10.6
15	4.07	4.9	5.5	6.0	6.4	6.7	7.1	7.3	7.5	7.8	8.0
20	3.32	3.8	4.3	4.6	4.9	5.1	5.3	5.5	5.6	5.8	5.9
30	2.63	3.0	3.3	3.4	3.6	3.7	3.8	3.9	4.0	4.1	4.2
60	1.96	2.2	2.3	2.4	2.4	2.5	2.5	2.6	2.6	2.7	2.7
∞	1.00	1.0	1.0	1.0	1.0	1.0	1.0	1.0	1.0	1.0	1.0

Note: s^2_{max} is the largest and s^2_{min} the smallest in a set of t independent mean squares, each based on $df_2 = n - 1$ degrees of freedom. Values in the column $t = 2$ and in the rows $df_2 = 2$ and ∞ are exact. Elsewhere, the third digit may be in error by a few units for the 5% points and several units for the 1% points. The third-digit figures in parentheses for $df_2 = 3$ are the most uncertain.

Source: From *Biometrika Tables for Statisticians*, 3rd ed., Vol. 1; edited by E. S. Pearson and H. O. Hartley (New York: Cambridge University Press, 1966), Table, p. 202. Reproduced by permission of the *Biometrika Trustees*.

TABLE 11

Percentage points of the Studentized range

Error df	α	t = Number of Treatment Means									
		2	**3**	**4**	**5**	**6**	**7**	**8**	**9**	**10**	**11**
5	.05	3.64	4.60	5.22	5.67	6.03	6.33	6.58	6.80	6.99	7.17
	.01	5.70	6.98	7.80	8.42	8.91	9.32	9.67	9.97	10.24	10.48
6	.05	3.46	4.34	4.90	5.30	5.63	5.90	6.12	6.32	6.49	6.65
	.01	5.24	6.33	7.03	7.56	7.97	8.32	8.61	8.87	9.10	9.30
7	.05	3.34	4.16	4.68	5.06	5.36	5.61	5.82	6.00	6.16	6.30
	.01	4.95	5.92	6.54	7.01	7.37	7.68	7.94	8.17	8.37	8.55
8	.05	3.26	4.04	4.53	4.89	5.17	5.40	5.60	5.77	5.92	6.05
	.01	4.75	5.64	6.20	6.62	6.96	7.24	7.47	7.68	7.86	8.03
9	.05	3.20	3.95	4.41	4.76	5.02	5.24	5.43	5.59	5.74	5.87
	.01	4.60	5.43	5.96	6.35	6.66	6.91	7.13	7.33	7.49	7.65
10	.05	3.15	3.88	4.33	4.65	4.91	5.12	5.30	5.46	5.60	5.72
	.01	4.48	5.27	5.77	6.14	6.43	6.67	6.87	7.05	7.21	7.36
11	.05	3.11	3.82	4.26	4.57	4.82	5.03	5.30	5.35	5.49	5.61
	.01	4.39	5.15	5.62	5.97	6.25	6.48	6.67	6.84	6.99	7.13
12	.05	3.08	3.77	4.20	4.52	4.75	4.95	5.12	5.27	5.39	5.51
	.01	4.32	5.05	5.50	5.84	6.10	6.32	6.51	6.67	6.81	6.94
13	.05	3.06	3.73	4.15	4.45	4.69	4.88	5.05	5.19	5.32	5.43
	.01	4.26	4.96	5.40	5.73	5.98	6.19	6.37	6.53	6.67	6.79
14	.05	3.03	3.70	4.11	4.41	4.64	4.83	4.99	5.13	5.25	5.36
	.01	4.21	4.89	5.32	5.63	5.88	6.08	6.26	6.41	6.54	6.66
15	.05	3.01	3.67	4.08	4.37	4.59	4.78	4.94	5.08	5.20	5.31
	.01	4.17	4.84	5.25	5.56	5.80	5.99	6.16	6.31	6.44	6.55
16	.05	3.00	3.65	4.05	4.33	4.56	4.74	4.90	5.03	5.15	5.26
	.01	4.13	4.79	5.19	5.49	5.72	5.92	6.08	6.22	6.35	6.46
17	.05	2.98	3.63	4.02	4.30	4.52	4.70	4.86	4.99	5.11	5.21
	.01	4.10	4.74	5.14	5.43	5.66	5.85	6.01	6.15	6.27	6.38
18	.05	2.97	3.61	4.00	4.28	4.49	4.67	4.82	4.96	5.07	5.17
	.01	4.07	4.70	5.09	5.38	5.60	5.79	5.94	6.08	6.20	6.31
19	.05	2.96	3.59	3.98	4.25	4.47	4.65	4.79	4.92	5.04	5.14
	.01	4.05	4.67	5.05	5.33	5.55	5.73	5.89	6.02	6.14	6.25
20	.05	2.95	3.58	3.96	4.23	4.45	4.62	4.77	4.90	5.01	5.11
	.01	4.02	4.64	5.02	5.29	5.51	5.69	5.84	5.97	6.09	6.19
24	.05	2.92	3.53	3.90	4.17	4.37	4.54	4.68	4.81	3.92	5.01
	.01	3.96	4.55	4.91	5.17	5.37	5.54	5.69	5.81	5.92	6.02
30	.05	2.89	3.49	3.85	4.10	4.30	4.46	4.60	4.72	4.82	4.92
	.01	3.89	4.45	4.80	5.05	5.24	5.40	5.54	5.65	5.76	5.85
40	.05	2.86	3.44	3.79	4.04	4.23	4.39	4.52	4.63	4.73	4.82
	.01	3.82	4.37	4.70	4.93	5.11	5.26	5.39	5.50	5.60	5.69
60	.05	2.83	3.40	3.74	3.98	4.16	4.31	4.44	4.55	4.65	4.73
	.01	3.76	4.28	4.59	4.82	4.99	5.13	5.25	5.36	5.45	5.53
120	.05	2.80	3.36	3.68	3.92	4.10	4.24	4.36	4.47	4.56	4.64
	.01	3.70	4.20	4.50	4.71	4.87	5.01	5.12	5.21	5.30	5.37
∞	.05	2.77	3.31	3.63	3.86	4.03	4.17	4.29	4.39	4.47	4.55
	.01	3.64	4.12	4.40	4.60	4.76	4.88	4.99	5.08	5.16	5.23

Source: Abridged from E. S. Pearson and H. O. Hartley, eds., *Biometrika Tables for Statisticians,* 2d ed., Vol 1 (New York: Cambridge University Press, 1958), Table 29. Reproduced with the permission of the editors and the trustees of *Biometrika*.

TABLE 11
(*continued*)

Error df	12	13	14	15	16	17	18	19	20	α
5	7.32	7.47	7.60	7.72	7.83	7.93	8.03	8.12	8.21	.05
	10.70	10.89	11.08	11.24	11.40	11.55	11.68	11.81	11.93	.01
6	6.79	6.92	7.03	7.14	7.24	7.34	7.43	7.51	7.59	.05
	9.48	9.65	9.81	9.95	10.08	10.21	10.32	10.43	10.54	.01
7	6.43	6.55	6.66	6.76	6.85	6.94	7.02	7.10	7.17	.05
	8.71	8.86	9.00	9.12	9.24	9.35	9.46	9.55	9.65	.01
8	6.18	6.29	6.39	6.48	6.57	6.65	6.73	6.80	6.87	.05
	8.18	8.31	8.44	8.55	8.66	8.76	8.85	8.94	9.03	.01
9	5.98	6.09	6.19	6.28	6.36	6.44	6.51	6.58	6.64	.05
	7.78	7.91	8.03	8.13	8.23	8.33	8.41	8.49	8.57	.01
10	5.83	5.93	6.03	6.11	6.19	6.27	6.34	6.40	6.47	.05
	7.49	7.60	7.71	7.81	7.91	7.99	8.08	8.15	8.23	.01
11	5.71	5.81	5.90	5.98	6.06	6.13	6.20	6.27	6.33	.05
	7.25	7.36	7.46	7.56	7.65	7.73	7.81	7.88	7.95	.01
12	5.61	5.71	5.80	5.86	5.95	6.02	6.09	6.15	6.21	.05
	7.06	7.17	7.26	7.36	7.44	7.52	7.59	7.66	7.73	.01
13	5.53	5.63	5.71	5.79	5.86	5.93	5.99	6.05	6.11	.05
	6.90	7.01	7.10	7.19	7.27	7.35	7.42	7.48	7.55	.01
14	5.46	5.55	5.64	5.71	5.79	5.85	5.91	5.97	6.03	.05
	6.77	6.87	6.96	7.05	7.13	7.20	7.27	7.33	7.39	.01
15	5.40	5.49	5.57	5.65	5.72	5.78	5.85	5.90	5.96	.05
	6.66	6.76	6.84	6.93	7.00	7.07	7.14	7.20	7.26	.01
16	5.35	5.44	5.52	5.59	5.66	5.73	5.79	5.84	5.90	.05
	6.56	6.66	6.74	6.82	6.90	6.97	7.03	7.09	7.15	.01
17	5.31	5.39	5.47	5.54	5.61	5.67	5.73	5.79	5.84	.05
	6.48	6.57	6.66	6.73	6.81	6.87	6.94	7.00	7.05	.01
18	5.27	5.35	5.43	5.50	5.57	5.63	5.69	5.74	5.79	.05
	6.41	6.50	6.58	6.65	6.73	6.79	6.85	6.91	6.97	.01
19	5.23	5.31	5.39	5.46	5.53	5.59	5.65	5.70	5.75	.05
	6.34	6.43	6.51	6.58	6.65	6.72	6.78	6.84	6.89	.01
20	5.20	5.28	5.36	5.43	5.49	5.55	5.61	5.66	5.71	.05
	6.28	6.37	6.45	6.52	6.59	6.65	6.71	6.77	6.82	.01
24	5.10	5.18	5.25	5.32	5.38	5.44	5.49	5.55	5.59	.05
	6.11	6.19	6.26	6.33	6.39	6.45	6.51	6.56	6.61	.01
30	5.00	5.08	5.15	5.21	5.27	5.33	5.38	5.43	5.47	.05
	5.93	6.01	6.08	6.14	6.20	6.26	6.31	6.36	6.41	.01
40	4.90	4.98	5.04	5.11	5.16	5.22	5.27	5.31	5.36	.05
	5.76	5.83	5.90	5.96	6.02	6.07	6.12	6.16	6.21	.01
60	4.81	4.88	4.94	5.00	5.06	5.11	5.15	5.20	5.24	.05
	5.60	5.67	5.73	5.78	5.84	5.89	5.93	5.97	6.01	.01
120	4.71	4.78	4.84	4.90	4.95	5.00	5.04	5.09	5.13	.05
	5.44	5.50	5.56	5.61	5.66	5.71	5.75	5.79	5.83	.01
∞	4.62	4.68	4.74	4.80	4.85	4.89	4.93	4.97	5.01	.05
	5.29	5.35	5.40	5.45	5.49	5.54	5.57	5.61	5.65	.01

t = Number of Treatment Means

Chapter 1: Statistics and the Scientific Method

1.1 **a.** The population of interest is the weight of shrimp maintained on the specific diet for a period of 6 months.
  b. The sample is the 100 shrimp selected from the pond and maintained on the specific diet for a period of 6 months.
  c. The weight gain of the shrimp over 6 months.
  d. Since the sample is only a small proportion of the whole population, it is necessary to evaluate what the mean weight may be for any other randomly selected 100 shrimp.

1.3 **a.** All households in the city that receive welfare support.
  b. The 400 households selected from the city welfare rolls.
  c. The number of children per household for those households in the city which receive welfare.
  d. In order to evaluate how closely the sample of 400 households matches the number of children in all households in the city receiving welfare.

1.5 **a.** All football helmets produced by the five companies over a given period of time.
  b. The 540 helmets selected from the output of the five companies.
  c. The amount of shock transmitted to the neck when the helmet's face mask is twisted.
  d. The neck strength of players is extremely variable for high school players. Hence, the amount of damage to the neck varies considerably from player to player for exactly the same amount of shock transmitted by the helmet.

Chapter 2: Collecting Data Using Surveys and Scientific Studies

2.1 The relative merits of the different types of sampling units depend on the availability of a sampling frame for individuals, the desired precision of the estimates from the sample to the population, and the budgetary and time constraints of the project.

2.3 The list of registered voters in the state could be used as the sampling frame for selecting the persons to be included in the sample.

2.5 **a.** Alumni that graduated from Yale in 1924.
  b. No. Alumni whose addresses were on file 25 years later would not necessarily be representative of their class.
  c. Alumni who responded to the mail survey would not necessarily be representative of those who were sent the questionnaires. Income figures may not be reported accurately (intentionally), or may be rounded off to the nearest $5,000, say, in a self-administered questionnaire.
  d. Rounded income responses would make the figure $25,111 highly unlikely. The fact that those with higher incomes would be more likely to respond and the fact that incomes are likely to be exaggerated would tend to make the estimate too high.

2.8 **a.** Factors: Temperature, Type of Seafood
  b. Factor Levels: Temperature ($0°C$, $5°C$, $10°C$), Type of Seafood (Oysters, Mussels)

c. Blocks: none
d. Experimental Unit: Package of Seafood
e. Measurement Unit: Sample from package
f. Replications: 3 packages per temperature
g. Treatments: (0°C, Oysters), (5°C, Oysters), (10°C, Oysters)
(0°C, Mussels), (5°C, Mussels), (10°C, Mussels)

2.9 a. Randomized complete block design with blocking variable (5 Farms) and 48 treatments in a $3 \times 4 \times 4$ factorial structure.
b. Completely randomized design with 10 treatments (Software Programs) and 3 replications of each treatment.
c. Latin square design with blocking variables (Position in Kiln, Day) each having 8 levels. The treatment structure is a 2×4 factorial structure (Type of Glaze, Thickness).

Chapter 3: Summarizing Data

3.5 Two separate bar graphs could be plotted, one with Lap Belt Only and the other with Lap and Shoulder Belt. A single bar graph with the Lap Belt Only value plotted next to the Lap and Shoulder Belt value for each value of Percentage of Use is probably the most effective plot.

3.7 The plot has a bimodal shape. This would be an indication that there are two separate populations. However, the evidence is not very convincing because the individual plots were similar in shape with the exception that the New Therapy had a few times that were somewhat larger than the survival times obtained under the Standard Therapy.

3.9 a. There appears to have been a dramatic drop in verbal scores for both sexes from 1970 to 1980, with a greater drop in female scores. There appears to have been a slight drop in math scores from 1970 to 1980, then a slow steady rise from 1980 to 1996.
b. Yes.
c. The female verbal scores had a larger decrease for the years 1970–1980 than the male scores.

3.13 The plots show an upward trend from year 1 to year 5. There is a strong seasonal (cyclic) effect; the number of units sold increases dramatically in the late summer and fall months.

3.15 Mean = 17.7, Median = 14.5, Mode = 18

3.17 a. Mean = 8.04, Median = 1.54
b. Terrestrial: Mean = 15.01, Median = 6.03
Aquatic: Mean = 0.38, Median = 0.375
c. The mean is more sensitive to extreme values than is the median.
d. Terrestrial: Median
Aquatic: Mean or median

3.19 a. If we use all 14 failure times, we obtain Mean > 173.7 and Median = 154. We know the mean is greater than 173.7 since the failure times for two of the engines are greater than the reported times of 300 days.
b. The median is unchanged. However, the mean is less than it would be if exact failure times were known.

3.21 Mean = 1.7707, Median = 1.7083, Mode = 1.273

3.23 a. $s = 7.95$
b. Because the magnitude of the racers' ages is larger than that of their experience.

3.25 a. Luxury: $\bar{y} = 145.0, s = 27.6$
Budget: $\bar{y} = 46.1, s = 5.13$
c. Luxury hotels vary in quality, location, and price, whereas budget hotels are more competitive for the low-end market so prices tend to be similar.

3.27 **a.** The stem-and-leaf plot is given here:

2	5
2	677
2	9
3	00111
3	223333
3	5
3	67
3	8

b. Min = 250, $Q_1 = 298$, $Q_2 = 317.5$, $Q_3 = (334 + 334)/2 = 334$, Max = 386
A boxplot showing the number of blood donors on Fridays is given here:

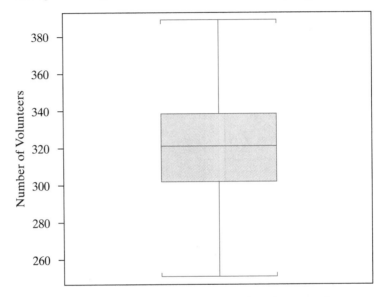

There are no outliers. The distribution is approximately symmetric.

3.29 **a.** A stacked bar graph showing the literacy levels of the three subsistence groups is given here:

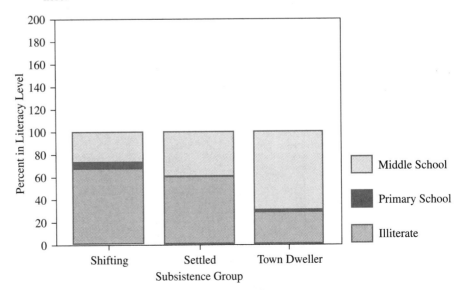

b. Illiterate: 46%, Primary Schooling: 4%, At Least Middle School: 50%, Shifting Cultivators: 27%, Settled Agriculturists: 21%, Town Dwellers: 51%

3.31 **a.** Yes

b. A scatterplot of M3 versus M2 (in trillions of dollars) is given here:

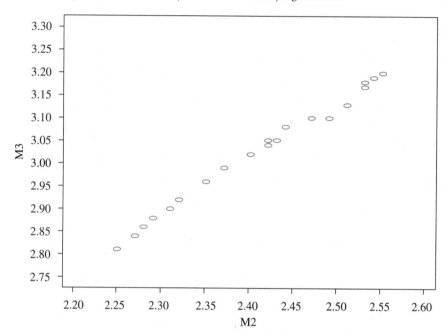

3.33 **a.** Mean = 57.5, Median = 34.0

b. Median

c. Range = 273, $s = 70.2$

d. Using the approximation, $s \approx range/4 = 273/4 = 68.3$.

e. $\bar{y} \pm s \Rightarrow (-12.7, 127.7)$; yields 82%
$\bar{y} \pm 2s \Rightarrow (-82.9, 197.0)$; yields 94%
$\bar{y} \pm 3s \Rightarrow (-153.1, 268.1)$; yields 97%

f. The Empirical Rule applies to data sets with a roughly "mound-shaped" histogram. The distribution of this data set is highly skewed right.

3.35 A scatterplot of the price per roll versus the number of sheets per roll for 24 brands of paper towels is given here:

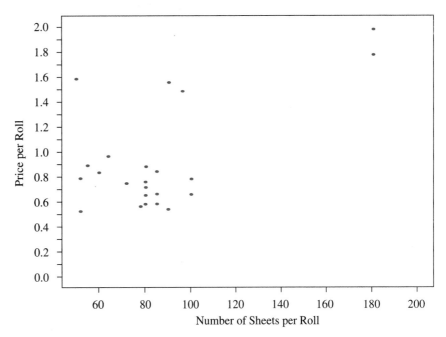

a. The points do not fall close to a straight line.

b. No

c. Paper towel sheets vary in thickness and size, which will affect the price.

3.37   a. Construct a relative frequency histogram.

b. Highly skewed to the right.

c. $1{,}424 \pm 3{,}488 \Rightarrow (-2{,}063, 4{,}912)$ contains $37/41 = 90.2\%$

$1{,}424 \pm (2)3{,}488 \Rightarrow (-5{,}551, 8{,}400)$ contains $38/41 = 92.7\%$

$1{,}424 \pm (3)3{,}488 \Rightarrow (-9{,}039, 11{,}888)$ contains $39/41 = 95.1\%$

These values do not match the percentages from the Empirical Rule: 68%, 95%, and 99.7%.

d. $1.48 \pm 1.54 \Rightarrow (-0.06, 3.02)$ contains $31/41 = 75.6\%$

$1.48 \pm (2)1.54 \Rightarrow (-1.60, 4.56)$ contains $40/41 = 97.6\%$

$1.48 \pm (3)1.54 \Rightarrow (-3.14, 6.10)$ contains $41/41 = 100\%$

These values more closely match the percentages from the Empirical Rule: 68%, 95%, and 99.7%.

3.39   a. There has been very little change from 1985 to 1996.

b. Yes

c. No

d. The cost of homes is very high.

3.41   a. Plot a relative frequency histogram.

b. 1,100

c. Mean = 1,091.96, Median = 1,039

d. Because the mean is slightly larger than the median, it is likely that the distribution is slightly skewed to the right.

3.43   The stem-and-leaf diagram is given here:

```
54        7
55
56
57
58
59
60
61
62        5
63        0
64
65        6
66        4 7 7
67        7 9
68        8 88 88
69        1 4 7 9
70        0 1 2 333 8
71        1
```

Yes, the distribution is left skewed.

3.45   **a.** New policy: $\bar{y} = 2.27$, $s = 3.26$
Old policy: $\bar{y} = 4.60$, $s = 2.61$
b. Both the average number of sick days and the variation in number of sick days have decreased with the new policy.

3.47   **a.** Average Price = 76.68
b. Range = 202.69
c. DJIA = 10,856.2
d. Yes; the stocks covered on the NYSE.
No; the companies selected are the leading companies in the different sectors of business.

3.49   **a.** The value 62 reflects the number of respondents in coal-producing states who preferred a national energy policy that encouraged coal production. The value 32.8 tells us that of those who favored a coal policy, 32.8% came from major coal-producing states. The value 41.3 tells us that 41.3% of those from coal states favored a coal policy. And the value 7.8 tells us that 7.8% of all the responses come from residents of major coal-producing states who were in favor of a national energy policy that encouraged coal production.
b. The column percentages because they display the distribution of opinions within each of the three types of states.
c. Yes. For both the coal and oil/gas states, the largest percentage of responses favored the type of energy produced in their own state.

Chapter 4: Probability and Probability Distributions

4.1   **a.** Subjective probability
b. Relative frequency
c. Classical
d. Relative frequency
e. Subjective probability
f. Subjective probability
g. Classical

4.3   0.021

4.5   HHH, HHT, HTH, THH, TTH, THT, HTT, TTT

4.7   **a.** $P(\overline{A}) = 1 - \dfrac{3}{8} = \dfrac{5}{8}$

$\qquad P(\overline{B}) = 1 - \dfrac{7}{8} = \dfrac{1}{8}$

$\qquad P(\overline{C}) = 1 - \dfrac{1}{8} = \dfrac{7}{8}$

    b. Events A and B are not mutually exclusive because B contains $A \Rightarrow A \cap B = A$, which is not the empty set.

4.9 A and B are not independent.
A and C are not independent.
B and C are not independent.

4.13   **a.** $S = \{F_1F_2, F_1F_3, F_1F_4, F_1F_5, F_2F_1, F_2F_3, F_2F_4, F_2F_5, F_3F_1, F_3F_2, F_3F_4, F_3F_5, F_4F_1, F_4F_2, F_4F_3, F_4F_5, F_5F_1, F_5F_2, F_5F_3, F_5F_4\}$

    b. Let T_1 be the event that the first firm chosen is stable and T_2 be the event that the second firm chosen is stable.

$\qquad P(T_1) = \dfrac{3}{5} \quad \text{and} \quad P(\overline{T}_1) = \dfrac{2}{5}$

$\qquad P(\text{Both Stable}) = P(T_1 \cap T_2) = P(T_2|T_1)P(T_1) = 0.30$

    c. $P(\text{One of two firms is Shaky}) = .60$

    d. $P(\text{Both Shaky}) = 0.10$
Alternatively, in the list of 20 outcomes there are 2 pairs in which both firms are Shaky (F_1 or F_2). Thus, the probability that both firms are Shaky is 2/20.

4.15   **a.** $P(A) = 0.63$
$\qquad P(B) = 0.36$
$\qquad P(C) = 0.14$

    b. $P(A|B) = 0.611$
$\qquad P(A|\overline{B}) = 0.64$
$\qquad P(\overline{B}|C) = 1$

    c. $P(A \cup B) = 0.77$
$\qquad P(A \cap C) = 0$
$\qquad P(B \cap C) = 0$

4.17   Let A = event customer pays first month's bill in full and B = event customer pays second month's bill in full. We are given that

$P(A) = 0.70, P(B|A) = 0.95, P(\overline{B}|\overline{A}) = 0.90, P(B|\overline{A}) = 0.10$

    a. $P(A \cap B) = 0.665$

    b. $P(\overline{B} \cap \overline{A}) = 0.270$

    c. $P(\text{pay exactly one month in full}) = 0.065$

4.19 $P(D|R_2) = \dfrac{P(R_2|D)P(D)}{P(R_2|D)P(D) + P(R_2|\overline{D})P(\overline{D})} = \dfrac{(0.40)(0.01)}{(0.40)(0.01) + (0.40)(0.99)} = 0.01$

The two probabilities are equal since the proportion of fair risk applicants is the same for both the defaulted and nondefaulted loans.

4.21   **a.** Sensitivity: $P(DA|C) = 50/53 = 0.943$
$\qquad$ Specificity: $P(DNA|RO) = 44/47 = 0.936$

    b. $P(C|DA) = \dfrac{(0.943)(0.00108)}{(0.943)(0.00108) + (0.021)(1 - 0.00108)} = 0.046$

    c. $P(RO|DA) = 1 - P(C|DA) = 1 - 0.046 = 0.954$

    d. $P(RO|DNA) = \dfrac{(0.936)(1 - 0.00108)}{(0.936)(1 - 0.00108) + (0.019)(0.00108)} = 0.99998$

4.23 **a.** $P(y \leq 4) = 0.8263$
b. $P(y > 4) = 0.1737$
c. $P(y \leq 7) = 0.9993$
d. $P(y > 6) = 0.0085$

4.25 **a.** $P(y \geq 3) = 0.64$
b. $P(2 \leq y \leq 6) = 0.60$
c. $P(y > 8) = 0.07$

4.27 No

4.29 **a.** $P(y = 0) = 0.0282$
$P(y = 6) = 0.0368$
$P(y \geq 6) = 0.0473$
$P(y = 10) = 0.000006$
b. $P(y \leq 100) = \sum_{i=0}^{100} \binom{1000}{i}(.3)^i(.7)^{1000-i}$

4.31 No

4.33 **a.** 0.4032
b. 0.4713

4.35 **a.** 0.4947
b. 0.2406

4.37 $z_0 = 1.96$

4.39 $z_0 = 1.645$

4.41 **a.** $P(500 < y < 696) = 0.475$
b. $P(y > 696) = 0.025$
c. $P(304 < y < 696) = 0.95$
d. $k = 84.5$

4.43 **a.** $z = 2.33$
b. $z = -1.28$

4.45 **a.** $\mu = 39; \sigma = 6; P(y > 50) = 0.0336$
b. Because 55 is $\dfrac{55 - 39}{6} = 2.67$ standard deviation above $\mu = 39$, thus $P(y > 55) = P(z > 2.67) = 0.0038$. We would then conclude that the voucher has been lost.

4.47 $\mu = 150; \sigma = 35$
a. $P(y > 200) = 0.0764$
b. $P(y > 220) = 0.0228$
c. $P(y < 120) = 0.1949$
d. $P(100 < y < 200) = 0.8472$

4.49 No. The sample would be biased toward homes for which the homeowner is at home much of the time. For example, the sample would tend to include more people who work at home and retired persons.

4.51 Starting at column 2, row 1, we obtain 150, 729, 611, 584, 255, 465, 143, 127, 323, 225, 483, 368, 213, 270, 062, 399, 695, 540, 330, 110, 069, 409, 539, 015, 564. These would be the women selected for the study.

4.53 The sampling distribution would have a mean of 60 and a standard deviation of $\dfrac{5}{\sqrt{16}} = 1.25$. Approximately 95% should fall in the interval (57.5, 62.5).

4.55 **a.** $P(800 < y < 1,100) = 0.7462$
b. $P(y < 800) = 0.1587$
c. $P(y > 1,200) = 0.0189$

4.57 **a.** $P(y > 2.7) = 0.0228$
    b. $P(z > 0.6745) = 2.30$
    c. Let μ_N be the new value of the mean; $\mu_N = 2.2065$

4.59 Individual baggage weight has $\mu = 95$; $\sigma = 35$; $P(y > 20{,}000) = 0.0217$

4.61 **a.** $\mu = 10$
    b. $\sigma = 3.16$
        $P(y < 5) = 0.0409$
    c. $P(y < 2) = 0.0036$

4.63 $\sigma_{\bar{y}} = 2.63$

4.65 No, there is strong evidence that the new fabric has a greater mean breaking strength.

4.67 One could use random sampling by first identifying all returns with income greater than $15,000. The next step would be to create a computer file with all such returns listed in alphabetical order by last name of client, then randomly select 1% of these returns using a systematic random sampling technique.

4.69 **a.** $P(y \leq 25) \approx 0$
    b. The ad is not successful.

4.71 **a, b.** The mean and standard deviation of the sampling distribution of $\bar{y}$ are given when the population distribution has values $\mu = 100$, $\sigma = 15$:

Sample Size	Mean	Standard Deviation
5	100	6.708
20	100	3.354
80	100	1.677

c. As the sample size increases, the sampling distribution of $\bar{y}$ concentrates about the true value of μ. For $n = 5$ and $n = 20$, the values of $\bar{y}$ could be a considerable distance from 100.

Chapter 5: Inferences about Population Central Values

5.1 **a.** All registered voters in the state.
    b. Simple random sample from a list of registered voters.

5.5 **a.** The width of the interval will be decreased.
    b. The width of the interval will be increased.

5.7 **a.** $10.4 \pm (2.58)\left(\dfrac{4.2}{\sqrt{400}}\right) = 10.4 \pm 0.54 = (9.86, 10.94)$
    b. No
    c. Yes. The 90% C.I. is $(10.05, 10.75)$.

5.9 $(824.7, 875.3)$

5.11 **a.** $n = 97$
    b. $n = 116$
    c. $n = 47$

5.13 $n = 125$

5.15 **a.** $n = 1{,}125$
    b. We would increase the odds of not containing the true average five-fold.

5.17 **a–c.** The power values, $PWR(\mu_a)$, are given here:

n	α	39	40	41	42	43	44
				μ_a			
50	0.05	0.3511	0.8107	0.9840	0.9997	1.0000	1.0000
50	0.025	0.2428	0.7141	0.9662	0.9990	0.9999	1.0000
20	0.05	0.1987	0.4809	0.7736	0.9394	0.9906	0.9992

5.19 $H_0: \mu \le 2$ vs. $H_a: \mu > 2$
$\bar{y} = 2.17, s = 1.05, n = 90$
 a. $z = 1.54 < 1.645 = z_{0.05} \Rightarrow$ Fail to reject H_0. The data does not support the hypothesis that the mean has been decreased from 2.
 b. $\beta(2.1) = 0.7704$

5.21 $n = 134$

5.23 $H_0: \mu \le 30$ vs. $H_a: \mu > 30$
$\alpha = 0.05, n = 37, \bar{y} = 37.24, s = 37.12$
 a. $z = 1.19 < 1.645 = z_{0.05} \Rightarrow$ Fail to reject H_0. There is not sufficient evidence to conclude that the mean lead concentration exceeds 30 mg kg^{-1} dry weight.
 b. $\beta(50) = 0.0513$
 c. No
 d. No

5.25 p-value $= 0.0359 > 0.025 = \alpha \Rightarrow$ No, there is not significant evidence that the mean is greater than 45.

5.27 $H_0: \mu = 1.6$ vs. $H_a: \mu \ne 1.6$; p-value $= < 0.0001 < 0.05 = \alpha \Rightarrow$ Yes, there is significant evidence that the mean time delay differs from 1.6 seconds.

5.29 **a.** Reject H_0 if $t \le -2.624$.
 b. Reject H_0 if $|t| \ge 2.819$.
 c. Reject H_0 if $t \ge 3.365$.

5.31 $H_0: \mu \le 80$ vs. $H_a: \mu > 80$; Reject H_0 if $t \ge 1.729$; $t = 0.84 \Rightarrow$ Fail to reject H_0 and conclude data does not support the hypothesis that the mean reading comprehension is greater than 80.
p-value $= P(t \ge 0.84) \approx 0.20$

5.33 **a.** Yes, the plotted points are all close to the straight line.
 b. Fairly close.
 c. Yes, because the confidence interval contains values which are greater than 35. However, the C.I. is a two-sided procedure that yields a one-sided α of 0.005, which would require greater evidence from the data to support the research hypothesis.

5.35 **a.** Untreated: (40.3, 46.9)
 Treated: (33.3, 38.9)
 b. The two intervals do not overlap. This would indicate that the average heights of the treated and untreated are significantly different.

5.37 Reject H_0 if $B \ge 20$.

5.39 **a.** The data set does not appear to be a sample from a normal distribution.
 b. The median would be a better choice than the mean.
 c. (208, 342)
 d. Reject $H_0: M \le 400$ if $B \ge 25 - 7 = 18$. We obtain $B = 4$; do not reject H_0.

5.41 **a.** Reject H_0: $M \leq 0.25$ in favor of H_a: $M > 0.25$ at level $\alpha = 0.01$ if $B \geq 25 - 6 = 19$. We obtain $B = 18$. Thus, we fail to reject H_0.
   b. Weight of driver, experience (age) of driver, amount of sleep in previous 24 hours, etc.

5.43 **a.** $\bar{y} = 1.466$
   b. 95% C.I.: $(1.26, 1.67)$
   c. H_a: $\mu > 1.20$; Reject H_0 if $t \geq 1.761$; $t = 2.74 \Rightarrow$ There is sufficient evidence that the mean mercury concentration has increased.
   d. Power values are given here:

μ_a	d	$PWR(\mu_a)$
1.28	0.250	0.235
1.32	0.375	0.396
1.36	0.500	0.578
1.40	0.625	0.744

5.45 H_0: $\mu \geq 300$ vs. H_a: $\mu < 300$
$n = 20$, $\bar{y} = 160$, $s = 90$, $\alpha = 0.05$
$p\text{-value} = P\left(t \leq \dfrac{160 - 300}{90/\sqrt{20}}\right) = P(t \leq -6.95) < 0.0001 < 0.05 = \alpha$
Yes, there is sufficient evidence to conclude that the average is less than $300.

5.47 **a.** $\bar{y} = 74.2$; 95% C.I.: $7.42 \pm (2.145)(44.2)/\sqrt{15} \Rightarrow (49.72, 98.68)$
   b. H_0: $\mu \leq 50$ vs. H_a: $\mu > 50$
   $n = 15$, $\alpha = 0.05$
   $p\text{-value} = P\left(t \geq \dfrac{74.2 - 50}{44.2/\sqrt{15}}\right) = P(t \geq 2.12) = 0.0262 < 0.05 = \alpha$
   Yes, there is sufficient evidence to conclude that the average daily output is greater than 50 tons of ore.

5.49 **a.** The summary statistics are given here:

Time	Mean	Standard Deviation	n	95% C.I.
6 A.M.	0.128	0.0355	15	(0.108, 0.148)
2 P.M.	0.116	0.0406	15	(0.094, 0.138)
10 P.M.	0.142	0.0428	15	(0.118, 0.166)
All Day	0.129	0.0403	45	(0.117, 0.141)

   b. No, the three C.I.s have a considerable overlap.
   c. H_0: $\mu \geq 0.145$ vs. H_a: $\mu < 0.145$
   $p\text{-value} = 0.0054$
   There is significant evidence (very small p-value) that the average SO_2 level using the new scrubber is less than 0.145.

5.51 $n = 587$

5.53 99% C.I. on μ: $(53.7, 62.3)$

Chapter 6: Inferences Comparing Two Population Central Values

6.1   **a.** Reject H_0 if $|t| \geq 2.064$.
 b. Reject H_0 if $t \geq 2.624$.
 c. Reject H_0 if $t \leq -1.860$.

6.3 p-value $= P(t \leq -2.6722) \Rightarrow 0.005 < p$-value < 0.01

6.5   **a.** $H_0: \mu_A = \mu_B$ vs. $H_a: \mu_A \neq \mu_B$; p-value $= 0.065 \Rightarrow$ Fail to reject H_0.
 b. The separate-variance t' test was used.
 d. The 95% C.I. estimate for the difference in means is $(-0.013, 0.378)$ ppm. The observed difference (0.183 ppm) is not significant.

6.7   We want to test $H_0: \mu_{No} \geq \mu_{Sub}$ vs. $H_a: \mu_{No} < \mu_{Sub}$; p-value $= 0.0049 \Rightarrow$ Reject H_0.

6.13   **a.** Plumber 1: $\bar{y}_1 = 88.81$, $s_1 = 7.89$
 Plumber 2: $\bar{y}_2 = 108.93$, $s_2 = 8.73$
 b. The boxplots are given here:

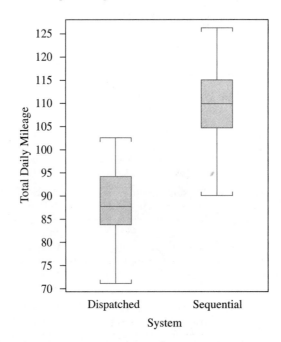

Since both graphs show a roughly symmetrical distribution with no outliers, a t test appears to be appropriate.

6.19   **a.** $H_0: \mu_d = 0$ vs. $H_a: \mu_d \neq 0$
 $t = 4.95$, df $= 29$, $\Rightarrow p$-value $= 2P(t \geq 4.91) < 0.001$
 b. $(2.23, 5.37)$

6.21 H_0: The distribution of differences (Benzedrine minus placebo) is symmetric about 0 vs.
 H_a: The differences tend to be larger than 0.
 With $n = 14$, $\alpha = 0.05$, $T = T_-$, reject H_0 if $T_- \leq 25$.
 From the data, we obtain $T_- = 16 < 25$, thus reject H_0 and conclude that the distribution of heart rates for dogs receiving Benzedrine is shifted to the right of the dogs receiving the placebo.

6.23 Let d = After − Before
   a. $H_0: \mu_d \leq 0$ vs. $H_a: \mu_d > 0$

   $$t = \frac{1.208 - 0}{1.077/\sqrt{12}} = 3.89 \Rightarrow \text{with df} = 11, 0.001 < p\text{-value} < 0.005 \Rightarrow$$

   Reject H_0 and conclude that the exposure has increased mean lung capacity.

   b. 95% C.I. on $\mu_{After} - \mu_{Before}$: $(0.52, 1.89)$

   c. If this was a well-controlled experiment with all factors except ozone exposure controlled, the experimenter would be justified in making the claim concerning the population of rats from which the rats in the study were randomly selected. However, no statement can be made about the effects of ozone on humans.

6.25 **a.** $H_0: \mu_{Within} = \mu_{Out}$ vs. $H_a: \mu_{Within} \neq \mu_{Out}$

   $$t' = \frac{3092 - 2450}{\sqrt{\frac{(1191)^2}{14} + \frac{(2229)^2}{12}}} = 0.89 \Rightarrow \text{with df} \approx 16, p\text{-value} \approx 0.384 \Rightarrow \text{Fail to reject } H_0.$$

   b. A 95% C.I. on $\mu_{Within} - \mu_{Out}$ is $(-879, 2164)$
   d. Construct plots.

6.27 **a.** Construct plots.
   b. 40 m: $H_0: \mu_{Within} = \mu_{Out}$ vs. $H_a: \mu_{Within} \neq \mu_{Out}$
   $t = 0.08 \Rightarrow \text{with df} \approx 6, p\text{-value} \approx 0.94 \Rightarrow \text{Fail to reject } H_0.$
   100 m: $H_0: \mu_{Within} = \mu_{Out}$ vs. $H_a: \mu_{Within} \neq \mu_{Out}$
   $t = 2.50 \Rightarrow \text{with df} \approx 7, p\text{-value} \approx 0.041 \Rightarrow \text{Reject } H_0.$

6.29 **a.** $H_0: \mu_{High} = \mu_{Con}$ vs. $H_a: \mu_{High} \neq \mu_{Con}$
   Separate-variance t test: $t' = 4.12$ with df ≈ 34, p-value $= 0.0002 \Rightarrow$ Reject H_0.
   b. 95% C.I. on $\mu_{High} - \mu_{Con}$: $(19.5, 57.6)$

6.31 **a.** $H_0: \mu_{Low} = \mu_{High}$ vs. $H_a: \mu_{Low} \neq \mu_{High}$
   Separate-variance t test: $t' = -5.73$ with df ≈ 29, p-value $< 0.0005 \Rightarrow$ Reject H_0.
   b. 95% C.I. on $\mu_{Low} - \mu_{High}$: $(-87.6, -41.5)$

6.33 **a.** $H_0: \mu_D \leq \mu_{RN}$ vs. $H_a: \mu_D > \mu_{RN}$
   Separate-variance t test: $t' = 9.04$ with df ≈ 76, p-value $< 0.0001 \Rightarrow$
   Reject H_0 and conclude there is significant evidence that the mean score of the Degreed nurses is higher than the mean scores of the Registered nurses.
   b. p-value < 0.0001
   c. 95% C.I. on $\mu_D - \mu_{RN}$: $(35.3, 55.2)$ points

Chapter 7: Inferences about Population Variances

7.1 **a.** 0.01   **b.** 0.90   **c.** 0.01   **d.** 0.98

7.3 **a.** Let y be the quantity in a randomly selected jar:

   Proportion = 2.28%

   b. The plot indicates that the distribution is approximately normal because the data values are reasonably close to the straight line.
   c. 95% C.I. on σ: $(0.113, 0.168)$
   d. $H_0: \sigma \leq 0.15$ vs. $H_a: \sigma > 0.15$

   Reject H_0 if $\dfrac{(n-1)(s)^2}{(.15)^2} \geq 66.34 \Rightarrow$ From the data, there is not sufficient evidence to reject H_0.

   e. $0.10 < p$-value < 0.90 (Using a computer program, p-value $= 0.8262$.)

7.5 **b.** 99% C.I. on σ: (8.290, 11.187)

c. $H_0: \sigma^2 \le 90$ vs. $H_a: \sigma^2 > 90$

With $\alpha = 0.05$, reject H_0 if $\dfrac{(n-1)(s)^2}{90} \ge 178.49$.

$$\frac{(n-1)(s)^2}{90} = \frac{(150-1)(9.537)^2}{90} = 150.58 < 178.49 \Rightarrow \text{Fail to reject } H_0.$$

7.7 **b.** $H_0: \sigma \ge 2$ vs. $H_a: \sigma < 2$

With $\alpha = 0.05$, reject H_0 if $\dfrac{(n-1)(s)^2}{(2)^2} \le 60.39$.

$$\frac{(81-1)(1.771)^2}{(2)^2} = 62.73 > 60.39 \Rightarrow \text{Fail to reject } H_0, \; p\text{-value} = 0.0772$$

c. 95% C.I. on σ: (1.534, 2.095)

7.9 $H_0: \sigma_A^2 \le \sigma_B^2$ vs. $H_a: \sigma_A^2 > \sigma_B^2$

With $\alpha = 0.05$, reject H_0 if $\dfrac{s_A^2}{s_B^2} \ge 3.79$.

$\dfrac{s_A^2}{s_B^2} = 3.15 < 3.79 \Rightarrow$ Fail to reject H_0 and conclude the data does not support σ_A^2 being greater than σ_B^2.

7.13 The data is summarized in the following table:

Method	n	Mean	95% C.I. on μ	Standard Deviation	95% C.I. on σ
L	10	5.90	(2.34, 9.46)	4.9766	(3.42, 9.09)
L/R	7	7.29	(2.31, 12.26)	5.3763	(3.46, 11.84)
L/C	9	16.00	(9.57, 22.43)	8.3666	(5.65, 16.03)
C	9	17.67	(5.45, 29.88)	15.8902	(10.73, 30.44)

Test $H_0: \sigma_L = \sigma_{L/R} = \sigma_{L/C} = \sigma_C$ vs. $H_a: \sigma$'s are different.
Reject H_0 at level $\alpha = 0.05$ if $L \ge F_{.05,3,31} = 2.91$.
From the data, $L = 2.345 < 2.91 \Rightarrow$ Fail to reject H_0.

7.15 **a.** $25 \times 90\% = 22.5$ and $25 \times 110\% = 27.5$ implies the limits are 22.5 to 27.5.

b. Construct plots.

c. $\hat{\sigma} = 5/4 = 1.25$

$H_0: \sigma = 1.25$ vs. $H_a: \sigma \ne 1.25$

With $\alpha = 0.05$, reject H_0 if $\dfrac{(n-1)s^2}{(1.25)^2} \le 16.05$ or $\dfrac{(n-1)s^2}{(1.25)^2} \ge 45.72$.

$$\frac{(n-1)s^2}{(1.25)^2} = \frac{(30-1)(1.4691)^2}{(1.25)^2} = 40.06 \Rightarrow 16.06 < 40.06 < 45.72 \Rightarrow \text{Fail to reject } H_0.$$

7.17 $H_0: \mu_1 = \mu_2$ vs. $H_a: \mu_1 \ne \mu_2$

$$t = \frac{131.60 - 147.20}{4.92\sqrt{\dfrac{1}{10} + \dfrac{1}{10}}} = -7.09 \text{ with df} = 18 \Rightarrow p\text{-value} < 0.0005 \Rightarrow \text{Reject } H_0.$$

7.19 We would now run a 1-tail test:

$H_0: \mu_A \ge \mu_B$ vs. $H_a: \mu_A < \mu_B$

$t = -2.40$ with df $= 15 \Rightarrow p$-value $= 0.0149 \Rightarrow$ Reject H_0.

Chapter 8: The Completely Randomized Design

8.1 **a.** Yes
   b. $H_0: \mu_A = \mu_B = \mu_C = \mu_D$ vs. H_a: Difference in μ's
   Reject H_0 if $F \geq F_{.05,3,20} = 3.10$.
   SSE = 0.8026
   $\bar{y}_{..} = 0.0826 \Rightarrow$
   SST = 0.5838 $\Rightarrow$
   $F = 4.85 > 3.10 \Rightarrow$
   Reject H_0.
   c. p-value $= P(F_{3,20} \geq 4.85) \Rightarrow 0.01 < p$-value < 0.025

8.5 **a.** The Kruskal–Wallis yields $H = 21.32 > 9.21$ with df $= 2 \Rightarrow p$-value < 0.001. Thus, reject H_0.
   b. Construct boxplots and a normal probability plot.
   c. The AOV table is given here:

Source	df	SS	MS	F	p-value
Supplier	2	10,723.8	5,361.9	161.09	0.000
Error	24	798.9	33.3		
Total	26	11,522.7			

   Reject H_0 if $F \geq 3.40$.
   $F = 161.09 > 3.40 \Rightarrow$ Reject H_0.
   d. 95% C.I. on μ_A: (185.26, 193.20)
   95% C.I. on μ_B: (152.31, 160.25)
   95% C.I. on μ_C: (199.97, 207.91)

8.11 **a.** The Kruskal–Wallis test yields $H' = 26.62$ with df $= 3 \Rightarrow p$-value $< 0.001 \Rightarrow$ Reject H_0.
   b. The two procedures yield equivalent conclusions.

8.13 **a.** $F = 54.70$ with df $= 3, 36 \Rightarrow p$-value $< 0.001 < 0.05$
   b. 95% C.I. on μ_A: (20.20, 26.54)   95% C.I. on μ_B: (5.41, 11.75)
   95% C.I. on μ_C: (11.76, 18.10)   95% C.I. on μ_D: (32.18, 38.52)
   c. $F = 2.10$ with df $= 3, 36 \Rightarrow 0.05 < 0.10 < p$-value $< 0.25 \Rightarrow$ Fail to reject H_0.

Chapter 9: More Complicated Experimental Designs

9.1 **a.** The F test from the AOV table tests 2-sided alternatives:

$$\text{Test } H_0: \mu_{Attend} = \mu_{DidNot} \text{ vs. } H_0: \mu_{Attend} \neq \mu_{DidNot}$$

   The AOV table is given here:

Source	df	SS	MS	F	p-value
Pair	5	1,319.42	263.88		
Treatment	1	420.08	420.08	71.40	0.0004
Error	5	29.42	5.88		
Total	11	1,768.92			

   p-value $= 0.0004 \Rightarrow$ Reject H_0.
   b. RE(RCB, CR) = 20.94

9.3 **a.** Blocks are Investigators and Treatments are Mixtures.
b. Randomly assign the four Mixtures to each of the Investigators.

9.5 **a.** $y_{ij} = \mu + \alpha_i + \beta_j + \varepsilon_{ij}$; $i = 1, 2, 3$; $j = 1, 2, 3, 4, 5, 6, 7$;
where y_{ij} is score on test of jth subject hearing the ith music type
α_i is the ith music type effect
β_j is the jth subject effect
$\hat{\mu} = 21.33$, $\hat{\alpha}_1 = -0.47$, $\hat{\alpha}_2 = -1.19$, $\hat{\alpha}_3 = 1.67$
$\hat{\beta}_1 = 0$, $\hat{\beta}_2 = -3$, $\hat{\beta}_3 = 3.33$, $\hat{\beta}_4 = -1.33$, $\hat{\beta}_5 = 1$, $\hat{\beta}_6 = 3.67$, $\hat{\beta}_7 = -3.67$
b. $F = 6.54$ with df $= 2, 12$
Therefore, p-value $= Pr(F_{2,12} \geq 6.54) = 0.0120 \Rightarrow$ Reject H_0: $\mu_1 = \mu_2 = \mu_3$.
c. Based on the interaction plot, the additive model may be inappropriate because there is some crossing of the three lines.
d. $t = 3$, $b = 7 \Rightarrow$ RE(RCB, CR) $= 3.86$

9.9 **a.** $y_{ijk} = \mu + \alpha_k + \beta_i + \gamma_j + \varepsilon_{ijk}$; $i, j, k = 1, 2, 3, 4$;
where y_{ijk} is the dry weight of a watermelon plant grown in Row i and Column j
α_k is the effect of the kth Treatment on dry weight
β_i is the effect of the ith Row on dry weight
γ_j is the effect of the jth Column on dry weight
b. The overall mean is $\bar{y}... = 1.5425$.
The parameter estimates are given here:

$$\hat{\mu} = 1.5425, \hat{\beta}_1 = -.0125, \hat{\beta}_2 = .005, \hat{\beta}_3 = .0025, \hat{\beta}_4 = .005$$
$$\hat{\gamma}_1 = .02, \hat{\gamma}_2 = .0325, \hat{\gamma}_3 = -.0375, \hat{\gamma}_4 = -.015$$
$$\hat{\alpha}_1 = .195, \hat{\alpha}_2 = .1425, \hat{\alpha}_3 = -.12, \hat{\alpha}_4 = -.2175$$

$F = 1280.4$ with df $= 3, 6 \Rightarrow p$-value $< .0001 \Rightarrow$
Reject H_0: $\mu_1 = \mu_2 - \mu_3 = \mu_4$.

9.12 **a.** Ten replications of a completely randomized design with a 3×2 factorial treatment structure.
b. $y_{ijk} = \mu + \alpha_i + \beta_j + \alpha\beta_{ij} + \varepsilon_{ijk}$; $i = 1, 2, 3$; $j = 1, 2$; $k = 1, \ldots, 10$;
where y_{ijk} is the attention span of the kth Child of Age i viewing Product j.
α_i is the effect of the ith Age on attention span
β_j is the effect of the jth Product on attention span
$\alpha\beta_{ij}$ is the interaction effect of the ith Age and jth Product on attention span
c. $\hat{\mu} = 27.27$, $\hat{\alpha}_1 = -4.27$, $\hat{\alpha}_2 = -2.22$, $\hat{\alpha}_3 = 6.48$
$\hat{\beta}_1 = -5.8$, $\hat{\beta}_2 = 5.8$
$\hat{\alpha}\beta_{11} = 5.7$, $\hat{\alpha}\beta_{21} = .35$, $\hat{\alpha}\beta_{31} = -6.05$
$\hat{\alpha}\beta_{12} = -5.7$, $\hat{\alpha}\beta_{22} = -.35$, $\hat{\alpha}\beta_{32} = 6.05$
d. The AOV table is given here:

Source	df	SS	MS	F	p-value
Age	2	1,303.0	651.5	4.43	0.017
Product	1	2,018.4	2,018.4	13.72	0.001
Interaction	2	1,384.3	692.1	4.70	0.013
Error	54	7,944.0	147.1		
Total	59	12,649.7			

9.13 **a.** The profile plot indicates an increasing effect of Product Type as Age increases.

9.15 The complete AOV table is given here:

Source	df	SS	MS	F
Treatment	14	SST	MST	
Factor A	2	SSA	MSA	MSA/MSE
Factor B	4	SSB	MSB	MSB/MSE
Interaction	8	SSAB	MSAB	MSAB/MSE
Blocks	2	SSBL	MSBL	
Error	28	SSE	MSE	
Total	44	TSS		

9.21 **a.** Using Tukey's W procedure with $\alpha = 0.05$, $s_\varepsilon^2 = $ MSE $= .0678$, $q_\alpha(t, \text{df}_{\text{error}}) = q_{.05}(3, 24) = 3.53 \Rightarrow$

$$W = (3.53)\sqrt{\frac{.0678}{3}} = 0.53 \Rightarrow$$

pH		Ca Rate 100	200	300
4	Mean	5.80	7.33	6.37
	Grouping	a	c	b
5	Mean	7.33	7.27	7.33
	Grouping	a	a	a
6	Mean	7.40	7.63	7.17
	Grouping	a	a	a
7	Mean	7.30	7.10	6.60
	Grouping	b	ab	a

9.25 **a.** The experiment is run as three replications of a completely randomized design with a 2×4 factorial treatment structure. A model for the experiment is given here:

$$y_{ijk} = \mu + \alpha_i + \beta_j + \alpha\beta_{ij} + \varepsilon_{ijk}; \quad i = 1, 2, 3, 4; \quad j = 1, 2; \quad k = 1, 2, 3;$$

where y_{ijk} is the amount of active ingredient (or pH) of the kth vial having the ith storage time in laboratory j

α_i is the effect of the ith storage time on amount of active ingredient (or pH)

β_j is the effect of the jth laboratory on amount of active ingredient (or pH)

$\alpha\beta_{ij}$ is the interaction effect of the ith storage time and jth laboratory on amount of active ingredient (or pH)

b. The complete AOV table is given here:

Source	df	SS	MS	F
Storage Time	3	SSA	SSA/3	MSA/MSE
Laboratory	1	SSB	SSB/1	MSB/MSE
Interaction	3	SSAB	SSAB/3	MSAB/MSE
Error	16	SSE	SSE/16	
Total	23	TSS		

9.27 **a.** With respect to Amount of Active Ingredient: The Time*Lab*Temp interaction (p-value $= .6914$), Lab*Temp interaction (p-value $= .3519$), and Time*Temp interaction (p-value $= .2817$) were not significant. However, there was significant evidence (p-value $= .0192$) of a difference in the means for Amount of Active Ingredient at the two temperatures. There is a strong interaction between Time and Lab (p-value $< .0001$).

9.29 **a.** Latin square design with blocking variables Farm and Fertility. The Treatment is the five types of fertilizers.
  b. There is significant evidence (p-value < 0.0001) the mean yields are different for the five fertilizers.

Chapter 10: Categorical Data

10.1 **a.** Yes
  b. $(0.15, 0.25)$ is a 90% C.I. for π.

10.3 **a.** $\hat{\pi} = 1{,}202/1{,}504 = 0.8 \Rightarrow$ 95% C.I. for π: $0.8 \pm 1.96\sqrt{(.8)(.2)/1{,}504} \Rightarrow (0.780, 0.820)$
  b. 90% C.I. for π: $0.8 \pm 1.645\sqrt{(.8)(.2)/1{,}500} \Rightarrow (0.783, 0.817)$

10.7 $\hat{\pi} = 88/254 = 0.346 \Rightarrow$ 90% C.I. for π: $(0.297, 0.395)$

10.11 **a.** $n\pi_0 = 76.8 > 5$ and $n(1 - \pi_0) = 723.2 > 5$
  b. $H_0: \pi \geq 0.096$ vs. $H_a: \pi < 0.096$
    $\hat{\pi} = 35/800 = 0.04375$, $z = -5.02 \Rightarrow$
    p-value $= P(z < -5.02) < 0.0001 \Rightarrow$ Reject H_0.

10.13 $\hat{\pi} = 10/24 = 0.417 \Rightarrow$ 95% C.I. on π: $(0.220, 0.614)$

10.15 $\hat{\pi}_1 = 109/200 = 0.545$ (Republicans) and $\hat{\pi}_2 = 86/200 = 0.43$ (Democrats)
  $z = 2.32 \Rightarrow p$-value $= 0.0102 \Rightarrow$ Reject H_0.

10.17 95% C.I. on $\pi_1 - \pi_2$: $(0.013, 0.167)$
Because 0 is not contained within the C.I., H_0 is rejected.

10.19 $H_0: \pi_1 = \pi_2$ vs. $H_a: \pi_1 \neq \pi_2$; p-value $= 0.0018 < 0.05 \Rightarrow$ Reject H_0.

10.21 **a.** $z = 9.54 \Rightarrow p$-value $= Pr(z > 9.54) < 0.0001 \Rightarrow$ Reject H_0.

10.23 $H_0: \pi_1 = \dfrac{1}{3}, \pi_2 = \dfrac{1}{3}, \pi_3 = \dfrac{1}{3}$

  H_a: at least one of the groups had probability of interning different from $\dfrac{1}{3}$
    $\chi^2 = 6.952$ with df $= 3 - 1 = 2 \Rightarrow 0.025 < p$-value $< 0.05 \Rightarrow$ Fail to reject H_0.

10.25 $H_0: \pi_1 = 0.50, \pi_2 = 0.40, \pi_3 = 0.10$
  H_a: at least one of the π_is differs from its hypothesized value
  $\chi^2 = 6.0$ with df $= 3 - 1 = 2 \Rightarrow 0.025 < p$-value $< 0.05 \Rightarrow$ Reject H_0.

10.31 $0.001 < p$-value < 0.005

10.33 **b.** $H_0: \pi_1 = 0.25, \pi_2 = 0.40, \pi_3 = 0.35$ vs. H_a: Specified proportions are not correct

$$\chi^2 = \sum_{i=1}^{3} \frac{(n_i - E_i)^2}{E_i} = 41.143 \quad \text{with df} = 2 \Rightarrow p\text{-value} < 0.0001 \Rightarrow$$

  Reject H_0.

10.35 H_0: Membership Status and Opinion are independent vs. H_a: Membership Status and Opinion are related

$$\chi^2 = \sum_{i,j} \frac{(n_{ij} - E_{ij})^2}{E_{ij}} = 162.80 \quad \text{with df} = (2 - 1)(3 - 1) = 2 \Rightarrow p\text{-value} < 0.0001 \Rightarrow$$

Reject H_0.

10.37 $\chi^2 = 72.521$ with df $= 8 \Rightarrow p$-value $= P(\chi^2 > 72.521) < 0.001 \Rightarrow$ Reject H_0.

10.41   **a.** Control: 10%; Low Dose: 14%; High Dose: 19%
    b. $H_0: \pi_1 = \pi_2 = \pi_3$ vs. H_a: The proportions are not all equal, where π_j is probability of a rat in Group j having One or More Tumors.
    $\chi^2 = 3.312$ with df $= (2 - 1)(3 - 1) = 2$ and p-value $= 0.191$
    c. No

10.43   **b.** $\chi^2 = 57.830$ with df $= (4 - 1)(4 - 1) = 9$
    c. p-value < 0.001
    d. There is significant evidence that Years of Education are related to Years on First Job.

Chapter 11: Linear Regression and Correlation

11.3   **a.** $\hat{y} = 4.698 + 1.97x$
    b. Yes, the plotted line is relatively close to all 10 data points.
    c. $\hat{y} = 4.698 + (1.97)(35) = 73.65$

11.5 $\hat{y} = 48.935 + 10.33(1.0) = 59.265$

11.7   **a.** Yes
    b. $\hat{y} = 3.10 + 2.76\sqrt{x}$

11.13   The estimated slope changed considerably from 1.80 to 2.46. This resulted from excluding the data value having high influence. Such a data point twists the fitted line toward itself and hence can greatly distort the value of the estimated slope.

11.15 p-value < 0.0001

11.17   **a.** $\hat{y} = 1.67 - 0.0159x$
    b. $s_\varepsilon = 0.1114$
    c. $SE(\hat{\beta}_0) = 0.05837 \qquad SE(\hat{\beta}_0) = 0.001651$

11.19   **a.** Yes
    b. $\hat{y} = 12.51 + 35.83x$

11.21   **b.** $\hat{y} = 99.777 + 5.1918x \qquad s = 12.2065$
    c. A 95% C.I. for the slope is given by $(5.072, 5.312)$.

11.23   **a.** $F = 7{,}837.26$ with p-value $= 0.0000$
    b. The F test and two-sided t test yield the same conclusion in this situation. In fact, for this type of hypothesis, $F = t^2$.

11.25   95% prediction interval is $(0.941, 1.449)$

11.27   **a.** 95% Confidence Intervals for $E(y)$ at selected values for x:

$x = 4 \Rightarrow (2.6679, 4.3987)$
$x = 5 \Rightarrow (4.2835, 5.4165)$
$x = 6 \Rightarrow (5.6001, 6.7332)$
$x = 7 \Rightarrow (6.6179, 8.3487)$

b. 95% Prediction Intervals for y at selected values for x:

$x = 4 \Rightarrow (1.5437, 5.5229)$
$x = 5 \Rightarrow (2.9710, 6.7290)$
$x = 6 \Rightarrow (4.2877, 8.0456)$
$x = 7 \Rightarrow (5.4937, 9.4729)$

11.29   No

11.31 The residual plot indicates that higher order terms in x may be needed in the model.

11.33 $\hat{y} = 1.47 + 0.797x$
$H_0: \beta_1 \le 0$ vs. $H_a: \beta_1 > 0$
$t = 7.53 \Rightarrow p\text{-value} = Pr(t_8 \ge 7.53) < 0.0005 \Rightarrow$
There is sufficient evidence in the data that the slope is positive.

11.35 **a.** $\hat{y} = 0.113 + 0.118x$
Dependent variable is Transformed CUMVOL and Independent variable is LOG(Dose).

b. $\hat{x} = (y - 0.11277)/0.11847 \qquad \hat{x}_L = 2.40 + \dfrac{1}{(1 - .2426)}(\hat{x} - 2.40 - d)$

$\hat{x}_U = 2.40 + \dfrac{1}{(1 - .2426)}(\hat{x} - 2.40 + d) \Rightarrow$

y	TRANS(y)	LOG(x)	$\hat{x}$	d	LOG(x)$_L$	LOG(x)$_U$	$\hat{x}_L$	$\hat{x}_U$
10	.322	1.764	5.84	.242289	1.24038	1.88018	3.46	6.55
14	.383	2.285	9.83	.198634	1.98616	2.51068	7.29	12.31
19	.451	2.855	17.38	.221359	2.70876	3.29328	15.01	26.93

11.37 $r = 0.9722$

11.43 **a.** $\hat{y} = 5.890659 + 0.0148652x$
b. $H_0: \beta_1 \le 0$ vs. $H_a: \beta_1 > 0$, from the output $t = 5.812$
The p-value on the output is for a two-sided test. Thus, $p\text{-value} = Pr(t_5 \ge 5.812) = 0.00106$, which indicates that there is significant evidence that the slope is greater than 0.

11.47 **a.** The prediction equation is $\hat{y} = 140.074 + 0.61896x$.
b. The coefficient of determination is $R^2 = 0.9420$ which implies that 94.20% of the variation in fuel usage is accounted for by its linear relationship with flight miles. Because the estimated slope is positive, the correlation coefficient is the positive square root of R^2, or $r = \sqrt{0.9402} = 0.97$.
c. The only point in testing $H_0: \beta_1 \le 0$ vs. $H_a: \beta_1 > 0$ would be in the situation where the flights were of essentially the same length and one wanted to determine whether there were other important factors that might affect fuel usage. Otherwise, it would be obvious that longer flights would be associated with greater fuel usage.

Chapter 12: Multiple Regression

12.1 **b.** $\hat{y} = 8.667 + 0.575x$

12.5 **a.** No, the two independent variables, Air Miles and Population, do not appear to be severely correlated, based on the correlation (-0.1502) and the scatterplot.
b. There are two potential leverage points in the Air Miles direction (around 300 and 350 miles). In addition, there is one possible leverage point in the population direction; this point has a value above 200.

12.7 **a.** For reduced model: $R^2 = 0.2049$
b. For complete model: $R^2 = 0.7973$
c. With INCOME as the only independent variable, there is a dramatic decrease in R^2 to a relatively small value. Thus, we can conclude that INCOME does not provide an adequate fit of the model to the data.

12.8 In the complete model, we want to test $H_0: \beta_1 = \beta_2 = 0$ vs. $H_a: \beta_1 \ne 0$ and/or $\beta_2 \ne 0$.
$F = 24.84$ with df $= 2, 17 \Rightarrow p\text{-value} = Pr(F_{2,17} \ge 24.84) < 0.0001 \Rightarrow$ Reject H_0.

12.12 **a.** $y = \beta_0 + \beta_1 x_1 + \beta_2 x_2 + \beta_3 x_3 + \beta_4 x_1 x_2 + \beta_5 x_1 x_3 + \varepsilon$, where

$x_1 = \text{LOG(dose)}$

$x_2 = \begin{cases} 1 & \text{if Product B} \\ 0 & \text{if Products A or C} \end{cases}$ $\qquad x_3 = \begin{cases} 1 & \text{if Product C} \\ 0 & \text{if Products A or B} \end{cases}$

$\beta_0 = y$-intercept for Product A regression line
$\beta_1 = $ slope for Product A regression line
$\beta_2 = $ difference in y-intercepts for Products A and B regression lines
$\beta_3 = $ difference in y-intercepts for Products A and C regression lines
$\beta_4 = $ difference in slopes for Products A and B regression lines
$\beta_5 = $ difference in slopes for Products A and C regression lines

b. $y = \beta_0 + \beta_1 x_1 + \beta_2 x_2 + \beta_3 x_3 + \varepsilon$

12.14 **a.** For testing $H_0: \beta_1 = 0$ vs. $H_a: \beta_1 \neq 0$, the p-value for the output is 0.0427. Thus, at the $\alpha = 0.05$ level we can reject H_0.

b. From the output, $\hat{p}(24) = 0.765$ with 95% C.I. (0.437, 0.932).

12.18 **a.** $\hat{y} = -1.320 + 5.550 \text{ Educ} + 0.885 \text{ Income} + 1.925 \text{ Popn} - 11.389 \text{ Famsize}$
$\qquad\quad$ (57.98) $\qquad$ (2.702) $\qquad\quad$ (1.308) $\qquad\qquad$ (1.371) $\qquad\qquad$ (6.669)

b. $R^2 = 96.2\%$ and $s_\varepsilon = 2.686$

12.21 **a.** The regression model is

$\hat{y} = -16.8198 + 1.47019 x_1 + .994778 x_2 - .0240071 x_1 x_2 - .01031 x_2^2 - .000249574 x_1 x_2^2$
$s_\varepsilon = 3.39011$

b. Test $H_0: \beta_3 = 0$ vs. $H_a: \beta_3 \neq 0$. From output, $t = -1.01$ with p-value $= 0.3243$. Thus, there is not substantial evidence that the variable $x_3 = x_1 x_2$ adds predictive value to a model that contains the other four independent variables.

12.23 **a.** $\hat{y} = 102.708 - 0.833 \text{ PROTEIN} - 4.000 \text{ ANTIBIO} - 1.375 \text{ SUPPLEM}$

b. $s_\varepsilon = 1.70956$

c. $R^2 = 90.07\%$

12.25 **a.** $\hat{y} = 89.8333 - 0.83333 \text{ PROTEIN}$

b. $R^2 = 0.5057$

c. In the complete model, we want to test
$H_0: \beta_2 = \beta_3 = 0$ vs. $H_a:$ at least one of $\beta_2, \beta_3 \neq 0$.

The F statistic has the form:
$F = 27.84$ with df $= 2, 14 \Rightarrow p$-value $= Pr(F_{2,14} \geq 27.84) < 0.0001 \Rightarrow$ Reject H_0.

There is substantial evidence to conclude that at least one of $\beta_2, \beta_3 \neq 0$.

REFERENCES

Agresti, A. (1990) *Categorical Data Analysis.* New York: Wiley.

Agresti, A., and B. A. Coull. (1998). "Approximate is better than 'exact' for interval estimation of binomial proportions," *The American Statistician* 52 (2), 119–126.

Appleton, D. R., J. M. French, and M. P. J. Vanderpump. (1996). "Ignoring a covariate: An example of Simpson's Paradox," *The American Statistician* 50 (4), 340–341.

Carmer, S., and M. Swanson. (1973). "An evaluation of ten pairwise multiple comparison procedures by Monte Carlo methods," *Journal of the American Statistical Association* 68, 66–74.

Carroll, R., R. Chen, E. George, T. Li, H. Newton, H. Schmiedliche, and N. Wang. (1997). "Ozone exposure and population density in Harris County, Texas," *Journal of the American Statistical Association* 92, 392–415.

Carter, R. (1981). "Restricted maximum likelihood estimation of bias and reliability in the comparison of several measuring methods," *Biometrics* 37, 733–741.

Cochran, W. (1954). "Some methods for strengthening the common χ^2 test," *Biometrics* 10, 417–451.

Cochran, W., and G. Cox. (1957). *Experimental Design,* 2nd ed. New York: Wiley.

Conover, J. (1998). *Practical Nonparametric Statistics,* 3rd ed. New York: Wiley.

Devore, J. (2000). *Probability and Statistics for Engineering and the Sciences.* 5th ed. Pacific Grove, Calif.: Duxbury Press.

Effron, B., and R. Tibshirani. (1993). *An Introduction to the Bootstrap.* London: Chapman and Hall.

Fisher, R. A. (1949). *The Design of Experiments.* Edinburgh: Oliver and Boyd.

Hartley, H. O. (1950). "The maximum *F*-ratio as a short-cut test for homogeneity variance," *Biometrika* 37, 308–312.

Hildebrand, D., and L. Ott. (1998). *Statistical Thinking for Managers,* 4th ed. Pacific Grove, Calif.: Duxbury Press.

Koehler, K. (1986). "Goodness-of-fit tests for log-linear models in sparse contingency tables," *Journal of the American Statistical Association* 81, 483–493.

Larntz, K. (1978). "Small-sample comparison of exact levels for chi-squared goodness-of-fit statistics," *Journal of the American Statistical Association* 73, 253–263.

Manly, B. (1998). *Randomization, Bootstrap and Monte Carlo Methods in Biology,* 2nd ed. London: Chapman and Hall.

Merriam-Webster's Collegiate Dictionary. (2001). 10th ed. Springfield, Mass.: Merriam-Webster.

Miller, B. F., and C. B. Keane. (1987). *Encyclopedia of Medicine, Nursing, and Allied Health.* W. B. Saunders: Philadelphia, 697–698.

Miller, R. (1981). *Simultaneous Statistical Inference,* 2nd ed. New York: Springer-Verlag.

Milliken, G. A., and D. E. Johnson. (1984). *Analysis of messy data.* Belmont, Calif.: Lifetime Learning Publications.

Morton, D., et al. (1982). "Lead absorption in children of employees in a lead industry," *American Journal of Epidemiology* 115, 549–555.

Newman, R. (1998). "Testing parallelism among the profiles after a certain time period." Unpublished PhD dissertation, Texas A&M University.

Ott, R. L., and M. Longnecker. (2001). *An Introduction to Statistical Methods and Data Analysis,* 5th ed. Pacific Grove, Calif.: Duxbury Press.

Pearson, E. S., and H. O. Hartley. (1966). *Biometrika Tables for Statisticians,* 3rd ed. Vol. 1. London: Cambridge University Press.

Raftery, A., and J. Zeh. (1998). "Estimating bowhead whale population size and rate of increase from the 1993 census," *Journal of the American Statistical Association* 93, 451–462.

Rosner, B., W. C. Willett, and D. Spiegelman. (1989). "Correction of logistic regression relative risk estimates and confidence intervals for systematic within-person measurement error," *Statistics in Medicine* 8, 1051–1070.

Scheaffer, R. L., W. Mendenhall, and L. Ott. (1996). *Elementary Survey Sampling,* 5th ed. Pacific Grove, Calif.: Duxbury Press.

Smith, A. F. (1967). "Diagnostic value of serum-creatinine-kinase in a coronary care unit," *Lancet* 2, 178.

Thall, P. F., and S. C. Vail. (1990). "Some covariance models for longitudinal count data with overdispersion," *Biometrics* 46, 657–672.

Tukey, J. (1953). "The problem of multiple comparisons." Princeton, N.J.: Princeton University. Mimeographed.

Tukey, J. (1977). *Exploratory Data Analysis.* Reading, Mass.: Addison-Wesley.

U.S. Bureau of Labor Statistics. (1982). *Handbook of Methods,* Vols. 1 and 2. Washington, D.C.: U.S. Department of Labor.

Welch, B. (1938). "The significance of the difference between two means when the population variances are unequal," *Biometrika* 29, 350–362.

Zelazo, P. R., N. A. Zelazo, and S. Kolb. (1972). "'Walking' in the newborn," *Science* 176, 314–315.

INDEX

Italic page numbers indicate material in tables or figures.